HIGH POLYMERS

CONDENSATION MONOMERS

EDITED BY

JOHN K. STILLE

Department of Chemistry
University of Iowa

AND

TOD W. CAMPBELL

WILEY-INTERSCIENCE

A Division of John Wiley & Sons, Inc.

New York · London · Sydney · Toronto

Library of Congress Cataloging in Publication Data:

Stille, John Kenneth, 1930–
    Condensation monomers.

    (High polymers, v. 27)
    Includes bibliographical references.
    1. Monomers. 2. Condensation products (Chemistry)
I. Campbell, Tod W., joint author. II. Title.
III. Series.

TP247.S72   547′.87   72-1260
ISBN 0-471-39370-3

Printed in the United States of America.

10 9 8 7 6 5 4 3 2 1

To

TOD W. CAMPBELL

(1919-1968)

CONTRIBUTORS TO VOLUME XXVII

W. B. ALSTON, *University of Iowa, Iowa City, Iowa*

H. C. BACH, *Chemstrand Research Center, Inc., Durham, North Carolina*

JOSEFINA T. BAKER, *Princeton Chemical Research, Inc., Princeton, New Jersey*

ROBERT BARCLAY, Jr., *Thiokol Chemical Corporation, Chemical Division, Trenton, New Jersey*

WILLIAM F. BRILL, *Halcon International, Inc., Little Ferry, New Jersey*

J. B. CLEMENTS, *Chemstrand Research Center, Inc., Durham, North Carolina*

STEVE GUST COTTIS, *Research and Development Division, The Carborundum Company, Niagara Falls, New York.*

J. C. COWAN, *Northern Regional Research Laboratory, Agricultural Research Service, U.S. Department of Agriculture, Peoria, Illinois*

JAMES ECONOMY, *Research and Development Division, The Carborundum Company, Niagara Falls, New York*

M. E. FREEBURGER, *University of Iowa, Iowa City, Iowa*

EUGENE V. HORT, *GAF Corporation, Wayne, New Jersey*

PETER T. KAN, *BASF Wyandotte Corporation, Wyandotte, Michigan*

E. L. MAINEN, *University of Iowa, Iowa City, Iowa*

J. PRESTON, *Chemstrand Research Center, Inc., Durham, North Carolina*

E. H. PRYDE, *Northern Regional Research Laboratory, Agricultural Research Service, U.S. Department of Agriculture, Peoria, Illinois*

ADRIAN A. R. SAYIGH, *The Upjohn Company, Research Laboratories, North Haven, Connecticut*

W. H. SHARKEY, *Central Research Department, E. I. du Pont de Nemours and Company, Wilmington, Delaware*

J. K. STILLE, *University of Iowa, Iowa City, Iowa*

THEODORE SULZBERG, *Sun Chemical Corporation, Corporate Research Laboratory, Corlstadt, New Jersey*

PREFACE

This book is one of several volumes of a series devoted to the synthesis, purification, and properties of monomers; its purpose is to provide chemists working in the field with an accurate—as far as possible—and practical reference source. We do not intend to make an extensive survey of the polymerization reactions of the monomers and the structure and the properties of the resulting polymers; instead, only a brief treatment of the kinds of polymerization reactions into which the monomers will enter is outlined.

Early in 1967, Tod W. Campbell took over the editorship of this volume on condensation-type monomers. Tod selected the group of authors and prepared a set of guidelines for them. Three years ago today, Tod passed away at his home in Tucson, Arizona, and shortly thereafter, I agreed to assume the responsibility for completing this volume.

This book originally contained 16 chapters, including discussions of inorganic monomers and silicone monomers and chapters on lactams, lactones, cyclic ethers, and carboanhydrides. Later it was decided that the cyclic monomers, which can undergo polymerization by typical addition-type or chain-growth polymerization, should be the subject of a separate volume.

The authors of each chapter have followed the format guidelines as closely as possible. Of course, differences in the commercial availability, the knowledge of the synthesis and properties each monomer, etc., have necessitated deviations from the format. In each chapter, however, the authors have attempted to provide the following information:

1. A historical introduction, including the commercial utility and production data.
2. A survey of the chemistry of all the practical synthetic routes to each monomer.
3. Details of the more important synthetic routes, including the commercial synthesis when feasible, or a practical laboratory synthesis in which pound quantities of the monomer can be prepared and purified.
4. The physical properties of the monomer.
5. Analytical procedures for determination of purity.
6. Storage and toxicology.
7. Polymerization and polymerizability.

In any book conceived in this way, featuring a number of contributors, the chapters are usually completed at widely different times. Unfortunately, this volume is no exception; the first chapters were received in the spring of 1968 and the last in the spring of 1971. Some of those chapters received early may not contain some of the more recent significant information.

Many thanks are due to the authors who have contributed these chapters; the burden of the work is theirs. I owe special gratitude to those many reviewers who so willingly gave excellent critiques of individual chapters.

JOHN K. STILLE

Iowa City, Iowa
May 9, 1971

CONTENTS

CONDENSATION MONOMERS

1. ALIPHATIC DIBASIC ACIDS

E. H. PRYDE AND J. C. COWAN, *Northern Regional Research Laboratory, Agricultural Research Service, U.S. Department of Agriculture, Peoria, Illinois*

Contents

I. THE ALIPHATIC DIBASIC ACIDS

A. Introduction

Aliphatic dibasic acids include straight-chain, branched, saturated, unsaturated, and, for the purposes of this chapter, alicyclic dicarboxylic acids. Of these, the most important commercially are the α-, ω-alkanedioic acids, especially adipic acid. Their importance arises chiefly from their bifunctionality, which permits them to undergo various polycondensation reactions for the preparation of certain polymers and plastics. Several of the most accessible acids along with pertinent references are listed in Table 1, and the most important of these are discussed in individual sections as indicated. The unsaturated dicarboxylic acids listed have been reviewed adequately elsewhere and, of the cyclic acids listed, only dimer acid is discussed here. The remaining cyclic acids were mentioned because of interest in them as monomers and their potential availability.

The alkanedioic acids constitute an interesting homologous series; yet major differences in their properties point out the fallacy of considering them to have complete uniformity. Carbonic acid is listed as the initial member of the homologous alkanedioic acids, being a dibasic although not a dicarboxylic acid. It is, of course, not usable in the free acid form because of its instability. However, it is available as a monomer in the form of the acid chloride (phosgene), the ester chloride (chloroformate), and the amide (urea).

B. Chemistry

Thermal stability is essential, both in the preparation of polymers by melt polymerization and in the polymer itself. Under comparable conditions, the following order of decreasing stability has been reported by Korshak and Rogozhin (22) for several acids (decarboxylation temperatures given in °C): sebacic (360) > suberic (350) > azelaic (330) > adipic (310) > succinic (300) > glutaric (285) > oxalic (170) > malonic (150). These decarboxylation temperatures are even lower under conditions of polymerization; that is,

TABLE 1
The Aliphatic Dibasic Acids as Monomers

Total number of carbon atoms	Common name	Geneva name	Formula	Reference	Comments
			Saturated Acids		
1	Carbonic	—	$HOC(O)OH$	1–3	Unstable, used as a monomer in the form of esters, acid chlorides, amides
2	Oxalic	Ethanedioic	$HOOCCOOH$	Section II	Polyoxamides not now commercial but have potential
3	Malonic	Propanedioic	$HOOCCH_2COOH$	Section III	Too unstable, both thermally and oxidatively
4	Succinic	Butanedioic	$HOOC(CH_2)_2COOH$	Section IV	Used to a minor extent
5	Glutaric	Pentanedioic	$HOOC(CH_2)_3COOH$	Section V	Not commercially important at present
6	Adipic	Hexanedioic	$HOOC(CH_2)_4COOH$	Section VI	The most important of all the dibasic acids; used for nylon-6/6
7	Pimelic	Heptanedioic	$HOOC(CH_2)_5COOH$	Section VII	Not commercially important
8	Suberic	Octanedioic	$HOOC(CH_2)_6COOH$	Section VIII	Not commercially important but may have potential
9	Azelaic	Nonanedioic	$HOOC(CH_2)_7COOH$	Section IX	Important commercially for uses other than polyamides
10	Sebacic	Decanedioic	$HOOC(CH_2)_8COOH$	Section X	Important commercially as in nylon-6/10

6

No.	Common name	Systematic name	Formula	Beilstein;[a] 4	
11	—	Undecanedioic	$HOOC(CH_2)_9COOH$	Section XI	Not commercially important
12	—	Dodecanedioic	$HOOC(CH_2)_{10}COOH$		Has recently become commercially important
13	Brassylic	Tridecanedioic	$HOOC(CH_2)_{11}COOH$	Section XII	Not commercially important but may have potential
14	—	Tetradecanedioic	$HOOC(CH_2)_{12}COOH$	Beilstein[a]	Not important
15	—	Pentadecanedioic	$HOOC(CH_2)_{13}COOH$	Beilstein[a]	Not important
16	Thapsic	Hexadecanedioic	$HOOC(CH_2)_{14}COOH$	Beilstein[a]	Not important
21	Japanic	Heneicosanedioic	$HOOC(CH_2)_{19}COOH$	Beilstein[a]	Not important

Unsaturated acids

No.	Common name	Systematic name	Formula		
4	Maleic	*Cis*-2-butenedioic	$HOOCCH{=}CHCOOH$	5, 6	Used mainly in unsaturated polyester resins
4	Fumaric	*Trans*-2-butenedioic	$HOOCCH{=}CHCOOH$	5, 6	Used mainly in unsaturated polyester resins
5	Citraconic	2-Methyl-*cis*-2-butenedioic	$HOOCC(CH_3){=}CHCOOH$	7	Used as a comonomer
5	Itaconic	2-Methylenesuccinic	$HOOCC({=}CH_2)CH_2COOH$	7, 8	Used mainly as a comonomer in latex polymers

Cyclic acids

No.	Common name	Systematic name	Structure		
6	—	1,1-Cyclobutanedicarboxylic	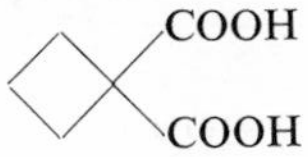	9, 10	Has potential value in polyesters and polyamides
7	Norcamphoric	1,3-Cyclopentanedicarboxylic	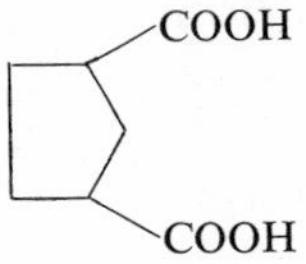	11–13	Has potential value in polyesters and polyamides
8	Tetrahydrophthalic	4-Cyclohexene-1,2-dicarboxylic	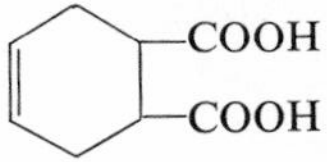	7	Used in alkyd resins

TABLE 1 (*continued*)

Total number of carbon atoms	Common name	Geneva name	Formula	Reference	Comments
8	—	1,1-Cyclohexanedicarboxylic	(cyclohexane with 1,1-diCOOH)	14	Has potential value in polyesters and polyamides
8	Hexahydrophthalic	1,2-Cyclohexanedicarboxylic	(cyclohexane with 1,2-diCOOH)	15	Has potential value in polyesters and polyamides
8	—	1,4-Cyclohexanedicarboxylic	HOOC–(cyclohexane)–COOH	16–18	Has potential value in polyesters and polyamides
9	Pinic	2,2-Dimethyl-3-carboxy-cyclobutylacetic	HOOC–(cyclobutane)–CH$_2$COOH	19	Has potential value in polyesters and polyamides
9	Chlorendic	1,4,5,6,7,7-Hexachloro-bicyclo[2.2.1]-5-heptene-2,3-dicarboxylic	(hexachlorobicycloheptene diCOOH structure)	20	Has potential in nonflammable polymer applications
10	—	1,4-Benzenediacetic	HOOCCH$_2$–(benzene)–CH$_2$COOH	21	Has potential value in polyesters and polyamides
36	Dimer	—	Mixture of acylic, mono-, and bicyclic compounds	Section XIII	Important for certain polyamides

[a] For acids that do not have specific references cited, the best source of general information is *Beilsteins Handbuch der Organischen Chemie*, 4th ed., 1918–to date, Beilstein-Institut für Literatur der Organischen Chemie, Springer Verlag, Berlin.

in the presence of glycols and diamines, which doubtless serve as catalysts. Decomposition temperatures as measured by derivatographic thermal analysis are somewhat at variance with those given by Korshak and Rogozhin (23): sebacic (330) > azelaic (320) = pimelic (320) > suberic (290) = adipic (290) = glutaric (290) > succinic (255) > oxalic (200) > malonic (185).

Anhydride formation is one example of the difference in chemical properties that can be displayed by alkanedioic acids. Oxalic and malonic acids do not form anhydrides. Succinic and glutaric acids form cyclic anhydrides when heated. When heated, adipic acid forms cyclopentanone, but when treated with acetic anhydride, it forms a polymeric anhydride. A cyclic anhydride also exists for adipic acid. Polymeric anhydrides are the normal products for the higher alkanedioic acids, but some produce fairly stable cyclic monomers or dimers when molecularly distilled (24). Pimelic acid also forms a cyclic ketone—cyclohexanone—when heated. This characteristic cyclization of succinic and glutaric acids to an anhydride and of adipic and pimelic acids to a ketone by application of heat is frequently expressed as Blanc's rule (25).

Oxidative degradation studies have been described for several reagents including chromium (VI) oxide (26,27), nitric acid (28), and hydrogen peroxide in the presence of a copper salt (29). Succinic acid is the most stable, with the higher acids progressively less stable according to number of carbon atoms. Mareš and Roček (27) give the oxidation rate constants, $k \times 10^{-3}$ in $1M^{-1}$ sec^{-1}, as follows: succinic, 0.0071; glutaric, 0.044; adipic, 0.32; pimelic, 0.80; suberic, 4.1; azelaic, 8.0; sebacic, 12; undecanedioic, 16; dodecanedioic, 24; and brassylic, 28. The effect of the carboxyl group in retarding oxidation of methylene groups decreases progressively up to the δ position.

Half-esters of the alkanedioic acids are useful chemical intermediates. Depending on the particular acid, they may be prepared by one of three different methods (30). For those acids forming cyclic anhydrides (succinic and glutaric), reaction of the anhydride with an alcohol forms the half-ester in good yield. For those acids between and including adipic and sebacic acids, partial esterification is used to best advantage. For the higher acids, the dimethyl ester is partially saponified with barium hydroxide; this method precipitates the barium salt of the half-ester and thereby removes the half-ester from further reaction. Half-esters tend to disproportionate with time, particularly when heated. The half-ester acid chlorides are readily prepared with thionyl chloride.

Cyclization reactions of the higher alkanedioic acids are characteristic and well known (31,32). For preparation of cyclic ketones having 5–6 ring atoms, the Dieckmann condensation of diesters is preferred. For those with 7–8 ring atoms, the Thorpe-Ziegler reaction of dinitriles is used; for those with 10–14 ring atoms, the acyloin condensation of the diester. None of these condensations is as satisfactory for the 4- or 9-carbon cyclic ketone as ring expansion of ketene or cyclooctanone with diazomethane. The acyloins (33) and the cyclic ketones obtained by treatment of the acyloins (34) have been reported for many of the higher alkanedioic acids.

Diesters have been converted to polymethylene diols by reduction with sodium and alcohol (35) or by hydrogenation over a barium chromate-stabilized copper-chromium oxide catalyst (36). Reduction of alkanedioic N-methylanilides with lithium aluminum hydride produces the dialdehydes in good yield (37).

C. Synthesis

A variety of methods and starting materials are used for producing the dibasic acids. Cyclic hydrocarbons from the petroleum industry are available for glutaric, adipic, suberic, and dodecanedioic acids; for these, the most likely commercial route is air oxidation followed by nitric acid oxidation of the resulting cyclic ketones and alcohols. Ozonolysis of unsaturated fatty acids is the preferred route for azelaic and brassylic acid, whereas alkaline cleavage of ricinoleic acid is used for sebacic acid.

An interesting, one-source reference for laboratory preparations of the alkanedioic acids from oxalic to brassylic acids is the work on heats of combustion by Verkade et al. (38).

Several general methods for the higher alkanedioic acids deserve mention, and many involve shorter chain α,ω-bifunctional compounds. Thus glycols may be converted to dibromides, which are converted to dicyanides, which in turn, when hydrolyzed to the dibasic acid, extend the carbon chain by two carbon atoms (39). Glycols may be converted directly to acids in high yields by the Reppe synthesis, based on the reaction of carbon monoxide under high pressure (40). Glycols may be oxidized by N_2O_4 in high yield to form acids with the same number of carbon atoms as the glycol (41).

An elegant synthesis for long-chain dibasic acids involves a ketene dimer as an intermediate in the following sequence (42–44):

$$ROOC(CH_2)_nCOCl \xrightarrow{Et_3N} ROOC(CH_2)_{n-1}CH\!\!=\!\!\overset{\displaystyle O-C=O}{\underset{\displaystyle |\quad\quad|}{C}}\!\!-CH(CH_2)_{n-1}COOR \xrightarrow[H^+ \text{ or } OH^-]{H_2O}$$

$$HOOC(CH_2)_nCO(CH_2)_nCOOH \xrightarrow{\text{reduction}} HOOC(CH_2)_{2n+1}COOH$$

The symmetrical keto dibasic acid is conveniently reduced with hydrazine and potassium hydroxide (4).

Electrolytic coupling of the half-ester of a dicarboxylic acid or a mixture of half-esters from two different dicarboxylic acids is a useful procedure (45–47):

$$2ROOC(CH_2)_xCOO^- \xrightarrow{-2e} ROOC(CH_2)_{2x}COOR + 2CO_2$$

A valuable synthesis extends the chain by 6 or 12 carbon atoms with 1-morpholino-1-cyclohexene (48,49). To extend the chain by six carbon atoms, the eneamine is acylated with the half-ester acid chloride of a dibasic acid:

$$HOOC(CH_2)_n CO(CH_2)_5COOH$$

The acylation with a diacid chloride succeeds best with suberic or sebacic acids to lengthen the chain by 12 carbon atoms. Acylation of the eneamine from cyclopentanone or cyclohexanone with 10-undecenoyl chloride yields a terminally unsaturated keto acid from which penta- or hexadecanedioic acid may be prepared (50).

Condensation of ethyl ω-bromoalkanoates with diethyl acetonedicarboxylate also provides a convenient, high-yielding route to higher keto-alkanedioic acids (51)

The malonic ester synthesis has been successful. The chain is extended by two carbon atoms when an ω-haloalkanoate is selected, or by four atoms with α,ω-dihaloalkane.

Another versatile procedure is the hydrogenolysis of various thiophene derivatives (52). Thus thapsic acid was prepared by acylation of 2,2'-methylene-dithiophene:

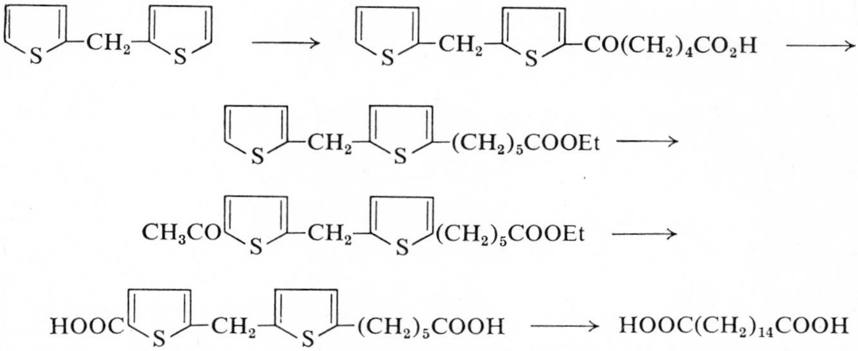

Chain extension by 6 carbon atoms is carried out with 1,3-cyclohexane-dione (dihydroresorcinol) alkylated in the 2-position by means of an alkyl ω-haloalkanoate (53). The alkylated product is submitted to reductive acid cleavage with hydrazine hydrate in alkaline solution.

The preceding methods are generally clear-cut and yield products free of homologs. Mixtures of dibasic acids may be obtained by nitric acid oxidation of air-oxidized petroleum waxes (54), of fatty acids (55, 56), and of fatty esters (57).

D. Physical Properties

Almost all alkanedioic acids have two crystalline modifications, but the second form has not yet been noted for some of the even-numbered acids, such as C_{10} and C_{12}. Glutaric acid is the first truly representative member of the series of odd-numbered acids; and adipic acid probably typifies the even-numbered acids. The differences in crystallographic and other properties can be explained in part by the molecular structure. In the even-numbered series, the plane of the carboxyl group is near the plane of the central carbon atoms, but for the odd-numbered series, the carboxyl plane may be as much as 40–50° from the central plane. Furthermore, the intermolecular distance in the strong hydrogen bridges between carboxyl groups of two glutaric acid molecules is on the order of 2.69 Å, whereas the lateral distance between two molecules is on the order of 3.59 Å, representing a relatively weak van der Waals attraction (58). Accordingly, these structural differences on the molecular level probably account for the softer crystal texture, the lower melting points, and greater solubilities for the odd-numbered acids.

The "saw-tooth" or alternating pattern of high and low melting points for the dibasic acids has excited frequent comment since Baeyer first made the observation in 1877. Erickson (59) has reported on an empirical expression, which takes the form of a mathematical sequence for predicting melting points. Agreement between observed and calculated values was good for the C_5–C_{34} dibasic acids. The convergence value for the series is 123°C. Esters of these acids also show an alternating effect (60, 61).

Solubilities of these acids also demonstrate an alternating or oscillating effect (62). Saracco and Marchetti (63) write the equations:

$$\ln S = \ln 4 - 1.05 \times n \qquad \text{(even)}$$
$$\ln S = \ln 6000 - 1.6 \times n \qquad \text{(odd)}$$

where S = moles per liter of solvent and n = number of carbon atoms. Breusch and Ulusoy (64) observed this effect in acetone and ethyl acetate as well as in water.

An alternating pattern is again evident in a plot of β index of refraction or of $(\alpha\beta\gamma)^{1/3}$ against carbon number (65,66).

Ionization constants have been reported for the alkanedioic acids through brassylic acid; there is apparently no alternating effect (67,68). Neither is there such an effect for densities and dielectric constants (64).

E. Analytical Procedures

Melting points of several derivatives for many of the dibasic acids are given in Table 2. The alternating effect previously mentioned has carried over to the derivatives.

TABLE 2

Melting Points (°C) of Derivatives of Dibasic Acids From Various Reagents

Dibasic acid	Ethanolamine[a] (69)	Benzylamine[a] (70)	p-Phenylphenacyl bromide[b] (71,72)	p-Phenylazophenacyl bromide[b] (73)
		Saturated		
Oxalic	169.0–169.5	222–223	165.5 dec.	153.5–156.5
Malonic	127.0–127.5	141.5–142.5	175	145.0–147.0
Succinic	156.2–156.7	205–206	208	215.5–216.5
Glutaric	119.6–120.0	169.5–170	152	166.0–167.5
Adipic	130.2–130.7	188–189	148	173.0–174.5
Pimelic	—	153–154	145–148 dec.	156.5–157.5
Suberic	138.5–138.9	—	151	169.0–171.0
Azelaic	125.5–125.9	—	141	166.0–167.5
Sebacic	144.5–145.0	166.0–167.5	140	159.0–160.0
		Unsaturated		
Maleic	—	149–150	168	—
Fumaric	—	203.5–205	—	—

[a] Forming the amide.
[b] Forming the ester.

F. Toxicology

Except for oxalic, malonic, and possibly glutaric acids, the alkanedioic acids may be considered as relatively nontoxic materials. Oxalic and malonic acids are hazardous in that they are strong acids and precipitants of blood calcium.

G. Polymer Applications

Reviews on polyesters (74) and polyamides (75) from many of these dicarboxylic acids have been published.

II. OXALIC ACID

A. Introduction

Oxalic acid exists as an acid salt in many different plants, such as saltwort, glasswort, wood sorrel or *Saureklee*, sour dock, and other plants of the *Oxalis* or *Rumex* genus, rhubarb, and beets. Some of these salts were observed as early as 1688 by Duclos. The acid was characterized by several investigators: Savary in 1773, Scheele and Bergman in 1776, and Wiegleb in 1779. Scheele recognized in 1784 the identity of *Kleesaure* with the *Zuckersaure* he had obtained earlier (1776) by the nitric acid oxidation of sugar. The latter process was further investigated by Thompson in 1847; and Blondeau in 1864 found that starch, dextrin, and cellulose also produced oxalic acid. A German patent obtained in 1907 by Naumann, Moeser, and Lindenbaum describes the use of vanadium pentoxide as a catalyst. Another route to oxalic acid is the caustic fusion process (76), first described by Vauquelin and further developed by Gay-Lussac in 1829 and Possoz in 1858; sugar, starch, sawdust, and fatty acids were among the starting materials used.

Many other preparations have been described: Wöhler hydrolyzed cyanogen in 1825; Dumas and Stas thermally decomposed sodium formate in 1840; Wurtz oxidized ethylene glycol with nitric acid in 1857; and Berthelot oxidized ethylene, propylene, or fatty acids with permanganate in 1867.

Commercial manufacture was initiated by Roberts, Dale & Co. in England as early as 1856, employing caustic fusion of sawdust (77). Annual production in the United States has averaged about 20 million lb for the past 25 years and was 22,854,000 lb in 1966 at a value of $0.21/lb (78). In 1968 producers were Allied Chemical Corporation and Charles Pfizer & Company, Inc.— each with an annual capacity of 10 million lb or more (79).

Oxalic acid has a number of industrial applications as follows (percentage of total market in parentheses): metal cleaning (27%); chemicals (25%); textile finishing, stripping, and cleaning (23%); leather tanning (4%), and other (21%) (79). It is used in commercial laundries as an acid rinse and as a remover of rust and ink stains; it is a component of radiator cleaners, and it serves as a bleaching agent for wood. Various iron salts are used for blueprint paper. Potassium oxalate serves as an analytical standard.

Good summaries of the various methods for producing oxalic acid, its properties, and its uses have been published (80,81).

B. Chemistry

Oxalic acid has anhydrous and dihydrated crystalline forms. The dihydrate can be converted to the anhydrous form by oven drying (82), by azeotropic removal of water (83–85), or by long standing over or crystallization from concentrated sulfuric acid (86).

It does not decompose at temperatures below 100°C (87), but at higher temperatures it decomposes to formic acid and carbon dioxide; above 160°C the formic acid also decomposes, yielding carbon monoxide and water (88). The decomposition in the vapor state at 0.9 mm of pressure and at 127–157°C is a first-order reaction with the Arrhenius parameters $E = 30.0 \pm 1.3$ kcal/mole and $\log A = 11.9 \pm 0.7$ sec^{-1} (89). In 100% sulfuric acid, oxalic acid rapidly decomposes even at room temperature to carbon monoxide, carbon dioxide, and water (90,91).

Russian investigators have pointed out the importance of knowing decomposition temperatures in connection with polycondensation reactions (92–94). The decomposition temperature is in the range 160–180°C for oxalic acid. However, in mixture with ethylene glycol, oxalic acid loses 25% of its carboxyl groups in a matter of 5 hr at 110°C and almost half in 2 hr at 130–140°C. Lowering of the decomposition temperature is less pronounced with diethylene glycol than with ethylene glycol. Clark has shown that apparent first-order rate constants for the decomposition in propylene glycol are 2.67×10^{-6} sec^{-1} at 118.38°C and 7.68×10^{-6} sec^{-1} at 148.65°C (95). The decomposition has been studied kinetically in a variety of solvents including glycerol (96) and other glycols (95).

The photochemical decomposition of oxalic acid in either the solid anhydrous state (88) or in aqueous solution (97) is a process with very low quantum efficiency.

Oxalic acid does not form an anhydride but does undergo readily various reactions characteristic of the carboxylic group. Although its half-esters are thermally unstable, the diesters are quite stable and may be distilled without change at temperatures as high at 245°C, the boiling point of the dibutyl ester at atmospheric pressure. Diethyl oxalate may be conveniently prepared from oxalic acid dihydrate by removal of product water and water of hydration by azeotropic distillation either as a ternary azeotrope with carbon tetrachloride or as the binary azeotrope with ethanol with continuous drying and recycling (98,99).

Oxidation of oxalic acid to carbon dioxide can be effected by a variety of oxidizing agents including hydrogen peroxide, permanganates, chromates, and cerium(IV) salts. Oxalic acid is relatively stable to nitric acid, which can be used in its preparation from carbohydrates or fatty materials. Electrolytic reduction of oxalic acid produces glyoxalic and glycolic acids.

Oxalyl chloride may be prepared by the action of phosphorus pentachloride on the acid (100,101). Thionyl chloride is without effect. Oxalyl chloride may also be prepared from phosgene and carbon monoxide (102) and by the catalytic decomposition of an ester of tetrachloroethylene glycol (103) or of bis(trichloromethyl) oxalate, obtained by chlorination of the dimethyl ester (104).

C. Synthesis

Several routes to oxalic acid have commercial potential. Many cheap or waste materials—including sawdust, waste cellulose liquors, corncobs, oat or rice hulls, lignin, peat, and lignite—can be used in some of these routes, but transportation costs or costs of other reagents may make the process uneconomic.

1. CAUSTIC FUSION

Sawdust or other waste cellulosic material is mixed with caustic soda and a small amount of water, and the mixture is heated to about 200–210°C in the presence of air. The oxalate is precipitated as the insoluble calcium salt, which is converted to the acid by reaction with sulfuric acid. During caustic fusion, carbon monoxide and dioxide are also produced. The caustic soda and sulfuric acid are obviously more expensive than sawdust, and recovery and recycle of the caustic soda are necessary. Yields of oxalic acid are on the order of 50–80 wt. % based on the sawdust used (105–114).

Othmer et al. (115–117) have investigated on a pilot scale continuous production of oxalic acid by the caustic fusion method and have reported oxalic yields as high as 79%. Projected yields were 65% for a commercial plant designed in accordance with the flow diagram suggested. Acetic acid (in yields up to 18%) and formic acid (yields up to 3.8%) were also recovered.

2. NITRIC ACID OXIDATION OF CARBOHYDRATES

The oxidation of carbohydrates in nitric acid is carried out preferably with a monosaccharide formed by preliminary acid hydrolysis of the carbohydrate (118,119). The reaction of nitric acid on the monosaccharide is

$$C_6H_{12}O_6 + 6HNO_3 \rightarrow 3HOOC-COOH + 6NO + 6H_2O$$

Studies have been reported for such starting materials as corncobs (111), sawdust (120), hydrolytic lignin (121), and furfural (122). Ethylene glycol can also be used (123), and yields of oxalic acid as high as 93% have been obtained from it (124).

In addition to vanadium compounds, molybdenum (125), manganese (125), and iron (126,127) compounds have been used as catalysts, with mixtures of vanadium and iron compounds most frequently mentioned (128). Vanadium pentoxide also catalyzes the oxidation of oxalic acid. Consequently, reaction conditions must be carefully delineated; e.g., a reaction temperature of 70°C (118). Tartaric acid is also formed, particularly at lower reaction temperatures, and an isomeric mixture can be recovered by precipitation as zinc or other metal salts (129,130) or by crystallization at −10°C (131).

3. SODIUM FORMATE DECOMPOSITION

For some decades the most important commercial route to oxalic acid has been the decomposition of sodium formate, which is synthesized in more than 90% yields by the reaction of carbon monoxide with sodium hydroxide under pressure:

$$CO + NaOH \rightarrow NaOOCH$$

If the reaction is continued by rapid heating to 360–420°C, sodium oxalate is formed in about 90% yield in a vigorous, exothermic reaction

$$2NaOOCH \rightarrow NaOOC-COONa + H_2$$

If the sodium formate is heated slowly to lower temperatures (300°C), then decomposition to sodium carbonate is the predominant reaction

$$2NaOOCH \rightarrow Na_2CO_3 + CO + H_2$$

An early German patent describes the use of small amounts of sodium hydroxide as a catalyst that moderates and promotes a smooth reaction at 360°C (132). In large amounts, sodium hydroxide reacts with sodium oxalate, giving sodium carbonate and hydrogen (133). Small amounts of calcium salts or magnesium oxide (134) also promote carbonate formation at the expense of oxalate yields.

The sodium oxalate product—a light, fluffy powder—is removed from the reactor with exclusion of air so that occluded hydrogen will not cause an explosion (114,135). Treatment with water, in which sodium oxalate is relatively insoluble, dissolves sodium carbonate and leaves a relatively pure oxalate product. To obtain pure oxalic acid, the sodium oxalate is treated with lime and the resulting calcium oxalate, with sulfuric acid as in the methods previously described.

Many descriptions of the sodium formate route were reported in the late nineteenth and early twentieth centuries (80,114); more recently, Russian and Eastern European workers have been active (136,137). Potassium formate apparently gives somewhat better yields than the sodium salt (138,139).

4. FERMENTATION

Oxalic acid is also formed in fermentation processes by a variety of organisms; commercially, it is produced as a by-product in citric acid manufacture.

5. PROPYLENE OXIDATION

Although oxidation of propylene or ethylene to oxalic acid was described by Berthelot in 1867, only recently have these starting materials received much attention. Propylene can be oxidized by "mixed acid" (nitric and sulfuric acids) (140) or by nitrogen dioxide and then mixed acid in a two-stage process (141). Oxygen may be used to oxidize evolved nitric oxide back to nitrogen dioxide, which is then recycled (142). Use of such catalysts as iron, chromium, aluminum, tin, and bismuth salts or iodine is said to improve yields (143). In the presence of palladium or mercury salts, ethylene is also oxidized to oxalic acid (144). In the absence of such catalysts but with careful control of reaction conditions, particularly the concentration of nitric and sulfuric acids, yields as high as 82% based on ethylene consumption are possible (145). Acetylene may also be used (146–150).

6. CARBONYLATION

Synthesis of oxalic acid in 29% yield has been accomplished in the reversible reaction of carbon monoxide with potassium carbonate at 243 atm and 470°C (151). The reaction of carbon monoxide in ethanol containing ethyl orthoformate and palladium chloride along with ferric or cupric chloride reportedly gives diethyl oxalate (152).

D. Physical Properties

Oxalic acid has three forms: a monoclinic and tabular or prismatic dihydrate (space group $P2_1/n$); an orthorhombic, dipyramidal, and anhydrous form (α-oxalic acid, space group P_{cab}); and another anhydrous form that is monoclinic and prismatic (β-oxalic acid, space group $P2_1/c$) (153–170). When crystallized from water, the dihydrate is tabular and polycrystalline with occluded mother liquor; single crystals of high purity are grown best from acetone-water solutions, and the habit is then prismatic (154). α-Oxalic acid is formed by dehydration of the hydrate below 97°C and by recrystallization from warm nitric or acetic acid; β-oxalic acid is formed by crystallization from nitric acid solution at room temperature, by sublimation, and by extraction of the dihydrate with boiling benzene.

Anhydrous oxalic acid is so volatile that it may be purified on a small scale by sublimation at reduced pressures. The vapor pressure at 60°C is 0.01 mm and at 105°C, about 0.53 mm (171).

Oxalic acid is moderately soluble in cold water and quite soluble in hot water. Its solubility may be expressed by the following equations

$$S, g/100 \text{ g of water } (0.5\text{–}70°C) = 3.543 + 0.1759016(t - 0.5)$$
$$+ 0.002721258(t - 0.5)^2 + 0.0001039899(t - 0.5)^3 \quad (172)$$
$$S', g/100 \text{ g of solution } (0\text{–}60°C) = 3.42 + 0.168t + 0.0048t^2 \quad (81)$$
$$S', g/100 \text{ g of solution } (50\text{–}90°C) = 0.333t + 0.003t^2 \quad (81)$$

Solubilities in representative organic solvents in grams per 100 g of solvent are 90% ethanol, 14.7 g at 15°C; 100% ethanol, 23.73 g at 15°C; ether, 1.47 g of the dihydrate and 23.59 g of the anhydrous acid. Johnson and Talbot (173) described the separation of oxalic and succinic acids in a countercurrent system consisting of 1-butanol and water.

Oxalic acid is a strong acid, about comparable to dichloroacetic acid and somewhat stronger than phosphoric, sulfurous, or the second acidic hydrogen of sulfuric acid. Oxalate ions in aqueous solution have been characterized by both Raman and infrared spectra (174).

Some useful properties of oxalic acid are presented in Table 3.

E. Analytical Procedures

Specifications and test methods have been published for the commercial product, which has a purity of about 99.5% (183–185).

Oxalic acid is both a strong acid and a reducing agent and can be assayed by either acidimetric or oxidimetric methods. However, forethought should be given to the sample history before a method is selected, since impurities also may be acidic, reducing, or both, as with malonic acid. Permanganate, bromate (in the presence of mercuric bromide), or cerium(IV) solutions are used in oxidimetric titrations. Pure sodium oxalate, available from the Bureau of Standards, is a primary standard and may be used conveniently for standardizing the titrant, preferably permanganate. Identical procedures for standardization and assay should be used, since variations will give erratic results. The preferred procedure involves addition of 90% of the permanganate at room temperature followed by the remainder at a temperature between 55 and 60°C (186,187).

Precise differential potentiometric (188) and coulometric methods (155) have been used with oxalic acid. The conductometric titration of oxalic acid in nonaqueous solutions has been described (189).

Mixtures of dibasic acids may be effectively resolved by gas-liquid chromatography (190). The acids are converted to their methyl esters, and then injected into a column of diethylene glycol succinate on Chromosorb W.

Thin-layer chromatography on silica gel successfully separated oxalic acid from homologous and other acids used in plasticizers (191). The solvent was

TABLE 3

Physical Properties of Oxalic Acid

Property	Oxalic acid dihydrate	Anhydrous oxalic acid
Molecular weight	126.07	90.04
Melting point, °C	101.5	187 (dec.)
Differential thermal analyses endotherms, °C (175)	110, 120, 125, 150, 195	—
Vapor pressure, $\log p$ (mm Hg)	$-(2741.9)(1/T) + 9.603$ (20–97°C) (164)	$-(4726.9499)(1/T) + 12.22292$ (60–105°C) (171)
Density, g/cc	1.650 (20°C)	α, 1.900; β 1.895
Specific volume, ml/g (87)	—	$0.5758 + 2.137t \times 10^{-4} - 0.705t^2 \times 10^{-6}$
Refractive index	$N_x = 1.417, N_y = 1.505, N_z = 1.550$ (65)	$\alpha: N_x = 1.445, N_y = 1.540, N_z = 1.638$ (167) $\beta: N_x = 1.445, N_y = 1.523, N_z = 1.631$ (167)
Heat capacity, c_p, joules/g (176)		
-200 to $+50$°C	—	$1.084 + 0.00319t$
0 to $+99.6$°C	—	1.31
Thermal conductivity, 10^{-4} Watts/cm deg -190 to 0°C (177)	—	$90 (1 - 58t \times 10^{-4})$
Coefficient of expansion, ml/g deg at 25°C (87)	—	1.784×10^{-4}
Coefficient of internal energy, cal/g atm (87)	—	-1.29×10^{-3}
Heat of combustion, cal/g (87)	—	654.7
Heat of formation, kcal/mole (87)	—	196.2
Dipole moment, in dioxane at 25°C (178)	—	2.63
Solubility, g/100 g H_2O (179)		
0°C	—	3.5
20°C	—	9.5
40°C	—	21.5
60°C	—	44.3
80°C	—	84.7
Heat of solution in water, kJ/mole (180)	-35.5	-9.58
Ionization constants in water at 25°C		
K_1 (181)	5.36×10^{-2}	
K_2 (182)	5.42×10^{-5}	

a mixture of 96% ethanol, water, and 25% ammonia in the volume ratio of 100:12:16. Oxalic acid has been separated from malonic, maleic, succinic, and tartaric acids also on silica gel plates (192). Separation was stated to be more effective with acidic than with alkaline solvents; butanol, 85% formic acid, and water in the ratio 7:3:12 was the preferred solvent.

Paper chromatographic analyses have been described (193,194).

F. Toxicology

Oxalic acid is a hazardous chemical on two counts: as a strong acid and as a precipitant of blood calcium. Casual, brief contact of the solid acid with the skin under dry conditions may be without effect, but prolonged exposure should be avoided, because dermatitis may result. Lengthy skin contact without adequate first aid causes ulcerous lesions to form. Contact of mucous membranes with the solid, dust, or vapor results in irritation of the eyes and upper respiratory tract, ulceration of the nose and throat, nosebleed, headache, irritability, and nervousness. Severe exposure may bring about chronic coughing, vomiting, and emaciation. Ingestion of oxalic acid causes severe corrosion of the mucous membranes, vomiting, burning abdominal pain, and collapse, followed by death. Obstruction of the renal tubules with calcium oxalate in a profound kidney disturbance is the result of precipitation of blood calcium. For further details see Sax (195), Steere (196), and Patty et al. (197).

G. Polymer Applications

1. POLYESTERS

Oxalic acid has developed no large-volume outlets in the polymer and plastics industry because of its thermal instablity, as well as certain undesirable properties of its polymers. Its polyesters are characterized by high melting points but poor hydrolytic and thermal stabilities. Carothers and co-workers have characterized several polyoxalates (198). Reaction of ethylene glycol with diethyl oxalate gives a polymeric product from which monomeric ethylene oxalate, a six-membered heterocyclic compound, can be recovered in about 50% yield by distillation *in vacuo*. Neither monomer nor polymer is stable at room temperature, and they are mutually interconvertible. Polymers from trimethylene and higher glycols do not depolymerize spontaneously.

Reaction of the glycol with an ester of oxalic acid is preferable to the reaction with the free acid, since oxalic acid decarboxylates at 170°C and at even lower temperatures in the presence of glycols. Organometallic aluminum compounds are suitable catalysts (199).

2. POLYAMIDES

Poly(hexamethylene oxamide) has certain desirable properties—including a high melting point, stiffness, and low moisture sensitivity. The property of low moisture sensitivity is anomalous in the series of homologous polyamides, but it is not too surprising in view of the low water solubility of oxamide itself in water (0.04 g/100 g). At its melting point (320°C), however, poly(hexamethylene oxamide) undergoes decomposition and gas evolution. Consequently, a satisfactory polymer is not obtainable by conventional methods. The polyamide may be prepared by the reaction of oxalyl chloride in solution (200) or gaseous form (201) with aqueous diamine and by ionic polymerization of the cyclic monomeric oxamide (202). It is reported that antimony(III) fluoride and arsenic(III) oxide catalyze the condensation of dibutyl oxalate and hexamethylene diamine in either solid-phase or melt polymerization and thereby obviate long heating (203).

Polyoxamides are generating considerable interest either as copolyamides or as a modifying agent for other polyamides.

The problem of too high a melting point can be overcome by the use of longer chain diamines, branched chain diamines, and certain aromatic diamines, and by the use of diacid or diamine mixtures as in copolyamides.

One patent describes some simple polyoxamides that are said to compare favorably with the best polyadipamides for spinnability and fiber strength (204). The yarn from the polyoxamide with 2-methylhexamethylenediamine is relatively insensitive to changes in temperature and humidity and has superior light and heat stability compared with other polyamides. The dibutyl oxalate used must be free of acidic impurities. Impurities catalyze decomposition, and for this reason, a polyoxamide prepared by gas-liquid interfacial polycondensation is not so stable as the polymer obtained by melt polymerization (205).

III. MALONIC ACID

A. Introduction

Malonic acid was not isolated and characterized until 1858, when Dessaignes prepared it by the dichromate oxidation of malic acid,

$$HOOCCH(OH)CH_2COOH,$$

and gave it its name in consideration of its source. Malonic acid occurs in beets and is deposited as the calcium salt in the evaporation of beet juice.

Several early syntheses are of interest, since they pertain to present or potential commercial syntheses. Kolbe hydrolyzed cyanoacetic acid, Henry used ethyl cyanoacetate, and Baeyer, barbituric acid, all in 1864. Cold permanganate oxidation of allene or propylene was reported by Berthelot in 1867; hydrolysis of malonitrile, again by Henry in 1886; Diels and Wolf in 1906 hydrated carbon suboxide, C_3O_2; Endemann obtained it in the oxidation of abietic acid a year later.

Production figures are available only for diethyl malonate: 1.2 million lb in 1951, but 0.59 million lb in 1959 (the last year for which figures were reported). In 1964, sales of diethyl malonate were 607,000 lb at a unit value of \$0.73/lb, up from 162,000 lb in 1953. The U.S. Tariff Commission has not reported sales figures since 1964 and lists the present manufacturers as Abbott Laboratories, Eli Lilly, and Kay-Fries Chemicals. The major markets for malonic acid, mainly as the diethyl ester, lie in the pharmaceutical industry.

B. Chemistry

In addition to the usual reactions of carboxylic acids, malonic acid undergoes several exceptional reactions as the result of activation of the methylene group by two carboxyl groups.

Malonic acid is the least stable of all the dicarboxylic acids, and a few degrees below the melting point it decomposes in the solid state to carbon dioxide and acetic acid. Since it is sufficiently stable at 95–100°C, it can be purified by vacuum sublimation. Decomposition in the melt is self-catalyzed and an apparent first-order reaction $k = 3.87 \times 10^{-4}$ (139.6°C), $\Delta H^* = 35.8$ kcal/mole, $\Delta S^* = 11.9$ (206). The intermolecular decomposition mechanism is apparently similar to that occurring in basic organic (207) and in carboxylic acid solvent media (208). A number of substances (aluminum and magnesium powders, zinc chloride, calcium oxide, potassium acid sulfate, and water) catalyze decomposition at 115–116°C (209).

Aqueous solutions of malonic acid decompose in an apparent first-order reaction. The rate of decomposition is 2.12×10^{-6} sec^{-1} at 80°C and 7.11×10^{-6} sec^{-1} at 90°C (210). The acid malonate ion is somewhat more stable, having the rate constants 0.24×10^{-6} sec^{-1} at 80°C and 0.71×10^{-6} sec^{-1} at 90°C. Aqueous solutions of neutral sodium malonate are essentially undecomposed at 125°C (211). The quantum yield in the photochemical decomposition of malonic acid solutions (ca. 0.5 molecule/quantum) (212) is higher than that for oxalic acid solutions (0.001–0.01 molecule/quantum) (97).

Dehydration of malonic acid with phosphorus pentoxide does not produce malonic anhydride, which is unknown, but carbon suboxide instead (213). However, anhydrides of alkyl substituted malonic acids are known (214).

Malonic acid undergoes oxidation to mesoxalic acid under mild conditions; under more stringent conditions, it may oxidize to oxalic and formic acids and carbon dioxide.

Aliphatic aldehydes react with malonic acid to form alkylidene malonic acids at low temperatures; under decarboxylating conditions, they form mixtures of α,β- and β,γ-unsaturated acids (215–217).

Bromine and chlorine react with malonic acid to produce the respective dibromo- and dichloromalonic acids. The monochloro acid may be prepared by reaction with sulfuryl chloride in ether (218).

Most reactions of the active methylene group are carried out with malonic diesters. One of the most important is the formation of either sodio- or disodiomalonic ester, from which a variety of acyclic or cyclic mono- and dicarboxylic acids can be prepared (219). An important reaction, the Michael condensation, is the addition to α,β-unsaturated compounds (220). Aldehydes react with malonic ester as well as with the acid; formaldehyde produces either the bismethylol derivative under mild conditions (221) or methylene-dimalonic ester under strongly basic conditions (222).

C. Synthesis

1. FROM SODIUM CHLOROACETATE AND SODIUM CYANIDE

Malonic acid is conveniently prepared from chloroacetic acid (223) as illustrated by the reactions

$$2ClCH_2COOH + Na_2CO_3 \rightarrow 2ClCH_2COONa + CO_2 + H_2O$$

$$ClCH_2COONa + NaCN \rightarrow NCCH_2COONa + NaCl$$

$$NCCH_2COONa + H_2O + NaOH \rightarrow NaOOCCH_2COONa + NH_3$$

The cyanation reaction (the second) is strongly exothermic, and the reaction temperature must be kept below about 95°C. If the reaction is allowed to proceed unchecked, hydrogen cyanide is evolved, and sodium chloroacetate is converted to glycolate. After hydrolysis of the nitrile group, two courses are open, depending on whether the acid or diethyl malonate is desired. The most important commercial process involves conversion of the dry mixture of sodium malonate and chloride to diethyl malonate

$$NaOOCCH_2COONa + H_2SO_4 \rightarrow HOOCCH_2COOH + Na_2SO_4$$

$$HOOCCH_2COOH + 2C_2H_5OH \rightarrow C_2H_5OOCCH_2COOC_2H_5 + 2H_2O$$

If malonic acid itself is desired, it may be separated from sodium chloride by precipitation as calcium malonate followed by regeneration to the acid. Alternatively, malonic acid may be made more conveniently by the hydrolysis of diethyl malonate (224).

At least three descriptions of pilot processes for diethyl malonate have been recorded, the major difference being in the way intermediate malonic acid is treated. In one process, malonic acid is not isolated from solution but is converted directly to the ester (225). In the other processes, malonic acid is removed from solution as a crude sodium malonate-sodium chloride solid (226) or as calcium malonate (227). Laboratory synthesis on a 1-lb scale is described in *Organic Syntheses* (228).

2. FROM POTASSIUM ACETATE AND CARBON DIOXIDE

Potassium acetate can be carboxylated in the presence of potassium carbonate, high surface silica, and a catalyst (229–231)

$$CH_3COOK + CO_2 \rightarrow [HOOCCH_2COOK]$$

$$[HOOCCH_2COOK] + K_2CO_3 \rightarrow KOOCCH_2COOK + KHCO_3$$

The reaction proceeds at 500 atm of carbon dioxide pressure and at 300°C with a catalyst such as iron, iron oxide, or cadmium oxide. Yields vary between 40 and 75%, the maximum being with iron powder.

3. MALONYL DICHLORIDE

Malonyl dichloride, a liquid distilling at 58–60°C at 28 mm, $n_D^{29} = 1.4572$, is a potential monomer having the advantage of high reactivity, so that polycondensation can be carried out at moderate temperatures rather than at the high temperatures required by the less reactive and thermally unstable acid or ester. It may be obtained by the action of thionyl chloride (232–234) or phosphorus pentachloride (235) on malonic acid, or by the action of hydrogen chloride on carbon suboxide (213).

4. MISCELLANEOUS SYNTHESES

Sodium malonate has been prepared by the action of carbon dioxide on sodium α-sodioacetate (236,237)

$$NaCH_2COONa + CO_2 \rightarrow NaOOCCH_2COONa$$

Reaction of ethyl α-bromoacetate with zinc and carbon dioxide at 80°C and 25 atm in tetrahydrofuran has given malonic half-ester in 16.5% yield (238). Under the same conditions, methyl α-bromoisobutyrate gives dimethylmalonic half ester in 50% yield. Although dioxan also promotes carboxylation, acyclic aliphatic ethers do not. The α-chloroesters are remarkably less reactive.

By slow electrochemical oxidation in alkaline solution, propiolactone may be converted to malonic acid (239).

Another potential route is the hydrolysis of malononitrile. Malononitrile has been prepared in a hot tube by the reaction of cyanogen chloride with acetonitrile; conversion was on the order of 16–18% (240,241). Yields as high as 65–69% have been reported when the reaction is carried out at 840°C for 6.5 sec in the presence of chlorine as a catalyst (242). Malononitrile may also be prepared by the reaction of acetonitrile with hydrogen cyanide

(243), cyanogen (244), or the crude reaction product of hydrogen cyanide and chlorine (245), and by the reaction of ammonia with chloroacetylene (246).

D. Physical Properties

Malonic acid is dimorphic, having a rhombic form (α-malonic acid) at temperatures above 80°C and a triclinic form (β-malonic acid) stable at room temperature (247). β-Malonic acid crystals are lath-shaped to acicular (65). The space group is $P_{\bar{1}}$. The three carbon atoms are in one plane with one carboxyl group only slightly out of plane (13°). However, the second carboxyl group is 90° out of the carbon atom plane. Malonic acid cannot be planar as a free molecule, because the oxygen atoms of the two carboxyl groups would then be at a distance of about 2.2 Å, less than the distance required by the van der Waals radii (248).

Infrared (249,250), near-infrared (161), and Raman (251) spectra for β-malonic acid have been described.

Malonic acid is extremely soluble in water. The solubility may be expressed from 0 to 50°C by the equation

$$\log \frac{1}{x} = 486.26 \frac{1}{T} - 0.978$$

where x is the mole fraction of solute and T is the absolute temperature (252). Solubilities in representative organic solvents are, in g/100 g of solution: methanol (19°C), 52.5; ethanol (19°C), 40.1; 1-propanol (19°C), 29.5; isobutyl alcohol (19°C), 21.2; ethyl ether, 6.25 at 0°C, 10.5 at 30°C; pyridine (26°C), 14.6; benzene (25°C), 0.0014 (253).

Malonic acid is a strong acid. Although weaker than oxalic, it is stronger than formic and about comparable to chloroacetic acid. Its aqueous solutions are corrosive, as would be expected. Corrosion of various metals by 0.1 N aqueous solutions is reported to occur in the following order: Al < Cu < Fe < Sn < Pb (254). Some of the more useful characteristics of malonic acid are given in Table 4.

E. Analytical Procedures

Malonic acid, as well as oxalic, may be assayed by standard titrimetric methods, either acidimetric or oxidimetric. Potentiometric (260,261) and conductometric (189) methods have been described.

Gas-liquid chromatography of malonic esters is useful for detecting homologous compounds (262). The quantitative aspects in the analysis, including conversion to the methyl esters, have been examined (190). Relative responses of a thermal conductivity detector have been determined for several

TABLE 4
Physical Properties of Malonic Acid

Property	Value
Molecular weight	104.06
Melting point, °C	134–136 (dec.)
Differential thermal analysis, endotherms, °C (175)	105, 160, 190, 290
Density at 16°C, g/cc (255)	1.619
Specific volume, ml/g (87)	$0.6250 + 0.900t \times 10^{-4} + 0.446t^2 \times 10^{-6}$
Refractive index of β form (65)	$N_x = 1.448,\ N_y = 1.488,\ N_z = 1.578$
Heat capacity, c_p, joules/g at 20°C (256)	1.15
Coefficient of expansion, ml/g deg at 25°C (87)	1.123
Coefficient of internal energy, cal/g atm (87)	-0.81
Heat of combustion, cal/g (87)	1983.5
Heat of formation, kcal/mole (87)	212.97
Dipole moment, in dioxane at 25°C (178)	2.57
Solubility, g/100 g H_2O (257)[a]	
0°C	108
20°C	152
40°C	212
60°C	292
80°C	455
Heat of solution in water, kJ/mole (180)	-18.8
Ionization constants in water at 25°C	
K_1 (258)[b]	1.4236×10^{-3}
K_2 (259)	2.014×10^{-6}

[a] Data recalculated from g/100 g of solution to g/100 g of water.

[b] There are discrepancies in reported values for K_1. See also Refs. 67 and 260.

dimethyl esters, including oxalate, malonate, and succinate esters (263). Molar responses of a radium ionization detector were investigated for homologous dimethyl esters from the oxalate through the sebacate (264). An alternation in molar responses occurred as a function of the number of carbon atoms up to the pimelate ester.

Paper (193,194) and thin-layer (191,265) chromatographic analyses have been reported. The chromatographic behavior of oxalic, malonic, succinic, and higher dibasic acids on ion-exchange resins has been described (266).

F. Toxicology

Malonic acid must be considered a hazardous chemical because it is a strong acid and because its calcium salt is insoluble; however, these properties are less extreme in malonic acid than in oxalic. Large doses of sodium malonate administered subcutaneously to rabbits inhibited succinoxidase activity of the heart muscle, and respiratory distress was caused by an accumulation of lactic acid (267). The LD_{50} in mole/kg is 0.0064 for rabbits if fed intravenously (268).

G. Polymer Applications

The activating effect of two carbonyl groups adjacent to the methylene group in malonic acid, its derivatives, and its polymers results in poor thermal and oxidative stabilities. Consequently, there is little application for malonic acid in the polymer field. The poly(ethylene malonate) prepared by Carothers and Arvin (269) was a liquid at room temperature. However, the linear polymer from dimethylmalonic acid and neopentyl glycol is a high-melting (261–268°C), microcrystalline polymer capable of being drawn into fibers (270). A prepolymer was prepared as a melt at 160°C, and the prepolymer was further polymerized in the solid state. The improved properties of this polymer as related to poly(ethylene malonate) result from the greater stability of the monomers and the increased rigidity of the chain shown in the structure

$$\left[\!\!\begin{array}{c} \quad\quad CH_3 \quad\quad\quad CH_3 \\ OCH_2\overset{|}{C}CH_2OC(O)\overset{|}{C}C(O) \\ \quad\quad CH_3 \quad\quad\quad CH_3 \end{array}\!\!\right]$$

IV. SUCCINIC ACID

A. Introduction

The dibasic acid most widely distributed in nature is probably succinic acid, found in many varieties of plant life. Agricola described it in 1546 as one of the products obtained by destructive distillation of amber (*sucinum* in the Latin). Lémery was first, in 1675, to recognize its nature as an acid, and in 1774 Wieglieb proved that it was present as such in amber and was not an artifact. Berzelius in 1815 and Liebig and Wöhler in 1830 established its elemental composition.

Its production during nitric acid oxidation of several fatty acids has been described: of stearic by Bromeis in 1840, of butyric by Dessaignes in 1850, and of sebacic or azelaic by Arppe in 1855. In 1847 Schmidt reported its formation during fermentation of sugar. In 1861 Kekulé synthesized it by reduction of maleic or fumaric acids with sodium amalgam, and Simpson accomplished the same result with the hydrolysis of succinonitrile. Its synthesis by the Kolbe electrolytic method was reported by Brown and Walker in 1891, and in 1899 Moritz and Wolffenstein reported its formation in the oxidation of acetic acid with potassium persulfate. Paal and Gerum hydrogenated aqueous sodium maleate with colloidal palladium in 1908, and a year later Vavon hydrogenated maleic acid in alcohol with platinum black.

Production of succinic acid was 60,000 lb in 1947, and 266,000 lb of succinic anhydride was produced in 1953 (the last years for which production was reported by the U.S. Tariff Commission). In the 1966 report, J. T. Baker Chemical Corporation and Allied Chemical Corporation were listed as producers of the acid, and Allied Chemical as the only producer of the anhydride. The major uses for succinic acid and its derivatives are in pharmaceuticals. The acid is available at $0.62/lb and the anhydride at $0.51.

B. Chemistry

Succinic acid has considerably higher thermal stability than its two lower homologs. According to Korshak and Rogozhin (93), the decomposition temperature is 290–310°C. However, during condensation of succinic acid with ethylene glycol, the acid loses three times as much carbon dioxide at 270°C as it does at 240°C, and the resultant polyester is low in terminal carboxyl groups. Heating succinic acid in an inert liquid medium for 5 hr at 255–260°C produces the dilactone of γ-ketopimelic acid plus a resin (271,272).

In contrast to its two lower homologs, succinic acid has an anhydride. This cyclic anhydride is formed when the acid is heated under reduced pressure or in the presence of a dehydrating agent such as acetyl chloride (273), phosphorus oxychloride (274), or acetic anhydride.

Esters of succinic acid are more stable to heat than is the acid and are prepared readily by the usual procedures. Half-esters are conveniently prepared by the action of an alcohol on the anhydride. The remaining free carboxyl group of the half-ester is an acid strong enough to cause slow disproportionation to the diester and diacid. Both the acid chloride of the half-ester (275) and succinyl chloride (276) are prepared through the use of thionyl chloride. A catalytic amount of pyridine is required for succinyl dichloride; otherwise the anhydride is formed.

A reaction generally unique to succinic esters is the Stobbe condensation, in which an aldehyde or a ketone forms an alkylidene succinate in the presence of an equimolar amount of base (277).

In addition to the usual amides, succinic acid forms the cyclic succinimide by reaction of ammonia with either the anhydride or the acid with application of heat. Succinimide can be halogenated to give an N-halosuccinimide, which is a mild halogenating agent as well as a disinfectant.

Succinic acid is considerably more stable to oxidizing agents than malonic and somewhat more stable than glutaric acid. For example, succinic acid is inert to neutral permanganate at room temperature and can be recrystallized from nitric acid. Its oxidation by hydrogen peroxide in the presence of catalysts has been studied extensively as a consequence of the biochemical importance of this oxidation.

C. Synthesis

1. FROM MALEIC ACID OR ANHYDRIDE

Because of its availability and low price, maleic anhydride is probably the most important source for the commercial synthesis of succinic acid or anhydride. Maleic anhydride is hydrogenated readily in either the melt or vapor state or in solution over a nickel catalyst (278–281). Other catalysts include palladium (282) and copper molybdate (283).

Both maleic acid and its sodium salt also have been used (284–288). Fel'dman and Troyanova (289) studied the reaction with Raney nickel in the temperature range of 40–140°C and pressure range of 15–30 atm. Rhodium, either in colloidal form (290) or deposited on charcoal (291), gives good results. Either palladium or rhodium on charcoal causes hydrogenation to proceed in 94–97% yields at lower temperatures (15–60°C) and pressures (7 atm) than are reportedly required with nickel (291).

Various electrolytic reduction methods have been described (292–296); mercury cathodes produced succinic acid in the best yields (297).

Reductions with various active metals, usually zinc, have also been reported (298–302).

2. CYANATION

Several routes to succinic acid exist through cyanation. Succinonitrile can be prepared by reaction of hydrogen cyanide with acetylene (303) or with acrylonitrile (304) and by the reaction of sodium cyanide with ethylene dichloride (305). The reaction of sodium cyanide with propiolactone (306) or with sodium acrylate (307) produces the half-nitrile. The nitrile group can be hydrolyzed in good yield under either strongly acidic or basic conditions.

3. CARBONYLATION

Reppe has described the reaction of carbon monoxide with sodium acrylate at 150°C and 150–200 atm or with acetylene at 100°C and 25–30 atm in the presence of $K_2[Ni(CN)_4]$ and in aqueous solution (308), and with ethylene glycol at 250°C and 200 atm with nickel carbonyl as catalyst in the presence of potassium carbonate (40); reported yields were well below 50% because of incomplete reaction or because of side reactions. These methods require an additional step to convert the succinate salt to the acid. On the other hand, succinic acid may be prepared directly in 80% yield by the reaction of carbon monoxide with acetylene in an organic solvent, such as acetone or cyclopentanone, in the presence of cobalt or other metal carbonyl (309). Dialkyl succinates are formed when the solvent is an alcohol. In methanol and with cobalt carbonyl as catalyst, dimethyl succinate is formed along with methyl acrylate and other by-products (310). In ethanol and with cobalt carbonyl hydride or a complex carbonylate salt as catalyst, diethyl succinate is formed (311).

Oxidation of an ethylene–carbon monoxide polymer with nitric acid also produces succinic acid (312).

Succinic acid has been produced directly by the reaction of carbon monoxide with acrylic acid in oleum at 40°C and 1200 psi for 12 hr (313).

In 1969 the use of palladium salts was reported for the carbonylation of acetylene at 75°C and atmospheric pressure (314). In n-butyl alcohol as solvent, various amounts of butyl acrylate, propionate, fumarate, and maleate were formed in addition to dibutyl succinate, depending on experimental conditions.

4. OXIDATION METHODS

Many methods and starting materials are available. For example, 4-carbon compounds of different degrees of oxidation have been used in the following methods: nitric acid (315–318) or electrolytic (319,320) oxidation of tetrahydrofuran; ammonium sulfide oxidation of thiophene, tetrahydrothiophene, or tetrahydrofuran (321); dinitrogen tetroxide (Maurer oxidation) (41) or nitric acid (50) oxidation of 1,4-butanediol; and nitric acid (322,323), catalytic (323) or noncatalytic (325) air, or electrolytic (326) oxidation of butyrolactone.

Numerous compounds having more than four carbon atoms have also been used; and because of its relative stability to oxidation, succinic acid is a frequent product of degradative oxidation of long-chain aliphatic compounds. In nitric acid oxidation of the mixture of cyclopentanol and cyclopentanone resulting from air oxidation of cyclopentane, large amounts of succinic as well as glutaric acids are formed (327). Cobalt-catalyzed oxidation of sodium α-hydroxyglutarate or α,α'-dihydroxyadipate with hypochlorite has been described (328), as have electrolytic oxidation of tetrahydrofurfuryl alcohol (329), catalytic air oxidation of levulinic acid (330), nitric acid oxidation of dichlorocyclohexane (331), persulfuric acid oxidation of furfural (332), nitric acid oxidation of acrolein dimer (333), and nitric acid oxidation of 6-aminocaproic acid, caprolactam, or its oligomers (334). Succinic acid is a by-product in the manufacture of adipic acid from cyclohexane (335–337). A major product in the alkaline permanganate oxidation of the unsaturated, short-chain nitriles formed by pyrolysis of stearonitrile is β-cyanopropionic acid (338).

Succinic acid is one of many products formed from coal by permanganate (339), alkali (340), or nitric acid (341) oxidation.

Mixtures of dibasic acids with succinic predominating are formed by air oxidation of paraffin waxes (342–344), nitric acid oxidation of asphaltic oils or paraffin wax treated with sulfur (345), or oxidation of paraffins in two stages—with air initially, then with nitric acid (54,346). Succinic anhydride can be separated from a crude mixture of dibasic acids by codistillation with o-dichlorobenzene (347). The mixture of acids has the typical composition: succinic, 33%; glutaric, 19%; adipic, 15%; pimelic, 9%; suberic and higher, 11%.

5. MISCELLANEOUS

Succinic esters are successfully prepared by the application of the Kolbe electrolysis reaction to malonic half-esters (348–351). Although impractical for succinic acid itself, this reaction may be useful for the synthesis of dialkyl- and tetraalkyl-substituted succinates (352).

Kharasch (353,354) prepared substituted succinic acids by reacting acetyl peroxide on appropriately substituted acetic acids, as well as succinic acid itself from acetic acid. Thus propionic acid produced α,α'-dimethylsuccinic acid; isobutyric acid, tetramethylsuccinic acid; and chloroacetic acid, meso-dichlorosuccinic acid. α,α'-Dimethylsuccinic ester has been produced by the alkylation of t-butyl acetate with t-butyl α-bromoisobutyrate by lithium amide in liquid ammonia (355). Salmon-Legagneur (356) reviewed syntheses for α,α-diphenylsuccinic acid, as well as other similarly substituted dibasic acids.

Succinic acid, with an mp of 187.8°C, has been obtained in high purity by zone refining (357).

The manufacture and properties of succinic acid have been reviewed (358).

D. Physical Properties

Succinic acid is dimorphic (359). α-Succinic acid is the first crystal modification formed upon cooling of its melt and is stable only above the transition temperature of about 137°C; after several weeks at room temperature, it changes to the β form. α-Succinic acid is triclinic (pseudomonoclinic because of twinning) with space group $P\bar{1}$ (360). The β form, stable at room temperature, is monoclinic and prismatic with space group $P2_1/a(C_{2h}^5)$ (361). In β-succinic acid, the distance between the central carbon atoms is normal at 1.533 Å, but the distance between a central carbon and its neighboring carboxyl carbon has contracted to 1.485 Å. The four carbons are only slightly nonplanar, and the angle between the carbon plane and the plane of the carboxyl group is 11°27'.

Infrared spectra for both crystalline and molten succinic acid have been described (362). The band at 1205 cm^{-1} is ascribed to the vibration of the —CH$_2$COOH group (363). Raman (251) and ultraviolet (364) spectra have also been reported. The infrared spectra of the anhydride (365,366) and of the acid or neutral potassium salts (367) are also available.

Succinic acid sublimes at reduced pressure; at 128°C, the vapor pressure is about 0.06 mm Hg and the enthalpy of sublimation is 28.1 cal/mole °K (368).

Succinic acid is moderately soluble in water. The solubility from 0 to 75°C may be expressed approximately by the equation

$$\log \frac{1}{x} = 1691.2 \, \frac{1}{T} - 3.778$$

where x is the mole fraction of solute and T is the absolute temperature (252). Solubilities in representative organic solvents are, in g/100 g of solvent: methanol (21.5°C), 19.4; ethanol (21.5°C), 9.49; and ether (15°C), 1.27 (369).

The corrosion of stainless steel by molten (370) and aqueous (371) succinic acid has been investigated.

Some of the more useful characteristics of succinic acid are given in Table 5.

TABLE 5
Physical Properties of Succinic Acid

Property	Value
Molecular weight	118.09
Melting point, °C (87)	187.6–187.9
Sublimation point, °C (87)	130–140
Differential thermal analysis, endotherms °C (175)	195, 255
Vapor pressure, log p (mm Hg), 99–128°C (368)	$14.048 - 6132/T$
Density at 25°C, g/cc (372)	1.572
Specific volume, ml/g (87)	$0.6349 + 1.097t \times 10^{-4} + 0.398t^2 \times 10^{-6}$
Refractive index of β form (65)	$N_x = 1.448, N_y = 1.531, N_z = 1.610$
Heat capacity from 0 to 160°C, joules/g (256)	$1.039 + 0.00641t$
Coefficient of expansion, ml/g deg at 25°C (87)	1.296
Coefficient of internal energy, cal/g atm (87)	−0.94
Heat of combustion, cal/g (87)	3019.6
Heat of formation, kcal/mole (87)	224.87
Dipole moment, in dioxane at 25°C (178)	2.20
Solubility, g/100 g H_2O (373)	
0°C	2.75
25°C	8.35
50°C	23.83
75°C	60.37
Ionization constants in water at 25°C	
K_1 (374)	6.21×10^{-5}
K_2 (375)	2.31×10^{-6}

E. Analytical Procedures

Potentiometric (261,376) and conductometric (189) titration methods are the procedures for the analysis of succinic acid that have been described. Thin-layer (377), paper (194), and gas-liquid (190,264) chromatographic procedures are available. Analytical separation of mixtures of succinic, glutaric, and adipic acids may be carried out on silicic acid columns containing water as a stationary phase (378). Either extrusion of the column packing or gradient elution as the isolation method had an accuracy of $\pm 1\%$. Other column chromatographic procedures have been described (379,380). Succinic acid may be isolated by chromatography as the p-phenylazoanil, formed from the reaction of p-aminoazobenzene with the anhydride (381). Maleic acid in succinic acid may be analyzed by polarography (382).

F. Toxicology

In view of its greatly reduced acid strength in comparison with its lower homologs, and in view of its importance in biochemical processes, a low order of toxicity would be expected for succinic acid. Succinic acid injected intravenously is quickly eliminated by animals (383). Large doses of sodium succinate administered either intravenously or orally to cats produced vomiting and diarrhea, but these effects were nonspecific, and no signs of systemic toxicity were noted (384). In fact, succinic acid can be used as an acidulant for foods (385). Other pharmacological effects and pharmaceutical uses for succinic acid have been reviewed (385).

G. Polymer Applications

1. Polyesters

Early investigations (by Lourenço in 1863, Davidoff in 1886, and Vörlander in 1894) on reactions of succinic acid with ethylene glycol or derivatives of these showed the products to be oligomers or low polymers. Menshutkin made a kinetic study of the reaction in 1881. Carothers and his co-workers (386,387) further characterized poly(ethylene succinates) of various molecular weights. Monomeric or dimeric cyclic esters from succinic acid and a host of alkylene glycols have been prepared by distillation of their polyesters in the presence of a catalyst (388), and their conformations have been investigated (389). The effects of substituents on the melting points of linear poly(alkylene succinates) have been reported (390). X-Ray studies have established the crystalline nature and molecular configurations of oriented fibers for several poly(alkylene succinates) (391). The chain molecules are arranged parallel to the fiber axis, are essentially planar zigzag in configuration, and have typical paraffinic packing in the plane perpendicular to the chain. Poly-(ethylene succinate) is an exception to the general pattern of polyesters and has a shorter fiber period than expected, probably because of a coiled or helical form. Infrared spectra for aliphatic polyesters reveal that the approximately planar trans configuration of the acid portion characteristic of the crystalline polyester does not exist for the amorphous state (392).

Linear, saturated polyesters, or those containing a small amount of unsaturation, may be crosslinked with peroxide to make rubberlike products. Immediately preceding and during World War II, such products were investigated as potential synthetic rubbers. Poly(propylene succinate) rubbers, although less stable hydrolytically, are extremely resistant to hydrocarbon oils and gasoline and more stable to oxidative degradation compared to poly(propylene sebacate) rubbers (393). These materials were prepared on a pilot scale, but commercial production was not achieved, apparently because of cost and lack of hydrolytic stability.

A low-molecular-weight polymer of succinic acid with diethylene glycol has found application in gas-liquid chromatography as the familiar "DEGS" stationary phase.

2. POLYAMIDES

Because of its ability to form cyclic compounds, succinic acid is seldom selected for synthesis of polyamides. Polymerization is terminated by the formation of cyclic imides (394,395). Cyclization is inhibited by the use of a secondary amine such as N,N'-dimethyl-1,4-butanediamine. Several co-polyamides prepared with succinic acid have been reported (75). N,N'-Ethylenedisuccinimide and N,N'-hexamethylenedisuccinimide react with alkylene diamines in ring-opening polyaddition to form high melting and regularly alternating polyamides having specific viscosities in the range 0.21 to 0.50 dl/g (396).

V. GLUTARIC ACID

A. Introduction

Although the isomeric pyrotartaric acid was known as early as 1807, glutaric (originally called "normal pyrotartaric") acid was not discovered until 1872, when Dittmar converted glutamic acid to α-hydroxyglutaric acid and reduced the latter with hydrogen iodide. In 1876 Markownikow and Lermantow, as well as Reboul, reported its synthesis by hydrolysis of glutaronitrile, derived from trimethylene dibromide. Glutaric acid was formed also when, in 1878, Wislicenus and Limpach carried out acid cleavage of α-acetoglutaric acid. Over the next 30 years, a number of investigators reported its synthesis by decarboxylation of several compounds—α-carboxyglutaric acid and particularly methylene bis(malonic acid). In 1886

Carette found glutaric acid to be produced by nitric acid oxidation of sebacic, stearic, and oleic acids, and Nordlinger made it by oxidation of myristic acid. Hentzschel and Wislicenus oxidized cyclopentanone with nitric acid in 1893.

Glutaric acid exists naturally to a limited extent and has been found reportedly in sheep's wool washings and in the juice of unripened sugar beets. Glutaric acid has no major markets, and consequently no statistics on its production have been published, although it is available as a by-product in adipic acid manufacture (397).

B. Chemistry

Glutaric acid has slightly lower thermal stability than succinic acid, following the alternating pattern observed for several properties of dibasic acids in the sequence of odd-even carbon atoms in the chain. Its decomposition temperature ranges from 280 to 290°C (22).

Like succinic acid, glutaric acid forms a cyclic anhydride when heated with a dehydrating agent, such as acetyl chloride, phosphorus pentachloride, or thionyl chloride, or when heated alone under reflux and at reduced pressure. Under milder conditions than those typical of anhydride formation, glutaric acid is converted to glutaryl chloride with thionyl chloride (398) or with phosphorus pentachloride (399). Like succinic acid, glutaric acid forms a cyclic imide as well as anhydride. Snethlage (26) has shown that glutaric acid is oxidized by chromic acid about 13 times more rapidly than succinic acid but only about one-seventh as rapidly as adipic acid. Similarly, glutaric and adipic acids are more rapidly oxidized by hydrogen peroxide in the presence of a cupric salt, and succinic acid is a product of the oxidation (29).

C. Synthesis

Organic syntheses procedures are available for the preparation of glutaric acid from trimethylene cyanide (400), from methylene bis(malonic acid) (401), from γ-butyrolactone (402), and from dihydropyran (403).

As a readily available petrochemical, cyclopentane would be preferred as a starting material compared with those given previously. For example, cyclopentane can be oxidized with air at 140°C and 400 psi to a mixture of cyclopentanone and cyclopentanol, which mixture is then oxidized with nitric acid at 65–75°C (327). Concentration of nitric acid and temperature are important factors, since glutaric acid is more rapidly oxidized than succinic (404). In the range of 25–35% nitric acid and at 65–75°C, succinic and glutaric acids are formed in the ratio 15:23, but at 75–80°C the ratio is

21:29 (327). Relatively pure cyclopentanol (405), cyclopentanone (406), or cyclopentene (407) have been used for the nitric acid oxidation.

Boedtker (408) obtained glutaric acid in more than 70% yield from cyclopentanone by using 13% nitric acid instead of more concentrated acid. Although succinic acid is insoluble and glutaric soluble in benzene, attempts to separate them by this means were unsuccessful because of a cosolvent effect. Instead, separation was accomplished by means of the barium salts; barium glutarate is soluble, barium succinate insoluble in water. Cyclopentanone can also be oxidized by air (409) or by nitrogen dioxide (410).

Vanadium pentoxide may be used for a catalyst in the nitric acid oxidation of cyclopentanol (405) or cyclopentanone (406); but care is required in regulating reaction conditions, since the catalyst also oxidizes glutaric acid to lower dicarboxylic acids.

Other methods include nitric acid oxidation of 5-hydroxypentanal (411), of 1,5-pentanediol or tetrahydropyran (412), and of 2-cyanocyclopentanone (413); alkali treatment of tetrahydrofurfuryl alcohol (414); dinitrogen tetroxide oxidation of 1,5-pentanediol (41); oxidation of α-hydroxyadipaldehyde with hydrogen peroxide (415); air oxidation of glutaraldehyde (416); ozonolysis of cyclopentene in ethyl acetate (417); and carbonylation of butyrolactone (418).

D. Physical Properties

The transition temperature to convert α- to β-glutaric acid is about 74–75°C; the α form is not stable at room temperature (419). β-Succinic acid is monoclinic prismatic with space group C_{2h}^6 (58). Glutaric acid is the first truly representative member of the well-defined "odd" series of normal aliphatic dicarboxylic acids (58,420). The bond lengths are more closely normal and do not have the slight alternating variations detected in the even series. The carboxyl group is inclined about 32° to the plane of the central carbon atoms, an inclination considerably greater than that in the even series. The end-to-end hydrogen bridging between carboxyl groups of adjoining molecules is about the same as in the even series, but the lateral connections are much weaker at 3.59 Å minimum distance. These differences in structure are no doubt responsible for the low melting point and the soft texture of glutaric acid crystals.

Glutaric acid is readily soluble in water and in polar organic solvents. Solubilities in representative solvents are: 8.06 g in 100 ml of ethyl acetate solution at 20°C (64); 27.62 g in 100 ml of acetone solution at 20°C (64); and 0.016 g in 100 g of benzene at 25°C (62).

Infrared spectra for glutaric acid (250,421–423) and its salts (424) have been described, as well as its far-infrared spectra (161).

Corrosion rates for the molten acid at 225°C in stainless steel have been reported (370).

Some properties of glutaric acid are given in Table 6.

TABLE 6
Physical Properties of Glutaric Acid

Property	Value
Molecular weight	132.11
Melting point, °C	98–99
Differential thermal analysis, endotherms, °C (175)	75, 115, 255, 300
Density at 25°C, g/cc (255)	1.424
Specific volume, ml/g, 0–50°C (87)	$0.7137 + 1.965t \times 10^{-4} + 0.194t^2 \times 10^{-6}$
Refractive index of β form (65)	$N_x = 1.451, N_y = 1.502, N_z = 1.585$
Heat capacity, c_p, joules/g at 20°C (256)	1.25
Coefficient of expansion, ml/g deg at 25°C (87)	2.062
Coefficient of internal energy, cal/g atm at 25°C (87)	−1.49
Heat of combustion, cal/g (87)	3891.1
Heat of formation, kcal/mole at 25°C (87)	229.44
Dipole moment, in dioxane at 25°C (178)	2.64
Solubility, g/100 g H_2O (425)	
3.4°C	46.7
23.9°C	130
45.8°C	297
Heat of solution in water, kJ/mole (180)	−22.6
Ionization constants in water at 25°C (426)	
K_1	4.58×10^{-5}
K_2	3.89×10^{-6}

E. Analytical Procedures

Mixtures of adipic, glutaric, and succinic acids (378) or of sebacic, suberic, adipic, and glutaric (427) have been separated by column chromatography on silicic acid with water as a stationary phase. Other methods of analysis are described in the sections dealing with methods of analysis for succinic and adipic acids.

F. Toxicology

Glutaric acid is much more toxic than either succinic or adipic acid. Sodium glutarate is severely nephrotoxic to rabbits when administered subcutaneously and leads to marked retention of nitrogenous waste products and to renal failure (428). The nephrotoxic effect in rabbits was confirmed by others (429).

G. Polymer Applications

Little has been done with glutaric acid polymers, since they show no advantage over corresponding adipic acid polymers and since glutaric is less readily available on a commercial scale than adipic. Furthermore, glutaric acid has the same property of forming cyclic imides as does succinic acid. Summaries of glutarate (74) and glutaramide (75) polymers have been published, and X-ray investigations on several polyesters have been reported (430–432). Polymers of glutaric acid generally have depressed melting points compared with those of succinic and adipic acids. Mixed copolyamides of glutaric acid with adipic, azelaic, or sebacic acids have been described (433), and the interfacial polymerization of glutaryl chloride with aromatic diamines has been studied. (434–436).

VI. ADIPIC ACID

A. Introduction

Adipic acid was first characterized by Laurent, who reported in 1837 its formation by degradative oxidation of oleic acid with nitric acid. Many other fatty materials have been used; for example, tallow by Malagati in 1846 and castor oil by Dieterle and Hell in 1884. Boedtker, in 1862, oxidized sebacic acid, in turn derived from castor oil. The fatty origin of the name of adipic acid is indicated by its derivation—from the Latin *adeps* (fat) and *adipis* (of fat). Other early syntheses are: electrolysis of succinic half-esters as described by Brown and Walker in 1891; saponification of the dinitrile from 1,4-dibromo- or diiodo-butane, recorded by Hamonet in 1901; permanganate oxidation of cyclohexanol, reported by Rosenlew in 1906, and Mannich and Hâncu in 1908; nitric acid oxidation of cyclohexane by Markownikow in 1898 and again by Nametkin in 1909; nitric acid oxidation of cyclohexanol by Bouveault and Locquin in 1909; ozonization of cyclohexene, by Harries et al. between 1906 and 1915; and oxidation of cyclohexane with oxygen in the presence of osmium by Willstätter and Sonnenfeld in 1913.

Commercial production of adipic acid was initiated by the du Pont Company in 1937, 100 years after Laurent's original description. In 1968 production of this acid in the United States alone exceeded 1.16 billion lb (437), and it has become the most important of the aliphatic dibasic acids. Because of large volume production and because of its petrochemical origin, it is available for $0.18–0.25/lb, indeed a low price for a chemical having a multistep synthesis. Capacity in 1968 totaled 1,270 million lb distributed among several companies (capacity in million pounds annually) (438): Allied (20), Celanese (90), du Pont (570), El Paso-Beaunit (80), Monsanto (490), and Rohm and Haas (20).

The most important outlet (about 90%) for adipic acid is in the production of nylon 6/6, but it is used widely in making plasticizers and urethanes, and to a lesser extent in lubricants and as an acidulant for foods and beverages.

An economic analysis for the production of adipic acid is available on a private subscription basis from the Process Economics Program of the Stanford Research Institute, Menlo Park, California. Adipic acid production, physical and chemical properties, and uses were reviewed in 1963 (439).

B. Chemistry

According to Korshak and Rogozhin (22), the decomposition temperature of adipic acid is about 310°C. Consequently, adipic acid is more stable than any of its lower homologs and the next higher, pimelic acid, but not as stable as the homologs higher than pimelic. It decomposes about 0.4% at 250°C and 4% at 280°C after 9 hr of heating. Decomposition is first order with rate constants of 0.0276×10^{-2} hr^{-1} at 250°C and 0.0848×10^{-2} at 260°C and with an activation energy of 57,000 cal/mole (74). In the presence of ethylene glycol, however, evolution of carbon dioxide starts at 150°C, and decomposition amounts to 15% at 250°C and 30% at 280°C after 9 hr. Furthermore, in the presence of such salts as calcium or barium adipate and copper sulfate, decarboxylation also accelerates.

In contrast to the behavior of its lower homologs, which produce anhydrides or monocarboxylic acids upon thermal treatment, adipic acid condenses intramolecularly with decarboxylation to form cyclopentanone. This reaction may be carried out without a catalyst, with use of acetic anhydride, with calcium adipate, and with various catalysts. An *Organic Syntheses* preparation describes the use of barium hydroxide as the catalyst (440). Slow distillation of adipic acid from a quartz flask at 300°C in the absence of a catalyst reportedly gives a yield of greater than 98% (441).

Adipic acid has two anhydrides (442). A polymeric anhydride is formed upon treatment of adipic acid with acetic anhydride followed by removal of the volatile constituents. When heated above 200°C under reduced pressure, the polymeric form breaks down to yield a monomeric anhydride. This liquid freezes at about 20°C and slowly reverts to the polymeric form upon standing at room temperature.

The preparation of adipoyl chloride and diethyl 2,2'-dibromoadipate as intermediates for synthesis of muconic acid has been described (443), as has the preparation of ethyl adipate (444). Ethyl adipate undergoes the Dieckmann condensation to 2-carbethoxycyclopentanone with 1 g-atom of sodium for each mole of diester (32,445). With 4 g-atoms of sodium for each mole, both methyl adipate and methyl glutarate undergo cyclic acyloin condensation to 2-hydroxycyclohexanone (adipoin) and 2-hydroxycyclopentanone (glutaroin), respectively (446). Pyrolysis of diethyl adipate at 470°C over

Pyrex helices produces ethyl hydrogen adipate in 28% conversion or 78% yield (447). The preparation of vinyl alkyl (448,449) and diallyl (450,451) adipates has been described.

An important reaction of adipic acid is the formation of crystalline "nylon salts" from various diamines—a convenient method for determining the exact stoichiometry of the reactants required for preparing polyamides. The nylon salt solution is best purified by treatment with charcoal that has been preconditioned with adipic acid and diamine (452). Hexamethylene-diammonium adipate, mp 183°C, has a pH of 7.62 at the inflection point, is very soluble in water (49% at 25°C), but has limited solubility in ethanol (453). There are two molecules of $C_{12}H_{30}N_2O_3$ in a unit cell and the carboxyl group is twisted out of the plane of adipate carbon atoms by 69.8° (454).

Adipic acid is quite stable to oxidation and may be recrystallized from hot concentrated nitric acid. However, adipic acid is oxidized by chromic acid at 25°C at a rate that is seven times faster than that for glutaric (26). Succinic acid is a major product in the oxidation of adipic acid by hydrogen peroxide in the presence of a cupric salt and at 60°C (29).

C. Synthesis

A two-step process involving cyclohexane is the principal commercial means of producing adipic acid. In the first step, cyclohexane is air oxidized to a mixture of cyclohexanol and cyclohexanone; the second step entails nitric acid oxidation of the ketone-alcohol mixture. Most information on these processes is to be found in a voluminous patent literature. However, a discussion on the processes used by various companies has appeared (455), and Russian investigators have published reviews in book form on the liquid-phase oxidation of hydrocarbons (456), the oxidation of cyclohexane (457), and the production of cyclohexanone and adipic acid (458). Originally, cyclohexanol was derived by hydrogenation of phenol, but air oxidation of cyclohexane has proved to be more economical—so much so, that cyclo-hexanol may turn out to be a feasible intermediate for phenol. Methods of production and oxidation processes for cyclohexane were reviewed in 1969 (459).

1. AIR OXIDATION OF CYCLOHEXANE TO CYCLOHEXANOL/CYCLOHEXANONE

Air oxidation in liquid phase, at low conversion (5–15%) and in the presence of a catalyst, produces the mixture of cyclohexanol and cyclo-hexanone; at higher conversions, further oxidation produces adipic acid and lower homologs which cause processing difficulties as well as lower efficiencies. Some reactions involved in this radical process are illustrated by the following equations:

$$RCH_2 + R' \cdot \rightarrow RCH \cdot + R'H$$

$$RCH \cdot + O_2 \rightarrow RCHOO \cdot$$

$$2RCHOO \cdot \rightarrow RC{=}O + RCHOH + O_2$$

$$RCHOO \cdot + RCH_2 \rightarrow RCHOOH + RCH \cdot$$

$$RCHOOH \rightarrow RCHO \cdot + \cdot OH$$

$$RCHO \cdot + RCH_2 \rightarrow RCHOH + RCH \cdot$$

$$RCHOOH \rightarrow RC{=}O + H_2O$$

$$[R = {-}(CH_2)_5{-}]$$

The catalyst, usually cobalt or manganese naphthenate, is used along with an initiator (e.g., cyclohexanone) and serves to shorten the induction period and induce hydroperoxide decomposition (460,461). Ota and Tezuka (462) report the following naphthenates in order of decreasing effectiveness: Co, Cr > Mn > Fe, Al > Pb > no catalyst, Hg, U, Zn, Ni ≫ Cu. Heterogeneous catalysts have also been used (463,464). Or the reaction may be initiated in the absence of a catalyst by chlorine (465), hydrogen bromide and ethyl bromide (466), radiation (467), or radiation in the presence of nitric oxide and chlorine (468).

In the absence of a catalyst, cyclohexyl peroxide is formed and can then be hydrogenated to cyclohexanol (469). Addition of a sequestering agent may be necessary to prevent decomposition by trace metal impurities (470). Again in the absence of catalyst, cyclohexanol may be oxidized to cyclohexanone and hydrogen peroxide (471).

Reaction conditions require careful control. Reaction temperatures are usually in the range of 140 to 160°C, although at 5–6% conversion, a temperature of 170°C may be more efficient (472). Water, which is formed during the reaction and tends to remove desirable intermediates in a separate phase, is optimum at a concentration of 6% or less (473) and may be removed by azeotropic distillation (474–476). Cyclohexanol in recycled cyclohexane has an optimum concentration of about 9% (473).

Typically, the oxidation products have the following composition: cyclohexanol (29.6%), cyclohexanone (28.4%), cyclohexenyl cyclohexyl ether (0.6%), cyclohexyl esters (3.2%), and compounds not volatile with steam (22.2%) (477,478). The nonvolatile material includes adipaldehydic acid, 6-hydroxyhexanoic acid, and esters of these along with adipates and esters of lower dibasic acids. Other by-products include valeric and caproic acids (479).

Duynstee and Hennekens (480), on the basis of experiments with labeled cyclohexane, found that the by-products were not necessarily formed from cyclohexanol or cyclohexanone but that they could be formed directly from cyclohexane. The percentages for each route for several by-products were:

Acid	From cyclohexane	From cyclohexanone
Valeric	58	42
Caproic	43	57
6-Hydroxyhexanoic	49	51
Glutaric	14	86
Adipic	5	95

A newly identified by-product was 6-cyclohexyloxyhexanoic acid, which is formed from the addition product of cyclohexyl peroxide and cyclohexanone.

Some useful process details are given in a du Pont patent (481). Variations in process details appear in numerous other patents (482–508).

2. AIR OXIDATION OF CYCLOHEXANE IN THE PRESENCE OF BORIC ACID

Ideally, the air-oxidation process would stop with production of cyclohexanol; but in fact, further oxidation and degradation occur. Successful limitation of an analogous oxidation was achieved some years ago in the production of alcohols from straight-chain hydrocarbons. The oxidation was carried out in the presence of boric anhydride or acid; the borate esters of the alcohols produced were not so subject to further oxidation as the alcohol (509–514). Later, borate esters of cyclohexanol formed *in situ* were successfully used to limit and direct the oxidation of cyclohexane (515,516). Plants utilizing this process are said to be under construction in several countries, including the United States.

In one patented process, oxidation was carried out at 160°C and 9.4 atm in the presence of 50 ppm of cobalt naphthenate and 5% metaboric acid (517,518). Boric acid was then removed by filtration, cyclohexane was removed by distillation, and the residue was oxidized with nitric acid. The overall yield was 76%, compared with 71% when boric acid was absent. In another process, air oxidation was performed with an oxygen-nitrogen mixture containing 8% oxygen at 165°C and 125 psig in the presence of 8% metaboric acid and 0.1% benzene; the yields were 92.1 and 2.7% of cyclohexanol cyclohexanone, respectively (519). The partial pressure of water in the exit gases should not exceed the limit expressed in the equation $P = 0.0175T - 1.85$, where P is the partial pressure in psi (absolute) and T is the reaction temperature in degrees centigrade (520). Processing details and particularly those pertaining to the form and recovery of boric acid are described in several patents (491,521–533). Cyclohexyl borate itself is the subject of a composition of matter patent (534).

TABLE 7

Adipic Acid from Cyclohexanol/Cyclohexanone (481)

Feed	Description	Yield of acid, lb/lb of feed				Nitric acid consumption, lb/lb adipic acid produced
		Adipic	Glutaric	Succinic		
Whole crude KA[a]	Stripped of cyclohexane; contains water to keep acids in solution and 34.5% of organics not volatile with steam	0.921	0.119	0.071		1.28
KA[b]	Steam-distilled oxidation products; contains 65.5% of organics in whole crude KA and 10.7% water	1.205	0.049	0.019		1.02
Wet KA	KA as above but also contains steam-volatile organics from aqueous wash layer	0.901	0.076	0.032		1.08
KALL[c]	Aqueous wash layer containing dibasic acids	0.553	0.220	0.138		1.30
KALL extract	Organics extracted from KALL with chloroform	0.508	0.044	0.036		0.59

[a] KA = Ketone-alcohol; i.e., cyclohexanone/cyclohexanol.
[b] Composition: cyclohexanol, 29.2%; cyclohexanone, 26.5%; cyclohexyl formate, 3.49%; high-boiling esters as cyclohexyl valerate, 1.93%; 1,2-cyclohexanediol, 0.53%; monobasic acids as valeric acid, 3.29%.
[c] KALL = Ketone-alcohol lower layer; i.e., the aqueous extract containing dibasic acids and some steam-volatile organics.

44

3. NITRIC ACID OXIDATION OF CYCLOHEXANOL/CYCLOHEXANONE

An *Organic Syntheses* procedure describes preparation of adipic acid by nitric acid oxidation of cyclohexanol in the presence of ammonium vanadate as catalyst (535). Nitric acid, 50% strength, is used in a molar ratio of 3.3:1 and at a temperature of 55–60°C.

In the technical production of adipic acid, a mixture of cyclohexanol and cyclohexanone is used. Goldbeck and Johnson (481) discuss various treatments given technical cyclohexanol/cyclohexanone and their effects on adipic acid yield (Table 7).

Reports by Lindsay (536), Godt and Quinn (537), and Van Asselt and Van Krevelen (538) indicate that the main reactions in the oxidation of cyclohexanol by nitric acid are:

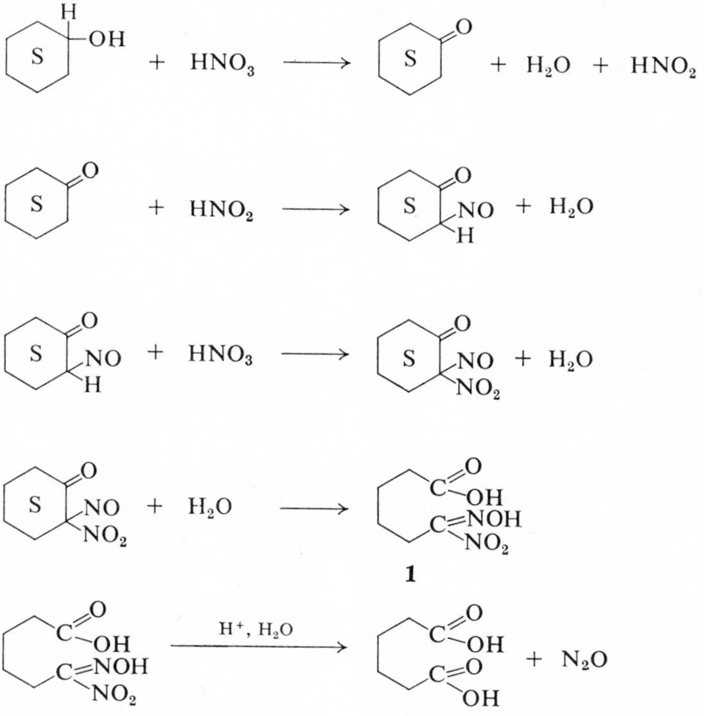

Gaseous products of the reaction include NO_2, NO, N_2O, N_2, and CO_2; by-products include glutaric, succinic, and adipic acids and cyclohexyl nitrite. When the reaction is carried out under milder conditions than used for adipic acid synthesis, either monoadiponitrolic acid (**1**) or a dimeric compound having structure **2** (539) can be isolated. Isolation of these

2

compounds suggests that they are intermediates. Treatment with nitric acid, under conditions that do not oxidatively degrade either adipic or glutaric acids, produces adipic and glutaric acids in the ratio 1:0.02 from **1** and adipic, glutaric, and succinic acids in the ratio 1:3:10 from **2** (537). Accordingly, **1** and **2** are probably the main sources for the by-product homologous acids, although there is no evidence for their existence under the conditions used for synthesis of adipic acid.

Catalysts make the nitric acid route a highly efficient process. Ammonium vanadate was mentioned in many of the early patents (540–542), and copper salts (543) were also used, singly. Today most processes combine copper and vanadium compounds, and no other catalyst or combination has been found

TABLE 8
Effect of Catalysts on Yield of Adipic Acid by Nitric Acid Oxidation (536)

Starting material	Catalyst	Temperature, °C	Adipic acid, %	Recoverable off-gases,[a] %
Cyclohexane[b]	None	—	33[c]	—
Cyclohexane	None	—	29[d]	—
Cyclohexene	None	—	70–80	—
Cyclohexylamine	None	—	55	—
Cyclohexanol	None	60	78	—
Cyclohexanol	NH_4VO_3	60	86	—
Cyclohexanol	NH_4VO_3	80	79	—
Cyclohexanol	Cu^{2+}	60	78–80	—
Cyclohexanol	Cu^{2+}	80	85	—
Cyclohexanol	Cu^{2+}, VO_3^-	55–85	92–93	15
Cyclohexanone	None	>80	~50	—
Cyclohexanone	VO_3^-	—	70	—
Cyclohexanone	Cu^{2+}	—	80	—
Cyclohexanone	Cu^{2+}, VO_3^-	—	90–95	25
Cyclohexanol/cyclohexanone	VO_3^-	55–80	74	—
Cyclohexanol/cyclohexanone	Cu^{2+}	—	85	—
Cyclohexanol/cyclohexanone	Cu^{2+}, VO_3^-	75–80	91[e]	17

[a] NO and NO_2 are easily recovered as nitric acid; N_2O and N_2 are not.
[b] Air oxidation.
[c] 65% at 5–12% conversion.
[d] In addition, 36% nitrocyclohexane.
[e] 92–94% in continuous, semitechnical scale.

to be as effective (478,481,544). Lindsay (536) reports the effects that these catalysts have on yields, and yields of 90–95% have been reached under optimum conditions (Table 8).

Van Asselt and Van Krevelen (538) propose that the function of the vanadate is the conversion of the dimer 2 to adipic acid, whereas copper has an inhibiting effect on side reactions, particularly formation of glutaric acid. They also suggest that oxidation of cyclohexanol is the fast reaction, that it is a chain reaction involving nitrous acid, and that the rate-determining step is the conversion of cyclohexanone. Compounds 1 and 2 are formed in a definite ratio, which is a function of temperature and of the concentrations of both nitric and nitrous acids.

The vanadium catalyst may serve to inhibit degradative oxidation by forming a cyclic intermediate with 1,2-cyclohexanediol (545).

Several patents describe various means for recovery and reuse of the catalysts (546–557).

Means of reducing the amount of unrecoverable N_2O include operation at supraatmospheric pressure, use of oxygen or air (558–560), and isolation and separate hydrolysis of monoadiponitrolic acid (561–565).

Russian workers have carried out several investigations on the nitric acid oxidation of cyclohexanol (566–574). Lubyanitskiĭ et al. (570) reported a 96% yield of adipic acid at 4 atm in a two-stage process, first at 55°C then at 90–100°C, with a 6:1 mole ratio of nitric acid to cyclohexanol and 0.01–0.03 mole/l of a 1:1 mole ratio of copper to ammonium vanadate.

Lindsay (536) discusses technical aspects, including corrosion, in pilot-plant scale. Many process details and variables are given in the patent literature (327,406,407,469,477,478,493,505,544,575–596).

Various other cyclohexyl derivatives used for the nitric acid oxidation include: cyclohexylamine (597), cyclohexyl mono- or disulfate (598), cyclohexene (580,599,600), chlorocyclohexane (601), cyclohexyl acetate (602), 1,2-cyclohexanediol (603), methyl 1-hydroxycyclohexane-1-carboxylate (604), 2-methylcyclohexanol (605), caprolactone (606), and 1,2-epoxycyclohexane (607). A mixture of 3- and 4-methylcyclohexanol gives substituted adipic acids (608).

4. NITRIC ACID OXIDATION OF CYCLOHEXANE

Various procedures have been described for the nitric acid oxidation of cyclohexane itself rather than an intermediate oxidation product of cyclohexane (580,582,586,609–611). The yields are usually low (34–70%) and compared with the oxidation of cyclohexanol-cyclohexanone, greater amounts of by-products, which include various nitrocyclohexanes and lower dibasic acids, are formed. However, yields up to 90% have been claimed when nitro- or dinitrocyclohexane serves as the reaction medium (612). Conversion of cyclohexane to nitrocyclohexanes is said to be prevented by complex formation between nitric oxide, the agent for nitrocyclohexane formation, and the solvent. Nitrocyclohexane is much more slowly oxidized than cyclohexane, and dinitrocyclohexanes represent a loss of nitric acid values.

5. AIR OXIDATION OF CYCLOHEXANE TO ADIPIC ACID

At conversions higher than those normally used for cyclohexanol/cyclohexanone production, cyclohexane can be oxidized to adipic acid as the main product. The reaction may be carried out in successive oxidation towers (613) or in a two-step process involving separation of the initial oxidation products (e.g., 614–617). The reaction proceeds in either absence (473, 618–622) or presence of a solvent, usually acetic or other carboxylic acid (e.g., 623–626). Acetone (627) and t-butyl alcohol (628) also are effective. For catalysts, cobalt (623), mixtures of manganese and copper (629–631), or combinations of all three (623,632) have been tried. Yamaguchi et al. (620) found that manganese naphthenate was superior to cobalt, chromium, nickel, or sodium salts. Optimum concentration was 0.003 wt.% as manganese; excess catalyst increased production of formic and acetic acids. Cyclohexanone (623), methyl ethyl ketone or acetaldehyde (624,625), butyraldehyde (626), or ozone (633,634) can be initiators. Various other patents discuss processing details (484,635–647).

According to Tanaka, Honda, and Inoue (648), catalytic activity was in the order $Co > Cr > Zn \geqq Hg > Ni > Pb$ as the acetates. Cobalt(III) acetylacetonate or acetate was particularly effective, since no induction period was required as with cobalt(II) acetate. The oxidation was first order in cyclohexane concentration and in cobalt(III) concentration. The initiating and rate-determining step was apparently

$$Co^{3+} + RH \rightarrow Co^{2+} + R\cdot + H^+$$
$$RH = cyclohexane$$

Optimum reaction conditions included acetic acid solvent and a reaction temperature of 80–90°C. Under these conditions, conversion of cyclohexane was 70–90% and adipic acid selectivity greater than 70%.

As shown by experiments in the presence of cyclohexanone tagged with ^{14}C, adipic acid is produced by oxidation of cyclohexanone and not directly from cyclohexane (649). Both monocarboxylic and homologous dicarboxylic acids are formed as by-products in rather large amounts by various decarboxylation reactions (650). However, oxidative decarboxylation of adipic acid represents only a very minor loss in selectivity (as found in experiments with labeled acids), and the origins of succinic and glutaric acid lie in some other route (651). The percentage per hour of decarboxylation is a function of temperature (651):

	90°C	150°C
Adipic	2.1	50
Glutaric	1.8	13
Succinic	1.3	13

6-Hydroxyhexanoic acid also is a product of the reaction (652–655) and is a potential industrial intermediate for caprolactam (653,656) or for hexamethylene diamine (656). Purification of adipic acid produced by this method presents problems because of the high proportion of by-products, and various extraction (484,636) or washing (638,647) techniques have been proposed.

Yields on the order of 50–80% are lower than with nitric acid oxidation.

6. Air Oxidation of Cyclohexanol or Cyclohexanone to Adipic Acid

According to an early patent (657), adipic acid may be obtained in 80% yield by air oxidation of cyclohexanone in acetic acid solution and in the presence of manganese acetate. With barium acetate as a modifier for the manganese catalyst, which also contained some cobalt (658), and at 40% conversions as high as 90% yields have been reported (659). Although cyclohexanol has been used (635,660), cyclohexanone apparently is the preferred starting material (631).

In one patented process, the cyclohexanol/cyclohexanone mixture prepared by the boric acid route is dehydrogenated to cyclohexanone over zinc oxide before carrying out further oxidation (661). Caprolactone and valeric acid are by-products in the oxidation of cyclohexanone (662); 1,12-dodecanedioic acid is a by-product in the oxidation of cyclohexanol (660,663). The peroxidic products obtained by oxidation of cyclohexanol in the absence of a catalyst may be converted to 6-hydroxycaproic acid in good yield by treatment with formic acid (664,665).

Process details and variations have been discussed in several publications (666–674). An interesting variation is the use of hexamethylphosphoramide and sodium methoxide to produce quantitative yields of adipic acid from cyclohexanone by oxidation with oxygen (675).

7. Oxidation with Nitrogen Dioxide

Without a catalyst, oxidation of cyclohexane by nitrogen dioxide can produce high yields of adipic acid, but so slowly that the process is impractical (676,677). Vanadium or silica catalysts inhibit formation of lower homologs, but the reaction still requires about two days at 50°C (678–681). With diatomaceous earth or silica gel as dehydrating agents and at low conversions, cyclohexyl nitrite and nitrocyclohexane are formed as the principal products (682). Cyclohexyl nitrite, but not nitrocyclohexane, is readily oxidized by nitric acid to adipic acid. Either cyclohexanol or cyclohexanone is also easily oxidized to adipic acid by nitrogen dioxide (683), particularly when air and metavanadate are present (684). Similar investigations have been reported (685,686).

Oxidation of 1,6-hexanediol in the Maurer reaction produces adipic acid in 96% yield (41).

Oxidation of cyclohexene by dinitrogen tetroxide and two subsequent steps have produced adipic acid in an overall yield of 76% (687):

$$\text{(cyclohexene)} + O_2 + N_2O_4 \longrightarrow \text{(cyclohexane with OONO}_2 \text{ and NO}_2)$$

$$\text{(cyclohexane with OONO}_2\text{, NO}_2) + (CH_3)_2N\overset{O}{\overset{\|}{C}}N(CH_3)_2 \longrightarrow$$

$$\text{(cyclohexanone with NO}_2) + (CH_3)_2N\overset{O}{\overset{\|}{C}}N(CH_2)_2 \cdot HNO_3$$

$$\text{(cyclohexanone with NO}_2) + H_2SO_4 + 2H_2O \longrightarrow$$

$$HOOC(CH_2)_4COOH + NH_2OH \cdot H_2SO_4$$

8. RECOVERY OF BY-PRODUCT DIBASIC ACIDS

Nitric acid and other mother liquors remaining after removal of adipic acid can be treated to recover glutaric, succinic, and oxalic acids (397,688–690). Separation may be accomplished by several methods: further evaporation precipitates oxalic and succinic acids first (691,692); distillation of the anhydrides of glutaric and succinic acids separates these from adipic (693–696); crystallization in water separates either oxalic or succinic acid from glutaric, which is much more soluble than the others (697); steam distillation removes succinic acid (698); and partial neutralization keeps the more strongly acidic succinic and glutaric acids in solution (699). Various extractants have been proposed: nitroparaffins (700), cyclohexanone-cumene mixtures (701,702), or 2-octanol (703–705) for adipic acid and diisopropyl ether for glutaric acid (706). Even-numbered dibasic acids, but not the odd-numbered ones, are said to crystallize from their solutions in 70% nitric acid and 1-nitropropane (701).

9. PURIFICATION OF ADIPIC ACID

Adipic acid in poly(hexamethylene adipamide) synthesis must be pure and free of the by-product dibasic acids formed during its synthesis. As little as 0.1 mole % succinic acid can reduce the relative viscosity of the polyamide by 0.4 unit and increase amine end groups by 1.3 moles per million grams (707,708) because succinic acid has less thermal stability than adipic.

The first crystallization of adipic acid from solution in 30–60% nitric acid should be conducted at a temperature above 40°C to avoid contamination with the byproduct dibasic acids. Under these conditions, recovery of adipic acid is only 70–85%. To recover the remaining adipic acid without contamination, the second crystal crop should be obtained from a 15% nitric acid solution. As a result, succinic acid purged from the system contains only 0.46 lb of adipic acid per pound of succinic acid instead of the normal 1.0 lb. Alternatively, adipic acid can be recovered by crystallization at 35°C from 38% nitric acid followed by centrifugation (709,710).

A high crystallization temperature (40–70°C) and the presence of dibasic acid impurities in a supersaturated solution suppress nucleation so that the adipic acid crystals are large and relatively pure (711,712).

During evaporative crystallization of adipic acid, excess foaming and buildup of crystals on walls and in lines can be prevented by addition of a silicone fluid (713). Other patented variations on crystallization procedures include filtration through an ultrafine filter (714), mixing a hot solution of adipic acid in 50% nitric acid with cold recycle crystal suspension (715), and recycling of mother and wash liquors to keep nitric acid losses to a minimum (716).

In recovering adipic acid from crude mixtures containing 50% or less of adipic acid, partial neutralization of a solution of the dibasic acids recovers 50–70% of relatively pure adipic acid upon crystallization (700,717). The increase in purity is the result of a favorable equilibrium, which arises from the greater acidities and water solubilities of glutaric and succinic acids compared with adipic acid. Partial neutralization accompanied by extraction with ether can separate sebacic from adipic acid (718–720).

The lower dibasic acids may be removed from crude adipic acid by trituration with hot glacial acetic acid (721).

Extraction of an aqueous ammoniacal solution of caproic and adipic acids with a hydrocarbon and a fatty alcohol or trialkylamine concentrates the caproic acid in the organic phase (722). Caproic acid may also be removed by steam distillation (723).

During recrystallization of adipic acid, growth rate and habit can be modified by addition of surfactants (724,725). Surfactants increase the supersaturation level at which adipic acid crystals can be grown from seeds without secondary nucleation, apparently by adsorbing on and poisoning the larger and more active embryos or heterogeneous nuclei. Crystals grown in the presence of these additives were better formed and had smoother faces than when grown without them. Anionic surfactants (e.g., sodium dodecyltetrapropylbenzene sulfonate) were selectively adsorbed on [110] end and [010] side faces, whereas cationic surfactants (e.g., trimethyloctadecyl-ammonium chloride) were selectively adsorbed to the extent of a monolayer on the hexagonal [001] face.

10. MISCELLANEOUS

A variety of oxidation methods other than those already described are available: ozonolysis of cyclohexene (417,726–732); ozonolysis of cyclo-hexane, cyclohexanol, or cyclohexanone in the vapor phase over a silica or alumina catalyst (733); oxidation of cyclohexane with air and dichromate (734); chromium(VI) oxide oxidation of cyclohexane (735); chromic acid oxidation of cyclohexanone (736); sodium hypobromite oxidation of cyclo-hexanone (737,738); performic acid treatment of cyclohexanone (739); nitric acid oxidation (606), alkali fusion (740), or electrochemical oxidation (741) of 6-hydroxycaproic acid; and air oxidation of cyclohexanesulfonic acid obtained by reaction of cyclohexane with air and sulfur dioxide (742). Adipic acid, along with other dibasic acids, is also formed by nitric acid oxidation of fatty acids or of paraffins.

1,1,1,5-Tetrachloropentane is one of several telomers produced in the reaction of ethylene with carbon tetrachloride, and it yields adipic acid by hydrolysis after conversion to 1,1,1-trichloro-5-cyanopentane (743–745).

Reppe and his co-workers have described the carbonylation of 1,4-butanediol (40,746,747) and of tetrahydrofuran (748,749). From 1,4-butanediol and with iodine and nickel carbonyl as catalysts, they prepared adipic acid in 69% yield after 4 hr at 200 atm pressure of carbon monoxide and at 260°C; they obtained about the same results with tetrahydrofuran. δ-Valerolactone is an intermediate (750). Thiophane has been used in place of tetrahydrofuran (751). Butadiene can also be carbonylated, but a mixture of isomeric acids is formed (752–755). Carbonylation of ethyl 2-pentenoate produces ethylsuccinic, methylglutaric, and adipic esters in the ratio 2:2–3:3–4 (756).

D. Physical Properties

The pattern of polymorphism of the even-numbered dibasic acids is not as regular as that of the odd-numbered acids. Adipic acid was thought to have only one crystalline form until 1961, when a second form was shown to exist at temperatures below $-150°C$ (757). The form stable at room temperature, α-adipic acid, normally crystallizes from aqueous solutions as flat, slightly elongated, hexagonal, monoclinic plates (724). Ethyl acetate and alcohol also may be used for recrystallization. Cell dimensions are $a = 10.07$ Å, $b = 5.16$ Å, and $c = 10.03$, and the space group is C_2^5h ($P2_1/a$) (758–760). The [001] face is predominant and is made up of carboxyl groups; the elongated [010] side faces and the [110] end faces contain both carboxylic and hydrocarbon groups. The carboxyl groups of the adipic acid molecule lie in a plane tilted about 6° from the plane of the central carbon chain. Thermal expansion of α-adipic acid occurs least in the c axis and greatest in the direction perpendicular to the [100] plane, and is explained by an increase in the angular vibration of the molecules (761).

Caking of adipic acid crystals can be prevented with a 25–200 ppm coating of stearic acid (762).

According to Davies and Thomas (368), the infrared spectrum of adipic acid is appreciably richer than that of succinic acid and shows a number of clear and continuous changes as the temperature rises. Other infrared studies have been recorded of adipic acid (423), of its —COOD derivative (249), of its solution in D_2O and quaternary ammonium salt solutions (763), of its thallium complexes (764), and of its disodium salt (424); the Raman spectrum has also been studied (765). Polarization of infrared active in-plane modes in adipic acid has been discussed (766).

Adipic acid is slightly soluble in cold water and quite soluble in hot water but is less soluble than its two neighboring homologs (Table 9) (425,767). It is very soluble in methanol but only slightly soluble in benzene (0.031 mmole/liter at 35°C and 0.250 mmole/liter at 65°C) (767). Solubilities in various mixtures of acetone-methanol and acetone-ethanol have been reported (773). Distribution coefficients ($k = C_W/C_{org}$) between water and several organic solvents at 26°C are (774): n-butanol, 0.31; cyclohexanone, 0.32; ethyl acetate, 0.91; ether, 2.2; isobutyl alcohol (at 22°C), 0.29 (775); methyl isobutyl ketone, 1.3 (776); and methyl isobutyl carbinol, 0.33 (776,777).

Adipic acid, either as the solid or in aqueous solution, is corrosive to mild steel but not to stainless steel.

E. Analytical Procedures

Adipic acid may be assayed by the usual titrimetric procedures, but exact purity is best determined by measurement of freezing-point depression (439,

TABLE 9
Physical Properties of Adipic Acid

Property	Value
Molecular weight	146.14
Melting point, °C (768)	153.0–153.1
Differential thermal analysis, endotherms, °C (769)	145, 310
Boiling point, °C/mm Hg (770)	265/100; 244.5/50; 216.5/15
Vapor pressure, $\log p$ (mm Hg), 86–133°C (368)	$15.463 - 6757/T$
Heat of sublimation, kcal/mole (368)	30.9
Density, g/cc (759)	1.345
Refractive index (65)	$N_x = 1.464,\ N_y = 1.506,\ N_z = 1.592$
Heat of combustion, cal/g (38)	4579.7
Dipole moment at 25°C (178,771)	2.30–2.60
Solubility, g/100 g H_2O (425)	
15°C	1.44
34.1°C	3.08
60°C	17.6
87.1°C	94.8
Heat of solution at 35–55°C, cal/mole (368)	16,400
Ionization constants in water at 25°C (772)	
K_1	3.85×10^{-5}
K_2	3.89×10^{-6}

778). Trace amounts of mono- and dicarboxylic acids can be separated and measured by column-partition chromatography. Procedures for this analysis and for foreign particles, heat stability (color-forming impurities), and metallic elements are given by Keller (778). Other partition-chromatographic procedures, chiefly on silicic acid columns with aqueous and buffered stationary phases, have been published (378,427,779).

Gas-liquid chromatographic procedures for the dimethyl ester (264,780), for the di-*n*-propyl ester (781), and for the free dicarboxylic acid (782) have been described. Various polyester columns may be used for the diesters, and carboxyl-terminated polyether column serves for the free acid.

The various solvent combinations used in thin-layer chromatographic analyses of adipic acid and its homologs include 96% ethanol, water, and 25% aqueous ammonia (100:12:16) (191); benzene, methanol, acetic acid (45:8:4) (265); diisopropyl ether, formic acid, water (90:7:3) (783); propanol, 28° Bé ammonia (70:30) (784); ethyl acetate, methanol, concentrated ammonia (20:20:10) (785); butanol, xylene, phenol, formic acid, water (10:70:30:8:2) (786); and the upper layer from an equilibrated mixture of benzene and 80% aqueous acetic acid (100:17.5) (377).

Paper chromatographic procedures are also available (193,194,787–790).

Melting-point–composition relations have been established for adipic acid mixtures with pimelic and suberic acids (791), and specifications for adipic acid have been published (439,778).

F. Toxicology

In contrast to the nephrotoxic nature of glutaric acid, adipic acid is only slightly irritating to the kidneys. Horn et al. (792) reviewed toxicological studies on adipic acid and found in their own tests that it is a safe food additive and comparable to citric and tartaric acids. Their tests were carried out by acute oral, acute intraperitoneal, and acute intravenous administration to mice, as well as by chronic feeding. Adipic acid was significantly less toxic than tartaric or citric acid upon intravenous administration. Diglyceride adipates, as well as polymeric glyceride adipates, have high digestibility coefficients in rats, and the stearic acid moiety is well absorbed (793). However, absorption is slow, particularly in the polyester. In rats fed a daily dose of adipic acid over a period of 33 weeks, a dose of below 400 mg caused no damage. At 400 mg, some damage occurred, and at 800 mg the effect was severe (794). Urea, glutamic acid, lactic acid, β-ketoadipic acid, and citric acid were among the oxidation products of adipic acid which along with the acid itself, were identified in the urine of rats fed labeled adipic acid (795). The urea and citric acid were probably not direct metabolites but were formed from labeled carbon dioxide.

Because of its nontoxic nature, adipic acid is used widely as a food acidulant (796).

G. Polymer Applications

The paramount use for adipic acid is in nylon 6/6 textile fibers, one of the most glamorous and publicized success stories of the chemical industry. Reviews are available on the subjects of nylon 6/6 synthesis and properties (75,797,798), textile fiber uses (799), and plastics uses (800). Procedures for the preparation of nylon salt and its polymerization are available (801,802).

Low-molecular-weight polyesters from adipic acid are used as the hydroxy-containing component for polyurethane foams and plastics (803, 804). Polyester preparation and properties (74), X-ray studies (391), and rubberlike properties (393) have been described.

VII. PIMELIC ACID

A. Introduction

Pimelic (Gr. *pimelē*, fat) acid was first reported and named in 1837 by Laurent, who believed he had prepared it along with several other dibasic acids by oxidation of oleic with nitric acid. Later investigators also obtained from fatty materials a product they thought to be pimelic acid, but they reported different melting points and crystalline forms. In 1862 Arppe pointed out that the material described by Laurent was probably a mixture of homologs and stated (erroneously) that pimelic acid is not formed in the oxidation of fats. Pure pimelic acid was not made until 1874, when Schorlemmer and Dale oxidized suberone (cycloheptanone) with nitric acid. In 1877 Baeyer synthesized it from 3-(2-furyl)propionic acid. Not until 1884 was an authentic sample prepared from a fatty material, when Gantter and Hell isolated it from the nitric acid oxidation products of castor oil. Bouveault, in 1898, also demonstrated that pimelic acid, as well as all the homologous dibasic acids from oxalic to sebacic, was formed during oxidation of fats. Six years later Hamonet reported its synthesis by saponification of 1,7-heptanedinitrile. Several early investigators employed malonic ester syntheses for pimelic acid preparation.

Pimelic acid is not made commercially and has no uses that apparently cannot be filled by other homologs. It is found in the urine of animals and humans.

B. Chemistry

Pimelic acid decomposes with decarboxylation at about 300°C (93). Dry distillation of the acid (or better, of the thorium salt, according to Ruzicka) produces cyclohexanone. Heating the acid with acetyl chloride or acetic anhydride yields the polymeric anhydride, mp 53–55°C. The polymeric anhydride gives a liquid monomer upon heating in a molecular still, but the monomer is unstable and rapidly converts to a polymer (24). Pimelic acid forms esters, the diamide, and dinitrile in the same manner as its homologs.

C. Synthesis

Organic Syntheses preparations are available for the synthesis of pimelic acid from (a) cyclohexanone via condensation with diethyl oxalate to ethyl 2-ketocyclohexylglyoxalate, decarbonylation to ethyl 2-ketohexahydro-benzoate, and acid cleavage of this β-keto ester in strong alkali (805); and (b) from salicylic acid by reduction and subsequent cleavage with sodium in isoamyl alcohol (806). The overall yield is about 50% by either method.

Simplified reaction steps have been reported for the conversion of penta-methylene glycol (derived from tetrahydropyran and furfural) to pimelic acid in an overall yield of about 75% by way of bromide and nitrile as intermediates (39,807).

Direct carboxylation of cyclohexanone with subsequent ring opening has reportedly been accomplished over difficultly reducible chromite catalyst containing manganese, zinc, and magnesium (808).

Carbonylation of cyclohexanone peroxides, particularly 1-hydroxy-1′-hydroperoxycyclohexyl peroxide, to pimelic acid is said to occur in a solvent, such as urea, and in the presence of iron(II) sulfate (809). Carbonylation of tetrahydropyran yields pimelic acid under conditions described by Reppe et al. (748).

Cleavage in strong alkali of various cyclohexane, and particularly cyclohexene, derivatives has attracted attention as a potential commercial route. For example, 2-cyanocyclohexanone (from 2-chlorocyclohexanone) produces pimelic acid in 85% yield upon treatment with strong alkali (810). However, the Diels-Alder adducts available from the reaction of butadiene with acrylonitrile, acrolein, acrylic acid, or maleic anhydride are more attractive starting materials (811–815). Yields of up to 88% are obtained. The product from alkali treatment of 3-cyclohexene-1-nitrile may be decolorized by treatment with ozone (815). Sulfur or an alkali-metal polysulfide added to the alkali improves the yield to more than 90% (816). 3-Cyclohexenecarboxaldehyde (1,2,3,6-tetrahydrobenzaldehyde) and its derivatives have received attention as starting materials (817–821). Alkali fusion of tetrahydrofuryl propanol gives pimelic acid in only 31% yield, whereas 1,7-heptanediol gave 79% (822).

Pimelic acid can also be obtained by a three-step synthesis from furfural in an overall yield of 86% by way of furylacrylic acid and diethyl 4-keto-pimelate (823). The furylacrylic acid was prepared from furfural and malonic acid, but it can be prepared by the Perkin condensation of furfural with acetic anhydride and sodium acetate (824a). 4-Ketopimelic acid is readily obtainable as the dilactone, which is formed by heating succinic acid or anhydride in an organic solvent (271). 4-Ketopimelic acid was reduced by hydrazine and potassium hydroxide (823), but it may also be hydrogenated as the sodium salt in aqueous solution (824b,825) or by electrolytic reduction (826).

Oxidation of 1,7-heptanediol with dinitrogen tetroxide produces pimelic acid in more than 90% yield (41). Pimelic acid completely substituted in both α positions has been prepared by the reaction sequence: acetone $\rightarrow$ phorone $\rightarrow$ phoronic acid dilactam and dilactone $\rightarrow$ 2,2,6,6-tetramethylpimelic acid (827).

D. Physical Properties

β-Pimelic acid is stable below the transition temperature of 74–75°C, and the α form is stable above it (828). However, α-pimelic acid is also stable at room temperature for at least several weeks before converting to the β form, unless it is pulverized or heat treated. Both forms are monoclinic, the β form having a tabular to lamellar habit. Apparently the α form has the space group $P2_1/c$ and the β form, space group $C2/c$ (420,760,829,830). The angle between the planes of each carboxyl group is about 60°, each about 30° out of the plane of the central carbon atoms in β-pimelic acid, but one at 44° and the other at 18° in α-pimelic acid (829).

Infrared spectra have been reported for the acid (423,831,832) and the sodium salt (424).

Pimelic acid is fairly soluble in water—100 g of saturated solution at 22° contains 4.965 g (810)—and easily soluble in alcohol, ether, and hot benzene. It is almost insoluble in cold benzene (0.0199 g/100 g of benzene at 25°C) (62). The distribution coefficients ($k = C_W/C_{org}$) at about 26°C for water and several solvents are (774): n-butanol, 0.15; ethyl acetate, 0.38; ether, 0.91; and isobutyl alcohol (at 20°C), 0.14 (775).

Houston and Van Sandt (791) have reported melting points and eutectic temperatures for a number of binary mixtures of pimelic and adipic, suberic, or azelaic acids. Some physical properties for pimelic acid are given in Table 10.

TABLE 10
Physical Properties of Pimelic Acid

Property	Value
Molecular weight	160.17
Melting point, °C (768)	105.7–105.8
Boiling point, °C/mm Hg	272/100; 251.5/50; 223/15
Density at 25°C, g/cc (372)	1.287
Refractive index (65)	$N_x = 1.458$, $N_y = 1.492$, $N_z = 1.579$
Heat of combustion, cal/g (38)	5169.5
Dipole moment (771)	2.36–2.47
Ionization constants in water (67,68)	
K_1 (833)	3.097×10^{-5}
K_2 (772)	3.74×10^{-6}

E. Analytical Procedures

Bouveault (834) separated pimelic from succinic, glutaric, and adipic acids. Calcium glutarate, being very soluble in water, is easily separated from calcium pimelate, which is slightly soluble in cold water but insoluble in boiling water.

F. Toxicology

Information on the toxicological or physiological effects of pimelic acid does not seen to be available. On the basis of the known properties of adipic acid, however, pimelic acid may be assumed to be nontoxic.

G. Polymer Applications

Korshak and Vinogradova (74) investigated various polyalkylene pimelates and have summarized their properties. Fuller, Frosch, and Pape (430) determined the fiber period (23.6 Å) for poly(trimethylene pimelate).

The polyamides, like the polyesters, have been investigated principally by the Russians (75), who also examined copolyamides of pimelic acid. The polyamide of pimelic acid formed with 4,4'-diamino-3,3'-dimethyldiphenyl-methane by either melt (435) or interfacial (434) polymerization has a melting point in the range 200–215°C and a decomposition temperature of 340–365°C.

VIII. SUBERIC ACID

A. Introduction

Suberic acid was obtained for the first time by the action of nitric acid on cork (*suber* in Latin) as reported by Brugnatelli in 1781. Later, in 1837, Laurent found it in the mixture of acids resulting from nitric acid oxidation of oleic or stearic acids. A better fatty acid source is castor oil, used first by Tilley in 1841 and then by Arppe in 1861. Castor oil produces a mixture of suberic and azelaic acids, the predominant one depending on reaction time, temperature, and acid strength. Verkade (835) investigated the nitric acid oxidation of ricinoleic acid and found that azelaic acid was more easily oxidized than suberic by nitric acid. Thus high temperatures, concentrated acid, and long reaction times favored higher proportions of suberic acid in the product mixture; but even under optimum conditions for each, only 24% yield of azelaic and 11% yield of suberic acids were obtained. Other early methods include: the electrolysis of ethyl hydrogen glutarate by Crum, Brown, and Walker in 1891; Hamonet's hydrolysis in 1903 of suberonitrile obtained from 1,6-dichlorohexane and potassium cyanide; and oxidation of cyclooctane either by nitric acid as described by Willstätter and Veraguth in 1907 or by chromic acid as reported the same year by Wallach.

At present suberic acid has no major uses but is available as a developmental chemical.

B. Chemistry

The decomposition temperature of suberic acid, according to Korshak and Rogozhin (93), is about 345°C; therefore, suberic acid is more stable than any of its lower homologs, as well as its next higher homolog, azelaic acid. Polymeric suberic anhydride (mp 65–66°C) is formed when the acid is refluxed with acetic anhydride (24). When heated in a molecular still, the polymer slowly yields a small amount of dimer stable below the melting point of 55–57°C. Dry distillation of the acid as the thorium or calcium salt, or in the presence of iron filings, produces suberone (cycloheptanone) in 35–40% yields. The acid chloride, ester, and amide are prepared according to standard procedures.

C. Synthesis

Cyclooctene and cyclooctane became commercially available in 1968 and probably are the most economically feasible starting materials for suberic acid. They are produced by hydrogenation of 1,5-cyclooctadiene, which is formed by cyclic oligomerization of butadiene. Several routes to suberic acid based on these materials have been developed, but one analogous to that developed for adipic acid appears the best; that is, air oxidation of cyclooctane to an alcohol/ketone mixture followed by nitric acid oxidation. See Section XI (dodecanedioic acid) for a discussion on a similar route from cyclododecane.

Air oxidation may be carried out to about 10% conversion in the absence of a catalyst, but the presence of either sodium bisulfate (836) or sodium dihydrogen phosphate (837) is required to minimize side reactions by controlling acid content. When the resultant hydroperoxide is reduced with sodium sulfide, the product contains 89% cyclooctanol and 6% cyclooctanone. Oxidation of this product with nitric acid then produces suberic acid in an overall yield of 76% from cyclooctanone.

Cyclooctane may also be oxidized by nitrogen dioxide in the presence of ammonium metavanadate (838).

Another route to cyclooctane is that developed by Reppe et al. (839,840), who prepared cyclooctatetraene from acetylene in the presence of nickel cyanide. Hydrogenation can be controlled to prepare either the monoene or the saturated hydrocarbon. Cyclooctene was readily cleaved to suberic acid by such oxidizing agents as chromic oxide, nitric acid, or potassium permanganate. Oxidation with nitric acid is usually accomplished in the presence of a vanadium compound, which is thought to be specific (841). A mixture of vanadium and manganese salts as catalysts reportedly gives higher yields than a vanadium compound alone (842).

An alternate route to suberic acid from butadiene takes the following course: 1,4- or 1,2-addition of chloromethyl methyl ether, simultaneous reaction with carbon monoxide and acetylene in the presence of nickel carbonyl, hydrogenation of the 8-methoxy-2,5- or 8-methoxy-2,4-octadiene-1-carboxylic acid, and oxidation of the ω-methoxy caprylic acid with nitric acid (843). Another sequence involves coupling of 1-methoxy-1-buten-3-yne (from diacetylene and methanol) to 1,8-dimethoxy-1,7-octadien-3,5-diyne followed by addition of methanol to the enol ether groups, hydrogenation, hydrolysis, and oxidation (844).

Reaction of carbon monoxide and acetylene with 1-chloro-4-cyano-2-butene in the presence of nickel carbonyl yields 7-cyano-2,5-heptadienoic acid, also a potential source of suberic acid (845).

An alternative to the oxidation of cyclooctane is the ozonolysis of cyclooctene. Fremery and Fields (730,731) ozonized cyclooctene in aqueous emulsion and in the presence of hydrogen peroxide and alkali to produce suberic acid in 63% yield in a single step. A surfactant helped to attain the small diameter of the dispersed phase droplets necessary to avoid the formation of peroxidic polymers, which would cause lower yields. 1,5-Cyclooctadiene was ozonized to 4-octenedioic acid in 52% yield. Several recent patents describe other solvents and operating conditions. Ozonolysis may be carried out in a carboxylic acid containing 1–10% water (846). Subsequent oxidation of the ozonolysis products by oxygen at reflux temperature in the presence of additional water (to a total water content of 25–35%) produces suberic acid in 60% yield. A preferred solvent is stated to be propionic acid; the ozonolysis products can be decomposed thermally at 100°C and in the absence of oxidant to produce a mixture of suberic and suberaldehydic acids, or in the presence of oxygen to produce mostly suberic acid (847). Perry (848) ozonized 1,5-cyclooctadiene in methanol-dichloromethane and oxidized the ozonolysis products with oxygen in the presence of silver oxide to make 4-octenedioic acid in 75% yield. Maggiolo (849) obtained suberic acid in 78% yield by ozonolysis of cyclooctene in propionic acid at 0°C followed by oxidation with oxygen beginning at 70°C and ending at 105–110°C in a careful time-temperature schedule.

Reaction of 2 moles of acetylene and 4 moles of carbon monoxide in the presence of cobalt carbonyl produces bifurandione, an unsaturated dilactone, which can be hydrogenated to suberic acid (850–852).

Telomerization of ethylene with carbon tetrachloride produces mixtures of homologous telomers, including 1,1,1,5-tetrachloroheptane. Cyanation followed by hydrolysis yields suberic acid (744,745).

Reppe et al. (40) carbonylated 1,6-hexanediol in the presence of nickel carbonyl to form suberic acid in 90% yield.

In the laboratory, suberic acid may be prepared in an overall 76% yield from 1,6-hexanediol by the sequence: diol → dibromide → dinitrile → suberic acid (39). It may also be prepared from tetrahydropyran by the sequence: tetrahydropyran → ω-acetoxypentyl chloride → ω-bromopentyl chloride → ω-halocapronitrile → diethyl ω-cyanopentylmalonate → suberic acid (807). The overall yield from tetrahydropyran is 30%. A generally useful synthesis is the condensation of 1,3-cyclohexanedione (dihydroresorcinol) with various organohalides (53). Suberic acid is obtained in 50% overall yield by condensation of ethyl bromoacetate with the dione followed by reductive cleavage of the product.

As we noted in the introduction, suberic mixed with azelaic acid can be prepared in low yields by nitric acid oxidation of ricinoleic acid. Separation of this mixture is probably effected best by distillation of the dimethyl esters (853). Separation may be accomplished also by crystallization of the less soluble suberic acid from benzene-ethanol (4:1 volume ratio) (854). Other feasible methods include (a) extraction of the solid mixture with ether to remove azelaic acid (solubility of azelaic 2.7 parts, of suberic 0.8 parts in 100 parts ether) then benzene (in which suberic acid is sparingly soluble) (855); (b) fractional crystallization from water from which azelaic acid crystallizes first (856); and (c) precipitation of the magnesium salts (100 parts of water at 18°C dissolve 3.63 parts magnesium azelate and 13.54 parts magnesium suberate) (857). Gantter and Hell (858) made the following observations:

1. Aqueous recrystallization is impractical.

2. Separation by ether could give pure suberic but not azelaic acid.

3. Fractional crystallization of the calcium, manganese, or magnesium salts yields pure azelaic but not suberic acid.

4. Complete separation is accomplished best by treatment with ether followed by crystallization of the magnesium salt.

D. Physical Properties

Suberic acid is dimorphic; the transition temperature apparently is about 90°C (859). The β form, stable at room temperature, has the space group $P2_1/C$ and is monoclinic with lamellar habit (760,860).

TABLE 11
Physical Properties of Suberic Acid

Property	Value
Molecular weight	174.19
Melting point, °C (368)	143.0–143.3
Boiling point, °C/mm Hg	279/100; 258.5/50; 230/15
Density, g/cc (860)	1.270
Refractive index (65)	$N_x = 1.469, N_y = 1.507, N_z = 1.587$
Heat of combustion, cal/g (38)	5645.7
Dipole moment at 25°C (771)	2.27–2.44
Solubility, g/100 cm³ aqueous solution (861)	
0°C	0.08
15°C	0.13
20°C	0.16
35°C	0.45
50°C	0.98
65°C	2.22
Ionization constants in water (772)	
K_1	3.05×10^{-5}
K_2	3.85×10^{-6}

Infrared spectra have been determined (368,423,832).

Suberic acid is less soluble in water than pimelic and even somewhat less soluble than azelaic acid (Table 11). Solubilities in benzene and dilute hydrochloric acid have been recorded (767). Solubilities in alcohols at 40°C are (g/100 g alcohol): methyl, 32.0; ethyl, 18.4; propyl, 13.9. Well-formed crystals for crystallographic studies are best obtained from ether in a sealed tube heated to 100°C and allowed to cool slowly (862). Marvel and Richards (774) have reported distribution coefficients ($k = C_W/C_{org}$) at 26°C for several solvents: n-butanol, 0.12; ethyl acetate, 0.16; ether, 0.34; chloroform, benzene, and hexane, all greater than 10.

E. Analytical Procedures

Suberic acid may be separated chromatographically from other dibasic acids on a silicic acid column with a stationary phase consisting of a citrate buffer at pH 5.4 (779). Composition–melting-point relations have been given for binary mixtures of suberic acid with adipic, pimelic, azelaic, and sebacic acids (791).

F. Toxicology

The sodium salt of suberic acid is only mildly nephropathic when injected subcutaneously in rabbits (863). Of the homologs in the even series, suberic acid is metabolized the least when injected as the sodium salt and is eliminated mostly unchanged (864–867).

G. Polymer Applications

X-Ray investigations have been reported for the polyesters of suberic acid with ethylene (868), trimethylene (432), and decamethylene (431) glycols.

A high melting (295°C) but thermally stable polymer has been prepared from suberic acid and *trans*-1,4-cyclohexanebis(methylamine) (869). This polyamide has excellent hydrolytic stability and has been tested in several molded plastics applications (870).

IX. AZELAIC ACID

A. Introduction

Laurent reported in 1837 an acid he found among products from the nitric acid oxidation of oleic acid and named it azelaic (from *azote* and *elaidic*). Bromeis was unable to confirm Laurent's results, and this acid was neglected for some years. In 1857 Wirz obtained an acid that he called lepargylic (cf. perlargonic), and Buckton obtained an acid from China wax that he called anchoic. Arppe, in 1862, identified both as azelaic acid and reported the best source to be the nitric acid oxidation of castor oil. This method was used for many years, although the yield was low. The simultaneous formation of suberic acid and the separation of these two acids are described in Section VIII.

Azelaic acid occurs naturally in pine bark oil, in rancid butter, in the urine of pregnant mares, and occasionally in the urine of cows.

Some early methods of synthesis include: reaction of pentamethylene bromide with sodiomalonic ester followed by decarboxylation (by Haworth and Perkin in 1894), oxidation of 9,10-dihydroxyoctadecanoic acid with permanganate (by Edmed in 1898) or in the presence of alkali (by Le Sueur in 1901), alkaline permanganate oxidation of potassium ricinoleate (by Maquenne in 1899), reduction of 5-hydroxynonanedioic acid with hydrogen iodide (by von Pechmann in 1904), ozonolysis of oleic acid (by Harries and Tank in 1907), carbonation of the magnesium derivative of 1,7-dibromo-heptane (by von Braun and Sobecki in 1911), and oxidation of oleic acid with N_2O_4 (by Jegerow in 1912) or with air (by Ciamician and Silber in 1914).

At present azelaic acid is made commercially by ozonolysis of oleic acid; Emery Industries, Inc., is the sole manufacturer in the United States. Production is probably on the order of 20 million lb annually, and the listed price in 1970 was $0.36/lb. The major outlet for azelaic acid is in azelate ester plasticizers for poly(vinyl chloride). According to the U.S. Tariff Commission Report, the 1967 production of azelate esters was 17.5 million lb having an average value of $0.29/lb. Azelate esters are used also as lubricants.

B.   Chemistry

Azelaic acid is somewhat less thermally stable than either suberic or sebacic acid and decomposes with decarboxylation at about 330°C [as determined by measurement of gas pressure variations (22)] or about 320°C [according to derivatographic thermogravimetric analysis (23)]. Mitskevich and Agabekov (871) found that oxidative decarboxylation starts as low as 130°C; oxidation is inhibited by cobalt or manganous bromide, and the inhibition is attributed to the bromide anion. Azelaic and suberic acids are less susceptible than their lower homologs to oxidation by hydrogen peroxide in the presence of a copper salt at 60°C (29).

Dry distillation of the thorium salt of azelaic acid produces cyclooctanone in 20% yield (872); the iron salt produces about 10% of the cyclic ketone (873). When treated with glacial acetic acid, azelaic acid forms a polymeric anhydride. Molecular distillation of the polymer yields an unstable liquid monomer that rapidly rearranges to a linear polymer (24).

Procedures have been described for the preparation of the following azelaic compounds: diethyl azelate, ethyl hydrogen azelate, and 8-carbethoxyoctanoyl chloride (874,875); azelayl dichloride (876); and azelanitrile (877).

C. Synthesis

1. OZONOLYSIS

Ozonolysis of oleic acid, the commercial method for producing azelaic acid, is one of the better routes for laboratory preparation:

$$CH_3(CH_2)_7CH{=}CH(CH_2)_7COOH \xrightarrow[\text{solvents}]{O_3} \text{ozonolysis products}$$

$$\text{ozonolysis products} \xrightarrow{(O)} CH_3(CH_2)_7COOH + HOOC(CH_2)_7COOH$$

pelargonic acid   azelaic acid

Several ozonolysis products other than an ozonide having the 1,2,4-trioxolane structure $-CH\overset{\displaystyle O-O}{\underset{\displaystyle O}{\diagdown\diagup}}CH-$ may be formed, depending on the solvent (878). Since the original work of Harries, who ozonized without a solvent and thermally decomposed the ozonolysis products in water, many modifications and improvements have been reported, mostly in the patent literature.

Rieche (879) treated oleic acid with ozone in the presence of acetic acid and water, and with (or without) hydrogen peroxide or chromic oxide to decompose the ozonolysis products; a catalyst, such as an iron or manganese salt, could be used. Silver oxide in aqueous alkali is a good method although expensive for oxidizing the ozonolysis products (880,881); a more practical method is the use of oxygen (882–884). A small amount of ozone (885) or manganous acetate (886) aids oxidation. Because an ozonolysis solvent is desirable, usually a carboxylic acid is selected—for example, acetic (884) or pelargonic (883) acid. Thermal decomposition of the ozonolysis products obtained in the absence of a solvent produces a mixture of carbonyl and carboxylic compounds—the maximum carbonyl at 100°C and a constant aldehyde/acid ratio at temperatures higher than 140°C (887). A mixture of capronitrile and pelargonitrile serves as the common solvent for oxidative ozonolysis of tall oil fatty acids, amidation, and dehydration steps to prepare azelanitrile (888).

The technical process is described in a patent (883). Oleic acid (1000 lb) is diluted with 500 lb of pelargonic acid and treated with oxygen containing 1.75% of ozone in a countercurrent absorber at 25–45°C. The ozonolysis products are then treated with oxygen at 75–120°C in three reactors in series for about 6 hr. Pelargonic acid (900 = 500 + 400 lb) is removed by distillation at 230°C and 25 mm Hg, and azelaic acid is distilled at 270°C and 3–4 mm. The latter is purified by treatment with water at 95°C, since at this temperature azelaic is soluble but the saturated fatty acids present in the original oleic acid are not. Water in the amount of about 100% of the total weight of oleic and caproic acids is said to serve as a reaction moderator and to improve yield of azelaic acid (889). Water may replace the organic solvent completely as, for example, with a 10% emulsion of oleic acid in water, the ozonolysis products are decomposed with sodium hydroxide (890). Chromogenic substances may be removed by hydrogenating the product acids with nickel catalyst in the presence of activated clay (891). Alternatively, the product acids can be purified by further treatment with ozone, extraction of monobasic acids with octane, crystallization, and finally distillation (892). Esterification of the residue from distillation of the pelargonic-containing phase further improves azelaic acid yields (893).

Numerous variants of this technical process exist. The reaction may be carried out in the liquid phase in the presence of a silent electrical discharge (894) or in the vapor phase at reduced pressure (10–150 mm Hg) (895).

A "reaction-promoting" reagent, such as orthophosphoric acid, reportedly improves azelaic yields when present during oxygen treatment in the presence of manganese acetate of thermally decomposed ozonolysis products (896).

A major problem when oxidation is carried out with air or oxygen is chain degradation, which forms lower homologs of pelargonic and azelaic acids. Pasero et al. (897) obtained only 60–70% yields of the expected acids and found that the major by-product was an ester. The by-product esters from oleonitrile, for example, were octyl ω-cyanooctanoate and ω-cyanoheptyl pelargonate (898). A possible mechanism is the oxidation of the aldehydes formed during ozonolysis to peracids, with subsequent formation of acyl peroxides and decomposition of the acyl peroxide to an ester (899). This reaction is essentially a Baeyer-Villiger type (900). Water present during ozonolysis increases yield of the desired product and reduces the by-product ester, perhaps acting as a free radical inhibitor (901).

Chain degradation is also minimized when performic acid is selected to oxidize ozonolysis products formed in methanol (902) or when reduced platinum oxide is the catalyst during oxidation with oxygen (901).

Manganous acetate has been used as a catalyst during oxidation with oxygen at 60–100°C for the ozonolysis products obtained from oleic acid in acetic acid (903).

2. NITRIC ACID OXIDATION

Nitric acid oxidation of oleic acid was one of the earliest methods investigated for the preparation of azelaic acid. Unfortunately, a large number of shorter chain fragments also formed, and elaborate separations were required to isolate azelaic acid. Thus Ellingboe (904), using 88% nitric acid at 30–35°C and with 0.1% ammonium vanadate catalyst, obtained a product in 60% yield said to contain 65% azelaic and 35% suberic acids. Kirjakka and Nieminen (905) preferred 95% acid at 25°C and found the following compositions of products (wt. %):

Product acid	Starting acid	
	Oleic	Linoleic
Sebacic	2	Trace
Azelaic	44–56	2
Suberic	16–17	36
Pimelic	7–12	12
Adipic	3–4	Trace
Glutaric⎫ Succinic⎭	6–10	38

With 85% nitric acid, product composition was about the same; with 75% acid, the reaction was very slow. When Kirjakka and Nieminen followed the procedure of Sprules and Griffith (906), which calls for 65% acid at 75–110°C with manganese dioxide catalyst, lower yields and relatively more suberic acid resulted.

Gut and Guyer (907) studied the reaction in considerable detail. At 105°C and 3 hr with 64% nitric acid and 0.1% vanadium pentoxide, nitro and nitroso derivatives of oleic acid decreased, and dicarboxylic acids increased when pressure was raised to 5 atm. From 10 to 20 atm, nitrogen compounds decreased and lower dicarboxylic acids increased with increasing pressure. With 32% nitric acid, the major products were nitrogen compounds at 80°C, and maximum yield of azelaic and other dicarboxylic acids was at 140°C. With 64% nitric acid, nitrogen compounds were much less at 80°C as well as at 140°C and total dicarboxylic acid production was much greater than with 32% nitric acid, but in neither case did azelaic acid yield exceed 10%. The major product under any condition was apparently glutaric acid. The highest yields of the higher (C_6–C_9) dicarboxylic acids at 90°C and with 85% nitric acid occurred when the mole proportion of nitric to oleic acid was at least 20:1. An adequate aqueous phase was essential for the best dicarboxylic acid yields; an inadequate amount of water led to increased nitrogen compounds.

Gut et al. (28) studied the oxidative cleavage of azelaic, sebacic, capric, and pelargonic acids by nitric acid. At 62% nitric acid, a mole ratio of 52:1, a temperature of 80°C, and a reaction time of 72 hr at 1 atm, the breakdown of azelaic acid depended on the presence of vanadium pentoxide catalyst as indicated:

	Catalyst, wt. %	
	0	2
Dicarboxylic acid	Product, wt. %	
Azelaic	92.0	63.0
Suberic	3.4	0.1
Pimelic	0.6	1.5
Adipic	1.0	4.9
Glutaric	1.0	13.5
Succinic	2.0	17.0

The organic nitrogen compounds may be removed from the dicarboxylic acids in the form of their esters by an alkaline wash (908).

Instead of oleic acid as such, the following materials have been used: dihydroxystearic acid (909) or sulfated oleic acid (910) in the presence of manganese dioxide, the product from air oxidation of oleic acid in the presence of cobalt acetate (911), a variety of waxy materials (912), and tall oil fatty acids containing 15% resin acids (913). Metallic compounds may be removed from the product acids by precipitation from a solution of the acids in o-dichlorobenzene (914). Optimum reaction conditions for cleavage of ricinoleic acid are said to be: nitric acid concentration, 65–70%; temperature, 60–70°C; catalyst, 1% of ammonium vanadate (915).

Variations in procedure have been described in several patents. Nitric acid of 8–15% strength supplements oxidation with air at 15–40 psi; pelargonic acid and water are continuously distilled from the reaction (916). Another patent cites omission of catalyst if the reaction is carried out with 18% nitric acid at 135°C and autogenous pressure (917).

The distillation residue remaining after removal of pelargonic acid can be extracted countercurrently with 50% aqueous methanol and petroleum hydrocarbons to separate azelaic acid from the saturated acids originally present (918).

Numerous extractants can separate pelargonic and azelaic acids. They include: hot water under pressure (919), a nonpolar solvent to extract pelargonic acid from the nonaqueous layer after nitric acid oxidation (920), a mixture consisting of 60% n-butyl acetate and 40% n-heptane to extract azelaic acid from the aqueous layer (921), aqueous ethylene glycol and a hydrocarbon for countercurrent extraction (922), water and naphthol mineral spirits at 85°C (923), and 1,1,2-trichloro-1,2,2-trifluoroethane (924).

Instead of removing nitric acid by distillation, it may be neutralized with ammonia; sulfuric acid is added, and the dibasic acids crystallize from the nonviscous mother liquor (925). Or, in an alternate procedure, after the oily layer is removed from the 63% nitric acid solution containing the dibasic acids, sulfuric acid is added, and the nitric acid removed by distillation (926). Ammonia and water are added to the residue, and suberic and azelaic acids crystallize out in a 4:6 ratio.

A 1968 patent describes the reaction of dinitrogen tetroxide with oleic acid to produce 9(10)-nitro-10(9)-peroxynitrato-octadecanoic acid followed by conversion to a nitroketone intermediate, which is then hydrolyzed to pelargonic and azelaic acid (927).

Nitric acid oxidation of 1,9-dimethoxynonane to produce azelaic acid has been carried out at 25°C with 100% nitric acid in the presence of vanadium catalyst (928). The dimethoxynonane was prepared from the coupled product from 1-methoxy-but-3-ene-1-yne and 1-chloro-5-methoxy-2-pentene.

3. PERMANGANATE OXIDATION

The permanganate oxidation of potassium ricinoleate is an *Organic Syntheses* preparation (929) and has been discussed in some detail by Cason and Rapoport (930). This method has the advantage over nitric acid oxidation, since only small amounts of suberic are formed. Azelaic acid melting at 104–106°C is obtained in about 35% yield.

Armstrong and Hilditch (931) reported an 83% yield when methyl oleate was oxidized with potassium permanganate in acetone solution.

4. AIR OXIDATION

Because air is the cheapest oxidizing agent, it has long been investigated, and continues to be investigated, for the preparation of azelaic acid. However, the reaction is not clean, yields are low, and a mixture of products is formed. Loder and Salzberg (932) carried out the oxidation for 4 hr in acetic acid at 30–40 atm and at 118–128°C in the presence of cobalt acetate to obtain a mixture of azelaic and suberic acid. Patrick and Emerson (933) studied: (a) air under pressure in the presence of chromium oxide in the absence of a solvent and (b) oxygen in acetic solvent with the mixed acetates of manganese, lead, and cobalt at atmospheric pressure. Suberic and azelaic acids were formed in no more than 10–15 wt.% yield. The reaction of oleic acid and air has been conducted in the vapor phase over various catalysts (934). Suitable catalysts include titanium vanadate–titanium dioxide on alumina at temperatures of 400–500°, and a mixture of vanadium, iron, and osmium oxides on pumice yields 22–32% azelaic acid (935,936).

Oxidation of oleic acid in acetic acid containing cobalt acetate at 50–100°C produces 9,10-dihydroxystearic acid in 10–15% yield, as well as azelaic acid in 65% yield (937). Oxidation of propyl oleate with uranium oleate as catalyst produces some 9,10-epoxystearic acid also (938). Oxidation of oleic acid in propanol at 86% in the presence of cobalt naphthenate is said to produce 9,10-dihydroxystearic acid in 19% yield and azelaic acid in 46% yield (939). In a two-stage process described by Morgan and Walker (911), the first step consists of air oxidation of oleic acid in acetic acid in the presence of acetaldehyde activator and with cobalt acetate catalyst, and the second step consists of nitric acid oxidation; a 92% yield of azelaic and suberic acids was reported.

Various other oxidation products have been described (940,941). Hydroperoxides from the autoxidation of methyl oleate occur in the order 10- > 11- > 8- > 9-hydroperoxido-oleates (942). Gamma-irradiation accelerates autoxidation of methyl oleate but does not affect the course of autoxidation; in this study the hydroperoxides occurred in the order 10 > 11 > 9 > 8 (941).

5. ALKALI FUSION

In a manner similar to the preparation of sebacic acid, alkaline cleavage of various oleic derivatives produces azelaic acid. Among the derivatives suitable are 9,10-dihydroxy and 9,10,12-trihydroxy stearic acid (the latter giving a mixture of azelaic and sebacic) (943,944), 9(10)-chloro-10(9)-hydroxystearic acid (945), 9(10)-cyano-10(9)-hydroxystearic acid (946), 9,10-epoxystearic acid, in low yield (947), cyclooctene-1 or 4-carboxylic acid (948), and chloropelargonic acid (949). Tall oil fatty acids are also adequate as starting materials when treated with aqueous caustic in the presence of group V salts (913).

6. DICHROMATE OXIDATION

Although chromium(VI) compounds are efficient oxidizing agents for cleavage of the double bond in oleic acid, they are expensive. Many efforts have been made to make the process less costly. These include employment of the reduced chromium in the form of $Cr(OH)SO_4$ as a tanning material (950,951) and electrolytic regeneration of the reduced chromium (952,953).

The technical process, used until development of the ozonolysis process, has been described in some detail (953). Pelargonic acid is removed from the reaction product by distillation at reduced pressure, and azelaic acid is recovered from the residue by recrystallization from hot water (954).

Other derivatives of oleic acid may be starting materials; for example, 9,10-dihydroxystearic acid (955–957), hydroxylated oleylacetate or methyl oleate (958), stearolic acid (959), and oleonitrile (960–962). The fatty acids from partially hydrogenated soybean or fish oil have also been used (963).

Applications for the cleavage products produced by this method have been described (961,962).

7. MISCELLANEOUS

Azelaic acid has been synthesized from chlorobromomethane-ethylene telomers (964). An interesting route is the dimerization of the ketene obtained from 4-carbomethoxybutyryl chloride and trimethyl amine to produce the symmetrical 5-ketoazelaic acid upon acidic hydrolysis (43). Korshak et al. (965) condensed acrylonitrile with cyclohexanone, then saponified and oxidized the product to get 4-oxononanedioic acid, which was reduced to azelaic acid.

Hydrogen peroxide also oxidizes 2-formylcyclooctanone (966) or methyl oleate (967).

Carmichael (855) carried out electrolysis of a mixture composed of potassium ethyl suberate and potassium ethyl malonate.

8. SEPARATION AND PURIFICATION

In addition to the methods for separating azelaic from suberic acid presented in Section VIII, a variety of methods have been described in the patent literature. When a solution of the sodium salts of a mixture of azelaic and sebacic acids is partially neutralized, sebacic acid is said to precipitate first, leaving sodium azelate in solution (968).

Either aqueous formic acid or dimethylformamide can extract azelaic from stearates or precipitate brassylic acid from a solution of azelaic and brassylic acid (969). A solution of 46% suberic and 54% azelaic acids in 89% acetic acid precipitates suberic acid in 93% purity; azelaic acid is recovered from a benzene solution of the remaining products (970). Fractional crystallization from dichlorobenzene (86% *ortho*) produces several fractions from which, after recrystallization from water, fairly pure dibasic acids are obtained (971,972). Homologous dibasic acids may be separated chromatographically on a column of silicic acid wet with buffered water by elution with increasingly polar chloroform-butanol solution (973). Fractional crystallization from 17.5% nitric acid has been used; suberic acid precipitates first (974).

Crude azelaic or sebacic acid may be purified by recrystallization from the aqueous solution, which has been first extracted at 98°C with a small amount of *sym*-tetrachloroethane to remove impurities less polar than the dibasic acids (975).

Fractional distillation offers a good method for separating the methyl esters of azelaic and sebacic acids (976), the propyl or butyl esters of the mixed acids from nitric acid oxidation of oleic acid (977), and the methyl esters of the products obtained by ozonolysis (893).

D. Physical Properties

Azelaic acid follows the pattern of the odd-numbered dibasic acids and is dimorphic. The β form, space group $C2/c$, is obtained as platelets by crystallization from acetone at room temperature; the α form, space group $P2_1/c$, as fine cylindrical needles from the melt (978,979). In the α form, one carboxyl group is 15° from the carbon plane and the other is 50° on the opposite side; in the β form, each carboxyl is 25° from the plane, on opposite sides (979). The two forms may crystallize side by side from the same solution. Elevation of temperature favors conversion of α to β forms (980a). Infrared spectra have been recorded for both crystalline (250) and molten (832) states.

Azelaic acid is much more soluble in water than sebacic and even somewhat more soluble than suberic (Table 12). Solubility in water in terms of grams per 100 cc of aqueous solution has been given by Lamouroux (861): 0°C, 0.10; 20°C, 0.24; 50°C, 0.82; and 65°C, 2.2. Molar solubilities in aqueous solutions containing various percentages of ethanol at 25°C has been given by Bonhomme (68): 0%, $1.12 \times 10^{-2}\,M$; 20%, $4.10 \times 10^{-2}\,M$; 40%, $32.1 \times 10^{-2}\,M$; 60%, $71.6 \times 10^{-2}\,M$; 80%, $100.3 \times 10^{-2}\,M$; 90%, $98.8 \times 10^{-2}\,M$; and 100%, $91.9 \times 10^{-2}\,M$. Solubility in ether is 1.88 g/100 g at 11°C and 2.68 g/100 g at 15°C (858); in benzene, 9.2 mg/100 g at 25°C (62). Distribution coefficients ($k = C_W/C_{org}$) at 26°C are: hexane, $\geqq 10$; chloroform, 4.6; ether, 0.14; ethyl acetate and n-butanol, $\leqq 0.1$ (774).

TABLE 12

Physical Properties of Azelaic Acid

Property	Value
Molecular weight	188.22
Melting point, °C (38)	107–108
Differential thermal analysis, endotherms, °C (23)	95, 310, 340
Boiling point, °C/mm Hg (770)	286.5/100, 265/50, 237/15
Density, g/cc (979)	1.235 (α)
	1.245 (β)
Refractive index (65)	$N_x = 1.466$, $N_y = 1.495$, $N_z = 1.582$
Heat of combustion, cal/g (38)	6066.7
Dipole moment at 25°C (771)	2.33
Solubility, g/100 cm³ aqueous solution (980b)	
15°C	0.212
22°C	0.214
44.5°C	0.817
55°C	1.648
Ionization constants in water at 18°C (68,772)	
K_1	2.88×10^{-5}
K_2	3.86×10^{-6}

E. Analytical Procedures

Composition–melting-point relations have been given for binary mixtures of azelaic with suberic, sebacic, and undecanedioic acids (791). Infrared, ultraviolet, and distillation characteristics for diethyl azelate have been assessed as a means of characterizing mixtures of acids in polymeric esters (981). Gas-liquid chromatographic analysis of dimethyl azelate has been reported (264,780). Thin-layer (377) and paper (788,789) as well as column (779,982) chromatographic methods have also been recorded. Mass spectral data for the dimethyl ester have been detailed (983).

F. Toxicology

Sodium azelate is mildly nephropathic when administered subcutaneously to rabbits (863). Azelaic acid is only slightly toxic when given in large single doses to rabbits or when fed daily to rats over long periods, after which it is excreted in the urine (984). An average of 60% of the ingested acid was recovered from the urine of dogs fed up to 45 g of azelaic acid dissolved in sodium carbonate and mixed with their food; none was recovered from the feces (985). Recovery in urine was 60–70% in humans fed 54–57 g of azelaic acid (867).

2-Ethylhexyl, cyclohexyl, and methyl isobutyl carbinol esters of azelaic acid were practically nontoxic upon intraperitoneal injection of animals (986). Rats showed growth retardation and elevated kidney weights when fed 15% but not when fed 0.5% of di-n-hexyl azelate in their diets for 2 years; dogs showed no changes (987).

G. Polymer Applications

The use of azelaic acid in the preparation of polyesters (74) and polyamides (75) has been reviewed. Crystallographic data for polyalkylene azelates have been reported (391). Polymer and other applications for azelaic acid were reviewed in 1954 (988). Nylon 6/9 is said to be equivalent to nylon 6/10 for many plastics (989). A bibliography of azelaic acid polymers and other uses is available (990). Polyazelamides from *trans*-1,4-cyclohexanebis-(methylamine) have considerably higher melting points than those from the cis isomer and somewhat higher than those from p-xylene-α,α'-diamine (869).

Nonamethylene diamine, derived from azelaic acid through the nitrile, has excited considerably more interest as a monomer than has the acid. The diamine is used to prepare poly(nonamethylene urea); this compound has reached semicommercial production in Japan, where it is known as "Urylon"

(991,992). A *Macromolecular Syntheses* procedure is available for this polyurea (993).

X. SEBACIC ACID

A. Introduction

Sebacic (L. *sebaceus*, made of tallow) acid has been called variously *Fettsäure*, *Brenzölsäure*, pyroleic, and ipomic acid. It was first described by Thenard in 1802, who obtained it by the dry distillation of hog fat. Berzelius believed it to be identical to benzoic acid, but the elemental analyses of Dumas and Peligot established its correct formula and proved its separate identity. Redtenbacher (1840) showed that other oleic-containing fats could be used, including olive oil and ox tallow. Bouis (1851) prepared it by distilling ricinoleic acid or castor oil in the presence of caustic—essentially the present commercial method. Mayer (1852) and Arppe (1865) showed that it could also be obtained by nitric acid oxidation of jalapinic acid or spermaceti. Brown and Walker (1891) reported its synthesis in 20% yield by the electrolysis of monoethyl adipate. Gauthier (1909) synthesized it by the action of carbon dioxide on the mixed product obtained by the action of magnesium on 1,4-dibromobutane. It has been prepared from undecylenic acid by ozonolysis as carried out by Noorduyn (1919) or by action of acid on its adduct with dinitrogen tetroxide as carried out by Jegorow (1912). A review on the manufacture and on the applications of sebacic acid was published in 1951 (994).

Sebacic acid is available commercially at a price of $0.70–0.74/lb and its dimethyl, dibutyl, and dioctyl esters at the respective prices of $1.25, $0.65, and $0.57/lb. Production figures are available only for the dibutyl and dioctyl esters, quoted at 5.0 and 5.1 million lb, respectively, for 1968 (437). Sebacic acid is used for the preparation of nylon 6/10, and its esters have value as low-temperature plasticizers for poly(vinyl chloride) and as lubricants. Manufacturers listed in the 1967 U.S. Tariff Commission Report were Rohm and Haas Company and Harchem Division, Pennwalt Corporation.

B. Chemistry

Sebacic acid has a higher degree of thermal stability than any of its lower homologs, since a rapid rise in carbon dioxide evolution occurs only at 360°C (22). Slow, dry distillation at 340°C produces chiefly pelargonic acid and minor amounts of other degradation products, including a nonenoic acid and a small amount of cyclononanone (995). In the presence of ethylene glycol, decomposition amounts to 0.02% after 9 hr of heating at 280°C (74). Thermal decomposition of its salts as in the Ruzicka large-ring synthesis results in only about 1% or less yield of the cyclic ketone (872,873).

When heated with acetic anhydride, sebacic acid forms a polymeric anhydride containing terminal acetyl groups (996). Molecular distillation of this anhydride produces a cyclic anhydride dimer and a residue of high molecular weight having the capability of being cold drawn.

Several reactions of sebacic acid and its derivatives have been described in *Organic Syntheses:* ethyl hydrogen sebacate can be isolated in 60–65% yield after a mixture of the diester and the diacid has been refluxed in the presence of hydrochloric acid (997), sebaconitrile and ω-cyanopelargonic acid are formed in the respective yields of about 50 and 30% by heating with urea (998), sebacoin is formed in 65% yield by treatment of the dimethyl ester with sodium (999), and electrolysis of methyl hydrogen sebacate produces dimethyl octadecanedioate in about 70% yield (1000). Methyl hydrogen sebacate may be prepared by esterification of the diacid in 72% yield when an ion-exchange catalyst is used (1001). Electrolysis of methyl hydrogen sebacate under conditions differing from those leading to the 18-carbon diester can lead to synthesis of capric (decanoic) acid in 75% yield (1002).

Sebacic acid is subject to oxidation by boiling nitric acid (1003), by chromium(VI) oxide (27), and by oxygen at 130–170°C accompanied by decarboxylation (1004).

Thermal stabilities (1005) and physical chemical properties (1006) have been reported for a number of sebacic esters.

C. Synthesis

1. ALKALI FUSION OF RICINOLEIC ACID

Ricinoleic (12-hydroxy-9-octadecenoic) acid is found in castor oil in amounts up to 90% of the fatty acids present; it is a unique starting material for a variety of cleavage products, with the specificity for any one pair of products depending on reaction conditions. Pyrolysis of methyl ricinoleate at 250°C in a quartz flask produces methyl undecylenate in about 70% yield and heptanal (1007). When ricinoleic acid is treated with excess caustic at 190–205°C 10-hydroxydecanoic acid and 2-octanone are produced, but at 250°C or higher sebacic acid and 2-octanol predominate (1008,1009). As high as 87% yields of sebacic acid have been claimed when the reaction is carried out at autogenous pressures of 1200–1500 psi to prevent loss of water vapor (1010). If the reaction at 190°C is carried out in a reducing medium, such as an unhindered primary or secondary alcohol, the hydroxy acid is formed in more than 80% yield (1011–1013).

The most likely reactions occurring in this sebacic acid synthesis are (1011,1014,1015):

Dehydrogenation of ricinoleic acid:

$$RCH(OH)CH_2CH{=}CHR' \xrightarrow{\text{OH}^-} RC(O)CH_2CH{=}CHR' + H_2$$

and predominantly

$$RCH(OH)CH_2CH{=}CHR' + OCHCH_2R' \xrightarrow{\text{OH}^-}$$
$$RC(O)CH_2CH{=}CHR' + HOCH_2CH_2R'$$

Double-bond shift:

$$RCH(OH)CH_2CH{=}CHR' \xrightarrow{\text{OH}^-} RCH(OH)CH{=}CHCH_2R'$$

and predominantly

$$RC(O)CH_2CH{=}CHR' \xrightarrow{\text{OH}^-} RC(O)CH{=}CHCH_2R'$$

retro-aldol condensation:

$$RC(O)CH{=}CHCH_2R' \xrightarrow{\text{H}_2\text{O,OH}^-} RC(O)CH_3 + OCHCH_2R'$$

reversible equilibration of ketone-alcohol:

$$RC(O)CH_3 + HOCH_2CH_2R' \xrightleftharpoons{\text{OH}^-} RCH(OH)CH_3 + OCHCH_2R'$$

irreversible oxidation:

$$R'CH_2CHO \xrightarrow{\text{H}_2\text{O,OH}^-} R'CH_2COOH + H_2$$
$$[R = CH_3(CH_2)_5, \; R' = (CH_2)_7COOH]$$

Since hydroxy acid is formed in high yield at 200°C or below, and since 2-octanol and sebacic acid are formed in high yield at higher temperature, ricinoleic acid must be dehydrogenated almost exclusively by the aldehydic acid. 2-Octanone and 10-hydroxydecanoic acid have been shown to react as indicated in the last two equations (1014).

Ricinoleic acid, castor oil (1010), alkyl ricinoleates, ricinoleic amide (1008,1009), or ricinoleyl alcohol (1016) are suitable starting materials.

A number of process variables have been patented. The reaction may be carried out continuously as well as by batch procedure under pressure; bleed-off of hydrogen without loss of water may be desirable to prevent too high a pressure (1010). Foaming due to hydrogen evolution in the presence of sodium soaps can be minimized by slow addition of starting material or by use of a high boiling hydrocarbon oil (1008,1009). A reaction medium of white mineral oil, having a boiling range of 300–400°C, avoids the need for pressure reactors and for return of water, which is usually necessary to reduce reaction viscosity (1017). A mixture of cresols has been used (1018). Steam at 325°C may take the place of external heating (1019). Lime has been used in place of sodium or potassium hydroxide (1020). Product sebacate salts are usually treated with acid to recover sebacic acid, but electrolytic conversion has also been described (1021). Sodium nitrate present during the reaction increases yields of both sebacic acid and 2-octanol (1022).

The continuous process can be carried out rapidly at normal pressure in simplified equipment with 25–30% instead of the usual 60–100% excess alkali (1023,1024).

Various catalyst systems have been described including calcium salts (1008,1009), cadmium soaps (1025,1026), barium salts (1027), lead oxide or salts (1028), lead salts in the presence of barium compounds (1029), nickel metal also in the presence of barium salts (1030), and zinc oxide (1031). Barium salts will precipitate barium sebacate, which then can be separated by filtration and avoid neutralizing the excess caustic in an alkali bath (1027).

Pilot-plant experiments in a 118-liter, copper-covered, iron kettle have been described (1032), as well as the pilot syntheses for dibutyl and dioctyl sebacates (1033).

Methods for recovering by-product acids from the alkaline fusion of castor oil have been devised (1034,1035).

Partial neutralization of crude, fatty-acid-containing sebacic acid can be used to remove the fatty acid impurities (1036). Charcoal treatment of the sebacic acid may be necessary to obtain a colorless product (1037).

2. ELECTROLYSIS

Adipic half-esters form sebacate esters in up to about 75% yield when electrolyzed in methanol containing sodium methoxide at 64°C (1038) or in methanol containing potassium formate (1039):

$$2ROOC(CH_2)_4COO^{\ominus} \xrightarrow{-2e} ROOC(CH_2)_8COOR + 2CO_2$$

Adipic acid monoamide may also be used as the starting material (1040). Loss of platinum from the anode can be minimized by use of a platinum-coated titanium or tantalum electrode; a graphite anode is more practical, even though yields based on starting material are somewhat lower (1041). Kinetics and mechanism have been discussed (1042). In a two-stage electrolysis process said to increase conversion (96.5%), with a yield of 80–85% and a current efficiency of 60–65% (1043), the second stage is performed with a mercury cathode from which sodium amalgam is continuously removed. Other anode materials have been examined (1044). Extraction of the electrolyte solution with isooctane permits isolation of dimethyl sebacate containing little adipic ester (1045).

An alternate synthesis is the hydrolysis of 1,1,1,5-tetrachloropentane to 5-chloropentanoic acid, followed by electrolysis, conversion of the dihalide to a dinitrile, and hydrolysis to sebacic acid (1046).

3. Carbonation of Disodiooctadiene

Under certain conditions, butadiene undergoes additive dimerization in the presence of sodium to form an isomeric mixture of disodiooctadiene (1047):

$$CH_2=CHCH=CH_2 + Na \longrightarrow Na^+[\bar{C}H_2CH=CHCH_2\cdot]$$

$$6[\bar{C}H_2CH=CHCH_2\cdot] \longrightarrow \begin{pmatrix} \bar{C}H_2CH=CHCH_2CH_2CH=CH\bar{C}H_2 \\ \updownarrow \\ CH_2=CH\bar{C}HCH_2CH_2CH=CH\bar{C}H_2 \\ \updownarrow \\ CH_2=CH\underline{C}HCH_2CH_2\underline{C}HCH=CH_2 \end{pmatrix}$$

Carbonation of this mixture followed by hydrogenation produced sebacic, 2-ethylsuberic, and 2,5-diethyladipic acids in the nonstatistical ratios of 3.5:5:1. The additive dimerization reaction was carried out with finely divided sodium in the presence of a sodium carrier (e.g., o-terphenyl) and in an active ether (e.g., ethylene glycol diethyl ether) solution at −40 to −50°C. Sebacic acid was recovered in about 35% yield by fractional crystallization from hot toluene. The remaining isomeric mixture, termed "isosebacic" acid by the developer, contains 72–80% 2-ethylsuberic, 12–18% diethyl adipic, and 5–10% sebacic acids. Although isosebacic acid reached pilot-plant development, it did not attain the status of commercial production. Numerous patents describe various processing details (1048–1072).

4. Miscellaneous

Ozonolysis has been used to prepare sebacic acid from undecylenic acid in 93% yield with oxidation of the ozonolysis products by alkaline silver oxide (1073). Thermal decomposition at 150°C in ethanol of the ozonolysis products from ethyl undecylenate obtained in ethanol resulted in about a 60% yield of diethyl sebacate (1074). Ozonolysis of 1,11-dodecadiene to sebacic dialdehyde has also been used, followed by oxidation with alkaline hydrogen peroxide or preferably with oxygen in the presence of permanganate (1075).

Sebacic acid has been produced by nitric acid oxidation of either cyclodecane (1076) or cyclodecanol (836), and by reaction of undecylenic acid with dinitrogen tetroxide followed by treatment with water, permanganate, or tin and hydrochloric acid (1077). Cyclodecanone, obtained from nonane-1,9-dicarboxylic acid, can be oxidized with chromic acid to give sebacic acid (1078).

Naphthalene is a cheap starting material but requires a multistep process to obtain sebacic acid. The steps include hydrogenation of naphthalene to decahydronaphthalene; autoxidation to the *trans*-9-hydroperoxide; decomposition of *trans*-9-hydroperoxydecahydronaphthalene or its benzoate ester to 6-hydroxycyclodecanone by acid treatment (1079–1084); dehydration to cyclodec-5-enone, followed by oxidative ozonolysis; and hydrogenolysis of the resultant 5-ketodecane-1,10-dioic acid (1085–1087). Alternatively, the cyclodec-5-enone can be hydrogenated to cyclodecanone (1088).

Reaction of ferrous sulfate with the peroxides produced by photoinitiated liquid-phase oxidation of cyclopentanol produced sebacic acid in 24% yield (1089). Thermal decomposition of the peracid from ethyl hydrogen adipate produced diethyl sebacate in 40% yield (1090).

4,7-Dioxodecan-1,10-dioic acid (dilevulinic acid) is available from furfural and levulinic acid and has been reduced to sebacic acid by hydrogenation over Raney nickel (1091), by electrolysis (826), and by amalgamated zinc and hydrochloric acid (1092). Other potential intermediates include 2,5-furan-dipropionaldehyde from condensation of furan with acrolein (1093); 5-methoxy-1-pentyne from tetrahydrofurfuryl chloride by reaction with sodium followed by methylation, bromination, and dehydrobromination (1094); the condensation product from 1,3-cyclohexadione (dihydroresorcinol) and γ-bromocrotonic acid (1095); and 1,8-dicyano-2,6-octadiene from the reductive dimerization of 5-chloro-3-pentenenitrile with iron powder (1096).

D. Physical Properties

Sebacic acid crystals are monoclinic with tabular habit and are best formed by recrystallization from warm concentrated nitric acid solution or from amyl acetate. The space group is $C_{2h}^5(P2_1/a)$ and there are 2 molecules per unit cell (862,1097). The carboxyl groups are tilted by approximately 3° from the plane of the central carbon atoms. The central carbon–carbon bond at the center of symmetry has the normal length of 1.54 Å, but the other bonds alternate between long and short. In general, the crystallographic properties are similar to those of adipic acid. Apparently a second crystalline form and a transition temperature have not been reported, although we might expect a transition at very low temperatures, as occurs for adipic acid.

Infrared absorption spectra of sebacic acid at various temperatures up to 124°C have almost complete coincidence, and the full pattern of absorption characterizing the higher homologs has become established in contrast to the changes shown by the lower homologs (368). The Raman (765) and infrared absorption spectra (250,423) have been recorded.

Sebacic acid is only slightly soluble in water, almost insoluble in benzene, but readily soluble in alcohol and ether. Quantitative data have been given for solubilities in benzene and in 0.0019 M hydrochloric acid at 35–65°C (767). At 20°C solubilities in grams per 100 ml of solution are 3.25 in acetone and 0.97 in ethyl acetate (64). Distribution coefficients ($k = C_W/C_{org}$) at 26°C are: hexane, 6.5; carbon tetrachloride, 4.5; benzene, 3.6; butyl acetate, 0.11; and ethyl acetate, methyl ethyl ketone, n-butanol ≤ 0.1 (774).

Physical properties for a number of dialkyl sebacates have been reported (1098), and useful physical properties for sebacic acid are given in Table 13.

TABLE 13
Physical Properties of Sebacic Acid

Property	Value
Molecular weight	202.24
Melting point, °C (368)	134.0–134.4
Differential thermal analysis, endotherms, °C	120 (769), 155 (175)
Differential thermal analysis, exotherms, °C (769)	270, 320, 415, 440
Boiling point, °C/mm Hg (770)	294.5/100, 273/50, 243.5/15
Vapor pressure, log p (mm Hg) (368)	$18.911 - 8395/T$
Enthalpy of sublimation, kcal/mole (368)	38.4
Density, g/cc (1097)	1.231
Refractive index (65)	$N_x = 1.470, N_y = 1.507, N_z = 1.589$
Heat of combustion, cal/g (38)	6415.2
Dipole moment at 25°C (771)=	2.38
Heat of solution in benzene at 35–55°C, cal/mole (767)	21,000
Solubility, g/100 cm³ aqueous solution (861)	
0°C	0.004
20°C	0.10
35°C	0.16
50°C	0.22
65°C	0.42
Ionization constants in water at 25°C (68)	
K_1	3.1×10^{-5}
K_2	3.6×10^{-6}

E. Analytical Procedures

Composition–melting-point relations have been given for binary mixtures of sebacic with suberic and dodecanedioic acid (791). Liquid-chromatographic separation of sebacic from other dibasic acids has been described (427,779), and paper (193,194,788,789), thin-layer (119,265,377,783,785), and gas-liquid (264) chromatographic procedures have been developed.

F. Toxicology

Sebacic acid is essentially nontoxic and is excreted 30–46% unchanged in the urine of humans (1099) and 30% unchanged in the urine of dogs (1100). It is similarly excreted by rats and ribbits (984). The acute oral LD_{50} in milligrams per kilogram for ethyl sebacate is 14,470 for the rat and 7280 for the guinea pig (1101).

G. Polymer Applications

Sebacic acid reacts with hexamethylene diamine to form the polyamide known as nylon 6/10:

$$x HOOC(CH_2)_8 COOH + x H_2N(CH_2)_6 NH_2 \rightarrow$$

$$HO [C(O)(CH_2)_8 C(O)NH(CH_2)_6 NH \overline{]_x} H + (2x - 1)H_2O$$

Nylon 6/10 absorbs less moisture than either nylon 6 or nylon 6/6 and, consequently, it possesses better dimensional stability and electrical properties. These properties have promoted nylon 6/10 for plastics where these properties are of special benefit. Because of the high cost of sebacic acid, it is often used for preparation of copolyamides rather than homopolymers.

Crystallographic data for polyalkylene sebacates have been reported (391). Among the polyester rubbers investigated during World War II, the sebacates had considerably greater hydrolytic stability than the succinates (393).

XI. DODECANEDIOIC ACID

A. Introduction

Dodecanedioic acid was first prepared, in low yield, by Noerdlinger in 1890 by adding hydrobromic acid to methyl 10-undecenoate, reacting the product with potassium cyanide and hydrolyzing the ester nitrile. Although he was unaware of the formation of isomeric bromo compounds, he apparently isolated the correct isomer. Walker and Lumsden, as well as Komppa in 1901, confirmed Noerdlinger's synthesis and independently carried out electrolysis of ethyl hydrogen pimelate to produce the diester. In 1909 von Braun reported its synthesis from 1,10-diiodododecane and potassium cyanide.

Dodecanedioic acid occurs naturally only as an ester in the urine of pregnant mares. However, its formation by oxidation of several natural products has been described: Barrowcliff and Power in 1907 reported the alkaline permanganate oxidation of hydnocarpic acid, and in 1910 Bougault presented his account of the chromic oxide oxidation of sabinic acid.

Dodecanedioic acid has only recently become an important industrial item, with the announcements by E. I. du Pont de Nemours & Company of the new textile fiber "Qiana" and of the new polyamide nylon 6/12 in 1968 and 1970, respectively. Both products have this acid as the dibasic acid component.

B. Chemistry

A polymeric anhydride is formed when dodecanedioic acid is refluxed in acetic anhydride (24). The polymeric anhydride, when heated at 110°C in a molecular still, forms a cyclic dimer stable up to its melting point of 76–78°C.

When heated gradually to 500°C, the thorium salt produces cycloundecanone in small yield (1078).

Mareš and Roček (27) found the oxidation rate constant to be $22 \times 10^{-3} \, lM^{-1}sec^{-1}$ when dodecanedioic acid is oxidized by chromic acid in acetic and sulfuric acid at 50°C.

The diester, acid chloride, and nitrile may be prepared in the usual manner. The half-ester methyl hydrogen dodecanedioate is best prepared by partial saponification of the diester in the presence of barium hydroxide (1102,1103). The acid chloride of the half-ester is readily obtained with thionyl chloride (1104). The acyloin condensation has been described for dimethyl dodecanedioate (33,34).

C. Synthesis

Cyclododecatriene and cyclododecane became commercially available in 1968 and are the most economically feasible starting materials for the synthesis of dodecanedioic acid. Cyclododecatriene is produced by the trimerization of butadiene (1105). The catalyst may be a Ziegler or a complex nickel type—one that permits use of a C_4 fraction rather than pure butadiene is a polyalkyltitanate-dialkyl-aluminum chloride (1106).

1. Oxidation of Cyclododecanol/one

Following the pattern of adipic acid synthesis, an intermediate stage of oxidation is usually required for the dodecanedioic acid synthesis. Cyclodecanol is formed by air oxidation of cyclododecane in the presence of a boron compound (1107–1115). A relatively small amount (8–10%) of cyclododecanone also forms as are various cyclododecanediols (1116). In the absence of a boron compound, air oxidation produces a mixture of cyclododecanol and cyclododecanone. Cobalt naphthenate is a suitable catalyst, and at 120–125°C the alcohol/ketone mixture appears in a ratio of about 1:2.5 (1117). If the oxidation is carried out at 135°C at lower conversion, in the absence of a redox catalyst, and if the hydroperoxide is reduced chemically, the alcohol/ketone ratio becomes 15:1 (836).

An alternate procedure to cyclododecanol involves epoxidation of cyclo-dodecatriene followed by hydrogenation of the monoepoxide (1107,1118–1120). Cyclodecanol may also be prepared via sulfation of cyclododecene (1121).

Cyclododecanol is preferred for the nitric acid oxidation to the dibasic acid (1122), but the ketone is necessary for making dodecalactam via the oxime (1123). 2,2-Dinitrocyclododecanone has been isolated from the reaction of cyclododecanone with 54% nitric acid at 65–70°C (1122).

Oxidation of cyclodecanol with 60% nitric acid in the presence of ammon-ium metavanadate forms dodecanedioic acid in about 90% yield (1107,1123). A copper-vanadium catalyst produces dodecanedioic acid in 92% yield from an alcohol/ketone mixture (1124). The product contains 97.1–98.6% of C_{12}, 0.9–2.5% of C_{11}, and 0.1–0.3% of C_{10} dibasic acid, along with 0.1–0.3% of unidentified material (1125). A two-stage system with the first reactor at 98°C and the second at 104°C may be used (1126).

A continuous pilot process in which 60% nitric acid is used at 60°C has been described (592).

Cyclododecene, formed by partial hydrogenation of cyclododecatriene, can also be oxidized with nitric acid to dodecanedioic acid. Oxidation may be carried out in acetic acid solution with 70% nitric acid (1119) or with 50% nitric acid with a vanadium-osmium catalyst (1127).

9,10-Dihydroxy-1,5-dodecadiene has been converted to dodecanedioic acid by hydrogenation followed by alkaline permanganate oxidation (1128).

Direct oxidation of cyclododecane to dodecanedioic acid has been accom-plished with dinitrogen tetroxide over a vanadium-on-silica catalyst, but the reaction is slow (1129).

Cyclododecanone may be oxidized by air in acetic acid solution containing 2–10% of water with a manganese acetate catalyst (1119). A Baeyer-Villiger reaction with a peracid and cyclododecanone also produces the dibasic acid in addition to the expected lactone (1130). A base-catalyzed oxidative cleavage of cyclododecanone occurs to the extent of 40–48% in hexamethyl-phosphoramide (675).

2. OZONOLYSIS

1,5,9-Cyclododecatriene has been ozonized selectively to the monoozonoly-sis product in either dichloromethane-ethanol at −78°C and 4% conversion (1131), or in aqueous emulsion in alkaline hydrogen peroxide and 50% conversion at 5°C (730,731). The aqueous ozonolysis products were oxidized *in situ* to give 4,8-dodecadienedioic acid in 60 mole % yield. The alcoholic ozonolysis products were oxidized by silver oxide to give the same unsaturated acid. The silver oxide is capable of being recycled, and less than stoichiometric amounts can be used if oxygen is bubbled continuously through the solution (848).

To avoid having to recycle large quantities of starting material as a result of low conversion, ozonolysis of the monocycloolefin is probably to be preferred. Cyclododecatriene can be reduced to cyclododecene by partial hydrogenation (1132–1134) or by reduction with lithium in ethyl amine (1135).

Thermal decomposition of cyclododecene ozonolysis products in propionic acid yields large amounts of 11-formylundecanoic acid, as well as dodecandioic acid (847). Oxidation of ozonolysis products is best accomplished with oxygen in steps of increasing temperature (849). Results are said to be good when water is absent from the ozonolysis step but present in the amount of 10% during oxidative decomposition (846).

3. PEROXIDIC SYNTHESES

An interesting reaction is the decomposition of cyclohexanone peroxide to dodecanedioic acid either thermally (1136) or by reduction with ferrous sulfate (1137,1138). Cyclohexanone peroxide is easily prepared by the reaction of cyclohexanone with hydrogen peroxide (1139) or by the oxidation of cyclohexanol with oxygen (663). The peroxide has a dimer structure, with the monomeric hydroxy hydroperoxide and other peroxides in equilibrium when in solution (663,1140,1141).

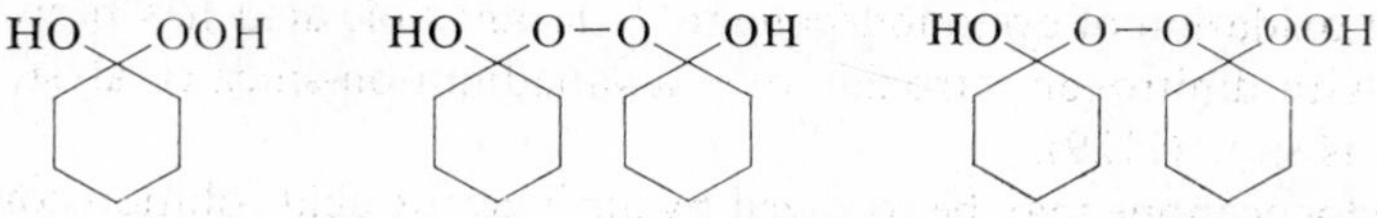

The formation of dodecanedioic acid occurs through the following sequence:

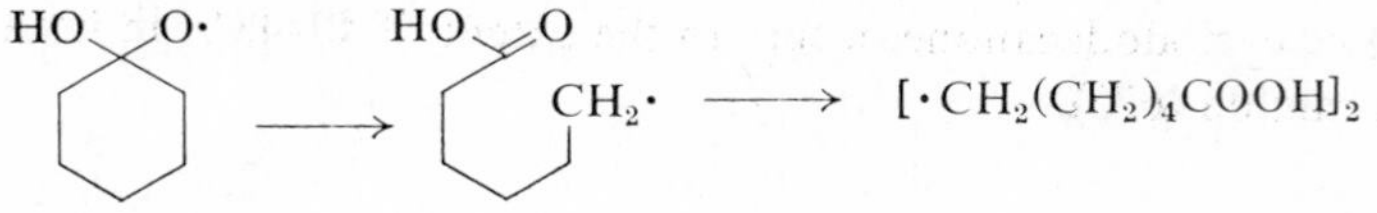

Yields were as high as 55% when the syrupy product first formed upon reaction of cyclohexanone with 90% hydrogen peroxide was treated with ferrous sulfate. Upon standing, the syrupy product gradually crystallized; the crystalline peroxide, similar to that obtained with hydrogen peroxide of lower concentration, formed dodecanedioic acid in low yield (1142). Alternatively, when the cyclohexanone peroxide was given a preliminary treatment with cyclohexanone and caused to react with ferrous caproate in toluene solution, it produced dodecanedioic acid in 64% yield (663).

Pimelic acid is formed if the decomposition of 1-hydroxycyclohexyl hydroperoxide is carried out in the presence of carbon monoxide (1143). Terminally substituted acids are formed at the expense of dodecanedioic acid if certain halogen compounds are present (1144).

Unsaturated dicarboxylic acids are also formed when certain peroxides are decomposed in the presence of butadiene (1145,1146). Thus decomposition of the ketone peroxide from ethyl acetoacetate produces crude unsaturated esters in 50–70% yields, but the straight-chain ester of dodecanedioic acid forms in less than 20% yield (1146).

Peroxidic addition of chloroform to methyl undecylenate (10-undecenoate) yields methyl 12,12,12-trichlorododecanoate, which forms dodecanedioic acid upon hydrolysis (1147).

Electrolysis of the half-nitrile of pimelic acid has produced the nitrile of dodecanedioic acid (1148). Electrolysis of potassium ethyl malonate in the presence of butadiene produced 4,8-dodecadienedioic acid (1149). Under certain conditions, electrolysis of the half-ester of glutaric acid in the presence of butadiene forms dimethyl 6-dodecenedioate (1150).

4. ALKALINE CLEAVAGE

Alkaline cleavage of ricinoleic acid yields sebacic acid (Section X), but if the starting material is first hydrogenated to remove unsaturation, the product is then a mixture of undecanedioic and dodecanedioic acids (1151–1155). Steadman and Peterson (1155) found that the presence of cadmium oxide during the reaction at 325–330°C doubled the yield of mixed acids to 61% compared with that with sodium hydroxide alone. The cadmium oxide also inhibited corrosion of the steel autoclave.

Bagby (1156) found that alkaline cleavage of lesquerolic acid, 14-hydroxy-11-eicosenoic acid, produced 2-octanol and dodecanedioic acid.

5. NITRIC ACID OXIDATION

A mixture of undecanedioic and dodecanedioic acids is also produced by the nitric acid oxidation of 12-hydroxyoctadecanoic acid (1155,1157–1160). Yields as high as 80% result when the oxidation is carried out with concentrated nitric acid at about 90°C in the presence of copper and vanadium catalysts (1155). Caproic and heptanoic acids can be removed from the reaction product by steam distillation, and the nonvolatile acids can be purified by recrystallization from 30% aqueous acetic acid. The dicarboxylic acids can be separated as the ethyl esters by fractional distillation (1159).

6. MISCELLANEOUS

Dodecanedioic acid has been prepared from decamethylene glycol with the dibromide and the dicyanide as intermediates (39). Bromination, cyanation, and hydrolysis were accomplished in the respective yields of 90, 91, and 98 % for an overall yield of about 80 %.

Carboxylation at the double bond in undecylenic acid by carbon monoxide and water can occur at either the 10- or 11-carbon to produce a mixture of straight-chain and branched acids. Thus Reppe and Kröper (1161) reacted carbon monoxide and water with undecylenic acid in the presence of nickel carbonyl. Conversions were from 45 to 55 % at 170–280°C, 200 atm, and 15–16 hr. Under similar conditions, 1,10-decanediol may be converted to dodecanedioic acid in 75 % yield (40). Conditions apparently can be milder if a solvent is present; Ercoli (1162) reported 76 % conversion of undecylenic acid at 145°C, 200–210 atm, and a 1-hr reaction time when acetone was the solvent.

In the Koch reaction, carboxylation occurs in the presence of sulfuric acid. Koch and Schauerte (1163) found that the newly formed carboxyl group could be tertiary, as well as primary or secondary, as a result of extensive rearrangement of the carbon skeleton. Various alkyl substituted azelaic and sebacic acids were formed for the most part.

Traumatic acid, 1-dodecenedioic acid, a plant hormone, has been synthesized by the reaction of sebacaldehydic acid with malonic acid in the presence of pyridine (1164).

7. PURIFICATION

Dodecanedioic acid can be purified by recrystallization from water, then toluene (1165). By performing the recrystallization in water at 100–108°C, advantage could be taken of a specific crystal modification. Overall yield of dodecanedioic acid of 99.99 % purity was 98 % when the crude acid contained 1 % undecanedioic acid and 0.2 % sebacic acid. Recrystallization from diethylene glycol dimethyl ether is also said to be effective (1166). Suspension of dodecanedioic acid containing 10 % undecanedioic acid in 1 % sodium hydroxide solution at 100°C reportedly removes the undecanedioic acid (1167).

D. Physical Properties

Dodecanedioic acid crystals are monoclinic, have the space group $P2_1/c$, and have two molecules per unit cell (1168). The coefficient of expansion of the a axis has been reported (1169). The small, but measurable, solubility of dodecanedioic acid in water at various temperatures is given in Table 14. Its solubility in benzene ranges from 0.0049 g/100 ml at 35°C to 0.123 g/100 ml at 65°C (767).

Infrared spectra have been recorded for the solid at various temperatures, as well as for the melt (368).

TABLE 14
Physical Properties of Dodecanedioic Acid

Property	Value
Molecular weight	230.30
Melting point, °C (791)	128.7–129.0
Boiling point, °C/mm Hg (1170)	254/15, 245/10
Heat of combustion, cal/g (38)	6994
Vapor pressure, log p (mm Hg) (368)	$17.728 - 8006/T$
Enthalpy of sublimation, kcal/mole (368)	36.6
Heat of solution in benzene at 35–45°C (767)	21,200
Solubility, g/100 g H_2O (1171)	
23°C	0.003
28°C	0.005
54°C	0.027
84°C	0.120
98°C	0.306
100°C	0.368
Ionization constants at 25°C in 40% aqueous ethanol (68)	
K_1	2.0×10^{-6}
K_2	2.5×10^{-7}

E. Analytical Procedures

Composition–melting-point relations have been given for binary mixtures of dodecanedioic with sebacic and undecanedioic acids (791).

F. Toxicology

Dogs, fed at the daily rate of 0.48 g dodecanedioic acid/kg excreted 6.5% of the acid, with sebacic, suberic, and adipic acids also present in the urine (1172). Men, fed at the daily rate of 0.074 g/kg excreted 7.1% of the acid, but adipic acid was the only other dibasic acid present in the urine.

G. Polymer Applications

The composition of Qiana, the textile fiber introduced by du Pont in 1968, includes the polyamide from dodecanedioic acid and bis(*para*-aminocyclohexyl)methane. The preparation of the nylon salt (1173) and of the polymer (1174) have been described. Use of this polyamide is anticipated for plastics (1174) and tire cord (1175), as well as for textile applications. Incorporation of chlorinated biphenyl imparts flame retardancy to the polymer, which is then 100% self-extinguishable (1176).

Properties have been reported for the polyamide from dodecanedioic acid and hexamethylene diamine (1177), as well as for the polyamide made with 1,4-cyclohexane bis(methylamine) (868,869).

Polyethylene dodecanedioate can be depolymerized to form a cyclic, monomeric ester having 16 atoms in the ring (388). X-Ray fiber patterns have been determined for the ethylene glycol (1178) and the trimethylene glycol (430).

XII. BRASSYLIC ACID

A. Introduction

Brassylic acid was first described and named in 1867 by Haussknecht, who oxidized behenolic acid $[CH_3(CH_2)_7C\equiv C(CH_2)_{11}COOH]$ and erucic acid $[CH_3(CH_2)_7CH\equiv CH(CH_2)_{11}COOH]$ with nitric acid. He incorrectly assigned the formula of a C_{11} dibasic acid to his product and believed the unsaturation of either starting material was in the 11,12 position. In 1893 Grossmann further investigated the oxidation of behenolic acid, and Fileti and Ponzio studied erucic acid oxidation. Grossmann showed that Haussknecht's product was impure and that the correct formula was that of the C_{13} dibasic acid. In 1900 Krafft and Seldis described the synthesis of brassylic acid from ethyl 11-bromoundecanoate and sodiomalonic ester. Komppa repeated the malonic ester synthesis in 1901 and obtained a product with different characteristics, apparently unaware that he probably was working with an isomer of ethyl 11-bromoundecanoate. Walker and Lumsden in the same year repeated the sodiomalonic ester synthesis with the correct, 11-bromoderivative to confirm the synthesis of Krafft and Seldis. von Braun and Danziger carried out a synthesis in 1912 from 1,11-dibromo- or 1,11-diiodoundecane via the dinitrile.

Brassylic acid is not now commercially available. Should polyamides (1179), lubricants (1180,1181), or plasticizers (1182–1184) from brassylic acid develop commercial utility, erucic acid is available as a starting material for brassylic acid synthesis from *Crambe abyssinica*, a new oilseeds crop first planted commercially in 1965. Erucic acid is also present, to a lesser extent, in most rapeseed oils.

B. Chemistry

Brassylic acid forms a polymeric anhydride when refluxed with acetic anhydride. Molecular distillation of the polymer produces an unstable, monomeric anhydride (24). Most if not all the reactions of brassylic acid are similar to those of its homologs.

C. Synthesis

1. OZONOLYSIS

Ozonolysis appears to be the synthesis method of choice by reason of the absence of homologous acids as obtained in the nitric acid method. Holde and Zadeck (1185) ozonized erucic acid in chloroform and thermally decomposed the ozonolysis products in water to get a mixture of aldehydes, pelargonic acid, and brassylaldehydic acid; the same products resulted from brassidic acid, the trans isomer of erucic acid. Verkade et al. (38), following the same procedure, oxidized the aldehydes with permanganate. Acetic acid has been used as the solvent (1186). Blackmore and Szatkowski (890) ozonized erucic acid in aqueous emulsion, decomposed the ozonolysis products with sodium hydroxide, and oxidized the aldehydes with ozonized air. Greiner (1187) and Grynberg et al. (1188) ozonized in pelargonic acid.

In laboratory studies preliminary to a continuous pilot-plant experiment, yield potential, product stability, thermal effects to be anticipated, and preferred separation and recovery procedures were evaluated (1189). Brassylic acid of 98% purity was obtained in 70% of the theoretical yield after recrystallization. Alternatively, an 88% yield of dimethyl brassylate in 95% purity was achieved by removal of methyl pelargonate by distillation. The pilot-plant process also provided brassylic acid in 95% purity, but the yield was lower. The ozonization was carried out in an emulsion of propionic acid and water; the oxidation, with air at temperatures ranging from 60 to 97°C in the sieve plate tower; and the purification, by recrystallization from naphthol spirits.

2. MISCELLANEOUS

Reaction of erucic acid with hydrogen peroxide in acetic acid and the subsequent oxidation of the dihydroxy derivative with chromic acid (956, 958), periodic acid (1190,1191), or permanganate (1192) have been described.

It has also been discovered that fuming nitric acid may be used on a small laboratory scale to oxidize erucic acid (1193). Furthermore, oxidation of erucic acid with permanganate in acetone solution has produced brassylic acid in 78% yield (1192,1194).

Brassylic acid has been prepared from cyclododecanone in a several-step, but high-yield, process following the sequence: conversion to the cyanohydrin with hydrogen cyanide, dehydration with thionyl chloride, and oxidative cleavage with sodium hydroxide at 300–350°C (1195).

"Acid cleavage" under strongly basic conditions of methylene bisdihydroresorcinol followed by reduction of the 5,9-diketotridecanedioic acid with hydrazine produced brassylic acid in excellent yield (1196).

$$\longrightarrow \ HOOC(CH_2)_3CO(CH_2)_3CO(CH_2)_3COOH$$

4,7,10-Triketobrassylic acid (from acid cleavage of difurfurylideneacetone) may be reduced with hydrogen over nickel (825). The symmetrical keto acid has been prepared from the condensation of ethyl 5-iodovalerate with diethyl acetonedicarboxylate (51) and by the reaction of the acid chloride of pimelic acid with 1-morpholino-1-cyclohexene (49).

The reaction of methyl 10-undecenoate and ethyl bromoacetate has produced ethyl methyl γ-bromobrassylate in 66% yield; a zinc-copper couple removed bromine (1147). Condensation of methyl 10-undecenoate with ethyl cyanoacetate is a similar route (1197).

The sodiomalonic ester synthesis may be carried out with 11-bromoundecanoic acid in place of the ester to form brassylic acid in 87% yield (1198). The malonic ester synthesis has also been performed with 1,9-dibromononane (35,1199).

Other methods, mostly producing mixtures, include telomerization (964); alkaline cleavage of 13,14-dihydroxybehenic acid (943,944), of 13-chlorobehenic acid (1200), and of 13,14-epoxybehenic acid (1201); oxidation of methyl erucate in a stream of air at 96°C (1202); reaction of methyl 10-undecenoate with acetic anhydride in the presence of peroxide (1203); oxidation of 2-formyl-cyclododecanone with 28% hydrogen peroxide (1204); and electrolysis of a mixture of the half-esters of adipic and azelaic acids (1205).

3. PURIFICATION

Brassylic acid may be removed from saturated fatty acids (e.g., stearic acid) by solution in 30% formic acid in water (1206) or in water alone at elevated temperatures and pressures (1207).

Extraction of brassylic acid can also be accomplished by contacting the crude product containing it with a solution of disodium brassylate; when the resulting solution of the monosodium brassylate is cold, it separates out brassylic acid of high purity (1208).

Pelargonic acid can be extracted from brassylic acid by contacting the mixture with 1,1,2-trichloro-1,2,2-trifluoroethane (Freon-113) or similar halogenated hydrocarbon (924,1209).

Charcoal treatment in ethanol solution followed by crystallization from ethanol may be necessary to remove color bodies, which discolor brassylic acid polyamides (1209).

D. Physical Properties

Dupré la Tour (419) observed two crystal forms for brassylic acid; but the transition from one to the other was incomplete, even at temperatures slightly below the melting point. Housty (1210) found that the α form was stable at room temperature. As expected, α-brassylic acid belongs in the space group $P2_1/c$, has four molecules per unit cell, and has one carboxyl group at an angle of 50° from the chain of carbon atoms with the other at an angle of 15°30′ on the other side of the plane.

Brassylic acid is only slightly soluble in cold water (Table 15) but is

TABLE 15
Physical Properties of Brassylic Acid

Property	Value
Molecular weight	244.32
Melting point, °C (1211)	114
Density, g/cc (1210)	1.16
Heat of combustion, cal/g (38)	7238.9
Dipole moment (1212)	2.68
Solubility, g/100 g H_2O at 24°C (1213)	0.004
Ionization constants at 25°C in 40% aqueous ethanol (68)	
K_1	1.6×10^{-6}
K_2	2.9×10^{-7}

readily soluble in alcohol, ether, and chloroform. This acid is soluble in benzene only when hot, and it is insoluble in petroleum ether.

E. Analytical Procedures

Data have been reported for thin-layer (786) and gas-liquid (780) chromatography.

F. Toxicology

Dogs injected with brassylic acid excrete in the urine C_7, C_9, and C_{11} but none of the C_{13} dibasic acid (865,1100).

G. Polymer Applications

Nylons 13, 13/13, and 6/13 have been prepared on a laboratory scale. The lower water absorption of these nylons makes them suitable for uses requiring retention of strength, toughness, abrasion resistance, and electrical properties under varying conditions of humidity. These nylons can be compression or injection molded, and extruded to form rods, films, or filaments (1179,1214,1215).

XIII. DIMER ACID

A. Introduction

Dimer acids are the 36-carbon dicarboxylic acids formed when 18-carbon unsaturated fatty acids are given various treatments. The first published report on dimer acids came from DeNordiske Fabriker (DE-NO-FA) in a patent issued in 1919 for polymerization of the alkaline soaps of poly-unsaturated fatty acids (1216). Scheiber (1217) wrote in 1929 that the 1,4-pentadiene systems of polyunsaturated fatty esters shifted to conjugation before dimerization during thermal polymerization, and Kappelmeier (1218) in 1933 suggested that the conjugated esters reacted with nonconjugated esters to form dimers. Wheeler (1219) extended this hypothesis and established that the reaction does occur but that other reactions are important and lead to aromatic, bicyclic, and tricyclic dimers, depending on the catalyst and conditions. In 1939 and 1940 Bradley (1220) studied dimeric and trimeric fatty esters from thermal polymerization of the methyl ester of poly-unsaturated fatty acids and subsequent molecular distillation.

Commercial use came as a result of industry's applying information developed at the Northern Regional Research Laboratory between 1941 and 1948 for the preparation of polyester rubber and polyamide resins (1221–1224). Goebel (1225–1228) learned how to polymerize fatty acids—first, under pressure with added water at 290°C or higher temperatures and, later, with clays and comparatively low temperatures of 210–220°C. Whereas Goebel used acid-activated clays, others found that alkaline clays also worked (1229).

Most of the dimer acid produced in the United States involves either acidic or alkaline types of clay. Gradually production has increased, and estimates suggest the total may be more than 30–35 million lb. Substantial but smaller amounts are manufactured in Europe. Tall oil fatty acids are the main source of raw material. Castor fatty acids are being used in England. The water of dehydration is retained in the reactor to minimize decarboxyla-tion and polyesterification (1230). A somewhat similar acid is available in Germany from the condensation of one molecule of styrene with two of conjugated fatty ester (1231), but the main part of the styrene adduct prob-ably has three fatty acids for two styrene molecules.

B.   Chemistry

Dimer acids are a unique commercial chemical and have been recognized and used in part because of their uniqueness (1232). They contain olefinic as well as aromatic unsaturation if made with clay catalyst and, in contrast with most other dibasic acids, this unsaturation does not shift readily to conjuga-tion with either the carboxyl group or other unsaturation. The unsaturation is reactive, but activity is readily controlled.

Dimer acids form simple salts with monovalent metallic ions such as sodium, polymeric salts with divalent such as calcium, and polymeric ammonia complex salts with zinc (1233). Dimer acids form polymeric an-hydrides when reacted with acetic anhydride, and acid chlorides with thionyl chloride. Heating with ammonia or primary and secondary amines gives the expected amide. The amides will dehydrate at higher temperature to the nitriles, the nitriles can be reduced to dimer diamine, and the latter reacted with phosgene to give a diisocyanate (1234). If a diamine is used in excess with dimer acids and the resulting amide heated to 285°C, imidazolines are formed. With a triamine, such as diethylenetriamine, the product still retains an active amine group for epoxy reaction (1235,1236).

$$R—CO_2H + HN(CH_2CH_2NH_2)_2 \longrightarrow RCON(CH_2CH_2NH_2)_2 \longrightarrow$$

$$\begin{array}{c} RC\text{——}NCH_2CH_2NH_2 \\ \| \quad\quad | \\ NCH_2CH_2 \end{array}$$

The high molecular weight of 560+ for dimer acids permits ready condensation to useful polymers. With ethylene glycol or ethylene diamine, only 5–10 moles of water need to be removed to reach molecular weights of 3000–6000. For ethylene glycol polyesters of more than 3000–5000, it is necessary to have an excess of glycol and to carry out a glycolysis reaction (1237). Ethylene glycol tends to decompose slowly above 160°C. Apparently when stoichiometric amounts of the glycol are present in polyesters, the polyesterification reaction ceases because hydroxyl groups are lost at the end of enough polymers to stop the reaction. If an excess of glycol is present, loss of some hydroxy groups is no longer critical, since polyesterification proceeds by a glycolysis reaction as follows:

$$x HOCH_2CH_2O[CORRCO_2CH_2CH_2O]H \rightarrow$$

$$(x - 1)HOCH_2CH_2OH + HOCH_2CH_2O[CORRCO_2CH_2CH_2O]_xH$$

C. Synthesis

Polyunsaturated fatty acids can be converted to dimeric and trimeric products by several different methods. These methods do not always give the same dimer acid or the same percentages of dimer and trimer. Esters of polyunsaturated acids form dimeric and trimeric esters when heated to 270–310°C (1220). The reaction is slow at the lower temperatures and fairly rapid at higher temperatures. The acids will dimerize at these temperatures, but they also undergo decarboxylation. If water is added to the reaction and if the reaction is kept under pressure, the decarboxylation is held to a minimum (1225). Such catalysts as activated clay, boron fluoride and its derivatives, hydrogen fluoride, hydrogen peroxide, and other free radical sources lower the temperature of dimerization.

Activated acidic (1226) or alkaline (1229) clays are heated with polyunsaturated fatty acids, such as linoleic, under autogeneous pressure to 220–230°C for 4–24 hr. The products from a basic clay (pH 8) catalyst are: monomer, 26.6%; dimer, 59.3%; and residual polymer, 14.1% (1238). The dimer fraction contains a monocyclic aromatic dibasic acid with both saturated and unsaturated side chains and bicyclic and tricyclic structures. From mass spectra data, we know that the mono-, bi-, and tricyclic structures are present in the ratios of 100, 78, and 19, respectively. Boron fluoride (1239) and hydrogen fluoride at 200°C give rapid conversion to trimeric or higher polymeric fatty esters. Only about 10% of the product is dimeric.

Free radical catalysts will convert methyl linoleate to dehydro dimers that contain at least one conjugated system and four double bonds (1240). When heated to 250°C, these dehydro dimers lose conjugation and two double bonds, giving a bicyclic product. Oleate will also dimerize with free radicals (1241). Oleate reacts with *t*-butylperoxide to give a dehydro dimer that retains unsaturation in each chain of the combined oleate radicals, and linoleate undergoes a similar reaction to give an intermediate similar to that shown in the equation for the bicyclic dimer below. When the linoleate dimer from peroxide treatment is heated, it forms an apparent bicyclic dimer. The monocyclic aromatic component of linoleate dimer can, in part, be oxidized to 1,2,3,4-benzene tetracarboxylic acid, sometimes called prehnitic acid but apparently incorrectly so (1232). Stearate esters with free radical catalysts will also give dimeric fatty acids (1242).

Fatty-acid-derived dibasic acids may also be obtained by the reaction of two conjugated fatty acids (1230), the reaction of styrene with conjugated fatty acids (1231), the addition of carbon monoxide and hydrogen by the oxo reaction (1243) and subsequent oxidation, and the Koch reaction (1244).

Much conjecture has been expressed about the mechanism of formation and structure of dimer acids. They do differ in structure somewhat, depending on their manner of formation. Probably the best explanation of the reactions of oleic and linoleic esters by thermal, free radical, and clay-catalyzed polymerization is that of Wheeler and co-workers (1219,1238,1240,1241, 1245,1246). Apparently clay catalyzes the shift to conjugation in one molecule as the initial step. It is followed by a Diels-Alder reaction of the conjugated diene with one double bond of the nonconjugated fatty acid that serves as the dienophile. Subsequent or other reactions form bicyclic and aromatic structures as the predominant species, with minor amounts of tricyclic dimeric products. The relative order of these different molecular species in the clay-catalyzed dimer were aromatic monocyclic, bicyclic, and tricyclic dimer. The bicyclic derivative is probably formed by a free radical mechanism involving the conversion of linoleic acid to a free radical, combination of these radicals to a dehydro dimer, and subsequent intramolecular Diels-Alder reaction of conjugated systems present.

$$RCH{=}CHCH_2CH{=}CHR' \longrightarrow RCHCH{=}CHCH{=}CHR' \longrightarrow$$

$$\begin{array}{l} RCHCH{=}CHCH{=}CHR' \longrightarrow \text{bicyclic dimer} \\ | \\ RCHCH{=}CHCH{=}CHR' \end{array}$$

D. Physical Properties

Dimer acids never crystallize, despite their high molecular weight of 560. They are comparatively nonvolatile and distill with difficulty, except in a molecular or alembic still. These unusual properties are summarized in Tables 16 and 17.

TABLE 16
Unusual Properties of Dimer Acid

High molecular weight of 560
Distills with difficulty
Unsaturation not conjugated with carboxyl group
Unsaturation reactive but controllable
Never crystallizes
Soluble in hydrocarbons

E. Analytical Procedures

Usually acid and saponification values are straightforward determinations on dimer acids. One commercial concern has published its methods (1247). Saponification and acid values do not usually agree, apparently because of the formation of some anhydrides that are difficult to obtain as free acids. Iodine value run by any method is a rough approximation of unsaturation, since aromatic or tertiary hydrogens, or both, will substitute readily with many iodine value reagents (1222). Kaufman's bromine number is a superior method. Determining hydrogen values with active catalyst seems to work very well with some dimer acids (1240).

Analysis for monomer, dimer, and trimer can be by alembic distillation (1248), molecular distillation (1220), micromolecular distillation using a glass helical spring to determine weights lost (1249), and chromatography on silicic acids by column or thin-layer techniques (1250). The presence of small amounts of "dimeric" products similar to dimer acids but decarboxylated may present some problems in analyses. Low values for saponification equivalents on distilled dimeric or other products will give a measure of this undesired by-product.

F. Toxicology

Since the polyamides and polyesters from dimer acids are not absorbed, they are generally believed to be nontoxic (1251). Dimer acid in 5% amounts in the diets of rats and fed for 32 weeks impairs growth, retards liver function, and lowers body temperature and metabolic rate (1252). Bottino (1253) reports that polymeric fatty esters from methyl linoleate are mildly toxic but

TABLE 17
Characteristics of Commercial Dimer Acids

Property	Standard dimer	Low monomer	Purified dimer	Hydrogenated dimer	Trimer
Acid value	188–197	188–197	194–198	191–197	183–191
Saponification value	191–201	189–201	197–201	193–200	192–200
Color, Gardner-1963 (max)	9	7–9	5	3	—
Monobasic acid, %	2–5	1	0–1	0	—
Dimer acid, %	73–75	75–79	95–97	97–99	20
Trimer acid, %	22–23	20–25	4	3–5	80
Flash point, ASTM, °F	530–540	530–580	530	585	595
Fire point, ASTM, °F	580–600	600–650	610	660	680
Viscosity at 25°C, Gardner-Holdt	Z-4	Z5-6	Z-4	Z-3	—
Viscosity at 25°C, centistokes	6200	9500	5600	5200	—
Specific gravity 25–20°C	0.95	0.95	0.95	0.94	—
Unsaponifiables, %	0.3–2.0	0.3–1.5	0.3	0.1	—

that their effects are counteracted by addition of a good-quality vegetable oil in the diet. The possible presence of toxic cyclic or otherwise modified monomeric fatty acids in these polymeric esters was not ruled out in these studies by Bottino (1253) and Griem (1252). Crampton et al. have shown that the thermally modified monomeric fatty acids are toxic (1254) and that the dimer fraction from thermally polymerized linseed oil is nontoxic (1255).

G. Polymer Applications

The first application of dimeric fatty acid or esters commercially was to manufacture rubber substitutes for static uses, such as jar rings and grommets, during World War II (1221,1222). Studies on their conversion to polyamides (1223,1224) also occurred during the war, with the ethylene diamine polyamide finding use as a heat-sealing agent for glassine paper (1224,1256).

The carboxylic acid groups of dimer acid undergo most if not all the polymerization reactions of the carboxyl group, so that a wide variety of diamines, amino alcohols, and glycols can be reacted (1222–1224,1232). In short- and long-oil alkyds, dimer acid can increase viscosity of the alkyd, influence its flexibility, permit better production control; however, it gives only slight improvement in film properties over bodied oil or bodying of alkyd *in situ* (1257).

When an excess of amine is present, a basic polyamide can be made that assists in control of corrosion. These basic polyamides have found extensive use as the amine component of epoxy adhesives (1236). When these amine polyamides or dimeric fatty acid amides from polyamines, such as ethylene diamine or diethylene triamine, are heated to 285°C, imidazoline formation occurs to give a more active catalyst for epoxy reaction. When a polyamine is chosen, a primary amine group remains on the end of the polyamide chain.

The many uses of polyamide resins include heat-sealing resins, epoxy glues, coatings, thixotropic agents for paints, and casting and laminating resins for a wide variety of end products. Two good sources for further details are a review article (1232) and Floyd's book (1258).

XIV. C$_{19}$ DICARBOXYLIC ACIDS

A. Introduction

The C$_{19}$ dicarboxylic acids are the isomeric mixtures obtained by the action of carbon monoxide on oleic acid by any one of several different methods. Apparently, the first report of these acids was by Mannes and Pack (1259), who prepared them by alkali fusion of formylstearyl alcohol or methyl formylstearate. A general structure may be written in the following way:

$$CH_3(CH_2)_x—\overset{\overset{\displaystyle (CH_2)_zH}{|}}{\underset{\underset{\displaystyle COOH}{|}}{C}}—(CH_2)_yCOOH \qquad \begin{array}{l} x + y + z = 15 \\[4pt] z = 0 \text{ or } 1 \end{array}$$

It is difficult to assign an unambiguous name to these acids. Specific names have been used frequently but incorrectly in the past including:

9(10)-carboxystearic acid
1,8(9)-heptadecanedicarboxylic acid
2-nonyldecanedioic acid
2-octylundecanedioic acid

The name 9(10)-carboxystearic acid ($z = 0$ in the general structure) can be used properly only for the product obtained under special circumstances. In the sections that follow, reference will be made to oxo-, Koch-, or Reppe-C_{19} dicarboxylic acids, since these have different properties.

The C_{19} dicarboxylic acids have aroused industrial interest because their properties should be intermediate between those of the straight-chain alkanedioic acids and those of dimer acids. In 1971, they were available as developmental products from at least two companies (Badische Anilin- & Soda-Fabrik and Union Camp Corporation). Previously, formylstearic acid had also been made in pilot quantities (Union Carbide Corporation) but apparently is not now available.

In addition to the preparation of polymers, the C_{19} dicarboxylic acids have potential value as intermediates for ester plasticizers (1260–1262) and lead or other metal salt stabilizers (1263) for poly(vinyl chloride), amide corrosion inhibitors (1264), and ester viscosity-index improvers (1265) for lubricating oils; and as an air-entraining agent for concrete (1266). The properties of carboxystearic acid as a corrosion inhibitor are reportedly inferior to those of 9(10)-phosphonostearic acid (1267). The diimidazoline derivative may be used as a curing agent for epoxy resins (1268).

B.   Chemistry

The oxo-C_{19} dicarboxylic acids can be esterified in the usual manner, for example, by reflux in toluene with p-toluenesulfonic acid as the catalyst with azeotropic removal of water (1260).

Dufek et al. (1269) studied the relative rates of esterification for the two carboxyl groups in 9(10)-carboxystearic acid with several alcohols. The rate was highly dependent upon the concentration of the sulfuric acid catalyst. These investigators found that the terminal carboxyl group was from 25 to 30 times more reactive than the internal carboxyl. The two ester groups showed similar rate differences upon transesterification so that preparation of mixed dialkyl esters (e.g., methyl 9(10)-carbobutoxystearate) was possible.

The Koch C_{19} (see below) dicarboxylic acids cannot be esterified completely by the usual methods within a reasonable reaction period because of steric hindrance at the internal carboxyl group. However, use of dimethyl sulfate permits complete esterification after about 42 hr of reflux (1244). The dibutyl ester can be prepared by transesterification of the dimethyl ester. Alternatively, the dibutyl ester can be obtained by reflux in toluene for 240 hr with azeotropic removal of water (1262). Neither the dimethyl nor the dibutyl ester can be hydrolyzed completely by reflux in alcoholic potassium hydroxide in 8 hr. The great difference in reactivity between the terminal and the branched ester groups permits hydrolysis of the diester to an acid ester, the branched ester group being essentially unaffected during the short time required for hydrolysis of the terminal ester group.

Liquid HF may serve as both a reaction solvent and a catalyst and has been used for preparing the dibutyl, dicetyl, and diisocetyl esters (1270). Reaction times required are on the order of 1–3 hr at temperatures of 10–20°C.

C. Synthesis

In 1933 and subsequently, a series of patents issued to Du Pont described the direct reaction of carbon monoxide and water with olefins (mainly ethylene) in the presence of an acidic catalyst to produce carboxylic acids (1271–1282):

$$2RCH = CH_2 + 2CO + 2H_2O \rightarrow RCH_2CH_2COOH + RCH(CH_3)COOH$$

High temperatures (325°C) and pressures (700 atm) were required. These extreme conditions, the corrosive nature of the reaction mixture requiring silver-lined reactors, and low yields precluded commercial application. One successful laboratory application with boron trifluoride as catalyst was the synthesis of 2,2,3-trimethylbutanoic acid in high yield from tetramethylethylene (1283). With boron trifluoride, alcohols as well as olefins produced acids (1284). With 73% sulfuric acid, Ford (1285) prepared acids by carbonylation of isoolefins at 100°C and 950 atm in fair yield (e.g., trimethylacetic acid from isobutylene).

Such was the situation until improvements in methods were effected by the Koch, Roelen, and Reppe processes. General reviews of these reactions are available (1286–1288), but the C_{19} dicarboxylic acids have not been reviewed in detail.

1. THE KOCH REACTION

In 1955, Koch (1289) described a technically significant two-step modification for the synthesis of acids by the carbonylation of olefins. The milder conditions employed and the higher yields obtained compared with the original Du Pont process have resulted in industrial application to a variety of olefins. For example, various *neo* acids are now available that have such structures as the following (1290,1291):

$$
\begin{array}{c}
CH_3 \\
| \\
R—C—COOH \\
| \\
CH_3
\end{array}
$$

Koch (1292–1301) has described conditions and catalysts in a number of patents and articles, and several reviews are available (1302–1305).

Catalysts include 96–97% sulfuric acid, hydrogen fluoride, the combination of hydrogen fluoride and boron trifluoride, and chlorosulfonic acid (1293, 1294); phosphoric acid (1301); boron trifluoride monohydrate and sulfuric or phosphoric acid (1297); and boron trifluoride dihydrate (1300). Others have described similar catalysts, including various boron trifluoride mixtures (1306–1308), hydrogen fluoride containing 3–10% water (1309,1310), phosphoric acid on kieselguler (1311), and alkyl or aryl sulfonic acids such as methanesulfonic acid (1312). The oxonium tetrafluoroborates are said to be especially effective for continuous operation and for olefins such as ethylene and propylene, which are not carbonylated under the usual conditions (1313).

The catalyst is usually present in large excess (100–200 mole %), but the reaction proceeds readily at 0–50°C and 1–50 atm of carbon monoxide pressure. Instead of gaseous carbon monoxide, formic acid may be used to generate carbon monoxide *in situ*, whereby mainly tertiary acids are formed (1294,1295). If an alcohol is present in place of the usual small amount of water, an ester is formed (1314,1315); alternatively, the alcohol may be added to the complex formed in the first step (1299).

An interesting variant of the Koch reaction is the synthesis of carboxylic acids from straight chain or cyclic paraffins in the presence of an olefin or alcohol (1316). A mixture of acids is formed from both the paraffin and the olefin as the result of intermolecular hydride exchange between the initial carbonium ion and the paraffin.

The products obtained by the Koch procedure consists of a mixture of positional and methyl-branched isomers formed by rearrangement of the intermediate carbonium ions. Such rearrangement was demonstrated by Koch and Schauerte (1163), who characterized the products formed from undecylenic acid by both the pressure and by the formic acid methods. They found the 2-methyl-2-alkylalkanedioic acids to be the only product of the formic-acid method and about 50% of the product of the pressure method. The relative amounts of methyl-branched, tertiary acids can be lowered by high pressures of carbon monoxide (1298) or, in the formic acid method, by reducing the degree of agitation to obtain a supersaturated solution of carbon monoxide (1317).

Detailed characterization of the Koch C$_{19}$ dicarboxylic acids has not been made. Frankel (1243) found by mass spectral evidence that the position of the carboxyl group was distributed along the chain with a maximum concentration at carbon 15. Position of methyl branching could not be determined by mass spectroscopy.

Koch (1292,1294) has described only briefly his studies with oleic acid. The C$_{19}$ dibasic acid was obtained in 83% yield when oleic acid in hexane solution was injected into a shaker autoclave containing 97% sulfuric acid with carbon monoxide at 45 atm and after a reaction period of 12 hr.

On the other hand, Roe and Swern have described in detail the conditions for preparing carboxystearic acid in good yield from oleic acid (1244,1318–1321). Optimum concentration of sulfuric acid is 97% and of water, 5 moles per mole of oleic acid. Oleic acid in formic acid is added by drops to the sulfuric acid at 10–20°C for a total reaction time of 3 hr. Alternatively, carbon monoxide from a cylinder may be used when introduced continuously through a gas dispersion tube. At the end of the reaction, the sulfuric acid solution is poured into ice water and the carboxystearic acid recovered. Interestingly, linoleic acid gave the same product, a lactone, as that obtained from ricinoleic acid. Undecylenic acid under these conditions produced a branched- and not a straight-chain dibasic acid. Oleyl alcohol formed carboxyoctadecanol.

Matsubara, Sasaki, and Ohtsuka (1322) obtained maximum yields of dibasic acid from oleic acid in a carbon monoxide atmosphere when the sulfuric acid concentration was 98%. Below a concentration of 90%, hydration of the double bond was the exclusive reaction.

In the formic acid method, sulfation of the double bond could be minimized, even at low concentrations of formic acid, if the reaction was carried out with phosphoric as well as sulfuric acid present (1323).

Miller et al. (1310) described the carbonylation of oleic, 10-undecenoic, and erucic acids in the presence of hydrogen fluoride in a Monel autoclave.

2. THE OXO (ROELEN) REACTION

The oxo reaction (also known as hydroformylation) was discovered by Roelen (1288,1324,1325) just prior to World War II. It consists of the addition of carbon monoxide and hydrogen (synthesis gas) to an olefin under anhydrous conditions to produce an aldehyde. The catalyst is a cobalt hydrocarbonyl, which is formed from various cobalt compounds under the conditions used, for example, pressures of 80–300 atm of synthesis gas at 100°C and above. To produce an acid, the product aldehyde then is oxidized by air or another oxidizing agent.

Application of the oxo reaction to various unsaturated fatty materials has been reported (Table 18). Yields have varied from 25 to 99%.

TABLE 18
Oxo Reactions on Unsaturated Fatty Derivatives

Starting material	Diluent	Catalyst	Temperature, °C	Pressure, psig	Yield, mole %	Reference
Methyl oleate	Methyl formate	Reduced, fused alkali-free cobalt	140–145	8000–11,000	72	1326
Methyl oleate, olive oil, grape seed oil	—	Basic cobalt carbonate + 2% basic iron carbonate on bentonite	100–110	220–300	—	1327,1328
Methyl oleate	Methanol	Cobalt carbonyl	150	3000	40	1329
Methyl oleate	—	Reduced cobalt	150	3680	—	1330
Rapeseed, soybean, linseed, whale oils	Benzene	Cobalt and thorium on kieselguler	110–170	1800–4000	75–85	1331
Oleyl alcohol	Benzene	Cobalt carbonyl	140–150	3000	—	1332
Oleyl alcohol	—	Cobalt, thorium, and magnesium oxides on kieselguler	180–190	2200	—	1333
Methyl esters of tall oil fatty acids	—	Cobalt tallate	170	3000–4000	92	1334
Ethyl oleate	Benzene	Cobalt carbonyl	133–144	3675–5100	25	1260
Tall oil fatty acids	—	Cobalt sulfate	—	3000	—	1268
Oleic acid	—	Cobalt sulfate + iron powder	180–182	2560–3560	—	1335

Substrate	Solvent	Catalyst	Temperature (°C)	Pressure	Yield (%)	Reference
Methyl 9,10-epoxystearate	Benzene	Cobalt carbonyl	120–150	1420–2600	—	1336
Methyl tallate	—	Cobalt tallate	165–175	3000–4000	—	1337
Methyl esters of tall oil fatty acids	—	Cobalt carbonyl	135[a]	2400	—	1338
			180[b]	2400	—	
Methyl oleate	—	Cobalt acetate	150[a]	2460	74	1339–1341
			190[b]	2460	90	
Methyl oleate, linoleate, linolenate, and soybean oil	Benzene	Cobalt carbonyl	100–110[a]	3500–4600	85	1342
			175–190[b]	3500–4600	80	—
Methyl oleate	Toluene	5% Rh/CaCO$_3$ + triphenyl phosphine	95–110	2000	90–99	1243
Soybean, linseed, and safflower oils and their methyl esters	Toluene	5% Rh/CaCO$_3$ + triphenyl phosphine	100	1000	c	1343

[a] Temperature used for formylstearate preparation.
[b] Temperature used for hydroxymethylstearate preparation.
c Mixture of polyformyl compounds.

103

Hydroformylation with conventional cobalt catalysts occurs exclusively at one double bond in polyunsaturated fatty acids, the other double bonds being hydrogenated (1331,1334). At higher temperatures than those normally used, reduction of the formyl group occurs to produce the hydroxymethyl derivative (1327,1328,1338):

$$-\underset{\underset{\text{CHO}}{|}}{\text{CH}}- + \text{H}_2 \longrightarrow -\underset{\underset{\text{CH}_2\text{OH}}{|}}{\text{CH}}-$$

Laï et al. (1339) studied the effects of pressure, temperature, and carbon monoxide/hydrogen ratio on the reaction when methyl oleate was the substrate and 8% of cobalt laurate was added as catalyst. If the aldehyde ester is the product desired, the reaction should be carried out at lower temperatures (150°C), at high pressure, and with an excess of carbon monoxide relative to hydrogen (2/1). If the hydroxymethyl derivative is desired, the reaction should be carried out at higher temperatures (190°C), lower pressure, and with an excess of hydrogen relative to carbon monoxide (2/1); estolide condensation products are major byproducts. Cobalt(II) acetate is the most desirable form of cobalt to use (1340). High proportions of catalyst (2% as cobalt) speed the reaction but favor hydroxymethylstearate formation; low proportions (0.5% as cobalt) favor formylstearate formation. Neither benzene nor ether was beneficial as a reaction solvent, the effect being to slow the reaction.

Frankel et al. (1342) studied the reaction at 3500–4600 psig using dicobalt octacarbonyl as the catalyst. At 100–110°C, the formyl esters predominated, and at 175–190°C, the hydroxymethyl esters did. Yields of distillable oxo products were greatest (85%) from methyl oleate at 100–110°C, least at 175–190°C, and less with linoleate and linolenate than with methyl oleate.

Usually, the product is described as a mixture of 9- and 10-formylstearates or, after oxidation, 9- and 10-carboxystearic acids. In reality, extensive double bond isomerization occurs before the hydroformylation step, so much so that isolation of the straight-chain isomer in the form of nonadecanedioic acid (1335) has been patented. In a more detailed study, Laï et al. (1344) have shown the extent to which isomerization occurs to the terminal position and have isolated dimethyl nonadecanedioate in 9% yield. The apparent catalyst for the isomerization is cobalt carbonyl or cobalt hydrocarbonyl. The isomerization also occurs in the direction of the carboxyl group, as shown by the isolation of methyl 2-octadecenoate from the product obtained when only carbon monoxide or nitrogen was present. The intermediate isomers between the 2- and the 17-positions also are present, presumably in greater concentration as the 9-position is approached.

Frankel et al. (1342) found 4–7% of linear oxo products at reaction temperatures of 100–110°C and 8–16% at 175–190°C. Tributylphosphine cobalt carbonyl catalyst increased the proportion of linear product but also increased the amount of nondistillable products. Distribution of isomers was determined by mass spectrometry. The dibasic acid methyl esters of the branched isomers provided the most distinct spectrum for location of the branch. Distribution occurred between the C-6 and C-13 positions, with a maximum in the C-11 position. Evidence for the branched isomers was provided also by thin-layer chromatography.

The extensive isomerization described in the preceding paragraphs can be avoided if desired. Frankel (1243) has reported that only methyl 9(10)-formylstearate is formed when hydroformylation of methyl oleate is carried out with rhodium-triphenylphosphine catalyst. Advantages of this catalyst system include not only the absence of double bond isomerization before the hydroformylation step but also higher conversions and yields (90–99%) and lower pressures (500–2000 psig) compared with cobalt carbonyl as a catalyst. Furthermore, hydroxymethylstearate formation is absent even at 180°C.

When the rhodium-triphenylphosphine catalyst system was applied to polyunsaturated fatty acid compounds (soybean, linseed, and safflower oils and their methyl esters), Frankel and Thomas (1343) found that diformylated products were formed. Hydroformylation on more than one double bond of the fatty acid chain represents another advantage of the rhodium-triphenylphosphine catalyst, since hydroformylation occurs at one double bond and hydrogenation at the other when cobalt carbonyl is used. At 100°C and 1000 psig, the product from methyl safflowerate was a mixture of formylstearate, formyoleate, and difomylstearate esters. At 150°C and 1500 psig, the product was a mixture of mono- and diformylstearates.

Ucciani et al. (1345) successfully applied the Roelen reaction to various other unsaturated fatty materials including oleyl alcohol and oleyl nitrile. Attempts to apply the Roelen reaction to oleyl amine were unsuccessful.

Hydroformylation of methyl 9,10-epoxystearate produces a hydroxy or an unsaturated formylstearate (1336):

$$\underset{\text{CHO}}{\overset{\overset{\displaystyle O}{\diagup\diagdown}}{-\text{CH}-\!\!-\text{CH}-}} + \text{CO} + \text{H}_2 \longrightarrow \underset{\underset{\text{CHO}}{|}}{\overset{\overset{\displaystyle OH}{|}}{-\text{CH}-\text{CH}-}} + \underset{\underset{\text{CHO}}{|}}{-\text{CH}=\text{C}-}$$

Compounds of the latter type (e.g., methyl formyloleate) were patented as novel compositions of matter (1336).

Procedures have been described for the reduction of methyl formylstearate by copper chromite catalyzed hydrogenation (1330) or by lithium aluminum hydride (1338) to nonadecanediol, said to be useful for preparing viscosity-index improvers for lubricants (1330) and hard, electrically resistant polyurethane resins of high-impact strength (1338).

The C_{19} dicarboxylic acids are readily obtained from the formyl or hydroxymethyl derivatives described in the preceding paragraphs by several methods. The first described preparation of these acids resulted when hydroformylated oleyl alcohol was fused in sodium hydroxide at 270–320°C (1259). Alm and Shepard (1334) used a slurry of potassium hydroxide in cetane at 250–280°C to convert methyl formylstearate. Temperatures for the alkali fusion have been reported as low as 200°C (1268) and as high as 335°C (1335). Rogier (1337) used potassium hydroxide at 260°C. The advantage of this method lies in its ability to oxidize alcohols as well as aldehydes as in the mixture obtained by the conventional oxo process.

Air oxidation of formylstearic acid has been accomplished (a) in the presence of residual cobalt ion from the hydroformylation step (1328,1342), (b) in the presence of manganese(II) acetate in acetic acid solution (1260), or (c) in the absence of a metal ion catalyst (1345). Schwab et al. (1346) found that calcium naphthenate was particularly effective. Conversions were on the order of 95% after treatment with this catalyst at 20°C for 24 hr with no solvent, and free radical decarbonylation and other side reactions were at a minimum. Catalyst activity based on disappearance of formylstearate was in the order cobalt > lead > manganese > cerium > iron > zirconium > calcium as the naphthenates, but yields of carboxystearate were obtained with the naphthenates in the following order: calcium > lead ≈ iron ≈ zirconium ≈ manganese > cobalt > cerium.

Various chemical oxidizing agents have been used including potassium permanganate (1243,1346) and potassium dichromate (1346). Alkaline hydrogen peroxide was not effective.

3. THE REPPE REACTION

Just prior to and during World War II, Reppe and his co-workers (1347) developed methods for carboxylating olefins and other compounds directly with carbon monoxide and water through the use of nickel carbonyl. The nickel carbonyl can be used either in stoichiometric amounts or in catalytic amounts, which require higher temperatures and pressures than do stoichiometric amounts. Catalysts other than nickel carbonyl have been used, including cobalt carbonyl (1162,1348,1349), rhodium trichloride, and palladium dichloride with or without a ligand such as triphenylphosphine (1350, 1351).

Reppe and Kröper (1352,1353) applied the reaction to oleic acid, using nickel carbonyl and nickel iodide at 270°C and 200 atm of carbon monoxide pressure. After a reaction period of 16 hr and after suitable work-up, the C_{19} dicarboxylic acids were obtained in 70% yield. When the reaction was carried out on methyl oleate in the presence of methanol in place of water, a 36% yield was obtained of methyl carbomethoxystearate. Similar results were obtained with ethyl and butyl oleates. Reppe et al. (1354) described a continuous pilot reactor which produced 64 kg of product from 60 kg of oleic acid over a period of 5 days. The product contained about 80% of the C$_{19}$ dicarboxylic acids (corresponding to an 80% yield), 10% of oleic acid, 5% of stearic acid, and 5% of hydroxystearic acid. In this system, one inlet stream consisted of a solution of nickel carbonyl in oleic acid, and another stream contained water, nickel acetate, and hydriodic acid; the temperature was 270°C and pressure, 70 to 215 atm.

Using dicobalt octacarbonyl as the catalyst, Levering (1355) carried out the reaction with oleic acid in much shorter time than required with nickel carbonyl. He obtained 60–80% yields of the dibasic ester by carrying out the reaction in methanol at 220°C and 6000 psig of carbon monoxide pressure.

Amines as well as water and alcohols can act as proton donors for this reaction. Crowe and Helsler (1356) carbonylated oleic acid with cobalt carbonyl at about 210°C and 5500 psig in the presence of aniline. Surprisingly, the major product isolated in fair to good yield was N-phenyl pentadecylsuccinimide:

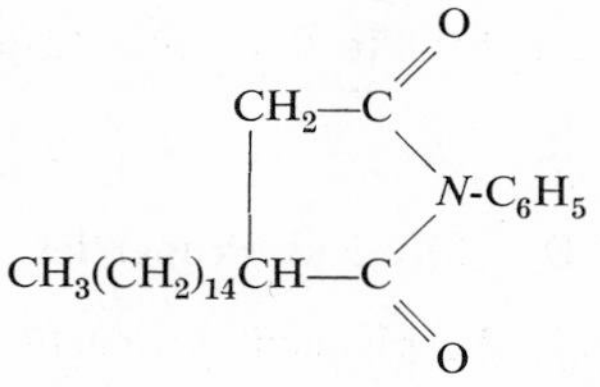

The isolation of this product illustrates the isomerization activity of the cobalt carbonyl catalyst, the shifting double bond having been trapped finally in the most stable position.

4.   MISCELLANEOUS

Elad and Rokach (1357) prepared methyl 9(10)-carbamoylstearate in 60% yield by the photochemical addition of formamide to methyl oleate in sunlight:

$$CH_3(CH_2)_7CH{=}CH(CH_2)_7COOCH_3 + HCONH_2 \xrightarrow{h\nu}$$

$$CH_3(CH_2)_8\underset{\underset{CONH_2}{|}}{C}H(CH_2)_7COOCH_3 + CH_3(CH_2)_7\underset{\underset{CONH_2}{|}}{C}H(CH_2)_8COOCH_3$$

Similarly, they prepared 9(10)-carbamoylstearamide in 74% yield from oleamide. Acetone was the initiator, and the reaction was carried out in dilute t-butyl alcohol solution at room temperature for 3–4 days with a mercury lamp or for 2–3 weeks when sunlight was used.

Reaction of sodium cyanide with methyl 9,10-epoxystearate produces a mixture of methyl 10-hydroxy-9-cyanostearate and methyl 9-hydroxy-10-cyanostearate (1358). Instead of the epoxy compound, methyl chloroformoxystearate can be used. The product may be considered as the half nitrile of the C_{19} dicarboxylic acids with an additional hydroxyl functionality. Alkali fusion of this compound results in chain scission producing a mixture of mono- and dibasic acids (946).

Homologs of the C_{19} dicarboxylic acids have been prepared by the free radical-initiated addition of acetic acid or anhydride to methyl oleate (1203,1359):

$$CH_3(CH_2)_7CH{=}CH(CH_2)_7COOCH_3 + CH_3COOH \xrightarrow[140°C]{\text{peroxide}}$$

$$CH_3(CH_2)_x\underset{\underset{CH_2COOH}{|}}{CH}(CH_2)_yCOOCH_3 \qquad x + y = 15$$

Reaction was about 80% complete after 90 hr at 140°C. The branched-chain carboxyl group in this compound is more reactive than that in the Koch C_{19} acid; the esterification rate is four to five times faster, and saponification of the ester is complete.

D. Physical Properties

As indicated in Table 19, the physical properties of the C_{19} dicarboxylic acids depend upon the method of preparation. The isomeric mixtures obtained by the Koch and cobalt-catalyzed oxo reactions are syrups or liquids at room temperature, whereas the more homogeneous product from the Reppe process or the selective oxo reaction with rhodium-triphenylphosphine catalyst melts above room temperature. Methyl 9(10)-carboxystearate may be recrystallized from hexane (1346).

E. Analytical Procedures

Frankel et al. (1243,1342) have described characterization of methyl carbomethoxystearate by gas–liquid and thin-layer chromatography, mass spectrometry, infrared spectroscopy, and nuclear magnetic resonance. Senn and Pine (1360) describe the nuclear magnetic spectra for α-methyl branched monocarboxylic acids prepared by the Koch reaction.

F. Toxicology

Apparently, no toxicological data for the C_{19} dicarboxylic acids have been reported. Until such information is available, these acids should be handled

TABLE 19
Physical Properties of Various C_{19} Dicarboxylic Acids and Methyl 9(10)-Carboxystearate

Property	Value				
	Koch	Oxo	Reppe	Nonadecanedioic	Methyl 9(10)-carboxystearate
Molecular weight	328.48	328.48	328.48	328.48	344.49
Melting point (°C)	[a]	[b]	46 (1353)	118–119.5 (1335)	[b]
Boiling point (°C/mm Hg)	200–201/0.45 (1244)	—	235–240/1.3 (1353)	—	201–205/0.07 (1346)
Refractive index, n_D^{30}	1.4615 (1244)	1.4614 (1345)	—	—	—
Density, d^{30}	0.9726 (1244)	—	—	—	—

[a] Syrup at room temperature.
[b] Liquid at room temperature.

with normal precautions, although no unusual toxicity problem would be expected.

G. Polymer Applications

Patents have issued for the use of certain polyamides and copolyamides from C_{19} dicarboxylic acids as finish coats for leather (1361,1362), curing agents for epoxy resins (1337,1363–1365), transparent moldings (1366,1367) and coatings (1368,1369), plasticizers (1370), and protective coatings (1371). The polyamide from 9,9-bis(aminopropyl)fluorene is said to be useful for films, printing inks, and hot melt adhesives (1367).

Copolymers of polyesters from the C_{19} dicarboxylic acids are said to be suitable for floor coverings and wire coatings (1372). Unsaturated polyester amide resins have also been made (1373).

The diisocyanate derived from aminomethylstearyl amine has value in the production of hard, flexible, and nontoxic polyurethane coatings and castings (1374).

References

1. J. H. Saunders, E. E. Hardy, and R. J. Slocombe, in "Encyclopedia of Chemical Technology," Vol. 10, R. E. Kirk and D. F. Othmer, Eds., The Interscience Encyclopedia, New York, pp. 391–403.
2. W. B. Tuemmler, in "Encyclopedia of Chemical Technology," 2nd ed., Vol. 4, H. F. Mark, Chairman of the Editorial Board, Interscience, New York, 1964, pp. 386–393.
3. E. Wygasch, in "Ullmanns Encyklopädie der technischen Chemie," Vol. 13, W. Foerst, Ed., Urban und Schwarzenberg, Munich, 1962, pp. 493–500.
4. L. J. Durham, D. J. McLeod, and J. Cason, *Org. Syn.*, **Coll. Vol. IV**, 510 (1963).
5. G. L. Brownell, in "Encylopedia of Polymer Science and Technology," Vol. 1, H. F. Mark, Chairman of the Editorial Board, Interscience, New York, 1964, pp. 67–95.
6. B. Dmuchovsky, and J. E. Franz, in "Encylopedia of Chemical Technology," Vol. 12, H. F. Mark, Chairman of the Editorial Board, Interscience, New York, 1967, pp. 819–837.
7. C. J. Knuth and A. M. Schiller, in Ref. 5, pp. 122–138.
8. C. J. Knuth, in Ref. 6, pp. 83–86.
9. R. P. Mariella and R. Raube, *Org. Syn.* **Coll. Vol. IV**, 288 (1963).
10. A. Armen, U.S. Patent 3,074,914 (to Dow Chemical Co.) (Jan. 22, 1963).
11. R. H. Perry, Jr. (to Esso Research and Engineering Co.), U.S. Patent 2,963,487 (Dec. 6, 1960).
12. C. A. Cohen (to Esso Research and Engineering Co.), U.S. Patent 2,991,308 (July 4, 1961).
13. N. P. Greco (to Koppers Co.), U.S. Patent 3,206,503 (Sept. 14, 1965).
14. A. W. Dox and L. Yoder, *J. Amer. Chem. Soc.*, **43**:1366 (1921).
15. H. R. Appell (to Koppers Company, Inc.), U.S. Patent 3,520,921 (July 21, 1970).
16. W. J. Bailey and R. Barclay, Jr., *J. Amer. Chem. Soc.*, **81**:5393 (1959).
17. W. L. Foohey (to. E. I. du Pont de Nemours & Co.), U.S. Patent 3,027,398 (Mar. 27, 1962).

18. H.-L. Huelsmann and G. Renckhoff (to Chemische Werke Witten, GmbH), U.S. Patent 3,428,668 (Feb. 18, 1969).
19. G. W. Hedrick and R. V. Lawrence, *Ind. Eng. Chem.*, **52**:853 (1960).
20. C. F. Baranauckas and A. L. Blackwell (to Hooker Chemical Corp.), U.S. Patent 2,903,463 (Sept. 8, 1959).
21. A. McLean, J. Habeshaw, and W. J. Oldham (to British Hydrocarbon Chemicals, Ltd.), British Patent 797,986 (July 9, 1958).
22. V. V. Korshak and S. V. Rogozhin, *Izv. Akad. Nauk SSSR, Otd. Khim. Nauk*, **1952**:531.
23. S. Gal, T. Meisel, and L. Erdey, *J. Therm. Anal.*, **1**:159 (1969).
24. J. W. Hill and W. H. Carothers, *J. Amer. Chem. Soc.*, **55**:5023 (1933).
25. H. G. Blanc, *C.R. Acad. Sci. (Paris)*, **144**:1356 (1907); *Bull. Soc. Chim. Fr.*, **3**:778 (1908).
26. H. C. S. Snethlage, *Rec. Trav. Chim. Pays-Bas*, **56**:873 (1937).
27. F. Mareš and J. Roček, *Coll. Chem. Commun. Czech.*, **26**:2389 (1961).
28. G. Gut, R. v. Falkenstein, and A. Guyer, *Helv. Chim. Acta.*, **49**:481 (1966).
29. A. P. Ponsford and I. Smedley-Maclean, *Biochem. J.*, **28**:892 (1934).
30. J. Cason and H. Rapoport, "Laboratory Text in Organic Chemistry," 2nd ed., Prentice-Hall, Englewood Cliffs, N.J., 1962, pp. 409–422.
31. K. Ziegler, in "Methoden der Organische Chemie (Houben-Weyl)," Vol. 4, Pt. 2, E. Muller, Ed., Georg Thieme Verlag, Stuttgart, 1955, pp. 731–811.
32. J. P. Schaefer and J. J. Bloomfield, in "Organic Reactions," Vol. 15, A. C. Cope, Ed.-in-chief, Wiley, New York, 1967, pp. 1–203.
33. M. Stoll and A. Rouvé, *Helv. Chim. Acta*, **30**:1822 (1947).
34. V. Prelog, L. Frenkiel, M. Kobelt, and P. Barman, *Helv. Chim. Acta*, **30**:1741 (1947).
35. P. Chuit and J. Hausser, *Helv. Chim. Acta*, **12**:850 (1929).
36. A. Guyer, A. Bieler, and M. Sommaruga, *Helv. Chim. Acta*, **38**:976 (1955).
37. F. Weygand, G. Eberhardt, H. Linden, F. Schäfer, and I. Eigen, *Angew. Chem.*, **65**:525 (1953).
38. P. E. Verkade, H. Hartman, and J. Coops, *Rec. Trav. Chim. Pays-Bas*, **45**:380 (1926).
39. K. E. Miller, D. I. Lusk, J. F. Marks, E. Blanc, and T. R. Fernandes, *J. Chem. Eng. Data*, **9**:227 (1964).
40. W. Reppe, H. Kröper, N. v. Kutepow, H. J. Pistor, and O. Weissbarth, *Justus Liebigs Ann. Chem.*, **582**:72 (1953).
41. W. Langenbeck and M. Richter, *Chem. Ber.* **89**:202 (1956).
42. J. C. Sauer, *J. Amer. Chem. Soc.*, **69**:2444 (1947).
43. A. T. Blomquist, J. R. Johnson, L. I. Diuguid, J. K. Shillington, and R. D. Spencer, *J. Amer. Chem. Soc.*, **74**:4203 (1952).
44. L. J. Durham, D. J. McLeod, and J. Cason, *Org. Syn.*, **Coll. Vol. IV**, 555 (1963).
45. B. C. L. Weedon, *Quart. Rev. (London)*, **6**:380 (1952).
46. S. Swann, Jr., in "Technique of Organic Chemistry," 2nd ed., Vol. 2, A. Weissberger, Ed., Interscience, New York, 1956, p. 385.
47. W. Fuchs and E. Dickersbach-Baronetzky, *Fette, Seifen, Anstrichm.* **57**:675 (1955).
48. S. Hünig, E. Lücke, and W. Brenninger, *Org. Syn.*, **41**:65 (1961).
49. S. Hünig and E. Lücke, *Chem. Ber.* **92**:652 (1959).
50. M. S. R. Nair, H. H. Mathur, and S. C. Bhattacharyya, *Tetrahedron*, **19**:905 (1963).
51. J. J. Leonard and W. E. Goode, *J. Amer. Chem. Soc.*, **72**:5404 (1950).
52. N. P. Buu-Hoï, M. Sy, and N. D. Xuong, *Bull. Soc. Chim. Fr.*, **1955**:1583.
53. H. Stetter, in "Newer Methods of Preparative Organic Chemistry," Vol. II, W. Foerst, Ed., Academic Press, New York, 1963, p. 79.

54. J. A. Patterson and S. M. Pier (to Texaco, Inc.), U.S. Patent 2,918,487 (Dec. 22, 1959).
55. H. v. Euler, H. Hasselquist, and Uno Lööv, *Ark. Kemi* **1**:307 (1949); *Chem. Abstr.* **44**:5813 (1950).
56. M. Kobayashi, *Abura Kagaku*, **4**:53 (1965); *Chem. Abstr.*, **54**:2787 (1960).
57. R. Perron and J. Petit, *J. Rech. Cent. natl. Rech. Sci., Lab. Bellevue* (*Paris*) No. **24**:122 (1953); *Chem. Abstr.*, **48**:6143 (1954).
58. J. D. Morrison and J. M. Robertson, *J. Chem. Soc.*, **1949**:1001.
59. J. G. Erickson, *J. Amer. Chem. Soc.*, **71**:307 (1949).
60. A. M. King and W. E. Garner, *J. Chem. Soc.*, **1934**:1449.
61. W. Schlenk, Jr., *Justus Liebigs Ann. Chem.*, **727**:1 (1969).
62. P. E. Verkade and J. Coops, Jr., *Rec. Trav. Chim. Pays-Bas*, **49**:578 (1930).
63. G. Saracco and E. S. Marchetti, *Ann. Chim.* (*Rome*), **48**:1357 (1958).
64. F. L. Breusch and E. Ulusoy, *Fette, Seifen, Anstrichm.*, **66**:739 (1964).
65. R. N. Castle, *Mikrochem. ver. Mikrochim. Acta*, **38**:92 (1951).
66. A. N. Winchell, "The Optical Properties of Organic Compounds," 2nd ed., Academic Press, New York, 1954.
67. G. Kortüm, W. Vogel, and K. Andrussow, *Pure Appl. Chem.*, **1**:190 (1961).
68. G. Bonhomme, *Bull. Soc. Chim. Fr.*, **1968**:60.
69. W. H. Rauscher and W. H. Clark, *J. Amer. Chem. Soc.*, **70**:438 (1948).
70. O. C. Dermer and J. King, *J. Org. Chem.*, **8**:168 (1943).
71. N. L. Drake and J. Bronitsky, *J. Amer. Chem. Soc.*, **52**:3715 (1930).
72. N. L. Drake and J. P. Sweeney, *J. Amer. Chem. Soc.*, **54**:2059 (1932).
73. E. Vioque and M. del P. de La Maza, *Grasas Aceites*, **8**:19 (1957); *Chem. Abstr.*, **51**:13795 (1957).
74. V. V. Korshak and S. V. Vinogradova, "Polyesters," B. J. Hazzard, Tr., Pergamon Press, Oxford, 1965.
75. V. V. Korshak and T. M. Frunze, "Synthetic Hetero-Chain Polyamides," N. Kaner, Tr., Davey, New York, 1964.
76. E. C. Jahn, in "Wood Chemistry," Vol. 2, 2nd ed., L. E. Wise and E. C. Jahn, Eds., Reinhold, New York, 1952, Chap. 22, pp. 939–944.
77. Anon., *Rev. Prod. Chim.*, **13**:259 (1916); *Chem. Abstr.*, **11**:870 (1917).
78. U.S. Tariff Commission, "Synthetic Organic Chemicals. United States Production and Sales, 1966," TC Publ. 248, Govt. Printing Office, Washington, D.C., 1968, p. 58.
79. *Oil, Paint, Drug Rep*, **April 1, 1968**.
80. Y. Mayor, *Rev. Chim. Ind.* (*Paris*), **46**:356 (1937); **47**:3 (1938); **47**:73 (1938).
81. P. A. Florio and G. R. Patel, in "Encyclopedia of Chemical Technology," 2nd ed., Vol. 14, H. F. Mark, Chairman of the Editorial Board, Interscience, New York, 1967, pp. 356–373.
82. E. Bowden, *Org. Syn.* (2nd ed.) **Coll. Vol. I.** 424 (1946).
83. H. T. Clarke and A. W. Davis, *Org. Syn.* (2nd ed.) **Coll. Vol. I**, 421 (1946).
84. E. I. Johnson and J. R. Partington, *J. Chem. Soc.*, **1930**:1510.
85. A. Hahn, (to Globus-Werke Fritz Schulz, Jr.), German Patent 825,838 (Dec. 20, 1951); *Chem. Abstr.*, **50**:1077 (1956).
86. W. W. Fisher, *Proc. Chem. Soc.*, **8**:186 (1892).
87. R. C. Wilhoit and D. Shiao, *J. Chem. Eng. Data*, **9**:595 (1964).
88. D. E. Wobbe and W. A. Noyes, Jr., *J. Amer. Chem. Soc.*, **48**:2856 (1926).
89. G. Lapidus, D. Barton, and P. E. Yankwich, *J. Phys. Chem.*, **68**:1863 (1964); **70**:407, 1575, 3135 (1966).
90. D. M. Lichty, *J. Phys. Chem.*, **11**:225 (1907).

91. M. Liler, *J. Chem. Soc.*, **1963**:3106.
92. V. V. Korshak and S. V. Rogozhin, *Dokl. Akad. Nauk SSSR*, **76**:539 (1951); *Chem. Abstr.*, **45**: 8455 (1951).
93. V. V. Korshak and S. V. Rogozhin, *Khim. Fiz.-Khim Vysokomol. Soedin. Dokl. Konf. Vysokomol. Soedin. 7-ya Konf.*, **1952**:11; *Chem. Abstr.*, **48**:3912 (1954).
94. V. V. Korshak and S. V. Vinogradova, "Polyesters," B. J. Hazzard, Tr., Pergamon Press, London, 1965, Chap. 2, p. 25.
95. L. W. Clark, *J. Phys. Chem.*, **70**:1597 (1966).
96. M. A. Haleem and P. E. Yankwich, *J. Phys. Chem.*, **69**:1729 (1965).
97. A. J. Allmand and L. Reeve, *J. Chem. Soc.*, **1926**:2834.
98. H. T. Clarke and A. W. Davis, *Org. Syn.* (2nd ed.) **Coll. Vol. I**, 261 (1946).
99. J. Kenyon, *Org. Syn.* (2nd ed.) **Coll. Vol. I**, 263 (1946).
100. H. Staudinger, *Chem. Ber.*, **41**:3563 (1908).
101. H. Biltz and E. Topp, *Chem. Ber.*, **46**:1387 (1913).
102. P. J. Wiezevich (to Standard Oil Development Co.), U.S. Patent 2,055,617 (Sept. 29, 1936).
103. E. K. Ellingboe and L. R. Melby (to E. I. du Pont de Nemours & Co.), U.S. Patent 2,816,140 (Dec. 10, 1957).
104. E. K. Ellingboe (to E. I. du Pont de Nemours & Co.), U.S. Patent 2,816,141 (Dec. 10, 1957).
105. B. L. Leonov, *Org. Chem. Ind. (USSR)*, **5**:489 (1938); *Chem. Abstr.*, **33**:843 (1939).
106. F. D. Leicester (to Imperial Chemical Industries Ltd.), U.S. Patent 2,193,337 (March 12, 1940).
107. L. J. Beckham (to Allied Chemical and Dye Corp.), U.S. Patent 2,687,433 (Aug. 24, 1954).
108. H. Klapproth (to Rudolph Koepp Chemische Fabrik A.G.), German Patent 1,014,095 (Aug. 28, 1957); *Chem. Abstr.*, **53**:15992 (1959).
109. K. Fay (to Rudolph Koepp & Co. Chemische Fabrik, gmbH), German Patent 1,015,420 (Sept. 12, 1957); *Chem. Abstr.*, **53**: 15992 (1959).
110. O. R. Sweeney, *Iowa Eng. Exp. Sta. Bull.*, **73**:7 (1924).
111. H. A. Webber, *Iowa Eng. Exp. Sta. Bull.*, **118**:55 pp (1934).
112. D. H. Grangaard (to Kimberly-Clark Corp.), U.S. Patent 2,928,868 (March 15, 1960).
113. D. H. Grangaard (to Kimberly-Clark Corp.), U.S. Patent 3,008,984 (Nov. 14, 1961).
114. F. Ullman, "Enzyklopaedie der technischen Chemie," 2nd ed., Vol. 8, Urban & Schwarzenberg, Berlin, 1931, pp. 217–226.
115. D. F. Othmer and R. H. Royer, *Ind. Eng. Chem.*, **34**:274 (1942).
116. D. F. Othmer, J. J. Jacobs, Jr., and A. C. Pabst, *Ind. Eng. Chem.*, **34**:268 (1942).
117. D. F. Othmer, C. H. Gamer, and J. J. Jacobs, Jr., *Ind. Eng. Chem.*, **34**:262(1942).
118. L. A. Price and D. I. Gleim, *Proc. Pa. Acad. Sci.*, **26**:45 (1952).
119. G. S. Simpson (to General Chemical Co.), U.S. Patent 2,057,119 (Oct. 13, 1936).
120. R. W. Bailey, *J. Appl. Chem.*, **4**:549 (1954).
121. A. P. Krasnova, E. A. Parshina, S. I. Sukhanovskii, and M. I. Chudakov, *Zh. Prikl Khim.*, **30**:802 (1957).
122. A. P. Salchinkin, *Tr. Kuban. Sel'skokhoz. Inst.*, **1959**:No. 2, 220.
123. K. M. Seymour, *J. Chem. Educ.*, **16**:285 (1939).
124. E. V. Obmornov, V. G. Karetnik, V. I. Koptelov, N. A. Dosovitskaya, Z. P. Koptelova, G. P. Masalova, E. I. Dosovitskii, and V. N. Ostrovskaya (to Novomoskovsk Aniline Dye Plant), British Patent 1,095,100 (Dec. 13, 1967); *Chem. Abstr.*, **68**:7495 (1968); U.S. Patent 3,531,520 (Sept. 29, 1970).
125. G. Kolsky, U.S. Patent 1,446,012 (Feb. 20, 1923).

126. A. Mittasch and O. Ball, U.S. Patent 1,518,597 (Dec. 9, 1925).

127. M. J. Brooks (to General Chemical Co.), U.S. Patent 2,322,915 (June 29, 1943).

128. G. H. Fuchs (to Allied Chemical Corp.), U.S. Patent 3,536,754 (Oct. 27, 1970).

129. W. E. Stokes and W. E. Burch (to Standard Brands Inc.), U.S. Patent 2,257,284 (Sept. 30, 1941).

130. S. Soltzberg (to Atlas Powder Co.), U.S. Patent 2,380,196 (July 10, 1945).

131. N. Makay, U.S. Patent 2,813,121 (Nov. 12, 1957).

132. Rudolph Koepp & Co., German Patent 161,512 (April 17, 1903); *Chem. Zentralbl.*, **1905**:II, 367.

133. A. A. Balandin and L. Kh. Freĭdlin, *J. Gen. Chem. (USSR) (Engl. transl.)*, **6**:868 (1936); *Chem. Abstr.*, **30**:6628 (1936).

134. S. Keimatsu and B. Ikeda, *J. Pharm. Soc. Jap.*, **1915**:399, 499; *Chem. Abstr.*, **9**:2232 (1915).

135. G. Laber, in "Ullmans Encyklopaedie der technischen Chemic," Vol. 13, 3rd ed., W. Foerst, Ed., Urban & Schwarzenberg, Munich, 1962, pp. 51–56.

136. V. M. Semenov, A. Yu. Shagalov, and P. I. Astrakhantzev, *J. Appl. Chem. (USSR)*, **8**:99 (1935); *Chem. Abstr.*, **29**:6882 (1935).

137. D. Condurache, *Rev. Chim. (Bucharest)*, **10**:146 (1959); *Chem. Abstr.*, **57**:5755 (1962).

138. M. Enderli (to Rudolph Koepp & Co., Chemische Fabrik A.G.), U.S. Patent 2,002,342 (May 21, 1935).

139. L. Kh. Freĭdlin, *Org. Chem. Ind. (USSR)*, **3**:681 (1937): *Chem. Abstr.*, **31**:7401 (1937).

140. P. Busse, German Patent 742,053 (Oct. 7, 1943); *Chem. Abstr.*, **40**:901 (1946).

141. E. J. Carlson and E. E. Gilbert (to Allied Chemical Corp.), U.S. Patent 3,081,345 (March 12, 1963).

142. S. A. Rhone-Poulenc, Netherlands Appl. 6,603,748 (Oct. 10, 1966); *Chem. Abstr.*, **66**:5183 (1967).

143. J. Duroux and L. Pichon (to Société Anonyme des Usines Chimiques Rhone-Poulenc), French Patent 1,487,446 (July 7, 1967); *Chem. Abstr.*, **68**:10082 (1968); U.S. Patent 3,549,696 (Dec. 22, 1970).

144. Rhone-Poulenc S. A., Netherlands Appl. 6,608,748 (Jan. 2, 1967); *Chem. Abstr.*, **67**:6872 (1967).

145. B. P. Brossard, J. Boichard, M. L. M. J. Gay, R. M. C. Janin, and L. M. E. Pichon (to Rhone-Poulenc S. A.), U.S. Patent 3,428,675 (Feb. 18, 1969).

146. A. Quilico and M. Freri, *Gazz. Chim. Ital.*, **59**:930 (1929).

147. S. N. Kazarnovskiĭ, *Org. Chem. Ind. (USSR)*, **2**:3 (1936).

148. S. Kakutani and T. Tashiro, *Rep. Imp. Ind. Res. Inst. Osaka*, **18**:13 (1938), 35 pp.

149. T. Abe, Japanese Patent 3609 ('50) (Oct. 19, 1950); *Chem. Abstr.*, **47**:3340 (1953).

150. Y. Koike, Japanese Patent 666 ('53) (Feb. 18, 1953); *Chem. Abstr.*, **48**:2090 (1954).

151. C. Matignon and C. Faurholt, *C.R. Acad. Sci. (Paris)*, **179**:271 (1924).

152. D. M. Fenton and P. J. Steinwand (to Union Oil Company of California), U.S. Pat. 3,393,136 (July 6, 1968).

153. J. L. Torgesen and J. Strassburger, *Science*, **146**:53 (1964).

154. J. K. Taylor and S. W. Smith, *J. Res. Natl. Bur. Std. (U.S.)*, **63A**:153 (1959).

155. R. E. Richards and J. A. S. Smith, *Trans. Faraday Soc.*, **47**:1261 (1951).

156. J. Itoh, R. Kusaka, R. Kiriyama, and S. Yabumoto, *J. Chem. Phys.*, **21**:1895 (1953).

157. G. E. Pringle, *Acta Crystallogr.*, **7**:716 (1954).

158. J. D. Dunitz and J. M. Robertson, *J. Chem. Soc.*, **1947**:142.

159. G. A. Jeffrey and G. S. Parry, *J. Chem. Soc.*, **1952**:4864.

160. L. J. Bellamy and R. J. Pace, *Spectrochim. Acta*, **19**:435 (1963).

161. V. Lorenzelli and A. Alemagna, *C.R. Acad. Sci. (Paris)*, **256**:3626 (1963).
162. R. Marignan, *Bull. Soc. Chim. Fr.*, **1948**:351.
163. A. Weil, *C.R. Acad. Sci. (Paris)*, **238**:576 (1954).
164. N. Gerard, A. Thrierr-Sorel, and G. Watelle-Marion, *C.R. Acid. Sci. (Paris)*, Ser. C, **262**:733 (1966).
165. S. Takagi and S. Oomi, *J. Soc. Chem. Ind. Jap.*, **42**: suppl. binding 302 (1939); *Chem. Abstr.*, **34**:2225 (1940).
166. J. Bell, *J. Chem. Soc.*, **1940**:72.
167. S. B. Hendricks, *Z. Kristallogr.*, **91**:48 (1935).
168. E. G. Cox, M. W. Dougill, and G. A. Jeffrey, *J. Chem. Soc.*, **1952**:4854.
169. J. Strassburger and J. L. Torgesen, *J. Res. Natl. Bur. Std. (U.S.)*, **67A**:347 (1963).
170. H. Murata and K. Kawai, *J. Chem. Phys.*, **25**:589 (1956).
171. W. A. Noyes, Jr., and D. E. Wobbe, *J. Amer. Chem. Soc.*, **48**:1882 (1926).
172. Z. N. Miczynski, *Monatsh. Chem.*, **7**:258 (1886).
173. J. D. A. Johnson and A. Talbot, *J. Chem. Soc.*, **1950**:1068.
174. K. Ito and H. J. Bernstein, *Can. J. Chem.*, **34**:170 (1956).
175. W. W. Wendlandt and J. A. Hoiberg, *Anal. Chim. Acta*, **28**:506 (1963).
176. H. T. Barnes, in "International Critical Tables," Vol. 5, E. W. Washburn, Ed., McGraw-Hill, New York, 1929, p. 108.
177. M. S. Van Dusen, in Ref. 176, p. 216.
178. M. T. Rogers, *J. Phys. Chem.*, **61**:1442 (1957).
179. M. Koppel and M. Cahn, *Z. Anorg. Allg. Chem.*, **60**:53 (1908).
180. E. Anderson, in Ref. 176, p. 148.
181. L. S. Darken, *J. Amer. Chem. Soc.*, **63**:1007 (1941).
182. G. D. Pinching and R. G. Bates, *J. Res. Natl. Bur. Std. (U.S.)*, **40**:405 (1948).
183. Federal Supply Classification, FSC 6810, "Oxalic Acid, Dihydrate, Technical," Spec. No. 0-0-690a, Govt. Printing Office, Washington, D.C., July 1, 1968.
184. American Chemical Society, Committee on Analytical Reagents, "Reagent Chemicals," 4th ed., American Chemical Society, Washington, D.C., 1968, p. 391.
185. Anon., *U.K. At. Energy Auth., Prod. Group PG Rep.* **92(W)**:(1960), 16 pp.
186. R. M. Fowler and H. A. Bright, *J. Res. Natl. Bur. Std. (U.S.)*, **15**:493 (1935).
187. H. A. Laitinen, "Chemical Analysis," McGraw-Hill, New York, 1960, p. 366.
188. R. G. Bates and E. Wichers, *J. Res. Natl. Bur. Std. (U.S.)*, **59**:9 (1957).
189. N. Van Meurs and E. A. M. F. Dahmen, *Anal. Chim. Acta*, **21**:10 (1959).
190. T. S. Rumsey and C. H. Noller, *J. Chromatogr.*, **24**:325 (1966).
191. D. Braun and H. Geenen, *J. Chromatogr.*, **7**:56 (1962).
192. P. G. Pifferi, *Boll. Lab. Chim. Prov. (Bologna)*, **17**:445 (1966); *Chem. Abstr.*, **66**:783 (1967).
193. H. Kalbe, *Hoppe-Seyler's Z. Physiol. Chem.*, **297**:19 (1954).
194. A. Seher, *Fette, Seifen, Anstrichm.*, **58**:401 (1956).
195. N. I. Sax, "Dangerous Properties of Industrial Materials," 3rd ed., Reinhold, New York, 1968, p. 987.
196. N. V. Steere, Ed., "Handbook of Laboratory Safety," The Chemical Rubber Co., Cleveland, Ohio, 1967.
197. F. A. Patty, D. Irish, and D. Fassett, Eds., "Industrial Hygiene and Toxicology," 2nd ed., Wiley, New York, 1963, p. 1773.
198. W. H. Carothers, J. A. Arvin, and G. L. Dorough, *J. Amer. Chem. Soc.*, **52**:3292 (1930).
199. J. R. Caldwell and D. D. Reynolds (to Eastman Kodak Co.), U.S. Patent 2,720,506 (Oct. 11, 1955).

200. E. E. Magat (to E. I. du Pont de Nemours & Co.), U.S. Patent 2,831,834 (Apr. 22, 1958).
201. L. B. Sokolov, *J. Polym. Sci.*, **58**:1253 (1962).
202. O. Vogl and A. C. Knight, *Macromolecules*, **1**:315 (1968).
203. S. D. Bruck, *Ind. Eng. Chem., Prod. Res. Develop.*, **2**:119 (1963).
204. G. S. Stamatoff and N. K. J. Symons (to E. I. du Pont de Nemours & Co.), U.S. Patent 3,247,168 (April 19, 1966).
205. Ye. P. Krasnov, L. B. Sokolov, and T. A. Polyakova, *Polym. Sci. (USSR) (Engl. trans.)*, **6**:1371 (1965).
206. L. W. Clark, *J. Phys. Chem.*, **67**:138 (1963).
207. G. Fraenkel, R. L. Belford, and P. E. Yankwich, *J. Amer. Chem. Soc.*, **76**:15 (1954).
208. L. W. Clark, *J. Phys. Chem.*, **64**:41, 692 (1960); **68**:3048 (1964).
209. J. F. Norris and H. F. Tucker, *J. Amer. Chem. Soc.*, **55**:4697 (1933).
210. G. A. Hall, Jr., *J. Amer. Chem. Soc.*, **71**:2691 (1949).
211. R. A. Fairclough, *J. Chem. Soc.*, **1938**:1186.
212. W. C. Pierce and G. Morey, *J. Amer. Chem. Soc.*, **54**:467 (1932).
213. O. Diels and B. Wolf, *Chem. Ber.*, **39**:689 (1906).
214. H. Staudinger and E. Ott, *Chem. Ber.*, **41**:2208 (1908).
215. A. I. Vogel, *J. Chem. Soc.*, **1927**:1985.
216. S. E. Boxer and R. P. Linstead, *J. Chem. Soc.*, **1931**:740.
217. J. R. Johnson, in "Organic Reactions," Vol. 1, R. Adams, Ed., Wiley, New York, 1942, p. 226.
218. M. Conrad and H. Reinbach, *Chem. Ber.*, **35**:1813 (1902).
219. A. C. Cope, H. L. Holmes, and H. O. House, in "Organic Reactions," Vol. 9, R. Adams, Ed., Wiley, New York, 1957, p. 107.
220. E. D. Bergmann, D. Ginsburg, and R. Pappo, in "Organic Reactions," Vol. 10, R. Adams, Ed., Wiley, New York, 1959, p. 179.
221. K. N. Welch, *J. Chem. Soc.*, **1930**:257.
222. K. N. Welch, *J. Chem. Soc.*, **1931**:673.
223. M. Conrad, and C. A. Bischoff, *Justus Liebigs Ann. Chem.*, **204**:126 (1880).
224. E. C. Britton and E. Monroe (to Dow Chemical Co.), U.S. Patent 2,373,011 (April 3, 1945).
225. A. A. Ross and F. E. Bibbins, *Ind. Eng. Chem.*, **29**:1341 (1937).
226. W. Wenner, in "Encyclopedia of Chemical Technology," Vol. 12, H. F. Mark, Chairman, Editorial Board, Interscience, New York, 1967, p. 854.
227. J. Schwyzer, "Die Fabrikation pharmazeutisches und chemischtechnischer Produkte," Julius Springer, Berlin, 1931, p. 102.
228. N. Weiner, *Org. Syn.* **Coll. Vol. II,** 376 (1946).
229. B. Raecke (to Henkel & Cie, GmbH), German Patent 1,185,602 (Jan. 21, 1965); *Chem. Abstr.*, **62**:9015 (1965).
230. B. Raecke, *Angew. Chem.*, **76**:892 (1964).
231. B. Raecke (to Henkel & Cie, GmbH), U.S. Patent 3,359,310 (Dec. 19, 1967).
232. T. Asher, *Chem. Ber.*, **30**:1023 (1897).
233. H. Staudinger and S. Bereza, *Chem. Ber.*, **41**:4461 (1908).
234. C. Raha, *Org. Syn.* **Coll, Vol. IV,** 263 (1963).
235. R. Black, H. Shaw, and T. K. Walker, *J. Chem. Soc.*, **1931**:276.
236. D. O. DePree and W. R. Eller (to Ethyl Corp.), U.S. Patent 2,852,559 (Sept. 16, 1958).
237. D. O. DePree and R. D. Closson, *J. Amer. Chem. Soc.*, **80**:2311 (1958).
238. G. Bottaccio and G. P. Chinsoli, *Chem. Ind. (Milan)*, **1966**:1457.

239. O. Heuse, M. Boldt, and R. Wirtz (to Farbwerke Hoechst Akt.), U.S. Patent 3,316,162 (April 25, 1967).
240. J. K. Dixon (to American Cyanamid Co.), U.S. Patent 2,553,406 (May 15, 1951).
241. J. K. Dixon (to American Cyanamid Co.), U.S. Patent 2,606,917 (Aug. 12, 1952).
242. M. Taguchi and S. Shoji (to Nissan Kagaku Kogyo Kabushiki Kaisha), U.S. Patent 3,417,126 (Dec. 17, 1968).
243. L. J. Krebaum (to Monsanto Chemical Co.), U.S. Patent 3,055,738 (Sept. 25, 1962).
244. P. L. Levine, D. E. Johnson, and W. L. Kranisch (to Arthur D. Little, Inc.), U.S. Patent 3,497,546 (Feb. 24, 1970).
245. D. E. Johnson, P. L. Levine, and W. L. Kranisch (to Arthur D. Little, Inc.), U.S. Patent 3,541,133 (Nov. 17, 1970).
246. K. Morita, N. Hashimoto, and T. Saraie (to Takeda Chemical Industries, Ltd.), U.S. Patent 3,502,709 (March 24, 1970).
247. F. Dupré la Tour, *C.R. Acad. Sci. (Paris)*, **193**:180 (1931).
248. J. A. Goedkoop and C. H. MacGillavry, *Acta Crystallogr.*, **10**:125 (1957).
249. D. Hadži and N. Sheppard, *Proc. Roy. Soc. (London), Ser A*, **216**:247 (1953).
250. J. Lecomte, *C.R. Acad. Sci. (Paris)*, **211**:776 (1940).
251. V. Ananthanarayanan, *Proc. Indian Acad Sci., Sect. A*, **51**:328 (1960).
252. N. W. Taylor, in "International Critical Tables," Vol. 4, E. W. Washburn, McGraw-Hill, New York, 1928, p. 251.
253. A. Seidell, "Solubilities of Organic Compounds," 3rd ed., Vol. 2, Van Nostrand, New York, 1941, p. 168.
254. V. D. Yakhontov, *J. Appl. Chem. USSR (Engl. transl.)*, **19**:761 (1946).
255. W. Biltz, W. Fischer, and E. Wunnenberg, *Z. Phys. Chem. (Leipzig) [A]*, **151**:25 (1930).
256. J. H. Awberry, in Ref. 176, p. 102.
257. A. Seidell, Ref. 253, p. 166.
258. S. N. Das and D. J. G. Ives, *Proc. Chem. Soc.*, **1961**:373.
259. W. J. Hamer, J. O. Burton, and S. F. Acree, *J. Res. Natl. Bur. Std. (U.S.)*, **24**:269 (1940).
260. J. O. Burton, W. J. Hamer, and S. F. Acree, *J. Res. Natl. Bur. Std. (U.S.)* **16**:575 (1936).
261. R. H. Gale and C. C. Lynch, *J. Amer. Chem. Soc.*, **64**:1153 (1942).
262. V. Procházková, J. Beneš, and K. Vereš, *J. Chromatogr.*, **21**:402 (1966).
263. D. T. Canvin, *Can. J. Biochem.*, **43**:1281 (1965).
264. N. E. Sharpless, *J. Chromatogr.*, **12**:401 (1963).
265. H. J. Petrowitz and G. Pastuska, *J. Chromatogr.*, **7**:128 (1962).
266. C. Davies, R. D. Hartley, and G. J. Lawson, *J. Chromatogr.*, **18**:47 (1965).
267. P. Handler, *J. Biol. Chem.*, **161**:53 (1945).
268. C. M. Gruber, Jr., C. de Berardinis, and L. A. Erf, *Arch. Int. Pharmacodyn.*, **79**:461 (1949).
269. W. H. Carothers and G. A. Arvin, *J. Amer. Chem. Soc.*, **51**:2560 (1929).
270. H. J. Hagemeyer (to Eastman Kodak Co.), U.S. Patent 3,043,808 (July 10, 1962).
271. H. Hopff and H. Griesshaber (vested in the Alien Property Custodian), U.S. Patent 2,302,321 (Nov. 17, 1942).
272. B. L. Moldavskiĭ and M. V. Blinova, *Neftekhimiya*, **5**:108 (1965); *Chem. Abstr.*, **62**:16043 (1965).
273. L. F. Fieser and E. L. Martin, *Org. Syn.* **Coll. Vol. II**, 560 (1943).
274. R. L. Shriner and H. C. Struck, *Org. Syn.* **Coll. Vol. II**, 560 (1943).
275. J. Cason, *Org. Syn.* **Coll. Vol. III**, 170 (1955).
276. J. Cason and E. J. Reist, *J. Org. Chem.*, **23**:1492 (1958).

277. W. S. Johnson and G. H. Daub, in "Organic Reactions," Vol. 6, R. Adams, Ed., Wiley, New York, 1951, pp. 1–73.
278. F. Webel (to I. G. Farbenindustrie A.G.), German Patent 441,002 (Feb. 19, 1927); *Chem. Zentralbl.*, **1927**:I, 2138.
279. K. W. Coons (to National Aniline and Chemical Co.), U.S. Patent 2,198,153 (April 23, 1940).
280. M. A. Kise and R. R. Wenner (to the Solvay Process Co.), U.S. Patent 2,245,404 (June 10, 1941).
281. J. H. Hahn (to Monsanto Chemical Co.), U.S. Patent 2,807,532 (Sept. 24, 1957).
282. O. W. Cass (to E. I. du Pont de Nemours & Co.), U.S. Patent 2,867,628 (Jan. 6, 1959).
283. A. Schulz (to Badische Anilin- & Soda-Fabrik A.G.), French Patent 1,360,486 (May 8, 1964).
284. A. E. Craver (to Barrett Co.), U.S. Patent 1,491,465 (April 22, 1922).
285. A. O. Jaeger (to The Selden Co.), U.S. Patent 1,844,394 (Feb. 9, 1932).
286. J. A. Bertsch and A. H. Krause (to Monsanto Chemical Co.), U.S. Patent 1,945,175 (Jan. 30, 1934).
287. B. B. Allen, B. W. Wyatt, and H. R. Henze, *J. Amer. Chem. Soc.*, **61**:843 (1939).
288. K. Ashida, *Mem. Inst. Ind. Res., Osaka Univ.*, **8**:193 (1951); *Chem. Abstr.*, **46**:7043 (1952).
289. I. Kh. Fel'dman, and E. S. Troyanova, *J. Appl. Chem. (USSR)*, **16**:15 (1943); *Chem. Abstr.*, **38**:2932 (1944).
290. C. Zenghelis and C. Stathis, *C.R. Acad. Sci. (Paris)*, **206**:682 (1938).
291. W. Schmitz (to Firma Carl Still), German Patent 1,259,869 (Feb. 1, 1968); *Chem. Abstr.*, **68**:6595 (1968).
292. J. F. Norris and E. O. Cummings, U.S. Patent 1,457,791 (June 5, 1923).
293. J. F. Norris and E. O. Cummings, *Ind. Eng. Chem.*, **17**:305 (1925).
294. L. P. Kyrides and J. A. Bertsh (to Monsanto Chemical Co.), U.S. Patent 1,927,289 (Sept. 19, 1933).
295. P. C. Condit (To California Research Corp.), U.S. Patent 2,537,304 (Jan. 9, 1951).
296. T. Yatani et al., Japanese Patent 3679 ('54) (June 23, 1954); *Chem. Abstr.*, **49**:8332 (1955).
297. S. Swann, Jr., K. H. Wanderer, H. J. Schaffer, and W. A. Streaker, *J. Electrochem. Soc.*, **96**:353 (1949).
298. E. E. Reid (to Hercules Powder Co.), U.S. Patent 2,141,406 (Dec. 27, 1938).
299. E. B. Punnett (to Allied Chemical and Dye Corp.), U.S. Patent 2,370,579 (Feb. 27, 1945).
300. V. Harlay, *C.R. Acad. Sci. (Paris)*, **213**:304 (1941).
301. Fabriques de Produits de Chimie Organique de Laire, French Patent 971,429 (Jan. 17, 1951).
302. E. Schwenk, D. Papa, B. Whitman, and H. F. Ginsberg, *J. Org. Chem.*, **9**:175 (1944).
303. W. M. Campbell (to Shawinigan Chemicals Ltd.), U.S. Patent 2,415,414 (Feb. 11, 1947).
304. P. Kurtz, *Justus Liebigs Ann. Chem.*, **572**:52 (1951).
305. A. O. Rogers (to E. I. du Pont de Nemours & Co.), U.S. Patent 2,415,261 (Feb. 4, 1947).
306. T. L. Gresham (to The B. F. Goodrich Co.), U.S. Patent 2,449,988 (Sept. 28, 1948).
307. R. T. Dean and E. O. Hook (to American Cyanamid Co.), U.S. Patent 2,351,667 (June 20, 1944).
308. W. Reppe, U.S. Patent 2,604,490 (July 22, 1952); Badische Anilin- & Soda-Fabrik A.G., British Patent 691,424 (May 13, 1953).

309. G. Natta and P. Pino (to Lonza Electric and Chemical Works Ltd.), U.S. Patent 2,851,486 (Sept. 9, 1958).

310. P. Pino, A. Miglierina, and E. Pietra, *Gazz. Chim. Ital.*, **84**:443 (1954).

311. W. Reppe, H. Albers, and H. H. Friederich (to Badische Anilin- & Soda-Fabrik A.G.), German Patent 1,040,526 (Oct. 9, 1958); *Chem. Abstr.*, **55**:6382 (1961).

312. S. I. Scott (to E. I. du Pont de Nemours & Co.), U.S. Patent 2,436,269 (Feb. 17, 1948).

313. L. Weintraub, J. F. Vitcha, and R. Limon, *Chem. Ind.* (*London*) **1965**:185.

314. C. B. Lines and R. Long, *Amer. Chem. Soc., Div. Petrol. Chem. Preprints*, **14**:No. 2, B159 (1969).

315. I. G. Farbeindustrie A.G., Belgian Patent 445,957 (July 31, 1942); *Chem. Abstr.*, **39**:712 (1945).

316. F. Ebel and F. Pyzik (to Alien Property Custodian), U.S. Patent 2,312,468 (March 2, 1943).

317. H. Heinze (to I. G. Farbenindustrie A.G.), German Patent 867,688 (Feb. 19, 1953); *Chem. Abstr.*, **48**:11487 (1954).

318. W. Reppe et al., *Justus Liebigs Ann. Chem.*, **596**:107 (1955).

319. British Celanese Ltd., British Patent 590,310 (July 14, 1947).

320. J. E. Bludworth, S. B. Jeffries, Jr., and M. O. Robeson (to Camille Dreyfuss), Canadian Patent 450,353 (Aug. 3, 1948); *Chem. Abstr.*, **42**:8096 (1948).

321. W. G. Toland, Jr. (to California Research Corp.), U.S. Patent 2,670,370 (Feb. 23, 1954).

322. Verein für Chem. Ind. A.G., German Patent 473,262 (Dec. 19, 1926); *Chem. Abstr.*, **23**:2989 (1929).

323. W. Reppe et al., *Justus Liebigs Ann. Chem.*, **596**:179 (1955).

324. R. M. Isham (to Danciger Oil & Refineries, Inc.), U.S. Patent 2,385,518 (Sept. 25, 1945).

325. K. Schmeidl and K. Scherf (to Badische Anilin- & Soda-Fabrik A.G.), German Patent 1,183,897 (Dec. 23, 1964); *Chem. Abstr.*, **62**:6397 (1965).

326. R. M. Isham (to Danciger Oil & Refineries, Inc.), U.S. Patent 2,420,954 (May 20, 1947).

327. H. W. Fleming (to Phillips Petroleum Co.), U.S. Patent 2,452,741 (Nov. 2, 1948).

328. W. Voss and R. Knopp (to VEB Filmfabrik Agfa Wolfen), British Patent 927,450 (May 29, 1963); *Chem. Abstr.*, **59**:11262 (1963).

329. Y. Takayama (to Alien Property Custodian), U.S. Patent 2,338,466 (Jan. 4, 1944).

330. A. P. Dunlop and S. Smith (to Quaker Oats Co.), U.S. Patent 2,676,186 (April 20, 1954).

331. E. Dietzel and K. Gieseler (to Vereinigte Glanzstoff-Fabriken A.G.), German Patent 765,011 (Aug. 9, 1951); *Chem. Abstr.*, **51**:17985 (1957).

332. N. Milas, *J. Amer. Chem. Soc.*, **49**:2005 (1927).

333. M. S. Konecky (to Esso Research and Engineering Co.), U.S. Patent 3,100,798 (Aug. 13, 1963).

334. H. Schade, F. W. Beckhaus, and J. Fiedler, (East) German Patent 23,813 (Oct. 1, 1962); *Chem. Abstr.*, **59**:6258 (1963).

335. A. F. Millidge, I. K. M. Robson, and A. Elce (to Distillers Co. Ltd.), British Patent 767,290 (Jan. 30, 1957).

336. R. F. Gilby, Jr., and C. E. Hoberg (to E. I. du Pont de Nemours & Co.), U.S. Patent 2,971,010 (Feb. 7, 1961).

337. Imperial Chemical Industries Ltd., Netherlands Appl. 6,516,174 (Feb. 25, 1966); *Chem. Abstr.*, **65**:3752 (1966).

338. G. W. Crosby and J. B. Braunworth (to Pure Oil Co.), U.S. Patent 2,862,028 (Nov. 25, 1958).
339. R. B. Randall, M. Benger, and C. M. Groocock, *Proc. Roy. Soc. (London) Ser. A*, **165**:432 (1938).
340. F. G. Parker, J. P. Fugassi, and H. C. Howard, *Ind. Eng. Chem.*, **47**:1586 (1955).
341. O. Grosskinsky and B. Jüttner (to Bergwerksverband zur Verwertung von Schutzrechten der Kohlentechnik GmbH), German Patent 1,021,349 (Dec. 27, 1957); *Chem. Abstr.*, **53**:22845 (1959).
342. O. L. Polly (to Union Oil Co. of California), U.S. Patent 2,533,620 (Dec. 12, 1950).
343. A. I. Smith (to Union Oil Co. of California), U.S. Patent 2,592,964 (April 15, 1952).
344. H. Chafetz and J. A. Patterson (to Texaco, Inc.), U.S. Patent 2,978,473 (April 4, 1961).
345. I. D. Elkins (to Kerr-McGee Oil Industries, Inc.), U.S. Patent 2,851,488 (Sept. 9, 1958).
346. B. L. Moldavskiĭ, M. V. Blinova, R. I. Rudakova, M. Sh. Usmanova, and E. I. Rubinshteĭn, *Zh. Prikl. Khim.*, **32**:2771 (1959); *Chem. Abstr.*, **54**:9754 (1960).
347. H. Chafetz (to Texaco, Inc.), U.S. Patent 3,036,127 (May 27, 1962).
348. A. C. Brown and J. Walker, *Justus Liebigs Ann. Chem.*, **261**:107 (1891).
349. J. B. Robertson, *J. Chem. Soc.*, **1925**:2057.
350. F. Fichter and J. Heer, *Helv. Chim. Acta*, **18**:704 (1935).
351. S. Glasstone and A. Hickling, *Chem. Rev.*, **25**:425 (1939).
352. L. Eberson, *Acta Chem. Scand.*, **13**:40 (1959).
353. M. S. Kharasch and M. T. Gladstone, *J. Amer. Chem. Soc.*, **65**:15 (1943).
354. M. S. Kharasch (to Eli Lilly and Co.), U.S. Patent 2,426,224 (Aug. 26, 1947).
355. K. Sisido, Y. Kazama, H. Kodama, and H. Nozaki, *J. Amer. Chem. Soc.*, **81**:5817 (1959).
356. F. Salmon-Legagneur, *Bull. Soc. Chim. Fr.*, **1956**:411.
357. E. A. Wynne, *Microchem. J.*, **5**:175 (1961).
358. P. Turi, in "Encyclopedia of Chemical Technology," 2nd ed., Vol. 19, H. F. Mark, Chairman of the Editorial Board, Interscience, New York, 1969, pp. 134–150.
359. F. Dupré la Tour, *C.R. Acad. Sci. (Paris)*, **191**:1348 (1930).
360. G. C. Rieck, *Rec. Trav. Chim. Pays-Bas*, **63**:170 (1944).
361. J. S. Broadley, D. W. J. Cruickshank, J. D. Morrison, J. M. Robertson, and H. M. M. Shearer, *Proc. Roy. Soc. (London) Ser. A*, **251**:441 (1959).
362. P. J. Corish and W. H. T. Davison, *J. Chem. Soc.*, **1955**: 2431.
363. T. Shimanouchi, M. Tsuobi, T. Takenishi, and N. Iwata, *Spectrochim. Acta*, **16**:1328 (1960).
364. J. Schurz, E. Treiber, and H. Toplak, *Z. Elektrochem.*, **60**:67 (1956).
365. L. J. Bellamy, B. R. Connelly, A. R. Philpotts, and R. L. Williams, *Z. Elektrochem.*, **64**:563 (1960).
366. W. G. Dauben and W. W. Epstein, *J. Org. Chem.*, **24**:1595 (1959).
367. D. Hadzĭ and A. Novak, *Nuovo Cimento*, **11**: Suppl. no. 3, 715 (1955); *Chem. Abstr.*, **55**:17215 (1961).
368. M. Davies and G. H. Thomas, *Trans. Faraday Soc.*, **56**:185 (1960).
369. A. Seidell, Ref. 253, p. 230.
370. A. I. Tsinman and V. S. Kuzub, *Zh. Prikl. Khim.*, **38**:1872 (1965); *Chem. Abstr.*, **63**:14473 (1965).
371. N. S. Novoshinskaya and N. G. Klyuchnikov, *Izv. Vyssh. Ucheb. Zaved. Pishch. Tekhnol.*, **1966**:36; *Chem. Abstr.*, **66**:5507 (1967).
372. W. Biltz, *Justus Liebigs Ann. Chem.*, **453**:278 (1927).
373. H. Marshall and D. Bain, *J. Chem. Soc.*, **97**:1074 (1910).

374. G. D. Pinching and R. G. Bates, *J. Res. Natl. Bur. Std. (U.S.)*, **45**:444 (1950).

375. G. D. Pinching and R. G. Bates, *J. Res. Natl. Bur. Std. (U.S.)*, **45**:322 (1950).

376. R. Gane and C. K. Ingold, *J. Chem. Soc.*, **1931**:2153.

377. S. Miyazaki, Y. Suhara, and T. Kobayashi, *J. Chromatogr.*, **39**:88 (1969).

378. A. I. Smith, *Anal. Chem.*, **31**:1621 (1959).

379. D. S. Kinnory, Y. Takeda, and D. M. Greenberg, *J. Biol. Chem.*, **212**:379 (1955).

380. R. W. Scott, *Anal. Chem.*, **27**:367 (1955).

381. H. B. Henbest and T. C. Owen, *J. Chem. Soc.*, **1955**:2968.

382. L. Silverman, *Chemist-Analyst*, **36**:57 (1947).

383. S. Forssman, *Acta Physiol. Scand.*, **Suppl. No. 5**:121 (1941); *Chem. Abstr.*, **36**:111 (1942).

384. V. L. Friend and H. Gold, *J. Amer. Pharm. Ass.*, **36**:50 (1947); *Chem. Abstr.*, **41**:7515 (1947).

385. W. H. Gardner, "Food Acidulants," Allied Chemical Corp., New York, 1966.

386. W. H. Carothers and G. L. Dorough, *J. Amer. Chem. Soc.*, **52**:711 (1930).

387. W. H. Carothers and J. W. Hill, *J. Amer. Chem. Soc.*, **54**:1559 (1932).

388. E. W. Spanagel and W. H. Carothers, *J. Amer. Chem. Soc.*, **57**:929 (1935).

389. J. Dale, *J. Chem. Soc.*, **1965**:72.

390. K. W. Doak and H. N. Campbell, *J. Polym. Sci.*, **18**:215 (1955).

391. C. S. Fuller et al., *J. Amer. Chem. Soc.*, **59**:344 (1937); **61**:2575 (1939); **64**:154 (1942); *J. Phys. Chem.*, **43**:323 (1939).

392. W. H. T. Davison and P. J. Corish, *J. Chem. Soc.*, **1955**:2428.

393. B. S. Biggs, R. H. Erickson, and C. S. Fuller, *Ind. Eng. Chem.*, **39**:1090 (1947).

394. V. Martello and V. Giolitti, *Gazz. Chim. Ital.*, **85**:1224 (1955).

395. V. V. Korshak, T. M. Frunze, and E. A. Krasnyanskaya, *Vysokomol. Soedin.*, **4**:1761 (1962); *Chem. Abstr.*, **59**:1762 (1963).

396. T. Kagiya, M. Izu, T. Matsuda, and K. Fukui, *J. Polym. Sci.*, *Pt A-1*, **5**:15 (1967); *Macromol. Syn.*, **3**:29 (1968).

397. *Chem. Eng. News*, **48**(30):50 (July 20, 1970).

398. S. Skraup and S. Guggenheimer, *Chem. Ber.*, **58**:2488 (1925).

399. K. v. Auwers and M. Schmidt, *Chem. Ber.*, **46**:457 (1913).

400. C. S. Marvel and W. F. Tuley, *Org. Syn* **Coll. Vol. I**, rev. ed., 289 (1946).

401. T. J. Otterbacher, *Org. Syn.* (2 nd ed.), **Coll. Vol. I**, 290 (1946).

402. G. Paris, L. Berlinguet, and R. Gaudry, *Org. Syn.* **Coll. Vol. IV**, 496 (1963).

403. J. English, Jr., and J. E. Dayan, *Org. Syn.* **Coll. Vol. IV**, 499 (1963).

404. E. N. Zil'berman and A. I. Kirillov, *Zh. Prikl. Khim.*, **30**:960 (1957); *Chem. Abstr.*, **52**:1077 (1958).

405. S. H. McAllister (to Shell Development Co.), U.S. Patent 2,193,562 (March 12, 1940).

406. S. H. McAllister (to Shell Development Co.), U.S. Patent 2,286,559 (June 16, 1942).

407. S. H. McAllister (to Shell Development Co.), U.S. Patent 2,285,601 (June 9, 1942).

408. E. Boedtker, *J. Pharm. Chim.*, **15**:225 (1932).

409. V. B. Fal'kovskii, T. A. Tyuricheva, E. M. Kalmykova, and S. V. L'vov, *Izv. Vyssh. Uchebn. Zaved., Khim. i Khim. Tekhnol.*, **6**:344 (1963); *Chem. Abstr.*, **59**:8584 (1963).

410. M. Asaka, *Yuki Gosei Kagaku Kyokai Shi*, **23**:589 (1965); *Chem. Abstr.*, **63**:8187 (1965).

411. J. G. M. Bremner, R. H. Stanley, D. G. Jones, and A. W. C. Taylor (to Imperial Chemical Industries Ltd.), U.S. Patent 2,389,950 (Nov. 27, 1945).

412. H. Ritter and W. Zerweck (to Cassella Farbwerke Mainkur A.G.), German Patent 848,038 (Sept. 1, 1952); *Chem. Abstr.*, **50**:16840 (1956).

413. A. Hrubesch and O. Schlichting (to Badische Anilin- & Soda-Fabrik A.G.), German Patent 887,943 (Aug. 27, 1953); *Chem. Abstr.*, **52**:13784 (1958).
414. W. Pack and M. Alsfeld (to Henkel & Cie, GmbH), German Patent 725,741 (Aug. 13, 1942).
415. R. R. Whetstone (to Shell Development Co.), U.S. Patent 2,513,766 (July 4, 1950).
416. H. R. Guest, H. A. Stansbury, Jr., and H. F. Lykins (to Union Carbide & Carbon Corp.), British Patent 767,416 (Feb. 6, 1957).
417. N. V. de Bataafsche Petroleum Moatschappij, British Patent 772,410 (April 10, 1957); *Chem. Abstr.*, **51**:12970 (1957).
418. N. v. Kutepow, W. Himmele, and I. Class (to Badische Anilin- & Soda-Fabrik A.G.), German Patent 1,026,297 (March 20, 1958).
419. F. Dupré la Tour, *C.R. Acad. Sci.* (*Paris*), **194**:622 (1932).
420. C. H. MacGillavry, Hoogschagen, and F. L. J. Sixma, *Rec. Trav. Chim. Pays-Bas*, **67**:869 (1948).
421. M. Wehrli and R. Fichter, *Helv. Phys. Acta*, **14**:189 (1941).
422. M. Wehrli, *Helv. Phys. Acta*, **14**:516 (1941).
423. R. Perron and J. Perichon, *Rev. Franc. Corps Gras*, **12**:381 (1965).
424. E. Childers and G. W. Struthers, *Anal. Chem.*, **27**:737 (1955).
425. E. C. Attané and T. F. Doumani, *Ind. Eng. Chem.*, **41**:2015 (1949).
426. I. Jones and F. G. Soper, *J. Chem. Soc.*, **1936**:133.
427. C. S. Marvel and R. D. Rands, Jr., *J. Amer. Chem. Soc.*, **72**:2642 (1950).
428. W. C. Rose, *J. Pharmacol.*, **24**:147 (1924).
429. V. J. Harding and T. F. Nicholson, *J. Pharmacol.*, **42**:373 (1931).
430. C. S. Fuller and C. L. Erickson, *J. Amer. Chem. Soc.*, **59**:344 (1937).
431. C. S. Fuller and C. J. Frosch, *J. Amer. Chem. Soc.*, **61**:2575 (1939).
432. C. S. Fuller, C. J. Frosch, and N. R. Pape, *J. Amer. Chem. Soc.*, **64**:154 (1942).
433. V. V. Korshak and T. M. Frunze, *Izv. Akad. Nauk SSSR, Otd. Khim. Nauk*, **1955**:934; *Chem. Abstr.*, **50**:9325 (1956).
434. O. Ya. Fedotova, S. A. Zakoshchikov, and I. P. Losev, *Vysokomol. Soedin.*, **5**:1671 (1963); *Polym. Sci.* (*USSR*) (*Engl. transl.*), **5**:783 (1964).
435. O. Ya. Fedotova, I. P. Losev, and S. A. Zakoshchikov, *Vysokomol. Soedin.*, **5**:531 (1963); *Polym. Sci.* (*USSR*) (*Engl. transl.*), **4**:1202 (1964).
436. O. Ya. Fedotova, S. A. Zakoshchikov, I. P. Losev, and T. R. Nakhodnova, *Izv. Vyssh. Uchebn. Zaved., Khim. i Khim. Tekhnol.*, **9**:764 (1966); *Chem. Abstr.*, **66**:6227 (1967).
437. U.S. Tariff Commission, "Synthetic Organic Chemicals, United States Production and Sales, 1968," Govt. Printing Office, Washington, D.C., 1970.
438. Anon., *Oil, Paint, Drug Rep.*, June 23, 1969, p. 9.
439. W. L. Standish and S. V. Abrams, in "Encyclopedia of Chemical Technology," Vol. 1, 2nd ed., H. F. Mark, Chairman of the Editorial Board, Interscience, New York, 1963, pp. 405–421.
440. J. F. Thorpe and G. A. R. Kon, *Org. Syn.* (2nd ed.) Coll. Vol. I, 192 (1946).
441. O. Neunhoeffer and P. Paschke, *Chem. Ber.*, **72**:919 (1939).
442. J. W. Hill, *J. Amer. Chem. Soc.*, **52**:4110 (1930).
443. P. C. Guha and D. K. Sankaran, *Org. Syn.* Coll. Vol. III, 623 (1955).
444. V. M. Mićović, *Org. Syn.* Coll. Vol. II, 264 (1943).
445. P. S. Pinckney, *Org. Syn.* Coll. Vol. II, 116 (1943).
446. J. C. Sheehan, R. C. O'Neill, and M. A. White, *J. Amer. Chem. Soc.*, **72**:3376 (1950).
447. W. J. Bailey and W. G. Carpenter, *J. Org. Chem.*, **29**:1252 (1964).
448. G. N. Freidlin, S. M. Zhenodarova, A. P. Chukur, and N. V. Fomina, *Zh. Obshch. Khim.*, **32**:792 (1962); *Chem. Abstr.*, **58**:1340 (1963).

449. G. N. Freidlin, A. A. Adamov, and P. M. Zaitsev, *Zh. Organ. Khim.*, **1**:666 (1965); *Chem. Abstr.*, **63**:6849 (1965).
450. V. I. Isagulyants and E. L. Markosyan, *Zh. Prikl. Khim.*, **35**:2109 (1962); *Chem. Abstr.*, **58**:5563 (1963).
451. D. M. Vinokurov and M. B. Khaikina, *Izv. Vyssh. Uchebn. Zaved. Khim. i Khim. Tekhnol.*, **6**:83 (1963); *Chem. Abstr.*, **59**:6250 (1963).
452. E. I. du Pont de Nemours & Co., British Patent 1,038,460 (Aug. 10, 1966).
453. D. D. Coffman, G. J. Berchet, W. R. Peterson, and E. W. Spanagel, *J. Polym. Sci.*, **2**:306 (1947).
454. C. J. Brown, *Acta Crystallogr.*, **21**:185 (1966).
455. Anon., *Chem. Week*, **89**(5):83 (1961).
456. N. M. Emanuel, E. T. Denisov, and Z. K. Maizus, "Liquid-Phase Oxidation of Hydrocarbons," B. J. Hazzard, Transl., Plenum Press, New York, 1967.
457. I. V. Berezin, E. T. Denisov, and N. M. Emanuel, "The Oxidation of Cyclohexane," K. A. Allen, Transl., Pergamon Press, New York, 1966.
458. M. Furman et al., "Production of Cyclohexanone and Adipic Acid (*Proizvodstvo Tsiklogeksanona i Adipinovoi Kisloty*)," Khimiya, Moscow, 1967, 240 pp.; *Chem. Abstr.*, **69**:203 (1968).
459. S. A. Miller, *Chem. Process Eng.*, **50**:63 (1969).
460. D. J. Loder (to E. I. du Pont de Nemours & Co.), U.S. Patent 2,223,494 (Dec. 3, 1940).
461. D. J. Loder (to E. I. du Pont de Nemours & Co.), U.S. Patent 2,321,551 (June 8, 1943).
462. N. Ota and T. Tezuka, *J. Chem. Soc. Jap. Ind. Chem. Sect.*, **57**:641, 723, 725 (1954); **58**:680 (1955); *Chem. Abstr.*, **49**:8675, 10201 (1955); **50**:11255 (1956).
463. H. L. Cates Jr., J. O. Penderson, R. W. Wheatcroft, and A. B. Stiles (to E. I. du Pont de Nemours & Co.), U.S. Patent 2,851,496 (Sept. 9, 1958).
464. J. Rouchard and P. Mulkay, *Bull. Soc. Chim. Belg.*, **76**:579 (1967).
465. G. J. Schmitt, J. Pisanchyn, and W. F. Chapman (to Allied Chemical Corp.), U.S. Patent 3,334,141 (Aug. 1, 1967).
466. J. H. Raley and F. F. Rust (to Shell Development Co.), U.S. Patent 2,391,740 (Dec. 25, 1945).
467. H. A. Dewhurst, *J. Phys. Chem.*, **63**:813 (1959).
468. A. J. Restaino and R. F. Hornbeck (to Atlas Chemical Ind., Inc.), U.S. Patent 3,368,955 (Feb. 13, 1968).
469. B. E. Kuiper (to E. I. du Pont de Nemours & Co.), U.S. Patent 3,035,092 (May 15, 1962).
470. C. Gardner, J. F. Prescott, and R. G. A. New (to Imperial Chemical Industries, Ltd.), U.S. Patent 3,428,690 (Feb. 18, 1969).
471. S. Ishimoto, T. Sasano, and K. Kawamura, *Ind. Eng. Chem.*, *Process Des. Develop.*, **7**:469 (1968).
472. M. S. Furman, V. V. Lipes, and N. A. Gol'tyaeva, *Inform. Soobshch. Gos. Nauch.-Issled. Proekt. Inst. Azotn. Prom. Prod. Org. Sin.*, **1966**:no. 17, pt. 1, 21–30; *Chem. Abstr.*, **69**:240 (1968).
473. J. G. D. Schulz and A. C. Whitaker (to Gulf Research and Development Corp.), U.S. Patent 3,340,304 (Sept. 5, 1967).
474. C. H. Hamblet and F. S. Chance (to E. I. du Pont de Nemours & Co.), U.S. Patent 2,557,281 (June 19, 1951).
475. A. P. Schueler and F. A. Wolff (to E. I. du Pont de Nemours & Co.), U.S. Patent 2,825,742 (March 4, 1958).
476. E. I. du Pont de Nemours & Co., British Patent 776,803 (June 12, 1957).

477. E. I. du Pont de Nemours & Co., British Patent 633,354 (Dec. 12, 1949).
478. C. H. Hamblet and A. McAlevy (to E. I. du Pont de Nemours & Co.), U.S. Patent 2,439,513 (April 13, 1948).
479. W. Pritzkow and K. A. Müller, *Chem. Ber.*, **89**:2321 (1956).
480. E. F. J. Duynstee and L. J. P. Hennekens, *Rec. Trav. Chim. Pays-Bas*, **89**:769 (1970).
481. M. Goldbeck, Jr., and F. C. Johnson (to E. I. du Pont de Nemours & Co.), U.S. Patent 2,703,331 (March 1, 1955).
482. F. Porter and J. N. Cosby (to Allied Chemical & Dye Corp.), U.S. Patent 2,565,087 (Aug. 21, 1951).
483. C. F. Dougherty, Jr., and C. C. Chapman (to Phillips Petroleum Co.), U.S. Patent 2,615,921 (Oct. 28, 2952).
484. A. D. Cyphers, Jr., and A. A. Gruber (to E. I. du Pont de Nemours & Co.), U.S. Patent 2,870,203 (Jan. 20, 1959).
485. W. W. Crouch and J. C. Hillyer (to Phillips Petroleum Co.), U.S. Patent 2,931,834 (April 5, 1960).
486. W. Simon, H. J. Waldmann, and E. Plauth (to Badische Anilin- & Soda-Fabrik A.G.), U.S. Patent 2,938,924 (May 31, 1960).
487. R. D. Chapman, C. R. Campbell, and R. Johnson (to Chemstrand Corp.), U.S. Patent 3,023,238 (Feb. 27, 1962).
488. J. W. M. Steeman (to Stamicarbon, N. V.), U.S. Patent 3,047,629 (July 31, 1962).
489. C. Brierley, N. S. Robson, J. C. Ruddell, and F. G. Webster (to Imperial Chemical Industries, Ltd.), British Patent 914,510 (Jan. 2, 1963).
490. W. Simon, H. J. Waldmann, R. Melan, and E. Plauth (to Badische Anilin- & Soda-Fabrik A.G.), U.S. Patent 3,093,686 (June 11, 1963).
491. S. N. Fox and J. W. Colton (to Halcon International, Inc.), U.S. Patent 3,109,864 (Nov. 5, 1963).
492. A. Buck (to Inventa A.G.), U.S. Patent 3,119,873 (Jan. 28, 1964).
493. W. B. Hogeman (to Monsanto Co.), French Patent 1,369,683 (Aug. 14, 1964).
494. W. B. Hogeman (to Monsanto Co.), French Patent 1,369,684 (Aug. 14, 1964).
495. W. B. Hogeman (to Monsanto Co.), French Patent 1,369,685 (Aug. 14, 1964).
496. G. Lemetre, R. Orlandi, and C. Alberto (to Bombrini Parodi-Delfino Società per Azioni), U.S. Patent 3,161,476 (Dec. 15, 1964).
497. H. J. Waldmann and H. Hoffmann (to Badische Anilin- & Soda-Fabrik, A.G.), U.S. Patent 3,179,699 (April 20, 1965).
498. W. B. Hogeman (to Monsanto Co.), U.S. Patent 3,260,742 (July 12, 1966).
499. W. B. Hogeman (to Monsanto Co.), U.S. Patent 3,260,743 (July 12, 1966).
500. W. L. Seddon (to Petrocarbon Developments, Ltd.), U.S. Patent 3,274,254 (Sept. 20, 1966).
501. J. W. M. Steeman and J. P. H. v.d. Hoff (to Stamicarbon, N.V.), U.S. Patent 3,316,302 (April 25, 1967).
502. C. Gardner and J. F. Prescott (to Imperial Chemical Industries, Ltd.), British Patent 1,077,374 (July 26, 1967).
503. S. M. Ciborowski and K. Z. Balcerzak (to Instytut Chemii Ogolnej, Warsaw), U.S. Patent 3,349,007 (Oct. 24, 1967).
504. J. H. Bonfield and A. W. Sogn (to Allied Chemical Corp.), U.S. Patent 3,350,444 (Oct. 31, 1967).
505. W. J. Arthur and L. S. Scott (to E. I. du Pont de Nemours & Co.), U.S. Patent 3,365,490 (Jan. 23, 1968).
506. C. Gardner and J. F. Prescott (to Imperial Chemical Industries, Ltd.), U.S. Patent 3,365,491 (Jan. 23, 1968).

507. A. Pounder and W. F. Sykens (to Imperial Chemical Industries, Ltd.), U.S. Patent 3,467,701 (Sept. 16, 1969).
508. K. Pugi (to E. I. du Pont de Nemours & Co.), U.S. Patent 3,530,185 (Sept. 22, 1970).
509. M. Luther and W. Dietrich (to I. G. Farbenindustrie, A.G.), U.S. Patent 1,931,501 (Oct. 24, 1933).
510. T. Hellthaler and E. Peter (to A. Riebeck'sche Montanwerke A.G.), U.S. Patent 1,947,989 (Feb. 20, 1934).
511. Société Anonyme d'Innovations Chimique Sinnova ou Sidac, French Patent 1,166,679 (Nov. 13, 1958); *Chem. Abstr.*, **55**:378 (1961).
512. A. N. Bashkirov, V. V. Kamzolkin, K. M. Sokova, and T. P. Andreeva, *Dokl. Akad. Nauk SSSR*, **118**:149 (1958); *Chem. Abstr.*, **54**:7531 (1960).
513. A. N. Bashkirov and S. Pal, *Dokl. Akad. Nauk SSSR*, **128**:1175 (1959); *Chem. Abstr.*, **54**:7531 (1960).
514. A. N. Bashkirov and A. I. Kistanova, *Dokl. Akad. Nauk SSSR*, **131**:827 (1960); *Chem. Abstr.*, **54**:16405 (1960).
515. Anon., *Chem. Eng. News*, **44**(23):17 (1966).
516. Anon., *J. Commer. (N. Y.)*, Feb. 27, 1969, p. 6.
517. Imperial Chemical Industries, Ltd., French Patent 1,351,666 (Feb. 7, 1964); *Chem. Abstr.*, **61**:8195 (1964).
518. Imperial Chemical Industries, Ltd., Netherlands Appl. 288,365 (March 10, 1965); *Chem. Abstr.*, **63**:9819 (1965).
519. J. B. Feder and J. H. Carroll (to Halcon International, Inc.), U.S. Patent 3,239,552 (March 8, 1966).
520. C. N. Winnick (to Halcon International, Inc.), U.S. Patent 3,243,449 (March 29, 1966).
521. J. E. Helbig, D. O. Nelsen, R. E. Pennington, and I. J. Satterfield (to Esso Research and Engineering Co.), U.S. Patent 3,232,704 (Feb. 1, 1966).
522. H. Olenberg and L. Kantrowitz (to Halcon International, Inc.), U.S. Patent 3,240,820 (March 15, 1966).
523. J. W. M. Steeman and J. P. H. v.d. Hoff (to Stamicarbon, N.V.), U.S. Patent 3,287,423 (Nov. 22, 1966).
524. M. Becker (to Halcon International, Inc.), U.S. Patent 3,317,581 (May 2, 1967).
525. H. Olenberg (to Halcon International, Inc.), U.S. Patent 3,324,186 (June 6, 1967).
526. D. O. Nelsen, R. E. Pennington, I. J. Satterfield, J. E. Motley, and M. Spielman (to Esso Research and Engineering Co.), U.S. Patent 3,336,390 (Aug. 15, 1967).
527. J. W. H. Steeman and J. P. H. v.d. Hoff (to Stamicarbon, N.V.), U.S. Patent 3,350,465 (Oct. 31, 1967).
528. J. L. Russell and H. Olenberg (to Halcon International, Inc.), U.S. Patent 3,420,897 (Jan. 7, 1969).
529. J. L. Russell and M. Becker (to Halcon International, Inc.), U.S. Patent 3,438,726 (April 15, 1969).
530. J. C. Brunie and N. Crenne (to Rhone-Poulenc S.A.), U.S. Patent 3,479,394 (Nov. 18, 1969).
531. H. Olenberg and J. B. Feder (to Halcon International, Inc.), U.S. Patent 3,492,355 (Jan. 27, 1970).
532. J. L. Russell (to Halcon International, Inc.), U.S. Patent 3,497,544 (Feb. 24, 1970).
533. J. Alagy, F. Defoor, and S. Frankowiak (to Institut Français du Pétrole des Carburants et Lubrifiants), U.S. Patent 3,510,530 (May 5, 1970).
534. M. M. Clark, U.S. Patent 2,613,219 (Oct. 7, 1952).
535. B. A. Ellis, *Org. Syn.* (2nd ed.) **Coll. Vol. I,** 18 (1941).
536. A. F. Lindsay, *Chem. Eng. Sci. Spec. Suppl.*, **3**:78 (1954).

537. H. C. Godt, Jr., and J. F. Quinn, *J. Amer. Chem. Soc.*, **78**:1461 (1956).
538. W. J. Van Asselt and D. W. Van Krevelen, *Rec. Trav. Chim. Pays-Bas*, **82**:51 (1963).
539. H. C. Godt, Jr. (to Monsanto Chemical Co.), U.S. Patent 2,881,215 (April 7, 1959).
540. J. D. Riedel, A.G., British Patent 265,959 (Dec. 8, 1927).
541. Hydrierwerke Deutsche, German Patent 473,960 (Feb. 13, 1926); *Chem. Abstr.*, **23**:2988 (1929).
542. W. Schrauth (to E. I. du Pont de Nemours & Co.), U.S. Patent 1,921,101 (Aug. 8, 1933).
543. R. P. Perkins and A. J. Dietzler (to Dow Chemical Co.), U.S. Patent 1,960,211 (May 22, 1934).
544. E. Harrison, J. M. Woolley, and R. May (to Imperial Chemical Industries, Ltd.), British Patent 572,260 (Sept. 28, 1945).
545. G. Gut, R. v. Falkenstein, and A. Guyer, *Chimia*, **19**:581 (1965).
546. F. G. Jeffers and F. G. Webster (to Imperial Chemical Industries, Ltd.), British Patent 756,679 (Sept. 5, 1956).
547. F. G. Jeffers (to Imperial Chemical Industries, Ltd.), U.S. Patent 2,791,566 (May 7, 1957).
548. D. W. Brubaker and D. E. Danly (to Chemstrand Corp.), Belgian Patent 609,530 (April 24, 1962); *Chem. Abstr.*, **58**:1373 (1963).
549. D. W. Brubaker and D. E. Danly (to Monsanto Chemical Co.), Belgian Patent 619,290 (Dec. 24, 1962); *Chem. Abstr.*, **59**:1135 (1963).
550. D. M. Leyshon and J. Stewart (to Imperial Chemical Industries, Ltd.), British Patent 937,547 (Sept. 25, 1963).
551. P. J. v.d. Berg (to Stamicarbon, N.V.), U.S. Patent 3,106,450 (Oct. 8, 1963).
552. Imperial Chemical Industries, Ltd., Belgian Patent 635,677 (Jan. 31, 1964); *Chem. Abstr.*, **61**:12976 (1964).
553. R. Johnson and C. R. Campbell (to Monsanto Co.), U.S. Patent 3,148,210 (Sept. 8, 1964).
554. D. M. Leyshon and J. Stewart (to Imperial Chemical Industries, Ltd.), U.S. Patent 3,161,603 (Dec. 15, 1964).
555. D. W. Brubaker and D. E. Danly (to Monsanto Co.), U.S. Patent 3,186,952 (June 1, 1965).
556. Imperial Chemical Industries, Ltd., Netherlands Appl. 6,600,039 (July 21, 1966); *Chem. Abstr.*, **65**:16868 (1966).
557. J. M. Connolly and C. J. Lowery (to Imperial Chemical Industries, Ltd.), U.S. Patent 3,459,512 (Aug. 5, 1969).
558. G. P. Brown, Jr., E. I. Crowley, and N. W. Franke (to Gulf Research & Development Co.), U.S. Patent 2,831,024 (April 15, 1958).
559. J. Kamlet (to The Goodyear Tire & Rubber Co.), U.S. Patent 2,844,626 (July 22, 1958).
560. E. Haarer, R. Plass, and P. Hornberger (to Badische Anilin- & Soda-Fabrik A.G.), German Patent 1,159,423 (Dec. 19, 1963).
561. J. J. Fuchs (to E. I. du Pont de Nemours & Co.), U.S. Patent 3,112,340 (Nov. 26, 1963).
562. J. O. White (to E. I. du Pont de Nemours & Co.), U.S. Patent 3,076,026 (Jan. 29, 1963).
563. J. O. White (to E. I. du Pont de Nemours & Co.), U.S. Patent 3,170,952 (Feb. 23, 1965).
564. E. I. du Pont de Nemours & Co., British Patent 908,757 (Oct. 24, 1962).
565. J. Hannin and P. Baumgartner (to Institut Français du Pétrole, des Carburants et Lubrifiants), French Patent 1,376,479 (Oct. 30, 1964).

566. E. N. Zil'berman, S. N. Suvorova, and Z. S. Smolyan, *Zh. Prikl. Khim.* (*Leningrad*), **29**:621 (1956); *Chem. Abstr.*, **50**:14546 (1956).

567. E. N. Zil'berman and N. M. Chernysheva, *Tr. Khim. Khim. Tekhnol.*, **1**:684 (1958); *Chem. Abstr.*, **54**:7583 (1960).

568. J. Putnik, *Chem. Listy* **41**:135 (1947); *Chem. Abstr.*, **45**:555 (1951).

569. I. Ya. Lubyanitskiĭ, R. V. Minati, and M. S. Furman, *Khim. Prom.* (*Kiev*), **1960**:453; *Chem. Abstr.*, **55**:10313 (1961).

570. I. Ya. Lubyanitskiĭ, R. V. Minati, and M. S. Furman, *Khim. Prom.* (*Kiev*), **1960**:529; *Chem. Abstr.*, **55**:13309 (1961).

571. I. Ya. Lubyanitskiĭ, G. I. Kostylev, and M. S. Furman, *Khim. Prom.* (*Kiev*.), **1960**: 533; *Chem. Abstr.*, **55**:13309 (1961).

572. A. M. Gol'dman, A. I. Zaitsev, G. I. Kostylev, L. S. Lakhmanchuk, I. Ya. Lubyanitskiĭ, V. A. Preobrazhenskii, and M. S. Furman, *Khim. Prom.* (*Kiev*), **1962**:323; *Chem. Abstr.*, **58**:5504 (1963).

573. A. M. Gol'dman, G. I. Kostylev, I. Ya. Lubyanitskiĭ, R. V. Minati, V. A. Preobrazhenskii, S. M. Sedova, V. I. Trubnikova, and M. S. Furman, *Poluprod. Sin. Poliamidov*, **1963**:17; *Chem. Abstr.*, **61**:4204 (1964).

574. I. Ya. Lubyanitskiĭ, *Zh. Prikl. Khim.* (*Leningrad*), **36**:860 (1963); *Chem. Abstr.*, **59**:6268 (1963).

575. A. H. Aronow, U.S. Patent 2,191,786 (Feb. 27, 1940).

576. J. F. Olin (to The Sharples Solvents Corp.), U.S. Patent 2,244,849 (June 10, 1941).

577. S. H. McAllister (to Shell Development Co.), British Patent 540,004 (Oct. 1, 1941).

578. R. M. Cavanaugh (to E. I. du Pont de Nemours & Co.), U.S. Patent 2,291,211 (July 28, 1942).

579. G. Meier (vested in the Alien Property Custodian), U.S. Patent 2,300,955 (Nov. 3, 1942).

580. C. N. Zellner (to Tide Water Associated Oil Co.), U.S. Patent 2,323,861 (July 6, 1943).

581. E. Harrison and R. May (to Imperial Chemical Industries, Ltd.), British Patent 567,525 (Feb. 19, 1945).

582. T. F. Doumani, C. S. Coe, and E. C. Attané, Jr. (to Union Oil Co. of California), U.S. Patent 2,459,690 (Jan. 18, 1949).

583. C. H. Hamblet and A. McAlvey (to E. I. du Pont de Nemours & Co.), U.S. Patent 2,557,282 (June 19, 1951).

584. Du Pont Co. of Canada, Ltd., British Patent 738,393 (Oct. 12, 1955).

585. C. H. Hamblet and D. B. Hanson (to E. I. du Pont de Nemours & Co.), U.S. Patent 2,750,415 (June 12, 1956).

586. J. B. O'Hara (to Olin Mathieson Chemical Corp.), U.S. Patent 2,789,136 (Apr. 16, 1957).

587. Hans. J. Zimmer Verfahrenstechnik, Belgian Patent 627,356 (May 15, 1963).

588. W. N. Baxter, U.S. Patent 3,139,423 (June 30, 1964).

589. I. Drimus, I. Velea, C. G. Matasa, and I. C. Cristescu (to Ministerul Industriel Petrolului si Chimiei, Bucharest), U.S. Patent 3,173,961 (March 16, 1965).

590. O. A. Sampson, Jr. (to E. I. du Pont de Nemours & Co.), French Patent 1,400,649 (May 28, 1965); *Chem. Abstr.*, **63**:11367 (1965).

591. Halcon International, Inc., Netherlands Appl. 6,413,619 (June 8, 1965); *Chem. Abstr.*, **63**:17906 (1965).

592. G. Riegelbauer, A. Wegerich, A. Kuerringer, and E. Haarer (to Badische Anilin- & Soda-Fabrik A.G.), Belgian Patent 660,937 (Sept. 13, 1965); *Chem. Abstr.*, **64**:1966 (1966).

593. Hans. J. Zimmer Verfahrenstechnik, French Patent 1,490,179 (July 28, 1967); *Chem. Abstr.*, **68**:7495 (1968).

594. O. A. Sampson, Jr. (to E. I. du Pont de Nemours & Co.), U.S. Patent 3,359,308 (Dec. 19, 1967).

595. C. Gardner and R. Peace (to Imperial Chemical Industries, Ltd.), British Patent 1,107,424 (March 27, 1968).

596. B. P. Brossard (to Rhone-Poulenc S.A., Paris), U.S. Patent 3,376,337 (April 2, 1968).

597. E. K. Ellingboe and J. E. Kirby (to E. I. du Pont de Nemours & Co.), U.S. Patent 2,196,357 (April 9, 1940).

598. J. F. Olin and F. P. Fritsch (to The Sharples Solvents Corp.), U.S. Patent 2,226,357 (Dec. 24, 1940).

599. A. Parant and J. Kapron (to Houillères du Bassin-du-Nord et du Pas-de-Calais), French Patent 981,609 (May 29, 1951).

600. M. G. R. Carter, R. G. A. New, and P. V. Youle (to Imperial Chemical Industries, Ltd.), British Patent 1,068,905 (May 17, 1967).

601. E. Dietzel and K. Gieseler (to Vereinigte Glanzstoff-Fabriken A.G.), German Patent 765,011 (Aug. 9, 1951); *Chem. Abstr.*, **51**:17985 (1957).

602. K. Rumscheidt (to I. G. Farbenindustrie A.G.), German Patent 767,846 (Feb. 8, 1954); *Chem. Abstr.*, **49**:11693 (1955).

603. M. Morita (to Noguchi Research Foundation), Japanese Patent 6506 (1962); *Chem. Abstr.*, **59**:13797 (1963).

604. J. J. Fuchs (to E. I. du Pont de Nemours & Co.), U.S. Patent 3,125,600 (May 17, 1964).

605. M. Matsumoto, Y. Maruse, and I. Yamazaki (to Hirata Chem. Ind. Co.), Japanese Patent 4371 (April 28, 1961); *Chem. Abstr.*, **58**:10084 (1963).

606. A. Isard and F. Weiss (to Société d'Electro-Chimie, d'Electro-Metallurgie et des Acieries Electriques d'Ugine), French Patent 1,400,437 (May 28, 1965); *Chem. Abstr.*, **63**:9819 (1965).

607. E. I. du Pont de Nemours & Co., Netherlands Appl. 6,601,148 (April 25, 1966); *Chem. Abstr.*, **65**:13550 (1966).

608. F. Kögler (to Badische Anilin- & Soda-Fabrik A.G.), German Patent 854,507 (Nov. 4, 1952).

609. E. K. Ellingboe (to E. I. du Pont de Nemours & Co.), U.S. Patent 2,228,261 (Jan. 14, 1941).

610. R. M. Cavanaugh and W. M. Nagle (to E. I. du Pont de Nemours & Co.), U.S. Patent 2,343,534 (March 7, 1944).

611. A. Benning and O. Grosskinsky (to Bergwerksverband GmbH), U.S. Patent 3,173,947 (March 16, 1965).

612. D. D. Davis (to E. I. du Pont de Nemours & Co.), U.S. Patent 3,306,932 (Feb. 28, 1967); Netherlands Appl. 6,412,492 (May 3, 1965).

613. Esso Research and Engineering Co., British Patent 738,808 (Oct. 19, 1955).

614. R. H. Carter (to Scientific Design Co., Inc.), Belgian Patent 613,853 (Aug. 13, 1962); *Chem. Abstr.*, **58**:3320 (1963).

615. R. L. Golden (to Scientific Design Co., Inc.), Belgian Patent 614,548 (Sept. 3, 1962); *Chem. Abstr.*, **59**:2653 (1963).

616. J. L. Russell and R. L. Golden (to Scientific Design Co., Inc.), Belgian Patent 614,820 (Sept. 10, 1962); *Chem. Abstr.*, **59**:2652 (1963).

617. R. E. Lidov (to Halcon International, Inc.), U.S. Patent 3,361,806 (Jan. 2, 1968).

618. E. Rindtorff, W. Ester, and A. Sommer (to Bergwerksgesellschaft Hibernia A.G.), German Patent 1,020,021 (Nov. 28, 1957); *Chem. Abstr.*, **54**:1363 (1960).

619. M. S. Furman, A. M. Gol'dman, V. M. Olevskiĭ, and V. R. Ruchinskiĭ, *Khim. Prom.* (*Kiev*), **1960**:265; *Chem. Abstr.*, **55**:3469 (1961).

620. T. Yamaguchi, N. Ohta, Y. Ishiwata, M. Takehara, and M. Tanaka, *Kôgyô Kagaku Zasshi*, **66**:44 (1963); *Chem. Abstr.*, **59**:6246 (1963).

621. Badische Anilin- & Soda-Fabrik A.G., German Patent 854,505 (Nov. 4, 1952); *Chem. Abstr.*, **50**:16841 (1956).

622. J. G. D. Schulz and A. C. Whitaker (to Gulf Research & Development Co.), U.S. Patent 3,390,174 (June 25, 1968).

623. D. J. Loder (to E. I. du Pont de Nemours & Co.), U.S. Patent 2,223,493 (Dec. 3, 1940).

624. Imperial Chemical Industries, Ltd., French Patent 1,365,852 (July 3, 1964).

625. L. A. Duncanson and H. G. Lawley (to Imperial Chemical Industries, Ltd.), U.S. Patent 3,361,807 (Jan. 2, 1968).

626. Imperial Chemical Industries, Ltd., Netherlands Appl. 6,415,031 (June 24, 1965); *Chem. Abstr.*, **63**:17932 (1965).

627. F. T. Wadsworth (to Pan American Refining Corp.), U.S. Patent 2,589,648 (Mar. 18, 1952).

628. S. G. Gallo and C. E. Morrell (to Standard Oil Development Co.), U.S. Patent 2,675,407 (April 13, 1954).

629. Scientific Design Co., Inc., British Patent 941,662 (Nov. 13, 1963).

630. R. E. Lidov (to Scientific Design Co., Inc.), French Patent 1,350,622 (Jan. 31, 1964).

631. R. S. Barker and M. A. Cohen (to Halcon International, Inc.), U.S. Patent 3,234,271 (Feb. 8, 1966).

632. Toa Gosei Chemical Industry Co., Ltd., French Patent 1,488,079 (July 7, 1967); *Chem. Abstr.*, **68**:6595 (1968).

633. J. Kollar (to Gulf Research & Development Co.), British Patent 1,007,987 (Oct. 22, 1965).

634. J. Kollar (to Gulf Research & Development Co.), U.S. Patent 3,231,608 (Jan. 25, 1966).

635. O. Drossbach (to E. I. du Pont de Nemours & Co., Inc.), U.S. Patent 2,285,914 (June 9, 1942).

636. E. I. du Pont de Nemours & Co., British Patent 806,660 (Dec. 31, 1958).

637. Lonza Electric and Chemical Works, Ltd., British Patent 870,105 (June 14, 1961).

638. G. H. Mock (to Chemstrand Corp.), Belgian Patent 615,895 (Oct. 2, 1962); *Chem. Abstr.*, **58**:11222 (1963).

639. R. L. Golden, J. L. Russell, and R. E. Lidov (to Scientific Design Co., Inc.), Belgian Patent 616,101 (Oct. 8, 1962).

640. R. E. Lidov (to Scientific Design Co., Inc.), Belgian Patent 616,102 (Oct. 8, 1962).

641. J. L. Russell and C. N. Winnick (to Scientific Design Co., Inc.), French Patent 1,330,137 (June 21, 1963).

642. Scientific Design Co., Inc., British Patent 939,798 (Oct. 16, 1963).

643. Halcon International, Inc., British Patent 956,779 (April 29, 1964).

644. Halcon International, Inc., British Patent 956,780 (April 29, 1964).

645. R. L. Golden (to Halcon International, Inc.), Belgian Patent 661,117 (Sept. 15, 1965).

646. R. H. Carter (to Halcon International, Inc.), U.S. Patent 3,207,783 (Sept. 21, 1965).

647. F. Jaffe (to W. R. Grace & Co.), U.S. Patent 3,383,413 (May 14, 1968).

648. K. Tanaka, M. Honda, and G. Inoue, *Kôgyô Kagaku Zasshi*, **72**:2587 (1969); *Chem. Abstr.*, **72**(19):296 (1970) Abstr. No. 99953; *idem.*, **72**:2590 (1969); *Chem. Abstr.*, **72**(19):283 (1970) Abstr. No. 99815r.

649. I. V. Berezin, B. G. Dzantiev, N. F. Kzaanskaya, L. N. Sinochkina, and N. M. Emanuel, *Zh. Fiz. Khim.*, **31**:554 (1957); *Chem. Abstr.*, **52**:58 (1958).

650. V. V. Lipes, L. K. Kazantseva, N. A. Gol'tyaeva, and M. S. Furman, *Khim. Prom.* (*Kiev*), **1964**(9):688; *Chem. Abstr.*, **61**:14544 (1964).

651. H. B. Tinker, *J. Catal.*, **19**:237 (1970).

652. Celanese Corp. of America, British Patent 935,029 (Aug. 28, 1963).

653. C. N. Winnick and S. Rudoff (to Scientific Design Co., Inc.), Belgian Patent 629,240 (Oct. 21, 1963).

654. Badische Anilin- & Soda-Fabrik A.G., Netherlands Appl. 6,413,762 (May 28, 1965); *Chem. Abstr.*, **63**:17906 (1965).

655. K. H. Koenig (to Badische Anilin- & Soda-Fabrik A.G.), U.S. Patent 3,277,168 (Oct. 4, 1966).

656. Halcon International, Inc., Netherlands Appl. 6,410,476 (April 2, 1965); *Chem. Abstr.*, **63**:11731 (1965).

657. W. Flemming and W. Speer (to I. G. Farbenindustrie A.G.), U.S. Patent 2,005,183 (June 18, 1935).

658. D. D. Lee and C. Sparacino (to E. I. du Pont de Nemours & Co.), U.S. Patent 2,511,475 (June 13, 1950).

659. W. J. Amend (to E. I. du Pont de Nemours & Co.), U.S. Patent 2,316,543 (April 13, 1943).

660. L. H. Buxbaum, *Justus Liebigs Ann. Chem.*, **706**:81 (1967).

661. C. N. Winnick (to Scientific Design Co.), French Patent 1,356,221 (April 7, 1964).

662. W. Pritzkow, *Chem. Ber.*, **87**:1668 (1954).

663. N. Brown, M. J. Hartig, M. J. Roedel, A. W. Anderson, and C. E. Schweitzer, *J. Amer. Chem. Soc.*, **77**:1756 (1955).

664. G. Sugerman and J. Kollar (to Halcon International, Inc.), U.S. Patent 3,405,173 (Oct. 8, 1968).

665. G. Sugerman and J. Kollar (to Halcon International, Inc.), U.S. Patent 3,405,174 (Oct. 8, 1968).

666. E. T. Crisp and G. H. Whitfield (to Imperial Chemical Industries, Ltd.), British Patent 824,116 (Nov. 25, 1959).

667. H. Prückner (to E. I. du Pont de Nemours & Co.), U.S. Patent 2,341,288 (Feb. 8, 1944).

668. K. Ohashi and K. Mizutani, *J. Soc. Chem. Ind. Japan.*, **44**:840 (1941); *Chem. Abstr.*, **44**:840 (1941).

669. I. V. Machinskaya and T. K. Veselovskaya, *J. Appl. Chem. USSR*, **17**:377 (1944); *Chem. Abstr.*, **39**:2738 (1945).

670. W. Speer (to I. G. Farbenindustrie A.G.), German Patent 767,813 (Oct. 12, 1953); *Chem. Abstr.*, **49**:11693 (1955).

671. K. A. Chervinskii and V. I. Karban, *Ukr. Khim. Zh.*, **28**:198 (1962); *Chem. Abstr.*, **58**:2335 (1963).

672. V. P. Ivanov, M. S. Furman, A. D. Shestakova, and I. L. Arest-Yakubovich, *Inform. Soobshch. Gos. Nauch.-Issled. Proekt. Inst. Azotn. Prom. Prod. Org. Sin.*, No. **17**:Pt. 1, 5 (1966) (Protsessy Proizyod); *Chem. Abstr.*, **68**:6860 (1968).

673. A. Brehm, *Z. Phys. Chem.* (*Leipzig*), **221**:1 (1962).

674. E. Yasui, T. Kawaguchi, T. Matsubara, and N. Hisanaga (to Toa Gosei Chemical Industry Co., Ltd.), U.S. Patent 3,513,194 (May 19, 1970).

675. T. J. Wallace and A. Schriesheim (to Esso Research and Engineering Co.), U.S. Patent 3,356,722 (Dec. 5, 1967).

676. W. F. Hoot and K. A. Kobe, *Ind. Eng. Chem.*, **47**:782 (1955).

677. A. Karl (to Heinrich Koppers GmbH), German Patent 1,012,296 (July 18, 1957).

678. F. Minisci and A. Quilico, *Atti Accad. Naz. Lincei, Rend., Cl. Sci. Fis., Mat. Nat.*, **31**:357 (1961); *Chem. Abstr.*, **58**:2380 (1963).

679. A. Quilico, F. Minisci, G. Belvedere, and M. Cecere (to Montecatini Società Generale per l'Industrià Mineraria e Chimica), Belgian Patent 611,138 (Dec. 29, 1961); *Chem. Abstr.*, **57**:14965 (1962).

680. Montecatini Società Generale per l'Industrià Mineraria e Chimica, Italian Patent 640,919 (June 8, 1962); *Chem. Abstr.*, **59**:11290 (1963).

681. F. Minisci, G. Belvedere, and A. Quilico (to Montecatini Società Generale per l'Industrià Mineraria e Chimica, French Patent 1,369,857 (Aug. 14, 1964); *Chem. Abstr.*, **62**:448 (1965).

682. T. F. Doumani, C. S. Coe, and E. C. Attané (to Union Oil Co. of California), U.S. Patent 2,465,984 (March 29, 1949).

683. W. O. Kenyon and G. V. Heyl (to Eastman Kodak Co.), U.S. Patent 2,298,387 (Oct. 13, 1942).

684. F. Minisci, G. Belvedere, M. Cecere, and A. Quilico (to Montecatini Edison S.p.A.), U.S. Patent 3,444,194 (May 13, 1969).

685. E. T. Denisov and N. M. Emanuel, *Zh. Fiz. Khim.*, **31**:1266 (1957); *Chem. Abstr.*, **52**:3485 (1958).

686. S. Ciberowski, *J. Phys. Chem.*, **67**:1375 (1963).

687. D. R. Lachowicz, T. S. Simmons, and K. L. Kreuz (to Texaco, Inc.), U.S. Patent 3,466,326 (Sept. 9, 1969).

688. F. Kögler (to Badische Anilin- & Soda-Fabrik A.G.), German Patent 848,805 (Sept. 8, 1952); *Chem. Abstr.*, **50**:10127 (1956).

689. F. Kögler (to Badische Anilin- & Soda-Fabrik A.G.), German Patent 871,601 (March 23, 1953).

690. J. H. Bonfield and R. H. Belden (to Allied Chemical Corp.), French Patent 1,347,525 (Dec. 27, 1963).

691. Badische Anilin- & Soda-Fabrik A.G.), British Patent 722,155 (Jan. 19, 1955).

692. E. Nebe and O. Boehm (to Badische Anilin- & Soda-Fabrik A.G.), U.S. Patent 2,719,172 (Sept. 27, 1955).

693. C. E. Hoberg and M. K. Phibbs (to E. I. du Pont de Nemours & Co. and du Pont Co. of Canada, Ltd.), U.S. Patent 2,971,010 (Feb. 7, 1961).

694. C. R. Campbell, J. J. Hicks, Jr., and R. Johnson (to Chemstrand Corp.), French Patent 1,316,914 (Feb. 1, 1963).

695. C. R. Campbell and R. Johnson (to Monsanto Co.), U.S. Patent 3,180,878 (April 27, 1965).

696. C. R. Campbell and R. Johnson (to Monsanto Co.), U.S. Patent 3,359,283 (Dec. 19, 1967).

697. J. H. Bonfield and R. H. Belden (to Allied Chemical Corp.), U.S. Patent 3,290,369 (Dec. 6, 1966).

698. Badische Anilin- & Soda-Fabrik A.G., Netherlands Appl .6,406,890 (Dec. 28, 1964); *Chem. Abstr.*, **62**:13050 (1965).

699. R. A. Hines (to E. I. du Pont de Nemours & Co.), U.S. Patent 2,776,990 (Jan. 8, 1957).

700. R. G. Hay and S. M. Hazen (to Gulf Research & Development Co.), U.S. Patent 2,804,475 (Aug. 27, 1957).

701. B. S. Kryuchkov, L. A. Serafinov, I. P. Strelets, Yu. F. Golynets, and S. V. L'vov, *Khim. Tekhnol. Topl. Masel*, **9**(4):6 (1964); *Chem. Abstr.*, **61**:2455 (1964).

702. B. S. Kryuchkov, L. A. Serafimov, Yu. F. Golynets, I. P. Strelets, and S. V. L'vov, Russian Patent 167,860 (Feb. 5, 1965); *Chem. Abstr.*, **62**:16064 (1965).

703. Monsanto Co., British Patent 982,751 (Feb. 10, 1965).

704. D. E. Danly and G. L. Whitesell (to Monsanto Co.), U.S. Patent 3,329,712 (July 4, 1967).
705. D. E. Danly (to Monsanto Co.), French Patent 1,374,694 (Oct. 9, 1964).
706. C. T. Sciance and L. S. Scott (to E. I. du Pont de Nemours & Co.), U.S. Patent 3,338,959 (Aug. 29, 1967).
707. C. H. Hamblet and R. E. Gee (to E. I. du Pont de Nemours & Co.), U.S. Patent 2,713,067 (July 12, 1955).
708. E. I. du Pont de Nemours & Co., British Patent 733,968 (July 20, 1955).
709. F. Bende, H. Vollinger, and K. Pohl (to Vickers-Zimmer A.G.), U.S. Patent 3,476,804 (Nov. 4, 1969).
710. H. Vollinger, K. Pohl, and F. Bende (to Vickers-Zimmer A.G.), U.S. Patent 3,476,805 (Nov. 4, 1969).
711. E. I. du Pont de Nemours & Co., British Patent 745,063 (Feb. 22, 1956).
712. W. B. Clark and R. E. Gee (to E. I. du Pont de Nemours & Co.), U.S. Patent 2,813,122 (Nov. 12, 1957).
713. M. Raynes (to Monsanto Co.), U.S. Patent 3,102,908 (Sept. 3, 1963).
714. Lonza Elektrizitätswerke und Chemisches Fabriken A.G., British Patent 874,407 (Aug. 10, 1961).
715. N. V. Stamicarbon, Dutch Patent 93,024 (Dec. 15, 1969); *Chem. Abstr.*, **55**:5351 (1961).
716. Scholven-Chemie A.G., Netherlands Appl. 6,411,575 (June 8, 1965); *Chem. Abstr.* **63**:17907 (1965).
717. E. I. du Pont de Nemours & Co., British Patent 745,034 (Feb. 15, 1956).
718. M. Clasper and J. Haslam, *Analyst* (*London*), **74**:224 (1949).
719. W. S. Wise, *Analyst* (*London*), **75**:219 (1950).
720. J. Haslam and M. Clasper, *Analyst* (*London*), **75**:688 (1950).
721. F. Becke, W. Flemming, W. Speer, and A. Woerner (to Badische Anilin- & Soda-Fabrik A.G.), German Patent 868,901 (March 2, 1953); *Chem. Abstr.*, **48**:11487 (1954).
722. J. B. Wilkes (to Chevron Research Corp.), U.S. Patent 3,433,830 (March 18, 1969).
723. F. G. Webster and D. J. Sutherland (to Imperial Chemical Industries, Ltd.), U.S. Patent 3,445,512 (May 20, 1969).
724. A. S. Michaels and A. R. Colville, Jr., *J. Phys. Chem.*, **64**:13 (1960).
725. A. S. Michaels and F. W. Tausch, Jr., *J. Phys. Chem.*, **65**:1730 (1961).
726. P. S. Bailey, *J. Org. Chem.*, **22**:1548 (1957).
727. P. S. Bailey, *Ind. Eng. Chem.*, **50**:993 (1958).
728. P. S. Bailey (to Esso Research and Engineering Co.), U.S. Patent 3,238,250 (March 1, 1966).
729. Cities Service Research and Development Co., British Patent 971,670 (Sept. 30, 1964).
730. M. I. Fremery and E. K. Fields, *J. Org. Chem.*, **28**:2537 (1963).
731. M. I. Fremery and E. K. Fields (to Standard Oil Co.), U.S. Patent 3,284,492 (Nov. 8, 1966).
732. A. T. Menyailo, I. E. Pokrovskaya, and A. K. Yakovleva, *Neftekhimiya*, **7**:70 (1967); *Chem. Abstr.*, **67**:247 (1967).
733. S. Frank and A. M. Feldman (to American Cyanamid Co.), U.S. Patent 3,026,353 (March 20, 1962).
734. W. H. Clingman, Jr. (to American Oil Co.), U.S. Patent 2,849,484 (Aug. 26, 1958).
735. F. Mareš, J. Roček, and J. Sicher, *Collect. Czech. Chem. Commun.*, **26**:2355 (1961).
736. J. Roček and A. Riehl, *J. Org. Chem.*, **32**:3569 (1967).
737. M. W. Farrar, *J. Org. Chem.*, **22**:1708 (1957).

738. M. W. Farrar (to Monsanto Chemical Co.), U.S. Patent 2,938,051 (May 24, 1960).
739. F. Weiss, J. Modiano, and A. Lakodey (to Société d'Electro-Chimie, d'Electro-Metallurgie et des Aciéries Electriques d'Ugine), French Addn. 86,281 (Jan. 7, 1966); *Chem. Abstr.*, **65**:10497 (1966).
740. F. Weiss, J. Modiano, and A. Lakodey (to Société d'Electro-Chimie, d'Electro-Metallurgic et des Aciéries Electriques d'Ugine), U.S. Patent 3,404,179 (Oct. 1, 1968).
741. Farbwerke Hoechst A.G., Belgian Patent 638,178 (April 3, 1964); *Chem. Abstr.*, **63**:16219 (1965).
742. S. F. Marrian and M. Marin (to British Hydrocarbon Chemicals, Ltd.), U.S. Patent 3,427,350 (Feb. 11, 1969).
743. Celanese Corp. of America, British Patent 765,536 (Jan. 9, 1957).
744. H. Oberrauch (to Farbwerke Hoechst A.G.), German Patent 1,021,347 (Dec. 27, 1957); *Chem. Abstr.*, **54**:1322 (1960).
745. I. Hirao, T. Fujimoto, T. Onizuka, and M. Maruyama, *Kôgyô Kagaku Zasshi*, **65**:1004 (1962); *Chem. Abstr.*, **58**:4413 (1963).
746. W. Reppe, N. v. Kutepow, W. Morsch, and P. Hornberger (to Badische Anilin- & Soda-Fabrik A.G.), German Patent 948,151 (Aug. 30, 1956); *Chem. Abstr.*, **52**:18221 (1958).
747. W. Reppe, H. Friederich, N. v. Kutepow, and W. Morsch (to Badische Anilin- & Soda-Fabrik A.G.), U.S. Patent 2,729,651 (Jan. 3, 1956).
748. W. Reppe, H. Kröper, H. J. Pistor, and O. Weissbarth, *Justus Liebigs Ann. Chem.*, **582**:87 (1953).
749. Badische Anilin- & Soda-Fabrik A.G., British Patent 742,038 (Dec. 21, 1955).
750. S. K. Bhattacharyya and D. K. Nandi, *Ind. Eng. Chem.*, **51**:143 (1959).
751. R. A. Hines (to E. I. du Pont de Nemours & Co.), U.S. Patent 2,809,991 (Oct. 15, 1957).
752. N. S. Imyanitov, B. E. Kuvaev, and D. M. Rudkovskii, *Zh. Prikl. Khim.* (*Leningrad*), **38**:2558 (1965); *Chem. Abstr.*, **64**:6484 (1966).
753. N. S. Imyanitov and D. M. Rudkovskii, *Zh. Org. Khim.*, **2**:231 (1966); *Chem. Abstr.*, **65**:2119 (1966).
754. D. M. Rudkovskii and N. S. Imyanitov (to All-Union Scientific-Research Institute of Petrochemical Processes), British Patent 1,092,694 (Nov. 29, 1967); *Chem. Abstr.*, **69**:3298 (1968).
755. D. M. Rudkovsky and N. Imjanitov (to Vsesojuzny Nauchno, Issledovateljsky Institute Neftekhimicheskikh Protsessov), U.S. Patent 3,481,975 (Dec. 2, 1969).
756. R. H. Hasek and E. U. Elam (to Eastman Kodak Co.), U.S. Patent 2,801,263 (July 30, 1957).
757. M. I. Arevalo and M. L. Canut, *Bol. Real Soc. Espan. Hist. Nat., Secc. Geol.*, **59**:37 (1961); *Chem. Abstr.*, **56**:2400 (1962).
758. C. H. MacGillavry, *Rec. Trav. Chim. Pays-Bas*, **60**:605 (1941).
759. J. D. Morrison and J. M. Robertson, *J. Chem. Soc.*, **1949**:987.
760. J. Housty, *Bull. Soc. Chim. Fr.*, **1967**:273.
761. J. L. Amoros, M. L. Canut, and E. Neira, *Proc. Roy. Soc.* (*London*) *A*, **285**:370 (1965).
762. J. T. Lassiter (to E. I. du Pont de Nemours & Co.), U.S. Patent 3,459,798 (Aug. 5, 1969).
763. J. Steigman, R. D. Iasi, H. Lilenfeld, and D. Sussman, *J. Phys. Chem.*, **72**:1132 (1968).
764. R. C. Aggarwal and A. K. Srivastava, *Indian J. Chem.*, **5**:627 (1967).
765. V. Ananthanarayanan, *Spectrochim. Acta*, **20**:197 (1964).
766. H. Susi, *Spectrochim. Acta*, **17**:1257 (1961).

767. M. Davies and D. M. L. Griffiths, *Trans. Faraday Soc.*, **49**:1405 (1953).
768. H. Serwy, *Bull. Soc. Chim. Belg.*, **42**:483 (1933).
769. S. Gal, J. Szammer, T. Meisel, L. Erdey, and L. Otvos, *Proc. Conf. Appl. Phys. Chem. Methods Chem. Anal.*, *Budapest 1966*, **3**:307; *Chem. Abstr.* **68**:8337 (1968).
770. F. Krafft and H. Noerdlinger, *Chem. Ber.*, **22**:816 (1889).
771. C. Béguin and T. Gäumann, *Helv. Chim. Acta*, **41**:1376 (1958).
772. B. Adell, *Z. Phys. Chem. (Leipzig)*, **185**:161 (1939).
773. W. D. Bancroft and F. J. C. Butler, *J. Phys. Chem.*, **36**:2515 (1932).
774. C. S. Marvel and J. C. Richards, *Anal. Chem.*, **21**:1480 (1949).
775. R. Collander, *Acta Chem. Scand.*, **4**:1085 (1950).
776. D. E. Pearson and M. Levine, *J. Org. Chem.*, **17**:1351 (1952).
777. M. R. Murty, M. R. Rao, and C. V. Rao, *J. Sci. Ind. Res. (New Delhi)*, **17B**:103 (1958).
778. R. E. Keller, in "Encyclopedia of Industrial Chemical Analysis," Vol. 4, F. D. Snell and C. L. Hilton, Eds., Interscience, New York, 1967, p. 408.
779. T. Higuchi, N. C. Hill, and G. B. Corcoran, *Anal. Chem.*, **24**:491 (1952).
780. T. K. Miwa, K. L. Mikolajczak, F. R. Earle, and I. A. Wolff, *Anal. Chem.*, **32**:1739 (1960).
781. A. J. Appleby and J. E. O. Mayne, *J. Gas. Chromatogr.*, **5**:266 (1967).
782. D. E. Gross and H. B. Tinker, *Anal. Chem.*, **40**:239 (1968).
783. E. Knappe and D. Peteri, *Fresenius Z. Anal. Chem.*, **188**:184 (1962).
784. C. Passera, A. Pedrotti, and G. Ferrari, *J. Chromatogr.*, **14**:289 (1964).
785. E. Bancher and H. Scherz, *Mikrochim. Acta*, **1964**:1159.
786. N. S. Rajagopal, P. K. Saraswathy, M. R. Subbaram, and K. T. Achaya, *J. Chromatogr.*, **24**:217 (1966).
787. G. Hammarberg and B. Wickberg, *Acta Chem. Scand.*, **14**:882 (1960).
788. J. L. Occolowitz, *J. Chromatogr.*, **5**:373 (1961).
789. R. D. Hartley and G. G. Lawson, *J. Chromatogr.*, **7**:69 (1962).
790. P. Fijolka, W. Radowitz, and F. Runge, *Plaste Kaut.*, **10**:521 (1963).
791. D. F. Houston and W. A. VanSandt, *Ind. Eng. Chem.*, **18**:538 (1946).
792. H. J. Horn, E. G. Holland, and L. W. Hazleton, *J. Agr. Food Chem.*, **5**:759 (1957).
793. R. L. Shull, L. A. Gayle, R. D. Coleman, R. B. Alfin-Slater, A. T. Gros, and R. O. Feuge, *J. Amer. Oil Chem. Soc.*, **38**:84 (1961).
794. K. Lang and A. R. Bartsch, *Biochem. Z.*, **323**:462 (1953); *Chem. Abstr.*, **47**:5502 (1953).
795. I. I. Rusoff, R. R. Baldwin, F. J. Domingues, C. Monder, W. J. Ohan, and R. Thiessen, *Toxicol. Appl. Pharmacol.*, **2**:316 (1960); *Chem. Abstr.*, **54**:21507 (1960).
796. W. H. Gardner, in "Handbook of Food Additives," T. E. Furia, Ed., The Chemical Rubber Co., Cleveland, Ohio, 1968, p. 255.
797. R. Vieweg and A. Müller, Eds., "Kunststoff-Handbuch Bd. VI, Polyamide," Carl Hanser, Münich, 1966.
798. W. Sweeny and J. Zimmerman, in "Encylopedia of Polymer Science and Technology," Vol. 10, H. F. Mark, Chairman of the Editorial Board, Interscience, New York, 1969, p. 483.
799. O. E. Snider and R. J. Richardson, in Ref. 798. p. 347.
800. E. C. Schule, in Ref. 798, p. 460.
801. W. R. Sorenson and T. W. Campbell, in "Preparative Methods of Polymer Chemistry," 2nd ed., Interscience, New York, 1968, p. 74.
802. P. E. Beck and E. E. Magat, *Macromol. Syn.*, **3**:101 (1969).
803. H. Nordt, in "Kunststoff-Handbuch. Bd. VII, Polyurethane," R. Vieweg and A. Höchtlen, Eds., Carl Hansen, Münich, 1966, pp. 45–60.

804. K. A. Pigott, in "Encyclopedia of Polymer Science and Technology," Vol. 11, H. F. Mark, Chairman of the Editorial Board, Interscience, New York, 1969, p. 506.
805. H. R. Snyder, C. A. Brooks, and S. H. Shapiro. *Org. Syn.*, **Coll. Vol. II**, rev. ed., 531 (1943).
806. A. Müller, *Org. Syn.*, **Coll. Vol. II**, rev. ed., 535 (1943).
807. J. Cason, L. Wallcave, and C. N. Whiteside, *J. Org. Chem.*, **14**:37 (1949).
808. M. W. Farlow and G. W. Whitman (to E. I. du Pont de Nemours & Co.), U.S. Patent 2,390,576 (Dec. 11, 1945).
809. G. Chiusoli (to Società Generale per l'Industrià mineraria e Chimica), Italian Patent 564,920 (July 5, 1957); *Chem. Abstr.*, **53**:14945 (1959).
810. R. E. Meyer, *Helv. Chim. Acta*, **16**:1291 (1933).
811. H. de V. Finch, S. A. Ballard, and T. W. Evans (to Shell Development Co.), U.S. Patent 2,454,047 (Nov. 16, 1948).
812. H.-J. Pistor and H. Plieninger, *Justus Liebigs Ann. Chem.*, **562**:239 (1949).
813. H.-J. Pistor (to Badische Anilin- & Soda-Fabrik A.G.), German Patent 844,145 (July 17, 1952); *Chem. Abstr.*, **52**:10161 (1958).
814. F. X. Werber, J. E. Jansen, and T. L. Gresham, *J. Amer. Chem. Soc.*, **74**:532 (1952).
815. R. L. Roberts and J. W. Lynn (to Union Carbide Corp.), U.S. Patent 3,043,872 (July 10, 1962).
816. J. Kamlet (to Goodyear Tire & Rubber Co.), U.S. Patent 2,826,609 (March 11, 1958).
817. F. W. Major and H. M. Stanley (to Distillers Co. Ltd.), U.S. Patent 2,673,219 (March 23, 1954).
818. E. G. E. Hawkins and D. J. G. Long (to Distillers Co. Ltd.), British Patent 710,002 (June 2, 1954).
819. E. G. E. Hawkins and E. S. Stern (to Distillers Co. Ltd.), U.S. Patent 2,698,339 (Dec. 28, 1954).
820. D. J. G. Long (to Distillers Co., Ltd.), U.S. Patent 2,800,507 (July 23, 1957).
821. R. B. Duke, Jr., and M. A. Perry (to Eastman Kodak Co.), U.S. Patent 3,468,927 (Sept. 23, 1969).
822. F. Runge, R. Hueter, and H.-D. Wulf, *Chem. Ber.*, **87**:1430 (1954).
823. P. D. Gardner, L. Rand, and G. R. Haynes, *J. Amer. Chem. Soc.*, **78**:3425 (1956).
824a. F. G. Singleton (to H. H. Robertson Co.), U.S. Patent 2,436,532 (Feb. 24, 1948).
824b. W. Reppe et al., *Justus Liebigs Ann. Chem.*, **596**:158 (1955).
825. W. Muench and G. Geist (to Thüringische Zellwolle A.G. and Zellwolle- und Kunstseide-Ring, GmbH), German Patent 765,970 (Nov. 4, 1952); *Chem. Abstr.*, **52**:3855 (1958).
826. R. Lukeš and F. Šorm, *Collect. Czech. Chem. Commun.*, **13**:585 (1948).
827. W. E. Wellman and A. R. Kittleson (to Esso Research and Engineering Co.), U.S. Patent 3,383,412 (May 14, 1968).
828. F. Dupré la Tour, *C.R. Acad. Sci. (Paris)*, **201**:479 (1935); **202**:1935 (1936); **208**:364 (1939).
829. M. I. Kay and L. Katz, *Acta Crystallogr.*, **11**:289 (1958).
830. J. Housty and M. Hospital, *Acta Crystallogr.*, **21**:29 (1966).
831. M. Wehrli and E. Schönmann, *Helv. Phys. Acta*, **15**:317 (1942).
832. E. Schönmann, *Helv. Phys. Acta*, **16**:343 (1943).
833. G. H. Jeffery and A. I. Vogel, *J. Chem. Soc.*, **1935**:21.
834. L. Bouveault, *Bull. Soc. Chim. Fr.*, [**3**]**19**:562 (1898).
835. P. E. Verkade, *Rec. Trav. Chim. Pays-Bas*, **46**:137 (1927).
836. Wallace & Tiernan, Inc., Netherlands Appl. 6,502,597 (Sept. 6, 1965); *Chem. Abstr.*, **64**:6500 (1966).

837. R. B. Judge, M. F. Levy, and J. H. Quinn (to Wallace & Tiernan, Inc.), Belgian Patent 660,522 (Sept. 2, 1965); *Chem. Abstr.*, **64**:9611 (1966).
838. F. Minisci, G. Belvedere, and A. Quilico (to Società Generale per l'Industrià Mineraria e Chimica), French Patent 1,379,466 (Nov. 20, 1964); *Chem. Abstr.*, **62**:9014 (1965).
839. W. Reppe, O. Schlichting, K. Klager, and T. Toepel, *Justus Liebigs Ann. Chem.*, **560**:1 (1948).
840. W. Reppe, K. Klager, and O. Schlichting (to Badische Anilin- & Soda-Fabrik A.G.), German Patent 890,950 (Sept. 24, 1953); *Chem. Abstr.*, **50**:16840 (1956).
841. J. E. Franz and W. S. Knowles, *Chem. Ind. (London)*, **1961**:250.
842. P. Baumgartner and P. Duhaut (to Institut Français du Pétrole des Carburants et Lubrifiants), U.S. Patent 3,200,144 (Aug. 10, 1965)
843. W. Ziegenbein (to Chemische Werke Huek A.G.), German Patent 1,133,711 (July 26, 1962); *Chem. Abstr.*, **58**:1354 (1963).
844. W. Stumpf, *Angew. Chem.*, **69**:727 (1957).
845. "Montecatini" Società Generale per l'Industria Mineraria e Chimica, Italian Patent 587,503 (Jan. 16, 1959); *Chem. Abstr.*, **55**:12304 (1961).
846. E. K. Baylis, W. Pickles, and K. D. Sparrow (to Geigy Chemical Corp.), U.S. Patent 3,441,604 (April 29, 1969); Netherlands Appl. 6,511,374 (March 2, 1966).
847. R. Seekircher (to Columbian Carbon Co.), U.S. Patent 3,219,675 (Nov. 23, 1965); British Patents 965,510 (July 29, 1964) and 971,670 (Sept. 30, 1964) (to Cities Service Research and Development Co.).
848. R. H. Perry, Jr. (to Esso Research and Engineering Co.), U.S. Patent 3,202,704 (Aug. 24, 1965).
849. A. Maggiolo (to Wallace & Tiernan, Inc.), U.S. Patent 3,280,183 (Oct. 18, 1966).
850. J. C. Sauer, R. D. Cramer, V. A. Engelhardt, T. A. Ford; H. E. Holmquist, and B. W. Howk, *J. Amer. Chem. Soc.*, **81**:3677 (1959).
851. J. C. Sauer (to E. I. du Pont de Nemours & Co.), U.S. Patent 2,840,570 (June 24, 1958).
852. H. E. Holmquist (to E. I. du Pont de Nemours & Co.), U.S. Patent 2,884,450 (Apr. 28, 1959).
853. N. A. Rozanov and Belikov, *J. Russ. Phys. Chem. Soc.*, **61**:2303 (1929); *Chem. Abstr.*, **24**:3765 (1930).
854. J. N. E. Day, G. A. R. Kon, and A. Stevenson, *J. Chem. Soc.*, **1920**:639.
855. M. Carmichael, *J. Chem. Soc.*, **1922**:2545.
856. H. Kiliani, *Chem. Ber.*, **54**:456 (1921).
857. K. G. H. Derlon, *Chem. Ber.*, **31**:1957 (1898).
858. F. Gantter and C. Hell, *Chem. Ber.*, **14**:1545 (1881).
859. G.-M. Schwab and M. Deffner, *Nature*, **167**:240 (1951).
860. J. Housty and M. Hospital, *Acta Crystallogr.*, **17**:1387 (1964); **18**:753 (1965).
861. F. Lamouroux, *C.R. Acad. Sci. (Paris)*, **128**:998 (1899).
862. W. A. Caspari, *J. Chem. Soc.*, **198**:3235.
863. W. C. Rose, C. J. Weber, R. C. Corley, and R. W. Jackson, *J. Pharmacol.*, **25**:59 (1925).
864. K. Bernhard and M. Andreae, *Z. Physiol. Chem.*, **245**:103 (1937).
865. P. E. Verkade, J. van der Lee, and A. J. S. van Alphen, *Z. Physiol. Chem.*, **252**:163 (1938).
866. K. Bernhard, *Helv. Chim. Acta*, **24**:1412 (1941).
867. G. Weitzel, *Hoppe-Seyler's Z. Physiol. Chem.*, **282**:185 (1947).
868. C. S. Fuller and C. J. Frosch, *J. Phys. Chem.*, **43**:323 (1939).

869. A. Bell, J. G. Smith, and C. J. Kibler (to Eastman Kodak Co.), U.S. Patent 3,012,994 (Dec. 12, 1961); *J. Polym. Sci., Pt. A*, **3**:19 (1965).
870. M. T. Watson and G. M. Armstrong, *SPE J.*, **21**:475 (1965).
871. N. I. Mitskevich and V. E. Agabekov, *Neftekhimiya*, **6**:867 (1966); *Chem. Abstr.*, **66**:8825 (1967).
872. L. Ruzicka and W. Brugger, *Helv. Chim. Acta*, **9**:339 (1926).
873. A. I. Vogel, *J. Chem. Soc.*, **1929**:721.
874. H. McKennis, Jr., and V. duVigneaud, *J. Amer. Chem. Soc.*, **68**:832 (1946).
875. G. A. Schmidt and D. A. Shirley, *J. Amer. Chem. Soc.*, **71**:3804 (1949).
876. H. Erlenmeyer and W. Büchler, *Helv. Chim. Acta*, **29**:1924 (1946).
877. A. C. Cope, R. J. Cotter, and L. L. Estes, *Org. Syn.*, **Coll. Vol. IV**:62 (1963).
878. E. H. Pryde and J. C. Cowan, in "Topics in Lipid Chemistry," Vol. 2, F. D. Gunstone, Ed., Logos Press, London, 1971, pp. 1–98.
879. A. Rieche, German Patent 565,158 (March 20, 1931); *Chem. Abstr.*, **27**:1008 (1933).
880. F. Asinger, *Chem. Ber.*, **75**:656 (1942).
881. Badische Anilin- & Soda-Fabrik A.G., German Patent 868,148 (Feb. 23, 1953); *Chem. Abstr.*, **50**:1894 (1956).
882. Emery Industries, Inc., British Patent 757,355 (Sept. 19, 1956).
883. C. G. Goebel, A. C. Brown, H. F. Oehlschlaeger, and R. P. Rolfes (to Emery Industries, Inc.), U.S. Patent 2,813,113 (Nov. 12, 1957).
884. G. Izumi, *Kôgyô Kagaku Zasshi*, **62**:219 (1959).
885. T. Kobayashi and S. Miyazaki (to Bureau of Industrial Technics), Japanese Patent 5079 (Aug. 14, 1954); *Chem. Abstr.*, **49**:13672 (1955).
886. S. Mihara, I. Miwa, K. Ueno, and S. Morita (to Oriental High Pressure Industries Co.), Japanese Patent 8172 (Sept. 25, 1957); *Chem. Abstr.*, **53**:1149 (1959).
887. G. Izumi, and K. Ando, *Kôgyô Kagaku Zasshi*, **66**:783 (1963); *Chem. Abstr.*, **60**:5328 (1964).
888. I. Miwa and S. Morita (to Toyo Koatsu Industries, Inc.), U.S. Patent 3,133,953 (May 19, 1964); British Patent 956,719 (April 29, 1964).
889. A. Maggiolo (to The Welsbach Corp.), U.S. Patent 2,865,937 (Dec. 23, 1958); British Patent 841,653 (July 20, 1960).
890. R. L. Blackmore and W. Szatkowski (to A. Boake, Roberts and Co.), British Patent 810,571 (March 18, 1959).
891. F. O. Barrett and C. G. Goebel (to Emery Industries, Inc.), U.S. Patent 3,207,784 (Sept. 21, 1965).
892. H. F. Oehlschlager and H. G. Rodenberg (to Emery Industries, Inc.), U.S. Patent 3,402,108 (Sept. 17, 1968).
893. S. J. Niegowski and A. Maggiolo (to The Welsbach Corp.), U.S. Patent 2,897,231 (July 28, 1959).
894. L. S. Moody (to General Electric Co.), U.S. Patent 2,732,338 (Jan. 24, 1956).
895. C. E. Thorp and A. J. Gaynor (to The Cudahy Packing Co.), U.S. Patent 2,857,410 (Oct. 21, 1958); British Patent 883,531 (Nov. 29, 1961).
896. H. Mihara, I. Miwa, K. Ueno, and S. Morita (to Toyo Koatsu Industries, Inc.), U.S. Patent 3,060,211 (Oct. 23, 1962); British Patent 884,438 (Dec. 13, 1961).
897. J. Pasero, J. Chouteau, and M. Naudet, *Bull. Soc. Chim. Fr.*, **1960**:1717.
898. J. Pasero, L. Comeau, and M. Naudet, *Bull. Soc. Chim. Fr.*, **1963**:1794.
899. J. Pasero, L. Comeau, and M. Naudet, *Bull. Soc. Chim. Fr.*, **1965**:493.
900. P. R. Story and J. R. Burgess, *Tetrahedron Lett.*, **1968**:1287.
901. D. G. M. Diaper, J. Pasero, and M. Naudet, *Can. J. Chem.*, **46**:2767 (1968).
902. R. G. Ackman, M. E. Retson, L. R. Gallay, and F. A. Vandenheuvel, *Can. J. Chem.*, **39**:1956 (1961).

903. S. Mihara, I. Miwa, K. Ueno, and S. Morita (to Oriental High Pressure Industries Co.), Japanese Patent 8171 (Sept. 25, 1957); *Chem. Abstr.*, **52**:14663 (1958).
904. E. K. Ellingboe (to E. I. du Pont de Nemours & Co.), U.S. Patent 2,203,680 (June 11, 1940).
905. P. Kirjakka and M. Nieminen, *Suom. Kemistil. A*, **28**:9 (1955).
906. F. J. Sprules and R. Griffith (to Nopco Chemical Co.), U.S. Patent 2,365,290 (Dec. 19, 1944).
907. G. Gut and A. Guyer, *Helv. Chim. Acta*, **47**:1673 (1964).
908. V. P. Kuceski (to C. P. Hall Co.), U.S. Patent 3,021,348 (Feb. 13, 1962).
909. D. Price and R. Griffith (to National Oil Products Co.), U.S. Patent 2,365,290 (Dec. 19, 1944).
910. F. J. Sprules and R. Griffith (to Nopco Chemical Co.), U.S. Patent 2,426,954 (Sept. 2, 1947); British Patent 585,315 (Feb. 4, 1947).
911. C. S. Morgan, Jr., and J. W. Walker (to Celanese Corp. of America), U.S. Patent 2,847,431 (Aug. 12, 1958); British Patent 813,842 (May 27, 1959).
912. Société générale d'entreprise du sud de la France, French Patent 891,134 (Feb. 28, 1944); *Chem. Abstr.*, **47**:3012 (1953).
913. S. U. K. A. Richter and B. S. Berndtsson (to Svenska Oljeslageri-Aktiebolaget), Swedish Patent 130,417 (Jan. 2, 1951); *Chem. Abstr.*, **45**:8552 (1951).
914. N. C. Hill (to C. P. Hall Co.), U.S. Patent 2,971,023 (Feb. 7, 1961).
915. M. F. Ishanina, Yu. A. Stepanov, and S. N. Danilov, *Tr. Leningrad. Tekhnol. Inst. im. Lensoveta*, **1957**:30–8; *Chem. Abstr.*, **53**:1138 (1958).
916. R. L. Logan (to C. P. Hall Co. of Illinois), U.S. Patent 2,662,908 (Dec. 15, 1953).
917. R. D. Englert and L. M. Richard (to Tallow Research, Inc.), U.S. Patent 2,773,095 (Dec. 4, 1956).
918. T. Kobayashi (to Bureau of Industrial Technics), Japanese Patent 6172 (Nov. 30, 1953); *Chem. Abstr.*, **49**:9028 (1955).
919. M. Imai and T. Wakabayashi (to Nippon Fats and Oils Co.), Japanese Patent 9867 (Nov. 25, 1957); *Chem. Abstr.*, **52**:17116 (1958).
920. N. C. Hill and T. Higuchi (to C. P. Hall Co.), U.S. Patent 2,841,601 (July 1, 1958).
921. Celanese Corp. of America, British Patent 792,487 (March 26, 1958).
922. R. R. Allen and A. A. Kiess (to Armour and Co.), U.S. Patent 2,916,502 (Dec. 8, 1959); British Patent 820,518 (Sept. 23, 1959).
923. T. C. Manley (to The Welsbach Corp.), U.S. Patent 2,998,439 (Aug. 29, 1961).
924. E. J. Bennett (to E. I. du Pont de Nemours & Co.), U.S. Patent 3,320,230 (May 16, 1967).
925. M. Miwa and K. Ueno (to Oriental High Pressure Industries Co.), Japanese Patent 5519 (Sept. 2, 1954); *Chem. Abstr.*, **49**:15953 (1955).
926. I. Miwa and K. Ueno (to Oriental High Pressure Industries Co.), Japanese Patent 5520 (Sept. 2, 1954); *Chem. Abstr.*, **50**:7126 (1956).
927. D. R. Lachowicz, T. S. Simmons, and K. L. Kreuz (to Texaco, Inc.), U.S. Patent 3,415,856 (Dec. 10, 1968).
928. W. Ziegenbein, W. Franke, and A. Striebeck (to Chemische Werke Hüls A.G.), U.S. Patent 2,937,201 (May 17, 1960).
929. J. W. Hill and W. L. McEwen, *Org. Syn.*, **Coll. Vol. II**, 53 (1943).
930. J. Cason and H. Rapoport, "Laboratory Text in Organic Chemistry," 2nd ed., Prentice-Hall, Englewood Cliffs, N.J., 1962, p. 107.
931. E. F. Armstrong and T. P. Hilditch, *J. Soc. Chem. Ind. Trans.*, **44**:43T (1925).
932. D. J. Loder and P. L. Salzberg (to E. I. du Pont de Nemours & Co.), U.S. Patent 2,292,950 (Aug. 11, 1942).
933. T. M. Patrick, Jr., and W. S. Emerson, *Ind. Eng. Chem.*, **41**:636 (1949).

934. K. G. Ölund, Swedish Patent 131,987-8 (June 19, 1951); *Chem. Abstr.*, **46**:5075 (1952).
935. Svenska Oljeslageri-Aktiebolaget, British Patent 652,355 (April 25, 1951); *Chem. Abstr.*, **46**:1585 (1952).
936. S. U. K. A. Richter and B. S. Berndtsson (to Svenska Oljeslageri-Aktiebolaget), Swedish Patent 133,160 (Oct. 9, 1951); Swedish Patent 134,616 (Feb. 26, 1952); *Chem. Abstr.*, **47**:3340 (1953).
937. D. Swern and H. B. Knight (to the United States of America, as represented by the Secretary of Agriculture), U.S. Patent 2,572,892 (Oct. 30, 1951).
938. A. J. Feuell and J. H. Skellon, *J. Chem. Soc.*, **1954**:3414.
939. J. S. Mackenzie and C. S. Morgan, Jr. (to Celanese Corp. of America), U.S. Patent 2,820,046 (Jan. 14, 1958).
940. G. King, *J. Chem. Soc.*, **1956**:587.
941. H. T. Slover and L. R. Dugan, Jr., *J. Amer. Oil Chem. Soc.*, **34**:333 (1957).
942. J. Ross, A. I. Gebhart, and J. F. Gerecht, *J. Amer. Chem. Soc.*, **71**:282 (1949).
943. R. L. Logan (to Kessler Chemical Co., Inc.), U.S. Patent 2,625,558 (Jan. 13, 1953).
944. Kessler Chemical Co., British Patent 731,061 (June 1, 1955).
945. J. L. Ohlson and L. H. Spitzmueller (to Swift & Co.), U.S. Patent 2,791,607 (May 7, 1957).
946. E. F. Riener (to Rohm & Haas Co.), U.S. Patent 2,811,538 (Oct. 29, 1957).
947. M. Kobayashi and Y. Nishino, *Kagaku Kôgyô*, **35**:91 (1961); *Chem. Abstr.*, **55**:25749 (1961).
948. L. I. Zakharkin and G. G. Zhigareva, *Izv. Akad. Nauk SSSR, Ser. Khim.*, **1965**:1497; *Chem. Abstr.*, **63**:16206 (1965).
949. W. Stein (to Henkel & Cie, GmbH), German Patent 1,138,385 (Oct. 25, 1962); *Chem. Abstr.*, **58**:8910 (1963).
950. L. R. B. Harvey (to Arthur D. Little, Inc.), U.S. Patent 2,178,874 (Nov. 7, 1939).
951. R. P. Follett and W. J. Murray (to Arthur D. Little, Inc.), U.S. Patent 2,580,417 (Jan. 1, 1952).
952. Emery Industries, Inc., British Patent 604,281 (July 1, 1948).
953. J. D. Fitzpatrick and L. D. Myers, U.S. Patent 2,450,858 (Oct. 5, 1948).
954. J. D. Fitzpatrick and L. D. Myers (to Emery Industries, Inc.), U.S. Patent 2,389,191 (Nov. 20, 1945).
955. Y. Asahina and Y. Ishida, *J. Pharm. Soc.*, **481**:171 (1922); *Chem. Abstr.*, **16**:1936 (1922).
956. G. M. Bennett and H. Gudgeon, *J. Chem. Soc.*, **1938**:1679.
957. T. Kobayashi and S. Miyazaki, *Rep. Govt. Chem. Ind. Res. Inst., Tokyo*, **49**:73 (1954); *Chem. Abstr.*, **50**:3221 (1956).
958. L. Canonica, Italian Patent 519,134 (March 11, 1955); *Chem. Abstr.*, **52**:1210 (1958).
959. A. Grün and F. Wittka, *Chem. Umsch. Geb. Fette Oele Wacshe Harze*, **32**:257 (1925); *Chem. Abstr.*, **20**:301 (1926).
960. J. D. Fitzpatrick and L. D. Myers (to Emery Industries, Inc.), U.S. Patent 2,468,436 (April 26, 1949).
961. L. D. Myers and J. D. Fitzpatrick (to Emery Industries, Inc.), U.S. Patent 2,470,515 (May 17, 1949).
962. L. D. Myers and J. D. Fitzpatrick (to Emery Industries, Inc.), U.S. Patent 2,474,010 (June 21, 1949).
963. Emery Industries, Inc., British Patent 618,412 (Feb. 22, 1949).
964. I. B. Afanas'ev, G. B. Ovakimyan, T. N. Eremina, I. B. Voronina, L. K. Smail's, and A. A. Beer, *Khim. Prom.*, **1962**:709; *Chem. Abstr.*, **60**:5331 (1964).

965. V. V. Korshak, S. L. Sosin, and E. M. Morozova, *Zh. Obshch. Khim.*, **30**:907 (1960); *Chem. Abstr.*, **55**:376 (1961).
966. S. I. Zav'yalov, L. P. Vinogradova, and G. V. Kondrat'eva, *Tetrahedron*, **20**:2745 (1964).
967. H. Kameoka, *Kôgyô Kagaku Zasshi*, **63**:741 (1960); *Chem. Abstr.*, **56**:6107 (1962).
968. W. Lehmann and R. Schröter (vested in the Alien Property Custodian), U.S. Patent 2,323,061 (June 29, 1943).
969. Henkel & Cie, GmbH, British Patent 783,369 (Sept. 25, 1957).
970. I. Miwa and K. Ueno (to Oriental High Pressure Industries Co.), Japanese Patent 5521 (Sept. 2, 1954); *Chem. Abstr.*, **50**:7127 (1956).
971. L. S. Pooler (to C. P. Hall Co. of Illinois), U.S. Patent 2,716,133 (Aug. 23, 1955).
972. N. C. Hill and V. P. Kuceski (to C. P. Hall Co. of Illinois), U.S. Patent 2,824,134 (Feb. 18, 1958).
973. T. Higuchi (to C. P. Hall Co. of Illinois), U.S. Patent 2,744,067 (May 1, 1956).
974. G. B. Corcoran (to C. P. Hall Co. of Illinois), U.S. Patent 2,824,135 (Feb. 18, 1958).
975. C. Paquot, R. Perron, and J. Petit, *Bull. Soc. Chim. Fr.*, **1959**:878.
976. F. Rennkamp, *Z. Physiol. Chem.*, **260**:276 (1939).
977. R. M. Cavanaugh and R. H. Weir (to E. I. du Pont de Nemours & Co.), U.S. Patent 2,560,156 (July 10, 1951).
978. W. A. Caspari, *J. Chem. Soc.*, **1929**:2709.
979. J. Housty and M. Hospital, *C.R. Acad. Sci.* (*Paris*), **258**:1551 (1964); *Acta Crystallogr.*, **22**:288 (1967).
980a. F. Dupré La Tour, *C.R. Acad. Sci.* (*Paris*), **202**:1935 (1936).
980b. E. Molinari and P. Fenaroli, *Chem. Ber.*, **41**:2789 (1908).
981. R. W. Stafford, J. F. Shay, and R. J. Francel, *Anal. Chem.*, **26**:656 (1954).
982. D. Chobanov, *C.R. Acad. Bulg. Sci.*, **14**:155 (1961).
983. R. Ryhage and E. Stenhagen, *Ark. Kemi*, **14**:483 (1959).
984. A. Enders, *Ark. Exp. Pathol. Pharmakol.*, **197**:706 (1941); *Chem. Abstr.*, **37**:3501 (1943).
985. H. G. Smith, *J. Biol. Chem.*, **103**:531 (1933).
986. F. S. Mallette and E. Von Haam, *Arch. Ind. Hyg. Occup. Med.*, **6**:231 (1952); *Chem. Abstr.*, **46**:11746 (1952).
987. H. C. Hodge, E. A. Maynard, W. L. Downs, L. Salerno, and J. T. Packer, *Toxicol. Appl. Pharmacol.*, **4**:247 (1962); *Chem. Abstr.*, **56**:14577 (1962).
988. R. W. Moncrieff, *Ind. Chem.*, **30**:259 (1954).
989. Emery Industries, Inc., Organic Chemicals Div., "Azelaic Acid in Polymer Applications," Tech. Bull. 450, Cincinnati, Ohio, 1967.
990. Emery Industries, Inc., Organic Chemicals Div., "Abstracts of Azelaic Acid Use— Patents and Journal References," Tech. Bull. 443, Cincinnati, Ohio, 1965.
991. H. G. J. Overmars, in "Encyclopedia of Polymer Science and Technology," Vol. 11, H. Mark, Chairman of the Editorial Board, Interscience, New York, 1969, pp. 464–495.
992. O. E. Snider and R. J. Richardson Ref. 991, pp. 495–506.
993. H. G. J. Overmars, *Macromol. Syn.*, **3**:8 (1968).
994. H. Jones, *Chimia*, **5**:169 (1951).
995. O. Aschan, *Chem. Ber.*, **45**:1603 (1912).
996. J. W. Hill and W. H. Carothers, *J. Amer. Chem. Soc.*, **54**:1569 (1932).
997. S. Swann, Jr., R. Oehler, and R. J. Buswell, *Org. Syn.*, **Coll. Vol. II**, 276 (1943).
998. B. S. Biggs and W. S. Bishop, *Org. Syn.*, **Coll. Vol. III**, 768 (1955).
999. N. L. Allinger, *Org. Syn.*, **Coll. Vol. IV**, 840 (1963).
1000. S. Swann, Jr., and W. E. Garrison, Jr., *Org. Syn.*, **41**:33 (1961).

1001. R. Perron and J. Perichon, *Fette Seifen, Anstrichm.*, **66**:750 (1964).
1002. G. Mullick and A. P. Chakravarty, *Labdev*, **3**:47 (1965); *Chem. Abstr.*, **62**:11680 (1965).
1003. E. Bödtker, *J. Prakt. Chem.*, **85**:221 (1912).
1004. V. E. Agabekov and N. I. Mitskevich, *Katal. Reakts. Zhidk. Faze, Tr. Vses. Konf.*, *2nd* (1966), **1967**:570; *Chem. Abstr.*, **69**:1716 (1968).
1005. R. LeMar, *U.S. Dep. Commer., Off. Tech. Serv.* AD, 277,080 (1962).
1006. K. Thinius and E. Kestner, *Plaste Kaut.*, **11**:393 (1964); *Chem. Abstr.*, **63**:16542 (1965).
1007. A. Grün and T. Wirth, *Chem. Ber.*, **55**:2206 (1922).
1008. A. G. Houpt (to American Cyanamid Co.), U.S. Patent 2,217,515 (Oct. 8, 1940).
1009. A. G. Houpt (to American Cyanamid Co.), U.S. Patent 2,217,516 (Oct. 8, 1940).
1010. H. A. Bruson and L. W. Covert (to Rohm & Haas Co.), U.S. Patent 2,182,056 (Dec. 5, 1939).
1011. M. J. Diamond, R. G. Binder, and T. H. Applewhite, *J. Amer. Oil Chem. Soc.*, **42**: 882 (1965).
1012. M. J. Diamond and T. H. Applewhite, *J. Amer. Oil Chem. Soc.*, **44**:656 (1967).
1013. M. J. Diamond (to the Secretary of Agriculture), U.S. Patent 3,466,310 (Sept. 9, 1969).
1014. G. H. Hargreaves and L. N. Owen, *J. Chem. Soc.*, **1947**:753.
1015. R. A. Dytham and B. C. L. Weedon, *Tetrahedron*, **8**:246 (1960).
1016. W. Schrauth and K. Hennig (to E. I. du Pont de Nemours & Co.), U.S. Patent 2,304,602 (Dec. 8, 1942).
1017. G. D. Davis and B. A. Dombrow (to National Oil Products Co.), U.S. Patent 2,318,762 (May 11, 1943).
1018. G. Dupont and O. Kostelitz (to Société Organico), U.S. Patent 2,674,608 (April 6, 1954).
1019. F. W. Lane (to E. I. du Pont de Nemours & Co.), U.S. Patent 2,580,931 (Jan. 1, 1952).
1020. V. E. Haury (to Simco, Inc.), U.S. Patent 2,693,480 (Nov. 2, 1954).
1021. Rohm & Haas Co., British Patent 738,516 (Oct. 12, 1955).
1022. Y. Bourgeois (to Société de Produits Chimiques de Bezons), U.S. Patent 2,935,530 (May 3, 1960).
1023. I. Gavat and I. Ciolan, *Rev. Chim. (Bucharest)*, **12**:708 (1961); *Chem. Abstr.*, **57**:2022 (1962).
1024. M. Klang and S. Marinescu, *Rev. Chim. (Bucharest*, **15**:542 (1964); *Chem. Abstr.*, **63**:17883 (1965).
1025. W. Stein (to Henkel & Cie; GmbH), U.S. Patent 2,696,500 (Dec. 7, 1954).
1026. W. Stein (to Henkel & Cie; GmbH), U.S. Patent 2,696,501 (Dec. 7, 1954).
1027. D. S. Bolley and F. C. Naughton (to The Baker Castor Oil Co.), U.S. Patent 2,734,916 (Feb. 14, 1956).
1028. F. C. Naughton and P. C. Daidone (to The Baker Castor Oil Co.), U.S. Patent 2,851,491 (Sept. 9, 1958).
1029. F. C. Naughton and P. C. Daidone (to The Baker Castor Oil Co.), U.S. Patent 2,851,492 (Sept. 9, 1958).
1030. F. C. Naughton (to The Baker Castor Oil Co.), U.S. Patent 2,851,493 (Sept. 9, 1958).
1031. Y. J. E. Bourgeois, P. C. Gosselin, and M. Roussos (to Société de Produits Chimiques et de Synthèse), French Patent 1,366,068 (June 1, 1964); *Chem. Abstr.*, **62**:2711 (1965).
1032. M. Kobayashi, T. Mochizuki, and A. Nakade, *Yukagaku*, **7**:133 (1958).
1033. M. Kobayashi, T. Mochizuki, and A. Nakade, *Yukagaku*, **7**:138 (1958).

1034. H. C. Cheetham and D. A. Rothrock (to The Resinous Products and Chemical Co.), U.S. Patent 2,267,269 (Dec. 23, 1941).

1035. W. E. Hanson (to Gulf Research and Development Co.), U.S. Patent 2,470,849 (May 24, 1949).

1036. R. S. Emslie (to E. I. du Pont de Nemours & Co.), U.S. Patent 2,731,495 (Jan. 17, 1956).

1037. N. Duinea and E. Popescu-Lascu, Romanian Patent 51,292 (Sept. 17, 1968); *Chem. Abstr.*, **70**(18): Abstr. No. 79386g (1969).

1038. H. A. Offe (to Badische Anilin- & Soda-Fabrik A.G.), German Patent 854,508 (Nov. 4, 1952); *Chem. Abstr.*, **52**:7914 (1958).

1039. H. A. Offe (to Badische Anilin- & Soda-Fabrik A.G.), German Patent 855,403 (Nov. 13, 1952); *Chem. Abstr.*, **49**:12163 (1955).

1040. H. A. Offe, *Z. Naturforsch. B*, **2**:185 (1947).

1041. A. I. Kamneva, M. Y. Fioshin, L. I. Kazakova, and S. M. Itenberg, *Neftekhimiya*, **2**:550 (1962).

1042. M. Y. Fioshin and L. I. Kazakova, *Dokl. Akad. Nauk SSSR*, **152**:1132 (1963); *Chem. Abstr.*, **60**:6481 (1964).

1043. M. Y. Fioshin and L. I. Kazakova, *Zashch. Metal. Oksidnye Pokrytiya, Korroz. Metal. Issled. Obled. Elektrokhim.*, **1965**:343; *Chem. Abstr.*, **65**:3343 (1966).

1044. Y. M. Tyurin, E. P. Kovsman, and E. A. Karavaeva, *Zh. Prikl. Khim.* (*Leningrad*), **38**:1818 (1965); *Chem. Abstr.*, **64**:3045 (1966).

1045. W. Himmele, N. v. Kutepow, and W. Schwab (to Badische Anilin- & Soda-Fabrik A.G.), German Patent 1,162,347 (Feb. 6, 1964); *Chem. Abstr.*, **60**:10554 (1964).

1046. K. Saotome, H. Komoto, and T. Yamazaki, *Bull. Chem. Soc. Jap.*, **39**:480 (1966); *Chem. Abstr.*, **65**:2120 (1966).

1047. C. E. Frank and W. E. Foster, *J. Org. Chem.*, **26**:303 (1961).

1048. J. F. Walker (to E. I. du Pont de Nemours & Co.), U.S. Patent 2,352,461 (June 27, 1944).

1049. H. Greenberg (to National Distillers Products Corp.), U.S. Patent 2,749,364 (June 5, 1956).

1050. D. R. Carley and W. E. Foster (to Ethyl Corp.), U.S. Patent 2,773,092 (Dec. 4, 1956).

1051. C. E. Frank and W. E. Foster (to National Distillers Products Corp.), U.S. Patent 2,790,002 (April 23, 1957).

1052. J. F. Nobis and R. E. Robinson (to National Distillers and Chemical Corp.), U.S. Patent 2,795,625 (June 11, 1957).

1053. C. E. Frank and J. R. Leebrick (to National Distillers and Chemical Corp.), U.S. Patent 2,816,913 (Dec. 17, 1957).

1054. C. E. Frank and J. R. Leebrick (to National Distillers and Chemical Corp.), U.S. Patent 2,816,914 (Dec. 17, 1957).

1055. C. E. Frank and W. E. Foster (to National Distillers and Chemical Corp.), U.S. Patent 2,816,916 (Dec. 17, 1957).

1056. V. L. Hansley and S. Schott (to National Distillers and Chemical Corp.), U.S. Patent 2,816,917 (Dec. 17, 1957).

1057. R. Wynkoop, J. F. Nobis, and J. J. Giachetto (to National Distillers and Chemical Corp.), U.S. Patent 2,816,918 (Dec. 17, 1957).

1058. R. Wynkoop and J. J. Giachetto (to National Distillers and Chemical Corp.), U.S. Patent 2,816,919 (Dec. 17, 1957).

1059. L. M. Watson and G. H. Slattery (to National Distillers and Chemical Corp.), U.S. Patent 2,816,935 (Dec. 17, 1957).

1060. V. L. Hansley and S. Schott (to National Distillers and Chemical Corp.), U.S. Patent 2,816,936 (Dec. 17, 1957).

1061. R. Wynkoop and J. J. Giachetto (to National Distillers and Chemical Corp.), U.S. Patent 2,822,389 (Feb. 4, 1958).

1062. C. E. Frank and J. F. Nobis (to National Distillers and Chemical Corp.), U.S. Patent 2,824,118 (Feb. 18, 1958).

1063. R. Wynkoop and J. J. Giachetto (to National Distillers and Chemical Corp.), U.S. Patent 2,837,564 (June 3, 1958).

1064. J. Feldman and O. D. Frampton (to National Distillers and Chemical Corp.), U.S. Patent 2,837,565 (June 3, 1958).

1065. J. Feldman and O. D. Frampton (to National Distillers and Chemical Corp.), U.S. Patent 2,837,566 (June 3, 1958).

1066. S. A. Mednick, R. Wynkoop, and J. Feldman (to National Distillers and Chemical Corp.), U.S. Patent 2,858,337 (Oct. 28, 1958).

1067. J. Feldman (to National Distillers and Chemical Corp.), U.S. Patent 2,862,027 (Nov. 25, 1958).

1068. S. Schott (to National Distillers and Chemical Corp.), U.S. Patent 2,865,969 (Dec. 23, 1958).

1069. R. E. Robinson and J. F. Nobis (to National Distillers and Chemical Corp.), U.S. Patent 2,867,656 (Jan. 6, 1959).

1070. F. G. Lum (to California Research Corp.), U.S. Patents 3,002,018–3,002,020 (Sept. 26, 1961).

1071. C. E. Frank and W. E. Foster (to National Distillers and Chemical Corp.), U.S. Patent 3,013,071 (Dec. 12, 1961).

1072. S. P. Rowland and E. G. Pritchett (to National Distillers and Chemical Corp.), U.S. Patent 3,222,394 (Dec. 7, 1965).

1073. F. Asinger and H. Eckoldt, *Chem. Ber.*, **76**:585 (1943).

1074. D. G. M. Diaper and D. L. Mitchell, *Can. J. Chem.*, **43**:319 (1965).

1075. J. v. Braun and W. Keller, *Chem. Ber.*, **66**:215 (1933).

1076. W. Hückel, A. Gercke, and A. Gross, *Chem. Ber.*, **66**:563 (1933).

1077. S. V. Vasil'ev, *Zh. Obshch. Khim.*, **26**:712 (1956); *Chem. Abstr.*, **50**:14545 (1956).

1078. L. Ruzicka, M. Stoll, and H. Schinz, *Helv. Chim. Acta*, **9**:249 (1926).

1079. H. E. Holmquist, H. S. Rothrock, C. W. Theobald, and B. E. Englund, *J. Amer. Chem. Soc.*, **78**:5339 (1956).

1080. R. Criegee, *Chem. Ber.*, **77**:722 (1944).

1081. W. J. Mijs, K. S. DeVries, J. G. Westra, H. A. A. Gaur, J. Smidt, and J, Vriend, *Rec. Trav. Chim. Pays-Bas*, **87**:580 (1968).

1082. Inventa A.G. für Forschung und Patentverwertung, Belgian Patent 628,945 (June 16, 1963); *Chem. Abstr.*, **60**:9166 (1964).

1083. Inventa A.G. für Forschung und Patentverwertung, British Patent 963,945 (July 15, 1964); *Chem. Abstr.*, **61**:9414 (1964).

1084. A. E. Schnider (to Inventa A.G. für Forschung und Patentverwertung), U.S. Patent 3,429,927 (Feb. 25, 1969).

1085. F. H. Howell (to Geigy, U.K., Ltd.), British Patent 1,196,598 (July 1, 1970).

1086. J. R. Geigy, Societé Anon., French Patent 1,545,224 (Nov. 8, 1968).

1087. F. H. Howell (to Geigy, U.K., Ltd.), British Patent 1,196,594 (July 1, 1970).

1088. J. Letort, J. Tiquet, J. Lahouste, P. Payen. C. Tamielian, and J. Van Overbeke, *Chim. Ind. Genie Chim.*, **103**:927 (1970).

1089. N. Brown, A. W. Anderson, and C. E. Schweitzer, *J. Amer. Chem. Soc.*, **77**:1760 (1955).

1090. F. Fichter and H. Buess, *Helv. Chim. Acta*, **18**:445 (1935).

1091. J. Kamlet (to Crown Zellerbach Corp.), U.S. Patent 2,815,375 (Dec. 3, 1957).
1092. O. Diels and K. Alder, *Justus Liebigs Ann. Chem.*, **486**:211 (1931).
1093. I. D. Webb and G. T. Borcherdt, *J. Amer. Chem. Soc.*, **73**:752 (1951).
1094. R. Paul and O. Riobé, *C.R. Acad. Sci. (Paris)*, **230**:1185 (1950).
1095. H. Stetter and W. Dierichs, *Chem. Ber.*, **86**:693 (1953).
1096. G. Boffa, D. Costabello, and A. Guilico (to "Montecatini" Società Generale per l'Industrià Mineraria e Chimica), U.S. Patent 2,956,075 (Oct. 11, 1960).
1097. J. D. Morrison and J. M. Robertson, *J. Chem. Soc.*, **1949**:993.
1098. W. H. Stahl and H. Pessen, *J. Amer. Chem. Soc.*, **74**:5487 (1952).
1099. R. Emmrich, *Deut. Arch. Klin. Med.*, **187**:504 (1941); *Chem. Abstr.*, **38**:5573 (1944).
1100. P. E. Verkade, J. van der Lee, and A. J. S. van Alphen, *Z. Physiol. Chem.*, **250**:47 (1937).
1101. P. M. Jenner, E. C. Hagan, J. M. Taylor, E. L. Cook, and O. G. Fitzhugh, *Food Cosmetol. Toxicol.*, **2**:327 (1964).
1102. R. Signer and P. Sprecher, *Helv. Chim. Acta*, **30**:1001 (1947).
1103. L. J. Durham, D. J. McLeod, and J. Cason, *Org. Syn.*, **Coll. Vol. IV**, 635 (1963).
1104. R. G. Jones, *J. Amer. Chem. Soc.*, **69**:2350 (1947).
1105. G. Wilke, *Angew. Chem. Int. Ed. Engl.*, **2**:105 (1963).
1106. J. Furukawa, H. Morikawa, and R. Yamamoto (to Mitsubishi Petrochemical Co., Ltd.), U.S. Patent 3,476,820 (Nov. 4, 1969).
1107. L. I. Zakharkin, V. V. Korneva, and G. M. Kunitskaya, *Izv. Akad. Nauk SSSR, Otd. Khim. Nauk*, **1961**:1908; *Chem. Abstr.*, **56**:8551 (1962).
1108. L. I. Zakharkin, A. N. Bashkirov, V. V. Kamzolkin, K. M. Sokova, V. V. Korneva, and T. P. Andreeva, USSR Patent 145,579 (Aug. 21, 1963); *Chem. Abstr.*, **60**:5343 (1964).
1109. V. V. Kamzolkin, A. N. Bashkirov, K. M. Sokova, and T. P. Andreeva, *Neftekhimiya*, **4**:96 (1964); *Chem. Abstr.*, **60**:14401 (1964).
1110. M. Martin and R. Champ (to Produits Chimiques Pechiney-Saint Gobain), French Patent 1,379,783 (Nov. 27, 1964); *Chem. Abstr.*, **62**:9013 (1965).
1111. E. I. du Pont de Nemours & Co., Netherlands Appl. 6,505,838 (Nov. 8, 1966); *Chem. Abstr.*, **66**:7093 (1967).
1112. Chemische Werke Hüls A.G., French Patent 1,486,391 (June 23, 1967); *Chem. Abstr.*, **68**:8362 (1968).
1113. F. Broich and H. Grasemann (to Chemische Werke Hüls A.G.), U.S. Patent 3,399,035 (Aug. 27, 1968).
1114. E. J. Inchalik, I. Kirschenbaum, and R. M. Hill (to Esso Research and Engineering Co.), U.S. Patent 3,419,615 (Dec. 31, 1968).
1115. H. Grasemann, *Erdoel Kohle, Erdgas, Petrochem.*, **22**:751 (1969); *Chem. Abstr.*, **72**: Abstr No. 42524r (1970).
1116. V. V. Kamzolkin, A. N. Bashkirov, K. M. Sokova, T. P. Andreeva, and G. A. Zelenaya, *Neftekhimiya*, **4**:599 (1964); *Chem. Abstr.*, **61**:13209 (1964).
1117. Inventa A.G. für Forschung und Patentverwertung, Belgian Patent 640,590 (Dec. 13, 1963); *Chem. Abstr.* **63**:515 (1965).
1118. Wises, H. K., and S. B. Lippincott (to Esso Research and Engineering Co.), U.S. Patent 2,978,464 (April 4, 1961).
1119. H. K. Wiese and S. B. Lippincott (to Esso Research and Engineering Co.), U.S. Patent 3,087,963 (April 30, 1963); British Patent 878,241; *Chem. Abstr.*, **56**:8570 (1962).
1120. C. N. Winnick (to Halcon International, Inc.), German Offen. 1,914,572 (Oct. 9, 1969); *Chem. Abstr.*, **72**: Abstr. No. 54773j (1970).

1121. Columbian Carbon Co., Netherlands Appl. 6,601,043 (Aug. 1, 1966); *Chem. Abstr.*, **66**:226 (1967).

1122. O. v. Schickh, F. Urbanek, and H. Metzger, *Z. Naturforsch.*, **18b**:980 (1963); *Chem. Abstr.*, **60**:5353 (1964).

1123. L. I. Zakharkin and V. V. Korneva, *Sin. Svoistva Monomerov, Akad. Nauk SSSR, Inst. Neftekhim. Sinteza, Sb. Rab. 12-oi (Dvenadtsatoi) Konf. Vysokomol. Soedin.*, **1962**:193; *Chem. Abstr.*, **62**:6389 (1965).

1124. C. M. White, V. L. Hughes, and W. E. Wellman (to Esso Research and Engineering Co.), French Patent 1,331,267 (June 28, 1963); *Chem. Abstr.*, **60**:2781 (1964).

1125. Esso Research and Engineering Co., Netherlands Appl. 6,408,320 (Jan. 25, 1965); *Chem. Abstr.*, **63**:499 (1965).

1126. J. O. White and D. D. Davis (to E. I. du Pont de Nemours & Co.), German Offen. 1,912,569 (Oct. 2, 1969); *Chem. Abstr.*, **72**: Abstr. No. 21327p (1970).

1127. E. D. Wilhoit (to E. I. du Pont de Nemours & Co.), U.S. Patent 3,461,160 (Aug. 12, 1969).

1128. R. J. Convery (to Sun Oil Co.), U.S. Patent 3,070,626 (Dec. 25, 1962).

1129. F. Minisci, G. Belvedere, R. Galli, and A. Quilico (to Montecatini Edison, S.p.A.), U.S. Patent 3,366,680 (Jan. 30, 1968).

1130. K. Kosswig, W. Stumpf, and W. Kirchhof, *Justus Liebigs Ann. Chem.*, **681**:28 (1965).

1131. R. H. Perry, Jr. (to Esso Research and Engineering Co.), U.S. Patent 3,059,028 (Oct. 16, 1962).

1132. C. G. McAlister (to Columbian Carbon Co.), U.S. Patent 3,400,164 (Sept. 3, 1968).

1133. R. Levine (to Columbian Carbon Co.), U.S. Patent 3,400,165 (Sept. 3, 1968).

1134. C. G. McAlister (to Columbian Carbon Co.), U.S. Patent 3,400,166 (Sept. 3, 1968).

1135. R. H. Perry, Jr. (to Esso Research and Engineering Co.), U.S. Patent 3,173,964 (Mar. 16, 1965).

1136. W. Cooper, *J. Chem. Soc.*, **1951**:1340.

1137. W. Cooper and W. H. T. Davison, *J. Chem. Soc.*, **1952**:1180.

1138. M. J. Roedel (to E. I. du Pont de Nemours and Co.), U.S. Patent 2,601,223 (June 24, 1952).

1139. N. A. Milas, S. A. Harris, and P. C. Panagiotakos, *J. Amer. Chem. Soc.*, **61**:2430 (1939).

1140. R. Criegee, *Fortschr. Chem. Forsch.*, **1**:508 (1950).

1141. M. S. Kharasch and G. Sosnovsky, *J. Org. Chem.*, **23**:1322 (1958).

1142. E. G. E. Hawkins, *J. Chem. Soc.*, **1955**:3463.

1143. D. D. Coffman, R. Cramer, and W. E. Mochel, *J. Amer. Chem. Soc.*, **80**:2882 (1958).

1144. J. B. Braunworth and G. W. Crosby, *J. Org. Chem.*, **27**:2064 (1962).

1145. M. S. Kharasch and W. Nudenberg, *J. Org. Chem.*, **19**:1921 (1954).

1146. D. D. Coffman and H. N. Cripps, *J. Amer. Chem. Soc.*, **80**:2880 (1958).

1147. G. Dupont, R. Dulou, and P. Quantin, *Bull. Soc. Chim. Fr.*, 1951, 59; French Patent 1,018,186 (Dec. 29, 1952).

1148. J. Kamlet (to Goodyear Tire & Rubber Co.), U.S. Patent 2,782,157 (Feb. 19, 1957).

1149. R. V. Lindsey, Jr., and M. L. Peterson (to E. I. du Pont de Nemours & Co.), U.S. Patent 2,680,713 (June 8, 1954).

1150. M. Y. Fioshin, A. I. Kamneva, L. A. Mirkind, A. G. Kormienko, and L. A. Salmin, *Khim. Prom.*, **42**:804 (1966); *Chem. Abstr.*, **66**:4095 (1967).

1151. L. A. Mikeska (to Standard Oil Development Co.), U.S. Patent 2,614,122 (Oct. 14, 1952).

1152. M. Kobayashi, *Yushi Kagaku Kyôkaishi*, **2**:183 (1953); *Chem. Abstr.*, **48**:9718 (1954).

1153. T. R. Steadman and J. O. H. Peterson, Jr. (to National Research Corp.), U.S. Patent 2,847,466 (Aug. 12, 1958).
1154. T. R. Steadman and J. O. H. Peterson, Jr. (to National Research Corp.), U.S. Patent 2,847,467 (Aug. 12, 1958).
1155. T. R. Steadman and J. O. H. Peterson, *Ind. Eng. Chem.*, **50**:59 (1958).
1156. M. O. Bagby, *J. Org. Chem.*, **26**:4735 (1961).
1157. Y. Asahina and H. Takimoto, *J. Pharm. Soc. Jap.*, **49**:1017 (1929); *Chem. Abstr.*, **24**:1346 (1930).
1158. E. P. Czerwin (to E. I. du Pont de Nemours & Co.), U.S. Patent 2,269,998 (Jan. 13, 1942).
1159. W. P. Hall and E. E. Reid, *J. Amer. Chem. Soc.*, **65**:1466 (1943).
1160. T. R. Steadman and J. O. H. Peterson (to National Research Corp.), U.S. Patent 2,847,432 (Aug. 12, 1958).
1161. W. Reppe and H. Kröper, *Justus Liebigs Ann. Chem.*, **582**:38 (1953).
1162. R. Ercoli, *Chim. Ind. (Milan)*, **37**:1029 (1955).
1163. H. Koch and K. Schauerte, *Brennstoff-Chem.*, **46**:392 (1965).
1164. J. English, Jr., J. Bonner, and A. J. Haagen-Smit, *J. Amer. Chem. Soc.*, **61**:3434 (1939).
1165. J. Imer (to Produits Chimiques Pechiney-Saint Gobain), French Patent 1,393,568-9 (March 26, 1965); *Chem. Abstr.*, **63**:8204 (1965).
1166. E. M. Amir and J. K. Nickerson (to Esso Research and Engineering Co.), U.S. Patent 3,417,138 (Dec. 17, 1968).
1167. R. Sailer (to Inventa A.G. für Forschung und Patentverwertung), German Offen. 1,903,571 (Oct. 16, 1969); *Chem. Abstr.*, **72**: Abstr. No. 21326n (1970).
1168. J. Housty and M. Hospital, *C.R. Acad. Sci. (Paris)*, **259**:2437 (1964).
1169. F. Dupré la Tour, and A. Riedberger, *C.R. Acad. Sci. (Paris)*, **199**:215 (1934).
1170. H. Noerdlinger, *Chem. Ber.*, **23**:2356 (1890).
1171. J. Walker and J. S. Lumsden, *J. Chem. Soc.*, **79**:1197 (1901).
1172. R. Emmrich and I. Emmrich-Glaser, *Hoppe-Seyler's Z. Physiol. Chem.*, **266**:183 (1940).
1173. C. B. Flack and R. H. Weir (to E. I. du Pont de Nemours & Co.), U.S. Patent **3,502,624** (March 24, 1970).
1174. **S. B. Speck** (to E. I. du Pont de Nemours & Co.) U.S. Patent 3,393,210 (July 16, **1968**).
1175. J. Zimmerman (to E. I. du Pont de Nemours & Co.), U.S. Patent 3,393,252 (July 16, **1968**).
1176. R. R. Lunt, Jr. (to E. I. du Pont de Nemours & Co.), U.S. Patent 3,440,211 (April 22, 1969).
1177. M. K. Dobrokhotova, K. N. Vlasova, E. K. Lyadysheva, and S. L. Kutuzova, *Sov. Plast.*, **1967**:26.
1178. C. S. Fuller, *J. Amer. Chem. Soc.*, **70**:421 (1948).
1179. R. B. Perkins, J. J. Roden, III, A. C. Tanquary, and I. A. Wolff, *Mod. Plast.*, **46**:136 (1969).
1180. H. Jones (to Geigy Co., Ltd.), British Patent 775,560 (May 29, 1957).
1181. S. W. Critchley, A. Hill, and I. Williamson (to Geigy Co., Ltd.), British Patent 787,467 (Dec. 11, 1957).
1182. H. J. Nieschlag, W. H. Tallent, I. A. Wolff, W. E. Palm, and L. P. Witnauer, *Polym Eng. Sci.*, **7**:51 (1967).
1183. H. J. Nieschlag, W. H. Tallent, I. A. Wolff, W. E. Palm, and L. P. Witnauer, *Ind. Eng. Chem. Prod. Res. Develop.*, **6**:201 (1967).

1184. H. J. Nieschlag, J. W. Hagemann, I. A. Wolff, W. E. Palm, and L. P. Witnauer, *Ind. Eng. Chem. Prod. Res. Develop.*, **3**:146 (1964).

1185. D. Holde and F. Zadeck, *Chem. Ber.*, **56**:2052 (1923).

1186. T. J. Mirchandani and J. L. Simonsen, *J. Chem. Soc.*, **1927**:371.

1187. A. Greiner, *Zesz. Probl. Postepow Nauk Roln.*, **1970**:351; *Chem. Abstr.*, **73**: Abstr. No. 76617j (1970).

1188. H. Grynberg, M. Beldowicz, and J. Cyganska, *Zesz. Probl. Postepow Nauk Roln.*, **1970**:359; *Chem. Abstr.*, **73**: Abstr. No. 76618k (1970).

1189. H. J. Nieschlag, I. A. Wolff, T. C. Manley, and R. J. Holland, *Ind. Eng. Chem. Prod. Res. Develop.*, **6**:120 (1967).

1190. J. Ross, A. I. Gebhart, and J. Fred Gerecht, *J. Amer. Chem. Soc.*, **67**:1275 (1945).

1191. B. B. Ghatgey, U. G. Nayak, K. K. Chakravarti, and S. C. Bhattacharyya (to Council of Scientific and Industrial Research), British Patent 857,163 (Dec. 29, 1960).

1192. S. Ohara and Y. Shinozaki, *Yukagaku*, **5**:222 (1956).

1193. A. Müller, *Chem. Ber.*, **67**:295 (1934).

1194. A. Fijita, Y. Hirose, S. Egami, K. Shioji, Y. Wake, and H. Nakamura, *J. Pharm. Soc. Jap.*, **74**:119 (1954); *Chem. Abstr.*, **49**:1565 (1955).

1195. L. I. Zakharkin and V. V. Korneva, *J. Org. Chem. USSR*, **2**:738 (1966).

1196. H. Stetter and W. Dierichs, *Chem. Ber.*, **85**:290 (1952); German Patent 915,085 (July 15, 1954).

1197. Henkel & Cie, GmbH, British Patent 792,486 (March 26, 1958).

1198. M. D. Potter and E. P. Taylor, *J. Chem. Soc.*, **1951**:3513.

1199. K. Mislow and I. V. Steinberg, *J. Amer. Chem. Soc.*, **77**:3807 (1955).

1200. R. L. Logan (to Kessler Chemical Co.), U.S. Patent 2,777,865 (Jan. 15, 1957).

1201. M. Kobayashi and Y. Nishino, *Kagaku To Kôgyô (Osako)*, **35**:91 (1961); *Chem. Abstr.*, **55**:25749 (1961).

1202. Y. Suzuki and T. Takeuchi, *Kôgyô Kagaku Zasshi*, **67**:902 (1964); *Chem. Abstr.*, **62**:1083 (1965).

1203. E. T. Roe, D. A. Konen, and D. Swern (to U.S. Department of Agriculture), U.S. Patent 3,308,140 (March 7, 1967).

1204. L. I. Zakharkin, L. P. Vinogradova, V. V. Korneva, and S. I. Zav'yalov, *Izv. Akad. Nauk SSSR, Otd. Khim. Nauk*, **1962**:1309; *Chem. Abstr.*, **58**:7824 (1963).

1205. T. Yumoto, *Rep. Govt. Ind. Res. Inst.*, Nagoya, **2**:56 (1953).

1206. H. Kosche (to Henkel & Cie., GmbH), German Patent 949,651 (Sept. 27, 1956); British Patent 783,369 (Sept. 25, 1957).

1207. B. Blaser and W. Stein (to Henkel & Cie., GmbH), German Patent 1,020,618 (Dec. 12, 1957).

1208. W. Stein (to Henkel & Cie., GmbH), U.S. Patent 2,798,093 (July 2, 1957).

1209. H. J. Nieschlag, unpublished results.

1210. J. Housty, *Acta Crystallogr.*, **B24**:486 (1968).

1211. M. Fileti and G. Ponzio, *J. Prakt. Chem.*, **48**:323 (1893); *Chem. Ber. Ref.*, **26**:811 (1893).

1212. R. J. R. Mohan Rao and S. R. Palit, *Indian J. Phys.*, **34**:55 (1960).

1213. J. Walker and J. S. Lumsden, *J. Chem. Soc.*, **79**:1191 (1901).

1214. J. L. Greene, Jr., E. L. Huffman, R. E. Burks, Jr., W. C. Sheehan, and I. A. Wolff, *J. Polym. Sci., Pt. A-1*, **5**:391 (1967).

1215. U.S. Department of Agriculture, Agricultural Research Service, CA-71-28, June, 1968, 5 pp.

1216. DeNordiske Fabriker (DE-NO-FA Aptieselskap), British Patent 127,814 (1919); *Chem. Abstr.*, **13**:2462 (1919); British Patent 166,236 (1921); *Chem. Abstr.*, **16**:848 (1922).

1217. J. Scheiber, *Farbe Lack*, **1929**:585.
1218. C. P. A. Kappelmeier, *Farben-Z.*, **38**:1018–1020, 1077–1079 (1933).
1219. D. H. Wheeler, *Off. Dig. Fed. Paint Varn. Prod. Clubs*, **322**:661 (1951)
1220. T. F. Bradley and W. B. Johnston, *Ind. Eng. Chem.*, **33**:86 (1941).
1221. J. C. Cowan, W. C. Ault, and H. M. Teeter, *Ind. Eng. Chem.*, **38**:1138 (1946).
1222. J. C. Cowan, D. H. Wheeler, H. M. Teeter, et al., *Ind. Eng. Chem.*, **41**:1647 (1949).
1223. L. B. Falkenburg, H. M. Teeter, P. S. Skell, and J. C. Cowan, *Oil Soap*, **22**:143 (1945).
1224. J. C. Cowan, A. J. Lewis, and L. B. Falkenburg, *Oil Soap*, **21**:101 (1944).
1225. C. G. Goebel, *J. Amer. Oil Chem. Soc.*, **24**:65 (1947).
1226. C. G. Goebel, U.S. Patent 2,482,761 (1949).
1227. F. O. Barrett, C. G. Goebel, and R. M. Peters, U.S. Patent 2,793,219 and 2,793,220 (1957).
1228. C. G. Goebel, U.S. Patent 2,664,429 (1953).
1229. S. E. Miller, U.S. Patent 3,412,039 (1968); *Chem. Abstr.*, **70**:2839y (1969).
1230. R. Rowe, *Paint Technol.*, **23**:257 (1959); British Patent 841,554 (July 20, 1960).
1231. J. Baltes, *Fette, Seifen, Anstrichm.*, **66**:942 (1964).
1232. J. C. Cowan, *J. Amer. Oil Chem. Soc.*, **39**:534 (1962).
1233. J. C. Cowan and H. M. Teeter, *Ind. Eng. Chem.*, **36**:148 (1944).
1234. M. R. Kamal and J. E. Weeklatz, preprint booklet "Organic Coatings and Plastic Chemistry," (Amer. Chem. Soc.), **26**:154 (1966).
1235. L. Peterson, U.S. Patent 3,002,941 (1961).
1236. R. W. Fulmer and D. W. Glaser, *Paint Ind. Mag.*, **74**:14–16, 18, 29 (1959).
1237. J. C. Cowan and D. H. Wheeler, *J. Amer. Chem. Soc.*, **66**:84 (1944).
1238. D. H. Wheeler, A. Melun, and F. Linn, *J. Amer. Oil Chem. Soc.*, **47**:242 (1970).
1239. C. B. Croston, I. L. Tubb, J. C. Cowan, and H. M. Teeter, *J. Amer. Oil Chem. Soc.*, **29**:331 (1952).
1240. S. A. Harrison and D. H. Wheeler, *J. Amer. Chem. Soc.*, **76**:2379 (1954).
1241. R. F. Paschke, L. E. Peterson, and D. H. Wheeler, *J. Amer. Oil Chem. Soc.*, **41**:56 (1964).
1242. S. A. Harrison, L. E. Peterson, and D. H. Wheeler, *J. Amer. Oil Chem. Soc.*, **42**:2 (1965).
1243. E. N. Frankel, *J. Amer. Oil Chem. Soc.*, **48**:248 (1971).
1244. E. T. Roe and D. Swern, *J. Amer. Oil Chem. Soc.*, **37**:661 (1960).
1245. D. H. Wheeler and J. White, *J. Amer. Oil Chem. Soc.*, **44**:298 (1967).
1246. R. F. Paschke, L. E. Peterson, and D. H. Wheeler, *J. Amer. Oil Chem. Soc.*, **41**:723 (1964).
1247. Emery Industries, "Analytical Methods for Dimer and Trimer Acid," Tech. Bull. 416, Cincinnati, Ohio, 1959.
1248. J. C. Cowan, L. B. Falkenburg, and H. M. Teeter, *Ind. Eng. Chem. Anal. Ed.*, **16**:90 (1944).
1249. R. F. Paschke, J. R. Kerns, and D. H. Wheeler, *J. Amer. Oil Chem. Soc.*, **31**:5 (1954).
1250. C. D. Evans, D. G. McConnell, E. N. Frankel, and J. C. Cowan, *J. Amer. Oil Chem. Soc.*, **42**:764 (1965).
1251. General Mills, Chemical Division, Kankakee, Illinois, Report on the Non-Toxicity of Versamid Polyamide Resins.
1252. W. Griem, *Z. Ernaehrungswiss.*, **7**:30 (1966).
1253. N. R. Bottino, *J. Amer. Oil Chem. Soc.*, **39**:25 (1962).
1254. E. W. Crampton, R. H. Common, E. T. Pritchard, and F. A. Farmer, *J. Nutr.*, **60**:13 (1956).

1255. E. W. Crampton, R. H. Common, F. A. Farmer, A. F. Wells, and D. Crawford, *J. Nutr.*, **49**:333 (1953).

1256. L. J. Gold, U.S. Patent 2,451,212 (1948); *Chem. Abstr.*, **43**:2037 (1949).

1257. R. Bult, *Off. Dig. Fed. Soc. Paint Technol.*, **33**:1594 (1961).

1258. D. E. Floyd, "Polyamide Resins," 2nd ed., Reinhold, New York, 1966.

1259. L. Mannes and W. Pack (to Henkel & Cie, GmbH), German Patent 745,265 (Mar. 1, 1944).

1260. G. H. Birum and T. M. Patrick (to Monsanto Chemical Company), U.S. Patent 2,965,598 (Dec. 20, 1960).

1261. D. H. Wheeler and E. R. Rogier (to General Mills, Inc.), U.S. Patent 3,016,359 (Jan. 9, 1962).

1262. E. T. Roe, G. R. Riser, and D. Swern, *J. Amer. Oil Chem. Soc.*, **38**:527 (1961).

1263. Société Anon. Bougies de la Cour, N.V. Netherlands Application 6511375 (Mar. 7, 1966).

1264. A. G. Rocchini (to Gulf Research & Development Co.), U.S. Patent 2,604,451 (July 22, 1952).

1265. Technochemie GmbH Verfahrenstechnik, Netherlands Application 6605281 (Oct. 21, 1966).

1266. D. M. Doty, *Riv. Ital. Sostanze Grasse*, **44**:417 (1967).

1267. Y. Machida, Y. Okamoto, and H. Sakurai, *Yukagaku*, **14**:563 (1965).

1268. S. A. Harrison (to General Mills, Inc.), U.S. Patent 2,966,478 (Dec. 27, 1960).

1269. E. J. Dufek, R. O. Butterfield, and E. N. Frankel, *J. Amer. Oil Chem. Soc.*, **48**(7): 325A (1971).

1270. Armour & Company, British Patent 1,077,547 (Aug. 2, 1967).

1271. G. B. Carpenter (to E. I. du Pont de Nemours & Co.), U.S. Patent 1,924,766 (Aug. 29, 1932).

1272. A. T. Larson (to E. I. du Pont de Nemours & Co.), U.S. Patent 2,022,244 (Nov. 26, 1935).

1273. J. C. Woodhouse (to E. I. du Pont de Nemours & Co.), U.S. Patent 1,924,762 (Aug. 29, 1932).

1274. G. B. Carpenter (to E. I. du Pont de Nemours & Co.), U.S. Patent 1,924,763 (Aug. 29, 1932).

1275. W. E. Vail (to E. I. du Pont de Nemours & Co.), U.S. Patent 1,924,764 (Aug. 29, 1932).

1276. A. T. Larson and W. E. Vail (to E. I. du Pont de Nemours & Co.), U.S. Patent 1,924,765 (Aug. 29, 1932).

1277. G. B. Carpenter (to E. I. du Pont de Nemours & Co.), U.S. Patent 1,924,767 (Aug. 29,1932).

1278. G. B. Carpenter (to E. I. du Pont de Nemours & Co.), U.S. Patent 1,957,939 (May 8, 1934).

1279. G. B. Carpenter (to E. I. du Pont de Nemours & Co.), U.S. Patent 1,924,768 (Aug. 29, 1932).

1280. G. B. Carpenter, Canadian Patent 340,503 (Apr. 3, 1934).

1281. A. T. Larson (to E. I. du Pont de Nemours & Co.), U.S. Patent 2,020,689 (Nov. 12, 1935).

1282. D. J. Loder (to E. I. du Pont de Nemours & Co.), U.S. Patents 2,135,456 and 2,135,459 (Nov. 1, 1938).

1283. T. A. Ford, H. W. Jacobson and F. C. McGrew, *J. Amer. Chem. Soc.*, **70**:3793 (1948).

1284. D. J. Loder (to E. I. du Pont de Nemours & Co.), U.S. Patents 2,135,448; 2,135,451; 2,135,452; 2,135,453; 2,135,457 (Nov. 1, 1938).

1285. T. A. Ford (to E. I. du Pont de Nemours & Co.), U.S. Patent 2,419,131 (Apr. 15, 1947).
1286. H. Kröper in "Methoden der Organischen Chemie (Houben-Weyl)," E. Müller, Ed., Vol. 4, pt. 2, Georg Thieme Verlag, Stuttgart, 1955, pp. 376–393.
1287. C. W. Bird, *Chem. Rev.*, **62**:283 (1962).
1288. J. Falbe, "Carbon Monoxide in Organic Synthesis," translated by C. R. Adams, Springer-Verlag, New York, 1970, Chapter 3.
1289. H. Kock, *Brennst.-Chem.*, **36**:321 (1955).
1290. J. van Dam and M. J. Waale, *Chim. Ind. (Paris)*, **90**:511 (1963).
1291. W. J. Ellis and C. Roming, Jr., *Hydrocarbon Process.*, **44**(6):139 (1965).
1292. H. Koch, *Fette, Seifen, Anstrichm.*, **59**:493 (1957).
1293. H. Koch (to Studiengesellschaft Kohle mbH), German Patent 942,987 (May 8, 1956).
1294. H. Koch (to Studiengesellschaft Kohle mbH), U.S. Patent 2,831,877 (Apr. 22, 1958); British Patent 743,597 (Jan. 18, 1956).
1295. H. Koch and W. Haaf, *Justus Liebigs Ann. Chem.*, **618**:251 (1958)
1296. H. Koch and W. Haaf, *Justus Liebigs Ann. Chem.*, **638**:111 (1960).
1297. H. Koch and W. Huisken (to Studiengesellschaft mbH), U.S. Patent 2,876,241 (Mar. 3, 1959); British Patent 798,065 (July 16, 1958); German Patents 972,291 and 973,077.
1298. H. Koch, W. Huisken, K. E. Möller, and K. Lohbeck (to Studiengesellschaft Kohle mbH), German Patent 972,315 (July 2, 1959).
1299. H. Koch and K. E. Moeller, German Patent 1,064,941 (Sept. 10, 1959); British Patent 880,788.
1300. H. Koch and K. E. Moeller (to Studiengesellschaft Kohle mbH), German Patent 1,095,802 (Nov. 13, 1958), addition to German Patent 973,077; British Patent 883,243.
1301. H. Koch and K. E. Moeller (to Studiengesellschaft Kohle mbH), U.S. Patent 3,061,621 (Oct. 30, 1962); German Patent 1,148,990.
1302. W. Haaf, *Brennst.-Chem.*, **45**:209 (1964).
1303. K. E. Möller, *Angew. Chem. Int. Ed. Engl.*, **2**:719 (1963).
1304. K. E. Möller, *Brennst.-Chem.*, **45**:129 (1964).
1305. D. Osteroth, *Seifen, Öle, Fette, Wachse*, **95**:741 (1969).
1306. S. D. Sumerford, H. G. Ellert, and R. C. Lohman (to Esso Research and Engineering Co.), French Patent 1,377,834 (Nov. 6, 1964); German Patent 1,202,265; *Chem. Abstr.*, **62**:10341 (1965).
1307. G. Rohlffs and S. Pawlenko (to Schering A.G.), U.S. Patent 3,099,687 (July 30, 1963); British Patent 908,497 (Oct. 17, 1962).
1308. M. J. Waale and J. M. Vos (to Shell Oil Co.), U.S. Patent 3,059,004 (Oct. 16, 1962).
1309. B. S. Friedman and S. M. Cotton (to Sinclair Refining Co.), U.S. Patent 2,975,199 (March 14, 1961).
1310. E. J. Miller, Jr., A. Mais, and D. Say (to Armour & Co.), U.S. Patent 3,481,977 (Dec. 2, 1969); French Patent 1,477,301 (Apr. 14, 1967).
1311. A. John, Jr. (to Hercules Powder Co.), U.S. Patent 3,036,124 (May 22, 1962).
1312. J. Devine and J. F. Davies (to Unilever Limited), British Patent 998,974 (July 21, 1965); French Patent 1,355,077 (Feb. 3, 1964).
1313. S. Pawlenko, *Chem.-Ing.-Tech. Z.*, **40**:52 (1968).
1314. B. S. Friedman and S. M. Cotton (to Sinclair Refining Co.), U.S. Patent 3,005,846 (Oct. 24, 1961).
1315. Schering A.G., Belgium Patent 619,229 (Dec. 21, 1962); French Patent 1,325,461.
1316. W. Haaf and H. Koch, *Justus Liebigs Ann. Chem.*, **638**:122 (1960).
1317. W. Haaf, *Chem. Ber.*, **99**:1149 (1966).

1318. E. T. Roe and D. Swern (to Secretary of Agriculture), U.S. Patent 3,169,140 (Feb. 9, 1965).

1319. E. T. Roe and D. Swern (to Secretary of Agriculture), U.S. Patent 3,170,939 (Feb. 23, 1965).

1320. E. T. Roe and D. Swern (to Secretary of Agriculture), U.S. Patent 3,270,035 (Aug. 30, 1966).

1321. Albright & Wilson Limited, British Patent 960,011 (June 10, 1964).

1322. M. Matsubara, M. Sasaki, and H. Ohtsuka, *Kogyo Kagaku Zasshi*, **71**:1179 (1968); *Chem. Abstr.*, **70**:Abstr. No. 28344j (1969).

1323. A. Garriloff and F. Dusart (to N. V. Bougies de la Cour et de Roubaix Oedenkoven), Belgium Patent 650,876 (Nov. 13, 1964); Netherlands Appl. 6409056 (Feb. 8, 1965); *Chem. Abstr.*, **63**:499 (1965).

1324. O. Roelen (to Alien Property Custodian), U.S. Patent 2,327,066 (Aug. 17, 1943).

1325. A. Landgraf and O. Roelen (to Alien Property Custodian), U.S. Patent 2,415,102 (Feb. 4, 1947).

1326. W. F. Gresham, R. E. Brooks, and W. M. Bruner (to E. I. du Pont de Nemours & Co.), U.S. Patent 2,437,600 (March 9, 1948); British Patent 614,010 (Dec. 8, 1948).

1327. G. Natta and E. Beati, *Chim. Ind. (Milan)*, **27**:84 (1945).

1328. G. Natta and E. Beati, British Patent 646,424 (Nov. 22, 1950).

1329. C. H. McKeever and G. H. Agnew (to Rohm and Haas Co.), U.S. Patent 2,533,276 (Dec. 12, 1950).

1330. C. H. McKeever (to Rohm and Haas Co.), U.S. Patent 2,599,468 (June 3, 1952).

1331. S. Kodama, I. Taniguchi, S. Yuasa, A. Watanabe, and I. Yoshida, *J. Chem. Soc. Jap. Ind. Chem. Sect.*, **57**:395 (1954); *Chem. Abstr.*, **49**:5005 (1955).

1332. J. C. LoCicero and C. L. Levesque (to Rohm and Haas Co.), U.S. Patent 2,686,200 (Aug. 10, 1954).

1333. B. Blaser and W. Stein (to Henkel & Cie, GmbH), German Patent 965,697 (June 13, 1957).

1334. R. M. Alm and J. W. Shepard (to Standard Oil Co.), U.S. Patent 2,891,084 (June 16, 1959).

1335. K. Büchner, O. Roelen, J. Meis, and H. Langwald (to Ruhrchemie Aktiengesellschaft), U.S. Patent 3,043,871 (July 10, 1962); British Patent 864,918 (Apr. 12, 1961); German Patent 1,109,661 (June 29, 1961).

1336. W. D. Niederhauser (to Rohm and Haas Co.), U.S. Patent 3,054,813 (Sept. 18, 1962).

1337. E. R. Rogier (to General Mills, Inc.), U.S. Patent 3,062,773 (Nov. 6, 1962); British Patent 904,543 (Aug. 29, 1962); German Patent 1,445,380.

1338. E. J. DeWitt and W. T. Murphy (to The B. F. Goodrich Company), U.S. Patent 3,243,414 (Mar. 29, 1966).

1339. R. Laï, M. Naudet, and E. Ucciani, *Rev. Fr. Corps Gras*, **13**:737 (1966).

1340. R. Laï, M. Naudet, and E. Ucciani, *Rev. Fr. Corps Gras*, **15**:15 (1968).

1341. R. Laï, *Rev. Fr. Corps Gras*, **17**:455 (1970).

1342. E. N. Frankel, A. Metlin, W. K. Rohwedder, and I. Wender, *J. Amer. Oil Chem. Soc.*, **46**:133 (1969).

1343. E. N. Frankel and F. L. Thomas, *J. Amer. Oil Chem. Soc.*, **49**:10 (1972).

1344. R. Laï, E. Ucciani, and M. Naudet, *Bull. Soc. Chim. Fr.*, **1969**:793.

1345. E. Ucciani, A. Bonfand, R. Laï, and M. Naudet, *Bull. Soc. Chim. Fr.*, **1969**:2826.

1346. A. W. Schwab, E. N. Frankel, E. J. Dufek, and J. C. Cowan, *J. Amer. Oil Chem. Soc.*, **49**:75 (1972).

1347. W. Reppe et al., *Justus Liebigs Ann. Chem.* **582**:1 (1953).

1348. G. Natta, P. Pino, and E. Mantica, *Chim. Ind.*, **32**:201 (1950).

1349. R. Ercoli (to Societa Generale per l'Industria Mineraria e Chimica), U.S. Patent 2,911,422 (Nov. 3, 1959).
1350. K. Bittler, N. v. Kutepow, D. Neubauer, and H. Reis, *Angew. Chem. Int. Ed. Engl.*, **7**:329 (1968).
1351. N. v. Kutepow, K. Bittler, and D. Neubauer (to Badische Anilin- & Soda-Fabrik A.G.), U.S. Patent 3,437,676 (April 8, 1969); Netherlands Application 6409121 (Feb. 10, 1965).
1352. W. Reppe and H. Kröper (to Badische Anilin- & Soda-Fabrik A.G.), German Patent 861,243 (Dec. 29, 1952).
1353. W. Reppe and H. Kröper, *Chem. Ber.*, **582**:38 (1953).
1354. W. Reppe, N. v. Kutepow and H. Detzer (to Badische Anilin- & Soda-Fabrik A.G.), German Patent 1,006,849 (April. 15, 1957).
1355. D. R. Levering, *J. Org. Chem.*, **24**:1833 (1959).
1356. B. F. Crowe and R. Y. Helsler (to Texaco Inc.), U.S. Patent 2,958,695 (Nov. 1, 1960).
1357. D. Elad and J. Rokach, *J. Org. Chem.*, **30**:3361 (1965).
1358. E. M. Beavers (to Rohm and Haas Co.), U.S. Patent 2,812,338 (Nov. 5, 1957).
1359. E. Roe, D. A. Konen and D. Swern, *J. Amer. Oil Chem. Soc.*, **42**:457 (1965).
1360. W. L. Senn, Jr. and L. A. Pine, *Anal. Chim. Acta*, **31**:441 (1964).
1361. K. Dachs, R. Schubert and H. Wilhelm (to Badische Anilin- & Soda-Fabrik A.G.), U.S. Patent 2,957,783 (Oct. 25, 1960).
1362. H. Wilhelm, E. W. Hann, K. Dachs, and L. Wuertele (to Badische Anilin- & Soda-Fabrik A.G.), Ger. Patent 1,242,858 (June 22, 1967).
1363. F. Bayerlin, K. Dachs, J. W. Hartmann, O. Lissner, F. Steden, and H. Wilhelm (to Badische Anilin- & Soda-Fabrik A.G.), Belgium Patent 629,013 (Aug. 28, 1963); *Chem. Abstr.*, **61**:10837 (1964).
1364. F. Steden (to Badische Anilin- & Soda-Fabrik A.G.), German Patent 1,212,299 (Mar. 10, 1966).
1365. Badische Anilin- & Soda-Fabrik A.G. Netherlands Application 6514333 (May 9, 1966); *Chem. Abstr.*, **65**:12375 (1966).
1366. G. Illing, H. Bittermann, and H. Wilhelm (to Badische Anilin- & Soda-Fabrik A.G.), German Patent 1,177,332 (Sept. 3, 1964).
1367. Schering A.G., Netherlands Application 6505071 (Oct. 29, 1965).
1368. K. Dachs, H. Fikentscher, H. Wilhelm, and H. Wolz (to Badische Anilin- & Soda-Fabrik A.G.), German Patent 1,028,328 (April 17, 1958).
1369. W. Reppe, N. v. Kutepow, H. Wilhelm, and K. Dachs (to Badische Anilin- & Soda-Fabrik A.G.), German Patent 1,050,053 (Feb. 5, 1959).
1370. E. Griebsch (to Schering A.G.), German Patent 1,210,870 (Feb. 17, 1966).
1371. W. Wagner, H. Oberfeuer, and W. Bruns (to Glasurit-Werke M. Winkelmann A.G.), German Offen. 1,918,235 (Jan. 14, 1971).
1372. H. Finkentscher, H. Willersinn, and H. Wilhelm (to Badische Anilin- & Soda-Fabrik A.G.), German Patent 1,031,962 (June 12, 1958).
1373. D. Helm and R. Janssen (to Schering A.G.), U.S. Patent 3,502,602 (March 24, 1970).
1374. A. J. Castro and L. F. Kinney (to Armour Industrial Chemical Co.), U.S. Patent 3,487,050 (Dec. 30, 1969).

Readings

Kürzinger, A., in "Ullmanns Encyklopädie der technischen Chemie," 3rd ed., Vol. 5, W. Foerst, Ed., Urban & Schwarzeberg, Münich, 1954, pp. 822–827.
Muir, W. M., in "Encylopedia of Chemical Technology," 2nd ed., Vol. 1, H. F. Mark, Chairman of the Editorial Board, Interscience, New York, 1963, pp. 240–254.

Oldham, J., in "Rodd's Chemistry of Carbon Compounds," 2nd ed., Vol. ID, S. Coffey, Ed., American Elsevier, New York, 1965.

Ralston, A. W., "Fatty Acids and Their Derivatives," Wiley, New York, 1948.

Schiller, A. M., in "Encyclopedia of Polymer Science and Technology," Vol. I., H. F. Mark, Chairman of the Editorial Board, Interscience, New York, 1964.

Sittig, M., "Dibasic Acids and Anhydrides," Noyes Development Corp., Park Ridge, N.J., 1966.

2. ALIPHATIC DIAMINES

PETER T. KAN, *BASF Wyandotte Corporation, Wyandotte, Michigan*

Contents

I. INTRODUCTION

Aliphatic diamines represent an important class of monomers for the preparation of condensation-type polymers. Historically, du Pont's Carothers and his co-workers initiated research on the synthesis of polyamides in 1928 and subsequently developed conditions for preparing high-molecular-weight polyamides from the condensation of aliphatic diamines with dibasic acids. This work has led to the commercial development of a wide variety of polyamide systems capable of forming useful filaments, fibers, molding plastics, and surface-coating materials. du Pont's commercial production of polyamides began in 1939. Since then tremendous growth has been realized— particularly in the past 20 years. Indeed, the world production of polyamide fibers reached 2.66 billion lb in 1966 in comparison to only 123 million lb in 1950. The growth of the polyamide resins continues at a rapid rate as new systems and new applications are continually being developed. The importance of aliphatic diamines in the synthetic polymer industry cannot be overstated inasmuch as the most significant commercial use for this class of materials in condensation polymers is in the manufacture of polyamides. Aliphatic diamines also serve as raw materials for the production of polyurethane and polyurea resins. Of lesser commercial importance are the polysulfonamides and polyphosphonamides derived from aliphatic diamines.

This chapter is principally concerned with the synthesis of a number of specific aliphatic diamines of commercial utility, present or potential. These include the straight-chain as well as cyclic derivatives. Major emphasis is placed on describing the experimental details of various available synthetic methods for these monomers. The commercial routes currently used by the manufacturers are first considered, but small-scale laboratory procedures are adopted when practical routes are unavailable. The physical properties, together with polymerization conditions for the diamines and the application of the resulting polymers, are also described. Other information to be discussed comprises of economic data, analytical procedures, toxicology, and storage of the aliphatic diamines.

II. COMMERCIAL DATA AND UTILITY

Ethylenediamine is commercially produced by the reaction of 1,2-dichloro-ethane and ammonia. This process results in a mixture of ethylene amines consisting of about 40–50% ethylenediamine, 25–30% diethylenetriamine, 12–20% triethylenetetramine, and 10–13% higher polyamines. Major producers are Union Carbide, Dow, and Jefferson, for an estimated total capacity of 215 million lb of ethylene amines in 1968 (1). Domestic consumption of ethylene amines in 1967 was about 75–80 million lb. Exports accounted for an additional 34–36 million lb, resulting in a total market of 110–115 million lb in 1967. With the estimated demand for 1968 at about 130 million lb, the current capacity exceeds demand significantly. Table 1

TABLE 1

Commercial Data For Aliphatic Diamines

Amine	Producer	Capacity, million lb/year	Total market, million lb/year	Price/lb, dollar
Ethylene amines[a]	Union Carbide	100	130	0.33
	Dow	65		
	Jefferson	50		
		215		
1,2-Propanediamine	Union Carbide	—	—	1.13
	Jefferson	—	—	
1,3-Propanediamine	Union Carbide	—	—	1.28
1,6-Hexanediamine[b]	du Pont	300	470	—
	Monsanto	145		
	Celanese	70		
	El Paso-Beaunit	25		
		540		
Piperazine[c]	Dow	0.5	3	1.43
	Jefferson	3		
	Union Carbide	0.5		
	Fleming Lab.	0.3		
		4.3		
2-Methylpiperazine	BASF Wyandotte	—	—	1.50
Xylylenediamines[d]	Showa Denko	2.8	—	—

[a] Including higher amines—diethylenetriamine, triethylenetetramine, and higher polyamines (1).

[b] Ref. 3.

[c] U.S. Tariff Commission, Synthetic Organic Chemicals, 1966, and Ref. 5.

[d] Ref. 7.

furnishes the available production and price information for the aliphatic diamines.

The major use for ethylenediamine is in the manufacture of carbamate insecticides; the higher ethylene amines are used for wet-strength paper, textile finishes, oil emulsification, and corrosion inhibitors. Two other outlets for ethylenediamine are polyamide resins and chelating agents. These markets have shown at least 10% or better yearly growth rates. Other applications for ethylenediamine include wet-strength resins, lubricating oil compositions, and textile crease-resistant resin formulations.

Both 1,2-propanediamine and 1,3-propanediamine are available from Union Carbide and the former is also produced by Jefferson. However, capacities and market information are not available. The disalicylaldehyde derivative of 1,2-propanediamine is used as a copper-sequestering agent in fuel oils and high octane gasoline (2). The chelate derived from the 5-methyl-salicylaldehyde derivative and cobalt can be used as an additive for gasoline and other hydrocarbons. 1,2-Propanediamine is also employed in rust and corrosion inhibitors and in textile softeners. 1,3-Propanediamine is reported to find utility as an intermediate for the manufacture of textile finishing agents, permanent press additive for textiles, ion-exchange resins, pharmaceuticals, rubber chemicals, insecticides, surfactants, dyestuffs, and corrosion inhibitors.

Of all the aliphatic amines manufactured in the United States, the production of 1,6-hexanediamine was estimated to account for more than half of the 695 million lb reported for 1965. The four producers are du Pont, Monsanto, Celanese, and El Paso-Beaunit. Several processes are available for the manufacture of 1,6-hexanediamine, and details are given in Section IV. The key intermediate is adiponitrile. du Pont is believed to make adiponitrile via butadiene, whereas Celanese, Monsanto, and El Paso-Beaunit all employ adipic acid as the raw material for adiponitrile. Monsanto also has an electrolytic process for making adiponitrile from acrylonitrile. The total estimated capacity for 1,6-hexanediamine in late 1967 was 540 million lb. Since 1,6-hexanediamine is produced captively almost exclusively for the manufacture of nylon 66, actual production data and price information are unavailable. An estimate that the 1,6-hexanediamine output in 1966 was at least 470 million lb is based on the production of 780 million lb of nylon 66 fiber and 60 million lb of nylon 66 plastics (3). Besides its major use as a nylon 66 intermediate, a small quantity of 1,6-hexanediamine is used to make nylon 6,10(1,6-hexanediamine and sebacic acid). The polymer derived from 1,6-hexanediamine and terephthalic acid is reported to show promise in laboratory tests for tire cord applications (4).

The leading producer of piperazine is Jefferson, with a capacity of 3 million lb/year. Other manufacturers of 0.5 million lb or less are Dow, Union

Carbide, and Fleming Laboratories; thus the total capacity in 1966 was about 4.3 million lb (5). Dow had announced plans for a multimillion pound per year piperazine plant in 1966 (6), but the proposal was subsequently deferred. About 93 % of the 3 million lb/year market of piperazine is used in animal deworming applications. There is no extensive use of piperazine in fibers and plastics. However, when the current price of $1.43/lb comes down, piperazine should show rapid growth and find applications in the fiber, plastic, textile, and drug industries.

2-Methylpiperazine, an alkyl-substituted piperazine derivative, is offered by BASF Wyandotte in commercial quantities at $1.50/lb. This cyclic diamine has potential applications as a urethane catalyst, a corrosion inhibitor, a specialized solvent, and an intermediate for polymers.

In 1965 Showa Denko K.K. of Japan developed the ammoxidation process, in which mixed xylenes (30 % para- and 70 % metaisomers) are converted to mixed phthalonitriles and finally to xylylenediamines by hydrogenation. Production was started in 1967, and by 1968 the Showa Denko capacity is 2.7 million lb/year (7). Xylylenediamine has shown promise in the area of polyamide fibers. High-molecular-weight polyamides have been obtained with dibasic acids. The polyamide with adipic acid possesses higher Young's modulus than that of conventional nylons, although its dyeability, moisture absorption, elongation, and strength properties are the same as those of conventional nylons (8). Xylylenediamine is readily converted to xylylene-diisocyanate by phosgenation to provide a nonyellowing polyurethane intermediate (9). Xylylenediamine also serves as epoxy curing agent and as rust inhibitor.

The other members of the aliphatic diamines included in this chapter have not yet reached the commercial development stage. Several amines, however, have shown potential as intermediates in fiber and coating applications. Bis(p-aminocyclohexyl)methane is reported to be the diamine component of du Pont's polyamide fiber "Qiana," announced in 1968 (10). The polyamide fiber derived from bis(p-aminocyclohexyl)methane and azelaic acid is described as silklike, and the resulting fabric has excellent wash-and-wear characteristics (11). However, the most likely diacid for "Qiana" is dodecane-dioic acid (12). Initially the fiber was offered in pilot plant quantities selling at $5–8/lb. Another potential application for bis(p-aminocyclohexyl)methane involves its conversion to 4,4'-methylene di(cyclohexylisocyanate). This alicyclic diisocyanate and methylcyclohexylene diisocyanate, another alicyclic diisocyanate derived from the phosgenation of methylcyclohexylenediamine, are intermediates for the preparation of polyurethane coatings having good gloss and color-retention properties (13). 4,4-Dimethyl-1,7-heptanediamine is used in reactions with pyromellitic acid esters to form polypyromellitimides with excellent injection-molding properties (14). These polypyromellitimides can be reinforced with glass to result in strong, moisture-resistant thermo-plastic resins (15).

The aliphatic diamines, being chemically related, all undergo typical polymerization reactions via the amino sites. The variations in the structures of the individual members impart different physical and chemical properties to the resulting polymers. Products with specific properties can be obtained by a proper choice of monomers and polymerization conditions. Although some of the aliphatic diamines described here have not yet reached commercial status, their polymerization conditions and potential application of the resulting polymers have been studied. The preparation and application of the various classes of polymers derived from aliphatic diamines are covered in a later section.

III. SURVEY OF CHEMISTRY

The representative aliphatic diamines useful in the formation of condensa-tion polymers include straight-chain α,ω-diamines, alicyclic diamines, piperazines, and the xylylenediamines. The survey of their general synthetic methods is divided and each is discussed separately, in view of the basic structural differences among these four classes of compounds. The specific synthetic routes to each monomer are extensively reviewed in Section IV.

A. α,ω-Straight-Chain Diamines

1. REDUCTION OF DINITRILES

$$NC(CH_2)_nCN + 2H_2 \rightarrow H_2NCH_2(CH_2)_nCH_2NH_2$$

The reduction of α,ω-straight-chain dinitriles yields readily the corre-sponding diamines. The reduction may be carried out chemically with metal hydrides such as lithium aluminum hydride (16) or with sodium in alcohol (17). Although high conversions are achieved in some cases, the handling of large quantities of the reactive reducing agents and the high cost of the metal hydrides render these processes of little commercial utility.

Dinitriles are also reduced catalytically to the diamines with hydrogen. The catalysts employed include nickel, cobalt, platinum, and palladium (18). The reduction is normally carried out at elevated temperatures under super-atmospheric conditions. The reduction product usually gives rise to a mixture of primary and secondary amines. The unwanted secondary amine is formed either by the coupling of the primary amine and the loss of ammonia, or by addition of the primary amine to the intermediate imine followed by deamination.

$$2RNH_2 \rightarrow R_2NH + NH_3$$

or

$$RCH_2NH_2 + RCH{=}NH \longrightarrow \underset{\underset{NH_2}{|}}{RCHNHCH_2R} \xrightarrow{H_2} RCH_2NHCH_2R + NH_3$$

The extent of this side reaction is somewhat dependent on the chain length of the dinitriles, that is, increasing chain length reduces secondary amine formation. The lower homologs of the aliphatic dinitriles yield cyclic secondary amines as by-products. For example, in the catalytic hydrogenation of adiponitrile to 1,6-hexanediamine, hexamethyleneimine is obtained as a by-product. Equal quantities of piperidine and 1,5-pentanediamine are produced from hydrogenation of glutaronitrile. Similarly, succinonitrile is converted to pyrrolidine and 1,4-butanediamine, with the former being the major product.

$$NC(CH_2)_4CN \longrightarrow H_2N(CH_2)_6NH_2 + CH_2(CH_2)_4CH_2NH$$

$$NC(CH_2)_3CN \longrightarrow H_2N(CH_2)_5NH_2 + CH_2(CH_2)_3CH_2NH$$

$$NC(CH_2)_2CN \longrightarrow H_2N(CH_2)_4NH_2 + CH_2(CH_2)_2CH_2NH$$

The secondary amine formation is also dependent on reaction temperature, inasmuch as the elimination of ammonia is favored at temperatures above 150°C. When nickel or cobalt catalysts are used, this side reaction may be suppressed by the addition of ammonia to the system. If the hydrogenation is catalyzed by platinum, the reaction can be performed in acetic anhydride, which acetylates the primary amine and thus avoids its addition to the intermediate imine (19). However, a hydrolysis step after the hydrogenation is needed to recover the free amine.

The catalytic hydrogenation of dinitriles with nickel or cobalt is a commercially important method for the manufacture of aliphatic diamines. Both slurried-batch and fixed bed-continuous processes have been developed. A typical example is the manufacture of 1,6-hexanediamine from the hydrogenation of adiponitrile. The detailed conditions of this process, together with the conditions for a number of other diamines, appear in Section IV.

2. AMMONOLYSIS OF DIHALIDES

Aliphatic diamines can be prepared from the reaction of the corresponding dihalide with ammonia. However, the reaction generally does not stop at the primary amine stage and a mixture of primary, secondary, and tertiary amines is usually obtained. The extent of the side reaction may be limited by using bulky alkyl halides as well as excess ammonia. Among the alkyl halides, alkyl iodides are most reactive and chloride the least reactive. For economic reasons, the alkyl chlorides are most often employed for commercial processes (20).

$$Cl(CH_2)_nCl + 2NH_3 \rightarrow H_2N(CH_2)_nNH_2 \cdot 2HCl$$

If secondary and tertiary amine by-products are readily separable from the desired primary amine, their formation is not necessarily a deterrent factor for a successful commercial process, since a demand for the higher amine by-products might exist to justify the use of such a process. This route is currently employed for the commercial production of ethylenediamine (via the ammonolysis of 1,2-dichloroethane). Uses have been developed for the polyamine by-products which are formed, constituting up to 50% of the total amine products in this process. The reaction is carried out at temperatures of above 100°C. Superatmospheric pressure is desirable to maintain a high ammonia concentration. The amine products are obtained as hydrochloride salts and a neutralization step with strong alkali is needed to liberate the free bases.

$$Cl(CH_2)_2Cl \xrightarrow[\text{2) NaOH}]{\text{1) NH}_3} H_2N(CH_2)_2NH_2 + H_2N(CH_2)_2NH(CH_2)_2NH_2 +$$
$$H_2N(CH_2)_2NH(CH_2)_2NH(CH_2)_2NH_2, \text{ etc.}$$

Catalysts can be used to facilitate the reaction of alkyl halides with ammonia. The conversion of 1,4-dibromobutane to 1,4-butanediamine is performed in the presence of an aluminum catalyst (21).

$$Br(CH_2)_4Br \xrightarrow[\text{Al cat.}]{\text{NH}_3} H_2N(CH_2)_4NH_2$$

Magnesium oxide is said to catalyze the reaction of ammonia with alkyl halides to a mixture of primary, secondary, and tertiary amines (22).

Reagents other than ammonia may be used with alkyl halides to form amines. The action of sodium amide in liquid ammonia on alkyl chlorides and bromides has been investigated by Shreve and co-workers (23,24). However, application of the sodium amide system to dihalides has not been reported. A somewhat related preparation of diamines involves the conversion of dibromoalkanes with magnesium to a Grignard intermediate followed by treatment with O-methylhydroxylamine. 1,5-Pentanediamine, 1,6-hexanediamine, and 1,10-decanediamine have been prepared in this manner from the corresponding dibromides uncontaminated by secondary and tertiary amines (25).

$$Br(CH_2)_nBr \xrightarrow[\text{2) CH}_3\text{ONH}_2]{\text{1) Mg}} H_2N(CH_2)_nNH_2 \qquad n = 5, 6, \text{ and } 10$$

3. Amination of Glycols and Aminoalcohols

The conversion of primary alcohols to amines requires high temperature and high pressure conditions. This transformation can be performed with ammonia over a dehydrating catalyst such as alumina, or in the presence of a hydrogenation catalyst such as nickel. Similar to the ammonolysis of the alkyl halides, a mixture of primary, secondary, and tertiary amines is obtained (26).

$$ROH \xrightarrow{\text{NH}_3} RNH_2 + R_2NH + R_3N$$

Because of the relative severity of reaction conditions and the generation of mixture of products, the amination of alcohol is seldom used in laboratory-scale synthesis of amines. However, this method is considered quite practical commercially, inasmuch as the production of specific amines can be controlled by varying the reactant ratio, the reaction conditions, and the catalysts. Although most of the work reported on the manufacture of amines by this amination process has been done on monohydric alcohols, extension of this method to the preparation of aliphatic diamines from glycols and amino-alcohols has been successful.

1,6-Hexanediamine is obtained from the amination of 1,6-hexanediol, using Raney nickel as catalyst (27). Conditions have been developed in which the side product hexamethyleneimine is recycled to suppress further formation of the secondary amine.

$$HO(CH_2)_6OH \xrightarrow{NH_3,\ Ni} H_2N(CH_2)_6NH_2 + \overline{CH_2(CH_2)_4CH_2NH}$$

Amination of glycols and aminoalcohols with ammonia and a ruthenium-on-carbon catalyst has been reported to furnish the corresponding diamines (28). Examples are 1,2-propanediol, 1,6-hexanediol, 1,7-heptanediol, 1,8-octanediol, 1,10-decanediol, and isopropanolamine. Compounds such as ethylene glycol, 1,4-butanediol, and 1,5-pentanediol, because of their tendency to yield cyclic derivatives, are considered unsuitable for this process. The amination is generally carried out under hydrogen pressure. The reaction conditions reported (225–232°C and 910 atm) are quite severe.

The amination of isopropanolamine can also be made with Raney nickel catalyst under considerably lower pressure conditions (1100 psig) to provide the desired 1,2-propanediamine in 65% yield (29). Since the starting iso-propanolamine is readily available from the reaction of ammonia and propylene oxide, this route is probably the process of choice for the commercial production of 1,2-propanediamine.

$$\underset{CH_2\!-\!\!-\!\!-CHCH_3}{\overset{O}{\triangle}} \xrightarrow{NH_3} \underset{H_2NCH_2\overset{\displaystyle |}{C}HCH_3}{\overset{OH}{}} \xrightarrow[Ni]{NH_3} \underset{H_2NCH_2\overset{\displaystyle |}{C}HCH_3}{\overset{NH_2}{}}$$

It should be pointed out that the amination of 1,2-propanediol under similar conditions does not furnish the 1,2-propanediamine. Instead, the partially aminated product, 2-amino-1-propanol, is obtained in low yield.

$$\underset{HOCH_2\overset{\displaystyle |}{C}HCH_3}{\overset{OH}{}} \xrightarrow[Ni]{NH_3} \underset{HOCH_2\overset{\displaystyle |}{C}HCH_3}{\overset{NH_2}{}}$$

When primary amine is used in place of ammonia for the amination of glycols, the corresponding secondary diamine is obtained. A French patent describes the preparation of N,N'-dimethyl-1,6-hexanediamine by the reaction of 1,6-hexanediol with methylamine in the presence of hydrogen and a cobalt oxide–cupric oxide catalyst (30).

$$HO(CH_2)_6OH + 2CH_3NH_2 \xrightarrow[\text{CoO–CuO}]{H_2} CH_3NH(CH_2)_6NHCH_3$$

4. HOFMANN DEGRADATION OF DIAMIDES

The reaction of amide with alkali hypobromite, prepared *in situ* from bromine and sodium hydroxide, leads to a primary amine containing one less carbon. The intermediate of this reaction is the corresponding isocyanate. This reaction has been reviewed in detail by Wallis and Lane (31).

$$\overset{\text{O}}{\overset{\|}{\text{R}\text{C}}}NH_2 + NaOBr \longrightarrow [RNCO] \xrightarrow{H_2O} RNH_2$$

The application of the Hofmann reaction to aliphatic diamides for the preparation of two carbon less diamines has been reported. The yield of diamines are usually in the 50–60% range. An exception is succinamide, which is not converted to ethylenediamine under the Hofmann conditions. The product isolated is dihydrouracil (32). Glutaramide and adipamide, on the other hand, furnish the corresponding 1,3-propanediamine and 1,4-

butanediamine in 54 (33) and 60% (34) yield, respectively. Higher aliphatic diamines, such as 1,6-hexanediamine and 1,7-heptanediamine, have also been prepared from the corresponding diamides (31). A National Distiller's patent describes the conversion of sebacic acid via the diamide to 1,8-octanediamine in two steps in 50% yield (35).

$$\overset{\text{O}\quad\text{O}}{\overset{\|\quad\|}{H_2N\text{C}(CH_2)_n\text{C}}}NH_2 \xrightarrow[\text{NaOH}]{\text{NaOBr}} H_2N(CH_2)_nNH_2 \qquad n = 3, 4, 6, 7, \text{ and } 8$$

The Hofmann reaction is considered an excellent general laboratory method for preparing amines. Its commercial utility for the manufacture of aliphatic diamines has not been fully demonstrated, however, probably because of the low yield and multistep operations required for the process.

5. SCHMIDT REACTION OF DICARBOXYLIC ACIDS AND HYDRAZOIC ACID

The Schmidt reaction represents a highly efficient laboratory procedure for the conversion of acids to amines via the intermediate azide. Concentrated sulfuric acid is most effective as catalyst. The reaction is normally carried out by contacting hydrazoic acid, generated *in situ* by the action of sulfuric acid on sodium azide, with a carboxylic acid. The formation of amine is evidenced by the evolution of carbon dioxide and nitrogen. Numerous modifications of the experimental conditions have been developed, and a review has been written (36).

$$RCOOH + NaN_3 \xrightarrow{H_2SO_4} RNH_2 + CO_2 + N_2$$

Aliphatic diamines are readily prepared by the Schmidt reaction from the corresponding dicarboxylic acids having two more carbon atoms. Poor to fair yields of diamines are obtained when succinic acid (8% ethylenediamine) (37) and glutaric acid (65% 1,3-propanediamine as the dipicrate) (33) are used. When diacids having six or more carbon atoms are employed, the yield of diamines is usually in the 80–90% range (36,38).

$$HOOC(CH_2)_nCOOH \xrightarrow{HN_3} H_2N(CH_2)_nNH_2 \qquad n = 4, 8, \text{ and } 12$$

The Schmidt reaction also is applicable to aminoacids. 1,5-Pentanediamine is obtained in 70% yield from ε-aminocaproic acid (39).

$$H_2N(CH_2)_5COOH \xrightarrow{HN_3} H_2N(CH_2)_5NH_2$$

The Schmidt reaction is superior to the Hofmann reaction as a laboratory preparative method in that the yields of diamines are generally higher. Furthermore, the one-step reaction of dicarboxylic acid to diamine for the Schmidt avoids the preparation of the diamide as in the case of the Hofmann. However, the high cost and potential hazard (toxicity and explosiveness) involving the use of hydrazoic acid render the former reaction impractical for commercial manufacture of aliphatic diamines.

6.   GABRIEL SYNTHESIS

The reaction of potassium phthalimide with an alkyl halide followed by hydrolysis of the resulting *N*-alkyl-substituted phthalimide provides a ready route to the primary amines. Unlike the amination of alkyl halides with ammonia, this method yields only primary amines uncontaminated by secondary and tertiary amines. The synthetic application of the Gabriel reaction has been summarized by Gibson and Bradshaw (40).

Excellent yields of aliphatic diamines are obtained from the reaction of potassium phthalimide and straight-chain α,ω-dibromoalkanes. The yields of a number of diamines prepared by this method are presented in Table 2.

TABLE 2
Synthesis of Aliphatic Diamines by the Gabriel Method

Amine	Yield, %	Reference
Ethylenediamine	90	41
1,3-Propanediamine	96	41
1,4-Butanediamine	74	42
1,5-Pentanediamine	"Nearly quantitative"	43
1,6-Hexanediamine	86	44

The Gabriel reaction represents an important laboratory synthetic method for primary amines. Currently this method is not used commercially, both because of the many steps involved and in view of the relatively high cost.

7. MISCELLANEOUS METHODS

The catalytic hydrogenation of a lactim ether in the presence of ammonia and a cobalt catalyst is reported to be a good general method of preparing straight-chain aliphatic diamines (45). The starting lactim ether is synthesized by the reaction of a dialkyl sulfate and a lactam.

$$\underset{\underset{\displaystyle NH}{\underline{\qquad\qquad}}}{CH_2(CH_2)_{n-1}C} = O + R_2SO_4 \longrightarrow \underset{\underset{\displaystyle N}{\underline{\qquad\qquad}}}{CH_2(CH_2)_{n-1}C} - OR \xrightarrow[NH_3]{H_2} H_2N(CH_2)_{n+1}NH_2$$

The final hydrogenation step is carried out under hydrogen pressure in a sealed pressure vessel, and the yield of diamine is in the 75–87% range. Diamines thus prepared include 1,4-butanediamine, 1,6-hexanediamine, 1,8-octanediamine, and 1,12-dodecanediamine.

Other methods of commercial significance developed for aliphatic diamines usually involve starting materials and reaction conditions aimed at the synthesis of a particular diamine rather than a general procedure applicable to the entire class of compounds. Some specific examples are the reductive amination of acrylonitrile or acrolein to yield 1,3-propanediamine, and similar treatment of hydroxyacetonitrile to yield ethylenediamine. The details of these examples, together with other specific routes, are discussed in Section IV.

B. Alicyclic Diamines

1. REDUCTION OF AROMATIC DIAMINES

The ease of catalytic hydrogenation of the aromatic nucleus depends on the type of substituents present on the aromatic ring. The hydrogenation of amino-substituted benzene systems is generally more difficult than that of the unsubstituted ring (46). Using catalysts such as nickel, rhodium, or ruthenium, hydrogenation of aromatic amines to alicyclic amines is carried out under high temperature and pressure conditions. Both slurried-batch and fixed bed-continuous processes have been developed commercially for the manufacture of alicyclic amines. Owing to the ready availability of the starting aromatic amines and the one-step hydrogenation reaction, this method is considered the most practical commercially.

The compound bis(*p*-aminocyclohexyl)methane is readily obtained in good yield from the hydrogenation of 4,4′-methylenedianiline in the presence of either an alkali-promoted cobalt oxide (47) or ruthenium dioxide (48).

$$H_2N-\langle\bigcirc\rangle-CH_2-\langle\bigcirc\rangle-NH_2 \xrightarrow{H_2} H_2N-\langle\bigcirc\rangle-CH_2-\langle\bigcirc\rangle-NH_2$$

The hydrogenation of phenylenediamine derivatives is performed with ruthenium catalysts such as ruthenium on carbon or ruthenium dioxide to provide the corresponding diaminocyclohexanes in excellent yields. Examples include 1,3-diaminocyclohexane (49), 1,4-diaminocyclohexane (49), and

R = H or CH$_3$

methyldiaminocyclohexanes (50).

The precursors to the starting phenylenediamines, such as the dinitrobenzenes or the nitroanilines, can be directly hydrogenated in one step to the corresponding diaminocyclohexanes. In fact, the preferred methods of preparing 1,3-diaminocyclohexane and 1,4-diaminocyclohexane are the hydrogenation of *m*-dinitrobenzene and *p*-nitroaniline, respectively (51).

2. CURTIUS REARRANGEMENT OF DIAZIDES

The Curtius reaction involves the rearrangement of an acid azide to an isocyanate, followed by hydrolysis to the amine. A review article has been presented for this reaction (52). There are a number of steps in the actual preparation, starting with the carboxylic acid derivative. The acid azide is prepared either from the acyl chloride and sodium azide or from the ester with hydrazine followed by diazotization with nitrous acid. The rearrangement of the azide to the isocyanate may be carried out in ethanol to yield the

$$\underset{RCN_3}{\overset{O}{\parallel}} \longrightarrow [RNCO] \xrightarrow{C_2H_5OH} \underset{RN\overset{O}{H}COC_2H_5}{} \xrightarrow{H_2O} RNH_2$$

corresponding urethane, which is then hydrolyzed to the amine.

The preparation of both 1,4-diaminocyclohexane (52) and 1,3-diaminocyclohexane (53) from the cyclohexane dicarboxylic acid esters by the Curtius method has been reported. However, because of the multistep operations involved and high cost of raw materials, the utility of this method remains in the laboratory.

C. Piperazines

Piperazine and its C-alkyl-substituted homologs are cyclic secondary diamines. Because of their structural difference as compared with the aliphatic and alicyclic diamines, the previously described general synthetic methods are not applicable to the preparation of piperazines. A general commercial route to the piperazines is the cycloamination of the appropriately substituted alkanolamines in the presence of a catalyst. Piperazine is manufactured by the reaction of monoethanolamine with hydrogen and ammonia. In this process, the reactants are passed through a fixed bed charged with a nickel-copper-chromia catalyst (54). Similarly, 2-methyl-

$$2H_2NCH_2CH_2OH \xrightarrow[NH_3]{H_2}$$

piperazine (55), 2,5-dimethylpiperazine (56), and 2,6-dimethylpiperazine (57) are prepared from the cyclization of N-(2-hydroxypropyl)ethylenediamine,

$$H_2NCH_2CH_2NHCH_2\overset{\overset{\displaystyle CH_3}{|}}{C}HOH \longrightarrow$$

$$2NH_2CH_2\overset{\overset{\displaystyle CH_3}{|}}{C}HOH \longrightarrow$$

$$HN(CH_2\overset{\overset{\displaystyle CH_3}{|}}{C}HOH)_2 \longrightarrow$$

isopropanolamine, and diisopropanolamine, respectively. Both slurried and fixed-bed conditions with nickel catalysts are used.

Other starting materials serving for the cycloamination reaction to prepare piperazines have been cited, and these are summarized in Section IV. There are also a number of specific routes to the individual piperazine derivatives. The dimerization of ethyleneimine (58) and the alkali hydrolysis of N,N'-di-(p-nitrosophenyl)piperazine, obtained from the condensation of 1,2-dibromo-ethane and p-nitrosoaniline (59), have been reported to furnish piperazine.

$$2 \overset{\overset{\displaystyle H}{\displaystyle N}}{CH_2 - CH_2} \longrightarrow \text{(piperazine)}$$

$$2 ON\!\!-\!\!\langle \rangle\!\!-\!\!NH_2 + 2 Br(CH_2)_2 Br \longrightarrow \overset{C_6H_4NO}{\underset{C_6H_4NO}{\text{(piperazine)}}} \xrightarrow{KOH} \text{(piperazine)}$$

Piperazine and 2,5-dialkyl-substituted piperazines may also be prepared by the reduction of the corresponding 2,5-diketopiperazines. The reaction sequence involves the condensation of two molecules of α-amino acids or their esters. The resulting 2,5-diketopiperazine is then readily converted by catalytic hydrogenation, electrolytic reduction, or sodium and alcohol to the corresponding piperazine. This route, however, is of academic interest only. A review on the preparation and reduction of the 2,5-diketopiperazines has been presented by Pratt (60).

$$2 H_2 N\overset{\overset{\displaystyle R}{|}}{CH}COOH \longrightarrow \text{(diketopiperazine)} \xrightarrow{H_2} \text{(piperazine)}$$

The preparation of 2,6-dimethylpiperazine from the corresponding 3,5-diketopiperazine derivative has been reported by Cignarella (61).

D. Xylylenediamines

Meta- and paraxylylenediamines are sometimes classified as aromatic diamines because the benzene nucleus is present. An early laboratory preparation of m-xylylenediamine comprises the reaction of m-xylenedibrom-ide and o-phthalimide (Gabriel synthesis) in 38% yield (62). The commercially practical process now employed for the manufacture of xylylenediamines is the catalytic hydrogenation of the corresponding phthalonitriles. The raw material is either terephthalic and isophthalic acids (63) or *p*- and *m*-xylenes (8).

The reduction of nitriles to amine as a general synthetic method has been discussed previously. The process details for the manufacture of xylylenediamine appear in Section IV-Q.

IV.   SYNTHESIS

In this section the specific syntheses of individual monomers are examined. A brief review is first made to summarize the various routes to each monomer. Since not all the monomers are currently manufactured in commercial quantities, the preferred routes to be discussed in detail include both commercially practical ones and small-scale laboratory procedures. The starting compounds in most cases are commercially available raw materials; otherwise, references are given indicating their sources or preparation.

A.   Ethylenediamine

The starting materials for the manufacture of ethylenediamine by various routes are 1,2-dichloroethane, ethylene glycol, formaldehyde, hydrogen cyanide, and ethanolamine. Ammonolysis of 1,2-dichloroethane is the most commercially practical route, and a number of processes based on this reaction have been reported in the patent literature (64–69). In this process ethylenediamine is formed as the hydrochloride salt, and a neutralization step is required to generate the free amine. Useful by-products, such as diethylenetriamine and triethylenetetramine, also occur.

$$ClCH_2CH_2Cl + NH_3 \rightarrow H_2NCH_2CH_2NH_2$$

The amination of ethylene glycol to furnish ethylenediamine is performed in the presence of hydrogen and a catalyst such as nickel–copper on alumina (70). The by-product is piperazine.

$$HOCH_2CH_2OH + NH_3 \rightarrow H_2NCH_2CH_2NH_2$$

A third method is the reductive amination of hydroxyacetonitrile (the cyanohydrin of formaldehyde) with ammonia, hydrogen, and a cobalt oxide catalyst (71). Ethylenediamine may also be produced directly by reacting formaldehyde and hydrogen cyanide with ammonia and hydrogen in the presence of a cobalt catalyst without isolation of the intermediate hydroxy-acetonitrile (72).

A further variation of this method is the hydrogenation of aminoaceto-nitrile, which is obtained from the amination of hydroxyacetonitrile (73–75). The foregoing routes starting from formaldehyde and hydrogen cyanide provide excellent yield of ethylenediamine. However, these routes are not yet economically competitive with the less efficient 1,2-dichloroethane route, since useful by-products are coproduced from the latter route. The instability of the intermediate acetonitrile derivatives also render the hydrogenation process less attractive.

$$HCHO + HCN \longrightarrow HOCH_2CN \xrightarrow{HN_3} H_2NCH_2CN \xrightarrow{H_2} H_2NCH_2CH_2NH_2$$

Ethylenediamine is also prepared by the amination of ethanolamine using Raney nickel as catalyst (76). The by-products are piperazine and diethylene-triamine. The conditions of this process may be varied to produce various ratios of amine products.

$$HOCH_2CH_2NH_2 + NH_3 \rightarrow H_2NCH_2CH_2NH_2$$

1. The 1,2-Dichloroethane Route

The reaction of 1,2-dichloroethane with ammonia is normally performed under aqueous conditions under pressure. Excess ammonia is required to minimize the formation of by-products. A process was reported by Nicolaisen (64) in 1957 in which ethylenediamine was produced by the reaction of 1,2-dichloroethane and aqueous ammonia in a fractionating tower. In this process the upper portion of the tower is maintained at 30–40°C, whereas the lower portion of the tower is kept at 100–200°C. Reaction pressures of from atmospheric to 300 psig may be used. 1,2-Dichloroethane, water, and ammonia are charged to the tower. The formation of ethylenediamine occurs in the middle portion of the tower. The excess ammonia is condensed overhead and can be returned totally to the reaction zone as reflux or can be recycled to the lower portions of the tower as well. The ethylenediamine hydrochloride in the lower portion of the tower is neutralized by passing a caustic solution into the tower at a point sufficiently below that at which appreciable quantities of 1,2-dichloroethane are present. Sufficient heat should be present at the lower portion of the tower to force the dissolved ammonia and 1,2-dichloro-ethane back to the middle portion of the tower for further reaction. When this process is operated continuously, the ratio of makeup ammonia to the

1,2-dichloroethane may be essentially theoretical, and equivalent amounts of ethylenediamine are removed from the lower portion of the tower.

Specifically, a tower of bubble cap or packed type, 35 ft high and 0.25 ft² in cross-sectional area may be used. 1,2-Dichloroethane is fed into the tower at a point just below the top plate at a rate of 100 lb/hr. Fresh anhydrous ammonia is charged at a rate of 35 lb/hr to the tower at a point one-third of the distance above the bottom of the tower. Caustic (40% solution) is charged at one-fifth of the height of the column from the bottom at 200 lb/hr and water is charged at about the level of the second plate from the top at 315 lb/hr. Heat is supplied by the reaction. The liquid and vapor in equilibrium on the top plate are essentially anhydrous ammonia. Under pressure of 225 psig ammonia gas is removed from the top of the tower at about 38°C. It is then taken overhead at a rate of about 568 lb/hr to a cooler, which reduces the temperature to about 35°C. About half the ammonia is returned to the top plate of the tower as reflux and about half is returned through a surge tank to the reaction zone (midpoint) of the tower. The ratio of ammonia to 1,2-dichloroethane is about 33:1. The liquid level on the bottom of the tower is just sufficient to maintain liquid feed to a reboiler, which returns vapors to the bottom of the tower. The bottoms, free of ammonia and 1,2-dichloroethane, are passed in part into the reboiler. The bottoms leave the tower at a temperature of about 140°C and comprise an aqueous solution of about 9% ethylenediamine and 18% sodium chloride. A 65% yield is reported for this process (64,77).

A patent issued to Jefferson in 1968 describes a similar process for the manufacture of ethylenediamine (69). Details are provided for the isolation of anhydrous ethylenediamine from an aqueous solution of ethylenediamine and sodium chloride. The purification step involves the distillation of the crude reaction effluent to remove a water-amine azeotrope. The azeotrope, after treatment with a 70% caustic solution, is distilled to furnish anhydrous ethylenediamine. To carry out the purification step, the reactor effluent containing amine products and sodium chloride is distilled at 180°C. The overhead consists of an azeotrope of amines and water. Some water and all the ammonia are removed from this azeotrope by further distillation. The bottoms from this distillation contain less than 30% water, the balance being amines. The by-product amines are removed as bottoms by a third distillation. The aqueous distillate, which is substantially free of by-product, is contacted countercurrently with a 70% caustic solution at 70–100°C in an extractor to reduce the water content of the aqueous ethylenediamine solution to less than 10%. This amine solution is then distilled under atmospheric pressure or under vacuum to isolate ethylenediamine containing only trace quantities of water. An example for the countercurrent extraction of water is described in which 100 parts of a 77 wt.% aqueous solution of ethylenediamine is brought into contact with 90 parts of a 73 wt. % caustic to result in 85 parts of extract fraction containing 92% ethylenediamine and about 105 parts of a raffinate fraction comprising about 62 wt. % solution of caustic.

2. THE ETHYLENE GLYCOL ROUTE

The amination of ethylene glycol with ammonia in the presence of hydrogen and a hydrogenation catalyst of nickel–copper on alumina to prepare ethylenediamine was reported by Fitz-William in 1964 (70). This liquid-phase, fixed-bed process consists of passing an aqueous solution of ethylene glycol, ammonia, and hydrogen to a preheater. The resulting mixture is heated to reaction temperature of 220–270°C at 3000–6000 psig. The preheated mixture is then passed through a reactor containing a hydrogenation catalyst of about 5–18% by weight of nickel and about 5–18% by weight of copper supported on activated alumina. The effluent is cooled and hydrogen and ammonia are vented and recycled. After the stripping of additional quantities of dissolved ammonia, the aqueous solution is distilled to isolate an aqueous ethylenediamine fraction, a piperazine fraction, an unreacted ethylene glycol fraction which is recycled, and a high boiling fraction which is discarded. The aqueous ethylenediamine fraction may be dried by azeotropic distillation with benzene. The following example illustrates this process in detail.

A quantity of 810 parts of a 50% aqueous solution of ethylene glycol, 7 parts of hydrogen, and 2975 parts of ammonia is charged to a preheater. The mole ratio of ethylene glycol to hydrogen to ammonia is 1:0.54:26.8. The preheater is set at 250°C under 4300 psig, and the preheated mixture is passed at a space velocity of 12.8 hr^{-1} through a reactor packed with 184 parts of a catalyst consisting of 12 wt. % copper and 16 wt. % nickel on activated alumina. The catalyst is prepared by melting a mixture of 475 parts of nickel nitrate and 396 parts of copper nitrate in a rotating gas-fired stainless steel drum; to this is added 408 parts of 10–20 mesh activated alumina. The resulting mixture is heated to a temperature of 400–800°C until the oxides of the metals are formed. The catalyst is screened to remove fines; and before the catalyst is used, the oxides are reduced to nickel and copper by heating with hydrogen at 400°C under atmospheric pressure. The reaction effluent, after a reaction period of 5 min, is cooled and passed through let-down valves to a storage tank where ammonia and hydrogen are vented. The liquid phase from the tank is stripped to remove ammonia. Distillation of the crude product mixture furnishes 494 parts of aqueous ethylenediamine, 15.6 parts of piperazine, and 250 parts of unreacted ethylene glycol. The aqueous ethylene-diamine fraction is dried by azeotropic distillation with benzene to yield 68.6 parts of relatively pure ethylenediamine. The yield is 46% based on the amount of ethylene glycol consumed.

3. The Formaldehyde–Hydrogen Cyanide Route

a. Via Formaldehyde and Hydrogen Cyanide

A one-step reaction of formaldehyde, hydrogen cyanide, ammonia, and hydrogen in the presence of a cobalt catalyst was reported by Nemec et al. in 1963 (72). This continuous process is carried out in a fixed bed under super-atmospheric conditions. Thus in a high-pressure tubular reactor (37.5 mm inner diameter) is charged 1 liter of a 60% cobalt on kieselguhr catalyst, 3 × 3 mm particle size, obtained from reduction of cobalt oxide with hydrogen at 120–400°C. To the reactor are fed hydrogen (60 g/hr) at 130°C and a liquid stream preheated at 110°C consisting of a mixture of ammonia (2100 g/hr) hydrogen cyanide (95 g/hr), and 37% formaldehyde (288 g/hr). The hydrogen stream contains about 75% recycled hydrogen, 25% fresh hydrogen, and some ammonia. The ammonia feed consists of about 97% recycled ammonia and 3% fresh ammonia. The reaction is carried out at 339 atm at 105–120°C in the reactor. The reactor effluent is passed into a stripper and cooling to 15°C, whereupon hydrogen is withdrawn and recycled to the hydrogen feed line. The liquid product is decompressed to 19.4 atm and distilled in such a manner that the vapor temperature is maintained at 50–55°C and the pot temperature at 210–220°C. The ammonia removed overhead is condensed and recycled. The bottoms containing the amine products are vented to atmospheric pressure. The crude product stream (430 g/hr) contains 40% ethylenediamine, 4% higher amine, and 56% water. The yield of ethylenediamine is 86% and the yield of total amine product is 95%, based on the hydrogen cyanide and formaldehyde charged.

b. Via Hydroxyacetonitrile

The conditions reported by Scholz et al. (71) for the conversion of hydroxy-acetonitrile to ethylenediamine are quite similar to those just described for the formaldehyde–hydrogen cyanide route. To carry out this process, 1.8 liters of a pilled, reduced cobalt oxide catalyst is charged into a tubular reactor 1.6 m long and 4.35 cm I.D. A 50% aqueous solution of technical grade hydroxyacetonitrile (240 g/hr), liquid ammonia (2 liters/hr), and hydrogen (1260 liters/hr) are passed through the reactor at 100°C under 315 atm of pressure. The reactor effluent is collected and decompressed. Hydrogen and ammonia may be recycled into the reaction. The liquid product is distilled at atmospheric pressure to provide 111.0 g of ethylene-diamine per hour (88% of theoretical yield). This is followed by 8 g/hr of higher amines, mainly diethylenetriamine and triethylenetetramine. An overall yield of amines of 94% is achieved by this process.

c. Via Aminoacetonitrile

The starting aminoacetonitrile is unstable on standing and therefore is freshly prepared either by the reaction of formaldehyde, hydrogen cyanide, and ammonia in water (78) or by ammination of an aqueous solution of hydroxyacetonitrile (75,79). Groggins describes a process for the preparation of glycine developed by the I. G. Farbenindustrie, in which an aqueous solution of aminoacetonitrile is prepared as an intermediate (80). In this process hydrogen cyanide is added to aqueous formaldehyde at 20–30°C and the solution is concentrated under vacuum to result in an 85% hydroxy-acetonitrile solution. A quantity of 20 kg of this solution is then added slowly to 40 liters of anhydrous ammonia in a water-cooled autoclave in 4 hr. The final pressure is 140–170 psig.

The du Pont process (75) of employing the crude aqueous solution of aminoacetonitrile to prepare ethylenediamine consists of first a flash distilla-tion to obtain a somewhat purified aqueous aminoacetonitrile, and this is followed by catalytic hydrogenation. In the purification step, the crude aqueous aminoacetonitrile containing ammonia is heated to 70°C to remove most of the ammonia, cooled to 5–10°C, filtered, and fed into the purification system. This system contains three separation stages. Each stage consists of a jacketed stainless tubular heater 6 ft × ½ in. I.D. followed by an 8 × 5 in. I.D. cyclone separator. The first and second stages are maintained at 40 mm of pressure and the third stage at 12–25 mm of pressure. The contact time in the heaters of each stage is less than 3 min. The filtrate is fed into the first separation stage. The vapor temperature is maintained at 75°C. The vapor from the first stage is condensed and fed into the second stage. The liquid from the first stage is fed to the third stage together with 0.08 lb of water per pound of the crude nitrile introduced into the purification system. The vapor from the third stage at 80°C is condensed and fed to the second stage, and the liquid fraction is discarded. The liquid from the second stage is recycled to the first stage. The vapor from the second stage at 72°C is condensed into a water-white aqueous solution of aminoacetonitrile. The purified aqueous solution thus obtained contains between 25 and 75% aminoacetonitrile. The preferred concentration for hydrogenation is 55–70% aminoacetonitrile. This solution is cooled to 5°C immediately, and within 30 min it is passed into the hydrogenation reactor for conversion to ethylenediamine.

The hydrogenation step is carried out in a vertically positioned, stainless steel cartridge type reactor of 6 ft × 2$\frac{1}{16}$ in. I.D. The reaction zone is charged with a reduced cobalt catalyst prepared by treating $\frac{3}{16}$-in. cobalt oxide pills with hydrogen at 250–450°C. A mixture of aqueous amino-acetonitrile containing 0.25 to 1.5 lb of water/lb of aminoacetonitrile, 8 lb of ammonia/lb of the nitrile, and 50–100 ft^3 (STP) of hydrogen is passed upward through the reactor. The space velocity for the nitrile is approximately 0.15. Hydrogenation is performed at 95–130°C under 300 atm of pressure. Operation is continued for 80 hr with a 92.5% conversion of aminoacetonitrile to ethylenediamine.

B. 1,2-Propanediamine

1,2-Propanediamine was prepared by Darzens in 1939 by the reaction of 1,2-dichloropropane and aqueous ammonia in high yield (81). However, the reaction was extremely slow (8 days at 78–80°C). Two other preparations involve the reductions of 1,2-dinitropropane with platinum oxide (82) and 1-nitro-2-aminopropane with Raney nickel (83); but these methods have no practical commercial application.

The method of choice for the manufacture of 1,2-propanediamine is the amination of isopropanolamine in the presence of ammonia and a hydrogenation catalyst (29). The isopropanolamine, prepared by the reaction of ammonia and propylene oxide (84), is commercially available.

$$\underset{\text{H}_2\text{NCH}_2\overset{\displaystyle|}{\text{C}}\text{HCH}_3}{\overset{\text{OH}}{}} \xrightarrow[\text{Ni}]{\text{NH}_3} \underset{\text{H}_2\text{NCH}_2\overset{\displaystyle|}{\text{C}}\text{HCH}_3}{\overset{\text{NH}_2}{}}$$

The amination reaction is carried out in an autoclave designed to operate at 2000 psig pressure. About equimolar quantities of ammonia and isopropanolamine are used. The usual hydrogenation catalysts such as nickel, cobalt, and copper chromite may be employed. Hydrogen may be used to provide the needed pressure and to some extent to reduce the color of the reaction mixture. The reaction is performed at 1000–1500 psig at about 180°C for several hours, and the 1,2-propanediamine is recovered by distillation of the crude product mixture.

In carrying out this reaction, the autoclave is charged with a mixture of 2640 g (35.2 moles) of isopropanolamine, 485 g (28.5 moles) of anhydrous ammonia, and 75 g of Raney nickel on a contained metal basis. The mixture is heated to 185°C for 8 hr with stirring and in the presence of hydrogen. The maximum pressure is 1100 psig. The autoclave is cooled and vented, and the crude product mixture is filtered and distilled to furnish a 65% yield of theory of 1,2-propanediamine, bp 120°C (81). The remainder comprises unreacted isopropanolamine and small amounts of higher boiling products.

C. 1,3-Propanediamine

The chief raw materials for the manufacture of 1,3-propanediamine are acrylonitrile and acrolein. Several processes consisting of the reaction of acrylonitrile with ammonia and hydrogen in the presence of a hydrogenation catalyst have been reported in the patent literature (85,86). These processes, which give good yield of product, are the methods of choice.

$$CH_2{=}CHCN + NH_3 \xrightarrow[\text{catalyst}]{H_2} H_2N(CH_2)_3NH_2$$

Other processes involve the reaction of acrolein with ammonia (87) or with urea, followed by hydrogenation of the resulting adduct (88). The

reaction of acrylonitrile with ethylene glycol and hydrogenation of the product ethylene bisoxydipropionitrile in the presence of ammonia has also been reported to yield 1,3-propanediamine (89).

$$CH_2{=}CHCN + HOCH_2CH_2OH \longrightarrow$$

$$NC(CH_2)_2O(CH_2)_2O(CH_2)_2CN \xrightarrow[\text{HN}_3]{H_2} H_2N(CH_2)_3NH_2$$

1. THE ACRYLONITRILE ROUTE

a. Continuous Process

Badische Anilin und Soda-Fabrik AG reports a continuous, fixed- or fluidized-bed process (86) for the condensation of acrylonitrile and ammonia. In addition to 1,3-propanediamine, this process gives di-(3-aminopropyl)-amine as a by-product, with a total of up to 97% yield of amine products. In this process, acrylonitrile and a large excess of liquid ammonia (about 12 parts ammonia per part of acrylonitrile) are intimately mixed and preheated to about 70°C. The large excess of ammonia is needed to minimize the formation of side products such as di-(2-cyanoethyl)amine and di-(3-aminopropyl)amine. The mixture and hydrogen are then fed through a tubular reactor containing a cobalt oxide catalyst prereduced by hydrogen in the reactor at 300–500°C under 50–350 atm. The amount of hydrogen used for the reaction is about 600–800 liters/kg of liquid feed mixture. The reaction is carried out at 90–120°C under 300 atm of hydrogen pressure. The liquid reactor effluent is then worked up by fractional distillation. The following example describes this process in detail.

A quantity of 8.2 parts of acrylonitrile per hour is fed together with 110 parts of liquid ammonia under 300 atm through a preheater in which the

reaction mixture is heated to 85°C. This mixture is then introduced continuously, with hydrogen, into the lower part of a vertical reactor containing
reduced cobalt oxide pellets. The reaction temperature is maintained between
90 and 110°C by superheating the hydrogen gas. The product mixture leaving
the top of the reactor is led through a heat exchanger into a separator, from
which the excess hydrogen is separated and recycled to the reactor. The
liquid product is distilled under pressure to recover excess ammonia in liquid
form. The remaining material is distilled at atmospheric pressure and then at
reduced pressure to isolate the amine products. The yield per hour is 7.15
parts of pure 1,3-propanediamine, bp 136–138°C and 3.1 parts of di-(3-
aminopropyl)amine, bp 95–97°C/3 mm. The total yield of amine mixture
is 93%, based on the acrylonitrile charged.

b.   Batch Process

A batch, slurried process using acrylonitrile and ammonia, but in the
presence of water and a high boiling solvent such as dipropylene glycol, is
described in a patent issued to the Standard Oil Company (85). In this process,
a mixture of the reactants and hydrogenation catalyst (e.g., Raney nickel,
acrylonitrile, ammonia, dipropylene glycol) and water is charged into an
autoclave equipped with stirrer. The molar ratios of reactants are 0.1 of
catalyst to acrylonitrile, 22–24 of ammonia to acrylonitrile, and 0.15–0.4
of water to ammonia. Sufficient dipropylene glycol is added to result in a
83–95% by volume of the glycol to total volume of glycol-water. Hydrogenation is performed at 25–50 atm in the temperature range of 75–155°C for 3 hr.
The reaction temperature is lowered quickly at the end of the reaction
period. The product is recovered by distillation, after stripping has been
performed to remove small amounts of water and ammonia. 1,3-Propane-
diamine is obtained in 78.5% conversion, together with up to 3% of the side
product n-propylamine.

2.   The Acrolein Route

The direct amination of acrolein with ammonia, hydrogen, and Raney
nickel in methanol-water has been reported to give about 34% yield of
1,3-propanediamine (87). The condensation of acrolein with urea, followed
by hydrogenation of the resulting 4-ureidotetrahydro-2-pyrimidone to
tetrahydropyrimidone and hydrolysis to 1,3-propanediamine provides a
somewhat better yield (52% of 1,3-propanediamine) (87). A detailed experimental procedure is given for this process.

To a solution prepared by dissolving 2 moles of urea and 6 g of concentrated nitric acid in 120 g of water, 1 mole of acrolein is added, drop by drop.
The temperature of the solution during addition is maintained at 65°C by
cooling. A 208-g portion (69%) of the resulting mixture is neutralized with

sodium hydroxide. To this solution is added 3 g of Raney nickel, and the mixture is hydrogenated at 150°C under 1000–1500 psig until 1 mole of hydrogen is absorbed per mole of acrolein charged. The catalyst is removed by filtration. After addition of 80 g of sodium hydroxide, the mixture is distilled at atmospheric pressure to hydrolyze the urea derivative. When solid appears in the distillation pot, steam is passed through the mixture to aid in distillation. All the distillates are combined, acidified, and evaporated to dryness. Treatment of the residue with 40% sodium hydroxide, followed by extraction with ether and distillation of the ether extracts, yields 18.3 g of 1,3-propanediamine, bp 136–138°C. Extraction of the steam distillation residue with ethanol and distillation of the extracts provides an additional 8.1 g of the diamine. The overall yield is 52%, based on the amount of acrolein charged.

D.   1,4-Butanediamine

The methods of preparation of 1,4-butanediamine include the hydrogenation of succinonitrile (90); the cleavage of tetrahydrofuran with hydrogen bromide to yield 1,4-dibromobutane, followed by ammonolysis with ammonia (21,91); and reaction of 2-pyrrolidone with dimethylsulfate to yield butyrolactim methyl ether, followed by catalytic hydrogenation in the presence of ammonia (45). The succinonitrile route, the 2-pyrrolidone route, and the 1,4-dibromobutane route are described in detail as follows.

1.   THE SUCCINONITRILE ROUTE

The starting succinonitrile may be obtained from the reaction of acrylonitrile and hydrogen cyanide (92). The reduction of succinonitrile with hydrogen and a catalyst such as nickel and without ammonia or other diluent usually provides only low yield of 1,4-butanediamine. The major side products are pyrrolidine and ammonia. However, the reported du Pont process (90) involves hydrogenation in the presence of excess ammonia to result in acceptable yield of product and virtually no undesirable cyclic by-products. The process can be either batch or continuous. Detailed descriptions for both the fixed-bed, continuous process and the batch process are given.

$$CH_2{=}CHCN \xrightarrow{\text{HCN}} NC(CH_2)_2CN \xrightarrow{\text{H}_2} H_2N(CH_2)_4NH_2$$

a.   Continuous Process

In a silver-lined vertically positioned reactor, $1\frac{3}{4}$ in. O.D. and 22 in. long, is charged as 8–14 mesh cobalt catalyst prepared by reduction of pure cobalt oxide granules by hydrogenation at 250–450°C. During the course of 5.67 hr, 1980 parts of a mixture consisting of 1570 parts of ammonia and 410 parts of succinonitrile, together with hydrogen, is passed upward through the reactor. The reaction is conducted under a pressure of approximately 700 atm supplied by hydrogen. The off gas is discharged from the reactor at the rate of about 3.55 ft^3/hr. Ammonia is removed by evaporation. The crude product, 220 parts, is distilled to yield 192 parts of pure 1,4-butanediamine, bp 158–159°C, mp 27°C.

b. Batch Process

In the batch process, 376 parts of succinonitrile, 7000 parts of ammonia, and 80 parts of reduced cobalt catalyst are charged into a silver-lined shaker tube. The reaction is performed under 700 atm of hydrogen pressure at 80°C during a 1-hr period. The combined reaction mixtures are filtered to remove catalyst, and the filtrate is distilled to furnish a 90.8% yield of product.

2. THE 2-PYRROLIDONE ROUTE

2-Pyrrolidone is prepared commercially from formaldehyde and acetylene by the following reaction sequence (93):

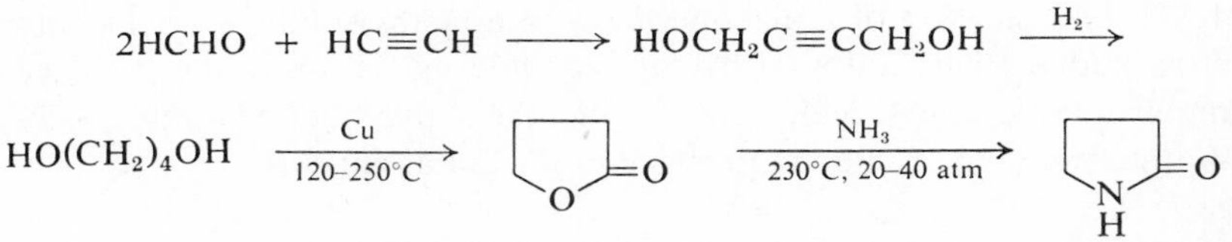

$$2HCHO \; + \; HC\equiv CH \; \longrightarrow \; HOCH_2C\equiv CCH_2OH \; \xrightarrow{\;H_2\;}$$

The intermediate butyrolactone may also be prepared from the air oxidation of tetrahydrofuran in the presence of a cobalt catalyst (94).

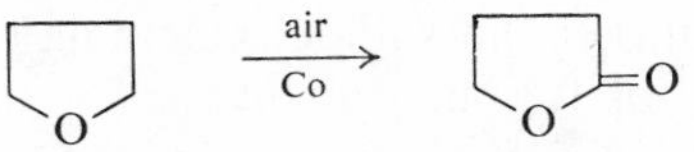

Interaction of 2-pyrrolidone with dimethylsulfate furnishes butyrolactim methyl ether (95), which in turn is hydrogenated in the presence of ammonia to 1,4-butanediamine (45).

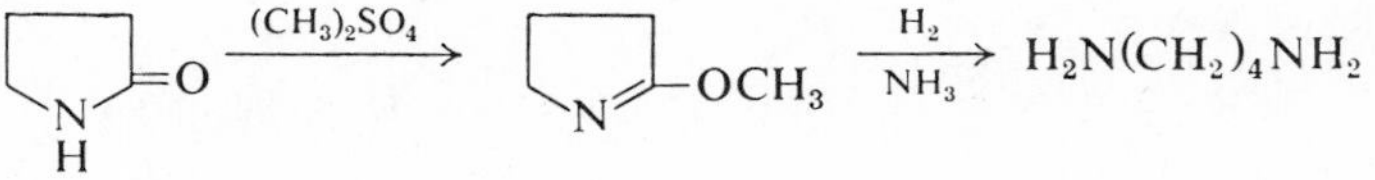

To carry out this process, 252 g (2 moles) of dimethylsulfate is added by drops with stirring to 170 g (2 moles) of 2-pyrrolidone in 100 ml of benzene at 60–70°C. The resulting mixture is refluxed for 8 hr. After cooling to 5°C, 150 g of potassium carbonate is added rapidly, followed by dropwise addition of 100 ml of water with cooling during 1 hr. Carbon dioxide is evolved during addition of water, and the temperature of the mixture rises to 20°C. The benzene layer is separated, dried, and distilled to yield 95 g (48%) of butyrolactim methyl ether, bp 118–121°C.

For the hydrogenation step, 50 g of butyrolactim methyl ether, 80 g of liquid ammonia, and 5 g of cobalt obtained by reduction of cobalt oxide are charged in a 1-liter autoclave. The autoclave is sealed and heated at 95°C for 4 hr. Then 70 atm of hydrogen is charged and hydrogenation is carried out at 150°C for 8 hr. The content is cooled, discharged, and taken up in ether. Then the ether solution is filtered and the filtrate distilled, providing a 73% yield of 1,4-butanediamine, bp 157–160°C.

3. The 1,4-Dibromobutane Route

Ammonolysis of 1,4-dibromobutane in the presence of an aluminum catalyst has been reported to provide good yield of 1,4-butanediamine (21).

$$Br(CH_2)_4Br \xrightarrow{NH_3} H_2N(CH_2)_4NH_2$$

An aluminum catalyst is prepared by boiling a mixture of 10 parts of aluminum powder, 10 parts of sodium carbonate, and 50 parts of water for 2 hr. To this is added 20 parts of a 50% sodium hydroxide solution, and the resulting mixture is heated at 90–100°C for 5–6 hr with agitation. The precipitate formed is allowed to settle and the supernatant liquid is decanted. The precipitate is washed with water to remove excess alkali, resulting in a muddy substance. A mixture of 1,4-dibromobutane and 5–8% by weight to the dibromo-compound of the aluminum catalyst prepared previously is charged into an autoclave. The autoclave is then pressurized with anhydrous ammonia to 3 atm. Ammonolysis is carried out by heating the reaction mixture at 105–120°C for 30 min. After cooling, excess ammonia is vented and the crude product mixture is washed with alcohol and recrystallized from water to provide an 82% yield of theory of 1,4-butanediamine.

E. 1,5-Pentanediamine

There are a number of preparations available for 1,5-pentanediamine, starting from various precursors. However, most of the reported procedures are only of academic interest. From the chemical reduction of glutaronitrile with sodium or sodium-potassium alloy in alcohol, up to 80% of 1,5-pentanediamine together with 18.5% piperidine has been obtained (96). Both the amination of 1,5-dichloropentane with anhydrous ammonia at 110°C and 10 atm (97) and reaction of a glycol with allylacetonitrile followed by catalytic hydrogenation of the adduct (98) have been reported to give 1,5-pentanediamine, but no yields are furnished. 1,5-Pentanediamine has also been prepared by the reaction of 1,5-dibromopentane with magnesium and O-methylhydroxylamine in 68% yield (25), and by the interaction of the same dibromo-derivative with N-potassium phthalimide in nearly theoretical yield (43).

Catalytic hydrogenation of glutaronitrile consists potentially of a ready route to 1,5-pentanediamine. The major complication is the formation of the cyclized product piperidine. In fact, piperidine is usually the major product obtained from the reduction of glutaronitrile. However, conditions are developed in which 1,5-pentanediamine and piperidine are each obtained in about 40% yield (99). The starting glutaronitrile is readily prepared from the reaction of sodium cyanide with 1,3-dichloropropane (100,101) or 1,3-dibromopropane (102). The experimental details using glutaronitrile are described.

$$X(CH_2)_3X \xrightarrow{\text{NaCN}} NC(CH_2)_3CN \xrightarrow{\text{H}_2} H_2N(CH_2)_5NH_2$$

$$X = Cl, Br$$

1. GLUTARONITRILE

a. Via 1,3-Dichloropropane (100,101)

A mixture of 30 g (0.61 mole) of sodium cyanide (dried at 110°C overnight) and 150 ml of dimethylsulfoxide (dried over calcium hydride) is placed in a flask fitted with stirrer, reflux condenser, dropping funnel, and thermometer. The thick slurry is heated to 90°C with a steam bath and then heating is removed. To the slurry is added slowly 28.3 g (0.25 mole) of 1,3-dichloro-propane. The addition is exothermic and the rate of addition is adjusted in such a manner that the temperature of the reaction does not rise beyond 160°C. The addition takes about 10 min and the mixture is stirred until the temperature drops below 50°C. Total reaction time is about 30 min. A quantity of 150 ml of chloroform is added to the flask, and the mixture is poured into saturated salt solution. Sufficient water is added to dissolve the precipitated salt. The chloroform fraction is separated and the aqueous fraction is extracted with chloroform. The combined chloroform solution is washed twice with salt solution, dried over calcium chloride, and distilled to provide 23.5 g (67%) of glutaronitrile, bp 101–102°C/1.5 mm, n_D^{25} 1.4339.

b. Via 1,3-Dibromopropane (102)

A mixture of 294 g (6 moles) of sodium cyanide and 300 ml of water is heated in a steam bath until most of the sodium cyanide is in solution (2–3 hr). A solution of 500 g (2.47 moles) of 1,3-dibromopropane in 1 liter of 95% alcohol is added over a 40–60 min period. The mixture is refluxed on a steam bath for 30–40 hr. The solvent is stripped under reduced pressure and the residue is extracted with 300–400 ml of ethyl acetate. The solution is filtered and the insoluble salts washed with 100 ml of ethyl acetate. The combined ethyl acetate solution is distilled at atmospheric pressure (to remove solvent) and then under reduced pressure, providing 180–200 g (77–86% yield) of glutaronitrile, bp 131–134°C/10 mm.

2. 1,5-PENTANEDIAMINE (99)

A mixture of 60 g (0.64 mole) of glutaronitrile, 12 g of methanol, 250 g of ammonia, and 12 g of cobalt metal on kieselguhr in an autoclave is stirred at 125°C under 3146 psig of hydrogen pressure for 12 hr. After cooling, the autoclave is vented and the contents discharged. The crude product mixture is washed with methanol and filtered. Fractionation of the filtrate up to 108°C furnishes 21.8 g (40%) of piperidine. Upon continued distillation, 26.8 g (41%) of 1,5-pentanediamine, bp 174–178°C, is obtained.

F. 1,6-Hexanediamine

Many commercially feasible processes are available for the manufacture of 1,6-hexanediamine, one of the most widely used aliphatic diamines (103,104). All the important routes involve the reduction of the key intermediate adiponitrile in the final step. The following brief summary describes the synthesis of adiponitrile starting from a wide variety of raw materials. Detailed conditions of some of these routes are discussed later.

The Monsanto electrohydrodimerization method of manufacturing adiponitrile from acrylonitrile is the most efficient of all the processes developed to date (105). The electrolytic reductive coupling of acrylonitrile provides nearly quantitative yield of adiponitrile at close to 100% current efficiency. The starting acrylonitrile is readily prepared in excellent yield by the ammoxidation of propylene, under the conditions reviewed by Sittig (106).

$$2CH_2{=}CHCN + H_2O \rightarrow NC(CH_2)_4CN + \tfrac{1}{2}O_2$$

Reaction of adipic acid with ammonia followed by dehydration of the resulting ammonium salt in the presence of a catalyst furnishes adiponitrile in good yield. The actual process involves a one-step gas-phase reaction of adipic acid and ammonia over a fixed-bed catalyst (104,107).

$$HOOC(CH_2)_4COOH + 2NH_3 \rightarrow NC(CH_2)_4CN + 4H_2O$$

The precursor adipic acid is made from either phenol or cyclohexane. Reduction of phenol to cyclohexanol or cyclohexanone with palladium catalysts followed by oxidation furnishes adipic acid. A mixture of cyclohexanol and cyclohexanone is obtained by the air oxidation of cyclohexane.

Further oxidation of this mixture by nitric acid also yields adipic acid. These various routes have been discussed in considerable detail by Hatch (104).

Another process of preparing adiponitrile involves the reaction of furfuraldehyde with calcium oxide to form furan. Hydrogenation of furan and interaction of the resulting tetrahydrofuran with hydrochloric acid in the presence of sulfuric acid yield 1,4-dichlorobutane. Treatment of this dichloroderivative with sodium cyanide provides the desired adiponitrile (94).

An alternate route to 1,4-dichlorobutane is the condensation of acetylene with formaldehyde and hydrogenation of the resulting 1,4-butynediol to 1,4-butanediol. Conversion of 1,4-butanediol to 1,4-dichlorobutane is accomplished with hydrochloric acid (107).

$$HC\equiv CH + 2CH_2O \longrightarrow HOCH_2C\equiv CCH_2OH \xrightarrow{H_2}$$
$$HO(CH_2)_4OH \xrightarrow{HCl} Cl(CH_2)_4Cl$$

Vapor-phase chlorination of butadiene provides high yields of a mixture of 3,4-dichloro-1-butene and 1,4-dichloro-2-butene. Reaction of the mixture of dichlorobutenes with sodium cyanide or hydrogen cyanide provides the same products, a mixture of cis- and trans-1,4-dicyano-2-butenes, which are hydrogenated to adiponitrile (104).

The hydrodimerization of acrylonitrile using sodium, potassium, or lithium amalgam to yield 69.5% of adiponitrile has been described (108). A chemical coupling of acrylonitrile performed in the presence of sodium dispersion to prepare adiponitrile (29%) has also been reported (109).

$$2H_2C{=}CHCN \rightarrow NC(CH_2)_4CN$$

We have just described the various routes to adiponitrile, which can be then readily hydrogenated to 1,6-hexanediamine. A number of other routes for 1,6-hexanediamine that do not require adiponitrile as the intermediate are summarized as follows.

The intermediate 1,4-dicyano-2-butene described earlier may be reduced in one step in high conversion (up to 91%) to 1,6-hexanediamine. A fixed-bed process using a reduced cobalt catalyst is employed (110).

$$NCCH_2CH{=}CHCH_2CN \xrightarrow{H_2} H_2N(CH_2)_6NH_2$$

Cyclohexene is reacted with ozone in methanol and the resulting oxidation product is hydrogenated in the presence of ammonia to yield 1,6-hexane-diamine (111).

$$\bigcirc \xrightarrow[\text{2) } NH_3, H_2]{\text{1) } O_3} H_2N(CH_2)_6NH_2$$

The amination of 1,6-hexanediol in dioxane in the presence of Raney nickel has been reported to furnish 1,6-hexanediamine. The side product, hexamethyleneimine, is recycled, and up to 71% conversion to 1,6-hexane-diamine is achieved (27).

$$HO(CH_2)_6OH \xrightarrow{NH_3, \, Ni} H_2N(CH_2)_6NH_2$$

Of the various routes reported for the preparation of 1,6-hexanediamine via adiponitrile, the ones using acrylonitrile, adipic acid, 1,4-dichlorobutane, and 1,4-dicyano-2-butene are all considered commercially significant. The electrohydrodimerization of acrylonitrile process, because of its low cost of starting material, simplicity and high efficiency, is the process of choice for adiponitrile. This process is discussed in detail, together with the final step of the other three processes mentioned previously.

1. ADIPONITRILE

a. Via Acrylonitrile

The Monsanto process of electrohydrodimerization of acrylonitrile represents the first large-scale commercial use of an electrochemical method for the manufacture of an organic compound (105,112–114). The apparatus consists of an electrolysis cell composed of a container, a cathode and an anode, and an electrolyte. The container is made of materials capable of resisting the corrosive action of the electrolyte, e.g., glass for laboratory-scale cell. For a batch process a diaphragm, such as a porous cup of unglazed porcelain, is placed in the container to separate the anode from the cathode. The platinum anode is immersed in an aqueous solution of tetraethylammonium p-toluenesulfonate (anolyte) contained in the porous cup. A quantity of mercury is added to the bottom of the container to serve as cathode. The porous cup and a solution (catholyte) of acrylonitrile in an aqueous concentrated tetraethylammonium p-toluenesulfonate are placed in the container. A quaternary ammonium electrolyte is used in order to increase the solubility of acrylonitrile in the electrolyte solution. This is necessary for high conversion to adiponitrile as well as for suppression of polymerization. A further advantage of using this electrolyte is that the reduction of acrylonitrile to the undesirable propionitrile is minimized. The electrolysis is carried out at 45–55°C by applying current to the cell. The quantity of current depends on the nature of the electrodes, the size of the cell, and the temperature. When a mercury cathode is used, the current density should be at least 5 A/dm² of cathode surface to provide good efficiency. The pH of the electrolyte is controlled by the addition of acetic acid, which serves to reduce alkalinity generated in the catholyte during electrolysis. Increasing alkalinity tends to favor the unwanted cyanoethylation of water to yield bis(cyanoethyl)ether. After electrolysis, the reaction mixture is neutralized and worked up by conventional methods. Details are given for both a batch process and a continuous process.

(*1*) *Batch Process.* The preparation of tetraethylammonium p-toluenesulfonate is as follows: a mixture of 200 g (1 mole) of ethyl p-toluenesulfonate, 101 g (1 mole) of triethylamine, and 100 ml of absolute alcohol is stirred at room temperature for 3.5 hr and then heated to 72°C in 40 min. The reaction becomes exothermic at this point and heating is discontinued. After it has been allowed to stand for 30 min, the mixture is heated to reflux and maintained at reflux for 6 hr. After cooling, solvent and unreacted starting materials are removed under aspirator vacuum. The remaining material (which solidified on standing) is washed with ether and dried to provide 296.8 g of the desired product, mp 103–104°C.

A 73.3% aqueous solution of tetraethylammonium p-toluenesulfonate is prepared by dissolving 586.7 g of the sulfonate in 215 g of water. The catholyte containing 40% acrylonitrile is made by mixing 33 g of water, 94.5 g of acrylonitrile, and 108 g of the 73.3% sulfonate solution. The anolyte is prepared by dissolving 20 ml of water to 20 ml of the 73.3% sulfonate solution. The platinum anode is placed in an Alundum cup containing the anolyte and immersed into a jacketed glass vessel containing the catholyte and 110 ml of mercury resting on the bottom as cathode. An electric current is passed through the cell for 7 hr at an average current of 2.0–3.2 A (a total of 22.3 A hr) and an EMF of 19–18 V during the first hour and 17.9–17.0 V during the last 6 hr. A 100% current efficiency is determined by copper coulombmeter measurement. A total of 4.40 ml of glacial acetic acid is intermittently added to the catholyte during the first 5 hr to maintain the catholyte just alkaline to phenol red. The temperature of the catholyte is kept at 23–25°C by water circulating in the cooling jacket of the cell container. After the 7-hr electrolysis period, the catholyte is neutralized, diluted with water, and extracted ten times with 50-ml portions of methylene dichloride. The combined extracts are washed with water and dried over potassium carbonate. After methylene dichloride has been removed, the concentrate may be distilled to recover unreacted acrylonitrile, leaving as residue the product adiponitrile. Analysis of the concentrate by vapor-phase chromatography (114) shows that 50% of the starting acrylonitrile is consumed, giving a 100% yield of adiponitrile based on unrecovered acrylonitrile. (adiponitrile, bp 127°C/1.4 mm, n_D^{25} 1.4378.)

(2) *Continuous Process.* A plate-and-frame type of cell is adopted for the continuous procedure. Lead plates separated by a cationic exchange membrane made from a sulfonated polystyrene resin serve as cathode and anode. The distance between the plates is less than ½ in. The catholyte and anolyte circulate separately to their respective electrodes, and means are provided for continuously adding acrylonitrile, separating adiponitrile from the catholyte and recycling the catholyte. Dilute mineral acid is used as the anolyte. The catholyte consists of 17.6% by weight acrylonitrile, 37.2% water, and 42.2% tetramethylammonium toluenesulfonate. A catholyte linear rate of circulation past the cathode of 1.2 ft/sec, a current density of 20 A/dm² of cathode surface, and a pH of 8.5–9.0 are employed for the hydrodimerization process. The yield of adiponitrile is 90%, based on acrylonitrile consumed together with 1% of propionitrile. This process may be carried out continuously for weeks with no lowering in yield.

b.   *Via Adipic Acid*

Adiponitrile is prepared in high yield by the vapor-phase reaction of adipic acid and ammonia in the presence of a dehydration catalyst such as boron phosphate (115) or silica gel (116). The process using boron phosphate is reported to be characterized by long catalyst life and the formation of minimum amounts of the by-product cyclopentanone. The boron phosphate catalyst is prepared by adding slowly and with stirring 824 g of powdered boric acid to 1730 g of 85% phosphoric acid. Stirring is continued until the mixture has become a homogeneous, viscous, toffeelike dough. After it has been allowed to stand at room temperature for 18–20 hr, the catalyst mixture sets into a firm gelatinous mass. The material is dried at 110°C and then baked at 350°C for 4 hr. The resulting hard, brittle white boron phosphate is broken up and screened to a particle size of 8–14 mesh.

The fixed-bed process is carried out by passing ammonia at the rate of 51.9 g/hr over 250 cc of the catalyst, maintained at 350°C. Adipic acid is vaporized and passed through the catalyst bed at a rate of 55.7 g/hr, together with the ammonia. The mole ratio of ammonia to adipic acid is 8:1. Space velocity of about 310 and contact time of approximately 5 sec are employed. The reaction is carried out for 49 hr and a total of 2730 g (18.7 moles) of adipic acid is charged. The reactor effluent, which consists of mainly water and adiponitrile, is cooled and separated into two layers. The aqueous layer is extracted with benzene. The combined benzene extract and oil layer are fractionated to yield 1763 g (87.5% yield based on adipic acid charged) of adiponitrile, bp 147–148°C/10 mm, 65 g (4.2%) of cyclopentanone, bp 128–130°C, and 51 g of distillation residue.

c.   Via 1,4-Dichlorobutane

The interaction of alkyl halides with sodium cyanide in dimethyl sulfoxide to prepare the corresponding nitrile has been reported by Smiley (100,101). The advantage of using dimethyl sulfoxide over the aqueous alcohol solvent normally employed for aliphatic chlorides is that a much shorter reaction time is realized. A slurry consisting of 110 g (2.24 moles) of sodium cyanide and 330 g of dimethyl sulfoxide is first stirred and heated to 80°C. Heat is removed and 127 g (1 mole) of 1,4-dichlorobutane is added by drops over a 5-min period. An exothermic reaction takes place during addition. The mixture is cooled in an ice bath when the temperature reaches 160°C. After addition is complete, the reaction mixture is stirred for 10 min and then cooled rapidly to below 50°C. To the crude product mixture is added 300 g of chloroform. The resulting mixture is poured into 800 g of water. The chloroform fraction is separated and the aqueous fraction is extracted two times with 150-g portions of chloroform. The combined chloroform solutions are washed once with a saturated sodium chloride solution and dried. The solution is then distilled at atmospheric pressure to remove chloroform. Fractionation of the remaining liquid yields 8 g of forerun, which is dimethyl-sulfoxide and 95 g of adiponitrile, bp 115°C/0.7 mm, n_D^{25} 1.4369. The yield of adiponitrile is 88%, based on the 1,4-dichlorobutane charged.

d.   Via 1,4-Dicyano-2-butene

The catalytic reduction of 1,4-dicyano-2-butene is performed either by a gas-phase or a liquid-phase hydrogenation using supported palladium catalysts (117). The process may be batchwise or continuous. The conditions for a fixed-bed, continuous process for the preparation of adiponitrile have been described (118).

A palladium-on-charcoal catalyst is prepared by saturating a sample of activated coconut charcoal, which has been preextracted with nitric acid, with a hot solution of palladium chloride. The quantity of palladium chloride used is calculated to result in a 2% palladium-on-charcoal mixture. The saturated charcoal mixture is evaporated to dryness in a steam bath, dried at 105–110°C, and then reduced in the presence of hydrogen at 200°C. The hydrogenation is carried out by passing upward a 5% solution of 1,4-dicyano-2-butene in methanol over a 2% palladium-on-activated-charcoal catalyst bed at 115–150°C. The 1,4-dicyano-2-butene charge is accompanied by hydrogen gas with a feed ratio of 50 moles of hydrogen per mole of dicyanobutene. The hydrogenation is performed at 400 psig pressure. The space velocity varies from 0.3 to 0.6 volume of dicyanobutene per volume of catalyst per hour. This run is carried on for 500 hr and the results are described in Table 3 (118). The catalyst may be regenerated by passing hydrogen over the catalyst bed at 500°C.

TABLE 3
Hydrogenation of Dicyanobutene

Space velocity	Temperature, °C	Amount conversion to adiponitrile, %
0.3	115	93
0.3	125	97
0.3	135	96–97
0.3	150	90
0.6	125	80
0.6	135	90
0.6	150	95

2. 1,6-HEXANEDIAMINE

a. Via Adiponitrile

Catalytic hydrogenation of adiponitrile is performed in the presence of ammonia and a cobalt catalyst. Excellent yield (over 90%) of 1,6-hexanediamine is obtained when a continuous fixed-bed process is employed for such a reduction (119,120). The starting adiponitrile may be purified by ion-exchange resins prior to hydrogenation to improve yield of the diamine and to extend catalyst life (121,122). The cobalt catalyst is usually prepared by decomposing freshly precipitated cobalt carbonate at 300–500°C in air to give cobalt oxide. The oxide is granulated, mixed with 2–4% of high-molecular-weight fatty acid ester, such as a stearic acid ester, and pelleted.

The pellets are sintered at 700–1100°C in an oxidizing atmosphere to form hard, dense pellets and then reduced in hydrogen at 400°C prior to use (123,124). Alloyed catalysts, such as cobalt-copper, cobalt-aluminum, and cobalt-chromium, have also shown good activity (120).

A fixed-bed process is reported by Imperial Chemical Industries to yield 97.3% of 1,6-hexanediamine (119). The experimental details provided are rather sketchy. In operation, adiponitrile at 1200 lb/hr and liquid ammonia at 2450 lb/hr are mixed and passed upward at 80°C into a hydrogenation reactor containing a cobalt catalyst. Excess hydrogen is admitted from the top of the reactor to a pressure of 230 atm. Hydrogenation is performed at 130–135°C. The reactor effluent collected from the top of the reactor is led into a separator unit. The gas phase containing hydrogen, ammonia, and nitrogen is cooled to 30–40°C and recycled to the reactor. The liquid phase of the separator is cooled and decompressed. After evaporation of dissolved ammonia, the product contains 97.3% 1,6-hexanediamine and 2.7% impurities.

The du Pont process of using a cobalt-copper catalyst is similar to that just described, but the patent (120) furnishes many more details. The cobalt-copper catalyst is prepared by mixing two parts by weight of cobalt oxide and one part of copper oxide and fusing the mixture in a furnace. The fused product is ground and screened to 8–14 mesh. The catalyst is then reduced to the metals in a hydrogen–carbon dioxide atmosphere by gradual heating to 375°C. The hydrogenation is carried out in a vertical fixed-bed reactor in which the catalyst is maintained at 125°C. A 20.1-wt.% solution of adiponitrile in liquid ammonia and hydrogen is passed upward through the catalyst bed under pressure. The reactants are preheated to 95°C before entering the reactor. The exothermic reaction developed in the reactor causes a 30° temperature rise, and the heat is partially dissipated by vaporization of excess ammonia into the gas phase. The feed rates of adiponitrile and ammonia correspond to 39 g and 152 g/hr/100 cc of catalyst, respectively. The reactor effluent discharged from the top of the reactor is led into a separator maintained at 45°C. Hydrogen is withdrawn from the top of the separator at a rate of 45 moles per mole of nitrile added and may be recycled to the reactor. The total pressure of the reactor system is kept at 600–630 atm. The liquid fraction in the separator is decompressed into a receiver containing methanol. After evaporation of ammonia and removal of methanol have taken place, the product 1,6-hexanediamine is obtained by fractionation. An average of 91% yield (based on adiponitrile charged) of 1,6-hexanediamine is obtained over a 160-hr hydrogenation period.

The 1,6-hexanediamine obtained from the hydrogenation of adiponitrile usually contains 600–2000 ppm of 1,2-diaminocyclohexane, which is not readily separable by simple fractional distillation. The presence of this impurity in 1,6-hexanediamine is undesirable in polyamide formation, since products of poor color stability and weak fiber strength are obtained. A purification procedure was developed by du Pont in which crude 1,6-hexanediamine is treated with water and 1,2-diaminocyclohexane is separated by azeotropic distillation through a series of four columns (125). Other impurities, such as hexamethyleneimine and ω-aminocapronitrile, are also removed by this process.

An alternate purification method reported by Monsanto is the formation of insoluble copper or nickel chelate of 1,2-diaminocyclohexane (126). The following example is illustrative. A quantity of 204 g of crude 1,6-hexanediamine containing 1135 ppm of 1,2-diaminocyclohexane is distilled with a 6-in. Vigreaux column to a pot temperature of 90°C and distillation head temperature of 44°C/70 mm in order to remove water and free ammonia. The system is allowed to reach atmospheric pressure and to the pot is added 6.5 ml of an aqueous solution of $CuSO_4 \cdot 5H_2O$ equivalent to 0.332 g of copper. The resulting mixture is agitated vigorously at 90°C for 10 min under a stream of nitrogen. The mole ratio of copper to 1,2-diaminocyclohexane is 2.5 and the water content in the mixture is approximately 3%. The 1,6-hexanediamine is then distilled at 125°C/66 mm. The purified product, after about 90% has been collected, is found to contain 20 ppm of 1,2-diamino-cyclohexane and no copper.

b. Via 1,4-Dicyano-2-butene

The direct hydrogenation of 1,4-dicyano-2-butene to 1,6-hexanediamine is made in the absence of ammonia to avoid extensive side reactions (110). Best results are obtained with a high-pressure, fixed-bed process. Thus, in a vertical tubular reactor is charged a layer of silica chips, 110 parts of 8–14 mesh cobalt catalyst prepared by the reduction of cobaltous oxide in hydrogen, and a zone of silica chips to serve as a preheating zone. A 5% solution of 1,4-dicyano-2-butene in methanol is charged into the top of the reactor at a rate of 240 parts/hr. Concurrently, a stream of hydrogen is passed downward through the reactor. The feed ratio of hydrogen and nitrile is adjusted to twentyfold excess of hydrogen to nitrile. The reaction pressure of 550–600 atm and the reaction temperature of 120°C in the catalyst zone are maintained. A total of 64.3 parts of 1,4-dicyano-2-butene is passed through the reactor during 5 hr at a space velocity of 0.1–0.5 g of the nitrile per cubic centimeter catalyst per hour. The reactor effluent is decompressed and worked up by fractional distillation to provide 64.5 parts of 1,6-hexane-diamine. This corresponds to 91.5% yield based on the 1,4-dicyano-2-butene charged. There are also isolated 1.9 parts of hexamethyleneimine and 4.0 parts of polymeric residue.

c. Via 1,6-Hexanediol

High yields of 1,6-hexanediamine are obtained from the amination of 1,6-hexanediol in the presence of hexamethyleneimine and Raney nickel (27). The process involves the reaction of 1,6-hexanediol with excess ammonia under superatmospheric conditions. The presence of about 30% by weight of hexamethyleneimine in the reaction mixture is necessary to suppress further formation of hexamethyleneimine and to serve as solvent. Thus in a multi-cycle run the hexamethyleneimine is recycled at about 30% concentration to result in an optimum yield of 1,6-hexanediamine, with the quantity of hexamethyleneimine remaining constant throughout the entire run. The amination is carried out in an autoclave. A quantity of 54 parts of hexamethyleneimine, 121 parts of 1,6-hexanediol, and 72 parts of Raney nickel is charged in the autoclave. To this mixture is added 530 parts of ammonia. The autoclave is sealed and agitated at 175°C for 5 hr. After cooling, the crude product mixture is analyzed to show a 30% conversion of the diol. The yield of 1,6-hexanediamine is 100% based on the diol consumed. No additional hexamethyleneimine is formed.

G. N,N'-Dimethyl-1,6-hexanediamine

The interaction of 1,6-dibromohexane with methylamine has been reported to yield N,N'-dimethyl-1,6-hexanediamine (127,128). Large excess of amine is required to minimize further alkylation of the resulting secondary amine, and no yields are mentioned. Low yields of N,N'-dimethyl-1,6-hexanediamine have been obtained from the interaction of 1,6-hexanediol and methylamine in the presence of hydrogen and a cobalt oxide–cupric oxide catalyst system (30), as well as from the reductive amination of the reaction product from cyclohexene, ozone, and isopropyl alcohol with methylamine (111). The methylation of 1,6-hexanediamine with methanol at 350°C in the presence of alumina to furnish a 90.4% yield of the desired product has also been recorded (129). The reaction of 1,6-hexanediamine with formaldehyde to yield a polymeric N,N'-bismethylene-1,6-hexanediamine followed by re-duction of its acetic acid salt with zinc dust to provide a 65–70% yield of N,N'-dimethyl-1,6-hexanediamine is a potentially attractive route. The experimental conditions given are rather sketchy, however (130).

An excellent three-step laboratory procedure for the synthesis of N,N'-dimethyl-1,6-hexanediamine via the p-toluenesulfonamide was described by Boon in 1947 (131). A similar route employing the benzenesulfonamide intermediate was also reported by Ried and Wesselborg in 1957 (132). The Boon route involves the reaction of 1,6-hexanediamine and p-toluenesulfonyl chloride to form N,N'-bis-p-toluenesulfonyl-1,6-hexanediamine; alkylation with methyl sulfate and hydrolysis of the N,N-dimethyl derivative with sulfuric acid follows. The yields of these steps are excellent, and the experimental details are:

$$H_2N(CH_2)_6NH_2 + p\text{-}CH_3C_6H_4SO_2Cl \longrightarrow p\text{-}CH_3C_6H_4SO_2NH(CH_2)_6NHSO_2C_6H_4\text{-}p\text{-}CH_3$$

$$\xrightarrow{(CH_3)_2SO_4} p\text{-}CH_3C_6H_4SO_2\underset{\underset{CH_3}{|}}{N}(CH_2)_6\underset{\underset{CH_3}{|}}{N}SO_2C_6H_4\text{-}p\text{-}CH_3$$

$$\xrightarrow{H_2SO_4} H_3C\underset{\underset{H}{|}}{N}(CH_2)_6\underset{\underset{H}{|}}{N}CH_3$$

1. N,N'-BIS-p-TOLUENESULFONYL-1,6-HEXANEDIAMINE (132,133)

N,N'-bis-p-toluenesulfonyl-1,6-hexanediamine is prepared under the normal Schotten-Baumann conditions. Thus to a mixture of 2:1 mole ratio of p-toluenesulfonyl chloride and 1,6-hexanediamine is added a quantity of 17% aqueous sodium hydroxide, calculated in slight excess to neutralize the hydrogen chloride formed from the reaction. The resulting mixture is heated on a steam bath for 1 hr. The solid product that separates out upon cooling is recrystallized from alcohol, mp 152°C. The yield is more than 90%.

2. N,N'-BIS-p-TOLUENESULFONYL-N,N'-DIMETHYL-1,6-HEXANEDIAMINE (131)

A solution of 424 g (1.0 mole) of N,N'-bis-p-toluenesulfonyl-1,6-hexane-diamine is prepared by dissolving the diamine in a mixture of methanol and a 32% sodium hydroxide solution containing 1.1 mole of sodium hydroxide. To this solution, cooled in an ice bath, is added 283.5 g (2.25 moles) of dimethylsulfate during 2 hr. The resulting reaction mixture is stirred at room temperature for 1 hr and then heated under reflux for 4 hr. Upon cooling, the product is filtered, washed with methanol and with water, and then dried. The yield of N,N'-bis-p-toluenesulfonyl-N,N'-dimethyl-1,6-hexanediamine is 84%, mp 140°C (from acetic acid).

3. N,N'-DIMETHYLHEXANEDIAMINE (131)

A quantity of 452 g (1 mole) of N,N'-bis-p-toluenesulfonyl-N,N'-dimethyl-1,6-hexanediamine is dissolved with stirring in 8.2 moles of 98% sulfuric acid. To this is added 162 g (9 moles) of water* and the resulting suspension is heated with stirring at 140–145°C for 7 hr. The mixture is cooled, diluted with water, and made alkaline with 32% sodium hydroxide. The resulting solution is steam distilled. The distillate is acidified with hydrochloric acid and evaporated to dryness. The crude diamine hydrochloride is dissolved in a 10% excess of 32% sodium hydroxide and the solution is distilled to dryness under reduced pressure. After the aqueous distillate has been saturated with sodium hydroxide, the oily layer is separated, dried over potassium hydroxide, and

---

* The author did not mention the manner of addition of water to the diamine–sulfuric acid solution. Precaution should be taken for this step.

then distilled to provide an 80% yield of N,N'-dimethylhexanediamine, bp 205°C.

H. 1,7-Heptanediamine

The route to 1,7-heptanediamine, as with the case of other long-chain aliphatic α,ω-alkanediamines, is via the reduction of the corresponding dinitrile. Pimelonitrile, the precursor for 1,7-heptanediamine, is obtained as one of the many possible intermediates starting from furfural. Hydrogenation of furfural and catalytic dehydration of the resulting tetrahydrofurfuryl alcohol yield the commercially available dihydropyran. Hydrogenolysis of dihydropyran in the presence of water furnishes the open-chain hydrated product, 1,5-pentanediol. Finally, the diol is converted to 1,5-dichloropentane with hydrogen chloride and then to pimelonitrile by sodium cyanide (94).

$$\text{(furyl)—CHO} \xrightarrow{\text{H}_2} \text{(furyl)—CH}_2\text{OH} \xrightarrow[-\text{H}_2\text{O}]{\text{Al}_2\text{O}_3} \text{(dihydropyran)} \xrightarrow[\text{H}_2]{\text{H}_2\text{O}}$$

$$\text{HO(CH}_2\text{)}_5\text{OH} \xrightarrow{\text{HCl}} \text{Cl(CH}_2\text{)}_5\text{Cl} \xrightarrow{\text{NaCN}} \text{NC(CH}_2\text{)}_5\text{CN} \xrightarrow{\text{H}_2} \text{H}_2\text{N(CH}_2\text{)}_7\text{NH}_2$$

The intermediate 1,5-pentanediol may also be prepared directly from tetrahydrofurfuryl alcohol by hydrogenolysis in the presence of a copper chromite catalyst (134). A somewhat similar route to pimelonitrile involves the hydrogenation of dihydropyran to tetrahydropyran and cleavage with hydrobromic acid to 1,5-dibromopentane. Pimelonitrile is then obtained by the reaction of 1,5-dibromopentane with sodium cyanide (135). Since dihydropyran and its precursors are all commercially available, the following processes, starting from dihydropyran via 1,5-pentanediol to prepare 1,7-heptanediamine, are described in detail.

1. 1,5-PENTANEDIOL

Dihydropyran may be hydrogenated directly under superatmospheric pressure in the presence of a catalyst such as foraminate copper or copper chromite (136). In the preferred method, however, the hydrolysis of dihydropyran with dilute mineral acid is followed by hydrogenation of the resulting di-2-tetrahydropyryl ether *in situ* with nickel, yielding 1,5-pentanediol (137). The presence of water in the hydrogenation step is essential, and a neutral aqueous medium is preferable. In this process, 300 g of dihydropyran is added to 200 g of 0.05N hydrochloric acid and the resulting mixture is heated to reflux. Initially the reaction is exothermic, and cooling is needed. When the hydrolysis is completed, the solution is neutralized with sodium bicarbonate and the resulting mixture is charged into an autoclave with 25 g of 30% nickel on kieselguhr. The mixture is heated to 150°C under 1000–1500 psig of hydrogen pressure with agitation. When hydrogen absorption ceases, the autoclave is cooled and vented, and the contents are discharged. The catalyst is removed by filtration. Distillation of the filtrate furnishes a 93% yield of pure 1,5-pentanediol, bp 105°C/4 mm.

2. 1,5-Dichloropentane

Conversion of 1,5-pentanediol to 1,5-dichloropentane may be performed with concentrated hydrochloric acid at 170° and under 10 atm of pressure (138), or at atmospheric pressure in the presence of zinc chloride (139). The atmospheric process is described as follows.

In a 10-liter reaction flask are placed 45 kg of 36% hydrochloric acid, 4.5 kg of anhydrous zinc chloride, and 3 kg of 1,5-pentanediol. The mixture is heated on a steam bath at 80–90°C for 4 hr, and a stream of hydrogen chloride is continuously bubbled into the mixture for the entire period. The mixture gradually separates into two layers. The top organic layer consists of 25–35% of the total volume. At this point, an additional quantity of 1,5-pentanediol together with hydrogen chloride is added slowly into the mixture by means of separate tubes extending into the heavier aqueous layer. The oily top layer, which comprises essentially all 1,5-dichloropentane, is allowed to overflow slowly by means of an overflow tube into a separate receiver. The 1,5-dichloropentane is then isolated by distillation (bp 84–85°C/34 mm). A small amount of side product dichloroamyl ether, which can be converted to 1,5-dichloropentane, is returned to the main reaction flask. The bottom layer in the reaction flask contains about 50 wt.% zinc chloride, 260–280 g of hydrogen chloride per liter, a little dissolved 1,5-pentanediol, and some tetrahydropyran. Since water is continuously formed in this reaction, the bottom aqueous layer is also siphoned slowly into a different receiver to remove water as hydrochloric acid. The tetrahydropyran and some hydrogen chloride are recycled from this receiver back into the main reactor. In this manner, the yield of 1,5-dichloropentane is 97.5% of theory, whereas the hourly conversion rate is 200–400 g of 1,5-dichloropentane.

3. Pimelonitrile

The conversion of 1,5-dichloropentane to pimelonitrile is performed in aqueous alcohol solvent with potassium cyanide (140). Thus a mixture of 141 g (1.0 mole) of 1,5-dichloropentane, 136.5 g (2.1 moles) of potassium cyanide, 6.6 g (0.04 mole) of potassium iodide, 160 ml of water, and 650 ml of 95% ethanol is heated under reflux for 10 hr. About 550 ml of the mixture is first removed by distillation. To the residue is added sufficient water to dissolve the precipitated salt. The oil phase is separated and the aqueous phase is extracted with benzene. The oil and benzene extracts are washed with 1N sodium hydroxide solution and then combined. Distillation of the remaining liquid, which occurs after stripping of solvent, gives 80–85% yield of pimelonitrile, bp 151–155°C/3 mm.

A similar procedure employing dimethylsulfoxide as solvent is reported to have shortened the reaction time to 30 min at 160°C. However, the reported yield is lowered to 75% (100,101).

4. 1,7-HEPTANEDIAMINE

The reduction of pimelonitrile to 1,7-heptanediamine can be carried out both chemically with sodium-alcohol (141) and catalytically with Raney nickel. The preferred route of catalytic reduction is usually performed in alcohol in the presence of ammonia to reduce the formation of secondary amines. The yield of diamine ranges from 85–90% (135,142–144).

To prepare 1,7-heptanediamine, a mixture of 61 g of pimelonitrile, 6 g of Raney nickel, 30 ml of ethanol, and 80 ml of liquid ammonia is placed in an autoclave. The autoclave is pressurized to 1500 psig with hydrogen. The mixture is heated at 130–150°C for 4–5 hr until hydrogen is no longer absorbed. The crude product mixture is filtered to remove catalyst and stripped to remove volatiles, whereupon it is fractionally distilled to obtain 55.5 g (85.5% yield) of 1,7-heptanediamine, bp 104–105°C/12 mm, mp 28–29°C.

I. 4,4-Dimethyl-1,7-heptanediamine

The raw materials for 4,4-dimethyl-1,7-heptanediamine may be either ethylene and 2,2-dichloropropane (145) or isoprene and hydrogen chloride (14,146). Both routes provide a common intermediate—3,3-dimethyl-1,5-dichloropentane.

$$CH_2{=}CH{-}\overset{\overset{\displaystyle CH_3}{|}}{C}{=}CH_2 \xrightarrow{2HCl} ClCH_2CH_2\overset{\overset{\displaystyle CH_3}{|}}{\underset{\underset{\displaystyle CH_3}{|}}{C}}Cl \xrightarrow[AlCl_3]{CH_2{=}CH_2} ClCH_2CH_2\overset{\overset{\displaystyle CH_3}{|}}{\underset{\underset{\displaystyle CH_3}{|}}{C}}CH_2CH_2Cl$$

$$\overset{\overset{\displaystyle CH_3}{|}}{\underset{\underset{\displaystyle CH_3}{|}}{Cl{\overset{|}{C}}Cl}} + 2CH_2{=}CH_2 + AlCl_3 \longrightarrow$$

Further reaction of 3,3-dimethyl-1,5-dichloropentane with sodium cyanide in refluxing tetrahydrofurfuryl alcohol furnishes 3,5-dimethyl-1,5-dicyanopentane, which is catalytically hydrogenated to 4,4-dimethyl-1,7-heptanediamine (145).

$$ClCH_2CH_2\overset{\overset{\displaystyle CH_3}{|}}{\underset{\underset{\displaystyle CH_3}{|}}{C}}CH_2CH_2Cl \xrightarrow{NaCN} NCCH_2CH_2\overset{\overset{\displaystyle CH_3}{|}}{\underset{\underset{\displaystyle CH_3}{|}}{C}}CH_2CH_2CN \xrightarrow{H_2}$$

$$H_2N(CH_2)_3\overset{\overset{\displaystyle CH_3}{|}}{\underset{\underset{\displaystyle CH_3}{|}}{C}}(CH_2)_3NH_2$$

1. 3,3-DIMETHYL-1,5-DICHLOROPENTANE

The reaction of ethylene and 2,2-dichloropropane in the presence of anhydrous aluminum chloride (145) is performed in a pressure vessel made of, or lined with, stainless steel, glass, noble metal, etc. High-speed agitation should be provided. The reaction may be carried out at temperatures between -40 and $+10°C$. Inert solvent may be employed, but it is preferable to perform the reaction in the absence of solvent. Exclusion of moisture from the reaction is essential.

Thus a mixture of 113 g (1 mole) of 2,2-dichloropropane and 11.3 g of reagent-grade anhydrous aluminum chloride is charged into a shaker tube and cooled to $-30°C$. Shaking is started and ethylene is admitted to a pressure of 300 psig at $-20°C$. The mixture is shaken for 1 hr at $-20°C$, and the pressure is maintained at 300 psig by replacing the ethylene used in the reaction. The crude product mixture is poured in 1 liter of cold 20% hydrochloric acid solution to decompose the catalyst. The organic layer is separated and fractionally distilled to furnish 22.6 g of starting 2,2-dichloropropane, bp 68–70°C, n_D^{25} 1.4120; 4.1 g of crude 3-methyl-1,3-dichlorobutane, bp 34–72°C/15 mm, n_D^{25} 1.4425; and 56.4 g of 3,3-dimethyl-1,5-dichloropentane, bp 90–100°C/13–15 mm, n_D^{25} 1.4640. The product fraction represents a 40% yield and an 80% conversion. The side product, 3-methyl-1,3-dichlorobutane, can be recycled with additional ethylene to yield more 3,3-dimethyl-1,5-dichloropentane.

An alternate route to the preparation of 3,3-dimethyl-1,5-dichloropentane is the reaction of 3-methyl-1,3-dichlorobutane with ethylene using aluminum chloride as catalyst (145). The starting 3-methyl-1,3-dichlorobutane is obtained in 86% yield by the reaction of isoprene and hydrogen chloride under 30 atm initial pressure at room temperature, bp 60°C/30 mm, n_D^{20} 1.4465. To carry out the condensation with ethylene, 252 g (1.8 moles) of 3-methyl-1,3-dichlorobutane and 15 g of aluminum chloride are placed in a three-necked flask immersed in a dry-ice–acetone bath at $-60°C$. Ethylene is bubbled into the stirred mixture while the temperature is permitted to rise to $-15°C$, and at this point absorption begins (determined by the difference of gas rates in the inlet and outlet bubblers). The temperature is maintained between -15 and $-10°C$ for the duration of the addition of ethylene (4 hr). The resulting crude product mixture is treated with water to decompose the catalyst. The organic fraction is washed with water, dried, and distilled to yield 205 g (68%) of 3,3-dimethyl-1,5-dichloropentane.

2. 3,3-DIMETHYL-1,5-DICYANOPENTANE (145)

A mixture of 75 g (0.44 mole) of 3,3-dimethyl-1,5-dichloropentane, 57.2 g (1.17 moles) of sodium cyanide, and 150 g of tetrahydrofurfuryl alcohol is heated under reflux for 3 hr. The mixture is cooled and filtered to remove sodium chloride, and the filtrate is distilled to furnish an 88% yield of 3,3-dimethyl-1,5-dicyanopentane, bp 195–196°C/14 mm, n_D^{23} 1.4551.

3.   4,4-DIMETHYL-1,7-HEPTANEDIAMINE (145)

The hydrogenation of 3,3-dimethyl-1,5-dicyanopentane is carried out by charging 80 g (0.53 mole) of the dinitrile and 8–10 g of a reduced cobalt oxide catalyst into an autoclave. A quantity of 60 g of anhydrous ammonia is introduced into the autoclave. Hydrogenation is performed at 135°C under 4500 psig of total pressure during a 2.5-hr period. The contents of the auto-clave is cooled, filtered, and then distilled to give a 92% yield of 4,4-dimethyl-1,7-heptanediamine, bp 72°C/0.3 mm, n_D^{20} 1.4640 (147).

The hydrogenation of 4,4-dimethyl-1,5-dicyanopentane may also be carried out with Raney nickel in ethanol to provide a 78% yield of 4,4-dimethyl-1,7-heptanediamine (147).

J.   1,8-Octanediamine

Many processes developed for 1,8-octanediamine are available, owing to the ease of synthesis of intermediates starting from acetylene or butadiene. The key intermediates for these processes are suberic acid, sebacic acid, and capryllactam. Both suberic acid and capryllactam are prepared from cyclo-octane, which is in turn obtained from acetylene or butadiene in two steps (104).

Suberic acid is then synthesized by the oxidation of cyclooctane with nitric acid (148).

An alternate route to suberic acid is the hydrogenation of cyclooctatetraene with Raney nickel (149) or palladium (150), followed by oxidation of the resulting cyclooctene (151).

Air oxidation of cyclooctane using cobalt naphthenate as the catalyst yields cyclooctanone. Cyclooctanone is converted to the oxime, followed by rearrangement to capryllactam with sulfuric acid (104).

$$\text{(cyclooctane)} \xrightarrow[\text{oxid.}]{\text{air}} \text{(cyclooctanone)}{=}O \xrightarrow[\text{2) H}_2\text{SO}_4]{\text{1) NH}_2\text{OH}} \begin{array}{c} CH_2-(CH_2)_6 \\ | \qquad\quad | \\ N \text{---} C{=}O \\ H \end{array}$$

The other intermediate, sebacic acid, is prepared by the fusion of sodium ricinoleate, which is obtained from castor oil. Acidification of the resulting disodium sebacate provides the free acid (104). Prior to the discussion of the

$$CH_3(CH_2)_5CHOHCH_2CH{=}CH(CH_2)_7COOH \xrightarrow{\text{NaOH}}$$

$$CH_3(CH_2)_5CHOHCH_3 + NaOOC(CH_2)_8COONa$$

$$NaOOC(CH_2)_8COONa \xrightarrow{\text{H}^+} HOOC(CH_2)_8COOH$$

detailed routes employing these key intermediates, it is appropriate to summarize a number of different approaches reported in the patent literature for the synthesis of 1,8-octanediamine.

Reaction of butadiene with the hydrochloride or sulfate salts of hydroxyl-amine in the presence of a titanous salt yields 1,8-diaminooctadiene-3,6. Hydrogenation of this unsaturated diamine gives the desired 1,8-octane-diamine (152,153).

$$2CH_2{=}CHCH{=}CH_2 \xrightarrow[\text{Ti}^{+3}]{\text{NH}_2\text{OH}} H_2NCH_2CH_2CH{=}CHCH_2CH{=}CHCH_2NH_2 \xrightarrow{\text{H}_2}$$

$$H_2N(CH_2)_8NH_2$$

Coupling of acrolein in acetic acid in the presence of zinc-copper couple yields 3,4-dihydroxy-1,5-hexadiene. Addition of hydrogen cyanide to the diene is accomplished with a soluble cuprous salt in aqueous acidic medium to form 1,6-dicyano-2,4-hexadiene. Hydrogenation of this dinitrile then furnishes 1,8-octanediamine (154).

$$CH_2{=}CHCHO \xrightarrow{\text{Zn--Cu}} \underset{\underset{OHOH}{|\ \ |}}{CH_2{=}CHCHCHCH{=}CH_2} \xrightarrow[\text{Cu}^+]{\text{HCN}}$$

$$NCCH_2CH{=}CHCH{=}CHCH_2CN \xrightarrow{\text{H}_2} H_2N(CH_2)_8NH_2$$

Amination of 1,8-octamethylene glycol using ruthenium has also been reported to give low yield of 1,8-octanediamine (28).

$$HO(CH_2)_8OH \xrightarrow[\text{Ru}]{\text{NH}_3} H_2N(CH_2)_8NH_2$$

Now that we have discussed briefly the various routes to 1,8-octanediamine, the preferred routes using suberic acid, sebacic acid, and capryllactam as intermediates may be indicated.

1. THE SUBERONITRILE ROUTE

Reaction of suberic acid with ammonia followed by dehydration of the ammonium salt furnishes suberonitrile. 1,8-Octanediamine is then obtained by the catalytic hydrogenation of suberonitrile (155). The detailed experimental conditions for this hydrogenation are described.

$$HOOC(CH_2)_6COOH \xrightarrow[\text{2) } -H_2O]{\text{1) } NH_3} NC(CH_2)_6CN \xrightarrow[NH_3]{H_2} H_2N(CH_2)_8NH_2$$

The preparation of the nickel catalyst for the hydrogenation is described in detail under the preparation of 1,10-decanediamine (see Section IV.K). Thus 100 g of suberonitrile, 50 ml of methanol, 50 g of anhydrous ammonia, and 10 g of nickel catalyst are charged into an autoclave and the resulting mixture is heated at 90–100°C. The autoclave is pressurized to 1000–2000 psig with hydrogen. Hydrogen absorption lasts 70 min. The autoclave is cooled and the crude product mixture is discharged and filtered to remove catalyst. The filtrate is fractionated to furnish 90 g (85%) of 1,8-octanediamine, bp 121–122°C/18 mm, which solidifies into a white crystalline solid.

2. THE CAPRYLLACTAM ROUTE

1,8-Octanediamine may be prepared by reaction of capryllactam and dimethyl sulfate to form caprylic lactim methyl ether, which is then hydrogenated (45).

$$\begin{array}{c} CH_2\!-\!(CH_2)_6 \\ | \quad\quad | \\ NH\!-\!\!-\!C\!=\!O \end{array} \xrightarrow{(CH_3)_2SO_4} \begin{array}{c} CH_2\!-\!(CH_2)_6 \\ | \quad\quad | \\ N\!=\!\!=\!C\!-\!OCH_3 \end{array} \xrightarrow[NH_3]{H_2} H_2N(CH_2)_8NH_2$$

The conversion of capryllactam to caprylic lactim methyl ether, bp 86°C/19 mm, in 75% yield, is effected by heating under reflux equimolar quantities of the lactam and dimethyl sulfate in benzene, shaking the reaction mixture with sodium carbonate solution, and distilling the product mixture (for experimental details, see preparation of butyrolactim methyl ether, Section IV.D.2). For the hydrogenation step, 39 g of caprylic lactim methyl ether, 5 g of cobalt obtained by reduction of cobalt oxide, and 120 g of ammonia are charged into a 1-liter autoclave. The autoclave is sealed and heated at 110°C for 4 hr. The autoclave is then pressurized to 70 atm with hydrogen and hydrogenation is carried on for 8 hr at 150°C. Upon cooling, the autoclave is vented and the contents are taken up in ether. Distillation of the ether extracts provides 28 g (77%) of 1,8-octanediamine, bp 112–113°C/11 mm, mp 56–57°C.

3. The Sebacic Acid Route

Two routes are available for the conversion of sebacic acid to 1,8-octane-diamine. In the first route the reaction of the diacid with ammonia or urea to sebacamide and conversion of the amide to 1,8-octanediamine with alkaline hypohalite is adopted (35). The second route employs the Schmidt reaction, in which sebacic acid is converted to the diazide and followed by rearrangement to 1,8-octanediamine (38). This route would not be commercially practical because of the high cost and potential hazards involving the use of sodium azide.

$$\text{HOOC(CH}_2)_8\text{COOH} \xrightarrow{\begin{array}{c}\text{NH}_3\end{array}} \text{H}_2\text{NC(CH}_2)_8\text{CNH}_2 \xrightarrow{\text{NaOBr}}$$

$$\xrightarrow[\text{H}_2\text{SO}_4]{\text{NaN}_3} \text{N}_3\text{C(CH}_2)_8\text{CN}_3 \longrightarrow \text{H}_2\text{N(CH}_2)_8\text{NH}_2$$

a. Via Sebacamide

In a reaction flask fitted with thermometer, stirrer, gas inlet tube, and vapor outlet leading to a water-cooled condenser, is placed 1010 g (5 moles) of sebacic acid. Heat is applied and anhydrous ammonia is introduced when the temperature reaches 150°C. Heating is continued until temperature reaches 200–220°C. At this point, the reaction temperature is held at 200–220°C for 7–15 hr, the flow of ammonia being maintained during this period. Water is continuously collected over the water-cooled condenser. When the amount of water collected reaches theoretical, the molten charge is transferred to a flat metal pan. The solidified crude amide is pulverized to about 20 mesh and extracted for several hours with 2% sodium hydroxide solution. Filtration, followed by washing with methanol and drying, results in the isolation of 80% yield of pure sebacamide, mp 210–212°C.

To prepare 1,8-octanediamine, the sebacamide obtained is used directly. In a reaction flask is placed a solution of 452 g (11.3 moles) of sodium hydroxide (40% excess) in 900 g of water and 1200 g of ice. To this is added dropwise 352 g (2.2 moles) of bromine with vigorous stirring at such a rate that little if any free bromine is present at any time. The temperature of the hypobromite solution is raised to 15°C, and 200 g (1 mole) of finely powdered sebacamide is added as rapidly as possible. The temperature of the resulting mixture rises gradually with the solution of all suspended matter. The final reaction temperature is about 60–65°C. When the resulting solution has cooled, it is extracted with ether by continuous liquid–liquid extraction technique for 8 hr to yield 71.5 g of 1,8-octanediamine (50%), bp 94°C/1 mm.

b. Via Sebacic Acid Azide

A quantity of 140 g of sodium azide is dissolved in 1400 g of concentrated sulfuric acid, the latter being covered with a layer of methylene chloride to retain any escaping hydrazoic acid. To this solution, 220 g of sebacic acid is slowly added, with cooling and stirring to maintain the reaction temperature at 25–30°C. When the evolution of gas has stopped, the mixture is diluted with water, resulting in a 30% sulfuric acid concentration. The methylene chloride layer is separated and the sulfuric acid solution is neutralized with soda. After the addition of an excess of sodium hydroxide, the crude 1,8-octanediamine, separating out as an oil, is dried over sodium hydroxide and purified by fractional distillation. Yield ranges from 85 to 90%.

K. 1,10-Decanediamine

The general method for the synthesis of long-chain aliphatic diamines is the reduction of the corresponding dinitrile. Amination of glycols has also been reported to yield diamines, but the yields are usually inferior (28,156). Thus 1,10-decanediamine has been obtained from the chemical reduction of sebaconitrile with sodium and methanol in 78% yield (157), and with lithium aluminum hydride in 50–58% yield (158,159). However, the best yield (80–90%) of 1,10-decanediamine is obtained from the catalytic hydrogenation of sebaconitrile in the presence of ammonia (160,161). The starting

$$NC(CH_2)_8CN \xrightarrow[\text{catalyst}]{H_2,\ NH_3} H_2N(CH_2)_{10}NH_2$$

sebaconitrile can be obtained from butadiene by the following sequence of reactions (160):

$$CH_2{=}CH{-}CH{=}CH_2 \xrightarrow{Cl_2} ClCH_2CH{=}CHCH_2Cl \xrightarrow[]{\text{partial cyanization}}$$

$$ClCH_2CH{=}CHCH_2CN \xrightarrow[Ni]{Fe}$$

$$NCCH_2CH{=}CHCH_2CH_2CH{=}CHCH_2CN \xrightarrow{H_2} NC(CH_2)_8CN$$

An alternate and preferred route to sebaconitrile (82% yield) starting from sebacic acid (obtained from castor oil) has been reported (162). This process involves the passing of ammonia through sebacic acid in the presence of a catalyst, such as phosphoric acid, at 250–300°C.

$$HOOC(CH_2)_8COOH \xrightarrow[H_3PO_4]{NH_3} NC(CH_2)_8CN$$

A similar path in which sebacic acid diamide is heated to 250°C in the presence of ammonium molybdate to yield sebaconitrile has also been described (163).

Finally, sebaconitrile can also be prepared starting from the telomerization of ethylene with carbon tetrachloride in the presence of benzoyl peroxide (164). The intermediate, 1,1,1,5-tetrachloropentane is obtained in 60% yield, together with lower and higher molecular weight telomers, which can be separated by distillation. Hydrolysis of 1,1,1,5-tetrachloropentane with sulfuric acid furnishes 5-chlorovaleric acid in 78% yield (164). The coupling of 5-chlorovaleric acid is performed under the Kolbe electrolysis conditions to yield 1,8-dichlorooctane (56%) (165). The desired sebaconitrile is then obtained by the reaction of 1,8-dichlorooctane with potassium cyanide in dimethyl formamide (165).

$$CH_2{=}CH_2 + CCl_4 \xrightarrow{\overset{\overset{\text{O} \quad \text{O}}{\underset{}{\|\quad\|}}}{C_6H_5COOCC_6H_5}} Cl(CH_2)_4CCl_3 \xrightarrow{H_2SO_4}$$

$$Cl(CH_2)_4COOH \xrightarrow{e^-} Cl(CH_2)_8Cl \xrightarrow{KCN} NC(CH_2)_8CN$$

A detailed procedure for the catalytic hydrogenation of sebaconitrile to 1,10-decanediamine is described in *Organic Syntheses* (161) and is summarized.

A quantity of 82 g (0.50 mole) of sebaconitrile is charged into a 1-liter autoclave. This is followed by 6 g of Raney nickel suspended in 50 ml of ethanol and 68 g (4 moles) of liquid ammonia. The autoclave is sealed and pressurized with hydrogen to 1500 psig, and heat is applied. Hydrogen absorption begins at 90°C and proceeds rapidly at 110–125°C. Hydrogenation lasts 1–2 hr. After the clave has been cooled, the excess hydrogen and ammonia are vented and the contents of the clave are discharged with the aid of ethanol. The ethanol solution is filtered through a layer of decolorizing carbon to remove the catalyst. Solvent is removed from the filtrate by atmospheric distillation and the remaining 1,10-decanediamine is distilled under reduced pressure. It boils at 143–146°C/14 mm and solidifies, on cooling, to a white solid that freezes at 60°C. The yield is 68–69 g (79–80%). On a fourfold scale-up the yield is 85–90%.

A large-scale reduction of sebaconitrile under conditions similar to those just described was reported in a British patent in 1938 (155). Detailed procedures for the preparation of nickel catalysts are also presented. An active nickel catalyst is prepared by dissolving 35 g of sodium metasilicate in 875 ml of water, and the solution is made slightly acidic by addition of 22 ml of concentrated nitric acid (73%). To this solution is added in succession 700 ml of nickel nitrate solution containing 70 g of nickel and 140 g of magnesia alba. After thorough mixing to produce a uniform suspension, 5600 ml of 5% sodium bicarbonate solution is added. The resulting slurry is coagulated by a few minutes of boiling. The solid is filtered, dried at 110°C, and calcined at 450°C for several hours to form a fine, pale green powder.

Finally, this powder is heated at 460–475°C under a stream of hydrogen. The resulting nickel catalyst is pyrophoric.

To carry out the reduction, 2000 g of sebaconitrile, 750 g of methanol, 700 g of ammonia, and 300 g of nickel catalyst are charged in a stirred autoclave. The clave is pressurized to 1000–1800 psig with hydrogen. The mixture is heated at 110–115°C until hydrogen absorption ceases (3–3.5 hr). The clave is cooled and vented. The crude reaction product mixture is filtered and fractionally distilled to obtain 1680 g (80% conversion) of 1,10-decanediamine, mp 60°C, bp 139–140°C/12 mm.

L.   1,12-Dodecanediamine

1,12-Dodecanediamine can be prepared by either one of the two reported methods for straight-chain polymethylene diamines. In the first method, a lactam is converted to the corresponding lactim ether and followed by catalytic hydrogenation to the diamine (45). The starting lauryllactam for

$$(CH_2)_{11}\!\!-\!\!C\!=\!\!O \xrightarrow{(CH_3)_2SO_4} (CH_2)_{11}\!\!-\!\!C\!-\!OCH_3 \xrightarrow[Co,\,NH_3]{H_2} H_2N(CH_2)_{12}NH_2$$

the foregoing process may be obtained from butadiene by the following sequence of reactions (104). Cyclization of 3 molecules of butadiene yields 1,5,9-cyclododecatriene. Oxidation of this triene with peracetic acid, which is then hydrogenated, furnishes the saturated epoxide. Cyclododecanone is obtained by rearrangement of the epoxide with magnesium iodide, and finally the desired lauryllactam is synthesized via the Beckmann rearrangement of the oxime.

The intermediate 1,5,9-cyclododecatriene is commercially available. However, in the final two steps—i.e., the conversion of the lauryllactam to the lactim ether and then to 1,12-dodecanediamine—the experimental conditions given are rather sketchy and no yields are mentioned (45).

The preferred method of preparing 1,12-dodecanediamine is the reduction of 1,10-dicyanodecane. This reduction can be carried out with either aluminum triisobutyl (166) or aluminum diisobutyl hydride in 78% yield (167), or by catalytic hydrogenation in the presence of Raney nickel (165). The starting 1,10-dicyanodecane may be prepared from 1,12-dodecanedioic acid via the ammonium salt (104). A recent reported synthesis of 1,10-dicyano-

$$HOOC(CH_2)_{10}COOH \xrightarrow{NH_3} H_4NOOC(CH_2)_{10}COONH_4 \xrightarrow{-H_2O} NC(CH_2)_{10}CN$$

decane employs the Kolbe electrolysis (165) of 6-chlorocaproic acid (168,169) to 1,10-dichlorodecane. The starting 6-chlorocaproic acid may be prepared from the reaction of caprolactam, hydrochloric acid, and sodium nitrite (169). The 1,10-dichlorodecane is converted to 1,10-dicyanodecane with potassium cyanide in dimethyl formamide (165).

$$2Cl(CH_2)_5COOH \xrightarrow{e^-} Cl(CH_2)_{10}Cl \xrightarrow{KCN} NC(CH_2)_{10}CN$$

The detailed experimental conditions for this process follow.

1. 1,10-DICHLORODECANE

In a glass cell ($10 \times 10 \times 1$ cm³) is fitted with two electrodes of titanium plates plated with platinum ($8 \times 8 \times 0.3$ cm³). The electrodes are 2 mm apart. The cell is placed in a cooling bath, and a solution consisting of 150 g (1 mole) of 6-chlorocaproic acid, 1.1 g of sodium, and 500 ml of absolute methanol is vigorously circulated through the cell by means of a circulation pump. The current density is maintained between 0.10 and 0.12 A/cm² throughout the reaction. The temperature in the cell is maintained below 50°C (30–40°C). The reaction is continued until the solution becomes slightly alkaline to litmus. The crude product mixture is then acidified with hydrochloric acid and most of the solvent is removed by distillation. The residue is extracted with ether. The ether extract is washed with 5% sodium carbonate solution to remove unreacted acid, whereupon it is dried with anhydrous sodium sulfate and then stripped. Fractional distillation of the remaining liquid furnishes 68 g (65% yield) of the desired 1,10-dichlorodecane, bp 101–102°C/1 mm, n_D^{25} 1.4580.

2. 1,10-DICYANODECANE

A mixture of 60 g (0.28 mole) of 1,10-dichlorodecane, 45 g of potassium cyanide, and 200 ml of dimethylformamide is heated at 110–115°C for 8 hr. The precipitated potassium chloride is removed by filtration and the filtrate is distilled to provide 40.5 g (75%) of 1,10-dicyanododecane, bp 145–146°C/ 2 mm, n_D^{16} 1.4550.

3. 1,12-DODECANEDIAMINE

The hydrogenation of 1,10-dicyanodecane is carried out in an autoclave rated at 110–130 atm operating pressure. Thus 20 g (1.04 moles) of 1,10-dicyanodecane, 8 g of Raney nickel (W-7) and 200 ml of methanol containing 20 g of ammonia are charged into the autoclave. Hydrogenation is performed at 100–110°C under 110–130 atm of hydrogen pressure for a 2-hr period. The catalyst is removed by filtration. After the volatile materials and solvent have been removed, the filtrate is distilled to furnish 13.8 g (66%) of 1,12-dodecanediamine, bp 145–148°C/2 mm, mp 65.5–66°C.

M. Bis(p-aminocyclohexyl)methane

The condensation of aniline and formaldehyde in the presence of hydrochloric acid provides the aromatic diamine, 4,4′-methylenedianiline. The conditions under which this condensation yields the diamine have been studied (170), and a number of commercial processes are available (171,172). Catalytic hydrogenation of 4,4′-methylenedianiline with either an alkali-promoted cobalt oxide catalyst system (47) or with ruthenium catalysts (47,48) furnishes the desired bis(p-aminocyclohexyl)methane in excellent yield.

$$\text{C}_6\text{H}_5\text{—NH}_2 \; + \; \text{HCHO} \; \xrightarrow{\text{HCl}} \; \text{H}_2\text{N—C}_6\text{H}_4\text{—CH}_2\text{—C}_6\text{H}_4\text{—NH}_2 \; \xrightarrow{\text{H}_2}$$

$$\text{H}_2\text{N—C}_6\text{H}_{10}\text{—CH}_2\text{—C}_6\text{H}_{10}\text{—NH}_2$$

An alternate route is the catalytic amination of 4,4′-methylenedicyclohexanol, which is prepared from the condensation of phenol and formaldehyde followed by hydrogenation of the resulting 4,4′-methylenediphenol (173).

$$\text{HO—C}_6\text{H}_{10}\text{—CH}_2\text{—C}_6\text{H}_{10}\text{—OH} \; \xrightarrow{\text{NH}_3} \; \text{H}_2\text{N—C}_6\text{H}_{10}\text{—CH}_2\text{—C}_6\text{H}_{10}\text{—NH}_2$$

Since 4,4′-methylenedianiline is readily available commercially, the routes employing this intermediate are considered the processes of choice. It should be pointed out that the product—bis(p-aminocyclohexyl)methane—exists in three geometrically isomeric forms; namely, trans-trans, cis-trans, and cis-cis. The isolation and identification of the three possible geometric isomers have been described by Barkdoll and co-workers (174). The bis(p-amino-cyclohexyl)methane product obtained from the catalytic hydrogenation of 4,4′-methylenedianiline invariably contains all three isomers, and the ratio of these isomers is dependent on the type of catalyst and on the reaction conditions (51,175). Separation techniques have been developed to change the ratio of the isomers in the product to suit application requirements (176–178).

1. BATCH HYDROGENATION USING COBALT OXIDE

The hydrogenation catalyst system of cobalt oxide–calcium oxide–sodium carbonate is employed (47,174). The cobalt oxide is prepared by heating cobaltous nitrate hexahydrate in the presence of air at temperatures ranging from 200 to 750°C for 15–24 hr. Calcium oxide is prepared by heating either reagent-grade samples of calcium hydroxide (500°C) or calcium carbonate (750°C) in a muffle furnace in air for 24 hr. Reagent-grade sodium carbonate is used as such without further purification. To carry out the hydrogenation, 100 g of 4,4′-methylenedianiline, 10 g of cobalt oxide, 15 g of calcium oxide, and 6.5 g of anhydrous sodium carbonate are charged into a rocking-type pressure vessel. Hydrogenation is performed at 215°C under 120–220 atm of pressure for 6 hr. After cooling, the contents of the vessel are discharged with the aid of methanol and then filtered through a bed of activated charcoal to remove the catalyst. In view of the pyrophoric nature of the catalyst, precautions should be taken. The filtrate is then stripped and distilled to provide an 86% yield of bis(p-aminocyclohexyl)methane.

2. BATCH HYDROGENATION USING RUTHENIUM DIOXIDE

The method described by Barkdoll et al. (47) serves in the batch hydrogenation of 4,4′-methylenedianiline using ruthenium dioxide. To a 1-gal stirred autoclave is charged 1250 g (6.3 moles) of 4,4′-methylenedianiline, 3125 ml of dioxane, and 25 g of ruthenium dioxide. After the clave has been purged with hydrogen to 200 psig several times to remove air, hydrogenation is made at 100–120°C and 166–200 atm of pressure for 4 hr. The reaction mixture is cooled and rinsed out with methanol. After the crude product has been filtered to remove catalyst and distilled at atmospheric pressure to remove solvent, it is fractionated under reduced pressure to provide 1223 g (92% based on 4,4′-methylenedianiline charged) of bis(p-aminocyclohexyl)methane, bp 120°C/0.8 mm–128°C/1.2 mm, n_D^{25} 1.5051; neutral equivalent 105.1 (calculated 105.2).

3. BATCH HYDROGENATION USING RUTHENIUM ON ALUMINA

Batch hydrogenation using ruthenium on alumina differs from the process using ruthenium dioxide in that the hydrogenation is performed in the presence of ammonia under higher pressure conditions (48). This process is reported to be characterized by a very short reaction time and a high trans-trans isomer content in the product. A high trans-trans isomer content is needed in the product for subsequent formation of polyamide having desirable properties. A wide variety of ruthenium catalysts may be used. Supported ruthenium catalysts may be prepared by fusing ruthenium with sodium peroxide and saturating the support with a solution of the fused salt followed by drying.

In operation 100 g of 4,4'-methylenedianiline, 100 ml of dioxane, 20 g of ammonia, and 12 g of finely divided catalyst of 5% ruthenium on γ-alumina are placed in a pressure vessel and hydrogenation is made at 4000 psig and 200°C for 3.75 min. Following an additional 15-min hold-up period at 5000 psig, the hydrogenation is stopped. After cooling and venting, the crude product mixture is filtered and solvent is removed by distillation. The desired bis(p-aminocyclohexyl)methane is obtained in 96% yield based on 4,4'-methylenedianiline by fractionation under reduced pressure. The distilled product has a freezing point of about 43.5°C and contains about 54% of the trans-trans isomer.

4. Continuous Hydrogenation Using Ruthenium on Alumina

The batch conditions using ruthenium on alumina just described may be adopted to a continuous process (48). A slurry consisting of 4 parts by weight of 5% ruthenium on alumina, 10 parts of ammonia, 50 parts of n-butyl ether, and 30 parts of 4,4'-methylenedianiline is continuously charged into the top of a vertical reactor fitted with a gas exit port at the top and a bottom exit for liquid. Hydrogen is introduced into the reactor near the bottom. The reaction is carried out at 5000 psig pressure and 225°C. The average residence time in the reactor is 6 min. The reactor effluent is decompressed and un-reacted ammonia is evaporated and recycled. Catalyst is recovered by centrifugation for reuse. The crude product mixture is distilled to recover solvent and the crude amine is further distilled to isolate product. From 30 parts of 4,4'-methylenedianiline, the following materials are obtained: 0.6 part of low boilers; 25.8 parts of bis(p-aminocyclohexyl)methane, freezing point 40°C; 3 parts of 4-(4-aminobenzyl)cyclohexylamine; 0.3 part of starting 4,4'-methylenedianiline; and 0.3 part of high boiling tar. The materials that are not fully hydrogenated are recycled as a portion of the feed.

N. 1,3-Diaminocyclohexane

1,3-Diaminocyclohexane may be prepared from the Curtius method (53) from dimethyl cyclohexane-1,3-dicarboxylate, which is obtained from the hydrogenation of the corresponding phthalate derivative (179). However, this complicated route consists of the conversion of the dimethyl ester to the dihydrazide with hydrazine and then to the diazide with sodium nitrite, followed by treatment with alcohol to the diurethane and finally hydrolysis with concentrated hydrochloric acid, furnishing the dihydrochloride salt of 1,3-diaminocyclohexane.

H_3COOC—⬡—$COOCH_3$ $\xrightarrow{H_2NNH_2}$ H_2NNHCO—⬡—$CONHNH_2$

$\xleftarrow{HNO_2}$

N_3CO—⬡—CON_3 $\xrightarrow{C_2H_5OH}$ H_5C_2OCONH—⬡—$NHCOOC_2H_5$

$\xleftarrow{HCl}$

H_2N—⬡—$NH_2 \cdot 2HCl$

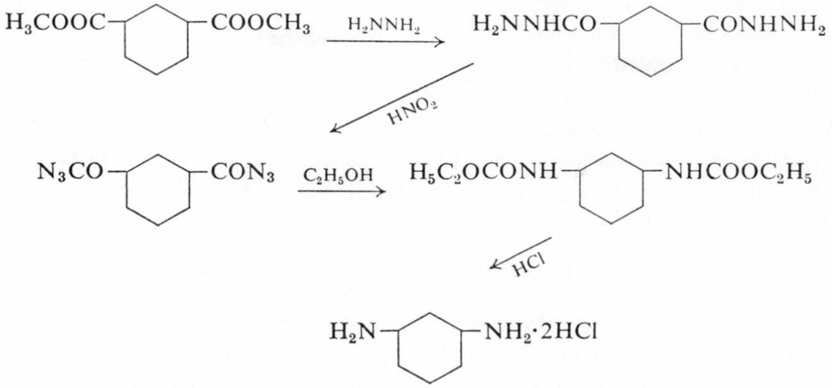

m-Dinitrobenzene is commercially prepared by the nitration of benzene in a stepwise manner (180). Reduction of *m*-dinitrobenzene to *m*-nitroaniline is readily accomplished with sodium sulfide (18). *m*-Nitroaniline is either hydrogenated to 1,3-diaminocyclohexane directly (53) or reduced first to *m*-phenylenediamine with a platinum–barium sulfate catalyst (53) followed by hydrogenation to 1,3-diaminocyclohexane in the presence of ruthenium dioxide (49).

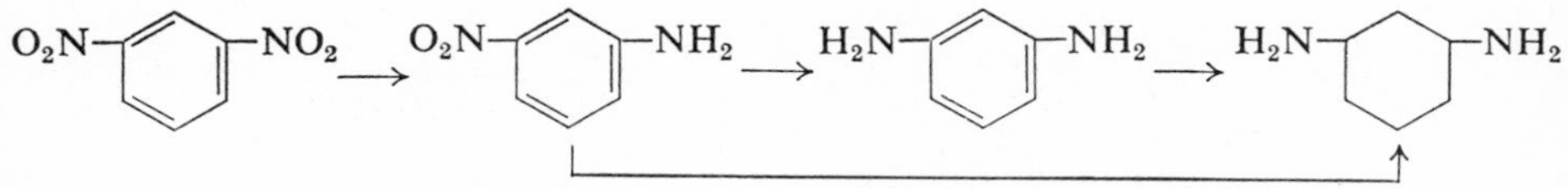

O_2N—⬡—NO_2 → O_2N—⬡—NH_2 → H_2N—⬡—NH_2 → H_2N—⬡—NH_2

The method of choice for the preparation of 1,3-diaminocyclohexane, however, is the direct hydrogenation of *m*-dinitrobenzene (51). The hydro-

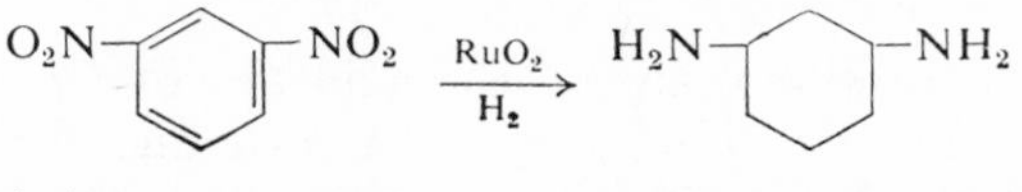

O_2N—⬡—NO_2 $\xrightarrow[H_2]{RuO_2}$ H_2N—⬡—NH_2

genation is carried out in the absence of ammonia with ruthenium dioxide as catalyst. The reduction of both the nitro groups and the aromatic ring is accomplished in one step under relatively low temperature conditions.

A mixture of 50.5 parts by weight of *m*-dinitrobenzene, 125 parts of dioxane, and 2.5 parts of ruthenium dioxide is heated and agitated in a pressure vessel under 125 atm of hydrogen pressure. Reduction of the nitro groups and the hydrogenation of the aromatic ring take place at 75 and 100°C, respectively. When hydrogenation is complete, the product is filtered and distilled to give a 75% of theoretical yield of 1,3-diaminocyclohexane, bp 95–97°C/30 mm, neutral equivalent 57.1 (calculated 57.1).

There are two possible geometric isomers for 1,3-diaminocyclohexane; namely, cis and trans. The product obtained from the hydrogenation of *m*-phenylenediamine consists of almost entirely the trans isomer (49). The high stereospecificity of this process is demonstrated by the absence of isomerization when a mixture of *cis*- and *trans*-1,3-diaminocyclohexanes is treated under the hydrogenation conditions. It is suggested that the 1,3-aminocyclohexane products obtained from the Curtius reaction, as well as from the reduction of *m*-nitroaniline, are in the cis form (53). The stereochemistry of the reduction product from *m*-dinitrobenzene is not described.

O.   1,4-Diaminocyclohexane

The synthetic routes developed for 1,4-diaminocyclohexane are quite similar to those described earlier for the preparation of 1,3-diaminocyclohexane. One method involves the Curtius reaction (52) starting from dimethyl cyclohexane-1,4-dicarboxylate, which is made from the catalytic hydrogenation of dimethyl terephthalate (182). Although the yield is quite good for each step of this four-step reaction, the multistep synthesis and the formation of the unstable diazide as an intermediate render this process of little commercial interest.

Nitration of chlorobenzene gives rise to a mixture of *o*- and *p*-chloronitrobenzene. The *p*-chloronitrobenzene is separated and aminated to yield *p*-nitroaniline (180). *p*-Nitroaniline is also produced by nitration of acetanilide and hydrolysis of the resulting *p*-nitroacetanilide (180). The preferred method of preparing 1,4-diaminocyclohexane is the hydrogenation of *p*-nitroaniline with ruthenium dioxide (51). 1,4-Diaminocyclohexane may also be prepared by a stepwise reduction of *p*-nitroaniline to *p*-phenylenediamine (180) and then to 1,4-diaminocyclohexane (49).

The direct reduction of *p*-nitroaniline to 1,4-diaminocyclohexane is performed in a pressure vessel (51). Thus a mixture of 50 parts by weight of *p*-nitroaniline, 125 parts of absolute ethanol, and 2.5 parts of ruthenium dioxide is shaken in the pressure vessel under 135 atm of hydrogen pressure. The reduction of the nitro group occurs at 80–85°C and the hydrogenation of the aromatic ring at 105–110°C. After the reaction is completed, the crude product mixture is worked up by distillation to give a 74% theoretical yield of 1,4-diaminocyclohexane, bp 90°C/22 mm, neutral equivalent 57.5 (calculated 57.1). Although the stereochemistry of the product thus obtained is not discussed, a similar reduction reported by Ponomarev and co-workers indicates that only the trans isomer of the two possible *trans-* and *cis*-1,4-diaminocyclohexanes is obtained (183).

The hydrogenation of *p*-phenylenediamine is performed with ruthenium catalysts, with 5% ruthenium on carbon being the catalyst of choice (49). In this reaction 27 g (0.25 mole) of *p*-phenylenediamine in three- to tenfold by weight of methanol and 4 g of 5% ruthenium on carbon in an autoclave are pressurized with hydrogen to 110 atm. Heating and agitation are initiated and hydrogenation is carried out at 160°C. When the required amounts of hydrogen are absorbed, the autoclave is discharged, catalyst is removed by filtration, and the product is isolated by distillation. The desired 1,4-diaminocyclohexane is obtained in 88% yield almost exclusively in the trans form.

P. Methyldiaminocyclohexanes

The ring methyl-substituted *m*-cyclohexylenediamine derivatives to be discussed are 1-methyl-2,4-diaminocyclohexane and 1-methyl-2,6-diaminocyclohexane. Stepwise dinitration of toluene yields either 2,4-dinitrotoluene after isolation of 4-nitrotoluene as the intermediate or a mixture of 2,4- and 2,6-dinitrotoluenes of different proportions (184). Hydrogenation of the mixed 2,4- and 2,6-dinitrotoluenes furnishes the corresponding tolylenediamines (184). The desired mixed methyldiaminocyclohexanes are obtained from catalytic hydrogenation of the aromatic amines (50).

The hydrogenation of tolylenediamine is carried out by charging 153.8 g (1.26 moles) of a tolylenediamine mixture containing 80% of the 2,4-isomer and 20% of the 2,6-isomer, 5 g of ruthenium dioxide, and 600 ml of dry dioxane in a 1-liter stirred autoclave. The air in the autoclave is displaced by hydrogen by purging five times with hydrogen to 200–250 psig. Hydrogenation is then performed at 120–140°C under 3390–4400 psig of pressure for 40 hr. The solvent is removed from the crude product mixture, which is then distilled under reduced pressure to provide 126 g of mixed methyldiaminocyclohexane, bp 73.5–75°C/3.0–3.1 mm, neutral equivalent 64.4 (calculated 64.2). The yield based on tolylenediamine charged is 78.2%.

Q. Piperazine

Piperazine is offered commercially by Jefferson, Union Carbide, and Dow, and a number of patents have been issued to these companies covering the synthesis of the product. Briefly, piperazine is prepared by cyclization of monoethanolamine (54,185–188) and diethanolamine (189), pyrolysis of monoethanolamine hydrochloride (190) and 2-(2-aminoethylamino)ethanol dihydrochloride (191,192), and reaction of 1-(2-aminoethyl)aziridine with hydrochloric acid followed by sodium hydroxide (193). Jefferson now employs the cyclization of monoethanolamine in the presence of a nickel-copper-chromia catalyst system and ammonia as the process yielding piperazine as the major product (54). Unlike the processes for alkyl-substituted piperazines, this process gives rise to a large number of other products, such as ethylenediamine, diethylene triamine, N-aminoethylpiperazine, aminoethylethanolamine, and hydroxyethylpiperazine. Subsequently, extensive distillations and recycling operations are needed to provide a good piperazine conversion. Some side products, such as 1-(2-aminoethyl)piperazine and triethylenediamine, are useful materials and can be produced along with piperazine with certain modifications of the basic process (186,187).

The basic process employs a fixed-bed reactor, and the reaction is carried out continuously under superatmospheric pressure. The catalyst for the fixed bed, which is prepared by pressing into pellets a mixture of 75 mole% nickel oxide, 23 mole% copper oxide, and 2 mole% chromium oxide, is prereduced with hydrogen at 200–400°C. The reduced catalyst charged to the reactor actually contains metallic nickel, metallic copper, and chromium oxide.

The general operation involves feeding to a Dowtherm heated reactor monoethanolamine, hydrogen, ammonia, and water. The reactor temperature is set at 220–230°C, and reaction pressure at 200 atm with hydrogen partial pressure amounts to between 60 and 80%, approximately, of the total pressure. The mole ratio of ammonia to monoethanolamine ranges from about 2.5 to 4.5, and water constituting between 15 and 100 wt.% of the total amine feed stock is used. Contact time corresponding to a feed rate of 0.5–5 lb/hr of monoethanolamine per pound of catalyst may be employed. Under these conditions, conversions of 75–95% of monoethanolamine to amine products can be realized. Table 4 summarizes the various amine

TABLE 4
Boiling Points of Amine Products from the
Cyclization of Monoethanolamine (54)

Compounds	bp, °C
N-Methylethylenediamine	115
Ethylenediamine	117
N-Ethylethylenediamine	129
N-Methylpiperazine	135
Piperazine	145
N-Ethylpiperazine	156
Monoethanolamine	170
Diethylenetriamine	207
N-Aminoethylpiperazine	221
N-Hydroxyethylpiperazine	242
2-(2-Aminoethylamino)ethanol	242

products and lists their boiling points.

For the workup procedure, the reactor effluent which comprises hydrogen, ammonia, water, and amines, is degassed, resulting in a mixture containing 10–50 wt.% water and 5–10 wt.% ammonia. The water content of this mixture is reduced to about 5–15% by distillation. Continued distillation gives a light distillation fraction, bp 102–145°C, consisting of 20–40 wt.% piperazine, all the remaining water, N-methylpiperazine, N-ethylpiperazine, and most of the ethylenediamine. This fraction may be redistilled to remove a mixture of water, N-alkylpiperazines, and ethylenediamine, which can be recycled to the monoethanolamine charge. The bottoms of the redistillation, containing most of the piperazine from the light distillation fraction, may be further purified to isolate piperazine by distillation. Continued distillation of the reactor effluent, after the light distillation fraction has been removed, affords a fraction, bp 145–152°C, which is almost pure piperazine. The next fraction is collected approximately between 152 and 180°C and comprises unconverted monoethanolamine and diethylenetriamine. This fraction may be returned to the reactor for further reaction. Finally, vacuum distillation of the remaining material yields substantially pure N-aminoethylpiperazine.

Table 5 summarizes the results of two runs performed in a reactor containing 15 gal of nickel-copper-chromia catalyst, under the reaction and workup conditions just described.

TABLE 5
Reaction Conditions and Products of Cyclization of
Monoethanolamine (54)

	Run 1	Run 2
Reaction temperature, °C	223	227
Hydrogen rate, SCFH[a]	800	800
Reaction pressure, psig	2800	2800
Feed rates, gal/hr		
Monoethanolamine (MEA)	158	158
Ammonia (anhydrous)	80	80
Water	140	140
NH_3/MEA mole ratio	1.1	1.1
Space velocity, g/hr, ml catalyst	3.1	3.1
Conversion of MEA, %	71.4	75.4
Yields (molar), %		
Ethylenediamine	12.0	15.0
Piperazine	47.6	58.0
Diethylenetriamine	9.0	3.3
N-aminoethylpiperazine	15.0	17.5
Hydroxyethylpiperazine	6.0	3.5
Residue	10.4	2.7

[a] Standard cubic feet per hour.

R. 2-Methylpiperazine

The synthesis of 2-methylpiperazine consists of the cyclization of N-(2-hydroxypropyl)ethylenediamine in the presence of a hydrogenation/dehydrogenation catalyst.

The starting ethylenediamine derivative is readily prepared by the reaction of excess ethylenediamine with propylene oxide (194). There are two processes reported in the patent literature for the synthesis of 2-methylpiperazine—a batch atmospheric process (55) and a superatmospheric, continuous process (195). Although the continuous process is more readily adaptable to scale-up operation, the atmospheric, batch process is more convenient for small-scale preparation. Therefore, both processes are described.

1. Batch Process

The batch process involves the heating of a mixture of N-(2-hydroxypropyl)ethylenediamine and Raney nickel catalyst. Any finely divided nickel hydrogenation/dehydrogenation catalyst can be used. A most convenient source is Raney Catalyst Company, Chattanooga, Tennessee, which supplies such a catalyst as a suspension in water (commercially sold as Raney nickel catalyst). Excess water is drained from the catalyst before use, and the actual weight of nickel (dry weight) in the "wet catalyst" after draining is about half the weight of the wet catalyst.

For this process, a mixture having a concentration of 4.3% catalyst (dry weight basis) of N-(2-hydroxypropyl)ethylenediamine and wet catalyst is heated and stirred in a three-necked glass flask equipped with stirrer, thermometer, and simple distillation head vented through a condenser. The mixture is heated to 100°C and maintained at this temperature for 4 hr. The presence of water, both from the catalyst and as a reaction product, serves to provide an effective upper temperature limit during the heating period. The heat applied to the reaction flask is then increased, and the crude 2-methylpiperazine is distilled until no further product is obtained. The reactor temperature ranges from 200 to 250°C at the end of distillation. The 2-methylpiperazine in the crude distillate corresponds to a yield of 88–91%.

Further purification is effected by fractional distillation in a 120 cm × 9 mm glass column, packed with a Nichrome spiral surrounded by an electrical heating jacket. The purified 2-methylpiperazine is obtained by collecting the fraction boiling between 151 and 156°C. The purity of the material is usually better than 98% by neutral equivalent determination.

2. Continuous Process

The main reactor for the continuous process consists of a stainless steel tube, 1 in. I.D. A 24-in.-long section is packed with 310 g of Harshaw catalyst Ni 0104-T, a nickel catalyst containing 60% nickel on Kieselguhr and supplied as $\frac{1}{8}$-in. pelleted tablets. A 12-in. section of the reactor immediately above the catalyst bed is filled with an inert packing, such as $\frac{1}{4}$-in. Raschig rings, to serve as a preheater for the feed to the reactor. The reactor is encased in a Dowtherm-heated jacket. The system is purged of air with hydrogen and heated to 130°C. Hydrogen is admitted to the system from the top to maintain a pressure of 50 psig throughout the run.

The process is carried out by pumping a mixture of N-(2-hydroxypropyl)-ethylenediamine and water (50–50 wt.%) as a feed solution, downward through the catalyst bed at a rate of 100 ml/hr. The condensed reactor effluent is then fractionally distilled and the product is collected at 151–156°C. A typical run employing these conditions is carried out for 225 hr and the overall yield, based on isolated 2-methylpiperazine, is 80 wt.%.

The variables studied for this process include reaction temperature and pressure. Hydrogen pressure and temperature are to some extent inter-dependent. At temperatures below about 160°C, increase in pressure beyond 50 psig gives decreased yield; whereas at temperatures above 160°C, increased hydrogen pressure in the range of 200–300 psig actually enhances the yield.

The effect on yield of amount of diamine passed over a given amount of catalyst per unit time (feed rate) is also studied. The best yield is obtained when a feed rate of 0.25–0.50 g diamine per gram of catalyst per hour is employed.

The use of water as diluent in the feed stock greatly improves the yield of product. This is exemplified by a run using anhydrous amine feed. An initial yield of 72% was decreased gradually to 55–59% at the end of 150 hr. When the anhydrous amine feed was replaced with an aqueous amine feed containing 50 wt.% water at this point, the yield for the next 12 hr was about 82%.

S. 2,5-Dimethylpiperazine

A number of methods have appeared in the patent literature since 1958 for the synthesis of 2,5-dimethylpiperazine. These include the following batch processes: cycloamination of isopropanolamine (56,196–198) and 2-amino-1-propanol (199), and catalytic hydrogenation of 2,5-dimethyl-pyrazine (200). A vapor-phase, continuous process using isopropanolamine has also been reported (56). Although none of the processes has reached the commercial stage, the bimolecular cycloamination of isopropanolamine is believed to be the commercially practical procedure.

$$2CH_3\overset{\displaystyle OH}{\underset{\displaystyle |}{CH}}CH_2NH_2 \longrightarrow$$

The starting isopropanolamine, which is obtained by the reaction of ammonia with propylene oxide (84), is currently available from Union Carbide. The condensation reaction is carried out under hydrogen pressure in the presence of Raney nickel as catalyst. The desired 2,5-dimethyl-piperazine is isolated as a mixture of trans and cis isomers, together with unreacted isopropanolamine and a side product, 2,5-dimethylpyrazine. *Trans*-2,5-dimethylpiperazine, mp 118°C, is the preferred isomer to serve as monomer in the preparation of condensation polymers. In the following procedure, therefore, the conditions for obtaining *trans*-2,5-dimethyl-piperazine as the major product, the separation of the trans isomer from the mixture, and conditions for the cis–trans conversion are described.

Isopropanolamine and wet Raney nickel containing about 50% water are charged into an autoclave equipped with an agitator and designed for operation at pressures up to 1200 psig. The ratio of catalyst to amine is 5 g of wet Raney nickel per mole of isopropanolamine. The agitator is started and air is removed by purging with nitrogen, and the nitrogen is then replaced by hydrogen. The autoclave is pressurized to 1200 psig and heated rapidly to 220°C, with occasional venting to compensate for the thermal expansion of hydrogen. The reaction is allowed to proceed for 4–8 hr. The contents is cooled to below 80°C and the pressure is gradually released. A quantity of water corresponding to about 10 wt.% of the isopropanolamine charge is added in order to prevent *trans*-2,5-dimethylpiperazine from crystallizing during the removal of the product from the clave. The aqueous mixture is filtered to remove Raney nickel, and the filtrate is worked up by fractional distillation through a 120 cm × 2.8 cm helix-packed column. The side product 2,5-dimethylpyrazine is first removed as a water azeotrope boiling at 98°C and containing about 30–35% 2,5-dimethylpyrazine; the yield of 2,5-dimethylpyrazine is about 1%, based on the isopropanolamine charged. Distillation is then continued to a head temperature of 155°C to remove the remaining water, which contains minor amounts of 2,5-dimethyl-pyrazine and 2,5-dimethylpiperazine. The dimethylpiperazine product fraction, together with unreacted isopropanolamine, is collected at 155–170°C. The unreacted isopropanolamine is removed from the dimethyl-piperazine product fraction by addition of 20–40 wt.% of the dimethyl-piperazine fraction of ethylbenzene and distillation of the ethylbenzene-isopropanolamine azeotrope at about 131°C. Toluene and chlorobenzene can also be employed as azeotroping agents for the removal of isopropanol-amine (201). The remaining material, a mixture of *trans*- and *cis*-2,5-dimethyl-piperazine, corresponds to a 77% yield of product. The composition of this product, as measured by infrared absorption procedures (198), is 80% trans and 20% cis.

Because of small differences among the boiling points of isopropanolamine and reaction products, direct fractional distillation is not a practical method of separation. The boiling points and melting points of the four components in the crude product mixture are listed in Table 6.

TABLE 6
Melting and Boiling Points of Cycloamination Products of
Isopropanolamine

	bp, °C	mp, °C
Trans-2,5-dimethylpiperazine	160	118
Cis-2,5-dimethylpiperazine	162	18
2,5-Dimethylpyrazine	153	15
Isopropanolamine	159	1

The following procedure (56) describes the separation of the *trans*-2,5-dimethylpiperazine from *cis*-2,5-dimethylpiperazine. A mixture consisting predominantly of trans isomer, such as the 80–20 mixture just described, is dissolved in 1.2 times its weight of heptane at 85–95°C and the solution is cooled to room temperature to obtain the crystalline *trans*-2,5-dimethylpiperazine. The cis isomer, being more soluble in heptane, remains in solution. The trans isomer is filtered and washed twice with heptane fractions weighing 0.4 times the weight of the original 2,5-dimethylpiperazine mixture. Further purification of the trans isomer is accomplished by repeat recrystallization from heptane. The original heptane filtrate and washings are combined and distilled to obtain a fraction containing predominantly the cis isomer and minor amounts of the trans isomer.

For the isomerization of *cis*-2,5-dimethylpiperazine to *trans*-2,5-dimethylpiperazine, the liquid cis isomer is heated in a sealed autoclave in the presence of hydrogen and a suitable catalyst. One method involves the use of a copper–barium chromite catalyst at 240°C at 1700 psig for 2.5 hr to give 80% cis–trans conversion (202). However, the isomerization can also be effected by employing the conditions described previously for the preparation of 2,5-dimethylpiperazine from isopropanolamine (56). Thus either a mixture of cis and trans isomers containing predominantly *cis*-2,5-dimethylpiperazine or a blend of the above-mentioned mixture with isopropanolamine is heated in an autoclave at 210°C under 1200 psig hydrogen pressure for 4 hr to yield 2,5-dimethylpiperazine rich in the trans isomer. This mixture can then be worked up by the separation procedure described earlier.

T.   2,6-Dimethylpiperazine

The process for the synthesis of 2,6-dimethylpiperazine involves the condensation of diisopropanolamine with ammonia in the presence of a nickel or cobalt hydrogenation/dehydrogenation catalyst (57). The starting diisopropanolamine, available from Union Carbide, is obtained from reaction of ammonia with propylene oxide (10).

This batch process is carried out by heating diisopropanolamine with about twofold excess of ammonia to about 170–200°C in the presence of a nickel or cobalt catalyst, under hydrogen pressure. The pressures range from 350 to 950 psig and a reaction time of 4–6 hr is used. The commercially available Raney nickel and Raney cobalt catalyst (Raney Catalyst Company,

Chattanooga, Tenn.) are found to be quite effective. A catalyst concentration of about 2 g/mole of diisopropanolamine will effect reaction, but best yields are obtained with the ratio of 10 g of catalyst per mole of amine. Both anhydrous or aqueous ammonia can be used, since water is an effective diluent for this process. Under these reaction conditions, yields of 60 to more than 70% of 2,6-dimethylpiperazine are generally obtained. The following is a detailed example of this process.

To a 1-gal stainless steel, stirred autoclave of 1200-psig operating pressure are charged 5 moles of diisopropanolamine and 50 g of Raney cobalt catalyst drained of excess water. The air in the autoclave is then replaced by nitrogen. A quantity of 10 moles of anhydrous ammonia from a tared cylinder is added to the autoclave by nitrogen pressure. The autoclave is heated at 190°C for 6 hr. The pressure of the system rises from 200 to 930 psig. At the end of the reaction period, the contents of the clave is cooled and filtered to remove catalyst. After ammonia and volatiles have been removed, the crude reaction product is fractionally distilled. The fraction boiling from 155–165°C is the desired 2,6-dimethylpiperazine, and the yield is 73% based on diisopropanol-amine charged. The product can be further purified by recrystallization from heptane, mp 113–114°C.

The product thus obtained corresponds to the *cis*-2,6-dimethylpiperazine reported by Pope and Read (203) and Cignarella (61). Pope and Read isolated minor quantities of *cis*-2,6-dimethylpiperazine from the reduction of isonitrosoacetone followed by hydrogenation of the resulting pyrazine derivative, the major product being 2,5-dimethylpiperazine. Cignarella subsequently reported the preparation of *cis*-2,6-dimethylpiperazine, mp 115–116°C, from the cyclization of 2,2′-iminodipropionic ethyl ester benzyl-amide to 4-benzyl-*cis*-2,6-dimethyl-3,5-diketopiperazine and lithium alu-minum hydride reduction of the diketopiperazine derivative, followed by catalytic debenzylation of the resulting 4-benzyl-*cis*-2,6-dimethylpiperazine (61). The trans isomer of 2,6-dimethylpiperazine has been reported by Ishiguro et al. (204) by the catalytic dehydrogenation of *N*-(2-hydroxypropyl)-2-aminopropylamine, bp 162–166°C; ditosylate, mp 199.5–200.5°C (hot ethanol).

U. Meta- and Paraxylylenediamines

Meta- and paraxylenes are obtained from either coal-tar or petroleum processing and are separated from a mixture of ortho, para, and meta isomers. Oxidation of *p*-xylene to terephthalic acid is accomplished with air or nitric acid (205). Similar oxidation conditions also apply to the con-version of *m*-xylene to isophthalic acid. The process developed by California Research Corporation (now Chevron Research Corp.) for the preparation of xylylenediamines involves the reaction of terephthalic and isophthalic acids with ammonia, followed by hydrogenation of the resulting phthalo-nitriles (63,206).

$$\text{HOOC}-\hspace{-4pt}\bigcirc\hspace{-4pt}-\text{COOH} \xrightarrow[-H_2O]{NH_3} \text{NC}-\hspace{-4pt}\bigcirc\hspace{-4pt}-\text{CN} \xrightarrow{H_2} \text{H}_2\text{NCH}_2-\hspace{-4pt}\bigcirc\hspace{-4pt}-\text{CH}_2\text{NH}_2$$

The development of a new and economical process for manufacturing mixed *p*- and *m*-xylylenediamines was announced in 1966 by Showa Denko of Japan (8,207). In this process a mixture of phthalonitriles is produced by the ammoxidation of a 70:30 mixture of *m*-xylene and *p*-xylene. The catalytic hydrogenation of the mixed dinitriles yields the desired xylylenediamines.

$$\text{H}_3\text{C}-\hspace{-4pt}\bigcirc\hspace{-4pt}-\text{CH}_3 \xrightarrow[O_2]{NH_3} \text{NC}-\hspace{-4pt}\bigcirc\hspace{-4pt}-\text{CN} \xrightarrow{H_2} \text{H}_2\text{NCH}_2-\hspace{-4pt}\bigcirc\hspace{-4pt}-\text{CH}_2\text{NH}_2$$

The ammoxidation process eliminates the oxidation of xylenes to the corresponding dicarboxylic acids. Furthermore, the cost of the mixed isomers of xylene is about one-fourth the cost of pure metaxylene (8). However, the detailed conditions of such process have not yet been fully disclosed (8,208). The ammoxidation of xylenes to phthalonitriles is described in a German patent by Showa Denko (209).

1. THE TEREPHTHALIC AND ISOPHTHALIC ACID ROUTE

The terephthalic-isophthalic acid process is applicable to the preparation of *m*-xylylenediamine, *p*-xylylenediamine, or a mixture of these two amines, depending on the starting acids used (63). Isophthalic acid, or terephthalic acid, or a mixture of the two acids in solid form, is charged continuously with the aid of nitrogen to a vaporizer maintained at 600–700°F. Ammonia gas, preheated to 800–900°F, is also passed continuously into the vaporizer at 20–60 psig in contact with the acid, forming a molten mass of partially reacted acid (ca. 20–30% acid groups reacted). Excess ammonia gas is used as a carrier for the vapor of the molten mass to a dehydration reactor packed with a dehydration catalyst, such as alumina. The dehydration reactor is maintained at 750–850°F and the unreacted and partially reacted acids are converted to the corresponding phthalonitriles. The reactor effluent is passed into a recovery tank containing xylene kept at 200–300°F. Ammonia and hydrogen are removed and the resulting solution, without further purification, is led into a catalytic hydrogenation reactor containing a hydrogenation catalyst such as reduced cobalt oxide or nickel. Hydrogenation is carried out at 2000–5000 psig at 220–325°F. Ammonia may be added in amounts of 10–50 wt.% of total feed. The reactor effluent is then distilled to isolate the xylylenediamines.

For a specific example, the preparation of *m*-xylenediamine is described. Solid isophthalic acid and nitrogen (mole ratio of 1:2.8) are fed continuously into the vaporizer held at 620–660°F under 18 psig of pressure. Simultaneously, ammonia gas preheated to 900–910°F is charged into the vaporizer with the feed ratio of 19–20 moles/mole of isophthalic acid. The resulting melt is vaporized and swept out of the vaporizer by the excess ammonia into the dehydration reactor at a space velocity of 63–65 lb of acid charged per cubic foot per hour. The reactor is packed with a dehydration type of alumina catalyst and the dehydration is carried out at 750–850°F under 1 psig pressure. The isophthalonitrile vapor, together with ammonia, nitrogen, water, benzonitrile, metacyanobenzamide, hydrogen, xylenes, and carbon dioxide leaves the reactor at 850–900°F and is passed into a recovery tank containing xylene at 235°F and under atmospheric pressure. The resulting xylene solution contains 31% of product, which consists of 97.9% isophthalonitrile, 1.1% benzonitrile, and 1.0% metacyanobenzamide. This xylene solution, together with a cobalt oxide hydrogenation catalyst and ammonia equivalent to 25 wt.% of total feed, is charged into a rocking autoclave. Hydrogenation is made at 250°F under 3000 psig of pressure. The rate of hydrogenation is 0.54 g of isophthalonitrile per cubic centimeter of catalyst per hour. After solvent has been removed by distillation, the resulting product contains 90.4% (92.3% mole yield) of metaxylylenediamine, 0.8% monoamines, and 8.8% tars.

2. THE AMMOXIDATION ROUTE

A general method of preparing aromatic nitriles from the corresponding alkyl-substituted benzene derivatives is the ammoxidation process. It consists of a gas-phase reaction of the alkyl-substituted benzenes (e.g., toluene) with ammonia and oxygen in the presence of a catalyst at elevated temperature. Studies have been made on the conversion of toluene, isopropylbenzene to cyanobenzene and three isomeric xylenes to terephthalonitrile, isophthalonitrile, and *o*-phthalonitrile under the ammoxidation conditions (208,210–214). The following conditions are given by Oga of Showa Denko in a German patent for the preparation of isophthalonitrile and a mixture of terephthalonitrile and isophthalonitrile from the corresponding xylenes (209). The reduction of phthalonitriles to the corresponding xylylenediamines has already been described.

The ammoxidation catalyst is prepared by dissolving 3.3 g of ammonium metavanadate in a solution of 4.2 g of arsenic oxide in 60 ml of water with warming. To this is added 2.5 g of potassium sulfate. The resulting solution is then added to 70 g of purified diatomaceous earth to form a paste, which is dried and then heated for 12 hr at 350°C in air. The catalyst thus prepared is charged to a reactor heated in a salt bath. A gas mixture of 1.5 vol.% of *m*-xylene, 5.5% of ammonia, and 93.0% of air is passed through the reactor at 383°C employing a 2.5-sec contact time. This results in the transformation of 75.8 mole% of *m*-xylene to pure white isophthalonitrile and 11.8 mole% of *m*-xylene to *m*-toluonitrile. The yield based on ammonia consumed is 73.8 mole%.

A somewhat different catalyst system may be prepared by dissolving 9.2 g of vanadyl sulfate in 50 ml of an aqueous solution containing 6.2 g of arsenic oxide and 4.2 g of lithium hydroxide. This is mixed with 55 g of titanium dioxide, and the resulting paste is dried and heated at 350°C in the presence of air for 16 hr. Ammoxidation is carried out by passing a blend of gases consisting of 1.5 vol.% of a mixture of 32.1% *p*-xylene and 67.9% *m*-xylene, 7.5% ammonia, and 91.0% air through the catalyst at 390°C, with a contact time of 2.5 sec. From this, 83.5 mole% of the xylene mixture is converted to phthalonitriles and 3.3 mole% to toluonitriles. The yield based on consumed ammonia is 75.8%.

V. PHYSICAL PROPERTIES

Some of the physical properties of the aliphatic diamines are listed in Tables 7–9. These compounds are strong bases, and the volatile members have an ammonical odor. They fume in the air and absorb carbon dioxide to form carbonate salts. All the aliphatic diamines discussed are soluble in alcohol. The lower molecular weight members are soluble in water, but the solubility decreases as molecular weight increases.

The characteristic infrared absorption bands of the aliphatic amines are summarized in Table 10. Additional information can be found in standard infrared reference books (215,216). Aliphatic primary and secondary amines display characteristic nitrogen–hydrogen overtone and combination bands in the near-infrared region. Primary amines exhibit absorption bands in both the combination and first overtone of the nitrogen–hydrogen stretching regions, whereas the secondary amines show absorption only in the first overtone of the nitrogen–hydrogen stretching region (217). The positions of absorption are included in Table 10.

Raman spectroscopy can be used for fingerprinting of compounds and detection of amino groups. This technique, however, has not seen extensive use (218).

The nuclear magnetic resonance (NMR) bands due to nitrogen–hydrogen protons of aliphatic and alicyclic amines are usually found in the δ 1–5 region. The position and shape of the bands vary, depending on the extent of hydrogen bonding and the fast proton exchange with solvents. The nitrogen–hydrogen protons of most aliphatic primary amines appear as a single sharp peak, owing to rapid proton exchange (219,220). This is unlike the situation involving the nitrogen–hydrogen protons of amines in acid solutions and amides, which show broadening and splitting of the nitrogen–hydrogen band into a triplet (221).

The alkyl hydrogens adjacent of nitrogen absorb near the δ 2.2–2.8 region. Some typical shifts are indicated in Table 11.

Inasmuch as the NMR spectra of the individual members of the aliphatic diamines are not readily available, the chemical shifts of a number of alkylamines are summarized in Table 12 for reference purposes.

Mass spectroscopy has been adopted for the structure determination of alkylamines. A high-intensity peak is produced by the scission of the carbon–carbon bond next to the nitrogen atom. The resulting fragment is stabilized by the electronegativity of the nitrogen atom (223,224).

$$R_1CH_2\ddot{N}\overset{R_2}{\underset{R_3}{\diagup}} \xrightarrow{-e^-} R_1\text{--}CH_2\overset{+}{N}\cdot\overset{R_2}{\underset{R_3}{\diagup}} \longrightarrow R_1\cdot + CH_2\overset{+}{=}N\overset{R_2}{\underset{R_3}{\diagup}}$$

and

$$CH_2\overset{+}{=}N\overset{R_2}{\underset{R_3}{\diagup}} \longrightarrow CH_2\overset{+}{=}N\overset{R_2}{\underset{H}{\diagup}} + R_4$$

$$R_1, R_2, R_3 = H, \text{ alkyl}$$

$$R_4 = R_3 - H$$

The most abundant ion from the cleavage of aliphatic primary amines is $CH_2\overset{+}{=}NH_2$ (m/e 30). For example, in the mass spectrum of ethylenediamine, the m/e 30 peak is by far the most intense one. The abundance of various fragments is also dependent on relative ion stabilities. For 1,2-propanediamine, the relative abundance of the $CH(CH_3)\overset{+}{=}NH_2$ ion (m/e 44) is 46% and of the $CH_2\overset{+}{=}NH_2$ ion (m/e 30) is 15% (225).

TABLE 7

Physical Properties of Aliphatic Diamines

Amine	Formula	mp, °C	bp, °C	d_4^{20}	n_D^{20}	Flash point, °F (open cup)
Ethylenediamine	$H_2N(CH_2)_2NH_2$	8.5	116.5	0.8994	1.4499	99
1,2-Propanediamine	$H_2NCH_2CH(NH_2)CH_3$	-37^d	121	0.8640_{20}	1.4455	92
1,3-Propanediamine	$H_2N(CH_2)_3NH_2$	-23.5	140	0.8881_{20}	1.4583	119
1,4-Butanediamine	$H_2N(CH_2)_4NH_2$	27–28	158–159	0.877^{25}		
1,5-Pentanediamine	$H_2N(CH_2)_5NH_2$	9	178–180	0.873^{25}		
1,6-Hexanediamine	$H_2N(CH_2)_6NH_2$	42	204–205			
N,N'-Dimethyl-1,6-hexanediamine	$CH_3NH(CH_2)_6NHCH_3$		205			
1,7-Heptanediamine	$H_2N(CH_2)_7NH_2$	28–29	223–225			
4,4-Dimethyl-1,7-heptanediamine	$H_2N(CH_2)_3C(CH_3)_2(CH_2)_3NH_2$		72/0.3 mm	0.8625	1.4640	
1,8-Octanediamine	$H_2N(CH_2)_8NH_2$	50–52	240–241			
1,10-Decanediamine	$H_2N(CH_2)_{10}NH_2$	61.5	140/12 mm			
1,12-Dodecanediamine	$H_2N(CH_2)_{12}NH_2$	66–67	187/16 mm			
Piperazine	$\overline{HN(CH_2)_2NHCH_2CH_2}$	104	148		1.446^{113}	190
2-Methylpiperazine	$\overline{HNCH(CH_3)CH_2NHCH_2CH_2}$	65–66	155.6			163
Cis-2,5-dimethylpiperazine	$\overline{HNCH(CH_3)CH_2NHCH(CH_3)CH_2}$	17–18	165	0.9195^{25}_{25}	1.4725	154

226

Trans-2,5-dimethylpiperazine	HNCH(CH₃)CH₂NHCH(CH₃)CH₂	117–118	162			235
Cis-2,6-dimethylpiperazine	HNCH(CH₃)CH₂NHCH₂CH(CH₃)	114	161			
Trans,trans-bis(4-amino-cyclohexyl)methane	H₂NC₆H₁₀CH₂C₆H₁₀NH₂	64–65.5	130–131/0.8 mm			
Cis,trans-bis(4-amino-cyclohexyl)methane	H₂NC₆H₁₀CH₂C₆H₁₀NH₂	35.7–36.9	127–128/1.2 mm	0.9608²⁵	1.5046	
Cis,cis-bis(4-amino-cyclohexyl)methane	H₂NC₆H₁₀CH₂C₆H₁₀NH₂	60.5–61.9	141/2 mm			
1,3-Diaminocyclohexane[a]	H₂NC₆H₁₀NH₂		95–97/30 mm	0.956¹⁵		
1,4-Diaminocyclohexane[a]	H₂NC₆H₁₀NH₂		90/22 mm			
Methyldiaminocyclohexane[b]	H₂N(CH₃)C₆H₉NH₂	72–73[e]	75/3.1 mm			
Metaxylylenediamine	H₂NCH₂C₆H₄CH₂NH₂	14[d]	105/2 mm			
Paraxylylenediamine	H₂NCH₂C₆H₄CH₂NH₂	35	153/22 mm			
Xylylenediamine[c]	H₂NCH₂C₆H₄CH₂NH₂	12[d]	247	1.051	1.5709	266

[a] Mixture of cis and trans isomers.
[b] Mixture of 80% 1-methyl-2,4-diaminocyclohexane and 20% 1-methyl-2,6-diaminocyclohexane.
[c] Mixture of 30% para and 70% metaisomers.
[d] Freezing point.
[e] Trans isomer.

TABLE 8

Further Physical Properties of Aliphatic Diamines

Amine	Solubility[d]					Absolute viscosity, cp at 20°C	Acid dissociation constants pK$_A$ at 20°C		Vapor density (air = 1)	Dipole moment at 25°C
	Water	Ethyl alcohol	Ether	Benzene	Chloroform		1	2		
Ethylenediamine	v	v	i	i	s	1.6	6.97	9.97	2.07	1.90
1,2-Propanediamine	v	s	i		v	1.6	6.85	9.90	2.56	
1,3-Propanediamine	v	v	v		s	1.9	8.58	10.65	2.56	1.94
1,4-Butanediamine	v	s					9.30	10.84	3.04	1.93
1,5-Pentanediamine	s	s	sl				9.74	11.05	3.52	1.91
1,6-Hexanediamine	v	s		s			9.83^{25}	10.93^{25}	4.01	1.91
N,N'-Dimethyl-1,6-hexanediamine	s	s								
1,7-Heptanediamine	s	s								
4,4-Dimethyl-1,7-heptanediamine		s								
1,8-Octanediamine	s	s	s	sl			10.10	10.99		1.99
1,10-Decanediamine		s	s							
1,12-Dodecanediamine		s	i							
Piperazine	v	s	i	s		0.666^{e}	5.66^{25}	10.03^{25}		
2-Methylpiperazine	v	s	s	s	s		5.46^{25}	9.90^{25}		

228

Compound								
Cis-2,5-dimethylpiperazine	v	v	v				5.23[25]	9.98[25]
Trans-2,5-dimethylpiperazine	v	v	sl	sl			5.34[25]	9.84[25]
Cis-2,6-dimethylpiperazine	s	s	i	sl	v	s	5.40[25]	9.86[25]
Trans,trans-bis(4-aminocyclohexyl)methane	s							
Cis-trans-bis(4-aminocyclohexyl)methane	s							
Cis-cis-bis(4-aminocyclohexyl)methane	s							
1,3-Diaminocyclohexane[a]	s	s	s					
1,4-Diaminocyclohexane[a]	s	s						
Methyldiaminocyclohexane[b]	s	s						
Metaxylylenediamine	s	s	s				8.49	9.64
Paraxylylenediamine	s	s	s				8.75	9.91
Xylylenediamine[c]	s	s	s				6.8	

[a] Mixture of cis and trans isomers.
[b] Mixture of 80% 1-methyl-2,4-diaminocyclohexane and 20% 1-methyl-2,6-diaminocyclohexane.
[c] Mixture of 30% para and 70% metaisomers.
[d] v = very soluble, s = soluble, sl = slightly soluble, i = insoluble.
[e] Centistokes at 138°C.

TABLE 9

Azeotropic Data of Aliphatic Diamines

Components	% Amine	bp, °C	d_{20}^{20}
Ethylenediamine/butanol	35.7	124.7	0.849
Ethylenediamine/isobutanol	50	120.5	0.856
Ethylenediamine/toluene	30	103	
Ethylenediamine/methyl cellosolve	31	130/733 mm	
Ethylenediamine/water	81.6	119.0	0.953
1,2-Propanediamine/butanol	49	126.5	0.843
1,2-Propanediamine/isobutanol	65	123	0.855
1,2-Propanediamine/toluene	32	105	0.865
2-Methylpiperazine/water	Nonazeotrope		
2,5-Dimethylpiperazine/amyl alcohol	Nonazeotrope		
2,5-Dimethylpiperazine/2-ethoxyethanol	Nonazeotrope		
2,5-Dimethylpiperazine/water	Nonazeotrope		

TABLE 10

Infrared and Near-Infrared Absorption Bands of Aliphatic Primary and
Secondary Amines (215,217)

Absorption	Region, cm^{-1}
C–N Vibrations	1220–1020
	1410
N–H Deformation	
Primary amines	1650–1590
Secondary amines	1650–1550
N–H Stretching (fundamental)	
Primary amines, 2 bands	3500–3300
Secondary amines, 1 band	3500–3300
N–H combination of stretching and deformation	
Primary amines	5000
N–H Stretching (first overtone)	
Primary and secondary amines	6700

TABLE 11

δ-Values of Hydrogens on Alkyl Groups Adjacent to Nitrogen (220)

N	CH$_3$N	–CH$_2$N	CHN
N (Acylic)	2.15	2.5	2.87
N (Cyclic secondary)		2.7	
N (Cyclic tertiary)	2.20	2.3	

TABLE 12
Nuclear Magnetic Resonance Bands of Alkylamines (222)

Structure	Chemical shift[a]					
	a	b	c	d	e	f
(a)　　　　(c) (b) $CH_3CH_2CH_2CH_2NH_2$	0.92	1.10	2.70			
(a)　　　(d) (b) $CH_3CH_2CHCH_3$ 　　　　| 　　　NH_2 　　　(c)	0.9	1.05	1.25	2.78		
(a) 　　CH_3 (a)　| CH_3CNH_2(b) 　　| 　　CH_3(a)	1.15	1.23				
(a) (b)　(c) (f) (e) $CH_3CH_2NCH_2CH_2OH$ 　　　　H 　　　(d)	1.12	2.68	2.73	3.53	3.53	3.65
(b)　　(a) 　CH_3　CH_3 (b) |　(d) | $CH_3C–CH_2–C–CH_3$(a) 　|　　　| 　NH_2　CH_3 　(c)　　(a)	1.02	1.18	1.25	1.45		
(piperidine) H (c) 　　　　CH_3 (a)	1.05	1.34	2.57			
(b) (a) H_3C—(piperazine)—CH_3 (a) (a)　　H (b)　(cis)	0.92	1.04				
(a) H (d) H_2—(piperazine)—H_2 (d) (c) H_2　　H_2 (c) CH_3 (b)	2.12	2.27	2.37	2.88		
NH_2 (b) 　　　| (c) H—C—CH_3 (a) (phenyl) (d)	1.38	1.58	4.10	7.30		

[a] In parts per million (ppm) of the δ scale, 0 for tetramethylsilane.

VI. ANALYTICAL PROCEDURES

The purity of aliphatic diamines can be determined by measuring specific gravity and boiling range. Specific gravity is determined at 20°C with a hydrometer calibrated to give the apparent specific gravity at 20/20°C. Alternatively, a calibrated pycnometer can be used (for extremely viscous materials it should have a large-bore plug). Distillation to determine boiling range should be performed according to ASTM procedures for lacquer solvents and diluents. The distillation is conducted with a 100-ml sample in a flask equipped with a condenser and a calibrated thermometer. Some commercial diamines contain water, which is measured by the Karl Fischer method. Other tests include color and odor. Color is determined by comparison to platinum-cobalt color standards or to Gardner color standards.

Aliphatic diamines are assayed by titration with either standard hydrochloric or perchloric acids. The hydrochloric acid titration is performed either in an aqueous medium using methyl red or mixed methyl red–bromocresol green as indicator, or in methanol with methyl orange and xylene cyanol FF as mixed indicator. The perchloric acid method is particularly suited for high-molecular-weight, water-insoluble amines. The titration is made in glacial acetic acid with crystal violet as the indicator. The potential fire and explosion hazards of using perchloric acid should be pointed out: aqueous solutions of 60–72% perchloric acid may form explosive mixtures with organic materials. Furthermore, strong dehydrating agents may convert the solution to the anhydrous acid, which is unstable at ordinary temperatures and explodes on contact with most organic materials (226). The detailed analytical procedures for the commercially available aliphatic diamines can usually be obtained from the producers or from the technical brochures.

A. Hydrochloric Acid–Water Titration (227)

The mixed-indicator solution used for hydrochloric acid–water titration is prepared by mixing 1 part of a 0.1% methyl red in methanol and 5 parts of a 0.1% bromocresol green in methanol. The separate indicator solutions should be prepared every two weeks and the mixed indicator solution is made daily. To each of two 250-ml glass-stoppered Erlenmeyer flasks is transferred 50 ml of water (for duplicate determinations). About 6–8 drops of methyl red–bromocresol green mixed-indicator solution is added, and the solution is neutralized by dropwise addition of $0.1N$ hydrochloric acid until the green color disappears. To each flask is added 3–4 meq of diamine, and the contents of the flask is titrated with standard $0.1N$ hydrochloric acid until the green color disappears. The amount of amine present is calculated by

$$\frac{AN \times \text{E.W.}}{\text{g sample} \times 10} = \text{amine, \% by weight}$$

where A = milliliters of N (normal) hydrochloric acid required and E.W. = equivalent weight of the amine.

B. Hydrochloric Acid–Methanol Titration (227)

For hydrochloric acid–methanol titration, the indicator solution is prepared by dissolving 0.15 g of methyl orange and 0.08 g of xylene cyanol FF in 100 ml of distilled water. To two 250-ml glass-stoppered Erlenmeyer flasks containing 25–50 ml of methanol each is added 3–5 drops of the indicator solution. The solution is made neutral by dropwise addition of 0.1N hydrochloric acid in methanol to an amber-brown end point. To each flask is added 3–4 meq of amine and the resulting solution is titrated with 0.1N hydrochloric acid in methanol to the original amber-brown color. The percentage of amine by weight is calculated in the same manner as that described in the hydrochloric acid–water titration method.

C. Perchloric Acid–Acetic Acid Titration (227)

Standard perchloric acid in acetic acid is prepared by dissolving 8 ml of 70–72% perchloric acid with sufficient glacial acetic acid in a 1000-ml volumetric flask and diluting to mark with glacial acetic acid. After the perchloric acid solution has been allowed to stand overnight, it is standardized against Bureau of Standards potassium phthalate using crystal violet indicator. To each of two 250-ml glass-stoppered Erlenmeyer flasks is added 50 ml of glacial acetic acid followed by 2 or 3 drops of a 0.1% solution of crystal violet indicator in glacial acetic acid. The solution is made neutral by dropwise addition of 0.1N perchloric acid in acetic acid to the first green color. To each flask is added 3–4 meq of amine. After thorough mixing, the solution is titrated with standard 0.1N perchloric acid to the green color of the neutral acetic acid. The amount of amine present is calculated by:

$$\frac{AN \times \text{E.W.}}{\text{g sample} \times 10} = \text{amine, \% by weight}$$

where A = milliliters of (N) normal perchloric acid required and E.W. = equivalent weight of amine.

VII. MONOMER PURITY BY POLYMERIZABILITY TEST

The high temperature, melt polycondensation technique described in Section IX.A represents an excellent method of testing the purity of diamines. Best results, such as high molecular weight and high yield of polymers, are obtained only with monomers of high purity and with the use of stoichiometric equivalence of reactants. Interfacial and solution polycondensation procedures require monomers of moderate to high purity. The polymerization is readily performed at low temperature and at atmospheric pressure. These procedures are also useful for adaptation as small scale, rapid polymerization tests (228).

A. Melt Polycondensation

High molecular weight polyamides are prepared by the reaction of aliphatic diamines and dicarboxylic acids under the melt polycondensation conditions. The preparation of the balanced diamine-dicarboxylic acid salt prior to polymerization is adopted to achieve 1 to 1 stoichiometry of reactants. The inherent viscosity, which is an indication of molecular weight of the polyamide, ranges from 0.5–2.0 for these products. As a typical example, the detailed experimental conditions for the preparation of poly(hexamethylene-adipamide) (243) are given.

1. 1,6-HEXANEDIAMINE-ADIPIC ACID SALT

To a solution of 14.60 g (0.100 mole) of adipic acid dissolved in 110 ml of absolute ethanol by warming followed by cooling to room temperature is added a solution of 11.83 g (0.12 mole) of 1,6-hexanediamine in 20 ml of absolute ethanol. The reaction mixture is allowed to stand overnight and the crystallized salt is filtered. After washing with cold absolute alcohol and air drying, the salt weighs 25.5 g (97%), mp 196–197°C. A 1% aqueous solution has a pH of about 7.6.

2. POLY(HEXAMETHYLENEADIPAMIDE)

To a polymer tube made with a constriction in the upper half of the neck and capable of withstanding high internal pressure is charged 20 g of the 1,6-hexanediamine-adipic acid salt. The tube is connected to a three-way stopcock. The air in the tube is removed by alternately evacuating by means of a vacuum pump to 0.5 mm and filling with nitrogen for three or four cycles. The constriction is sealed under vacuum by a torch. The tube is inserted into a steel jacket and heated in a sand bath at 215°C for 1.5–2 hr. After cooling to room temperature, the tube is opened cautiously. A new neck containing a side arm is sealed onto the polymer tube. The tube is clamped vertically and to the side arm are connected in sequence a trap, and a three-way stopcock leading both to a vacuum pump and a nitrogen source. A capillary tube is inserted from the top of the tube via a short section of rubber tubing and is extended to the bottom of the polymer tube. This capillary tube is attached to a source of low-pressure nitrogen. The air in the tube is removed by means of vacuum pumping and nitrogen flushing. The tube is placed in a 270°C vapor bath (60/40 of diphenylmethane/o-hydroxydiphenyl). The tube is heated for 30–60 minutes at atmospheric pressure and then evacuated gradually by means of the three-way stopcock. After heating at 0.2–1.5 mm for 1 hr the reaction is discontinued. The completion of reaction is evidenced by the rate of bubble rise which indicates the polymer has reached a maximum melt viscosity. After raising the capillary tube, the polymer is cooled under a stream of nitrogen entering from the side arm. The solid polymer (14 g, 80%) is recovered by breaking the polymer tube. The product has a melt temperature of 265°C. The inherent viscosity of a 0.5% concentration in *m*-cresol is about 1.0–1.4.

B. Interfacial Polycondensation

1. Polyterephthalamides

Aliphatic diamines react readily with terephthaloyl chloride under the interfacial condensation conditions to yield polyterephthalamides (229). The representative polyamides are described in Table 13. The yield and

TABLE 13
Polyterephthalamides by Interfacial Polycondensation

Diamine monomer	Polymer yield, %	Tm,[a] °C	η_{inh}[b]	Reference
Ethylenediamine	75	455	1.00	229
1,2-Propanediamine	89	—	1.70	229
1,3-Propanediamine	83	399	1.70	229
1,4-Butanediamine	80	436	1.20	229
1,5-Pentanediamine	88	353	2.00	229
1,6-Hexanediamine	85	371	0.9	229
N,N'-Dimethyl-1,6-hexanediamine	18[c]	—	0.3	229
Bis(4-aminocyclohexyl)methane	100	—	0.99[d]	230
1,3-Diamino-2-methylcyclohexane	—	350	0.78[d]	231
p-Xylylenediamine	—	360	0.53[e]	232

[a] Polymer melt temperature determined by noting the first point of the formation of a molten trail on a hot bar.
[b] Inherent viscosity ($[\ln \eta_{rel}]/C$) at $C = 0.5\%$ in concentrated sulfuric acid at 30°C.
[c] Yield of water-insoluble polymer.
[d] Inherent viscosity in *m*-cresol.
[e] Reduced specific viscosity η_{sp}/C determined at $C = 0.5\%$ in concentrated sulfuric acid at 20°C.

polymer melt temperatures are also given. The inherent viscosity, which is an indication of molecular weight of the polymer, is listed to serve as a standard for test of diamine purity.

The commercial terephthaloyl chloride is purified by refluxing a 200-g sample in 1 liter of hexane and 100 g of thionyl chloride for 15 hr. After most of the hexane and thionyl chloride have been removed by distillation, the terephthaloyl chloride crystallizes upon cooling. The material is filtered, washed with dry petroleum ether, and stored in a vacuum desiccator over phosphoric anhydride.

The following experimental conditions apply to the preparation of polymers derived from ethylenediamine to 1,6-hexanediamine. A solution of 0.063 mole of diamine and 0.126 mole of potassium hydroxide in 4.5 liters of distilled water is placed in an 8-liter stainless steel beaker. With high-speed mixing, a solution of 0.063 mole of terephthaloyl chloride dissolved in 1 liter of methylene chloride is added rapidly. The polymerization is run for 10 min at room temperature. The solid polymer is then isolated and boiled in distilled water to remove adsorbed methylene chloride. The product is dried at 80°C and 20 mm. The yield is 8–12 g.

For the polyterephthalamide of N,N'-dimethyl-1,6-hexanediamine, some modifications of the foregoing procedure are made. A solution of 0.05 mole of N,N'-dimethyl-1,6-hexanediamine and 10.6 g of sodium carbonate in 250 ml of water is stirred rapidly in a blender. A solution of 10.1 g (0.05 mole) of terephthaloyl chloride in 80 ml of chloroform is added rapidly, and the resulting mixture is stirred for 10 min. Chloroform is evaporated by heating to precipitate the polymer. The mixture is then dialyzed for 24 hr against distilled water, in order to remove salts and any unreacted materials. The product is isolated by filtration and drying. The yield of water-insoluble polymer portion is 18%.

The polyamide from bis(4-aminocyclohexyl)methane and terephthaloyl chloride is prepared with further modifications because the diamine is not soluble in water (230). Thus a dispersion is prepared by mixing 250 ml of water, 100 ml of chloroform, 2.0 g of sodium lauryl sulfate, 10.6 g (0.1 mole) of sodium carbonate, and 10.5 g (0.05 mole) of bis(4-aminocyclohexyl)-methane at room temperature in a blender. A solution of 10.15 g (0.05 mole) of terephthaloyl chloride in 100 ml of chloroform is added to the dispersion in 30 sec. After the mixture has been stirred for 5 min, an equal volume of hexane is added with moderate stirring. The product is washed and dried (17 g, 100% yield).

2. POLYADIPAMIDES

Reaction of diamines with adipyl chloride yields polyadipamides. A typical example is the preparation of the polyadipamide derived from 1,6-hexanediamine (230). A mixture of 3.93 g (0.070 mole) of potassium hydroxide, 3.95 g (0.034 mole) of 1,6-hexanediamine, and 200 ml of water is placed in an ice-jacketed home blender. Then, over a 5-min period, a solution of 6.22 g (0.034 mole) of adipyl chloride in 200-ml xylene is added with stirring. The stirring speed is gradually increased during addition to effect good mixing. The polyamide is collected by filtration, washed with water, and then dried. The yield is 5.6 g (73%), and the product has an inherent viscosity of 1.16 (*m*-cresol). Similarly prepared are the polyadipamides from 1,7-heptanediamine, 1,8-octanediamine, and 1,10-decanediamine (233). The reactions are performed using an aqueous diamine phase and adipyl chloride in carbon tetrachloride. Sodium hydroxide (in aqueous phase) is employed as the acid acceptor. After isolation by filtration, the products are washed with 50% aqueous ethanol and then acetone, and dried at 60°C in a vacuum oven. The inherent viscosities of these products are between 1 and 2 at 30°C (0.5 g/100 ml *m*-cresol).

3. POLYAMIDE FROM 4,4-DIMETHYL-1,7-HEPTANEDIAMINE AND 4,4'-SULFONYLDIBENZOIC ACID

A number of polyamides derived from 4,4'-sulfonyldibenzoic acid and aliphatic diamines have been reported by Stephens (234). The condensation of 4,4-dimethyl-1,7-heptanediamine and 4,4'-sulfonyldibenzoyl chloride is presented as a typical example. The reagent 4,4'-sulfonyldibenzoyl chloride, which is prepared by oxidation of *p*-tolylsulfone followed by reaction of the resulting 4,4'-sulfonyldibenzoic acid with phosphorus oxychloride, is doubly purified by recrystallization from trichloroethylene and distillation through a 16-in. Widmer column (234).

To prepare the polyamide, a solution of 4.29 g (0.0125 mole) of 4,4'-sulfonyldibenzoyl chloride in 50 ml of dichloromethane is added to 200 ml of distilled water containing 2.04 g (0.0129 mole) of 4,4-dimethyl-1,7-heptanediamine and 2.65 g (0.025 mole) of sodium carbonate under high-speed stirring. The resulting mixture is stirred for an additional 5 min, poured into 1 liter of distilled water, and boiled for 1 hr to remove dichloromethane and salts. The polyamide is isolated by filtration, washed successively with water and methanol, and dried at 60°C in a vacuum oven. The yield is 100%, inherent viscosity = 1.07 (determined at a concentration of 0.5 g of polymer per 100 ml of solution in 40/60 by weight of *sym*-tetrachloroethane/phenol at 30°C).

4. POLYURETHANE BASED ON ETHYLENE BIS(CHLOROFORMATE)

Polyurethane is prepared by reacting a diamine with a bichloroformate under interfacial polycondensation conditions. An emulsifying agent, such as Duponol ME (sodium lauryl sulfate, du Pont trademark), is used to obtain high inherent viscosity (235). Ethylene bis(chloroformate) is purified by distillation, bp 72°C at 2.2 mm, prior to use. The polymerization is performed by dissolving 0.05 mole of the diamine, 10.6 g (0.1 mole) of sodium carbonate, and 1.5 g of Duponol ME in 150 ml of water. The solution is cooled to 5°C and placed in a home blender. To this is added rapidly with stirring a solution of 9.35 g (0.05 mole) of ethylene bis(chloroformate) in

125 ml of benzene cooled to 10°C. The mixture is stirred for 5 min, whereupon the polymer is filtered, washed, and dried. The representative polyurethanes thus prepared are listed in Table 14.

TABLE 14

Polyurethanes Based on Ethylene Bis(Chloroformate) by
Interfacial Polycondensation

Diamine monomer	Polymer yield, %	Tm,[a] °C	η_{inh}[b]
1,6-Hexanediamine	72	180	1.19
1,4-Diaminocyclohexane	85	250	0.50
1,3-Diaminocyclohexane	55	200	0.62

[a] Polymer melt temperature (see Table 13 for definition).
[b] Inherent viscosity at 30°C, $C = 0.5$ g/100 ml of m-cresol.

5. POLYURETHANE-UREA BASED ON ω-ISOCYANATOALKYL CHLOROFORMATE

Polyurethane-ureas are obtained from the interaction of ω-isocyanatoalkyl chloroformates and aliphatic diamines under the interfacial polycondensation conditions (236). The starting ω-isocyanatoalkyl chloroformates are prepared by the phosgenation of the corresponding ω-aminoalkanols (232). For example, a solution of 12 g (0.117 mole) of 5-aminopentanol in 200 ml of dioxane is added during 30 min to a solution of 130 g (1.3 moles) of phosgene in 400 ml of dioxane cooled with an ice-water bath. The mixture stands overnight at room temperature, and then unreacted phosgene and solvent are stripped under vacuum. The remaining crude product mixture is distilled to yield a broad boiling fraction, 18 g, bp 80–120°C/5–25 mm. Redistillation of this fraction furnishes 6.5 g (38%) of 5-chloropentylisocyanate, bp 65–66°C/3 mm, and then 8.0 g (36%) of the desired 5-isocyanatopentyl chloroformate, bp 108–109°C/3 mm.

For a typical example, the preparation of polyurethane-urea from 5-isocyanatopentyl chloroformate and 1,6-hexanediamine is described. A solution consisting of 1.21 g (0.0104 mole) of 1,6-hexanediamine, 1.1 g (0.01 mole) of sodium carbonate, and 10 mg of sodium laurylsulfate in 70 ml of water is cooled at 2°C, and to it is added rapidly a solution of 2.03 g (0.0106 mole) of 5-isocyanatopentyl chloroformate in 25 ml of toluene with rapid stirring. An additional 5 ml of toluene is used to rinse the residual chloroformate into the reaction mixture. The mixture is stirred for 10 min and then filtered. The solid polymer is washed thoroughly with water and with methanol, and then dried. The yield is 2.5 g (89%) having an inherent viscosity of 1.39 (0.5 g/100 ml of m-cresol at 30°C). Table 15 describes a number of polyurethane-ureas derived from aliphatic diamines.

TABLE 15
Polyurethane-Ureas Based on ω-Isocyanatopentyl Chloroformate
by Interfacial Polycondensation (236)

Diamine monomer	Polymer yield, %	Tm,[a] °C	η_{inh}^b
Ethylenediamine	46	195	0.86
1,4-Butanediamine	75	197	0.59
1,6-Hexanediamine	89	189	1.39
m-Xylylenediamine	45	147	0.95

[a] Polymer melt temperature.
[b] Inherent viscosity, $C = 0\ 0.5$ g/100 ml of m-cresol at 30°C.

C. Solution Polycondensation

Low-temperature solution-polycondensation technique represents an excellent test for purity of monomers. The tolerance for impurity using the solution method is lower than that for the interfacial method. Therefore, all reagents and solvents must be purified and dried prior to use.

1. POLYTEREPHTHALAMIDES

Piperazine and C-alkyl-substituted piperazines are typical diamine monomers that form polyterephthalamides with terephthaloyl chloride under solution-polycondensation conditions (231). Two methods have been reported for their preparation, and they are described below using trans-2,5-dimethylpiperazine as the typical diamine monomer (237).

a. Method 1: In Benzene with Triethylamine as Acid Acceptor

In a quart-size blender is placed a solution of 2.28 g (0.02 mole) of trans-2,5-dimethylpiperazine and 5.6 ml of pure triethylamine in 100 ml of benzene. The blender top is covered with aluminum foil and over this is placed the plastic cap. A wide powder funnel is inserted through a ¾-in. hole in the center of the cap and foil. A quantity of 4.06 g of terephthaloyl chloride in 90 ml of benzene is added rapidly through the funnel, while stirring speed is increased.

Traces of acid chloride are rinsed immediately into the blender with 10 ml of benzene. A precipitate appears and the resulting mixture is stirred for 5 min at moderate speed. The mixture is then diluted with an equal volume of hexene. The precipitate is collected on a medium-pored, fritted glass funnel and washed thoroughly, first with water and then with acetone; after it has dried at 100°C, the yield of polyamide is 90%, and the inherent viscosity is 1.21 (m-cresol).

b. Method 2: In Chloroform with Calcium Hydroxide as Acid Acceptor

To a solution of 2.28 g (0.02 mole) of *trans*-2,5-dimethylpiperazine in 100 ml of chloroform in a blender is added 5.92 g (0.08 mole) of reagent-grade, powdered calcium hydroxide. During a 5-min period, 4.06 g of terephthaloyl chloride in 100 ml of chloroform is added. The mixture is stirred for 10 min, allowed to stand for an additional 15 min, and then diluted with acetone. Finally, the precipitate is collected and washed successively with 2% aqueous hydrogen chloride, water, and acetone. The yield of polyamide is 95% with an inherent viscosity of 3.14 (*m*-cresol).

Table 16 summarizes the preparation of polyterephthalamides from the piperazine monomers (237).

TABLE 16
Polyterephthalamides by Solution Polycondensation

Diamine monomer	Polymerization solvent	Method	Yield, %	Tm,[a] °C	η_{inh}^{b}
Piperazine	Chloroform	1	85	>400	1.35
2-Methylpiperazine	1,1,2-Trichloroethane	1	85	>375	1.92
Cis-2,5-dimethylpiperazine	Chloroform	2	98	335	0.86
Trans-2,5-dimethylpiperazine	Cis-1,2-dichloro-ethylene	1	91	>400	2.20
Cis-2,6-dimethylpiperazine	Dichloromethane	2	98	>400	1.09

[a] Polymer melt temperature (see Table 13 for definition).
[b] Inherent viscosity at 30°C, $C = 0.5$ g/100 ml of *m*-cresol.

2.   POLYUREAS BASED ON HEXAMETHYLENEDIISOCYANATE

Reaction of aliphatic diamines with diisocyanates produces polyureas. The use of hexamethylene diisocyanate under low-temperature solution-polycondensation conditions has been reported by Boenig et al. (238). The general procedure consists of placing a solution of 3 wt.% amine in 97 wt.% acetone in a 1-liter reaction flask fitted with a stirrer, a nitrogen gas-inlet tube, a reflux condenser, and a dropping funnel. The weighing of the amine is performed in a dry nitrogen atmosphere. The flask is placed in a constant temperature bath. With stirring, a solution of 8 wt.% hexamethylene diisocyanate in acetone is added in 30 min, and the mixture is stirred for an additional 30 min. The polyurea is isolated by filtration and washed three times with 200-ml portions of ether. The product is dried, first at 25°C for 3 hr under nitrogen and then at 70°C for 24 hr in a vacuum oven.

The intrinsic viscosity η is determined by extrapolating the viscosity values of polymer solutions containing 0.1, 0.2, and 0.3 g/100 ml of phenol to zero concentration. The polyureas prepared from a number of diamine monomers are listed in Table 17.

TABLE 17
Polyureas Based on Hexamethylene Diisocyanate by Solution Polycondensation (238)

Diamine monomer	Reaction Temperature, °C	mp, °C	η^a
Ethylenediamine	−50	293	0.40
1,3-Propanediamine	−65	266	0.40
1,4-Butanediamine	−55	283	0.61
1,5-Pentanediamine	−35	251	0.39
1,6-Hexanediamine	−45	274	0.48
1,7-Heptanediamine	−25	254	0.44
1,8-Octanediamine	−25	253	0.39
1,12-Dodecanediamine	0	238	0.46

[a] Intrinsic viscosity in phenol, $\eta = \left[\dfrac{\ln (t_x/t_0)}{C} \right]_{c \to 0}$ measured at 25°C.

VIII. TOXICITY, STORAGE AND HANDLING

Aliphatic diamines exhibit a relatively high order of toxicity. The toxicities of individual compounds are described in detail in standard reference books (239–242). Available toxicity data on some of the diamines are summarized in Table 18. In general, aliphatic diamines are strong local irritants. Exposure to liquids, solutions, and vapors of these materials will cause lung and mucous membrane irritation, skin and eye burns, and sensitization reactions (from some members). Standard industrial hygiene techniques should be observed in handling this class of materials. The aliphatic diamines are best stored under an inert atmosphere such as nitrogen. This helps to improve color stability and excludes atmospheric carbon dioxide and moisture. Copper or its alloys should not be used in contact with these amines. When iron pick-up is detrimental to their end use, delivery and storage should be made in aluminum or stainless steel containers. Some comments follow on safety, storage, and handling of the most common commercial compounds.

A. Ethylenediamine

Ethylenediamine has a relatively high order of toxicity. Precautions should be taken to prevent contact with the skin and eyes, since this compound is capable of causing severe burns. Human contact experience has demonstrated that ethylenediamine has sensitizing properties. Skin sensitization is common and is easily acquired. Usually repeated and prolonged exposure to vapors

TABLE 18
Toxicity of Aliphatic Diamines

Diamines	Oral LD$_{50}$, rats, g/kg	Subcutaneous LD$_{50}$, mice, g/kg	Cutaneous LD$_{50}$, rabbits, ml/kg	Intraperitoneal LD$_{50}$, mice, g/kg	Inhalation, concentrated vapor, rats mortality (8 hr)	Primary skin irritation, rabbits	Eye injury, rabbits
Ethylenediamine	1.16	0.42	0.73		0/6	Severe	Severe
1,2-Propanediamine	2.23		0.50		1/6	Severe	Severe
1,3-Propanediamine	0.35[a]		0.20		0/6	Severe	Severe
1,8-Octanediamine·2HCl				0.0035			
1,10-Decanediamine·2HCl				0.12–0.17			
Piperazine		1.10					
Piperazine citrate	11[b]						
2,5-Dimethylpiperazine	3.16						
Xylylenediamines (30–70 meta–para)	1.75[b]						

[a] Dosage in milliliters per kilogram.
[b] Mice were used instead of rats.

will cause sensitization of the respiratory tract, evidenced by asthmatic attacks. (Occasionally a single heavy exposure will have the same effects.) Once sensitized, the individual retains his sensitivity for many years. Therefore, it is necessary to transfer the person to a position where he will no longer come in contact with either the liquid or vapor. Dermatitis is reported to occur in a high proportion of exposed workers in the manufacturing of mixed ethylene amines. Voluntary inhalation of ethylenediamine vapor for 5–10 sec produces tingling of the face and irritation of the nasal mucosa at 200 ppm and severe nasal irritation at 400 ppm. The threshold limit value for this compound has been suggested at 10 ppm.

Ethylenediamine is sometimes stored and handled in plain carbon steel equipment, but it is preferable to use stainless steel or aluminum to preserve quality. Galvanized iron, and copper and its alloys, are to be avoided. Ethylenediamine is shipped in tin-lined drums and is usually stored in stainless steel or aluminum. Iron contamination from carbon steel causes discoloration. This material has a fairly high freezing point, so outside tanks and piping require heating and insulation. A centrifugal pump is suitable for transfer purposes. Asbestos is adequate for gaskets and packing.

B. 1,2-Propanediamine

The single, oral-dose toxicity of 1,2-propanediamine is said to be in the general range of a 10% solution of acetic acid. Breathing concentrated vapors at room temperature is slightly hazardous. Undiluted 1,2-propanediamine causes severe skin or eye burns. A 5% solution in water is the least concentration causing detectable injury in the eye. Therefore, eye protection is mandatory for those handling this material. In the event of accidental contact, the eye should be washed immediately with water for 15 min and a physician should be contacted as promptly as possible. 1,2-Propanediamine is a sensitizer for both the skin and the respiratory tract. Skin contact may cause severe dermatitis, and breathing vapors may cause bronchial asthma in sensitized persons.

1,2-Propanediamine should be stored and handled in stainless steel or aluminum equipment. It is mildly corrosive to steel and discolors rapidly when stored in a carbon steel container. Other metals that are attacked or cause discoloration are nickel, galvanized or tinned iron, and copper and copper alloys. Since this compound has a low freezing point and a relatively low viscosity, heated storage is not required.

C. 1,3-Propanediamine

1,3-Propanediamine has a higher order of oral and dermal toxicity than the above-mentioned diamines. It is rapidly adsorbed through the skin and is capable of causing severe skin and eye burns on direct contact. If spilled on clothing or shoes, the garment should be removed at once and the affected part of the body washed thoroughly. Storage precautions similar to those used with ethylenediamine should be observed.

D. Higher Aliphatic Diamines

Similar to the previously discussed low-molecular-weight aliphatic diamines, the higher molecular weight members (from 1,4-butanediamine to 1,12-dodecanediamine) are strong bases and they also exhibit skin and eye irritant properties. They are considered moderately toxic as acute local irritants and allergens. With respect to systemic toxicity, the shorter chain diamines, such as 1,4-butanediamine and 1,5-pentanediamine, cause animal blood pressure depression. The longer chain diamines may exhibit sympathomimetic activity—activity produced by epinephrine. The effects include blood pressure elevation, contraction of smooth muscle, salivation, and dilation of the pupil.

Aliphatic diamines are capable of causing the release of histamine when administered intravenously in man. Potent histamine releasors, as with histamine itself, cause decrease in blood pressure, headache, itching, erythema, and facial edema. This histamine-releasing ability increases with the increase of molecular weight of the diamines. 1,4-Butanediamine has only slight activity. Increasing activity is observed from 1,6-hexanediamine to a maximum at 1,10-decanediamine.

1,6-Hexanediamine is similar to ethylenediamine in that it can be absorbed through the skin and can exhibit sensitization properties. The acute percutaneous toxicity of the diamines resembles that of the corresponding monoamines. The renal tubular damage produced by intraperitoneal injections in rats of 1,2-propanediamine and 1,3-propanediamine is not observed with 1,4-butanediamine, 1,5-pentanediamine, 1,6-hexanediamine, and 1,10-decanediamine.

1,6-Hexanediamine is reported to cause anemia, weight loss, and degenerative microscopic changes in the kidneys and liver of guinea pigs after repeated doses. For workers handling 1,6-hexanediamine, conjunctival, and upper respiratory tract irritations have been observed. The results of one study showed that 1,6-hexanediamine was responsible for the development of acute hepatitis, followed by dermatitis, in one worker out of twenty.

E. Piperazine

Piperazine exhibits a low acute oral toxicity. However, the ingestion of large quantities may cause serious injury. Individuals taking 30–75 mg/kg/ day of various piperazine salts for medicinal purposes have shown side reactions. These include hives, headaches, nausea and vomiting, diarrhea, lethargy, tremor, incoordination, and muscular weakness. The symptoms are only temporary and disappear when the medication is stopped. Dermatitis as a result of skin contact has been reported. A 5% solution has caused severe injury of the rabbit eye. Therefore, goggles and protective clothing should be used in handling piperazine. Piperazine is marketed as the anhydrous, flaked material in sealed polyethylene containers enclosed in fibreboard drums. The material is hygroscopic. Storage should be made in a cool, dry place.

IX. POLYMERIZATION AND APPLICATIONS

The techniques used for the preparation of condensation polymers include both the melt polycondensation performed at high temperature and the interfacial and solution polycondensations made at low temperature. An excellent reference source summarizing the various experimental conditions for the small scale preparation of condensation polymers is the text *Preparative Methods of Polymer Chemistry*, by Sorenson and Campbell (243). In this section, we describe the polymerization techniques as well as the commercial applications of the various polymers derived from aliphatic diamines.

A. Polyamides

The melt polycondensation is the commercially practical method of preparing polyamides. The polymerization condition consists of heating stoichiometric quantities of a diamine and a dicarboxylic acid above the melting point of the resulting polyamide product. The preparation of a balanced diamine–dicarboxylic acid salt prior to the polymerization step is the preferred method of providing the desired 1:1 stoichiometry of reactants.

$$H_2NRNH_2 + HOOCR'COOH \longrightarrow [H_3\overset{+}{N}R\overset{+}{N}H_3\overset{-}{O}OCR'\overset{-}{C}OO] \overset{-H_2O}{\longrightarrow}$$

$$\underset{}{+NHRNHC\overset{O}{\overset{\|}{R'}}\overset{O}{\overset{\|}{C}}}_{\overline{n}}$$

The polymerization is usually carried out at temperatures above 200°C. The first heating cycle is made in a sealed reactor under pressure. This is followed by a second heating cycle in which the reactor is opened to release pressure and vacuum is applied to remove the by-product water. The molecular weight control is effected by varying the length of heating. The addition of a monofunctional compound, such as acetic acid, can also serve to stabilize the molecular weight of the polymer. Alkaline oxides and carbonates, and halogen salts of polyvalent metals and acids, may be used as catalysts. The polyamides thus prepared can form fibers and films by melt or solution methods.

A modification of the foregoing method is the heating directly of equimolar quantities of a diamine and a dicarboxylic acid ester without the salt formation step. The by-product in this case is alcohol.

$$H_2NRNH_2 + R''OOCR'COOR'' \xrightarrow{-R''OH} \left[NHRNHCR'\overset{\overset{O}{\|}}{C} \right]_{\overline{n}}$$

A different method of preparing polyamides is interfacial polycondensation. The polymer is made by the reaction of a diacid chloride in a water-immiscible organic solvent and a diamine in water at the interface between the two solutions. An acid acceptor, such as an alkali metal hydroxide or carbonate, is used to remove the by-product hydrogen chloride. In contrast to the melt polycondensation method, the interfacial polycondensation can be performed in open equipment at temperatures below 40°C. The reaction is extremely rapid and is normally over in several minutes. The resulting polymer possesses at least as high a molecular weight as that obtained by the melt method Some specific examples were given in Section VII.

$$H_2NRNH_2 + Cl\overset{\overset{O}{\|}}{C}R'\overset{\overset{O}{\|}}{C}Cl \xrightarrow{-HCl} \left[NHRNHC\overset{\overset{O}{\|}}{R'}\overset{\overset{O}{\|}}{C} \right]_{\overline{n}}$$

The interfacial polycondensation method incurs certain disadvantages upon scale-up. These are the formation of large quantities of by-product salt from hydrogen chloride and the acid acceptor, the presence of solvents, and the high cost of the diacid chloride.

Polyamides are probably the most widely known class of condensation polymers. Their applications in the field of textile fibers, molding compounds, coatings, and adhesives have shown tremendous growth in recent years. The term nylon, originally coined by du Pont to describe a specific class of polyamides, has now been universally adopted to describe the entire class of synthetic polyamide resins capable of forming useful filaments and fibers. This includes not only the polyamides prepared from the condensation of diamines and dicarboxylic acids (nylons 6,6 and 6,10), but also the ones obtained from the polymerizations of lactams (nylons 3–6, 8, 12) and ω-aminoacids (nylons 7, 9, 11) (104). A detailed discussion of the commercial application of polyamides has been made by Floyd (244).

Some physical properties of the polyamides, such as melting point, stability, and solubility, are important considerations in determining the use of these materials. The approximate melting points of a number of polyamides are listed in Table 19. For fiber applications, the strength, elasticity, toughness, flexibility, water-absorption characteristics, and abrasion resistance of the polyamide are also of importance. Nylon 6,6, the polyamide from 1,6-hexanediamine and adipic acid, is by far the largest single volume polyamide manufactured in the United States. Its applications in the fiber field include general apparel, tufted carpeting, rope, thread belting, and filter cloth. In the production of tire cord, nylon 6,6 fiber shows good growth potential and is competing with rayon. For use as molding resin, nylon 6,6 serves as a substitute for metal in bearings, gears, cams, rollers, and slides. Nylon 6,6 is also used as abrasion-resistant outer covers to protect primary insulation on electrical wires.

TABLE 19
Melting Points of Polyamides (244)

Diamine	Acid	Melting point of polyamide, °C
Ethylene	Sebacic	254
1,4-Butane	Adipic	278
1,4-Butane	Suberic	250
1,4-Butane	Azelaic	223
1,4-Butane	Sebacic	239
1,5-Pentane	Glutaric	198
1,5-Pentane	Adipic	223
1,5-Pentane	Pimelic	183
1,5-Pentane	Suberic	202
1,5-Pentane	Azelaic	178
1,5-Pentane	Sebacic	186
1,6-Hexane	Adipic	265
1,6-Hexane	Sebacic	209
1,8-Octane	Adipic	235
1,8-Octane	Sebacic	197
1,10-Decane	Carbonic	200
1,10-Decane	Oxalic	229
1,10-Decane	Sebacic	194
p-Xylylene	Sebacic	268
Piperazine	Sebacic	153

Nylon 6,10, the polyamide from 1,6-hexanediamine and sebacic acid, is used in many bristle applications. The brushes, sports equipment, and bristles made from nylon 6,10 monofilaments, because of their low moisture absorption capacity, are able to retain stiffness and mechanical properties when wet (245). A polyamide derived from bis(p-aminocyclohexyl)methane and an 8–12 carbon dicarboxylic acid is reported to be silklike. It possesses excellent wrinkle resistance, pleat retention, and dimensional stability. This fiber, named "Qiana" and introduced by du Pont in 1968 is currently aimed at the women's clothing market (10).

Other polyamides of commercial significance include high-molecular-weight polymers prepared by xylylenediamines and dicarboxylic acids. The product derived from adipic acid possesses a Young's modulus higher than that of conventional nylons. It shows promise in fiber and filament applications (8). A polyamide of piperazine and sebacic acid is reported to form a tough, waterproof resin that can be used to impregnate textile braid for covering electrical conductors (246). The reaction of 4,4-dimethyl-1,7-heptanediamine with pyromellitic anhydride under the melt polycondensation conditions forms a high molecular weight polyamide. This polymer exhibits excellent properties, such as thermal stability and toughness for injection molding applications (14) and as water-resistant glass-reinforced thermoplastic resin (15).

B. Polyurethanes

There are two commercially practical routes for the preparation of polyurethanes. The first is the reaction of a diol with a diisocyanate, which is obtained from reaction of a diamine and phosgene. High-temperature melt

$$H_2NRNH_2 + 2COCl_2 \longrightarrow OCNRNCO + 4HCl$$

$$OCNRNCO + HOR'OH \longrightarrow \underset{n}{+} \overset{O}{\underset{\parallel}{C}} NHRNH \overset{O}{\underset{\parallel}{C}} OR'O \underset{n}{+}$$

and solution techniques are most often adopted for the alcohol-isocyanate reaction (247). The melt method consists of adding a diisocyanate to a glycol while gradually increasing the reaction temperature. The reaction is complete when no more viscosity increase occurs. The product is ground, washed with methanol, and then dried. The polymer can be melt pressed and melt spun to form films and fibers. The condensation can be also performed in a solution such as 80/20 by volume mixture of chlorobenzene and o-dichlorobenzene (248).

The other route to polyurethanes is the reaction of a bischloroformate with a diamine. The bischloroformate is readily prepared from the reaction of a diol with phosgene. The urethane-formation step is performed under low-temperature interfacial and solution-polycondensation conditions. Under

$$HOROH + 2COCl_2 \longrightarrow Cl\overset{O}{\underset{\parallel}{C}}OROC\overset{O}{\underset{\parallel}{C}}Cl + 2HCl$$

$$Cl\overset{O}{\underset{\parallel}{C}}OROC\overset{O}{\underset{\parallel}{C}}Cl + H_2NR'NH_2 \xrightarrow{-HCl} \underset{n}{+} NHR'NHC\overset{O}{\underset{\parallel}{C}}OROC\overset{O}{\underset{\parallel}{C}} \underset{n}{+}$$

the interfacial conditions, the bischloroformate in an organic solvent is added with stirring to the aqueous phase containing the diamine, an emulsifying agent (e.g., sodium lauryl sulfate), and an acid acceptor such as sodium carbonate or sodium hydroxide. The reaction is performed at low temperature and is over in a few minutes. The product is isolated by filtration. A number of representative polyurethanes thus prepared have been described by Wittbecker and Katz (235). The experimental details appear in Section VII.

The solution technique is less frequently used than the interfacial method for the amine-chloroformate reaction. The urethanes derived from piperazines and ethylene bischloroformate form one example. They are synthesized in chloroform with excess diamine as the acid acceptor (249).

The polyurethanes, although considered a relatively new class of polymers, have grown tremendously in recent years. Flexible and rigid urethane foams are produced from the reaction of isocyanates and alcohols in the presence of a blowing agent and a catalyst. However, the urethane foam systems of commercial utility employ mostly aromatic diisocyanates, such as tolylene diisocyanate and 4,4'-diphenylmethane diisocyanate, which are derived from aromatic diamines. Polyurethanes using aliphatic diamines as raw materials serve in areas other than urethane foams.

Elastomers from 1,6-hexanediisocyanate and various polyesters have shown good tensile strength and elongation (250). Two-component coating systems are prepared from aliphatic diisocyanates and hydroxyl-terminated alkyd resins. 1,6-Hexanediisocyanate and adduct of 3 moles of 1,6-hexanediisocyanate with 1 mole of hexanetriol have been used as isocyanate components in combination with low-molecular-weight hydroxyl-terminated polyesters in various coating systems (251). Polyurethane coatings having good gloss and color retention properties have been prepared from 4,4'-methylenedi(cyclohexylisocyanate) and methylcyclohexylene diisocyanate with various polyols (13).

Aliphatic diamines used in the manufacture of a class of elastomeric fibers named "spandex" have found use in the apparel industry. The spandex fiber consists of alternating polyester- or polyester-urethane blocks and urea blocks. The polymer is produced by: (a) reaction of a dihydroxy polyether or polyester with a diisocyanate (e.g., tolylene diisocyanate) to form a hydroxy-terminated polyurethane, (b) reaction of the polyurethane with another diisocyanate (e.g., 4,4'-diphenylmethane diisocyanate) to form an isocyanate-terminated adduct, and (c) reaction of the adduct with an aliphatic diamine in dimethylformamide to yield the high-molecular-weight polymer, which can be spun in the form of highly viscous solutions into fibers (252).

C. Polyureas

Polyureas can be considered as polyamides of carbonic acid. Because of the high melt temperatures of this class of polymers and their thermal instability under the melt conditions, the polyureas are most frequently prepared by solution and interfacial polycondensation techniques (253).

Polyurea is formed by reaction of a diamine with a diisocyanate and, since no by-product is produced, the solution technique is most suitable for this

$$H_2NRNH_2 + OCNR'NCO \longrightarrow \left[NHRNH\overset{\displaystyle O}{\overset{\displaystyle \|}{C}}NHR'NH\overset{\displaystyle O}{\overset{\displaystyle \|}{C}} \right]_n$$

purpose. Phenols and alcohols can be used as solvent because of the much greater reactivity of isocyanates with amines. A typical example is poly-decamethyleneurea, which is prepared by the interaction of 1,10-decane-diamine and 1,10-decanediisocyanate in m-cresol. The reaction is carried out at 218°C for 5 hr. After the solution has cooled, it is poured into methanol to precipitate the polyurea (254). If the polyurea is soluble in the reaction medium, e.g., the polyurea of $trans$-2,5-dimethylpiperazine and methylene-bis(4-phenylisocyanate) in tetramethylene sulfone–chloroform, heating is unnecessary, the reaction being completed at room temperature after 7 min. The polymerization is stopped by the addition of aqueous n-butylamine (254).

Polyurea can be also prepared by the reaction of an aqueous solution of diamine and alkali and an organic solvent containing phosgene under the interfacial polycondensation conditions. The reaction is exothermic and

$$H_2NRNH_2 + COCl_2 \xrightarrow{-HCl} +NHRNH\overset{\overset{\displaystyle O}{\|}}{C}\!\!\xrightarrow{}_{\!n}$$

the polymerization is completed in 8–10 min. Carbon tetrachloride is a suitable solvent for use with 1,6-hexanediamine (255).

The third method of preparing polyurea involves the reaction of diamine with a biscarbamyl chloride of secondary diamines (256). The condensation

$$H_2NRNH_2 + Cl\overset{\overset{\displaystyle O}{\|}}{C}NR'N\overset{\overset{\displaystyle O}{\|}}{C}Cl \xrightarrow{-HCl} +NHRNH\overset{\overset{\displaystyle O}{\|}}{C}NR'N\overset{\overset{\displaystyle O}{\|}}{C}\!\!\xrightarrow{}_{\!n}$$

is carried out under the interfacial conditions. The biscarbamyl chlorides of piperazine and 1,4-bis(methylamino)hexane, etc., are used.

There are a number of other methods of preparing polyureas. These include the decomposition of the thiocarbamate salt of a diamine (257), exchange reaction of diamines with N,N'-carbonyldiimidazole (258), the reaction of diamines with N-methylnitrosourea (259), and the condensation of a diamine with a diurethane (260). The high-temperature, high-pressure reaction of diamines with carbon dioxide to yield polyureas has also been reported (261).

The polyurea can be prepared in linear form essentially free of cross-linked by-products for use as fibers (202,263,264). The polyureas obtained from the reaction of the carbamate of a number of aliphatic diamines with aliphatic diisocyanates are reportedly suitable for the manufacture of fibers, bristles, and foils (265). As a class, however, the polyureas are generally higher melting than the corresponding polyamides. This, plus the tendency of the polyureas to be thermally unstable at their melt temperatures, may have accounted for the limited commercial development of polyurea in the fiber field.

For nonfiber applications, polyureas derived from 1,6-hexanediisocyanate and primary amines containing additional secondary and tertiary amine groups for use as vehicles for textile printing pastes have been reported (262,266). Polyureas containing sulfonic acid substituents in the form of water-soluble ammonium salts possess cation-exchange properties (267).

D. Miscellaneous Polymers

Polysulfonamides are prepared by the reaction of aromatic disulfonyl chlorides and aliphatic diamines under the interfacial polycondensation conditions (268,269). A dispersing agent, such as sodium lauryl sulfate, is employed to improve the contact of the two phases. The resulting polymers are reported to form fibers by melt-spinning and films by melt-pressing techniques (268). The polyphosphonamides are similarly prepared by reacting phosphorus halides with aliphatic diamines (270). However, the polymers formed are usually low in molecular weight.

$$H_2NRNH_2 + ClSO_2ArSO_2Cl \xrightarrow{-HCl} \{NHRNHSO_2ArSO_2\}_n$$

$$H_2NRNH_2 + \overset{\overset{O}{\parallel}}{\underset{\underset{R'}{|}}{C}lPCl} \xrightarrow{-HCl} [\overset{\overset{O}{\parallel}}{\underset{\underset{R,}{|}}{NHRNHP}}]_n$$

Aliphatic diamines are used as cross-linking agents for epoxy resins. The reaction involves the epoxy end group of the resin from epichlorohydrin and bisphenol-A with the diamine to form hydroxy-substituted intermediates. The newly generated hydroxyl groups are then further reacted with other molecules to yield highly cross-linked polymeric system. A summary of the chemistry and applications of epoxy resins has been described by Simonds and Church (271).

$$CH_2\overset{}{\underset{O}{\diagdown\diagup}}CHCH_2R'CH_2CH\overset{}{\underset{O}{\diagdown\diagup}}CH_2 + H_2NRNH_2 \longrightarrow$$

$$\{NHRNHCH_2CHCH_2R'CH_2CHCH_2\}_n$$
$$\underset{NHRNH-}{|}$$

References

1. *Oil, Paint and Drug Reporter*, **193**:44 (April 8, 1968).
2. B. L. Moulthrop (to Socony Vacuum Oil Co.), U.S. Patent 2,651,595 (1953).
3. P. F. Lewis, "Amines," in "*Chem. Econ. Handbook*," 611.6030E (Oct. 1967).
4. *Chem. Week*, p. 101 (April 10, 1965).
5. *Chem. Week*, p. 47 (July 9, 1966).
6. *Oil, Paint and Drug Reporter*, **190**:41 (July 25, 1966).
7. *Chem. Week*, p. 66 (September 28, 1968).
8. T. Oga, *Hydrocarbon Process.*, **45**(11):174 (1966).
9. T. Kanzawa and K. Naito, *Jap. Chem. Quart.*, **3**(4):38 (1967).
10. *Chem. Week*, p. 17 (July 6, 1968).
11. F. A. Gadecki and S. B. Speck (to E. I. du Pont de Nemours & Co.), U.S. Patent 3,249,591 (1966).

12. *Chem. Week*, p. 20 (February 24, 1968); p. 56 (May 18, 1968); p. 57 (June 29, 1968).
13. K. A. Pigott, E. R. Wells, and G. A. Hudson, *Paint Varn. Prod.*, **58**(1):39 (1968.)
14. W. M. Edwards and I. M. Robinson (to E. I. du Pont de Nemours & Co.), U.S. Patent 2,710,853 (1955).
15. D. L. Brebner, W. M. Edwards, I. M. Robinson, E. N. Squire, and H. W. Starkweather, Jr. (to E. I. du Pont de Nemours & Co.), U.S. Patent 2,944,993 (1960).
16. W. G. Brown, "Reduction by Lithium Aluminum Hydride," in "Organic Reactions," Vol. VI, R. Adams, Ed., Wiley, New York, 1957, p. 469.
17. M. J. Astle, "Industrial Organic Nitrogen Compounds," Reinhold, New York, 1961, p. 21.
18. Ref. 17, pp. 22–25.
19. R. B. Wagner and H. D. Zook, "Synthetic Organic Chemistry," Wiley, New York, 1953, pp. 658–659.
20. P. H. Groggins and W. V. Wirth, "Amination by Ammonolysis," in "Unit Processes in Organic Synthesis," 5th ed., P. H. Groggins, Ed., McGraw-Hill, New York, 1958, pp. 397–403.
21. Gunshi Industrial Co., Japanese Patent 158,398 (1943).
22. Ref. 17, p. 6.
23. R. N. Shreve and L. W. Rothenburger, *Ind. Eng. Chem.*, **29**:1361 (1937).
24. R. N. Shreve and D. R. Burtsfield, *Ind. Eng. Chem.*, **33**:219 (1941).
25. R. Brown and W. E. Jones, *J. Chem. Soc.*, **781** (1946).
26. Ref. 17, pp. 7–10.
27. T. Horlenko and H. W. Tatum (to Celanese Corp. of America), U.S. Patent 3,215,742 (1965).
28. R. C. Schreyer (to E. I. du Pont de Nemours & Co.), U.S. Patent 2,754,330 (1956).
29. G. W. Fowler (to Union Carbide Corp.), U.S. Patent 2,519,560 (1950).
30. S. Winderl, E. Haarer, H. Corr, and P. Hornberger (to Badische Anilin- & Soda-Fabrik AG), French Patent 1,347,648 (1963).
31. E. S. Wallis and J. F. Lane, "The Hofmann Reaction," in "Organic Reactions," Vol. III, R. Adams, Ed., Wiley, New York, 1946, p. 267.
32. H. Weidel and E. Roithrer, *Monatsh.*, **17**:183 (1896); also see *J. Chem. Soc.*, **70**(A-1):470 (1896).
33. J. Crum and R. Robinson, *J. Chem. Soc.*, **561** (1943).
34. J. v. Braun and F. Jostes, *Ber.*, **59**:1091 (1926).
35. J. F. Nobis and H. Greenberg (to National Distillers and Chem. Corp.), U.S. Patent 2,865,940 (1958).
36. H. Wolff, "The Schmidt Reaction," in Ref. 31, p. 307.
37. M. Oesterlin, *Z. Angew. Chem.*, **45**:536 (1932).
38. Alfred Nobel and Co., British Patent 701,789 (1954).
39. K. F. Schmidt and S. Strzygowski (to Knoll AG), U.S. Patent 1,926,756 (1933).
40. M. S. Gibson and R. W. Bradshaw, *Angew. Chem. Int. Ed.*, **7**(12):919 (1968).
41. H. R. Ing and R. H. F. Manske, *J. Chem. Soc.*, **2348** (1926).
42. F. Chambret and D. Joly, *Bull. Soc. Chim. Fr.*, **14**(5):1023 (1947).
43. N. Putokhin, *Trans. Inst. Pure Chem. Reagents (Moscow)*, No. **6**:10 (1927); *Chem. Abstr.*, **23**:2938 (1929).
44. A. Muller and E. Feld, *Monatsh.*, **58**:12 (1931).
45. Badische Anilin- & Soda-Fabrik AG, British Patent 824,419 (1959).
46. R. L. Augustine, "Catalytic Hydrogenation," Marcel Dekker, New York, 1965, p. 71.
47. A. E. Barkdoll, D. C. England, H. W. Gray, W. Kirk, Jr., and G. M. Whitman, *J. Amer. Chem. Soc.*, **75**:1156 (1953).

48. W. J. Arthur (to E. I. du Pont de Nemours & Co.), U.S. Patent 3,347,917 (1967).
49. A. S. Chegolya, N. S. Smirnova, B. I. Zhizdyuk, L. M. Ryzhenko, G. I. Golub, and A. A. Panomarev, *Zh. Organ. Khim.*, **1**(10), 1900 (1965) (Eng. transl.).
50. J. M. Cross, S. H. Metzger, Jr., and C. D. Campbell (to Mobay Chem. Co.), U.S. Patent 3,351,650 (1967).
51. G. M. Whitman (to E. I. du Pont de Nemours & Co.), U.S. Patent 2,606,925 (1952).
52. P. A. S. Smith, "The Curtius Reaction," in Ref. 31, p. 386.
53. A. Skita and R. Rossler, *Ber.*, **72B**:461 (1939).
54. P. H. Moss and N. B. Godfrey (to Jefferson Chemicals Co., Inc.), U.S. Patent 3,151,115 (1964).
55. W. W. Levis, Jr., and W. K. Langdon (to Wyandotte Chemicals Corp.), U.S. Patent 2,875,206 (1959).
56. W. K. Langdon (to Wyandotte Chemical Corp.), U.S. Patent 3,067,199 (1962).
57. W. K. Langdon and W. W. Levis, Jr. (to Wyandotte Chemicals Corp.), U.S. Patent 2,911,407 (1959).
58. Ref. 17, p. 156.
59. G. Sonna, *Rend. Semin. Fac. Sci. Univ. Cagliari*, **10**:46 (1940); *Chem. Abstr.*, **37**:1718 (1943).
60. Y. T. Pratt, "The Pyrazines and Piperazines," in "Heterocyclic Compounds," Vol. VI, R. C. Edderfield, Ed., Wiley, New York, 1957, p. 435.
61. G. Cignarella, *J. Med. Chem.*, **7**(2):241 (1964).
62. P. Ruggli, E. Leupin, and H. Dahn, *Helv. Chim. Acta*, **30**:1845 (1947).
63. W. H. Lind (to California Research Corp.), U.S. Patent 2,970,170 (1961).
64. B. H. Nicolaisen (to Olin Mathieson Chemicals Corp.), U.S. Patent 2,805,254 (1957).
65. Farbenfabriken Bayer AG, British Patent 735,779 (1955).
66. G. O. Curma, Jr., and F. W. Lommen (to Union Carbide Corp.), U.S. Patent 1,832,534 (1931).
67. S. W. Dylewski, H. G. Dulude, and G. W. Warren (to Dow Chemical Co.), U.S. Patent 2,769,841 (1956).
68. F. C. Bersworth (to F. C. Bersworth Laboratories), U.S. Patents 2,028,041 (1936).
69. H. G. Muhlbauer (to Jefferson Chemicals Co.), U.S. Patent 3,394,186 (1968).
70. C. B. Fitz-William (to Allied Chemical Corp.), U.S. Patent 3,137,730 (1964).
71. H. Scholz and P. Guenthert (to Badische Anilin- & Soda-Fabrik AG), U.S. Patent 3,067,255 (1962).
72. J. W. Nemec, C. H. McKeever, and E. L. Wolffe (to Rohm and Haas Co.), German Patent 1,154,121 (1963).
73. W. F. Gresham (to E. I. du Pont de Nemours & Co.), U.S. Patent 2,429,876 (1947).
74. A. G. Weber and C. D. Bell (to E. I. du Pont de Nemours & Co.), U.S. Patent 2,436,368 (1948).
75. A. G. Weber and C. D. Bell (to E. I. du Pont de Nemours & Co.), U.S. Patent 2,519,803 (1950).
76. G. F. Mackenzie (to Dow Chemical Co.), U.S. Patent 2,861,995 (1958).
77. M. Sittig, "Organic Chemical Process Encyclopedia," Noyes Development Corp., Park Ridge, N.J., 1967, p. 259.
78. W. Gluud, W. Klempt, and E. Wiebeck, German Patent 659,771 (1938).
79. Gesellschaft für Kohlentechnik, British Patent 436,692 (1935).
80. Ref. 20, p. 407.
81. G. Darzens, *Compt. Rend.*, **208**:1503 (1939).
82. N. Levy and C. W. Scaife, *J. Chem. Soc.*, **1100** (1946).
83. R. L. Heath and J. D. Ross, *J. Chem. Soc.*, **1486** (1947).

84. J. N. Wickert (to Union Carbide Corp.), U.S. Patent 1,988,225 (1935).
85. A. F. Miller, M. Salehar, and J. Williams (to Standard Oil Co.), U.S. Patent 3,260,752 (1966).
86. H. Scholz and P. Guenthert (to Badische Anilin- & Soda-Fabrik AG), U.S. Patent 3,223,735 (1965).
87. Sharples Chemicals, Inc., British Patent 615,715 (1949).
88. C. W. Smith (to Shell Development Co.), U.S. Patent 2,662,080 (1953).
89. E. M. Smolin (to American Cyanamid Co.), U.S. Patent 3,331,877 (1967).
90. E. I. du Pont de Nemours & Co., British Patent 576,015 (1946).
91. F. Codignola and M. Piacenza, Italian Patent 450,090 (1949).
92. M. G. Gergel and M. Revelise, "Nitriles and Isocyanides," in Vol. 9, "Encyclopedia of Chemical Technology," R. E. Kirk and O. F. Othmer, Eds., Interscience, New York, 1952, p. 365.
93. A. Astle, "The Chemistry of Petrochemicals," Reinhold, New York, 1956, pp. 137–140.
94. O. W. Cass, *Ind. Eng. Chem.*, 40(2):216 (1948).
95. S. Petersen and E. Tietze, *Chem. Ber.*, 90:909 (1957).
96. A. P. Terent'ev and V. G. Yashunskii, *Zh. Obshch. Khim.*, 24:291 (1954); Engl. transl., *J. Gen. Chem. USSR*, 24:295 (1954).
97. F. Codignola and M. Piacenza, Italian Patent 422,763 (1947).
98. A. Hrubesch (to Badische Anilin- & Soda-Fabrik AG), German Patent 823,295 (1951).
99. G. A. Silverstone (to Imperial Chemical Industries, Ltd.), British Patent 768,257 (1957).
100. R. A. Smiley (to E. I. du Pont de Nemours & Co.), U.S. Patent 2,912,455 (1959).
101. R. A. Smiley and C. Arnold, *J. Org. Chem.*, 25:257 (1960).
102. C. S. Marvel and E. M. McColm, *Org. Syn.*, **Coll. Vol. 1**, p. 536 (1941).
103. J. D. Behun, "Amines," in Vol. 1, "Encyclopedia of Polymer Science and Technology," N. M. Bikales, Ed., Interscience, New York, 1964, p. 822.
104. L. F. Hatch, *Hydrocarbon Process. Pet. Refiner*, 42(4):157 (1963).
105. J. H. Prescott, *Chem. Eng.*, p. 238 (Nov. 8, 1965).
106. M. Sittig, "Combine Hydrocarbons and Nitrogen for Profit," Noyes Development Corp., Park Ridge, N.J., 1967, p. 14.
107. R. J. W. Reynolds, "Polyamides, Polyesters, and Polyurethanes," in "Fibers from Synthetic Polymers," R. Hill, Ed., Elsevier, New York, 1953, pp. 123–124.
108. Union Chimique-Chemische Bedrijven, S. A., Netherlands Appl. 6,609,241 (1967); *Chem. Abstr.* 67:53724 (1967); and Union Chimique-Chemische Bedrijven, S. A., Belgian Patent 692,153 (1968).
109. R. E. Robinson (to National Distillers and Chemicals Corp.), U.S. Patent 3,133,956 (1964).
110. B. W. Howk and G. M. Whitman (to E. I. du Pont de Nemours & Co.), U.S. Patent 2,504,024 (1950).
111. R. E. Foster and H. E. Schroeder (to E. I. du Pont de Nemours & Co.), U.S. Patent 2,657,240 (1953).
112. M. M. Baizer (to Monsanto Co.), U.S. Patent 3,193,477 (1965).
113. M. M. Baizer (to Monsanto Co.), U.S. Patent 3,193,481 (1965).
114. M. M. Baizer, *J. Electrochem. Soc.*, 111(2):215 (1964).
115. E. I. du Pont de Nemours & Co., British Patent 535,187 (1941).
116. E. I. du Pont de Nemours & Co., British Patent 494,236 (1938).
117. B. W. Howk and M. W. Farlow (to E. I. du Pont de Nemours & Co.), U.S. Patent 2,532,311 (1950).

118. L. E. Romilly (to E. I. du Pont de Nemours & Co.), U.S. Patent 2,532,312 (1950).

119. R. A. Williams (to Imperial Chemical Industries, Ltd.), U.S. Patent 3,398,195 (1968).

120. A. W. Larchar and H. S. Young (to E. I. du Pont de Nemours & Co.), U.S. Patent 2,284,525 (1942).

121. C. R. Campbell, R. Johnson, and R. R. Spiegelhalter (to Monsanto Co.), U.S. Patent 3,152,186 (1964).

122. Imperial Chemical Industries, Ltd., Netherlands Appl. 6,600,443 (July 21, 1966).

123. J. W. Conner, P. W. Evans, A. J. Isacks, Jr., and C. P. Neiswender, Jr. (to Monsanto Co.), French Patent 1,386,911 (1965).

124. E. I. du Pont de Nemours & Co., British Patent 728,599 (1955).

125. E. I. du Pont de Nemours & Co., British Patent 731,819 (1955).

126. D. C. Griffith, R. E. Jones, and R. L. Rose (to Monsanto Co.), U.S. Patent 3,254,126 (1966).

127. C. J. Cavallito, A. P. Gray, and E. E. Spinner, *J. Amer. Chem. Soc.*, **76**:1862 (1954).

128. Sterling Drug, Inc., British Patent 903,200 (1962).

129. A. S. Shpital'nyi and I. V. Kuznetsova, *Zh. Prikl. Khim.*, **30**:1848 (1957); *Chem. Abstr.*, **52**:9647 (1958).

130. H. Krassig, *Makromol. Chem.*, **8**:208 (1952).

131. W. R. Boon, *J. Chem. Soc.*, **307** (1947).

132. W. Ried and K. Wesselborg, *Ann.*, **611**:71 (1957).

133. T. S. Work, *J. Chem. Soc.*, **1315** (1940).

134. J. A. Robertson (to E. I. du Pont de Nemours & Co.) U.S. Patent 2,768,978 (1956).

135. R. Chretien, *Ann. Chim.* (*13*), **2**:682 (1957).

136. J. G. M. Bremner and F. Stankey (to Imperial Chemical Industries, Ltd.), U.S. Patent 2,440,929 (1948).

137. H. B. Copelin (to E. I. du Pont de Nemours & Co.), U.S. Patent 2,497,812 (1950).

138. W. Hentrich and A. Kirstahler (to Dentsche Hydrierwerke AG), German Patent 856,888 (1952).

139. H. Indest (to Vereinigte Glanzstoff-Fabriken AG), German Patent 926,186 (1955).

140. J. Cason, L. Wallcave, and C. N. Whiteside, *J. Org. Chem.*, **14**:37 (1949).

141. J. v. Braun and C. Muller, *Ber.*, **38**:2206 (1908).

142. E. A. Steck, J. S. Buck, and L. T. Fletcher, *J. Amer. Chem. Soc.*, **79**:4414 (1957).

143. H. J. Barber and K. Gaimster, *J. Appl. Chem.*, **2**:565 (1952).

144. I. B. Afanas'ev, G. B. Ovakimeyan, T. N. Eremina, I. B. Voronina, L. K. Smail's, and A. A. Beer, *Khim. Prom.*, **10**:702 (1962), *Chem. Abstr.*, **60**:5331 (1964).

145. A. W. Anderson (to E. I. du Pont de Nemours & Co.), U.S. Patent 2,769,848 (1956).

146. L. Schmerling and J. P. West, *J. Amer. Chem. Soc.*, **74**:2885 (1952).

147. A. J. Yu and R. D. Evans, *J. Polym. Sci.*, **42**:249 (1960).

148. Badische Anilin- & Soda-Fabrik AG, German Patent 857,376 (1952).

149. Badische Anilin- & Soda-Fabrik AG, British Patent 722,479 (1955).

150. Badische Anilin- & Soda-Fabrik AG, German Patent 860,490 (1952).

151. Badische Anilin- & Soda-Fabrik AG, German Patent 890,950 (1953).

152. National Distillers and Chemicals Corp., British Patent 821,984 (1959).

153. D. D. Coffman and E. L. Jener (to E. I. du Pont de Nemours & Co.), U.S. Patent 3,017,435 (1962).

154. I. D. Webb (to E. I. du Pont de Nemours & Co.), U.S. Patent 2,485,225 (1949).

155. E. C. G. Clark, British Patent 490,922 (1938).

156. J. B. Dickey and J. G. McNally (to Eastman Kodak Co.), U.S. Patent 2,412,209 (1946).

157. *Akad. Nauk SSSR*, Inst. Org. Khim. Sintezy Org. Soedin. I (1950); *Chem. Abstr.*, **47**:7999 (1953).

158. H. L. Yale, *J. Amer. Chem. Soc.*, **75**:675 (1953).

159. R. F. Nystrom, *J. Amer. Chem. Soc.*, **77**:2544 (1955).

160. G. Boffa, D. Costabello, F. Minisci, and A. Quilico, *Gazz. Chim. Ital.*, **89**:1390 (1959); *Chem. Abstr.*, **54**:22354 (1960).

161. B. S. Biggs and W. S. Bishop, *Org. Syn.*, **Coll. Vol. 3**, p. 229 (1955).

162. J. Schmidt and W. Schulzel (to Badische Anilin-Soda-Fabrik AG), German Patent 857,194 (1952).

163. C. H. Greenewalt and G. W. Rigby (to E. I. du Pont de Nemours & Co.), U.S. Patent 2,132,849 (1938).

164. R. M. Joyce, W. E. Hanford, and J. Harmon, *J. Amer. Chem. Soc.*, **70**:2529 (1948).

165. K. Saotome, H. Komoto, and T. Yamazaki, *Bull. Chem. Soc. Jap.*, **39**(3):480 (1966).

166. K. Ziegler, K. Schneider, and J. Schneider, *Ann.*, **623**:9 (1959).

167. K. Ziegler, British Patent 803,178 (1958).

168. F. L. M. Pattison, J. B. Stothers, and R. G. Woolford, *J. Amer. Chem. Soc.*, **78**:2255 (1956).

169. G. Nischk and E. Muller, *Ann.*, **576**:232 (1952).

170. A. A. Zalikin and Yu. A. Atrephikheev, *Tr. Mosk. Khim.-Tekhnol. Inst.*, **42**:114 (1963); *Chem. Abstr.*, **62**:463 (1965).

171. G. T. Perkins (to E. I. du Pont de Nemours & Co.), U.S. Patent 3,367,969 (1968).

172. E. Haarer and G. Wenner (to Badische Anilin- & Soda-Fabrik AG), German Patent 1,205,975 (1965).

173. L. D. Brake (to E. I. du Pont de Nemours & Co.), U.S. Patent 3,283,002 (1966).

174. A. E. Barkdoll, H. W. Gray, and W. Kirk, Jr., *J. Amer. Chem. Soc.*, **73**:741 (1951).

175. G. M. Whitman (to E. I. du Pont de Nemours & Co.), U.S. Patent 2,606,924 (1952).

176. J. R. Kuszewski (to E. I. du Pont de Nemours & Co.), U.S. Patent 3,393,236 (1968).

177. W. J. Arthur (to E. I. du Pont de Nemours & Co.), U.S. Patent 3,384,661 (1968).

178. W. J. Arthur (to E. I. du Pont de Nemours & Co.), U.S. Patent 3,153,088 (1964).

179. W. A. Pryor (to California Research Corp.), U.S. Patent 2,828,335 (1958).

180. L. P. Kuhn, W. J. Taylor, Jr., and P. H. Groggins, "Nitration," in "Unit Processes in Organic Synthesis," 5th ed., P. H. Groggins, Ed., McGraw-Hill, New York, 1958, p. 113.

181. J. Werner and P. H. Groggins, "Amination by Nitration," in Ref. 180, p. 188.

182. H. L. Yale (to Olin Mathieson Chemicals Corp.), U.S. Patent 2,672,472 (1954).

183. A. A. Ponomarev, A. S. Chegolya, and B. I. Zhizdyuk, Russian Patent 193,528 (1967); *Chem. Abstr.*, **69**:243 (1968).

184. J. H. Saunders and K. C. Frisch, "Polyurethanes, Chemistry and Technology," Part I, "Chemistry," Interscience, New York, 1962, p. 21.

185. P. H. Moss and N. B. Godfrey (to Jefferson Chemicals Co., Inc.), U.S. Patent 3,037,023 (1962).

186. H. G. Muhlbauer and M. Lichtenwalter (to Jefferson Chemicals Co., Inc.), U.S. Patent 3,285,920 (1966).

187. H. G. Muhlbauer and M. Lichtenwalter (to Jefferson Chemicals Co., Inc.), U.S. Patent 3,297,700 (1967).

188. R. C. Lemon, S. Depot, and R. C. Myerly (to Union Carbide Corp.), U.S. Patent 3,112,318 (1963).

189. M. W. Long, Jr. (to Dow Chemical Co.), U.S. Patent 2,910,477 (1959).

190. F. Poppelsdorf and R. C. Myerly (to Union Carbide Corp.), U.S. Patent 3,095,417 (1963).

191. F. Poppelsdorf and R. C. Myerly (to Union Carbide Corp.), U.S. Patent 3,016,558 (1963).
192. J. J. Scigliano and E. C. Britton (to Dow Chemical Co.), U.S. Patent 2,843,590 (1958).
193. G. E. Ham and P. M. Phillips (to Dow Chemical Co.), U.S. Patent 3,324,130 (1967).
194. L. J. Kitchen and C. B. Pollard, *J. Org. Chem.*, **8**:342 (1943).
195. W. K. Langdon (to Wyandotte Chemicals Corp.), U.S. Patent 2,835,673 (1958).
196. G. W. Fowler (to Union Carbide Corp.), U.S. Patent 2,980,682 (1961).
197. D. E. Trucker (to Wyandotte Chemicals Corp.), U.S. Patent 3,067,201 (1962).
198. W. K. Langdon, W. W. Levis, Jr., and D. R. Jackson, *I & EC Process Des. Develop.*, **1**:153 (1962).
199. W. P. Coker and G. W. Strother, Jr. (to Dow Chemical Co.), U.S. Patent 2,861,994 (1958).
200. G. W. Fowler, D. G. Crosby, and W. R. Proops (to Union Carbide Corp.), U.S. Patent 2,920,076 (1960).
201. J. T. Patton, Jr. (to Wyandotte Chemicals Corp.), U.S. Patent 2,940,973 (1960).
202. H. D. Williams (to E. I. du Pont de Nemours & Co.), U.S. Patent 3,074,949 (1963).
203. W. J. Pope and J. Read, *J. Chem. Soc.*, **105**:219 (1914).
204. T. Tishiguro, M. Matsumura, and M. Awamura, *Yakugaku Zasshi*, **78**:751 (1958); *Chem. Abstr.*, **52**:18453 (1958).
205. L. F. Marek, "Oxidation," in Ref. 180, p. 517.
206. C. D. Heaton (to California Research Corp.), U.S. Patent 2,773,902 (1956).
207. T. Oga, *Jap. Chem. Quart.*, **2**(3):55 (1966).
208. S. Sakuyama, K. Oda, and T. Ohara, *Chem. Eng. Progr.*, **60**(9):48 (1964).
209. T. Oga, H. Ichinokawa, and M. Ito (to Showa Denko), German Patent 1,236,494 (1967).
210. S. Saito and N. Ota, *Yuki Gosei Kagaku Shi*, **22**(9):730 (1964); *Chem. Abstr.*, **61**:11929 (1964).
211. S. Saito, H. Iwasaki, and N. Ota, *Yuki Gosei Kagkai Shi*, **22**(10):828 (1964); *Chem. Abstr.*, **61**:14578 (1964).
212. Y. Ogata and K. Sakanishi, *Kôgyô Kagaku Zasshi*, **69**(12):2294 (1966); *Chem. Abstr.*, **66**:9759 (1967).
213. Y. Ogata and K. Sakanishi, *Chem. Ind.* (*London*), **2055** (1966).
214. S. D. Mekhtiev, G. N. Suleimanov, Sh. F. Mamedova, and R. Yu Magerramova, *Azerb. Khim. Zh.*, **25** (1966); *Chem. Abstr.*, **65**:8815 (1966).
215. L. J. Bellamy, "The Infrared Spectra of Complex Molecules," 2nd ed., Wiley, New York, 1959.
216. R. G. White, "Handbook of Industrial Infrared Analysis," Plenum Press, New York, 1964.
217. F. H. Lohman and W. E. Norteman, Jr., *Anal. Chem.*, **35**(6):707 (1963).
218. J. Zabicky, "Detection, Determination, and Characterization of Amines," in "The Chemistry of Amino Groups," S. Patai, Ed., Interscience, New York, 1968, p. 111.
219. Ref. 218, p. 113.
220. L. M. Jackman," Applications of Nuclear Magnetic Resonance in Organic Chemistry," Pergamon Press, Oxford, 1962, p. 72.
221. J. B. Stothers, "Applications of Nuclear Magnetic Resonance Spectroscopy," in "Elucidation of Structures by Physical and Chemical Methods (Technique of Organic Chemistry, Vol. XI)," Pt. I, K. W. Bently, Ed., Interscience, New York, 1963, p. 211.
222. N. S. Bhacca, D. P. Hollis, L. F. Johnson, E. A. Pier, and J. N. Schoolery, 'NMR Spectra Catalog," Vols. I and II, Varian Associates, Palo Alto, Calif., 1962 and 1963.
223. Ref. 218, p. 127.

224. K. Biemann, "Mass Spectrometry, Organic Chemical Applications," McGraw-Hill, New York, 1962, p. 87.
225. F. W. McLafferty, "Interpretation of Mass Spectra," W. A. Benjamin, New York, 1966, p. 109.
226. "NFPA No. 49—Hazardous Chemicals Data—1966" International Fire Protection Association, Boston, 1966, p. 126.
227. E. F. Hillenbrand, Jr., and C. A. Pentz, "Determination of Amines and Amides," in "Organic Analysis," Vol. III, J. Mitchell, Jr., I. M. Kolthoff, E. S. Proskauer, and A. Weissberger, Eds., Interscience, New York, 1956, pp. 143–145.
228. P. W. Morgan, "Condensation Polymers: By Interfacial and Solution Methods," Interscience, New York, 1965, p. 12.
229. V. E. Shashova and W. M. Eareckson, III, *J. Polym. Sci.*, **40**:343 (1959).
230. R. G. Beaman, P. W. Morgan, C. R. Koller, E. L. Wittbecker, and E. E. Magat, *J. Polym. Sci.*, **40**:333 (1959).
231. C. R. Koller (to E. I. du Pont de Nemours & Co.), U.S. Patent 3,070,562 (1962).
232. Ref. 228, p. 173.
233. B. Ke and A. W. Sisko, *J. Polym. Sci.*, **50**:87 (1961).
234. C. W. Stephens, *J. Polym. Sci.*, **40**:359 (1959).
235. E. L. Wittbecker and M. Katz, *J. Polym. Sci.*, **40**:367 (1959).
236. K. Hayashi and Y. Iwakura, *Makromol. Chem.*, **94**:132 (1966).
237. P. W. Morgan and S. L. Kwolek, *J. Polym. Sci.*, **A2**:181 (1964).
238. H. V. Boening, N. Walker, and E. H. Myers, *J. Appl. Polym. Sci.*, **5**(16):384 (1961).
239. W. S. Spector, Ed., "Handbook of Toxicology," Vol. I, "Acute Toxicities," W. B. Saunders, Philadelphia, 1956.
240. M. N. Gleason, R. E. Gosselin, and H. C. Hodge, "Chemical Toxicity of Commercial Products," Williams and Wilkins, Baltimore, 1957.
241. N. I. Sax, "Dangerous Properties of Industrial Materials," 3rd ed., Reinhold, New York, 1968.
242. W. L. Sutton, "Aliphatic and Alicyclic Amines," in "Industrial Hygiene and Toxicology," Vol. II, "Toxicology," F. A. Patty, Ed., Interscience, New York, 1963, p. 2037.
243. W. R. Sorenson and T. W. Campbell, "Preparative Methods of Polymer Chemistry," 2nd ed., Interscience, New York, 1968.
244. D. E. Floyd, "Polyamide Resins," 2nd ed., Reinhold, New York, 1966.
245. L. F. Hatch, *Hydrocarbon Process. Pet. Refiner*, **42**(5):171 (1963).
246. C. S. Fuller and A. R. Kemp (to Bell Telephone Laboratories), U.S. Patent 2,349,951 (1944).
247. Ref. 243, p. 126.
248. C. S. Marvel and J. H. Johnson, *J. Amer. Chem. Soc.*, **72**:1674 (1950).
249. S. L. Kwolek and P. W. Morgan, *J. Polym. Sci.*, **A2**:2693 (1964).
250. O. Bayer, E. Mueller, S. Petersen, H. F. Piepenbrink, and E. Windemuth, *Angew. Chem.*, **62**:57 (1950).
251. J. H. Saunders and K. C. Frisch, "Polyurethanes, Chemistry and Technology," Pt. II, "Technology," Interscience, New York, 1964, p. 491.
252. Ref. 251, p. 694.
253. Ref. 228, p. 209.
254. Ref. 243, pp. 106–107.
255. E. L. Wittbecker (to E. I. du Pont de Nemours & Co.), U.S. Patent 2,816,879 (1957).
256. Ref. 243, p. 211.

257. G. J. M. Van der Kerk, H. G. J. Overmars, and G. M. Van der Want, *Rec. Trav. Chim. Pays-Bas*, **74**:1301 (1955).

258. W. R. Grace and Co., French Patent 1,299,698 (1962).

259. H. A. Walter (to Monsanto Chemical Co.), U.S. Patent 3,006,898 (1961).

260. E. I. du Pont de Nemours & Co., British Patent 528,437 (1940).

261. G. D. Buckley and N. H. Ray (to Imperial Chemical Co.), U.S. Patent 2,550,767 (1951).

262. W. Lehmann and H. Rinke (to Farbenfabriken Bayer AG), U.S. Patent 2,8524,94.

263. Y. Inaba and K. Kimoto (to Koyo Koatsu Industries, Inc.), U.S. Patent 2,973,342 (1961).

264. R. J. W. Reynolds, "Some Miscellaneous Fiber-forming Condensation Polymers," in "Fibers from Synthetic Polymers," R. Hill, Ed., Elsevier, New York, 1953, p. 179.

265. W. Lehmann and H. Rinke (to Farbenfabriken Bayer AG), U.S. Patent 2,855,384 (1958).

266. W. Lehmann and H. Rinke (to Farbenfabriken Bayer AG), U.S. Patent 2,761,852 (1956).

267. R. Neher (to Ciba AG), German Patent 1,046,309 (1958).

268. S. B. Speck (to E. I. du Pont de Nemours & Co.), U.S. Patent 2,808,394 (1957).

269. S. A. Sundet, W. A. Murphey, and S. B. Speck, *J. Polym. Sci.*, **40**:389 (1959).

270. D. M. Harris, R. L. Jenkins, and M. L. Nielson, *J. Polym. Sci.*, **35**:540 (1959).

271. H. R. Simonds and J. M. Church, "A Concise Guide to Plastics," 2nd ed., Reinhold, New York, 1963.

3. GLYCOLS AND BISCHLOROFORMATES

Eugene V. Hort

GAF Corporation,
Wayne, New Jersey

I. INTRODUCTION

In 1856* the eminent French chemist Adolphe Wurtz described the first member of a new class of compounds, the dibasic alcohols. He prepared this material by treating ethylene diiodide with silver acetate and saponifying the resulting diacetate. Because its properties were intermediate between those of glycerol and of alcohol, he proposed for this new compound the name "glycol" (1). He predicted that a similar route from propylene diiodide would give the next homolog, which he named "propyl-glycol"; and indeed in 1858 he described the synthesis and properties of this material too (2). The name glycol is still occasionally used to specifically designate the compound 1,2-ethanediol; but more frequently it appears as a generic name for any compound containing two hydroxyl groups located on aliphatic carbons.

A number of glycols have attained great industrial importance. These are covered in detail in Curme's ACS monograph, "Glycols" (3). A more recent review, which includes some of the less important glycols, is Mellan's "Polyhydric Alcohols" (4). There does not seem to be a good review of glycols from the standpoint of monomers, even though glycols have become an exceedingly important set of building blocks for polymers.

As early as 1863 (5) a resinous material, later shown to be a polyester, was prepared by heating ethylene glycol with succinic acid. In subsequent years many other polyesterifications were reported, but the structures of the products were not generally understood. Starting in 1929, Carothers and his

---

* The earliest references to glycols listed in Beilstein involve papers appearing in 1859. As a probably consequence, the ACS monographs dealing with glycols as well as various review articles have credited Wurtz with first describing glycols in 1859. Actually, he had presented a number of papers on glycols previously, starting in 1856.

co-workers elucidated the chemistry of polyesterification in a series of brilliant papers (6–11). Subsequently, an English group investigated polyesters of aromatic dicarboxylic acids and found them to be high melting and relatively stable to hydrolysis. This work culminated in the development of poly(ethylene terephthalate), introduced as Terylene by Imperial Chemical Industries in Great Britain (12) and Dacron by du Pont in the United States (13).

In the late 1930s German investigators studied the properties of polyurethanes obtained from the reacton of various diols with diisocyanates. It was soon discovered that equivalent polyurethanes could be prepared by the reaction of bischloroformates with diamines.

Of all the many glycol derivatives that have been prepared, one group, the bischloroformates, are unique in the interest they have aroused as monomers for condensation polymerization. Chloroformates, ClCOOR, may be considered either as derivatives of carbonic acid in which a chlorine atom has replaced a hydroxyl group or as derivatives of formic acid in which chlorine has replaced a hydrogen atom. The chlorocarbonate nomenclature is common in the old literature, but chloroformate is the standard *Chemical Abstracts* usage and is prevalent in recent literature. The proper nomenclature for the derivative of a diol is bis(chloroformate), but in common practice the parentheses are usually omitted. Although monochloroformates were described as early as 1833 (14), the first reported synthesis of a bischloroformate did not appear until 1925 (15). A recent article in *Chemical Reviews* (16) thoroughly covers the chemistry of chloroformates but does not deal with bischloroformates as monomers.

II. SURVEY OF CHEMISTRY

A. Glycols

We would expect the simplest member of the glycol series to be the one-carbon representative, methylene glycol. In 1859 Butlerov (17) treated methylene diiodide with silver acetate and saponified the resulting methylene diacetate. Instead of the expected methylene glycol, formaldehyde was obtained. It has subsequently been shown that dilute aqueous solutions of formaldehyde exist principally in the form of the monohydrate, methylene glycol. This is in equilibrium with a series of oligomeric hydrates and with a trace of unhydrated formaldehyde (18). Because of this lability, methylene glycol and other *gem*-diols are not usually grouped with the glycols.

Except when both hydroxyl groups are involved in a single reaction, the chemical behavior of glycols is exactly the same as that of alcohols. As with alcohols, glycol condensation reactions such as esterification proceed well with primary hydroxyls, fairly well with secondary hydroxyls, and poorly or not at all with tertiary hydroxyls. Furthermore, the tertiary hydroxyl group is easily split off, particularly at high temperatures or with acid catalysts. In consequence, glycols containing tertiary hydroxyls are of little interest as monomers for condensation polymers. Most of the glycols of interest for polymerization have two primary hydroxyls, a few have one primary and one secondary, and a few have two secondaries.

The chemical behavior of a glycol is greatly affected by the relative locations of the hydroxyl groups. When condensed with a difunctional comonomer, glycols can either give linear chains with functional groups at either end, or they can cyclize to rings. A general rule was developed by Carothers (19) based on a study of the esterification of glycols with dibasic acids. In its original form, this rule states that the nature of the ester is completely determined by the number of atoms in the repeating group backbone: if this number is five, a five-membered-ring will be obtained; if the number is six, interconvertible six-membered-ring and linear polymeric forms can be obtained; if the number is more than six, only the linear polymer is obtained. Such behavior is illustrated by the products obtained upon transesterification of various glycols with diethyl carbonate. Subsequently it was shown that

$$HO(CH_2)_2OH \longrightarrow$$

$$
\begin{array}{c}
CH_2-CH_2 \\
/ \quad \backslash \\
O \quad O \\
\backslash \quad / \\
C \\
\parallel \\
O
\end{array}
$$

5-ring

$$HO(CH_2)_3OH \longrightarrow$$

$$
\begin{array}{c}
CH_2 \\
CH_2 \quad CH_2 \\
| \quad \quad | \\
O \quad \quad O \\
\backslash \quad / \\
C \\
\parallel \\
O
\end{array}
$$

6-ring
+
linear polymer

$$HO(CH_2)_4OH \qquad HO+(CH_2)_4OCOO+_n(CH_2)_4OH \qquad \text{linear polymer}$$

Carothers's rules could not be applied rigidly, even for the glycol carbonate esters used as illustration. For example, it was found that the five-membered ring, ethylene carbonate, could be polymerized by heating under pressure with a little glycol and a trace of catalyst (20). It was also discovered that linear polyesters could be broken down into monomeric and dimeric cyclic esters by heating at a high temperature and distilling out the cyclic products (21).

Furthermore, with acetals, the optimum ring size has been shown to be one unit larger than with esters. With acetals containing a repeating group of

five or six, cyclic products are formed, and with a repeating group of seven,
a mixture of cyclic and linear polymeric acetals is obtained (22–24). Cyclic

$$HO(CH_2)_2OH \longrightarrow$$

CH₂—CH₂ ring with O, O and CH₂ (5-ring)

$$HO(CH_2)_3OH \longrightarrow$$

6-ring structure

$$HO(CH_2)_4OH \longrightarrow$$

7-ring + linear polymer

$$HO(CH_2)_5OH \longrightarrow HO \{(CH_2)_5OCH_2O\}_{\overline{n}} (CH_2)_5OH \qquad \text{linear polymer}$$

formals, such as the product from ethylene glycol (1,3-dioxolane), can be
polymerized with suitable cationic catalysts to high-molecular-weight polymers (25,26).

Another important respect in which the behavior of glycols varies with
chain length is in ease and mode of dehydration. Ethylene glycol is dehydrated
only slowly when heated with strong acids at high temperatures, giving 1,4-
dioxane and, under certain conditions, moderate conversions to lower
polyethers such as diethylene glycol and triethylene glycol (27). Trimethylene
glycol, on the other hand, cannot give a five- or six-membered ring by dehydration. Here, dehydration proceeds more rapidly than with ethylene glycol,
but it still requires acid catalysts and high temperatures. The products are
the expected hydroxyl-terminated polyether oligomers (28). 1,4-Butanediol
is very easily dehydrated to the five-membered cyclic ether, tetrahydrofuran;
this reaction proceeds rapidly in the presence of strong acid catalysts even
at comparatively low temperatures. 1,5-Pentanediol is dehydrated to the
six-membered cyclic ether, although its dehydration is slower than that of
1,4-butanediol. Finally, 1,6-hexanediol and higher homologs show little tendency to cyclic dehydration. Like trimethylene glycol, they are dehydrated by
acid catalysts and high temperatures to linear hydroxyl-teminated polyether
oligomers (29). Under very severe conditions (prolonged heating with 50%
sulfuric acid) 1,6-hexanediol has been reported to give cyclic dehydration to
a mixture of cyclic oxides, consisting of about 10% seven-membered, 25%
six-membered and 65% five-membered oxide rings (30).

The only glycol reactions potentially capable of forming polymers are
those in which each of the hydroxyl groups behaves independently and links

with a different site. Hydroxyl groups undergo a multitude of different reactions, but only four types of reaction have attracted much interest for polymerization:

1. Esterification

$$ROH + R'COOH \text{ (or acyl halide or anhydride)} \rightarrow R'COOR$$

2. Urethane formation

$$ROH + R'NCO \rightarrow R'NHCOOR$$

3. Acetal formation

$$2ROH + R'CHO \rightarrow ROCH(R')OR$$

4. Ether formation

$$2ROH \rightarrow ROR$$

Each of these reactions (except with isocyanate) involves condensation; water or some other small molecule is split out. The reaction with isocyanate involves addition rather than condensation.* However, the same product can be obtained by means of the condensation reaction between a chloroformate and an amine. On a commercial basis, only esterification (and to a lesser extent urethane formation) have utilized substantial volumes of glycols as monomers. Houben-Weyl (31) gives detailed laboratory procedures for the preparation of each of these types of polymers. Gaylord (32) has written an excellent review on the preparation of polyacetals and polyethers by means of condensation reactions.

B. Bischloroformates

The chloroformate group undergoes the typical reactions of an acyl chloride, although it is much less reactive than the usual acyl chloride and, in the absence of catalysts, is hydrolyzed by cold water at an extremely slow rate. The relative reactivity of several acyl halides has been shown to be (33):

$$CH_3COCl > C_6H_5COCl > C_2H_5OCOCl$$

The products of hydrolysis of a chloroformate are usually an alcohol, carbon dioxide, and hydrochloric acid. Carbonates, resulting from reaction of the alcohol with unhydrolyzed chloroformate, are found as by-products. Chloroformates react with alcohols or phenols to give carbonate esters and hydrochloric acid:

$$ROCOCl + R'OH \rightarrow ROCOOR' + HCl$$

With alcohols, the reaction frequently proceeds well even without a catalyst or acid acceptor, whereas with phenols a catalyst (usually a base) is necessary

---

* A book by P. W. Morgan on condensation polymers classifies the reaction of isocyanate with hydroxyl groups as a "hydrogen transfer reaction" and includes it in the categories covered (34).

(35). Thiols react more slowly than alcohols when not catalyzed but more rapidly than alcohols in the presence of base. Hence 2-mercaptoethanol gives the mercaptocarbonate when uncatalyzed and the hydroxythiolcarbonate when sodium hydroxide is added (36).

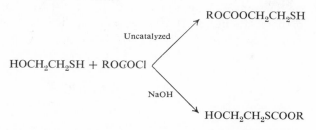

Chloroformates react readily with ammonia or amines to give carbamic esters, usually called urethanes (*Chemical Abstracts* spells it "urethans").

$$ROCOCl + R'NH_2 \rightarrow ROCONHR'$$

An excess of ammonia or amine can act as an acid acceptor for the hydrochloric acid produced, or an inorganic base or tertiary amine can be added. The unsubstituted carbamates obtained from ammonia are high melting crystalline compounds which traditionally have been used as derivatives for identification of chloroformates. The reaction with primary or secondary amines is extremely rapid, and it is faster with aliphatic amines than with aromatic amines. Urethanes derived from ammonia or primary amines still have hydrogen atoms on the nitrogen atoms. These are very low in reactivity, but under extreme conditions they can be further reacted with chloroformates (37). Tertiary amines react with chloroformates to form complexes, which presumably are quaternary ammonium salts of the structure $R'OCONR_3^+Cl^-$. With trialkyl amines, refluxing these complexes in benzene solution gives good yields of $R'OCONR_2$ and RCl (38). Aminoalcohols or aminophenols react at the amine group first. If a second mole of chloroformate is present, it can subsequently react with the hydroxyl group (39,40). Hydrazine (41) and hydroxylamine (42) react at both ends in stepwise fashion. With hydroxylamine, the amino group reacts first.

Carboxylic acids react with chloroformates, giving as the primary products mixed anhydrides of carboxylic and carbonic acids. These mixed anhydrides are limited in stability, and can be thermally decomposed to esters and carbon dioxide in high yields (43):

$$R'COOCOOR \rightarrow R'COOR + CO_2$$

In most cases, the mixed anhydride is so unstable that the only product isolated after treating an acid with a chloroformate is the ester (44).

Any of the previously described reactions of chloroformates can give a polymer by condensing a bischloroformate with a polyfunctional compound. The most important polycondensation is with diamines to form polyure-thanes. This often permits the preparation of urethanes when the diisocyanate required for the alternative synthesis (by condensation with a diol) is extremely difficult to prepare. Using interfacial techniques, high-molecular-weight polyurethanes can be prepared very easily from bischloroformates and diamines.

III. SYNTHETIC PROCEDURES

A. α-Diols

Only two α-diols have elicited much interest as monomers, but these are the most important of all from the commercial point of view. The first is ethylene glycol, manufactured in greater volume than all other glycols combined. The second is propylene glycol, which leads the rest of the pack in volume by a wide margin. Although both were formerly prepared by other methods, the sole surviving important method of manufacture is via hydrolysis of the corresponding epoxides.

1. HYDROLYSIS OF EPOXIDES

Epoxides can be most readily prepared by dehydrohalogenation of halo-hydrins or by oxidation of olefins. Upon hydrolysis, high yields of α-diols are obtained. Dittus (45) has written an excellent review on the preparation and reactions of epoxides; such glycols as 3-chloro-1,2-propanediol, 5-chloro-1,2-pentanediol, 3-butene-1,2-diol, and trans-1,2-cyclohexanediol, serve as illustrations.

a. Ethylene Glycol

In 1859 Wurtz prepared ethylene oxide by treating ethylene chlorohydrin with alkali (46). Upon hydrolyzing ethylene oxide with water, he obtained ethylene glycol and polyethylene glycols (47). This chlorohydrin route to ethylene glycol was introduced into commercial manufacture in Germany during World War I and in the United States in 1925 (48). Although in recent years, manufacture of ethylene oxide via direct oxidation has become dominant, many plants using the chlorohydrin process are still operating. In a typical commerical plant, water, chlorine, and ethylene are allowed to react continuously in a packed tower. The yield of ethylene chlorohydrin is 87–90%, and the principal by-product is 6–9% of ethylene dichloride. Product from the chlorohydrin tower is heated with a lime slurry, and the

crude ethylene oxide distillate is purified by fractional distillation. Flow
charts are given by Faith et al. (49).

$$Cl_2 + H_2O \longrightarrow HOCl + HCl$$

$$CH_2{=}CH_2 + HOCl \longrightarrow ClCH_2CH_2OH$$

$$ClCH_2CH_2OH + OH^- \longrightarrow CH_2\overset{O}{\overgroup{\qquad}}CH_2 + Cl^- + H_2O$$

In 1931 Lefort (50) disclosed the direct oxidation of ethylene to ethylene
oxide using silver as catalyst. The first direct oxidation plant went into operation in the United States in 1937. All recent ethylene oxide plants have used
direct oxidation. Voge and Adams (51) have written an excellent review on
catalytic oxidation of ethylene to ethylene oxide. Ethylene and a large excess
of air or oxygen plus inert gas are passed over a catalyst of silver oxide
deposited on an inert carrier. In typical plants, temperatures from 250 to
290°C and pressures from atmospheric to 200 psi are used. Typical yields are
60–70%. Ethylene oxide is extracted from the effluent gases by absorption in
water under pressure. The solution is stripped and the ethylene oxide is
fractionally distilled.

$$CH_2{=}CH_2 + \tfrac{1}{2}O_2 \longrightarrow CH_2\overset{O}{\overgroup{\qquad}}CH_2$$

Ethylene oxide can be hydrolyzed by heating with water alone at high
temperature and pressure, such as 1 hr at 195°C and 185 psi. In the presence
of 0.5–1.0% sulfuric acid, 30 min at 50–70°C is sufficient. Diethylene glycol
and triethylene glycol are formed as by-products. The smaller the excess of
water used for hydrolysis, the greater the proportion of higher oligomers.
Diethylene glycol, triethylene glycol, and higher polyglycols are produced by
using high ratios of ethylene oxide to water or by treating ethylene glycol with
ethylene oxide.

$$CH_2\overset{O}{\overgroup{\quad}}CH_2 + H_2O \longrightarrow HOCH_2CH_2OH$$

$$CH_2\overset{O}{\overgroup{\quad}}CH_2 + HOCH_2CH_2OH \longrightarrow HOCH_2CH_2OCH_2CH_2OH$$

The glycol oligomers are important commercial products. Diethylene
glycol has the third largest sales volume of the glycols, surpassed only by
ethylene and propylene glycols.

b. Propylene Glycol

Unlike ethylene oxide, the two-step (chlorohydrin) route from propylene
is still the dominant manufacturing process for propylene oxide, although
recently there has been considerable activity in direct oxidation (52–54).

The treatment of propylene with hypochlorous acid gives a mixture of chloro-hydrins, about 90% 1-chloropropan-2-ol and 10% 2-chloropropan-1-ol (55). Both isomers are converted to propylene oxide by treatment with a lime slurry. The propylene oxide is preferably distilled out as fast as it is prepared:

$$CH_3CH{=}CH_2 + HOCl \longrightarrow CH_3CHOHCH_2Cl$$

$$CH_3CHOHCH_2Cl + OH^- \longrightarrow CH_3\overset{O}{\overset{\diagup\diagdown}{CH{-}\!{-}CH_2}} + Cl^- + H_2O$$

Direct oxidation of propylene with air or oxygen has not given worthwhile yields of propylene oxide. However, a two-step oxidation involving a hydro-peroxide as oxidizing agent is feasible, and new plants involving this type of processing have been announced.

$$RH + O_2 \longrightarrow ROOH$$

$$ROOH + CH_3CH{=}CH_2 \longrightarrow ROH + CH_3\overset{O}{\overset{\diagup\diagdown}{CH{-}\!{-}CH_2}}$$

With ethylbenzene hydroperoxide, *t*-butyl alcohol, and propylene, using 0.006 mole of molybdenum (as the naphthenate) per mole of hydroperoxide as catalyst, the following results were reported: 92% conversion of hydro-peroxide and 83% yield of propylene oxide, based on hydroperoxide consumed (56). The 1-phenylethanol by-product may be dehydrated to styrene.

Propylene oxide can be hydrolyzed to propylene glycol either uncatalyzed at relatively severe conditions or catalyzed with acid under milder conditions. Dipropylene glycol is a by-product of propylene glycol manufacture, just as diethylene glycol is a by-product of ethylene glycol manufacture. Unlike ethylene glycol, propylene glycol has nonequivalent hydroxyl groups, and therefore three diglycols are possible. The commercial product is a mixture of all three, in approximately the following proportions (57):

$CH_3CHOHCH_2OCH(CH_3)CH_2OH$	primary-secondary glycol, 53%
$CH_3CHOHCH_2OCH_2CHOHCH_3$	disecondary glycol, 43%
$HOCH_2CH(CH_3)OCH(CH_3)CH_2OH$	diprimary glycol, 4%

2. HYDROLYSIS OF OTHER FUNCTIONAL GROUPS

The first syntheses of glycols ever described were by saponification of diacetates (1,58), and many subsequent investigators prepared α-diols by hydrolysis of many derivatives. Although glycols have been prepared by hydrolysis of sundry other compounds, such as *N,N'*-dinitroethylenediamine (59) and ethanolamine (60), the hydrolysis of compounds containing chloride groups is the only reaction in this category that has achieved importance as a synthetic method. The hydrolysis of ethylene dichloride has been extensively studied (61) and was at one time used in Germany for limited-scale manufacture of ethylene glycol. For many years, hydrolysis of the chlorohydrin was the principal method of manufacture of ethylene glycol. An aqueous solution of ethylene chlorohydrin was heated with an equivalent amount of sodium bicarbonate at 70–80°C for 4–6 hr (49). This method was replaced by indirect hydrolysis via ethylene oxide, principally because the glycol was separated only with difficulty from the by-product salt.

3. HYDROGENATION AND HYDROGENOLYSIS

α-Diols have been prepared by the hydrogenation or hydrogenolysis of many different substrates. Ethylene glycol has been obtained by hydrogenation of glyoxal (62), glycolic esters (63), glycolic acid or oxalic acid (64), and 2,3-diphenyl-1,4-dioxane (65). Similarly, propylene glycol has been obtained from lactate esters (66), glycidol (67), acetol (68), and glycerol (69). A small amount of propylene glycol is produced as a by-product in the hydrogenation of fatty acid glycerides. Various polyols such as lignin (70), xylitol (71), and sugars (72) give both ethylene glycol and propylene glycol in the mixtures of products obtained by hydrogenolysis. The combined hydrogenation and hydrolysis of aldehyde starch gives high yields of ethylene glycol and erythritol (73).

Until the plant using this method closed in 1968, ethylene glycol had been produced by one manufacturer through hydrogenation of glycolic acid, prepared from the reaction of formaldehyde, carbon monoxide, and water:

$$HCHO + CO + H_2O \rightarrow HOCH_2COOH$$

$$HOCH_2COOH + 2H_2 \rightarrow HOCH_2CH_2OH + H_2O$$

4. OTHER PROCEDURES

A route that has been widely investigated, although no commercial processes have been developed from it, is the direct oxidation of the double bond to the diol. The addition of hydrogen peroxide across double bonds has been shown to be catalyzed by various metal oxides and even by ultraviolet light. With osmium tetroxide as catalyst, oxidation of various olefins with hydrogen peroxide has given good yields of the corresponding diols (74).

$$CH_2{=}CH_2 + H_2O_2 \rightarrow HOCH_2CH_2OH$$

Other investigators have bis-hydroxylated double bonds with permanganate (75), with chlorate (76), with thallium(III) salts (77), electrolytically (78), and with air or oxygen (79). A particularly good general method for bis-hydroxylating double bonds is by treatment with organic peroxyacids. This is critically discussed (together with epoxidation which yields intermediates for α-diols) in an "Organic Reactions" chapter (80). The organic peroxyacids are prepared as needed from aqueous hydrogen peroxide and carboxylic acids or anhydrides. The following preparation of 1,2-tetradecanediol illustrates hydroxylation by peroxyacids (81).

At room temperature, 35 g of 25.6% hydrogen peroxide (0.263 mole) is added in one portion to a well-stirred mixture of 49.2 g (0.25 mole) of 1-tetradecene and 295 ml of 98–100% formic acid. The mixture is heated and stirred for about 24 hr at 40°C, when an analysis should reveal that slightly more than the theoretical 0.25 mole of hydrogen peroxide has been consumed. Two phases are present throughout the reaction. The formic acid is removed at reduced pressure and the residue heated at reflux for one hour with excess 3N alcoholic potassium hydroxide. After removal of most of the alcohol on a steam bath, a large quantity of hot water is added leaving the glycol as an oil. On cooling, the glycol crystallizes and is separated. It is then washed again with hot water as before and allowed to resolidify. The water washes are extracted with ether to recover a small amount of dissolved glycol. The crude glycol, including the residue from evaporation of the ether, weighs about 55 g (95%) and melts at about 65°C. By recrystallization from 8 volumes of methanol at 0°C, about 40 g (69%) of pure 1,2-tetradecanediol, melting at 68–68.5°C, is obtained.

Propylene glycol has been prepared by fermentation of glycerol (82). It can also be produced by the destructive distillation of the monosodium salt of glycerol (83). A variation of this synthesis involving destructive distillation of glycerol and sodium hydroxide (84) was frequently used as a laboratory preparation in the days before propylene glycol became commercially available. Ethylene glycol has been prepared by destructive distillation of choline (85). Ethylene glycol has also been prepared by treating methanol with free radicals (86) or with ultraviolet light (87). Irradiation of methanol by fission fragments has been described as giving a 65% chemical conversion to ethylene glycol with 1 lb produced for every 10 kW-hr consumed (88). One of the products obtained by heating ethanol, methanol and di-*tert*-butyl peroxide is propylene glycol (89).

B. β-Diols

β-Diols may be prepared by many of the same types of reactions used for diols with greater or lesser hydroxyl group separation, but the most important synthetic routes are those which lead exclusively to 1,3-diols. Foremost is the aldol condensation, which is a step in the industrial manufacture of three principal β-diols: 1,3-butylene glycol, neopentyl glycol, and 2,2,4-trimethyl-1,3-pentanediol. Another procedure applicable only to β-diols is the Prins reaction.

1. ALDOL CONDENSATION

The most important general method for the preparation of β-diols is by reduction of aldols, prepared from aldehydes by means of aldol condensations.

When a single aldehyde is used, the reaction is defined as self-condensation:

$$2RR'CHCHO \longrightarrow RR'CHCHOH\overset{\overset{\displaystyle R}{|}}{\underset{\underset{\displaystyle R'}{|}}{C}}CHO$$

When more than one aldehyde is used, the reaction is called mixed condensation. Whereas in self-condensations the presence of a hydrogen atom in the α position of the aldehyde is necessary, in mixed condensations only one of the aldehydes need have an α-hydrogen:

$$RR'CHCHO + HCHO \longrightarrow RR'\overset{\overset{\displaystyle }{}}{\underset{\underset{\displaystyle CH_2OH}{|}}{C}}CHO$$

When formaldehyde is used as one of the aldehydes, it is very difficult to stop at the monomethylol stage with an aldehyde containing more than one α-hydrogen. Also with formaldehyde, there is usually partial or total reduction of the aldol aldehyde group by means of a crossed Cannizzaro reaction:

$$RCHO + HCHO + H_2O \rightarrow RCH_2OH + HCOOH$$

The aldol condensation is catalyzed by either acids or bases, usually the latter. An entire volume of "Organic Reactions" is devoted solely to the aldol condensation (90).

a.   1,3-Butylene Glycol

The first syntheses of 1,3-butylene glycol, reported in 1872 and 1873 (91,92), involved sodium amalgam reduction of acetaldol (3-hydroxybutyraldehyde). The present principal method of manufacture is still based on acetaldol, although catalytic hydrogenation is now the method of reduction. The following procedure of Kyriakides (93) is suitable for laboratory preparation of acetaldol.

$$2CH_3CHO \rightarrow CH_3CHOHCH_2CHO$$

$$CH_3CHOHCH_2CHO \rightarrow CH_3CHOHCH_2CH_2OH$$

One kilogram (22.7 moles) of acetaldehyde is cooled below 5°C in a 2-liter flask. Over a 20-min period, 25 ml of 10% sodium hydroxide solution is added dropwise with vigorous stirring, while the temperature of the mixture is maintained between 4 and 5°C. After the solution has been stirred for 1 hr, it is acidified slightly with tartaric acid. The mixture is filtered to remove sodium tartrate (addition of ethyl ether facilitates the filtration). The filtrate is distilled under reduced pressure, raising the pot temperature very slowly. Acetaldehyde and aldol distill together slowly, as the mixed cyclic acetal (aldoxan) of acetaldehyde and aldol slowly dissociates. Redistillation gives about 500 g (50%) of aldol, bp 72°C/12 mm. Aldol should be distilled immediately before use because it dimerizes to paraldol on standing.

b.   Neopentyl Glycol

Neopentyl glycol was first prepared in 1894 (94) by what is still the commercial route—aldol condensation of isobutyraldehyde and formaldehyde, followed by a crossed Cannizzaro reaction:

$$
\begin{array}{ccc}
\overset{\displaystyle CH_3}{\underset{\displaystyle |}{}} & & \overset{\displaystyle CH_3}{\underset{\displaystyle |}{}} \\
CH_3CHCHO + HCHO & \longrightarrow & CH_3CCHO \\
& & \underset{\displaystyle |}{} \\
& & CH_2OH
\end{array}
$$

$$
\begin{array}{ccc}
\overset{\displaystyle CH_3}{\underset{\displaystyle |}{}} & & \overset{\displaystyle CH_3}{\underset{\displaystyle |}{}} \\
CH_3CCHO + HCHO + H_2O & \longrightarrow & CH_3CCH_2OH + HCOOH \\
\underset{\displaystyle |}{} & & \underset{\displaystyle |}{} \\
CH_2OH & & CH_2OH
\end{array}
$$

A good laboratory procedure for the preparation of neopentyl glycol is found in Organic Syntheses (95).

c.   2,2,4-Trimethyl-1,3-pentanediol

Trimethylpentanediol, also called octaglycol, although only recently commercially available, was first described in 1883 (96); its structure was elucidated in 1896 (97). Numerous workers (98–100) subsequently have reported the preparation of trimethylpentanediol and its isobutyrate monoester by treatment of isobutyraldehyde with alkalis. Hagemeyer and Wright (101) describe the effect of various factors on the proportions of free diol, monoisobutyrate, and diisobutyrate produced. Under optimum conditions their yield of monoisobutyrate approaches 90%.

$$
\begin{array}{ccc}
\overset{\displaystyle CH_3}{\underset{\displaystyle |}{}} & & \overset{\displaystyle CH_3}{} \quad \overset{\displaystyle CH_3}{} \\
2CH_3CHCHO & \longrightarrow & CH_3CHCHOHCCHO \\
& & \underset{\displaystyle |}{} \\
& & CH_3
\end{array}
$$

$$
\begin{array}{ccc}
\overset{\displaystyle CH_3}{} \; \overset{\displaystyle CH_3}{} & \overset{\displaystyle CH_3}{} & \overset{\displaystyle CH_3}{} \; \overset{\displaystyle CH_3}{} \; \overset{\displaystyle CH_3}{} \\
CH_3CHCHOHCCHO + CH_3CHCHO & \longrightarrow & CH_3CHCHOHCCH_2OOCCHCH_3 \\
\underset{\displaystyle |}{CH_3} & & \underset{\displaystyle |}{CH_3}
\end{array}
$$

The monoisobutyrate is formed as a result of esterification with the isobutyric acid formed by a crossed Cannizzaro reaction or directly by a Tischtschenko-Claisen rearrangement. Such a mixture (principally monoester) is offered commercially by Eastman Kodak. High yields of the free diol can best be prepared by a two-step process (102): first, preparation of the aldol (keeping the temperature low to avoid disproportionation reactions), and then hydrogenation over a nickel catalyst.

2. HYDROGENATION AND HYDROGENOLYSIS

β-Diols have been prepared by the hydrogenation of many different difunctional compounds.

As mentioned in the previous section, several commercial processes are based on the reduction of 3-hydroxyaldehydes prepared by means of aldol condensations. Another 3-hydroxyaldehyde of commercial interest is 3-hydroxypropionaldehyde, prepared by hydration of acrolein and used to make trimethylene glycol (103,104). β-Ketoalcohols such as 4-hydroxy-2-butanone (105) can also be hydrogenated to β-diols.

Hydroxyl-groups beta to a carbonyl group are extremely labile, and dehydration or hydrogenolysis frequently occurs during hydrogenation, giving rise to alcohols as by-products. Chemical reducing agents such as aluminum amalgam (106) and lithium aluminum hydride (107) can be used to circumvent this difficulty, but hydrogenation at relatively low temperatures is usually the preferred route. The following procedure utilizes platinum oxide promoted with ferrous chloride (108). Ketones do not require a promoter. Ruthenium appears to be even better than platinum as a catalyst, but it has not been extensively used for this purpose.

To a solution of 17.6 g (0.20 moles) of aldol in 50 ml of 95% ethanol is added 0.5 ml of 0.2M ferrous chloride solution and 0.1725 g of platinum oxide catalyst. Hydrogenated in a rocker at ambient temperature and 3 atm (total) of hydrogen pressure, it takes up 93% of the theoretical hydrogen in about 2 hr. Distillation at reduced pressure yields 14.4 g (80%) of 1,3-butanediol, boiling range 103–104°C/8 mm.

Hydrogenolysis of ester groups, usually at high temperatures and pressures with copper chromite catalysts, can also be used as a route to β-diols (109). Thus diethyl malonate gives trimethylene glycol. Both hydrogenolysis of an ester and reduction of a carbonyl take place when acetoacetic ester is reduced to 1,3-butylene glycol:

$$EtOOCCH_2COOEt \rightarrow HOCH_2CH_2CH_2OH$$

$$CH_3COCH_2COOEt \rightarrow CH_3CHOHCH_2CH_2OH$$

Hydrogenolysis of sugars gives, among other fragments, trimethylene glycol (110,111).

3. PRINS REACTION

Under the influence of acid catalysts, aqueous solutions of formaldehyde add (as methylene glycol) across substituted double bonds. The β-glycol product frequently reacts with additional formaldehyde to form a cyclic formal (1,3-dioxan). A formerly used commercial route to 1,3-butylene glycol involved the addition of formaldehyde to propylene. In addition to low yields of 1,3-butylene glycol, high yields of 4-methyl-1,3-dioxan were obtained (112). This, in turn, could be hydrolyzed to give additional butylene glycol (113).

$$CH_3CH{=}CH_2 + HOCH_2OH \longrightarrow \underset{\underset{\displaystyle CH_3CH-CH_2}{|\quad\quad|}}{OH\ \ CH_2OH}$$

$$\underset{\underset{\displaystyle CH_3CH-CH_2}{|\quad\quad|}}{OH\ \ CH_2OH} + HCHO \underset{\longleftarrow}{\longrightarrow} \underset{\underset{\displaystyle CH_3CH-CH_2}{|\quad\quad|}}{\overset{\overset{\displaystyle CH_2-O}{|\quad\quad|}}{O\quad\ \ CH_2}}$$

4. OTHER PROCEDURES

Although the method is not of great present interest, many of the β-diols were first prepared by saponification of other functional groups. For example, trimethylene glycol was first prepared by saponification of the diacetate obtained on treatment of 1,3-dibromopropane with potassium acetate (114). Direct saponification of the dihalide gives the same glycol (115). Similarly saponification of 1,3-dichlorobutane in one step (116) or via the diacetate (117) gives 1,3-butylene glycol:

$$\underset{\underset{\displaystyle |\quad|}{|\quad|}}{XC-CX} \xrightarrow{\quad\quad\quad\quad} \underset{\underset{\displaystyle |\quad|}{|\quad|}}{HOC-COH}$$

$$AcOC-COAc$$

β-Glycols have also been prepared by hydrolysis of β-oxides (118) and of such ethers as ditrimethylene glycol (119).

In addition to 1,4-butanediol, which is the principal product (65%), hydroboration of 1,3-butadiene has been reported to give 31% of 1,3-butylene glycol (120). Treatment of allene with diborane gives trimethylene glycol (121,122). Trimethylene glycol, a normal constituent of glycerol solutions from rancid fats, can be prepared by fermentation of glycerol (123). Optically active butylene glycol has been obtained by biological reduction of aldol with certain yeasts (124). Dechlorination of glycerol-β-monochlorohydrin with sodium amalgam also gives trimethylene glycol (125).

C. γ-Diols and Higher

For γ-diols and higher, the two hydroxyl groups have little influence on each other, except in cyclization reactions. Reduction of suitable functional groups such as carbonyls or carboxylic esters usually proceeds well with few complications. Certain syntheses are particularly well suited for the preparation of classes of long-chain diols: γ-diols are readily available through ethynylation reactions and δ-diols through reduction of alkoxydihydropyrans.

1. REDUCTION

All the presently important ways of synthesizing γ-diols and higher involve reduction, usually catalytic hydrogenation, either to create hydroxyl groups or to alter the backbone of the molecule when hydroxyls are introduced in some other fashion.

a. 1,4-Butanediol via 1,4-Butynediol

A unique method of preparing γ-diols features ethynylation, the addition of acetylene across a carbonyl double bond. Reaction with 1 molecule of aldehyde or ketone gives an acetylenic alcohol; reaction with 2 molecules an acetylenic glycol:

$$HC{\equiv}CH + RCOR' \longrightarrow HC{\equiv}C\overset{\displaystyle R}{\underset{\displaystyle R'}{C}}OH$$

$$HC{\equiv}C\overset{\displaystyle R}{\underset{\displaystyle R'}{C}}OH + RCOR' \longrightarrow HO\overset{\displaystyle R}{\underset{\displaystyle R'}{C}}C{\equiv}C\overset{\displaystyle R}{\underset{\displaystyle R'}{C}}OH$$

The ethynylation of ketones, usually carried out with alkaline catalysts (126), gives tertiary alcohols and glycols which are of little interest for condensation polymerization. The ethynylation of aldehydes, other than formaldehyde, has also been most often accomplished with alkaline catalysts; although a copper acetylide catalyst can be used, the rate is slow and the alcohol rather than the glycol is the principal product (127).

The preparation of 1,4-butynediol by the reaction of acetylene and formaldehyde is an important commercial process. Acetylene (diluted with nitrogen) and aqueous formaldehyde are passed over a supported copper acetylide catalyst at about 100°C and 5 atm of gage pressure. Butynediol is the principal product, accompanied by small amounts of propargyl alcohol.

As early as 1910, Lespieau (128) described the preparation of butanediol by catalytic hydrogenation of 1,4-butynediol over a platinum black catalyst. Much later, after the development of a practical process for butynediol from acetylene and formaldehyde, hydrogenation of 1,4-butynediol became the commercial source of 1,4-butanediol. Many hydrogenation schemes have been described for the preparation of butanediol. Reppe gives the following procedure (129).

A 30% aqueous butynediol solution is hydrogenated with about 11% of Raney nickel as catalyst at 40–60°C and 3000 psig of hydrogen pressure. After filtration from the catalyst, the solution is distilled giving a nearly quantitative yield of butanediol, boiling at 106°C/0.7 mm.

$$HOCH_2C{\equiv}CCH_2OH + 2H_2 \rightarrow HOCH_2CH_2CH_2CH_2OH$$

b. 1,5-Pentanediol via 2-Alkoxy-3,4-dihydro-2H-pyrans

A unique general method for the preparation of 1,5-diols involves the hydrolysis and hydrogenation of 2-alkoxy-3,4-dihydro-2H-pyrans. This is now the commercial route to 1,5-pentanediol.

Vinyl ethers react very readily with α,β-unsaturated carbonyl compounds in a manner that is empirically similar to the Diels-Alder reaction (131). This may be accomplished simply by heating the two reactants together in an

autoclave, with a small amount (0.1–1.0%) of hydroquinone to inhibit polymerization. Thus, for example, heating 1·35 moles of methyl vinyl ether with 1 mole of acrolein for 12 hr at 135°C gives 81% of 2-methoxy-3,4-dihydro-2H-pyran. Similarly, ethyl vinyl ether plus crotonaldehyde gives 87% of the 2-ethoxy-4-methyl analog, and ethyl vinyl ether plus methyl vinyl ketone yields 50% of the 2-ethoxy-6-methyl analog.

1,5-Diols can be obtained either by hydrolyzing the dihydropyran and then hydrogenating, or alternately, by simultaneous hydrolysis and hydrogenation. The following examples illustrate both types of procedure (132).

(1) 3-Methyl-1,5-Pentanediol (Two-Step Procedure). A mixture of 425 g (3.32 moles) 2-methoxy-4-methyl-3,4-dihydro-2H-pyran, 800 ml of water, and 30 ml of concentrated hydrochloric acid is stirred at 25–40°C for 2 hr, when it becomes homogeneous. After neutralization with sodium bicarbonate it is hydrogenated with 50 g of Raney nickel for 4 hr at 125°C and 1625 psig of pressure. After it is filtered from the catalyst, the solution is distilled at reduced pressure. The yield of 3-methyl-1,5-pentanediol, boiling at 134–137°C/6 mm is 357 g (91%).

(2) 1,5-Pentanediol (One-Step Procedure). A mixture of 325 g (2.85 moles) of 2-methoxy-3,4-dihydro-2H-pyran, 100 ml of water, and 40 g of copper chromite catalyst is hydrogenated 7.5 hr at 175°C and 2000 psig. After filtering and distilling, 257 g (87%) of 1,5-pentanediol is obtained boiling at 145–147°C/17 mm.

c. Reduction of Carboxylic Acid Derivatives

Dicarboxylic acids and their derivatives, particularly esters, have long been the favorite precursors for laboratory preparation of the corresponding diols. A number of different techniques have been widely used for reduction of acids and derivatives. These include sodium-and-alcohol, lithium aluminum hydride, and catalytic hydrogenation. For industrial purposes, catalytic hydrogenation, which is by far the least expensive, is the only procedure of importance. Even in the laboratory, hydrogenation is usually preferable, since chemical procedures produce large amounts of salts as by-products, and separation from these salts makes work-up difficult. The principal area in which chemical methods are preferred is in selective reduction: chemical methods will reduce unsaturated esters to unsaturated alcohols, which is very difficult to achieve cleanly by means of catalytic hydrogenation.

(*1*) *Reductions with Sodium and Alcohol.* As early as 1903 Bouveault and Blank (133) reported the reduction of diethyl adipate to hexanediol using sodium and alcohol. This procedure is generally applicable and has been frequently utilized for the preparation of long-chain glycols:

$$RCOOR^1 + 4Na + 2R''OH \rightarrow RCH_2ONa + R^1ONa + 2R''ONa$$

Diols are produced not only by the reduction of diesters, but also of hydroxy-esters (134).

Following is a summary of an excellent *Organic Syntheses* procedure for the reduction of diethyl sebacate to 1,10-decanediol (135).

A solution of 65 g (0.25 moles) of diethyl sebacate in 800 ml of anhydrous ethanol is treated with 70 g (3.0 gram-atoms) of sodium in large pieces. After it has been cooled to moderate the reaction, the mixture is heated at reflux until all the sodium dissolves. Water is added and the ethanol is distilled off. The organic layer plus evaporated benzene extracts of the aqueous layer are recrystallized from alcohol-benzene, giving 32–33 g (73–75%) of 1,10-decanediol, mp 72–74°C.

The same procedure was found by the checkers to give the following yields of α,ω-glycols, starting with other dicarboxylic esters.

Number of carbons	Yield, %
7	88
9	71
11	57
13	88
14	61
18	54

(*2*) *Lithium Aluminum Hydride Reductions.* Solutions of lithium aluminum hydride in ethers have been widely used to reduce carboxylic acids and their derivatives to alcohols. Yields are usually high. For example, diethyl adipate

$$4RCOOH + 3LiAlH_4 \rightarrow (RCH_2O)_4LiAl + 2LiAlO_2 + 4H_2$$

$$2RCOOR^1 + LiAlH_4 \rightarrow (RCH_2O)_2(R^1O)_2LiAl$$

$$(RCH_2O)_2(R^1O)_2LiAl + 2H_2O \rightarrow 2RCH_2OH + 2R^1OH + LiAlO_2$$

and sebacic acid have been reported to give 83% hexanediol (136) and 97% decanediol (137), respectively.

(3) *Catalytic Hydrogenation.* The most commonly used procedure for the preparation of diols by reduction of dibasic acid derivatives is catalytic hydrogenation of the esters. Although nickel and cobalt catalysts have been used in certain instances, copper chromite catalysts are most useful for hydrogenolysis of esters (109). The following *Organic Syntheses* procedure illustrates a typical hydrogenation (138).

A suspension of 20 g of copper chromite catalyst in 252 g (1.25 moles) of diethyl adipate is hydrogenated at 255°C and 2000–3000 psig of pressure until hydrogen absorption ceases (ca. 6–12 hr). After cooling, venting, and discharging, the solution is filtered from the catalyst. A 50 ml portion of 40% sodium hydroxide solution is added, and the system is heated at reflux for 2 hr to saponify any esters. Low boiling alcohols are removed by distilling to a pot temperature of 95°C. The aqueous residue is exhaustively extracted with ether or benzene and the extracts distilled. About 125–132 g (85–90%) of 1,6-hexanediol is obtained, boiling at 143–144°C/4 mm and melting at 41–42°C.

Attempts to hydrogenate unesterified dicarboxylic acids have generally been rather unsuccessful. It is claimed in patents that pelleted cobalt and nickel catalysts, prepared by reduction of the oxides, serve to hydrogenate aqueous solutions of diacids to diols in high yields (139). Ruthenium catalysts have been described as suitable for reduction of diacids to diols, with yields approximating 50% (64).

d. Other Hydrogenations

Long-chain α,ω-diols have been prepared by the reduction of a multitude of difunctional or cyclic compounds. For example, each of the following gives 1,6-hexanediol through suitable catalytic hydrogenation: caprolactone (140), 2-hydroxymethyltetrahydropyran (141), 2,5-tetrahydrofurandimethanol (142), 5-hydroxymethylfurfural (143), and acrolein dimer (3,4-dihydro-2H-pyran-2-carboxaldehyde (144). Similarly, hydrolysis of dihydropyran gives 5-hydroxyvaleraldehyde, which can be hydrogenated to pentanediol (145), and addition of formaldehyde to crotonaldehyde gives 5-hydroxy-2-penten-1-al, which can be hydrogenated in a like manner (146). Hydrogenations of various furan derivatives including furfural (147) and furfuryl alcohol (148) give, among other products, 1,5-pentanediol. Hydrogenation of tetrahydrofurfuryl alcohol is described in *Organic Syntheses* (149) and gives 40–47% conversion, with recovery of about 20% of the starting material. Hydrogenolysis of esters of 2,5-dicarboxyglutaric acid and 2-acetylglutaric acid have also been reported to give 1,5-pentanediol (150).

2. OTHER PROCEDURES

For γ-diols and higher, where the hydroxyl groups are well separated and behave largely as separate entities, virtually any reaction capable of preparing an alcohol can also be used to prepare a glycol. A synthesis commonly used by early investigators was saponification of a dihalide, either directly (151,152), or by first forming the diacetate (153,154). A similar synthesis is through hydrolysis of suitable ethers (155). Treatment of pentanediamine with nitrous acid yields pentanediol (156), and treatment of N,N'-dinitro-1,4-butanediamine with acid gives butanediol (157).

Hydroboration of dienes gives good yields of mixture of diols, rather than of single diol species. Oxidation of the polymer obtained from diborane and 1,3-butadiene with alkaline hydrogen peroxide gives 65–80% yields of a mixture containing about 65% 1,4-butanediol, most of the remainder being 1,3-butanediol with traces of the 1,2-isomer (120). Similarly, 1,4-pentadiene gives a diol composition of 62% 1,4- and 35% 1,5-pentanediol.

1,4-Butanediol has been obtained by treatment of ethylene chlorohydrin with excess copper powder, using an amine as catalyst (158). 1,6-Hexanediol has also been reported as one of the products obtained by treating butadiene with carbon monoxide, formaldehyde, and hydrogen in the presence of a suitable catalyst (159).

D. Cycloaliphatic Diols

All the cycloaliphatic diols which have aroused much interest as monomers are obtained by reduction of aromatic rings or alicyclic diones. Such syntheses give mixtures of *cis* and *trans* diols, and if a single isomer is required, difficult separations are necessary. Fortunately, the diol mixtures are frequently suitable for use as monomers without separation of isomers.

1. HYDROGENATION AND HYDROGENOLYSIS

Each of the three principal classes of cycloaliphatic diols is obtained by hydrogenation of a class of substrate: cyclohexanediols from aromatic diols, cyclohexanedimethanols from aromatic dicarboxylic acid derivatives, and cyclobutanediols from cyclobutanediones.

a. Cyclohexanediols via Aromatic Diols

In 1908 Sabatier and Mailhe (160) prepared 1,4-cyclohexanediol from hydroquinone and 1,3-cyclohexanediol from resorcinol, by means of catalytic hydrogenation over a nickel catalyst. Other metal catalysts [e.g., rhodium

(161)] have been shown to have certain advantages, but nickel catalysts are still the standard for this hydrogenation.

Hydrogenation of hydroquinone with 10% of either nickel-on-kieselguhr or Raney nickel as catalyst has been reported to give almost quantitative yields of mixed 1,4-cyclohexanediol isomers (162–164). Ethanol and methanol have been reported useful as solvents; other workers do equally well with no solvent at all. Typical hydrogenation pressures and temperatures are 1000–2000 psig and 130–200°C. The hydrogenated product may be dissolved in a low boiling alcohol or boiling acetone and filtered from the catalyst. After distillation, using an air condenser to prevent blocking, the mixed diols are collected at 148–149°C/10 mm (172–175°C/42 mm). Such a product, melting at 97–100°C, can be used as a monomer when an approximately 50:50 mixture of *cis* and *trans*-1,4-cyclohexanediol is suitable.

The cis and trans isomers can be separated by recrystallization of the diacetates, although other derivatives such as the trityl ethers or the bischloroformates have also been separated by recrystallization. The higher-melting trans diacetate is easily separated in pure form, mp 104.5°C, and can be hydrolyzed to the trans diol, mp 143°C. The lower-melting cis diacetate, however, is very difficult to crystallize totally free of trans diacetate, and in consequence, early workers isolated and reported as the cis diacetate a eutectic containing considerable trans isomer. The "cis" diol from the eutectic melts at about 102°C (165). Two papers have presented phase diagrams of the *cis* and *trans*-1,4-cyclohexanediol system, but the reported results are widely divergent. The first presents a relatively simple picture with the cis diol melting at 107°C and the trans diol at 142°C, and a eutectic containing 58% trans diol melting at 97°C (164). A later phase diagram is more complex, showing the pure cis diol melting at 112°C falling to a eutectic of 102°C at 16% trans diol, rising gradually to an inflection at about 104°C and 50% trans diol, then rising rapidly to 143°C for pure trans diol (163). This behavior is explained in terms of pure *cis*-1,4-cyclohexanediol experiencing a transition to a liquid crystal at 101.4°C and melting at 112–113°C

(166). Later workers (167) through an elaborate purification procedure have obtained pure cis-1,4-cyclohexanediol mp 112.4–112.8°C, tending to confirm the later phase diagram. To obtain the pure trans diol on a laboratory scale, crystallization of the diacetate followed by hydrolysis (164) or ester interchange (166) is recommended. When the pure cis diol is required, the procedure of first recrystallizing and then hydrolyzing the trityl ether is probably the simplest (163).

b. Cyclohexanedimethanols via Aromatic Dicarboxylic Esters

The 1,4-cyclohexanedimethanol of commerce is apparently prepared by a two-step hydrogenation of dimethyl terephthalate: first hydrogenation of the benzene ring over nickel or palladium and then hydrogenolysis of the carbomethoxy groups over copper chromite (168,169). It has been claimed

$$COOCH_3 \quad\quad COOCH_3 \quad\quad CH_2OH$$

(ring structures) $\longrightarrow$ (S) $\longrightarrow$ (S)

$$COOCH_3 \quad\quad COOCH_3 \quad\quad CH_2OH$$

that a copper chromite catalyst that has been exhaustively extracted with water before development gives superior results (170). Overall yields for the two-step hydrogenation have been claimed to be as high as 97%.

Although hydrogenation of dimethyl terephthalate to the alicyclic diester gives an isomer mixture consisting principally of the cis isomer, the diol isomer mixture is richer in the trans isomer. Early workers reported that hydrogenation over copper chromite did not change the cis–trans ratio of 1,4-cyclohexanedicarboxylic diesters, and they ascribed an isomerization to the diester which gives, after hydrogenation, a diol enriched in trans isomer (171). Later workers stated that only the trans isomer hydrogenated without isomerization and that the cis isomer equilibrated during hydrogenation to a mixture containing 60–70% trans-1,4-cyclohexanedimethanol (172). Still more recently it has been claimed that with either cis or trans-1,4-cyclohexanedicarboxylic diesters, the same isomer ratio, about 70% trans-1,4-cyclohexanedimethanol, is obtained (173). It may be noted that 1,4-cyclohexanediol is commercially available as a mixture containing about 70% trans and 30% cis diol. The preparation of pure trans isomer has been reported by crystallization from a cis–trans mixture. The residual diols can be reequilibrated (by heating with alkaline reagents) and additional trans isomer crystallized (174).

Attempts to prepare 1,2-cyclohexanedimethanol by catalytic hydrogenations analogous to those used for the 1,4-diol have produced poor results.

Although a two-step hydrogenation gives better results than a one-step hydrogenation, the principal product is 2-methylcyclohexanemethanol (175,176):

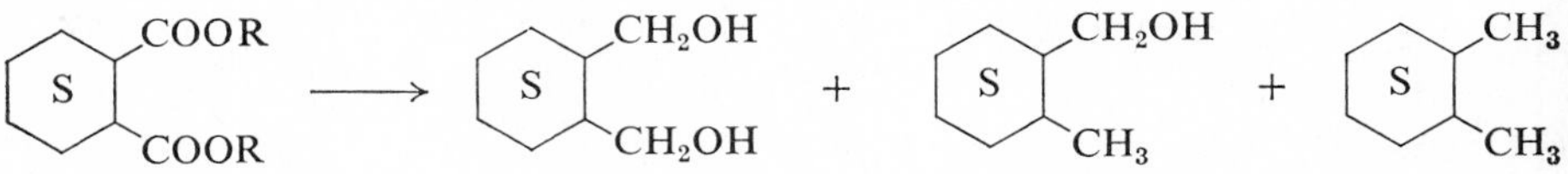

principal product
(70–94%)

Reduction of 1,2-cyclohexanedicarboxylic esters with sodium-and-alcohol has been reported to give a 68% yield, and lithium aluminum hydride reduction a 94% yield of 1,2-cyclohexanedimethanol (177). Although the sodium-and-alcohol reduction causes cis–trans isomerization, the hydride reduction proceeds with retention of configuration. 1,3-Cyclohexanedimethanol (178) and 1,4-cyclohexanedimethanol (179) have also been prepared in high yields by lithium aluminum hydride reductions of the cyclohexanedicarboxylic esters.

c. Cyclobutanediols via Cyclobutanediones

Ketenes are dimerized spontaneously upon standing, more rapidly when heated. Ketene itself tends to be dimerized unsymmetrically with a carbonyl added to an olefinic bond, but substituted ketenes are dimerized symmetrically through the olefinic bonds:

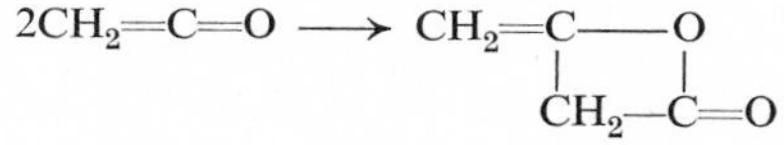

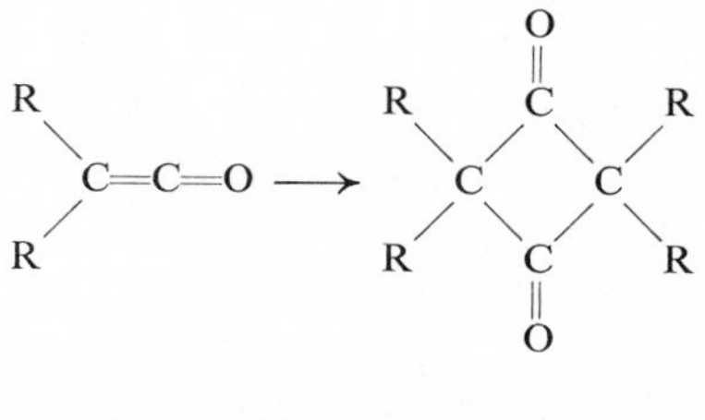

Such substituted 1,3-cyclobutanediones can serve as starting materials for the preparation of cyclic diols, and one such product, 2,2,4,4-tetramethyl-1,3-cyclobutanediol has recently become commercially available.

Dimethylketene has been prepared by a variety of different methods. Some give the dimer directly, whereas other methods produce the ketene, which may be dimerized by heating. On an industrial scale, a preferred route involves pyrolysis of isobutyric anhydride. Short exposure to 625°C at 25 mm of total pressure has been reported to give more than 95% yield at 80% conversion (180).

$$\left(\begin{array}{c}CH_3\\ \diagdown\\ \diagup CHCO\\ CH_3\end{array}\right)_2 O \longrightarrow \begin{array}{c}CH_3\\ \diagdown\\ \diagup C=C=O\\ CH_3\end{array}$$

For laboratory preparations, other procedures are useful; *Organic Syntheses* (18) suggests one involving bromination and debromination of isobutyric acid, with a 46–54% yield:

$$\begin{array}{c}CH_3\\ \diagdown\\ \diagup CHCOOH + Br_2\\ CH_3\end{array} \xrightarrow{P} \begin{array}{c}CH_3\\ \diagdown\\ \diagup CBrCOBr\\ CH_3\end{array} \xrightarrow{Zn} \begin{array}{c}CH_3\\ \diagdown\\ \diagup C=C=O\\ CH_3\end{array}$$

Treatment of isobutyryl chloride with triethylamine yields 57% of the dimer (182):

$$\begin{array}{c}CH_3\\ \diagdown\\ \diagup CHCOCl + (Et)_3N\\ CH_3\end{array} \longrightarrow \left[\begin{array}{c}CH_3\\ \diagdown\\ \diagup C=C=O\\ CH_3\end{array}\right] \longrightarrow \begin{array}{c}\quad O\\ \quad \parallel\\ CH_3\ \ C\ \ CH_3\\ \diagdown\diagup \diagup\diagdown\\ \quad C\quad C\\ \diagup\diagdown\diagdown\diagup\\ CH_3\ \ C\ \ CH_3\\ \quad \parallel\\ \quad O\end{array}$$

Heating dimethylmalonic anhydride in a bomb at 120°C gives an 80–90% yield of the dimer (183):

$$\begin{array}{c}\quad O\\ \quad \parallel\\ CH_3\ \ C\\ \diagdown\diagup\diagdown\\ \quad C\quad O\\ \diagup\diagdown\diagup\\ CH_3\ \ C\\ \quad \parallel\\ \quad O\end{array} \longrightarrow \left[\begin{array}{c}CH_3\\ \diagdown\\ \diagup C=C=O\\ CH_3\end{array}\right] \longrightarrow \begin{array}{c}\quad O\\ \quad \parallel\\ CH_3\ \ C\ \ CH_3\\ \diagdown\diagup \diagup\diagdown\\ \quad C\quad C\\ \diagup\diagdown\diagdown\diagup\\ CH_3\ \ C\ \ CH_3\\ \quad \parallel\\ \quad O\end{array}$$

Hydrogenation of tetramethylbutanedione can be carried out with ruthenium-on-carbon catalyst at 125°C and 1000–1500 psi of hydrogen pressure (184,185) or with nickel-on-kieselguhr at 130°C and 750 psi (186). Yields are better than 90%. In each case the product is an approximately 50:50 mixture of the cis and trans diols. These isomers have been separated by crystallization of their diformates, followed by transesterification with methanol. The cis isomer melts at 162.5–163.5°C and the trans isomer at 148°C.

Although cyclobutanediols are the only class of alicyclic diols in which the principal route to synthesis involves reduction of the dione, a similar procedure may also be applied to other cyclic diols. For example, the original synthesis (in 1894) of 1,4-cyclohexanediol involved reduction of 1,4-cyclo-hexanedione (165).

2.   OTHER PROCEDURES

Although we know that hydrogenation of catechol under many different sets of conditions produces 1,2-cyclohexanediol, this method of synthesis is not as attractive as the hydroxylation of cyclohexene, first accomplished (using potassium permanganate) in 1898 (187). *Organic Syntheses* (188) furnishes a good procedure involving hydroxylation with aqueous hydrogen peroxide and formic acid:

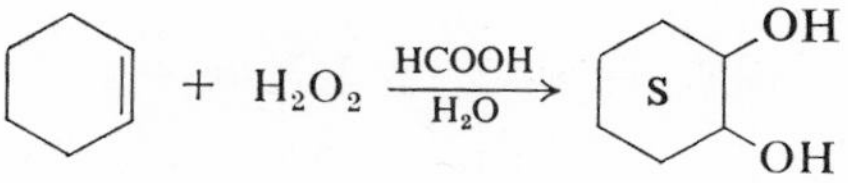

As described earlier, catalytic hydrogenation of 1,2-cyclohexanedicarboxy-lic esters leads to extensive hydrogenolysis with loss of hydroxyl groups. Because of this, a phthalic ester is hydrogenated for synthesis of 1,2-cyclo-hexanedimethanol (e.g., with Raney nickel at 170°C) to the cycloaliphatic diester, which is then chemically reduced to the diol. This can be done with sodium and alcohol (189) or with lithium aluminum hydride (190):

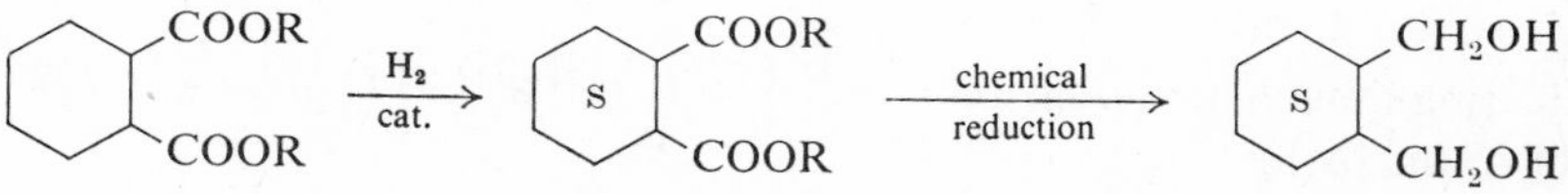

Unlike the 1,2-, 1,3-, and 1,4-isomers, 1,1-cyclohexanedimethanol cannot be prepared by reduction of an aromatic precursor. It is best prepared via aldol condensation of cyclohexanecarboxaldehyde with formaldehyde followed by a crossed Cannizzaro reaction (191) or by using the corre-sponding cyclohexene derivative and reducing the double bond as well as the aldehyde (192):

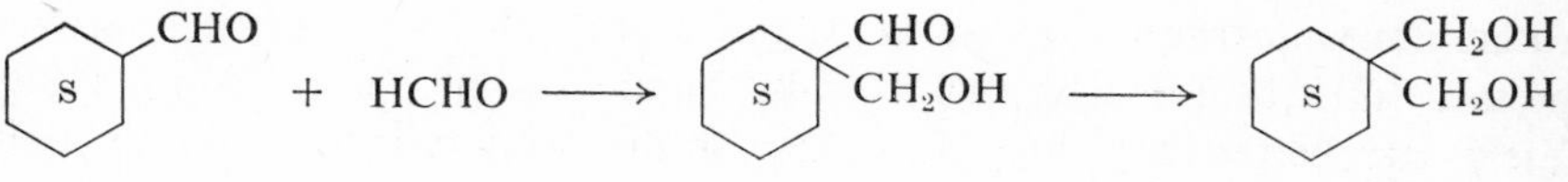

E.   Aralkyl Diols

The most important class of aralkyl diols is the xylenediols. In particular, *p*-xylene-α,α′-diol has attracted interest as a monomer, and it was available for several years as a semicommercial material. Two principal routes have been used for the preparation of xylenediols: chemical or electrolytic reduction of aromatic dicarboxylic derivatives, and side-chain halogenation of xylenes followed by hydrolysis:

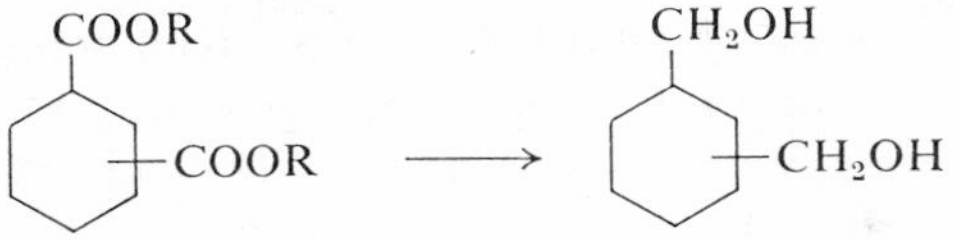

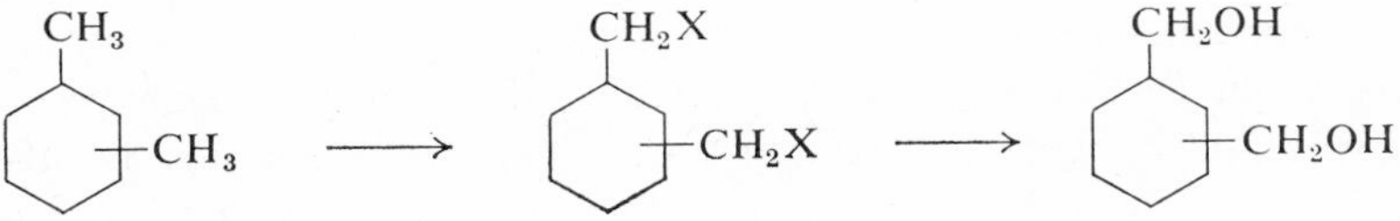

Catalytic hydrogenolysis of aromatic diesters has given principally cleavage to methyl rather than to methylol groups (175). Air oxidation of xylenes, catalyzed by t-alkyl boric esters, has been claimed to give xylene-α,α'-diols, among other products (193).

Lithium aluminum hydride reduction of phthalic anhydride gives an 87% yield of o-xylenediol, and yields with other phthalic acid derivatives are even higher (136). Reduction of dimethyl terephthalate with sodium borohydride in the presence of aluminum chloride gives a 72.5% yield of p-xylenediol (194). Aluminum amalgam in ethanol is also suitable for reduction of phthalic esters (195).

Probably the most practical method for preparation of xylenediols on an industrial scale is by chlorination of xylenes followed by hydrolysis. In one process for selective chlorination of xylenes, the separation and purification of the α,α'-dichloroxylenes are also outlined (196). However, on a moderate laboratory scale, bromination is probably simpler than chlorination. An *Organic Syntheses* preparation makes use of the ultraviolet-catalyzed bromination of o-xylene to dibromoxylenes in an overall yield of 48–53% (197). The brominations of m–xylene and of p-xylene have also been described (198–200). The dihaloxylenes have been hydrolyzed with alkali hydroxides, alkali carbonates, and even with water alone. A particularly suitable procedure is treatment with alkali acetate followed by hydrolysis of the acetate ester; yields in excess of 90% are claimed (201).

F.   Bischloroformates

The only method of preparing chloroformates having general preparative importance is the reaction of hydroxyl compounds with phosgene:

$$ROH + COCl_2 \rightarrow ROCOCl + HCl$$

Although diols with hydroxyl groups separated by at least four carbon atoms behave like monohydric alcohols in phosgenation, α-diols and β-diols tend to form cyclic five-membered or six-membered carbonates in a competitive reaction. The first synthesis of a bischloroformate, described in 1925 (15), was accomplished by slowly adding a tertiary amine (acid acceptor) to a nonaqueous solution of 2 moles of phosgene and 1 mole of ethylene glycol or trimethylene glycol. Previous investigators had obtained only cyclic ethylene carbonate when they treated ethylene glycol with phosgene on an equimolar basis (202,203). A process has been described for preparing bischloroformates of α-glycols with a minimum of cyclization by simultaneously but separately introducing streams of phosgene and glycol into a sump of liquid phosgene containing a trace of a tertiary amine as catalyst (204). The rate of addition is such that there is always a very high ratio of phosgene to diol in the reactor. In this manner a high yield of bischloroformate is obtained, which contains very little cyclic carbonate. A recent patent (205) claims that the use of an ether as solvent for the phosgenation and tetramethylurea as catalyst gives a particularly pure ethylene bischloroformate. An earlier paper (206) describes the preparation of the bischloroformate of neopentyl glycol by slow addition of the diol in a cyclic ether solvent to a solution of phosgene in the same solvent at about 0°C. Whereas yields of 90 and 89% were obtained when dioxane and tetrahydrofuran, respectively, were used as solvents, under similar conditions the yields from benzene and chloroform were only 25 and 21%, respectively. Although 1,2- and 1,3-diol bischloroformates may be purified by removing cyclic carbonates through washing with water, repeated washings are required because the partition ratio is not very high (204). A preferred procedure is by fractional distillation, using a low temperature and pressure to avoid decomposition.

For glycols with hydroxyls separated by at least four carbons, the conditions are less critical. Simple, uncatalyzed, slow addition of the glycol to a large stoichiometric excess of liquid phosgene (207), or alternately to a benzene solution of phosgene (208) at temperatures not exceeding about 25°C gives excellent yields. One author states that 1,3-cyclohexanediol, which ordinarily reacts very slowly with phosgene, reacts rapidly and in high yields without a catalyst, when acetone is used as solvent (209). The following examples illustrate the preparation and purification of bischloroformates. All reactions involving the highly toxic phosgene must be carried out in an efficient hood.

1. 1,4-CYCLOHEXYL BISCHLOROFORMATE (210)

In a three-necked flask fitted with an agitator and a dry-ice-cooled condenser is charged 116 g (1.0 mole) of cis-trans-1,4-cyclohexanediol and 100 ml of benzene. After stirring to disperse, about 200 ml (274 g, 2.77 mole) of phosgene is added, and the mix is stirred until a clear solution is obtained. Evaporation of the excess phosgene and benzene leaves the crystalline bischloroformate. If pure trans isomer is desired, it may be prepared by dissolving the crude mix in one volume of boiling benzene. On cooling to room temperature, pure trans isomer melting at 113–114°C is obtained. After concentrating the residue, crops of the cis isomer may be obtained, melting at 38–42°C.

2. 2,2,4,4-TETRAMETHYL-1,3-CYCLOBUTYL BISCHLOROFORMATE (211)

To a stirred mix of 72 g (0.50 mole) of 2,2,4,4-tetramethyl-1,3-cyclo-butanediol and 300 ml of dry dioxane under a dry-ice-cooled condenser is added 216 g (2.18 mole) of phosgene. The mix is heated at gentle reflux for 10 hr; then the hydrogen chloride and excess phosgene are removed using a stream of dry air as carrier. After evaporation of the residual solvent in a rotary vacuum evaporator at 50°C, the bischloroformate is left as a crystalline residue. Upon recrystallization from hexane it melts at 85–88°C.

Various other routes to chloroformates have been described. Direct chlorination of formates gives chloroformic esters; however, extensive chlorination of the ester moiety usually occurs simultaneously (212). The reaction of alcohols with diphosgene ($ClCOOCCl_3$) or with triphosgene ($CCl_3OCOOCCl_3$) also gives chloroformates (213,214). Phosgene adds to an epoxide group, producing a 2-chloroalkyl chloroformate (215). Alkyl carbonates are cleaved by phosphorus pentachloride to chloroformates and chlorides: with a cyclic carbonate the product is an ω-chloroalkyl chloroformate (216).

IV. PHYSICAL PROPERTIES

A. Selection of Values

The values listed in Tables 1–7 have been selected from the literature, with a single exception: the melting point of trimethylene glycol was determined in the author's laboratory. For all but the most important commerical glycols, the constants reported in the literature have been infrequent and varied. In general, when no single literature value appears more precise than the others, this compilation presents an average value.

B. Discussion of Tabulated Constants

Except for melting points, the physical constants of glycols are simple additive properties that change in a predictable fashion with changes in structure or molecular weight. Mathematical expressions have been derived which give calculated boiling points in good agreement with the measured values (217). Melting points, on the other hand, are much more complex: a number of the cyclic diols are even reported to crystallize in more than one

TABLE 1

Physical Properties of the Linear α,ω-Diols

| Glycol | mp, °C | bp, °C | | d_4^{20} | n_D^{20} |
		760 mm	10 mm		
Ethylene glycol	−13.0	197.6	89	1.109	1.4316
Trimethylene glycol	−25	214	104	1.054	1.4395
1,4-Butanediol	20.1	229	118	1.019	1.4454
1,5-Pentanediol	−18	242	128	0.992	1.4498
1,6-Hexanediol	42.8	250	135	—	—
1,7-Heptanediol	20	—	144	0.953	1.4570
1,8-Octanediol	62	—	152	—	—
1,9-Nonanediol	46	—	160	—	—
1,10-Decanediol	74	—	167	—	—
1,11-Undecanediol	63	—	175	—	—
1,12-Dodecanediol	84	—	183	—	—

modification. A glance at Table 1 would seem to confirm that melting points, even of linear α,ω-diols, vary haphazardly with chain length. However, closer examination shows that a zigzag alternation takes place: the even-numbered glycols form a higher melting series and the odd-numbered glycols a lower melting series. Both series gradually converge as the lengths increase.

As can be seen by comparison of the isomeric butanediols (Tables 1 and 2), for a given hydrocarbon skeleton, the values for boiling point, density, and refractive index all increase progressively with increased separation between the hydroxyl groups. Neopentyl glycol and trimethylpentanediol are relatively compact and hence are lower boiling than their linear isomers. The highly symmetrical neopentyl glycol is comparatively high melting.

As illustrated by Tables 1 and 3, the polyethylene glycols boil in the same general area as the α,ω-diols of equal chain length: e.g., diethylene glycol

TABLE 2

Physical Properties of Other Aliphatic Glycols

| Glycol | mp, °C | bp, °C | | d_4^{20} | n_D^{20} |
		760 mm	10 mm		
Propylene glycol	—	187.3	85	1.036	1.4326
Threo-2,3-butanediol	7	179	78	0.997	1.4315
Erythro (*meso*)-2,3-butanediol	34.4	182	82	0.999	1.4317
1,2-Butanediol	—	190.5	96.5	1.002	1.4378
1,3-Butanediol	−77	207.5	98	1.005	1.4412
Neopentyl glycol	130	207	160(100 mm)	—	—
2,2,4-Trimethyl-1,3-pentanediol	51.5	233	130	—	—

TABLE 3

Physical Properties of Glycol Oligomers

Glycol	mp, °C	bp, °C		d_4^{20}	n_D^{20}
		760 mm	10 mm		
Diethylene glycol	−8	245	128	1.116	1.4472
Triethylene glycol	−7.2	287.4	162	1.123	1.4559
Tetraethylene glycol	−4.1	327	190	1.123	1.4598
Dipropylene glycol[a]	—	232	116	1.023	1.4440

[a] Mixture of isomers.

TABLE 4

Physical Properties of Cyclic Glycols

Glycol	mp, °C	bp, °C	
		760 mm	10 mm
Cis-1,2-cyclohexanediol	99.6[a]	232	110
Trans-1,2-cyclohexanediol	104.4	236	113
Cis-1,3-cyclohexanediol	85.5	—	—
Trans-1,3-cyclohexanediol	115.5	—	—
Mixed 1,3-cyclohexanediols	—	240–245	130–135
Cis-1,4-cyclohexanediol	112.8[b]	—	—
Trans-1,4-cyclohexanediol	143	—	—
Mixed 1,4-cyclohexanediols	—	247–252	140–145
1,1-Cyclohexanedimethanol	98.5	—	143–144
Cis-1,2-cyclohexanedimethanol	43.5[c]	—	—
Trans-1,2-cyclohexanedimethanol	57[c]	—	—
Mixed 1,2-cyclohexanedimethanols	—	—	155
Cis-1,3-cyclohexanedimethanol	55	—	—
Trans-1,3-cyclohexanedimethanol	(liq.) n_D^{24} 1.4941	—	—
Cis-1,4-cyclohexanedimethanol	43	—	165
Trans-1,4-cyclohexanedimethanol	67	—	167
Cis-2,2,4,4-tetramethyl-1,3-cyclobutanediol	163.5	—	—
Trans-2,2,4,4-tetramethyl-1,3-cyclobutanediol	148	—	—
Mixed 2,2,4,4-tetramethyl-1,3-cyclobutanediols	129–150	239–244	—
o-Xylenediol	66	—	—
m-Xylenediol	47	—	150–155
p-Xylenediol	119	—	—

[a] Transitions between three enantiotropic crystal forms at 80.5 and 87.3°C.

[b] Transition to a liquid crystal form at 101.4°C.

[c] Although cis and trans-1,2-cyclohexanedimethanols have been frequently reported as melting at about 43.5 and 57°C respectively, there have also been reports that the cis compound melts at 49–50°C and the trans (in two crystal modifications) at 48.8 and 51.2°C.

boils close to the same point as pentanediol. At least for the first few members of the polyglycol series, density and refractive index increase with chain length.

The alicyclic diols introduce the additional complication of cis and trans isomerization. In general, cyclic diols are high melting, with the melting/ point frequently increasing with increasing symmetry. The boiling point increases as the hydroxyl group separation increases (see Table 4).

Glycols above their melting points are viscous liquids. Viscosity increases with increased chain length, with hydrocarbon chain segments much more effective than polyether segments. Surface tensions, while lower than for water, are much higher than for most organic liquids. The surface tension falls only slowly with increased hydrocarbon chain length. Glycols with one secondary hydroxyl group have lower surface tensions than corresponding diprimary glycols (Table 5).

TABLE 5

Viscosity and Surface Tension

Glycol	Viscosity at 20°C, cp	Surface tension at 20°, dynes/cm
Ethylene glycol	20.9	48.4
Diethylene glycol	35	48.5[a]
Triethylene glycol	48	45.2
Propylene glycol	56	36.5[a]
Trimethylene glycol	39[a]	—
1,3-Butanediol	65[a]	37.8[a]
1,4-Butanediol	72[a]	—
1,5-Pentanediol	128	43.2

[a] At 25°C.

C. Solubility of Glycols

The lower aliphatic glycols, up to 1,6-hexanediol, are reported to be miscible with water at 25°C, 1,8-octanediol is moderately soluble, and 1,10-decanediol is only slightly soluble. Solubility in lower alcohols parallels water solubility except that it falls off less rapidly as the chain is lengthened. Solubility, even in organic solvents, tends to drop as the chain length of diprimary linear glycols is increased. At temperatures below the melting point, solubility usually decreases as the melting point is increased. Glycols with a secondary hydroxyl group are much more soluble in most solvents than are diprimary glycols. Chain branching also increases solubility (Table 6).

The different classes of cycloaliphatic glycols exhibit similar solubility behavior: the cyclohexanediols and cyclohexanedimethanols are all very soluble in water or lower alcohols and sparingly soluble in aromatic

TABLE 6

Solubility of Glycols

Solvent	Solubility at 25°C, g/100 g of solvent[a]				
	EG	DEG	PG	BD	PD
Water	Misc.	Misc.	Misc.	Misc.	Misc.
Ethanol	Misc.	Misc.	Misc.	Misc.	Misc.
Acetone	Misc.	Misc.	Misc.	Misc.	Misc.
Ether	8.9	19.5	Misc.	3.1	6.0
Benzene	6.0	45.5	23.8	0.3	0.6
Carbon tetrachloride	6.6	35.5	30.5	0.4	<0.1

[a] EG = ethylene glycol, DEG = diethylene glycol, PG = propylene glycol, BD = 1,4-butanediol, and PD = 1,5-pentanediol.

solvents such as toluene, which has frequently been used as a recrystallization solvent. The xylenediols behave like the cycloaliphatic glycols. In general the lower melting cyclic diols are moderately soluble in ether, acetone, and chlorinated aliphatics; the higher melting isomers much less so.

D. Physical Properties of Bischloroformates

Bischloroformates are colorless crystalline solids or viscous liquids. Table 7 lists the physical properties of some of the more important members of this class.

TABLE 7

Physical Properties of Bischloroformates

ClCOOROCOCl R Group	bp, °C/mm Hg or (where indicated) mp	n_D^{25}	mp of Carbamate, °C	Reference
$(CH_2)_2$	109–14/24–25	1.4481	163–165	208
$(CH_2)_3$	130/28	1.4528	167–169.5	208
$(CH_2)_4$	96–97/4	1.4513	198.5–199.5	208
$(CH_2)_5$	114/3	1.4508	164–167	208
$(CH_2)_6$	119–120/3.5	1.4518	188.5–190	208
$(CH_2)_7$	129/3.5	1.4518	169–171	208
$(CH_2)_8$	107–10/0.1	1.4524	178–182	208
$(CH_2)_9$	127/0.2	1.4532	154.5–156	208
$(CH_2)_{10}$	149–155/3, mp −25°C		171.2	208
Trans-1,4-cyclohexyl	mp 113.5–115			208
Cis-1,4-cyclohexyl	mp 38–40			208
Trans-1,4-cyclohexanedimethyl	mp 68–71			218
Cis-1,4-cyclohexanedimethyl	Liquid			218
1,1-Cyclohexanedimethyl	110–113/5			206
2,2,2,4-Tetramethyl-1,3-cyclo-butyl	mp 85–88			211

V. ANALYTICAL PROCEDURES

Glycols are usually determined quantitatively by esterification with acetic anhydride.

$$2(CH_3CO)_2O + HOROH \rightarrow CH_3COOROCOCH_3 + 2CH_3COOH$$

After hydrolysis of excess anhydride, the acetic acid is titrated with sodium hydroxide solution. In addition to this general method, vicinal glycols may be determined by oxidation with periodic acid:

$$RCHOHCHOHR' + HIO_4 \rightarrow RCHO + R'CHO + H_2O + HIO_3$$

After the oxidation, potassium iodide is added and the liberated iodine is titrated with thiosulfate solution. Procedures for these and many other types of hydroxyl determinations are given by Siggia (219a). Glycols containing other functional groups can be determined by methods suitable for those groups—as, for example, bromination for unsaturated glycols.

Gas chromatography is particularly valuable for determining the purity of glycols. In most cases, infrared spectroscopy and NMR are not sensitive enough to measure small amounts of impurities.

Gas chromatography is also useful in ascertaining the purity of bischloroformates. Siggia gives methods (219b) suitable for the determination of hydrogen chloride and free carboxylic acid in acid chlorides such as bischloroformates. This is accomplished by potentiometric titration.

VI. STORAGE, HANDLING, AND TOXICITY

A. Glycols

Glycols are stable viscous liquids or crystalline solids. They are generally considered to be noncorrosive, and even the lowest boiling glycol (propylene glycol, bp 187.3°C) has a flash point (open cup) of 225°F. On an industrial scale, they are usually stored in mild steel vessels. For long-term storage, particularly when trace iron contamination is objectionable, stainless steel, aluminum, and glass containers have been used. A more economical alternative is the use of a storage vessel lined with a baked phenolic, epoxy-phenolic, or vinyl resin. It is not general practice to store glycols under nitrogen, although in some industrial installations where very low water tolerance is required, a dry nitrogen purge is used.

Glycols as a class have a low order of toxicity. Because of their low vapor pressures, they do not ordinarily present a hazard from inhalation except when heated or in the form of spray or (for solid glycols) dust. Tests on rabbits indicate that there is very little absorption through the skin, and, in general, saturated glycols are not active skin irritants. Oral toxicity values for rats, mice, or guinea pigs are available for most of the commercially available glycols. At least in some cases, the toxicity to humans appears to be higher than is found for small mammals. Whereas ethylene glycol has a single oral dose LD_{50} of 5.5–8.5 g/kg for rats, the lethal dose to humans is estimated to be as low as 1.4 g/kg. The toxicity falls off sharply for the polyethylene glycols. Propylene glycol is practically nontoxic. Most other glycols that have been studied display LD_{50} values for rats in the range of 2 to 20 g/kg.

B. Bischloroformates

Unlike many of the glycols that have become important industrial products, bischloroformates are still specialty chemicals and laboratory products. Again unlike the glycols, bischloroformates are reactive, corrosive, and toxic.

They are readily hydrolyzed and must be rigorously protected from water. Since the products of the hydrolysis are glycols plus 2 moles each of carbon dioxide and hydrogen chloride, precautions should be taken to prevent a dangerous pressure buildup. Metal containers are generally unsuitable, and the most desirable storage and reaction vessels are made of glass or glass-lined. Precautions must be taken to avoid inadvertent contamination not only with water but with any material that reacts with acid chlorides.

Although they are all high boiling, the vapors of bischloroformates, and of the hydrogen chloride resulting from hydrolysis in moist air, are irritating and must be avoided. All members of this group must be treated as potent skin irritants. The eyes and mucous membranes are particularly sensitive.

VII. POLYMERIZATION AND POLYMER APPLICATIONS

A. Polymerization of Glycols

Glycols have aroused considerable interest as monomers for four types of condensation polymers: polyesters, polyurethanes, polyacetals, and polyethers. Of these, only the first two have achieved commercial importance; the polyacetals and polyethers of commerce are prepared from monomers other than glycols.

1. POLYESTERS

Polyesters are the most important condensation polymers prepared from glycols. Among the polyester types that have been prepared are alkyds and unsaturated polyesters, fiber-formers, polycarbonates, and polyester oligomers for polyurethane chain segments.

a. Alkyds and Unsaturated Polyesters

The original alkyds were crosslinked polyesters based on glycerol and phthalic anhydride. In the 1920s, ethylene glycol was used as a minor ingredient of alkyds (220). By decreasing the amount of crosslinking, this gave improved solubility and retarded gelation. Later, pentaerythritol came into extensive use in alkyds as a replacement for glycerol, and glycols became even more necessary to decrease the amount of crosslinking.

Although ethylene glycol is important in alkyds, it has been replaced to a considerable extent by other gylcols that impart desired properties. The compatibility of alkyds with organic solvents is improved by substituting part of the ethylene glycol with propylene glycol or with 1,3-butylene glycol (221). It has been well established that compounds containing hydrogen in positions gamma to unsaturation readily undergo thermal cleavage by a cyclic mechanism (222–224). Polyesters of neopentyl glycol have no such hydrogens in the alcohol moiety and display high thermal stability, plus hydrolytic stability due to steric hindrance. Many compositions and formulations use neopentyl glycol as a replacement for all or part of the ethylene glycol or propylene glycol (225). Esters of the highly hindered 2,2,4,4-tetramethyl-1,3-cyclobutanediol were found to be exceptionally stable to hydrolysis, closely followed by those prepared from 2,2,4-trimethyl-1,3-pentanediol (226).

It was early discovered that the incorporation of maleic acid into an alkyd enhanced its drying properties, and that crosslinking could be accomplished by heating or by exposure to air after addition of a cobalt drier. Soon it was learned that linear polyesters of glycols and maleic acid functioned as drying oils and could be readily crosslinked. Even with proportions as low as 10% maleic anhydride to 90% succinic anhydride, a polyester can be fully crosslinked (227). It was next demonstrated that unsaturated polyesters, such as from diethylene glycol and maleic anhydride, could be copolymerized with such monomers as vinyl acetate, styrene, and methyl methacrylate (228). "Polyester resins," usually solutions of unsaturated polyesters in a monomer (principally styrene), have become an important class of laminating polymers.

Propylene glycol is the principal glycol used in the preparation of polyester resins because it gives better compatibility with styrene than do analogs from ethylene glycol. Diethylene glycol is often used, and longer chain diols, such as 1,5-pentanediol (229) and 1,6-hexanediol (230), are describing as imparting flexibility. 1,3-Butylene glycol imparts particularly good compatibility; indeed, it finds an important use in saturated polyesters for use as polymeric plasticizers (231,232).

b. Fiber-Forming Polyesters

Polyesters of linear α,ω-diols and linear dicarboxylic acids may be considered to be analogs of polyethylene in which carboxyl groups have been substituted at intervals for methylene groups in the polymer backbone. Measurements of solutions have revealed that this substitution increases the stiffness of the chain, and this stiffness reportedly increases regularly with increasing number of ester groups per unit chain length (233). Characteristically, for each dicarboxylic acid a glycol of a particular length gives a minimum melting point in the homologous polyester series. As the glycol length is decreased past this minimum, the melting point rises; for most high melting polyester series, the ethylene glycol polyester melts highest. As the glycol length is increased past the lowest melting product, the melting point rises, slowly approaching the limiting value (137°C) for a linear hydrocarbon of infinite length. Another characteristic phenomenon is that even-numbered and odd-numbered glycols form two separate melting-point progressions. For even-numbered dicarboxylic acids (including ortho and para aromatic diacids) the series with even-numbered glycols melts much higher. This leads to zigzag alternation of high and low melting points. For odd-numbered dicarboxylic acids (including meta aromatic diacids) the alternating effect is less pronounced, and may even be reversed. This alternating effect is gradually damped out as the length of the glycol increases. Table 8 presents the melting points of some representative homologous polyester series.

TABLE 8

Melting Points of Some Homologous Polyester Series

	mp, °C			
α,ω-Diol length	Oxalate (262)	Diglycolate (263)	Terephthalate (264)	Isophthalate (264)
2	173	17–20	256	103–108
3	87–88	29–32	217–18	92–96
4	96–98	67–70	222–23	88–94
5	—	30–33	134–140	76–82
6	73–76	47–51	148–154	75–80
7	31–33	—	—	—
10	77–79	61–64	123–127	34–36
20	—	86–89	108–113	47–49

When the diol is held constant and the chain length of the dicarboxylic acid is varied, similar phenomena are observed, and the melting points pass through a minimum at some particular dicarboxylic acid (usually at malonic acid). For even-numbered diols, the polyesters from even-numbered diacids melt much higher than the odd-numbered homologs. For odd-numbered diols, this is not necessarily true (see Table 9).

Because of their higher polarity, polyesters will crystallize, attain their maximum melting points, and become fiber forming at lower molecular weights than will polyethylene. This property is, of course, affected by the ength of the intervals between the ester linkages. As these intervals grow, the

TABLE 9

Melting Points of Some Homologous Polyester Series

	mp, °C		
Acid	Ethylene glycol (265)	Trimethylene glycol (266)	1,4-Butanediol (267)
Oxalic	159–161	89	103–105
Malonic	−22	33	−20 to −24
Succinic	102–103	52	113–114
Glutaric	−19	53	36–38
Adipic	47	46	58–60
Pimelic	25	51	38–41
Suberic	63	52	—
Azelaic	44	60	49–51
Sebacic	72	58	64–67

polymer behavior gradually approaches that of linear polyethylene (Table 10).

Although the polyesters of the linear α,ω-diols and linear dicarboxylic acids form strong fibers, they are too low melting and too susceptible to hydrolysis for practical commercial use. Ethylene glycol, which is less expensive than its higher homologs and gives higher melting polyesters, is the only linear glycol of importance in polyester fibers. An aromatic acid, and in particular terephthalic acid, is used to give the high melting point and hydrolytic stability that are needed. Another commercial polyester is prepared from mixed cis and trans isomers of 1,4-cyclohexanedimethanol and terephthalic acid. Although there are no commercial polyester fibers based on linear aliphatic dicarboxylic acids (such as are used in polyamide fibers), high melting polyesters may be prepared by introducing rigidity in the glycol rather than the diacid moiety. This can be done by use of cyclic diols, particularly those with high symmetry and maximum separation between hydroxyl groups. Thus poly(*trans*-1,4-cyclohexyl adipate) is reported to melt at 225°C (234,235) as compared to 58°C (236) for poly(1,6-hexyl adipate).

TABLE 10

Fiber-Forming Ability as a Function of Chain Length (268)

No. of atoms in repeating unit	Glycol	Acid	Shortest chain length that forms fibers
12	1,6-Hexanediol	Succinic	500
14	1,6-Hexanediol	Adipic	580
18	1,6-Hexanediol	Sebacic	760
28	1,16-Hexadecanediol	Sebacic	1120

c. Polycarbonates

Polycarbonates can be prepared by ester interchange of glycols with dialkyl carbonates (7,237) or by reaction of a glycol with equimolar amounts of phosgene. The interaction of a bischloroformate and a glycol, however is a particularly good method for laboratory preparation (see Section VII.B).

d. Polyesters for Polyurethane Chain Segments

Diisocyanates can be used to link together lower molecular weight polymers that have reactive groups at each end. The first polyesters to be used in such a manner as "prepolymers" were terminated by carboxyl as well as by hydroxyl groups. These foamed in use as carbon dioxide was emitted by decomposition of the first product—a mixed anhydride.

$$RCOOH + R'NCO \rightarrow RCOOCONHR' \rightarrow RCONHR' + CO_2$$

Later, hydroxyl-terminated polyesters with low acid numbers were used. Foaming was controlled by the addition of a measured amount of water. Such hydroxyl-terminated polyesters were also useful for polyurethane elastomers and coatings. Ethylene glycol and various other glycols were used in such polyesters, together with triols when crosslinking was desired. Poly(ethylene adipate) gives polyurethanes with good properties (238). Polyesters for modification with diisocyanates also frequently use propylene glycol, usually in combination with ethylene glycol. Many such polyesters have been claimed for this purpose. Hydroxyl-terminated polyesters of adipic acid with excess neopentyl glycol have been described as polyurethane pre-polymers (239,240). Polyurethane foams derived from hydroxyl-terminated neopentyl glycol polyesters are tough, have good resistance to discoloration, and possess superior hydrolytic and thermal stability. They are generally less resilient than foams based on unbranched-chain glycols. Coatings prepared from such hydroxyl-terminated polyesters and commercial diisocyanates have improved chemical resistance, light resistance, and weatherability. Low-molecular-weight hydroxyl-terminated polyesters of butanediol have been used by many investigators as ingredients of polyurethanes. An adipic acid polyester treated with methylene bis(phenylene isocyanate) (241) has been described as having the properties of a vulcanized rubber without being crosslinked. An adipic acid–terephthalic acid–butanediol terpolyester is described as giving a good flooring resin with toluene diisocyanate (242).

In recent years polyethers, which are lower in cost and frequently provide better properties, have displaced the polyesters from most of their applications in the polyurethane field. Polyethylene glycols have been of little interest here because of their water sensitivity. Polypropylene glycols see much service as prepolymers, particularly when containing other components such as triols to give crosslinking, or polyethylene glycol blocks, which provide a higher incidence of primary hydroxyl end-groups. Poly(1,4-butanediols) have proved particularly suitable as soft segments for segmented elastic polyurethanes.

2. POLYURETHANES

Substantially identical polyurethanes may be prepared in two different ways. The method of commercial importance involves reaction of diisocyanates with diols.

$$OCN\!-\!R\!-\!NCO + HO\!-\!R'\!-\!OH \rightarrow OCN\!\!\left[\!RNHCOOR'O\right]_{\overline{n}} H$$

A method of little commercial value, but excellent for laboratory preparation of polyurethanes, is the reaction of bischloroformates with diamines (see Section VII.B).

$$ClCOOROCOCl + NH_2R'NH_2 \longrightarrow Cl\!\!\left[\!COOROCONHR'NH\right]_{\overline{n}} H$$

A group led by Otto Bayer in Germany was first to investigate the reaction of linear diisocyanates with various diols. Trimethylene glycol and its higher homologs were found to give polyurethanes suitable for films or fibers, whereas polyurethanes from ethylene glycol were found to be thermally unstable, decomposing on melting with evolution of gas (243). In a patent based on a German application of 1937 (244), the reaction of 1,4-butanediol with linear diisocyanates of from five to eight carbons was disclosed. Fiber-forming polyurethanes prepared from butanediol and hexamethylene diisocyanate were commercially available in Germany as Igamid U and Perlon U (245), although apparently they have been displaced by other materials.

In Japan, considerable work was done on the polyurethane from hexanediol and tetramethylene diisocyanate, which was introduced as "Poluran" in 1940. Poluran was reported to melt at 189°C (comparable to the isomeric German product, Perlon U, from butanediol and hexamethylene diisocyanate, which is reported to melt at 184°C. This polyurethane was shown to have substantially identical properties, whether prepared from hexanediol and the diisocyanate or from hexanediol bischloroformate and the diamine (246). Unlike Perlon U, Poluran does not seem to have achieved commercial production.

The condensation of diisocyanates with hydroxyl-terminated prepolymers has had much greater commercial importance than direct condensation of diisocyanates with monomeric diols. In many such formulations, residual isocyanate end groups can be made to react further by treating with a glycol, which serves as a chain extender or crosslinker. 1,4-Butanediol has proved to be particularly desirable for this purpose and has long been used as a curing agent, starting with the German Vulkollan polyurethane (247), a product based on an ethylene glycol–adipic acid polyester. In curing a casting resin prepared from a linear dihydroxyl caprolactone polyester reacted with excess *p*-phenylene diisocyanate, the use of butanediol is claimed to give good pot life as well as good physical and electrical properties (248).

3. POLYACETALS

Polyacetals of the linear glycols are too susceptible to hydrolysis to be of much interest, although the polyformal of 1,4-butanediol has been claimed to be useful as a hydroxyl-terminated prepolymer for reaction with diisocyanates (249). Cyclic glycols give polyacetals that are much more stable. 1,4-Cyclohexanedimethanol yields polymeric formals with interesting properties (250,251), and condensation of this glycol with 2,2-dimethoxypropane produces a comparatively stable crystalline polyketal (252).

The polyformal of 2,2,4,4-tetramethyl-1,3-cyclobutanediol appears to be of special interest. Not only is it extremely stable to alkaline hydrolysis, it is even fairly resistant to acid hydrolysis (253). The cis, trans, and 1:1 cis–trans mixture give polyformals with softening ranges of 284–289, 275–280, and 283–291°C, respectively, which are all much higher than the ranges for polyformals of *trans*-1,4-cyclohexanediol, *trans*-1,4-cyclohexanedimethanol, and 2,5 (or 2,6)-norbornanediol. These tetramethylcyclobutanediol polymers give strong orientable films and fibers.

4. POLYETHERS

Except for ethylene glycol (which gives principally 1,4-dioxane), and 1,4-butanediol and 1,5-pentanediol (which give cyclic ethers), linear diprimary glycols can be readily dehydrated to give oligomers. Direct dehydration is of little importance, however, in the preparation of useful polyethers.

The commercially significant polyethers are prepared by opening the rings of cyclic ethers. Diethylene glycol and dipropylene glycol may be considered to be low-molecular-weight polyethers, and higher molecular weight products are prepared in the same way by treating an epoxide with less than the stoichiometric amount of water or with a glycol. 1,4-Butanediol can easily be dehydrated to tetrahydrofuran, and the latter can be polymerized to a hydroxy-terminated polyether (254).

B. Polymerization of Bischloroformates

1. POLYURETHANES

Although identical polyurethanes may be prepared by reaction of diisocyanates with glycols or by reaction of bischloroformates with amines, the latter are frequently preferred for laboratory preparations. First, the required bischloroformate is often readily synthesized when the diisocyanate is not. Second, it is far easier to prepare high polymers from bischloroformates (via interfacial techniques).

One type of polyurethane easily prepared from bischloroformates but inaccessible through diisocyanates is the group of products obtained by condensing bischloroformates with secondary diamines. The polyurethanes from piperazine appear to be especially attractive because of their high melting points, which are listed in Table 11 (255,256).

TABLE 11

Polyurethanes from Piperazine

Glycol bischloroformate	Inherent viscosity	Polymer melting temperature, °C
Trans-1,4-cyclohexanediol	1.01	375
Cis-trans-1,4-cyclohexanediol	1.08	275
Cis-1,4-cyclohexanediol	0.30	225
Neopentyl glycol	1.6	230
Ethylene glycol	1.6	245

$$\text{ClCOOROCOCl} + \text{HN}\underset{}{\overset{}{\bigcirc}}\text{NH} \longrightarrow \text{Cl}\left[\text{COOROCON}\underset{}{\overset{}{\bigcirc}}\text{N}\right]_n\text{H}$$

An important technique which has been developed for polycondensations using bischloroformates or other diacid chlorides is that of interfacial polycondensation (257). In this technique, a rapid irreversible polycondensation occurs at or near the interface of two immiscible phases. In the customary interfacial polymerization, a diamine and an acid acceptor are dissolved in an aqueous solution, while the bischloroformate is dissolved in an immiscible organic phase. Stirring is not necessary for demonstration purposes, and a continuous layer of polymer can be removed from the phase interface. For high yields and reasonable reaction times, the two phases may be contacted by means of vigorous agitation. In contrast to the usual experience with bulk polymerization, relatively impure bischloroformates can often be used in interfacial polymerization and still give high molecular weight polymers (257). The whole area of condensation polymerization by interfacial techniques is discussed in a recent *Polymer Review* (258).

2.   POLYCARBONATES

Polycarbonates are easily prepared from bischloroformates by reaction with polyhydroxy compounds. The condensation with aliphatic diols usually proceeds well, merely by heating under reduced pressure (259). Unbranched linear aliphatic polycarbonates have aroused little interest because they are low melting and hydrolytically and thermally low in stability, although certain hindered (and particularly cyclic) diols give much higher melting points. The dependence of melting point upon structure is illustrated in Table 12. Aliphatic bischloroformates can also be condensed with aromatic

TABLE 12

Melting Points of Aliphatic and Alicyclic Polycarbonates

Glycol	mp of polycarbonate, °C (269)
Trimethylene glycol	38–45
Neopentyl glycol	107–119
Cis-2,2,4,4-tetramethyl-1,3-cyclobutanediol	253
Trans-2,2,4,4-tetramethyl-1,3-cyclobutanediol	>360

diols, such as bisphenol-A, in the presence of an acid acceptor. The ethylene–bisphenol-A polycarbonate has been shown to decompose at 280°C to a mixture of cyclic ethylene carbonate and bisphenol-A polycarbonate (260). These mixed polycarbonates have not found practical applications. Only the fully aromatic polycarbonates have attained commercial importance.

3. POLYESTERS

Polyesters may be prepared by condensation of bischloroformates with dicarboxylic acids. Under certain conditions, the intermediate mixed polyanhydride can be isolated and, if desired, converted to the polyester by heating at a suitable temperature (261).

References

1. A. Wurtz, *Ann.*, **100**:111 (1856).
2. A. Wurtz, *Ann.*, **105**:202 (1858).
3. G. O. Curme, "Glycols," Reinhold, New York, 1952.
4. I. Mellan, "Polyhydric Alcohols," Spartan Books, Washington, D.C., 1962.
5. A. Lourenço, *Ann. Chim. Phys.* [3], **67**:296 (1863).
6. W. H. Carothers and J. A. Arvin, *J. Amer. Chem. Soc.*, **51**:2560 (1929).
7. W. H. Carothers and F. J. v. Natta, *J. Amer. Chem. Soc.*, **52**:314 (1930).
8. W. H. Carothers and G. L. Dorough, *J. Amer. Chem. Soc.*, **52**:711 (1930).
9. W. H. Carothers, J. A. Arvin, and G. L. Dorough, *J. Amer. Chem. Soc.*, **52**:3292 (1930).
10. W. H. Carothers and J. W. Hill, *J. Amer. Chem. Soc.*, **54**:1559 (1932).
11. W. H. Carothers and J. W. Hill, *J. Amer. Chem. Soc.*, **54**:1579 (1932).
12. J. R. Whinfield and J. T. Jackson (to Imperial Chemical Industries), British Patent 578,079 (June 14, 1946). *Chem. Abstr.* **41**:1495 (1947).
13. J. R. Whinfield and J. T. Jackson (to du Pont), U.S. Patent 2,465,319 (March 22, 1949).
14. J. Dumas, *Ann. Chim. Phys.* [2], **54**:226 (1833).
15. R. E. Oesper, W. Broker, and W. A. Cook, *J. Amer. Chem. Soc.*, **47**:2609 (1925).

16. M. Matzner, R. P. Kurkjy, and R. J. Cotter, *Chem. Rev.*, **64**:645–687 (1964).
17. A. Butlerov, *Ann.*, **111**:242 (1859).
18. J. F. Walker, "Formaldehyde" 3rd ed., Reinhold, New York-London, 1964, pp. 52–82.
19. W. H. Carothers, *Chem. Rev.*, **8**:361–362 (1931).
20. H. C. Stevens (to Columbia-Southern), British Patent 820,603 (Sept. 23, 1959); *Chem. Abstr.*, **54**:4060 (1960).
21. E. W. Spanagel and W. H. Carothers, *J. Amer. Chem. Soc.*, **57**:929 (1935).
22. H. S. Hill and H. Hibbert, *J. Amer. Chem. Soc.*, **45**:3117, 3124 (1923).
23. J. W. Hill and W. H. Carothers, *J. Amer. Chem. Soc.*, **57**:925 (1935).
24. D. B. Pattison, *J. Org. Chem.*, **22**:662 (1957).
25. W. F. Gresham (to du Pont), U.S. Patent 2,395,265 (Feb. 19, 1946).
26. W. F. Gresham (to du Pont), U.S. Patent 2,457,224 (Dec. 28, 1948).
27. G. H. Coleman and G. V. Moore (to Dow), U.S. Patent 2,056,830 (Oct. 6, 1936).
28. R. C. Morris and A. V. Snider (to Shell), U.S. Patent 2,520,733 (Aug. 29, 1950).
29. S. A. Ballard, R. C. Morris, and J. L. Van Winkle (to Shell), U.S. Patent 2,492,955 (Jan. 3, 1950).
30. A. Franke, A. Kroupa, F. Schweitzer, M. Winischhofer, H. Klein-Lehr, M. Just, M. Hackl, I. V. Reyher, and R. Bader., *Monatsh.*, **69**:167 (1936).
31. E. Mueller, "Methoden der Organischen Chemie (Houben-Weyl)" Vol. XIV/2, Georg Theime, Stuttgart, 1963, pp. 1–98, 580–586.
32. N. Gaylord, "Polyethers," Pt 1, Interscience, New York-London, 1963, pp. 409–453.
33. H. K. Hall, Jr., *J. Amer. Chem. Soc.*, **77**:5993 (1955).
34. P. W. Morgan, "Condensation Polymers: By Interfacial and Solution Methods," Interscience, New York-London, 1965, p. 466.
35. H. M. Fatianov, *Jahresber. Fortschr. Chem.*, **1864**:477.
36. D. D. Reynolds, D. L. Fields, and D. L. Johnson, *J. Org. Chem.*, **26**:5121 (1961).
37. O. Diels, *Ber.*, **36**:736 (1903).
38. J. A. Campbell, *J. Org. Chem.*, **22**:1259 (1957).
39. H. Najer, P. Chabrier, and R. Guidicelli, *Bull. Soc. Chim. Fr.*, **1959**:611.
40. S. R. Newman, R. Y. Heisler, and N. Alpert (to Texaco), U.S. Patent 2,921,955 (Jan. 19, 1960).
41. T. Curtius and K. Heidenreich, *Ber.*, **27**:773 (1894).
42. A. Hantzsch, *Ber.*, **27**:1254 (1894).
43. T. B. Windholz, *J. Org. Chem.*, **25**:1706 (1960).
44. A. Einhorn, *Ber.*, **42**:2772 (1909).
45. G. Dittus, "Three-Membered Ring Ethers (1,2-Epoxides)," in "Methoden der Organischen Chemie (Houben-Weyl)," Georg Theime, Stuttgart, 1965, pp. 367–488.
46. A. Wurtz, *Ann.*, **110**:126 (1859).
47. A. Wurtz, *Ann. Chim. Phys.* [3], **69**:329 (1863).
48. H. C. Schultze, "Ethylene Oxide" in "Encyclopedia of Chemical Technology," 2nd ed., Vol. 8, R. E. Kirk and D. F. Othmer, Eds., Interscience, New York-London, 1965, p. 531.
49. W. L. Faith, D. B. Keyes, and R. L. Clark, "Industrial Chemicals," 3rd Ed., Wiley, New York-London, 1965, pp. 372, 380.
50. T. E. Lefort (to Société Française de Catalyse Générale), French Patent 729,952 (March 27, 1931); *Chem. Abstr.*, **27**:312 (1933); T. E. Lefort (to Union Carbide) U.S. Patent 1,998,878 (April 23, 1935).
51. H. H. Voge and C. R. Adams, "Catalytic Oxidation of Olefins," in "Advances in Catalysis," Vol. XVII, Academic Press, New York-London, 1967, pp. 151–172.

52. A. L. Stautzenberger and A. H. Richey (to Celanese), U.S. Patent 3,341,556 (Sept 12, 1967).
53. V. W. Gash (to Monsanto), U.S. Patent 3,275,662 (Sept. 27, 1966).
54. R. Lanthier (to Shawinigan), U.S. Patent 3,222,382 (Dec. 7, 1965).
55. L. Smith, Z. Phys. Chem., 93:59 (1918).
56. R. Landau; D. Brown, J. L. Russell, and J. Kollar, Proc. 7th World Pet. Congr. Mexico 1967, 5:67.
57. P. H. Miller, "Glycols," in Kirk-Othmer," Encyclopedia of Chemical Technology, 2nd ed., Vol. 10, R. E. Kirk and D. F. Othmer, Eds., Interscience, New York–London, 1966, p. 651.
58. A. Wurtz, Ann. Chim. Phys. [3], 55:438 (1859).
59. W. R. Tomlinson, Jr., J. Org. Chem., 17:648 (1952).
60. G. N. Cohen, B. Nisman, and M. Raynaud, Compt. Rend., 225:647 (1947).
61. H. Bahr and H. Zieler, Angew. Chem., 43:286 (1930).
62. A. Wohl and K. Braünig, Chem-Z., 44:157 (1920); Chem. Abstr., 14:2174 (1920).
63. D. J. Loder (to du Pont), U.S. Patent 2,285,448, June 9, 1942.
64. J. E. Carnahan, T. A. Ford, W. F. Gresham, W. E. Grigsby, and G. F. Hager, J. Amer. Chem. Soc., 77:3766 (1955).
65. W. Stumpf, Z. Elektrochem., 57:690–691 (1953); Chem. Abstr., 49:1430 (1955).
66. K. Folkers and H. Adkins, J. Amer. Chem. Soc., 54:1146 (1932).
67. A. Kötz and K. Richter, J. Prakt. Chem. [2], 111:397 (1925).
68. R. Weidenhagen and H. Wegner, Ber., 71:2715 (1938).
69. R. Connor and H. Adkins, J. Amer. Chem. Soc., 54:4680 (1932).
70. J. F. Saeman and E. E. Harris, J. Amer. Chem. Soc., 68:2509 (1946).
71. A. A. Balandin, N. A. Vasyunina, and Y. Mamatov, Uzb. Khim. Zh., 6:3, 64–72 (1962); Chem. Abstr., 57:13657 (1962).
72. W. H. Zartman and H. Adkins, J. Amer. Chem. Soc., 55:4561 (1933).
73. F. H. Otey, J. W. Sloan, C. A. Wilham, and C. L. Mehltretter, Ind. Eng. Chem., 53:267 (1961).
74. N. A. Milas and S. Sussman, J. Amer. Chem. Soc., 59:2345 (1937).
75. G. Wagner, Ber., 21:1234 (1889).
76. K. A. Hofmann, O. Ehrhart, and O. Schneider, Ber., 46:1666 (1913).
77. R. R. Grinstead (to Dow), U.S. Patent 3,048,636 (Aug. 7, 1962).
78. S. K. Bhattacharyya, M. S. Muthanna, and A. D. Patankar, J. Sci. Ind. Res., 11B:369–370 (1952); Chem. Abstr., 47:2064 (1953).
79. D. M. Newitt and P. S. Mene, J. Chem. Soc., 1946:99.
80. D. Swern, in "Organic Reactions," Vol. 7, R. Adams, Ed., Wiley, New York–London, 1953, p. 378.
81. D. Swern, G. N. Billen, and J. T. Scanlan, J. Amer. Chem. Soc., 68:1504 (1946).
82. L. M. Utkin and R. E. Topshteïn, Mikrobiologiya, 19:116–120 (1950); Chem. Abstrs., 44:8047 (1950).
83. A. Belohoubek, Ber., 12:1873 (1879).
84. F. Raschig and W. Prahl, Ber., 61:185 (1928).
85. K. H. Meyer and H. Hopff, Ber., 54:2279 (1921).
86. K. Schwetlick and H. Hartmann, Z. Chem., 1:375–6 (1961); Chem. Abstr., 57:14979 (1962).
87. M. K. Phibbs and B. deB. Darwent, J. Chem. Phys., 18:495 (1950).
88. A. K. Wiebe, W. P. Conner, and G. W. Kinzer, Nucleonics, 19:50 (1961).
89. K. Schwetlick, W. Geyer, and H. Hartmann, Angew. Chem., 72:779 (1960).

90. A. T. Nielsen and W. J. Houlihan, "Organic Reactions," Vol. 16, John Wiley, New York, 1968.
91. A. Kekulé, *Ann.* **162**:310 (1872).
92. A. Wurtz, *Jahresber. Fortschr. Chem.*, **1873**:474.
93. L. P. Kyriakides, *J. Am. Chem. Soc.*, **36**:530 (1914).
94. M. Apel and B. Tollens, *Ber.*, **27**:1087 (1894).
95. H. B. J. Schurink, *Org. Syn.*, *Collective Vol. 1*, 425 (1941).
96. W. Fossek, *Monatsh.*, **4**:664 (1883).
97. A. Lieben, *Monatsh.*, **17**:68–75 (1896).
98. A. Franke, *Monatsh.*, **17**:85 (1896).
99. M. Haüsermann, *Helv. Chim. Acta.*, **34**:1211 (1951).
100. E. Elkik, *Bull. Soc. Chim. Fr.*, **1959**:933.
101. H. J. Hagemeyer, Jr. and H. N. Wright, Jr. (to Eastman Kodak) U.S. Patent 3,091,632, May 28, 1963.
102. H. J. Hagemeyer, Jr. (to Eastman Kodak) U.S. Patent 2,829,169, Apr. 1, 1958.
103. L. F. Hatch and T. W. Evans (to Shell) U.S. Patent 2,434,110, Jan. 6, 1948.
104. R. F. Fisher and C. W. Smith (to Shell) U.S. Patent 2,888,492, May 26, 1959.
105. J. Lincoln and J. G. N. Drewitt (to British Celanese) U.S. Patent 2,395,414, Feb. 26, 1946.
106. J. H. Halpern, *Monatsh.*, **22**:63 (1901).
107. L. A. Pohoryles, S. Sarel, and R. Ben-Shoshan, *J. Org. Chem.*, **24**:1878 (1959).
108. W. H. Carothers and R. Adams, *J. Amer. Chem. Soc.*, **46**:1682 (1924).
109. H. Adkins and H. R. Billica, *J. Amer. Chem. Soc.*, **70**:3121 (1948).
110. T. Tanno, *Bull. Chem. Soc. Jap.*, **11**:204–207 (1936); *Chem. Abstr.*, **30**:5946 (1936).
111. F. Conradin, G. Bertossa, and J. Giesen (to Inventa), U.S. Patent 2,852,570 (Sept. 16, 1958).
112. M. I. Farberov, *Dokl. Akad. Nauk SSSR*, **110**:1005–1008 (1956); *Chem. Abstr.*, **51**:8102 (1957).
113. M. I. Farberov and N. K. Shemyakina, *Zh. Obshch. Khim.*, **26**:2749–54 (1956); *Chem Abstr.*, **51**:7376 (1957).
114. E. Reboul, *Ann. Chim. Phys.* [5], **14**:491 (1878).
115. F. Beilstein and E. Wiegand, *Ber.*, **15**:1497 (1882).
116. D. V. Tishchenko and A. Churbakov, *J. Gen. Chem. (USSR)*, **7**:665 (1937); *Chem. Abstr.*, **31**:5754 (1937).
117. E. G. Bainbridge, *J. Chem. Soc.*, **105**:2301 (1914).
118. Celanese Corp. of America, British Patent 585,245 (Feb. 3, 1947); *Chem. Abstr.* **41**:4167 (1947).
119. C. A. Rojahn, *Ber.*, **54**:3120 (1921).
120. G. Zweifel, K. Nagase, and H. C. Brown, *J. Amer. Chem. Soc.*, **84**:183 (1962).
121. K. A. Saegebarth (to du Pont), U.S. Patent 3,060,244 (Oct. 23, 1962).
122. S. Corsano, *Atti Accad. Nazl. Lincei. Rend.*, *Classe Sci. Fis. Mat. Nat.*, **34**:(4), 430–433 (1963); *Chem. Abstr.*, **60**:3993 (1964).
123. A. Freund, *Monatsh.*, **2**:638 (1881).
124. P. A. Levene, A. Walti, and H. L. Haller, *J. Biol. Chem.*, **71**:466 (1926).
125. L. Henry, *Rec. Trav. Chim. Pays-Bas*, **16**:208 (1897).
126. T. F. Rutledge, "Acetylenic Compounds," Reinhold, New York, 1968, pp. 146–196.
127. W. Reppe et al., *Ann.*, **596**:25 (1955).
128. Lespieau, *Compt. Rend.*, **150**:1761 (1910).
129. W. Reppe et al., *Ann.*, **596**:58 (1955).

130. C. McKinley and J. P. Brusie (to GAF Corporation) U.S. Patent 2,768,214 (Oct. 23, 1956).
131. R. I. Longley, Jr., and W. S. Emerson, *J. Amer. Chem. Soc.*, **72**:3079 (1950).
132. R. I. Longley, Jr., W. S. Emerson, and T. C. Shafer, *J. Amer. Chem. Soc.*, **74**:2012 (1952).
133. L. Bouveault and G. Blank, *Compt. Rend.*, **137**:329 (1903).
134. R. Robinson and L. H. Smith, *J. Chem. Soc.*, **1937**:373.
135. R. H. Manske, *Org. Syn.*, Coll. Vol. 2, 154 (1943).
136. R. F. Nystrom and W. G. Brown, *J. Amer. Chem. Soc.*, **69**:1198 (1947).
137. R. F. Nystrom and W. G. Brown, *J. Amer. Chem. Soc.*, **69**:2158 (1947).
138. W. A. Lazier, J. W. Hill, and W. J. Amend, *Org. Syn.*, Coll. Vol. 2, 325 (1943).
139. H. Corr, E. Haarer, and P. Hornberger (to Badische Anilin- U. Soda-Fabrik A.G.) British Patent 921,477 (March 20, 1963); *Chem. Abstr.*, **59**:7430 (1963).
140. C. H. Hood and H. W. Tatum (to Celanese) French Patent 1,374,807 (Oct. 9, 1964); *Chem. Abstr.*, **62**:7638 (1965).
141. J. Colonge and P. Corbet, *Bull. Soc. Chim. Fr.*, **1960**:287.
142. T. Utne, R. E. Jones, and J. D. Garber (to Merck), U.S. Patent 3,070,633 (Dec. 25, 1962).
143. T. Utne, J. D. Garber, and R. E. Jones (to Merck), U.S. Patent 3,083,236 (March 26, 1963).
144. G. Kimura, T. Uematsu, and K. Miyake, *Yuke Gosei Kagaku Kyokai Shi*, **22**(5):372–379 (1964); *Chem. Abstr.*, **61**:2967 (1964).
145. G. F. Woods and H. Sanders, *J. Amer. Chem. Soc.*, **68**:2111 (1946).
146. J. C. Milligan (to Jefferson), U.S. Patent 3,046,311 (July 24, 1962).
147. K. Sugino and J. Midzuguchi, *J. Chem. Soc. Jap.*, **64**:1385–1392 (1943); *Chem. Abstr.*, **41**:4483 (1947).
148. H. Adkins and R. Connor, *J. Amer. Chem. Soc.*, **53**:1093 (1931).
149. D. Kaufman and W. Reeve, *Org. Syn.*, Coll. Vol. III, 693 (1955).
150. B. Wojcik and H. Adkins, *J. Amer. Chem. Soc.*, **55**:4939 (1933).
151. A. Franke, F. Lieben, and S. Glaubach, *Monatsh.*, **43**:228 (1922).
152. E. Haworth and W. H. Perkin, Jr., *J. Chem. Soc.*, **65**:598 (1894).
153. G. M. Bennett and F. Heathcoat, *J. Chem. Soc.*, **1929**:271.
154. J. Hamonet, *Bull. Soc. Chim. Fr.* [3], **33**:538 (1905).
155. R. Dionneau, *Ann. Chim. Phys.* [9], **3**:228 (1915).
156. G. Gustavson and N. Demjanov, *J. Prakt. Chem.* [2], **39**:542 (1889).
157. P. J. Dekkers, *Rec. Trav. Chim. Pays-Bas*, **9**:101 (1890).
158. H. Dreyfus (to Celanese), U.S. Patent 2,389,347 (Nov. 20, 1945).
159. T. Anderson and R. V. Lindsey (to du Pont), U.S. Patent 3,081,357 (March 12, 1963).
160. P. Sabatier and A. Mailhe, *Compt. Rend.*, **146**:1194 (1908).
161. A. Roebuck and H. Adkins, *Org. Syn.*, **28**:35 (1948).
162. H. Staudinger and F. Staiger, *Ann.*, **517**:72 (1935).
163. J. Coops, J. W. Dienske, and A. Aten, *Rec. Trav. Chim. Pays-Bas*, **57**:307 (1938).
164. R. C. Olberg, H. Pines, and V. N. Ipatieff, *J. Amer. Chem. Soc.*, **66**:1097 (1944).
165. A. Baeyer, *Ann.*, **278**:92 (1894).
166. J. Coops, J. W. Dienske, and W. M. Smit, *Rec. Trav. Chim. Pays-Bas*, **57**:637 (1938).
167. T. D. Perrine and W. C. White, *J. Amer. Chem. Soc.*, **69**:1543 (1947).
168. S. Z. Levin, I. S. Diner, and G. S. Gurevich, *Neftekhimiya* **2**:566–572 (1962). *Chem. Abstr.*, **58**:13834 (1963).

169. G. A. Akin, H. J. Lewis, and T. F. Reid (to Eastman Kodak), British Patent 988,012 (March 31, 1965); *Chem. Abstr.*, **63**:514 (1965).
170. R. H. Hasek and E. U. Elam (to Eastman Kodak), German Patent 1,144,703 (March 7, 1963); *Chem. Abstr.*, **60**:4028 (1964).
171. P. Smith and A. H. Jubb (to Imperial Chemical Industries), British Patent 902,372 (Aug. 1, 1962); *Chem. Abstr.*, **58**:458 (1963).
172. G. S. Gurevich, S. Z. Levin, and I. S. Diner, *Zh. Obshch. Khim.*, **34**(2):696–699 (1964); *Chem. Abstr.*, **61**:5556 (1964).
173. H. J. Naumann and H. Schmidt, *J. Prakt. Chem.*, **29**(3–6):230 (1965).
174. R. H. Hasek and M. B. Knowles (to Eastman Kodak), U.S. Patent 2,917,549 (Dec. 15, 1959).
175. H. Adkins, B. Wojcik, and L. W. Covert, *J. Amer. Chem. Soc.*, **55**:1673 (1933).
176. W. A. Lazier (to du Pont), U.S. Patent 2,105,664 (Jan. 18, 1938).
177. G. A. Haggis and L. N. Owen, *J. Chem. Soc.*, **1953**:393.
178. G. A. Haggis and L. N. Owen, *J. Chem. Soc.*, **1953**:399.
179. G. A. Haggis and L. N. Owen, *J. Chem. Soc.*, **1953**:406.
180. M. Mugno and M. Bornengo, *Chim. Ind. (Milan)*, **46**(1):5–9 (1964); *Chem. Abstr.*, **60**:9143 (1964).
181. C. W. Smith and D. G. Norton, *Org. Syn.*, Coll. Vol. 4, 348 (1963).
182. L. L. Miller and J. R. Johnson, *J. Org. Chem.*, **1**:138 (1936).
183. F. Felix, P. Meyer, and H. Harder, *Helv. Chim. Acta* **8**:325 (1925).
184. R. H. Hasek, E. U. Elam, J. C. Martin, and R. G. Nations, *J. Org. Chem.*, **26**:700 (1961).
185. R. H. Hasek and E. U. Elam (to Eastman Kodak), U.S. Patent 2.936,324 (May 10, 1960).
186. E. U. Elam and R. H. Hasek (to Eastman Kodak), U.S. Patent 3,190,928 (June 22, 1965).
187. W. Markownikoff, *Ann.*, **302**:22 (1898).
188. A. Roebuck and H. Adkins, *Org. Syn.*, Coll. Vol. 3, 217 (1955).
189. H. Wieland, O. Schlichting, and W. v. Langsdorff, *Z. Physiol., Chem.*, **161**:77 (1926).
190. A. T. Blomquist and D. T. Longone, *J. Amer. Chem. Soc.*, **79**:3916 (1957).
191. A. Franke and F. Sigmund, *Monatsh.*, **46**:70 (1925).
192. R. W. Shortridge, R. A. Craig, K. W. Greenlee, J. M. Derfer, and C. E. Boord, *J. Amer. Chem. Soc.*, **70**:949 (1948).
193. Esso, British Patent 944,110 (Dec. 11, 1963); *Chem. Abstr.*, **61**:6878 (1964).
194. H. C. Brown and B. C. Subba Rao, *J. Amer. Chem. Soc.*, **78**:2582 (1956).
195. J. N. Ray, A. Mukherji, and N. D. Gupta, *J. Indian Chem. Soc.*, **38**:705 (1961); *Chem. Abstr.*, **56**:11476 (1962).
196. J. E. Pritchard (to Phillips Petroleum), U.S. Patent 2,814,649 (Nov. 26, 1957).
197. E. F. M. Stephenson, *Org. Syn.*, Coll. Vol. 4, 984 (1963).
198. P. Ruggli and W. Theilheimer, *Helv. Chim. Acta*, **24**:906 (1941).
199. P. Ruggli, B B. Bussemaker, and W. Müller, *Helv. Chim. Acta*, **18**:616, 619 (1935).
200. P. Ruggli and B. Prijs, *Helv. Chim. Acta*, **28**:688 (1945).
201. J. E. Pritchard and B. Franzus (to Phillips Petroleum), U.S. Patent 2,939,886 (June 7, 1960).
202. J. Nemirowsky, *J. Prakt. Chem.* [2], **28**:439 (1882).
203. D. Vorländer, *Ann.*, **280**:187 (1894).
204. L. Spiegler (to du Pont), U.S. Patent 2,873,291 (Feb. 10, 1959).
205. M. Brown (to du Pont), U.S. Patent 3,334,128 (Aug. 1, 1967).

206. M. Katz and E. L. Wittbecker (to du Pont), U.S. Patent 2,787,630 (April 2, 1957).
207. N. Rabjohn, *J. Amer. Chem. Soc.*, **70**:1181–1183 (1948).
208. Y. Iwakura, M. Sakamoto, and H. Yasuda, *Nippon Kagaku Zasshi*, **82**:606–613 (1961); *Chem. Abstr.*, **56**:8534 (1962).
209. F. H. Norton (to Dow), U.S. Patent 2,923,732 (Feb. 2, 1960).
210. E. L. Wittbecker and M. Katz, *J. Polym. Sci.*, **40**:373 (1959).
211. J. R. Caldwell and W. J. Jackson, Jr. (to Eastman Kodak), U.S. Patent 3,096,359 (July 2, 1963).
212. W. Hentschel, *J. Prakt. Chem.* [2], **36**:213 (1887).
213. S. Cloëz, *Ann. Chim. Phys.* [3], **17**:302 (1846).
214. A. Cahours, *Ann. Chim. Phys.* [3], **19**:346 (1847).
215. J. I. Jones, *J. Chem. Soc.*, **1957**:2735.
216. H. Gross, A. Rieche, and E. Höft, *Chem. Ber.*, **94**:544 (1961).
217. H. J. Bernstein, *J. Amer. Chem. Soc.*, **74**:2674 (1952).
218. Chemische Werke Huels, Belgian Patent 635,368 (Nov. 18, 1963); *Chem. Abstr.*, **61**:9413 (1964)
219a. S. Siggia, "Quantitative Organic Analysis via Functional Groups," 3rd ed., Wiley, New York-London, 1963, pp. 8–72.
219b. Ref. 219a, pp. 179–186.
220. R. H. Kienle and C. S. Ferguson, *Ind. Eng. Chem.*, **21**:349 (1929).
221. E. F. Carlston, *Amer. Paint J.*, **42**:1, 50 (1957).
222. C. D. Hurd and F. H. Blunck, *J. Amer. Chem. Soc.*, **60**:2419 (1938).
223. R. T. Arnold, G. G. Smith, and R. M. Dodson, *J. Org. Chem.*, **15**:1256 (1950).
224. W. J. Bailey, *SPE Trans.*, **5**(1):59–64 (1965).
225. "Neopentyl Glycol," Bull. No. N-115 Eastman Kodak Co., pp. 9–31, 35–47, 1963.
226. P. Morison and J. E. Hutchins, *Amer. Chem. Soc.*, *Div. Org. Coatings Plastics, Chem., Preprints*, **21**:1, 159–170 (1961).
227. H. L. Vincent, *Ind. Eng. Chem.*, **29**:1267 (1937).
228. J. B. Rust, *Ind. Eng. Chem.*, **32**:64 (1940).
229. L. Gallacher and F. A. Bettelheim, *J. Polym. Sci.*, **58**:697 (1962).
230. W. Hinz and G. Solow, *Silik. Tech.*, **8**:178–185 (1957); *Chem. Abstr.*, **51**:17225 (1957).
231. R. M. Brice, J. M. Eakman, and D. M. Kaufer, *SPE J.*, **19**:984 (1963).
232. A. H. Selker, *Mod. Plast.*, **40**:1, 172 (1962).
233. E. A. Zavaglia and F. W. Billmeyer, Jr., *Off. Dig. J. Paint Technol. Eng.*, **36**(470):221 (1964).
234. M. Takahashi, *Sen-i Gakkaishi*, **14**:374–377 (1958); *Chem. Abstr.*, **52**:21217 (1958).
235. M. Takahashi, *Kobunshi Kagaku*, **15**:273–278 (1958); *Chem. Abstr.*, **54**:2802 (1960).
236. V. V. Korshak, S. V. Vinogradova, and E. S. Vlasova, *Izv. Akad. Nauk USSR, Otd. Khim. Nauk*, **1954**:1089–1096; *Chem. Abstr.*, **50**:252 (1956).
237. J. W. Hill and W. H. Carothers, *J. Amer. Chem. Soc.*, **55**:5031 (1933).
238. J. H. Saunders and K. C. Frisch, "Polyurethanes: Chemistry and Technology," Pt. 1, Interscience, New York–London, 1962, pp. 276–293.
239. B. F. Cinadr and E. G. Bobalek, *J. Appl. Polym. Sci.*, **6**:32 (1962).
240. C. S. Schollenberger (to B. F. Goodrich) U.S. Patent 3,015,650 (Jan. 2, 1962).
241. C. S. Schollenberger, H. Scott, and G. R. Moore, *Rubber World*, **137**:549 (1958).
242. G. E. Graham and J. A. Parker (to Armstrong Cork), U.S. Patent 2,893,969 (July 7, 1959).
243. O. Bayer, *Angew. Chem.*, **A49**:257 (1947).

244. H. Rinke, H. Schild, and W. Siefken (vested U.S. Attorney General), U.S. Patent 2,511,544 (June 13, 1950).

245. A. Höchtlen, *Kunststoffe*, **40**:221–232 (1950).

246. T. Hoshino and I. Ichikizaki, *Chem. High Polym.* (*Jap.*), **2**:328–330 (1945); *Chem. Abstr.*, **44**:5150 (1950).

247. O. Bayer and E. Müller, *Angew. Chem.*, **72**:934 (1960).

248. C. H. Smith, *Ind. Eng. Chem.*, *Prod. Res. Develop.*, **4**(1):9 (1965).

249. D. B. Pattison (to du Pont), U.S. Patent 2,870,097 (Jan. 20, 1959).

250. J. R. Caldwell and W. J. Jackson, Jr. (to Eastman Kodak), U.S. Patent 2,968,646 (Jan. 17, 1961).

251. W. J. Jackson, Jr., and J. R. Caldwell, *Adv. Chem. Ser.*, No. **34**:200–207 (1962).

252. Chemische Werke Huels, French Patent 1,361,204 (May 15, 1964); *Chem. Abstr.*, **62**:6592 (1965).

253. W. J. Jackson, Jr., and J. R. Caldwell, *J. Appl. Polym. Sci.*, **7**:1975 (1963).

254. Y. Hachihama and T. Shono, *Technol. Rep. Osaka Univ.*, **9**:361, 229–235 (1959); *Chem. Abstr.*, **54**:9354 (1960).

255. E. L. Wittbecker (to du Pont), U.S. Patent 2,731,446 (Jan. 17, 1956).

256. E. L. Wittbecker and M. Katz, *J. Polym. Sci.*, **40**:367–375 (1959).

257. E. L. Wittbecker and P. W. Morgan, *J. Polym. Sci.*, **40**:289–296 (1959).

258. P. W. Morgan, "Condensation Polymers: By Interfacial and Solution Methods," Interscience, New York–London, 1965.

259. H. Krzikalla and E. Bauer (to I. G. Farbenindustrie), German Patent 801,989, Jan. 29, 1951; *Chem. Abstr.*, **45**:3622 (1951).

260. W. Sweeny, *J. Appl. Polym. Sci.*, **5**:16, S15 (1961).

261. T. B. Windholz (to Celanese), U.S. Patent 3,200,097 (Aug. 10, 1965).

262. H. Batzer and H. Lang, *Makromol. Chem.*, **15**:217 (1955).

263. V. V. Korshak and S. V. Vinogradova, *Izv. Akad. Nauk SSSR, Otd. Khim. Nauk*, **1957**:866–870; *Chem. Abstr.*, **52**:2803 (1958).

264. V. V. Korshak, S. V. Vinogradova, and V. M. Belyakov, *Izv. Akad. Nauk SSSR, Otd. Khim. Nauk*, **1957**:730–736; *Chem. Abstr.*, **52**:2799 (1958).

265. V. V. Korshak, S. V. Vinogradova, and E. S. Vlasova, *Dokl. Akad. Nauk SSSR*, **94**:61–64 (1954); *Chem. Abstr.*, **49**:3823 (1955).

266. K. W. Doak and H. N. Campbell, *J. Polym. Sci.*, **18**:215 (1955).

267. V. V. Korshak and S. V. Vinogradova, *Zh. Obshch. Khim.*, **26**:539–544 (1956); *Chem. Abstr.*, **50**:13814 (1956).

268. H. Batzer, *Makromol. Chem.*, **10**:19 (1953).

269. H. Schnell, "Chemistry and Physics of Polycarbonates," Interscience, New York–London, 1964, p. 21.

4. HYDROXY ACIDS

STEVE GUST COTTIS AND JAMES ECONOMY

*Research and Development Division, The Carborundum Company,
Niagara Falls New York*

Contents

I. INTRODUCTION

Hydroxycarboxylic acids were among the first organic compounds to be characterized and synthesized. The hydroxy acids that occur in nature or are formed as intermediate products in biochemical processes include glycolic, lactic, β-hydroxybutyric, ricinoleic, malic, tartaric, citric, and salicylic. A great many hydroxycarboxylic acids have been synthesized; however, polyesters have been prepared from relatively few of them. This chapter deals mainly with those hydroxy acids whose condensation polymerization has been studied and from which polyesters have been prepared.

As we discuss in the section on polymerization (Section V), polyesters are not always produced when an attempt is made to polymerize certain hydroxycarboxylic acids; e.g., β-hydroxy acids tend to dehydrate, whereas the γ-hydroxy acids form lactones exclusively. Thanks to more recent work on the ring-opening polymerization of lactones, the AB-type polyesters of β-hydroxy acids are known. Because of the close relation of the lactones to the hydroxycarboxylic acids, some mention of the synthesis and polymerization of the more important lactones is made.

In general, the aliphatic polyesters derived from hydroxy acids have found little commercial utility because of their low melting points. Polymers derived from aromatic hydroxy acids would appear to be of greater commercial interest because of their potentially high melting points and relatively low cost. However, polymers from salicyclic acid tend to form cyclic structures, and the polymer from *m*-hydroxybenzoic acid is low melting. Perhaps the most interesting polymers are those derived from *p*-hydroxybenzoic acid. The fiber-forming polyesters, poly(oxymethylenebenzoyl) and poly(oxyethoxybenzoyl) have found commercial applications, particularly abroad. The homopolyester of *p*-hydroxybenzoic acid, has recently acquired the trade name "EKONOL." The polymer does not melt even at 550°C, and yet it can be fabricated by compression sintering at 420°C and 10,000 psi. This polymer combines self-lubricating character along with the highest reported elastic modulus, thermal conductivity, electrical insulation, and solvent resistance of any available polymer.

In this chapter the general synthesis of aliphatic and aromatic hydroxy acids is covered first. Then follows a more detailed discussion of the synthesis or manufacture and properties of the more important monomers. The chapter concludes with the types of polymers derived from the hydroxy acids, the methods of synthesis of the polyesters, and some of their applications.

II. GENERAL SYNTHETIC ROUTES TO HYDROXY ACIDS

The general syntheses of hydroxycarboxylic acids fall into four categories: (a) introduction of the carboxyl group into alcohols or phenols, (b) introduction of the hydroxyl group into carboxylic acids, (c) simultaneous introduction of hydroxyl- and carboxyl-forming groups into the molecule, and (d) decarboxylation of polyacids.

A. Introduction of the Carboxyl Group into Alcohols or Phenols

1. OXIDATION REACTIONS

Glycols may be oxidized by dilute nitric acid or platinum black and air to the corresponding hydroxy acid. For example, ethylene glycol oxidizes to form glycolic acid (1).

$$HOCH_2CH_2OH \xrightarrow{[O]} HOCH_2COOH$$

Hydroxy aldehydes can be oxidized to form hydroxy acids. The oxidation of aldol (β-hydroxybutyraldehyde) gives β-hydroxybutyric acid (2).

$$CH_3\overset{\overset{\displaystyle OH}{|}}{C}HCH_2CHO \xrightarrow[\text{H}_2\text{O}]{\text{Hg}_2\text{O}} CH_3\overset{\overset{\displaystyle OH}{|}}{C}HCH_2COOH$$

Similarly, phenolic aldehydes can be oxidized to aromatic acids.

2. ALKALINE FUSION OF SUBSTITUTED PHENOLS

The alkyl group attached to the aromatic nucleus of phenols is oxidized to a carboxyl group by alkali fusion. When, for example, o-cresol is heated with caustic soda at 260–270°C in the presence of copper oxide (3), salicyclic acid is obtained after acidification:

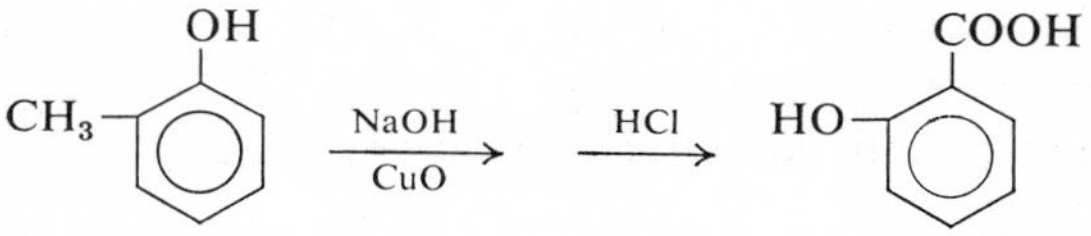

In a related reaction, p-hydroxybenzoic acid and some salicyclic acid can be obtained from phenol, carbon tetrachloride, and potassium hydroxide (4).

3. ADDITION OF CARBON DIOXIDE TO PHENOLS

A method for the synthesis of one common hydroxy acid, developed by Kolbe and Schmitt, starts with the alkali salt of the phenolic compound. The addition of carbon dioxide followed by acidification yields the acid. The action of carbon dioxide on the sodium salt of β-naphthol under pressure at 280–290°C yields 3-hydroxy-2-naphthoic acid (5).

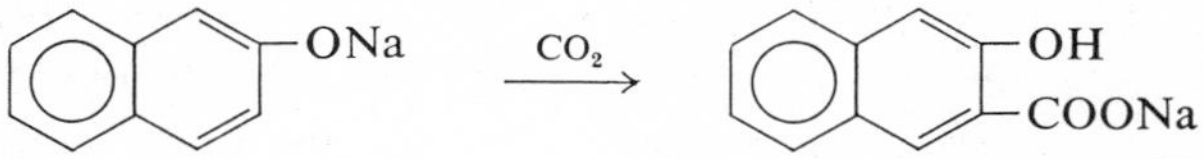

4. HYDROLYSIS OF HYDROXY NITRILES

Base or acid hydrolysis of hydroxy nitriles has long been used as a method for producing hydroxy acids. Hydracrylic acid (β-hydroxypropionic acid) is obtained by the hydrolysis of ethylene cyanohydrin (6,7).

$$HOCH_2CH_2CN \xrightarrow[\text{or acid}]{\text{base}} HOCH_2CH_2COOH$$

5. USE OF HYDROXY NITRO COMPOUNDS

A method (8) recently developed for the synthesis of α-hydroxy acids yields 65–75% of the following hydroxy acids: propionic, butyric, valeric, isovaleric, caproic, and isocaproic. The method depends on the conversion of hydroxy nitro compounds into hydroxy acids and hydroxylamine by refluxing the nitro compound in hydrochloric acid. Thus from 1-nitro-2-propanol, α-hydroxypropionic acid is obtained:

$$NO_2CH_2\underset{\underset{OH}{|}}{C}HCH_3 \xrightarrow{HCl} HOOC—\underset{\underset{OH}{|}}{C}HCH_3 + HONH_2·HCl$$

B. Introduction of the Hydroxyl Group into Carboxylic Acids

1. Reduction of Aldo Acids, Keto Acids, and Dicarboxylic Acid Monoesters

The reduction of aldo acids and keto acids with sodium amalgam or with zinc and mineral acids gives hydroxy acids. Levulinic acid yields γ-valerolactone by reduction with sodium amalgam (9).

$$CH_3\overset{\overset{O}{\|}}{C}CH_2CH_2COOH \xrightarrow{Na(Hg)} CH_3\underset{\underset{\llcorner—O—\lrcorner}{|}}{C}HCH_2CHC{=}O$$

Glycolic acid may be obtained by the reduction of oxalic acid with zinc (10).

$$HOOCCOOH \xrightarrow[HCl]{Zn} HOCH_2COOH$$

A general synthesis of ω-hydroxy acids starting with monoesters of dicarboxylic acids in overall yields of 55–76% is due to Dale. First the acid chloride is prepared by treating the acid with thionyl chloride. The chloride is then reduced with sodium borohydride in dioxane to the ω-hydroxy acid (11).

$$HOOC(CH_2)_nCO_2Et \xrightarrow{SOCl_2} \xrightarrow{NaBH_4} HOCH_2(CH_2)_nCO_2Et$$

2. Replacement of Halogens in Halocarboxylic Acids

Perhaps one of the most widely used syntheses of hydroxy acids is the replacement of halogens of halo acids by hydroxyl groups. Usually a catalyst such as a metal oxide or hydroxide is necessary, but with labile halogens, water alone can be sufficient to replace the halogen atom (12).

$$BrCH_2COOH \xrightarrow{H_2O} HOCH_2COOH$$

3. REPLACEMENT OF AMINO GROUPS FROM AMINOCARBOXYLIC ACIDS WITH HYDROXYL GROUPS

Treatment of amino acids with nitrous acid, followed by warming in the case of aromatic amines, replaces the amino group by hydroxyl group (13).

$$NH_2CH_2COOH \xrightarrow{HNO_2} HOCH_2COOH$$

4. ALKALI FUSION OF SULFO ACIDS

Aromatic acids that can be sulfonated can then be fused to yield hydroxy acids. For example, m-hydroxybenzoic acid is manufactured by the sulfonation of benzoic acid with oleum. The monosodium salt of m-sulfobenzoic acid is salted out and fused with alkali at 210–220°C.

5. THE REFORMATSKY REACTION

The Reformatsky reaction can be used to produce β-hydroxy acids by the reaction of α-haloesters with aldehydes or ketones using a zinc catalyst in an anhydrous medium (14).

$$BrCH_2COOEt + Zn \longrightarrow BrZnCH_2COOEt$$

$$\Big\downarrow R_2C=O$$

$$R_2C(OH)CH_2CO_2Et \xleftarrow{H_2O} R_2C(OZnBr)CH_2COOEt$$

6. ADDITION OF WATER TO UNSATURATED ACIDS

Heating unsaturated acids with alkali or sometimes sulfuric acid causes the addition of water to occur, yielding the hydroxy acid. For example, acrylic acid forms hydracrylic acid in the presence of sodium hydroxide (15).

$$CH_2{=}CHCOOH \xrightarrow[NaOH]{H_2O} \xrightarrow{H^+} HOCH_2CH_2COOH$$

C. Simultaneous Introduction of Hydroxyl- and Carboxyl-Forming Groups into the Molecule

1. ADDITION OF HYDROGEN CYANIDE

The addition of hydrogen cyanide to aldehydes, ketones, and ethylene oxides provides one of the most general methods of synthesizing hydroxy acids. Examples are the syntheses of lactic acid from acetaldehyde (16) and hydracrylic acid from ethylene oxide (17):

$$CH_3CHO + HCN \longrightarrow CH_3CH(OH)CN \xrightarrow{H_2O} CH_3CH(OH)COOH$$

$$CH_2\!\!-\!\!CH_2 + HCN \longrightarrow CNCH_2CH_2OH \xrightarrow{H_2O} HOOCCH_2CH_2OH$$
$$\underset{O}{\diagdown\diagup}$$

2. ALKALI FUSION OF ALKYLARYLSULFONIC ACIDS

Salicyclic acid can be prepared by the alkali fusion of o-toluenesulfonic acid (18).

3. OXIDATION OF PARAFFINS

A method of commercial importance is the oxidation of paraffins (19). This generally yields mixtures of hydroxy acids containing 13–35 carbon atoms.

D. Decarboxylation of Polyacids

Ethyl tartaric acid, for example, yields α-hydroxybutyric acid on heating to 180°C (20).

$$C_2H_5C(COOH)_2 \xrightarrow[-CO_2]{180°C} C_2H_5CHCOOH$$
$$\underset{OH}{|} \qquad\qquad\qquad \underset{OH}{|}$$

Also, 2-methyl-1,1-cyclopropanedicarboxylic acid yields γ-valerolactone, together with 2-methylcyclopropanecarboxylic acid (21):

III. MONOMER SYNTHESES

This section is intended to acquaint the reader with the most practical synthetic routes to the various monomers. Reaction conditions and the methods of isolation and purification of the monomer are given, as well as the yields. In some cases, particularly where the syntheses have been described in recent patent literature, lack of experimental details has necessitated some interpretation by the authors. Where possible, analyses of the monomer, toxicological data, and problems such as storage are described. Polyesters are not reported for several of the hydroxy acids, but their syntheses are included as a point of reference. Monomers containing dihydroxy or dicarboxylic acid groups and monomers containing other functional groups, such as halo groups or olefinic groups, are not considered in this chapter.

Only a limited number of hydroxy acids have been prepared on a commercial scale. Lactic acid and salicylic acid have been commercially available for many years. A few of the syntheses have been described in detail in "Organic Syntheses" and "Organic Reactions." In many cases, the isolation of the pure monomer presents a problem. The low melting points make purification by recrystallization difficult, especially when the product is contaminated with starting materials and with by-products (lactones) having similar solubilities and melting points. Distillation of the crude monomers often leads to lactone formation or polymerization (see Section V). Frequently, isolation of the lactone is the only practical method of isolation of a product from the reaction mixture. For the foregoing reasons, the syntheses of the appropriate lactones are described in addition to, or instead of, the corresponding hydroxy acid.

The hydroxy acids discussed here are arranged according to chemical formula.

A. $C_2H_4O_3$

1. GLYCOLIC ACID [HYDROXYACETIC ACID, HYDROXYETHANOIC ACID (IUPAC)], $HOCH_2COOH$

Before 1940 glycolic acid was available commercially on a limited scale by the hydrolysis of monochloroacetic acid with sodium hydroxide. Later, manufacture of this monomer by the electrolytic reduction of oxalic acid was introduced. The commercial preparation of glycolic acid in the United States today is based on the reaction of formaldehyde and carbon monoxide in the presence of a catalyst. One process (22) for the preparation of the acid entails the reaction of formaldehyde and carbon monoxide using as a catalyst sulfuric acid, dissolved in at least 0.5 mole of an organic acid, such as acetic, per mole of formaldehyde, preferably at 160–170°C for 1 hr. The reaction of carbon monoxide and formaldehyde can also be carried out above atmospheric pressure in the presence of water (23). Hydrochloric acid, phosphoric acid, formic acid, glycolic acid, and boron trifluoride are other suitable catalysts for the reaction.

Glycolic acid is offered in only one form, as a 70% aqueous solution. Of this, the free acid represents 62% and the balance is mainly polyglycolides that are in equilibrium with the monomer. The maximum amount of formic acid present is 0.45%; the ash content amounts to 0.35%. Inorganics, such as iron, chlorine, and copper, are present in amounts less than 10 ppm. Hydroxyacetic acid solutions can be assayed by direct titration with standard alkali if other acids and saponifiable materials are absent. To determine the free acid, care should be taken not to use excess alkali or permit the temperature to rise above 5°C, since the hydrolysis of any polyacids present tends to give high results. The total acid content is determined by adding excess alkali, boiling for a few minutes and back-titrating with acid.

Pure glycolic acid is mildly toxic if taken internally. It has an irritating effect if it comes in contact with broken skin. Short contact with normal skin is not likely to be harmful, although a warning label is carried on 55-gal drums. In addition to the drums, hydroxyacetic acid is available in specially lined tank cars.

Although commercial glycolic acid can be used to form the polyester by removal of water, poly(oxyacetyl) polymers are best obtained by heating salts of chloroacetic acid (92–96% yields) (24) or as with lactic acid, from the dimeric lactone, in this case glycolide.

2. GLYCOLIDE (p-DIOXANE-2,5-DIONE), $C_4H_4O_4$

The glycolide was first prepared but not characterized by von Heintz in 1859 (25). Preparation of glycolide from sodium chloroacetate generally gives yields of only 20% when heated alone in an oil bath at temperatures as high as 320°C and pressures of 7 mm (24). However, vacuum distillation of dry sodium chloroacetate in the presence of copper gives yields of 72% of glycolide, mp 82–83°C (24).

A current method for the manufacture of glycolide (26) consists of heating an alkali or alkaline earth monochloroacetate with silica in a ratio of 1:3 to 1:6 at 200–240°C. The monochloroacetate is first dried *in vacuo* and heated at 210–215°C for 2 hr *in vacuo* with activated silica. Distillation *in vacuo* gives 83% of glycolide, mp 80°C.

B. $C_3H_6O_3$

1. LACTIC ACID (2-HYDROXYPROPANOIC ACID), $CH_3CH(OH)COOH$

The manufacture and chief uses of lactic acid were well established in the United States before the manufacture of the hydroxy acid was undertaken in other countries. The first lactic acid factory using a fermentation process was established in 1881. Alternate routes such as oxidation of propylene (foreign) and hydrolysis of lactonitrile (U.S.) were not developed until after 1962. Both fermentation and synthetic processes are in use today (27).

2. FERMENTATION PROCESS

In practice, lactic acid is obtained by fermenting starch, dextrose, or a mash from corn or potatoes. The equation for the conversion is $C_6H_{12}O_6 \rightarrow 2CH_3CH(OH)COOH$, or 100% theoretical conversion of the available hexose substrate. Numerous strains of the bacterial genus *Lactobacillus* are available as well as some of their molds. By use of a thermophilic strain of *L. delbrueckii*, which operates most efficiently at 50°C at a pH of 5.0–5.5, most contamination and sterilization problems are eliminated. The reaction is carried out in the presence of calcium carbonate, which serves to neutralize the lactic acid (so that the pH does not become too low) and to facilitate isolation of the product. Sulfuric acid serves to precipitate the calcium as calcium sulfate and to liberate the free acid.

The technology of the recovery of the lactic acid depends on the grade of the acid desired. Grades of lactic acid are quoted as 50 and 80% edible grade, 50 and 80% plastic grade, and 22 and 44% technical grade, all in barrels. An 85% grade is sold in carboys. A brief review of the microbiological syntheses and manufacture and uses of lactic acid has been compiled by Sabin (28).

3. SYNTHETIC LACTIC ACID

The hydroxy acid is obtained as a by-product in the manufacture of acrylonitrile from lactonitrile (acetaldehyde cyanohydrin). An aqueous solution of crude lactic acid is first prepared by the acid hydrolysis of lactonitrile, and this solution is refined and adjusted to provide either a 50% or an 88% solution. In the commercial manufacture of lactic acid, greater than 80% yields of water-white, heat-stable grade, free from residual sugars, are obtained by the hydrolysis of lactonitrile in the conventional manner (29). In other countries, propylene is oxidized to lactic acid. In one process (30), propylene is passed through nitric acid (or nitrogen(V) oxide or oxygen and nitric acid or a combination of these) and the mixture kept one hour at 10°C, cooled to 0°C, degassed, and treated with water followed by sodium carbonate to give a pH of 1. After heating at 100°C for 20 hr, extraction with isopropyl ether produces 73.8% lactic acid. In another process (31), the hydrolysis of lactic acid nitrate, obtained from propylene, nitrogen tetroxide, and oxygen, gives lactic acid.

When lactic acid solutions are concentrated, the removal of water is accompanied by self-esterification of lactic acid, first to lactyllactic acid and then to polylactic acid. Equilibrium occurs among the several components when a concentrated solution is allowed to stand for several weeks at room temperature. The standard method (32) for determination of lactic acid is to oxidize the acid under carefully controlled conditions to acetaldehyde, which, in turn, is absorbed in sodium bisulfite and titrated iodometrically. For a quantitative titration of total available lactic acid, it is first necessary to saponify any polylactic acids by boiling in water or standard alkali. Optical activity can be used as a measure of purity for the optically active forms of lactic acid. The toxic hazard rating of lactic acid is moderate to slight.

4.   LACTIDE (2,5-DIMETHYL-p-DIOXANE-3,6-DIONE), $C_6H_8O_4$

As with glycolic acid, attempts to purify lactic acid by distillation at reduced pressures lead to the formation of the dimer, lactide. The cyclic ester has been prepared and characterized by Carothers et al. (33). Lactide can be prepared commercially (34) in good yields by heating lactic acid first to about 135°C, whereby the acid is polymerized. When the temperature is raised to about 200°C, the polymer reorganizes to the lactide, which can be distilled *in vacuo* in the presence or absence of a suitable liquid or gaseous vehicle. Preparation of optically active lactide has been accomplished (35) from 1(+) lactic acid (80% commercial grade) by polymerization followed by depolymerization of the polylactide *in vacuo* in the presence of zinc oxide at 210–250°C, $[\alpha]_D$ (20°C)–255°C. Lactide is available commercially from the Clinton Corn Company, Clinton, Iowa. The toxicity is unknown, but lactide emits acrid fumes upon heating.

5.   HYDRACRYLIC ACID (3-HYDROXYPROPANOIC ACID), $HOCH_2CH_2COOH$

Hydracrylic acid is not a suitable monomer for polymerization by conventional method, since it dehydrates to form acrylic acid rather than a lactone or a polyester. Its preparation by the alkaline hydrolysis of ethylene chlorohydrin is described in "Organic Syntheses" (6). The hydroxy acid is obtained only in low yields (28–31%) and is found to be an uncrystallizable, hygroscopic syrup. Other methods of preparation include hydrolysis of β-bromopropionic acid and the action of alkali on acrylic acid.

6. β-PROPIOLACTONE (OXETANE-2-ONE), $C_3H_4O_2$

Poly(3-oxypropionyl) can be prepared by the polymerization of β-propiolactone, which is formed by reacting ketene and formaldehyde at temperatures below 25°C in the presence of a Friedel-Crafts catalyst (36–38). Gaseous ketene is mixed with a stream of formaldehyde at a rate of 0.5 mole/hr below the surface of a stirred solution of aluminum chloride–zinc chloride dissolved in previously obtained β-propiolactone. The temperature of the reaction is maintained between 5–20°C. After 6 hr the mixture is distilled at 10 mm, producing the lactone, bp 47–51°C, in excellent yields. Purification of mixtures of β-propiolactone and aliphatic carboxylic anhydrides has been carried out (39) by first hydrolyzing the material with 4–6 moles of water per mole of anhydride. The products are fractionated in a tray distilling column at an overhead pressure of 50 mm and a base temperature of 110°C. The β-propiolactone is recovered in the fifth plate from the bottom (98.5% pure) from a mixture containing 83% lactone, 15% acetic anhydride, and 2% acetic acid.

The toxic hazard rating of the lactone is high. It may cause death or permanent injury after very short exposures to small quantities. Therefore, the lactone should be handled with extreme caution. In one study of the carcinogenic activity of β-lactones (40), it was found that β-propiolactone leads to sarcomas at the injection site at a dose level of 0.1 mg/injection (twice weekly, for 34 weeks). In another study (41), the β-lactone is stated as having no carcinogenic activity. The fire hazard is moderate when the lactone is exposed to heat or flame, and the substance can react with oxidizing materials. Storage of the lactone is recommended at 5–10°C, the temperatures at which the lactone is the most stable. Today β-propiolactone is available commercially from the Celanese Chemical Company, New York.

C. $C_4H_8O_3$

1. α-HYDROXYBUTYRIC ACID (2-HYDROXYBUTANOIC ACID), $CH_3CH_2CH(OH)COOH$

In an early preparation (42) of α-hydroxybutyric acid, the hydroxy acid was obtained by heating α-bromobutyric acid with aqueous potassium carbonate for 5–6 hr. The carbonate was decomposed with hydrochloric acid, the potassium chloride and bromide filtered, and the filtrate extracted with ether. The acid was purified by vacuum distillation. Gilman and Abbott (43) prepared the acid in 46% yield starting with trichloropropylene oxide. The oxide is dissolved in anhydrous ether, and methyllithium is added to the solution at 75°C. The product is hydrolyzed with acid and water. Distillation at 44–46°C at 3 mm gives an 85% yield of 1-trichloro-2-butanol. Refluxing the trichlorobutanol in a sodium carbonate–water–alcohol mixture hydrolyzes the trichloro compound to the acid, which is isolated as the zinc salt. Acidification of the salt followed by distillation at 138–141°C/11 mm gives the pure acid, mp 41–43°C. Moderate yields (65–75%) of the hydroxy acid have been obtained (8) by refluxing 1-nitro-2-butanol with 2 parts 1:1 hydrochloric acid solution for 10 hr, followed by filtration and evaporation of the filtrate. The manufacture of this hydroxy acid starting with acetone cyanohydrin has been described (44).

2. β-Hydroxybutyric Acid (3-Hydroxybutanoic Acid), $CH_3CH(OH)CH_2COOH$

In 1869 Wislicenus (45) described a synthesis for β-hydroxybutyric acid by the reduction of ethyl acetoacetate with sodium amalgam. More recently (46), the hydroxy acid has been synthesized by the oxidation of freshly distilled β-hydroxybutyraldehyde. The oxidation is carried out by using vanadic acid in acetic acid at a temperature of 55°C and at a pressure of 1200 mm for a duration of 24 hr. A yield of 25 g of β-hydroxybutyric acid, bp 125–130°C/8 mm is obtained from 54 g of aldehyde. Under similar conditions, using calcium acetate, 65 g of the aldehyde gives 35 g of the acid.

3. β-Butyrolactone (2-Methyloxetane-3-one), $C_4H_6O_2$

The synthesis of β-butyrolactone, which is commercially available, is described in "Organic Reactions" (47). Acetaldehyde dissolved in dry ether and gaseous ketene are added simultaneously to a solution of boron trifluoride–ethyl ether complex in dry ether. The temperature is maintained between 10 and 15°C. A 50% solution of sodium hydroxide is added to decompose the catalyst. The reaction mixture is then distilled, first at atmospheric pressure, then at reduced pressure. Redistillation of the crude product yields 70% of the lactone, bp 54–56°C/10 mm. The butyrolactone is also prepared by the reduction of diketene (48). Freshly distilled diketene containing 4% acetic anhydride and palladium black diluted with ethyl acetate is hydrogenated with agitation at 0°C. When the hydrogen absorption has ceased, the mixture is filtered and the butyrolactone is distilled. The conversion of diketene to the lactone, which is practically free of butyric acid, is 93%. In one study (41), β-butyrolactone has been shown to possess carcinogenic activity. Of the β and γ lactones studied, it was the only one having this activity.

4. γ-Hydroxybutyric Acid (4-Hydroxybutanoic Acid), $HOCH_2CH_2CHCOOH$

Isolation of the pure γ-hydroxybutyric acid is difficult because the lactone forms so readily. Poly(4-oxybutyroyl) has only been recently obtained (see Section V) by the polymerization of γ-butyrolactone.

5. γ-Butyrolactone (Tetrahydrofuran-2-one), $C_4H_6O_2$

γ-Butyrolactone can be obtained in up to 50% yield along with tetrahydrofuran and 1,4-dihydroxybutane from the hydrogenation of maleic anhydride (5–20% solutions) at temperatures of 225–275°C and pressures of 3000–10,000 psi (49). At 25°C and 2000 psi, a 20% maleic anhydride solution yields 74% of the butyrolactone. A solution of γ-bromobutyric acid and sodium ethoxide dissolved in absolute alcohol was refluxed for 5 hr. During this time sodium bromide separated. After distillation of the ethanol, the lactone separated from the cake after extraction in ether. The ether was evaporated and the lactone distilled at ordinary pressures. The yield is 67% of the lactone, bp 202–206°C, specific gravity (28°C) 1.1054, η_D (26.5°C), 1.4343 (50). The toxic hazard rating for ingestion of γ-butyrolactone is slight. It presents a moderate fire hazard when exposed to heat or flame, and it can react with oxidizing materials. The lactone is commercially available from G.A.F. Corporation, New York.

D. $C_5H_{10}O_3$

1. α-HYDROXYVALERIC ACID (2-HYDROXYPENTANOIC ACID), $CH_3CH_2CH_2CH(OH)COOH$

Free α-hydroxyvaleric acid is obtained by the reduction of the corresponding α-keto acid with sodium amalgam (51). The acid is purified by recrystallization of its zinc salt. The acid is highly hygroscopic and has a relatively low melting point (34°C). Alternatively, the hydroxy acid can be prepared by hydrolysis of the cyanohydrin of butyraldehyde (52) or by refluxing 1-nitro-2-pentanol with hydrochloric acid solution (8).

2. β-HYDROXYVALERIC ACID (3-HYDROXYPENTANOIC ACID), $CH_3CH_2CH(OH)CH_2COOH$

To produce β-hydroxyvaleric acid, propionaldehyde and bromoacetic acid are combined (53) according to the Reformatsky reaction (see page 319). The sodium salt, which is obtained after hydrolysis with caustic, is acidified with dilute sulfuric acid. The resulting hydroxy acid is extracted with ether and the ether layer refrigerated for a long time. Eventually, a low yield of almost pure β-hydroxyvaleric acid is obtained. A melting point of 43–44°C is obtained after repeated recrystallization.

3. γ-VALEROLACTONE (2-METHYLTETRAHYDROFURAN-5-ONE), $C_5H_8O_2$

The manufacture of γ-valerolactone is accomplished in quantitative yields by the hydrogenation of levulinic acid (4-oxopentanoic acid) in the gas or liquid phase. Levulinic acid is readily obtained by the hydrolysis of sucrose with concentrated hydrochloric acid. Vapor-phase hydrogenation of levulinic acid to γ-valerolactone (54) is carried out using reduced copper(II) oxide on silica or copper chromite at atmospheric pressure. A quantitative yield is obtained with the copper chromite catalyst when a feed rate of 0.82–1.2 g/min of levulinic acid and 5 liters/min of hydrogen at 200°C is used. When anhydrous ether serves as a solvent (55) and levulinic acid is shaken over platinum oxide under a pressure of 2.3–3.0 atm of hydrogen, a theoretical yield of the lactone is obtained. The toxicity of the lactone is probably low, but details are unknown. The lactone can be obtained commercially.

4. δ-VALEROLACTONE (TETRAHYDROPYRAN-2-ONE), $C_5H_8O_2$

According to an early preparation (56) of δ-valerolactone, purified δ-iodovaleric acid is heated with sodium ethoxide in absolute alcohol. The crude material is distilled at 218–220°C and the lactone is obtained in 58% yield—specific gravity (20°C), 1.1130; η_D (20°C), 1.4600. The lactone has been synthesized by the reaction of tetrahydrofuran and carbon monoxide in a pressure bomb (57). Adipic acid and its lactone are by-products. The highest yields of δ-valerolactone are obtained when nickel iodide is used as a catalyst at a temperature of 200°C and a pressure of 7200 psi. Total adipic acid yields are 28.6%, and the yield of δ-valerolactone is 32%. Other syntheses of the lactone from tetrahydrofuran using slightly different conditions and catalysts are reported as well (58). The lactone is also prepared by the oxidation of cyclopentanone with peracetic acid. After heating the ketone for 8 hr at 40°C, 84% of δ-valerolactone is obtained, bp 83°C/4 mm, η_D (20°C), 1.4540 (59). The lactone is available commercially.

5. α-HYDROXYISOVALERIC ACID (2-HYDROXY-3-METHYLBUTANOIC ACID), $(CH_3)_2CHCH(OH)COOH$

The procedure for making α-hydroxyisovaleric acid begins when a 2% calcium carbonate solution and α-bromoisovaleric acid are boiled together for 10 min. The mixture is then filtered and the filtrate concentrated. Crystallization of the calcium salt occurs when alcohol is added. After filtration, a second crop is obtained by adding zinc chloride and isolating the product as the precipitated zinc salt. The total yield of the acid in the form of its salts is 70% (60).

6. β-HYDROXYISOVALERIC ACID (3-HYDROXY-3-METHYLBUTANOIC ACID), $(CH_3)_2C(OH)CH_2COOH$

Oxidation of diacetone alcohol readily gives β-hydroxyisovaleric acid (61). Diacetone alcohol is added to a cold solution (10°C) of an alkaline solution of sodium hypochlorite, prepared from sodium hydroxide, chlorine, and water. Reaction begins immediately, and within a few minutes the mixture refluxes vigorously. The temperature is maintained at or near reflux. The mixture is stirred for a total of 3 hr and then allowed to stand overnight. The product is extracted from the acidified mixture with methyl ethyl ketone. Distillation of the acid gives 70% of β-hydroxyisovaleric acid.

7. β-ISOVALEROLACTONE (3-DIMETHYLOXETANE-4-ONE), $C_5H_8O_2$

To obtain β-isovalerolactone, ketene gas is passed through a mixture of acetone and zinc thiocyanate (about 0.2%) while stirring at 25–30°C. The catalyst is neutralized with aqueous sodium carbonate, and most of the acetone is removed under vacuum at 25°C. The residue is distilled to give 75% of β-isovalerolactone, bp 54–55°C/10 mm; η_D (20°C), 1.4126 (47).

8. HYDROXYPIVALIC ACID (2,2-DIMETHYL-3-HYDROXYPROPANOIC ACID), $HOCH_2C(CH_3)_2COOH$

Hydroxypivalic acid can be prepared by the oxidation of 2,2-dimethyl-3-hydroxypropanal. The aldehyde and water are placed in a rocking autoclave and oxygen is introduced. The mixture is agitated at 80°C and 100 atm for 2 hr. After cooling, the acid is extracted with ether giving 75% of the hydroxy acid, mp 123°C. Unreacted aldehyde amounts to 11% (62).

E. $C_6H_{12}O_3$

1. α-HYDROXYCAPROIC ACID (2-HYDROXYHEXANOIC ACID), $CH_3(CH_2)_3CH(OH)COOH$

The first step in producing α-hydroxycaproic acid is to boil a solution of α-bromocaproic acid and aqueous sodium carbonate for 1 hr. To the hot solution is added copper acetate, which serves to precipitate the copper salt of the hydroxy acid as a light green solid. After cooling, the salt is collected by filtration, washed with water, and dried in air to yield 60% of the copper salt. A suspension of the copper salt in hot water is treated with hydrogen sulfide and the copper sulfide filtered off. After the filtrate has been concentrated, the hydroxy acid is extracted with ether. Removal of the ether and recrystallization from petroleum ether gives the pure acid, mp 60–62°C. Considerable loss of product occurs in the conversion of the salt to the free acid (63).

The hydroxy acid is also obtained (64) by treatment of 2-thienylglycolic acid with Raney nickel in 20% sodium hydroxide or by heating butylhydroxymalonic acid above 123°C (65).

2. β-HYDROXYCAPROIC ACID (3-HYDROXYHEXANOIC ACID), $CH_3CH_2CH_2CH(OH)CH_2COOH$

The desired acid salt is obtained after hydrolysis and neutralization of the product from the Reformatsky reaction of butyraldehyde and ethyl bromoacetate (53). β-Hydroxycaproic acid can be obtained by crystallization from an acetone–carbon dioxide mixture which is refrigerated, or by allowing the oil to cool without a solvent. The melting point is 13°C.

3. γ-CAPROLACTONE (2-ETHYLTETRAHYDROFURAN-5-ONE), $C_6H_{10}O_2$

Condensation of propionaldehyde and malonic acid in the presence of a tertiary amine in benzene gives a 67% yield of a mixture of hexenoic acids;

from this, 74% of γ-caprolactone is formed by heating the acid mixture in 80% sulfuric acid for 1 hr at 80°C (66). The lactone is isolated by neutralization in sodium carbonate solution, followed by extraction with ether. Alternatively (67), a mixture of propionaldehyde, malonic acid, and pyridine is heated for 1 hr on a boiling water bath and poured into cold, dilute hydrochloric acid. The mixture is extracted with ether and the ether solution is extracted with 10% sodium hydroxide. The resulting aqueous solution is neutralized and extracted with ether, the ether evaporated, and the hexenoic acids distilled. Lactonization to γ-lactone is carried out as described previously. By treatment of *trans*-δ-hexenoic acid with polyphosphoric acid (68), the γ-lactone can be isolated in 50% yield (bp 103°C/14 mm).

4. δ-CAPROLACTONE (2-METHYLTETRAHYDROPYRAN-6-ONE), $C_6H_{10}O_2$

δ-Caprolactone production is begun when resorcinol and aqueous sodium hydroxide are hydrogenated over Raney nickel for 8 hr at 150°C and 50 atm of (initial) pressure in the presence of a silicone antifoam agent (69). The mixture is filtered, acidified with concentrated hydrochloric acid, boiled, and extracted for 18 hr with ether. Distillation gives 80% of the lactone, bp 110–111°C/11 mm. Reduction of 5-oxohexanoic acid over Raney nickel catalyst at 65–100°C and 90–100 atm in ethanol or acetic acid gives 81% of the lactone, bp 110–112°C/15 mm, η_D (20°C), 1.4402 (70). Reduction of 5-oxohexanoic acid with sodium borohydride in dioxane also yields the lactone in 46% yield, bp 97°C/8 mm η_D (20°C), 1.4424. Reduction of γ-acetobutyric acid (bp 146–148°C/7 mm) with sodium amalgam in sodium hydroxide solution gives 21% yield of δ-caprolactone, bp 107°C/14 mm (71).

5. ε-HYDROXYCAPROIC ACID (6-HYDROXYHEXANOIC ACID), $HO(CH_2)_5COOH$

Reaction of cyclohexanone with hydrogen peroxide in the presence of fluoroboric acid gives good yields of ε-hydroxycaproic acid, a useful precursor in the preparation of ε-caprolactam (72). To a mixture of cyclohexanone and dioxane is added 22.5N fluoroboric acid. Then 30% hydrogen peroxide is added during 1 hr at 60°C, and the mixture is heated at 600°C for an additional 5 hr. After the mixture has been allowed to cool, it is extracted with ether to give 80% of ε-hydroxycaproic acid.

6. ε-CAPROLACTONE, OXEPANE-2-ONE (OXACYCLOHEPTANE-2-ONE), $C_6H_{10}O_2$

ε-Caprolactone (available commercially from the Union Carbide Corp.) is manufactured by the oxidation of cyclohexanone (59). Cyclohexanone is heated to 40°C with stirring, and a 25.5% solution of peracetic acid in ethyl acetate is added dropwise over a period of 4 hr, with occasional cooling, to maintain a temperature of 40°C. After an additional 2.5-hr reaction period, analysis for peracetic acid indicates a 95% conversion. Distillation of the reaction mixture under reduced pressure gives 88% of monomeric ε-caprolactone and 9% of the corresponding polyester.

Synthesis of the lactone from the polyester is accomplished by heating the polyester together with magnesium chloride in a distilling apparatus to 270°C at 1 mm for 30 mins followed by heating to 310°C for 1 hr (73). The yield of the lactone is 86.5%. Similarly, when a cyclic dimer of caprolactone, mp 114°C, and magnesium chloride is heated to 250°C at 50 mm then at 1 mm, 36% of the monomer results.

In a process described in a Belgian patent (74), a mixture of formic acid and 83.5% hydrogen peroxide is allowed to stand 2 hr at 20°C and then is treated with cyclohexanone at 20°C. Extraction with benzene, washing with sodium carbonate, and evaporation follow, and there is an 86% yield of ε-caprolactone.

Stabilization of the ε-caprolactone with triorganic phosphites, alone or in mixtures with alkyl phenols, has been reported (75). For example, 100 g of ε-caprolactone containing less than 0.05% water is treated with 0.1 g of tridecyl phosphite at 25°C under nitrogen. At a concentration of tridecyl phosphite of 1190 ppm, ε-caprolactone shows a color of 10 after 6 hr at 95°C (ASTM D-1209-54) compared with a color of 150, under the same conditions, when no stabilizer is used.

7. γ-ISOCAPROLACTONE (2,2-DIMETHYLTETRAHYDROFURAN-5-ONE), $C_6H_{10}O_2$

Isocaproic acid is heated in alcoholic potassium permanganate at 50–60°C until the color disappears. The manganese dioxide is filtered off and the filtrate plus the washings from the manganese dioxide cake are concentrated and acidified with dilute sulfuric acid. Isolation of the acid followed by distillation at 208°C gives 35–40% yields of γ-isocaprolactone (76).

In another preparation (77), ethyl levulate is treated with methyl magnesium iodide in cold ether. After about 1 hr, the magnesium complex is decomposed with dilute sulfuric acid and the ether layer is separated. After the ether layer has been washed with sodium hydrogen sulfite and dried, the solvent is removed. The yellow oil is hydrolyzed with alcoholic potash, water is added, and the alcohol is removed. The mixture is concentrated and then acidified with hydrochloric acid and heated for 15 min, which converts the hydroxy acid to the lactone. After extraction with ether, washing, and drying, distillation gives the pure lactone, bp 205–207°C.

F. $C_7H_6O_3$

1.   SALICYLIC ACID (o-HYDROXYBENZOIC ACID, 2-HYDROXYBENZOIC ACID), 2-HO—C_6H_4-COOH

Salicylic acid (78), especially in the form of its methyl ester, is widely distributed in nature. The ester has been isolated from wintergreen leaves and sweet birch bark and from the blossoms of *Acacia farnesiana*. Saponification of the methyl ester isolated from natural sources served as the only commercial source of the acid for many years.

In 1859 Kolbe prepared the acid from phenol and carbon dioxide in the presence of sodium. He later modified his synthesis, heating sodium phenoxide and carbon dioxide under pressure at 183–200°C, and in 1874 this became the first suitable synthetic commercial process. Even today a modified version of the Kolbe reaction serves as the chief method for the manufacture of salicyclic acid.

Carbon dioxide reacts with sodium phenoxide to give almost exclusively the *ortho*-carboxylic acid when the temperature is not above 200°C. The original Kolbe process at 183–200°C gives disodium salicylide and phenol as the products. Schmitt modified the process, first combining dry sodium phenoxide and carbon dioxide in the cold and then heating the mixture under pressure at 120–140°C to give a quantitative yield of the monosodium salt of the hydroxy acid. Siefert later found that the important factor in obtaining complete conversion of the reactants is to keep the temperature below 140°C.

The commercial process used today consists of mixing hot aqueous sodium hydroxide (50%) and phenol in a thermocoil autoclave. When the mixture has been heated to 130°C, it is evaporated to dryness, first at atmospheric pressure, then *in vacuo*. By carrying out this process on heated ball mills, a dry, powdered sodium phenoxide is obtained. The charge is then cooled to 100°C and an excess of carbon dioxide introduced to the autoclave at a pressure of 5–6 atm. After the desired amount of carbon dioxide has been absorbed, the mixture is heated to 140°C and held at 140–170°C for several hours. The pressure is released and any regenerated phenol is recovered by vacuum distillation. The carboxylation reaction can also be carried out in a fluidized-solids reactor system with the addition of agents to insure free flow, or the carboxylation reaction can be carried out in the presence of an aliphatic alcohol under anhydrous conditions.

Water is added to the cooled autoclave product, and the hexahydrate of sodium salicylate crystallizes out at temperatures below 20°C. The hexahydrate can then be recrystallized from water to provide U.S.P. sodium salicylate, or it can be dissolved in water and the salicylic acid precipitated out with acid. Further purification of the free acid to U.S.P. salicylic acid is accomplished by sublimation of the dried crude product. The products obtained are U.S.P. salicylic acid and a colored technical grade. Yields of technical acid run around 88%. Sublimation to U.S.P. grade gives a recovery of around 95%. Salicylic acid is packed in tightly closed 100, 200, and 300 lb fiber drums. No special shipping precautions are necessary.

Other methods of synthesis of the hydroxy acid include the microbial oxidation of naphthalene (79), treatment of anhydrous cupric benzoate with air at 175–210°C (75% yield) (80), and oxidation of o-cresol at temperatures above 230°C in the presence of sodium and potassium hydroxide and a copper catalyst (81). None of these processes is practiced commercially today.

In 1970 U.S.P. salicylic acid was available in the form of white needles ($0.55/lb in drums at 1000 lb or more) or as a white, odorless powder ($0.62/lb, fiber drums, 1000 lb or more). The technical grade sells for $0.425/lb. Specifications for the U.S.P. grade are listed below. Loss on drying over silica gel should not exceed 0.5%, and the residue on ignition should be not more than 0.05%. Chloride content should not be more than 140 ppm, sulfate not more than 200 ppm, and heavy metals not more than 20 ppm. The purity of salicylic acid can be assayed by titration with standard alkali. A colorometric test for the acid is based on the intense violet color developed with ferric chloride solution. Free acids, especially hydrochloric acid and acetic acid, interfere with the test.

Salicylic acid is not especially toxic, but systemic poisoning may occur when it is applied to a large area of the skin.

2. m-Hydroxybenzoic Acid (3-Hydroxybenzoic Acid), 3-HO—C_6H_4-COOH

The manufacture of m-hydroxybenzoic acid is carried out by the sulfonation of benzoic acid with oleum, followed by salting out the monosodium salt of m-sulfobenzoic acid and fusing it with alkali at 210–220°C (78). After the fusion product has been dissolved with water, the free acid is precipitated from solution with hydrochloric acid. Recrystallization from water gives the pure acid, mp 201.5–202°C; specific gravity (25°C), 1.473.

Other methods of preparation of the hydroxy acid include: fusion of m-cresol with alkali in the presence of lead oxide, copper oxide, manganese dioxide, or iron oxide (82); fusion of m-hydroxybenzaldehyde with alkali; fusion of m-chlorobenzoic acid with potassium hydroxide; decarboxylation of 4-hydroxyphthalic acid with hydrochloric acid at 180°C; or heating of 2-hydroxyterephthalic acid with hydrochloric acid at 220°C in a closed system (83).

The metahydroxy acid is more stable at temperatures above 200°C than the para- or orthoisomers. Only at 300°C does it begin to decompose. The acid is packed in 100-lb drums and is available commercially.

3. p-HYDROXYBENZOIC ACID (4-HYDROXYBENZOIC ACID), $4\text{-HO}—C_6H_4\text{-COOH}$

Like salicylic acid, p-hydroxybenzoic acid is found in the leaves and blossoms of certain plants. It has been discovered, too, in the urine of horses. Early syntheses of this hydroxy acid include the reaction of carbon tetrachloride with phenol in aqueous alcoholic potassium or sodium hydroxide (Riemer and Tiemann) and heating 4-bromobenzoic acid with barium hydroxide and water in a copper vessel at 160°C.

Today p-hydroxybenzoic acid is manufactured mainly by the Kolbe-Schmitt synthesis, whereby carbon dioxide is passed over dry potassium phenoxide (78). The resulting intermediate undergoes rearrangement at a temperature of about 200°C to give the potassium salt of the parahydroxy acid. Preferably, the reaction mixture is heated under pressures of 20 or more atm and temperatures from 180–250°C. It is also possible to start with the dipotassium salt of salicylic acid, which is first dehydrated *in vacuo* and then heated in a carbon dioxide atmosphere, to effect almost complete rearrangement to the paraisomer. Potassium salicylate itself, when heated to 240°C for 1.5 hr, gives high yields of the paraisomer. The resulting potassium salt of p-hydroxybenzoic acid is converted to the free acid by treatment with a mineral acid. Yields above 95% of the crude acid have been obtained starting with the dipotassium salt of salicylic acid (84).

Purification of the hydroxy acid starting from the potassium salt from the reactor has been reported as follows (85). The reactor liquor is first diluted and acidified to a pH of 5.8 by the addition of concentrated hydrochloric acid. Sodium bisulfite (0.8%), zinc dust (1.2%), and activated charcoal (2%) are added to the filtrate after filtration. This mixture is stirred for 0.5 hr and then filtered. The pH of the filtrate is brought to 3.4 by the addition of concentrated hydrochloric acid. Cooling the solution to 5–10°C yields fine, off-white crystals, which are collected and dried at 55–60°C. This yields an acid of high quality, suitable for the preparation of the esters of p-hydroxybenzoic acid. The hydroxy acid can be purified further by recrystallization from water-free acetone, ethyl acetate, or carbon tetrachloride, yielding the acid in the form of water-free monoclinic prisms, mp 213–214°C; specific gravity (20°C), 1.497. In the synthesis of p-hydroxybenzoic acid described in "Organic Syntheses" (86), yields of 70–80% of the acid, mp 211–212°C, are obtained after recrystallization from water.

In another method of synthesis, the alkali salt of p-cresol is converted at temperatures of 260–270°C over metallic oxides to yield the hydroxy acid.

The monomer can also be obtained starting with potassium phenoxide, treating this with carbon monoxide, potassium carbonate, and potassium formate at temperatures above 210°C. This process gives almost a quantitative yield of the potassium salt of p-hydroxybenzoic acid (87). The role of the potassium formate is not clearly understood, but the high yields depend on its use. A 93% yield of crude p-hydroxybenzoic acid has been obtained from sodium phenoxide (88) when the sodium salt is used instead of the corresponding potassium salt.

Specifications for p-hydroxybenzoic acid suitable for the preparation of poly(p-oxybenzoyl) are as follows: the ash content should be 0.05% or less, the moisture content 0.4% or less, mp 214–215°C, color 7 APHA, and salicylic acid 0.02% or less.

The hydroxy acid is nontoxic; it has a diuretic effect when taken internally. In the bodies of animals, the hydroxy acid is combined and found in the urine, converted partly to 4-hydroxyhippuric acid and partly into 4-carboxyphenylsulfuric acid. It does not act as an antiseptic, although many of its esters do. Under the influence of pancreatic juices, p-hydroxybenzoic acid is converted into phenol. The hydroxy acid is available in 50 and 100 lb drums and in 50-kg bags.

G. $C_7H_{12}O_3$

1.   1-Hydroxycyclohexanecarboxylic Acid,

$$\begin{array}{c} CH_2CH_2C(OH)COOH \\ | \quad\quad | \\ CH_2CH_2CH_2 \end{array}$$

Cyclohexanone is converted to its cyanohydrin, which, in turn, is hydrolyzed with hydrochloric acid and the solution concentrated at low temperatures. 1-hydroxycyclohexanecarboxylic acid is extracted with ether and recrystallized from benzene (89). In a similar preparation (90), cyclohexanone cyanohydrin is reacted for 1 hr with a 10% hydrochloric acid solution to give 67% of the acid, mp 106–107°C. Alternatively, cyclohexanecarboxylic acid and its ethyl ester can be converted to the corresponding 1-hydroxy derivative (bp 90–100°C/0.5 mm, mp 106–107°C) by hydroxylation (oxidation) with peracetic acid in ethyl acetate (91).

2.   Hexahydrosalicylic Acid (2-Hydroxycyclohexanecarboxylic Acid),

$$\begin{array}{c} CH_2CH(OH)CHCOOH \\ | \quad\quad\quad / \\ CH_2CH_2CH_2 \end{array}$$

Ethyl hexahydrosalicylate is prepared by the high-pressure hydrogenation of an ethanol solution of methyl salicylate in the presence of a Raney nickel catalyst (92). The ester boils at 110–111°C at 13 mm. Free hexahydrosalicylic acid is obtained by shaking the ester at room temperature with a slight excess of 10% aqueous sodium hydroxide for 5 min. The resulting solution is allowed to stand for 1.5 hr, concentrated under vacuum, washed with ether, acidified with 10% ammonium sulfate, and thoroughly extracted with ether. Removal of the ether gives a quantitative yield of a mixture of cis–trans isomers. Repeated recrystallization from dry ether and ethyl acetate gives an acid that melts at 76–78°C and an acid that melts at 109–110°C.

3. 3-Hydroxycyclohexanecarboxylic Acid,

$$\text{HOCHCH}_2\text{CHCOOH}$$
$$\overset{|}{\text{CH}_2\text{CH}_2}\overset{\diagdown}{\text{CH}_2}$$

Benzoic acid is converted to m-hydroxybenzoic acid in a 90% yield by sulfonation, followed by fusion with potassium hydroxide (93). Esterification of the acid with ethanol and sulfuric acid gives 89% yield of the ethyl ester, (mp 72°C, from petroleum ether). The ethyl ester is then stirred in ethanol in the presence of Raney nickel with hydrogen at 140°C/130 atm. Absorption ceases after 30 hr, and the filtrate solution is concentrated and fractionally distilled. Ethyl hexahydrobenzoate is collected at 84–90°C/16 mm and the product, ethyl 3-hydroxyhexahydrobenzoate is collected at a boiling point of 117–122°C/1 mm in a 70% yield. The cis-3-hydroxyhexahydrobenzoic acid is obtained by alkaline hydrolysis, acidification, extraction with ether, and repeated recrystallization from ethyl acetate (mp 131–132°C).

4. 4-Hydroxycyclohexanecarboxylic Acid,

$$\text{CH}_2\text{CH}_2\text{CHCOOH}$$
$$\overset{|}{\text{HOCHCH}_2}\overset{|}{\text{CH}_2}$$

The reduction of 4-ketocyclohexanecarboxylic acid and ethyl p-hydroxybenzoate yields mixtures of the cis and trans forms of 4-hydroxycyclohexanecarboxylic acid, whereas hydrogenation of p-hydroxybenzoic acid gives mainly the cis form of the acid.

5. *cis-trans*-4-Hydroxycyclohexanecarboxylic Acid

Reduction of 4-ketocyclohexanecarboxylic acid was first carried out by Perkin (94), who obtained *cis-trans*-4-hydroxycyclohexanecarboxylic acid, mp 121°C. More recently Owens and Robbins (95) hydrogenated ethyl p-hydroxybenzoate over Raney nickel for 20 hr at 150°C/85 atm to obtain the ethyl ester. After saponification, 80% of the crude acid is obtained. Recrystallization gives a mixture of the isomers, mp 121°C.

Campbell and Hunt (96) reduced ethyl p-hydroxybenzoate by shaking a dioxane solution in the presence of 2% palladized strontium carbonate catalyst and hydrogen, initially at 150°C/118 atm and finally at 157°C/90 atm.

The yield of the crude ethyl ester is quantitative. Purified free acid melts at 120–121°C. Hydrogenation of methyl *p*-hydroxybenzoate or 4-keto-cyclohexanecarboxylic acid at 150°/130 atm, using a palladium-on-strontium carbonate catalyst, gives the same product, mp 120–121°C.

6. PURE *trans*-4-HYDROXYCYCLOHEXANECARBOXYLIC ACID, MP 148°C

The mixture of *cis-trans* isomers of 4-hydroxycyclohexanecarboxylic acid, mp 120–121°C (probably in equal portions) is refluxed for 4 hr in a mixture of acetic acid, acetic anhydride, and acetyl chloride. Under reduced pressure, the excess of the reagents is distilled off and the residual oil recrystallized several times with water to yield an acetoxy derivative, mp 139°C. The combined aqueous mother-liquors from the preparation of the acetoxy compound are saturated with ammonium sulfate and extracted with ether. After evaporation of the ether, the residue is distilled at 146–152°C/0.4 mm. The crude product has a melting point of 131°C; after recrystallization from chloroform (three times) and ethyl acetate, however, the pure *trans*-4-hydroxycyclohexanecarboxylic acid melts at 148°C (97).

7. PURE *cis*-4-HYDROXYCYCLOHEXANECARBOXYLIC ACID, MP 152°C

Hydrogenation of *p*-hydroxybenzoic acid in ethanol over Adams platinum catalyst for 28 hr followed by crystallization from acetone gives 53% yield of pure *cis*-4-hydroxycyclohexanecarboxylic acid, mp 152°C. A slight amount of the isomeric mixture, mp 120°C, is also obtained (98).

8. LACTONE OF 4-HYDROXYCYCLOHEXANECARBOXYLIC ACID

Distillation of the cis acid at atmospheric pressure and collection of the distillate boiling at 250–270°C gives a solid that, unlike the acid, is soluble in ether. The lactone of 4-hydroxycyclohexanecarboxylic acid separates from the solution in time in the form of leaflets, mp 108–112°C Saponification of the lactone with potassium hydroxide solution followed by acidification gives back the cis acid, mp 152–153°C.

H. $C_7H_{14}O_3$

1. α-HYDROXYENANTHIC ACID (2-HYDROXYHEPTANOIC ACID), $CH_3(CH_2)_4CH(OH)COOH$

α-Hydroxyenanthic acid has been prepared by the hydrolysis of 2,2-dibromoheptanol with lead hydroxide (99) and by hydrolysis of the corresponding α-amino acid with silver nitrate and hydrochloric acid.

2. β-Hydroxyenanthic Acid (3-Hydroxyheptanoic Acid), $CH_3(CH_2)_3CH(OH)CH_2COOH$

Coupling of *n*-valeraldehyde and ethyl bromoacetic acid via the Reformatsky reaction gives pure β-hydroxyenanthic acid, mp 40–41°C in low yields (53).

3. γ-Enantholactone (4-Heptanolactone, 5-*n*-Propyltetrahydrofuran-2-one),

$$n\text{-}C_3H_7CHCH_2CH_2C{=}O$$
$$\underset{\text{O}}{\rule{1.2cm}{0.4pt}}$$

Hydrogenation of β-furylpropionic acid at 250–265°C over a carbon catalyst gives γ-enantholactone, bp 112–113°C/14 mm η_D (20°C), 1.14385, in 60% yield (100). Some hexanoic acid is also isolated from the reaction. Alternatively, the lactone can be prepared from *trans*-heptenoic acid (68).

4. ε-Enantholactone, ε-Methyl-ε-caprolactone, 6-Heptanolactone, 2-Methyloxacycloheptene-7-one),

$$CH_3CH(CH_2)_4C{=}O$$
$$\underset{\text{O}}{\rule{1.2cm}{0.4pt}}$$

Oxidation of 2-methylcyclohexanone with peracetic acid at 40°C for 8.5 hr gives 92% of ε-enantholactone, bp 94°C/5 mm, η_D (20°C), 1.4588 (59).

5. ζ-Hydroxyenanthic Acid (7-Hydroxyheptanoic Acid), $HO(CH_2)_6COOH$

The monoethyl ester of pimelic acid (heptanedioic acid) is refluxed with thionyl chloride until no more hydrogen chloride is evolved, and the resulting acid chloride distilled under reduced pressure. The acid chloride is reduced with sodium borohydride in dioxane to yield crude ethyl 7-hydroxyheptanoate. Hydrolysis of this ester with 10% sodium hydroxide, followed by ether extraction and neutralization of the aqueous layer, gives crude ζ-hydroxyenanthic acid in 55% yield (11).

The acid can also be obtained in 49% yield by the hydrolysis of 7-chloroheptanoic acid (101) and from the reduction of poly(3-ethoxy-7-oxyheptanoyl) (102).

6. ζ-Enantholactone (Oxacyclooctane-2-one),

$$CH_2(CH_2)_5C{=}O$$
$$\underset{\text{O}}{\rule{1.2cm}{0.4pt}}$$

Cycloheptanone is oxidized with trifluoroperacetic acid in methylene chloride in the presence of disodium hydrogen phosphate to yield 68% of

7-heptanolactone, bp 80–81°C/11 mm (103). Cycloheptanone can also be oxidized with peracetic acid to yield 29% of ζ-enantholactone, bp 70°C/5 mm; η_D (20°C), 1.4688 (57). The major by-product in the latter method is 33% pimelic acid, mp 102°C.

7. A Dimeric Lactone of 7-Hydroxyheptanoic Acid,

$$
\begin{array}{c}
\text{O}-(\text{CH}_2)_6-\text{C}=\text{O} \\
| \qquad\qquad | \\
\text{O}=\text{C}-(\text{CH}_2)_6-\text{O}
\end{array}
$$

The polyester formed from heating 7-hydroxyheptanoic acid *in vacuo* is heated at 270–300°C *in vacuo* with magnesium chloride as the catalyst to give 16% of 7-heptanolactone and 65% of the dilactone, bp 138–140°C/1 mm, mp 46°C (from methanol). The dilactone has a musklike odor (104).

I. $C_8H_8O_3$

1. Mandelic Acid (2-Phenyl-2-hydroxyethanoic Acid), $C_6H_5CH(OH)COOH$

Originally produced by heating an extract of bitter almonds with hydrochloric acid, mandelic acid is produced today by the hydrolysis of benzaldehyde cyanohydrin. It is used in medicine for the treatment of urinary infections and is available (NF grade) in drums for use in synthetic organic chemistry. Mandelic acid is sold by a number of companies. The synthesis of mandelic acid from benzaldehyde (52%) using the bisulfite addition product as an intermediate to form mandelonitrile is described in "Organic Syntheses" (105).

Mandelic acid can also be synthesized in high yields (87%) from acetophenone (106). Acetophenone is first chlorinated to yield 90–97% of dichloroacetophenone, which, in turn, is hydrolyzed with base followed by acidification to yield the hydroxy acid. Other methods of preparation include: the hydrolysis of α-bromophenylacetic acid; synthesis from benzaldehyde, chloroform, and potassium carbonate; and the hydrolysis of ethyl mandelate (prepared by the catalytic reduction of phenylglyoxalate).

2. 2-Hydroxyphenylacetic Acid, $2\text{-HOC}_6H_4CH_2COOH$

To prepare 2-hydroxyphenylacetic acid, the methyl ether of salicylaldehyde is first melted and treated with a concentrated sodium bisulfite solution. The mixture is shaken vigorously for some time until the entire contents of the flask solidifies to the bisulfite addition compound. After filtering and washing, the material is placed in a vessel together with a saturated potassium cyanide solution, where the addition compound quickly goes into solution, forming the nitrile of *o*-methoxymandelic acid, which separates as an oil. The oil is solidified by treatment with ether and recrystallization from benzene. In one operation, the purified material is converted into o-hydroxyphenylacetic acid by boiling in an eightfold quantity of hydroiodic acid (specific gravity = 1.96). The hydroxy acid is recrystallized from chloroform, mp 144–145°C (107).

3. LACTONE (2-COUMARANONE, 2,3-DIHYDROBENZOFURAN-2-ONE)

The lactone of 2-hydroxyphenylacetic acid is prepared by first drying an ether solution of the acid over calcium chloride. Evaporation of the ether and distillation of the residue at 246°C/726 mm gives a colorless oil, which solidifies on cooling. The melting point of the lactone is 29°C.

4. 3-HYDROXYPHENYLACETIC ACID, $3\text{-HOC}_6\text{H}_4\text{CH}_2\text{COOH}$

Treatment of 3-methoxybenzaldehyde with potassium cyanide gives a yellow oil which, when boiled in hydroiodic acid, gives 3-hydroxyphenylacetic acid, mp 129°C (107).

5. 4-HYDROXYPHENYLACETIC ACID, $4\text{-HOC}_6\text{H}_4\text{CH}_2\text{COOH}$

First p-cresol is converted in 93% yield to p-tolyl methanesulfonate by use of methanesulfuryl chloride in pyridine solution. The p-tolyl sulfonate is oxidized to p-methanesulfonoxybenzoic acid by potassium permanganate in the cold (30% yield), by potassium permanganate in dilute sulfuric acid (35% yield), or—best—by chromium trioxide in 88% sulfuric acid (44% yield). The crude sulfonoxybenzoic acid melts at 214–221°C. After conversion of this acid to its ethyl ester, refluxing in 10% sodium hydroxide followed by neutralization, an 82% yield of the 4-hydroxyphenylacetic acid, mp 147–148°C, is obtained (108).

J. $C_8H_{16}O_3$

1. α-HYDROXYCAPRYLIC ACID (2-HYDROXYOCTANOIC ACID), $CH_3(CH_2)_5CH(OH)COOH$

The sodium bisulfite addition compound of enanthaldehyde is converted with potassium cyanide to the hydroxycaprylonitrile, which upon hydrolysis gives α-hydroxycaprylic acid (109). Purification of the acid is accomplished by recrystallization from dry benzene, followed by dissolving in dry ether and evaporating the solvent, mp 67°C.

2. β-HYDROXYCAPRYLIC ACID (3-HYDROXYOCTANOIC ACID), $CH_3(CH_2)_4CHOHCH_2COOH$

To produce β-hydroxycaprylic acid, ethyl bromoacetate and n-hexanal are treated with zinc activated with iodine or copper. The complex is hydrolyzed with base, and the resulting mixture is acidified. Recrystallization of the hydroxy acid from petroleum ether results in a 44% yield, mp 38–38.5°C (53).

3. γ-CAPRYLOLACTONE (2-n-BUTYLTETRAHYDROFURAN-5-ONE),

$$n\text{-}C_4H_9CHCH_2CH_2C{=}O$$
$$\underset{\displaystyle \quad\quad O\quad\quad}{\rule{0pt}{0pt}}$$

The first step in obtaining γ-capryolactone is the reduction of β-n-valeryl-propionic acid in solution with about 10 times the theoretical quantity of sodium amalgam (4%). After neutralization of the mixture and dilution with water, the mixture is extracted with ether and the product distilled. The yield of the lactone, bp 132–133°C/20 mm is 40% (110).

4. ε-HYDROXYCAPRYLIC ACID (6-HYDROXYOCTANOIC ACID), $CH_3CH_2CH(OH)(CH_2)_4COOH$

Reduction of propionyl-n-valeric acid with sodium amalgam, zinc-potash or zinc-potash-platinum chloride gives ε-hydroxycaprylic acid (110). The acid is isolated via its semicarbizide and is obtained as an odorless, colorless liquid, which forms the lactone at temperatures above 100°C.

5. LACTONE, $C_8H_{14}O_2$

Distillation of ε-hydroxycaprylic acid at 148–150°C/10 mm yields 50% of the lactone in the form of a colorless liquid, boiling without decomposition at 114–115°C/10 mm (110).

6. η-HYDROXYCAPRYLIC ACID (8-HYDROXYOCTANOIC ACID), $HO(CH_2)_7COOH$

Suberic acid is converted to its monoethyl ester and treated with thionyl chloride to form the acid chloride. Reduction of the acid chloride with sodium borohydride gives η-hydroxycaprylic acid in 50% yield (11).

7. η-CAPRYLOLACTONE (OXACYCLONONANE-2-ONE),

$$CH_2(CH_2)_6C{=}O$$
$$\underset{\displaystyle \quad\quad O\quad\quad}{\rule{0pt}{0pt}}$$

Whereas oxidation of cyclooctanone with peracetic acid gives only 6% of η-caprylolactone and 59% suberic acid (57), cyclooctanone gives 72% of the lactone (bp 72–73°C/11 mm) when oxidized with trifluoroperacetic acid in methylene chloride in the presence of disodium hydrogen phosphate (103).

K. $C_9H_{10}O_3$

1. ATROLACTIC ACID (2-HYDROXY-2-PHENYLPROPIONIC ACID),

$$CH_3C(OH)COOH$$
$$|$$
$$C_6H_5$$

The synthesis of atrolactic acid is accomplished by hydrolysis of the cyanohydrin of acetophenone (111). Crude yields of the hydroxy acid, mp 88–90°C, run around 50%. Recrystallization from boiling water gives the colorless crystals of atrolactic acid hemihydrate, mp 88–90°C (softening at 75°C). The anhydrous pure acid can be obtained by drying the hemihydrate at 55°C/1–2 mm, mp 94.5–95°C.

Other methods of preparation include: hydrolysis of α-halohydrotropic acids (112), hydrolysis of dibromopropiophenone (113), and hydrolysis of the reaction product of ethyl phenylglyoxylate with methyl magnesium iodide (114).

2. β(p-HYDROXYPHENYL)PROPIONIC ACID [(4-HYDROXYPHENYL)-3-PROPIONIC ACID], $HOC_6H_4CH_2CH_2COOH$

Anisaldehyde is condensed with ethyl acetate with sodium to give 82% of ethyl p-methoxycinnamate, which is hydrogenated over Raney nickel to ethyl β(p-methoxyphenyl)propionate. This ester is treated with hydroiodic acid, and β(p-hydroxyphenyl)propionic acid, which melts at 122°C, is then recrystallized from water. The hydroxy acid is also prepared by condensing anisaldehyde with dimethyl malonate, followed by hydrogenation of the resulting alkene group, hydrolysis, and decarboxylation (115).

L. $C_9H_{18}O_3$

1. ε-ISOPELARGONOLACTONE, ε-ISOPROPYL-ε-CAPROLACTONE (2-ISOPROPYLOXACYCLOHEPTANE-7-ONE),

$$(CH_3)_2CHCH(CH_2)_4C=O$$

Treatment of 2-isopropylcyclohexanone with peracetic acid at 50°C for 9 hr yields 84.5% of ε-isopelargonolactone, bp 107°C/5 mm; η_D (20°C), 1.4604 (57).

2. ω-HYDROXYPELARGONIC ACID (9-HYDROXYNONANOIC ACID), $HO(CH_2)_8COOH$

Reduction of 8-carbethoxycapryloyl chloride (from the monoethyl ester of azelaic acid and thionyl chloride) with sodium borohydride followed by hydrolysis gives 71% of ω-hydroxypelargonic acid (11). This hydroxy acid can also be prepared starting with the methyl ester of oleic acid (116). Ozonization of the oleate gives methyl aldehydocaprylate, which is reduced to methyl 9-hydroxyperlargonate in yields of 80–90%, bp 137–139°C/3 mm; η_D (20°C), 1.4438. Hydrolysis gives the free acid which, when recrystallized from ethyl acetate, has a melting point of 53–54°C.

M. $C_{10}H_{20}O_3$

1.   α-HYDROXYCAPRIC ACID (2-HYDROXYDECANOIC ACID), $CH_3(CH_2)_7CH(OH)COOH$

Capric acid is treated with bromine and phosphorus trichloride to yield the α-bromo acid, which is a liquid at ordinary temperatures. The bromo acid is transformed into α-hydroxycapric acid by dissolving in dilute potassium hydroxide solution and boiling. The hydroxy acid is purified by a series of recrystallizations (ether, followed by chloroform). The acid is also recovered from the mother-liquors via the potassium salt. From 418 g of capric acid, 335 g of the α-hydroxy acid is obtained in the pure form (117).

2.   β-HYDROXYCAPRIC ACID (3-HYDROXYDECANOIC ACID), $CH_3(CH_2)_6CH(OH)CH_2COOH$

β-Hydroxycapric acid can be prepared by reaction of ethyl bromoacetate with octylaldehyde by the Reformatsky reaction (53,118).

3.   γ-CAPRILACTONE (2-n-HEXYLTETRAHYDROFURAN-5-ONE),

$$n\text{-}C_6H_{13}CH(CH_2)_2CO$$
$$\underset{\quad\quad O \quad\quad}{\lfloor\text{———}\rfloor}$$

Ethyl enanthylsuccinate is hydrolyzed and decarboxylated to γ-oxocapric acid (119). Hydrogenation of the keto acid over Raney nickel in ethanol at 135°C/1000 psi gives 83% of γ-caprilactone, bp 109–110°C/2.5 mm; η_D (25°C), 1.4470. The overall yield from the succinate is 65%.

4.   ζ-METHYL-ε-PELARGONOLACTONE (2-sec-BUTYLOXACYCLOHEPTANE-7-ONE,

$$CH_3CH_2CH(CH_3)CH(CH_2)_4C=O$$
$$\underset{\quad\quad O \quad\quad}{\lfloor\text{———}\rfloor}$$

By heating 2-sec-butylcyclohexanone with peracetic acid at 50°C for 13 hr, a 91.6% yield of ζ-methyl-ε-pelargonolactone is produced, bp 113°C/4 mm; η_D (20°C), 1.4625 (57).

5. ω-Hydroxycapric Acid (10-Hydroxydecanoic Acid), $HO(CH_2)_9COOH$

The monoethyl ester of sebacic acid is treated with thionyl chloride and the product reduced with sodium borohydride according to the method of Dale (11). After hydrolysis, a 76% yield of the crude ester is obtained. The method of Lycan and Adams (116)—starting with methyl undecenate, followed by ozonization to the aldehyde, reduction, and hydrolysis—gives ω-hydroxycapric acid, bp 145–147°C/3 mm; η_D (20°C), 1.4471. [See also the similar method of Benton and Kiess (120), who obtained 12 g of the hydroxy acid, mp 75–76°C, from 20 g of undecenoic acid.]

Synthesis of the acid from a natural source first involves the isolation of ricinoleic acid (12-hydroxy-9-octadecenoic acid) from castor oil, which contains 80–85% of the unsaturated acid. Pyrolysis of ricinoleic acid gives n-hexylaldehyde and undecenoic acid (121), which can be converted by the previous methods to the ω-hydroxy acid. Alkali scission of ricinoleic acid at 200°C gives 10-hydroxydecenoic acid and methyl n-hexylketone; alkali scission at 240°C gives the hydroxy acid, hydrogen, and methyl-n-hexylcarbinol (121).

N. $C_{11}H_{22}O_3$—ω-Hydroxyundecanoic Acid (11-Hydroxyundecanoic Acid), $HO(CH_2)_{10}COOH$

Treatment of 10-bromodecanoic acid with potassium cyanide gives good yields of the corresponding nitrile, bp 186–187°C/13 mm; this in turn is hydrolyzed to give ω-hydroxyundecanoic acid (122). Alternatively, dimethyl malonate and 9-bromononane-1-ol are condensed together (in the presence of sodium) and the resulting ester saponified. The resulting salt is acidified, acetylated, and decomposed at 150°C. A yield of 50 g of 11-acetoxyundecanoic acid is obtained, bp 184–185°C/8 mm, mp 34°C, from 45 g of dimethyl malonate. Saponification of the acetoxy acid gives 40 g of the 11-hydroxy acid. Employing the method of Lycan and Adams (116), methyl dodecenoate is converted by ozonization to the aldehyde, which is reduced to the methyl ester of 11-hydroxyundecanoic acid, bp 156–157°C/3 mm. The free acid, mp 65.5–66°C, is obtained by hydrolysis of the ester.

O. $C_{12}H_{24}O_3$—Sabinic Acid, ω-Hydroxylauric Acid (12-Hydroxydodecanoic Acid), $HO(CH_2)_{11}COOH$

Dimethyl malonate is condensed with 10-bromodecanoic acid (122) and the resulting ester saponified. The sodium salt is acidified and the acid acetylated and decarboxylated to give the ω-acetoxy acid. Acetylsabinic acid, bp 174–176°C/0.5 mm, mp 45°C, is also obtained by the oxidation of the ester, Δ^{12}-tridecenyl acetate. The free sabinic acid is obtained by hydrolysis of the acetyl derivative.

The methyl ester of sabinic acid is obtained from methyl tridecenoate, after ozonization and reduction of the resulting aldehyde. The methyl ester, bp 164–166°C/3 mm, mp 34–36°C, is hydrolyzed with 10% aqueous potassium hydroxide solution. Recrystallization of the potassium salt from absolute ethanol, followed by treatment of the ethanol-free aqueous solution of the purified salt with 1N hydrochloric acid liberates the free acid. Extraction with ether, evaporation, and recrystallization from ethyl acetate gives pure ω-hydroxylauric acid, mp 83–84°C (116). Undecanol (32 g, from the reduction of ethyl undecenoate) is acetylated and brominated to give 11-acetoxy-undecyl bromide (46 g), η_D (30°C), 1.4647 (123). Refluxing the bromide with potassium cyanide in 75% ethanol followed by hydrolysis with alcoholic potassium hydroxide gives sabinic acid (35 g), mp 83–83.5°C (benzene). The hydroxy acid can also be isolated from the plant waxes of *Chamaecyporis obtusa* (124).

P. $C_{13}H_{26}O_3$—13-Hydroxytridecanoic Acid, $HO(CH_2)_{12}COOH$

Electrolysis of ω-acetoxydecanoic acid and methyl hydrogen succinate containing sodium for 4 hr at 1.1 A, followed by distillation *in vacuo*, saponification, and treatment of the acid fraction with benzene, gives 13-hydroxytridecanoic acid, mp 78–79°C (ethyl acetate) (123). The potassium salt of the monoethyl ester of tridecanedicarboxylic acid is reduced with sodium-alcohol and the product acidified and acetylated to give 13-acet-oxytridecanoic acid, bp 202–203°C/1 mm, mp 49°C. Saponification gives the free acid, mp 79–79.5°C (122). In another approach, methyl erucate is converted to methyl 11-hydroxytridecenoate, bp 170–173°C/3 mm, mp 41°C, by ozonization, followed by reduction of the aldehyde with sodium borohydride. Saponification gives the free acid, mp 77–78°C (116).

R. $C_{14}H_{28}O_3$—14-Hydroxytetradecanoic Acid, $HO(CH_2)_{13}COOH$

Conversion of 13-bromotridecanol to the nitrile (122), bp 205°C/4 mm, mp 53°C, followed by hydrolysis, gives 14-hydroxytetradecanoic acid (10 g of the bromo compound gives 7 g of the nitrile). When hydrolysis occurs after the reduction of 1,12-dicarbomethoxydodecane, followed by saponification and conversion to 14-acetoxytetradecanoic acid (mp 54°C), the hydroxy acid is produced.

S. $C_{15}H_{28}O_3$ (also $C_{16}H_{30}O_3$ and $C_{17}H_{32}O_3$)—
7-Hydroxy-7-Cyclohexylnonanoic Acid, $C_6H_{11}CH(OH)(CH_2)_7COOH$

The synthesis of ω-cyclohexylhydroxy acids is described by Hiers and Adams (125). Methyl-n-aldehydooctanoate and cyclohexyl magnesium bromide

gives 23% of methyl-7-cyclohexylnonanoate, bp 186–192°C/5 mm. Saponification gives 7-hydroxy-7-cyclohexylnonanoic acid, mp 78–79°C. In a similar
manner, 8-cyclohexyl-8-hydroxydecanoic acid and 9-cyclohexyl-9-hydroxy-
undecanoic acid are prepared.

T. $C_{15}H_{30}O_3$

1. CONVOLVULINOLIC ACID (11-HYDROXYPENTADECANOIC ACID),
$CH_3(CH_2)_3CH(OH)(CH_2)_9COOH$

Methyl 11-hydroxypentadecanoate is prepared by condensing methyl
10-aldehydododecanoate with n-butyl magnesium bromide (126), mp 29–32°C,
bp 166°C/2 mm. Hydrolysis gives the free convolvulinolic acid which, when
recrystallized four times from ethyl acetate, melts at 63.4–64°C.

2. 15-HYDROXYPENTADECANOIC ACID, $HO(CH_2)_{14}COOH$

Refluxing 10,16-diphenylhexadec-15-en-1-ol with acetic anhydride and
pyridine gives the corresponding acetate, which is oxidized in acetone–
acetic acid by heating with potassium permanganate. The acetone is removed,
the residue is treated with an aqueous suspension of sulfur dioxide, and the
product is extracted into ether. Saponification with potassium hydroxide is
carried out and the alcohol is removed. Benzophenone is extracted from the
residue with petroleum ether and the remaining solution is acidified. Crude
15-hydroxypentadecanoic acid (35%) is crystallized from benzene, mp
84–85°C (127).

A yield of 12% of the acid is obtained by reduction of 1,13-dicarbo-
methoxytridecane (122).

3. EXALTOLIDE

$CH_2(CH_2)_{13}C{=}O$
$\underline{\qquad O \qquad}$

The 15-hydroxy acid is polymerized to the polyester by heating for 3 hr
at 200–230°C, followed by heating at 240°C/1 mm for 1.5 hr. Magnesium
oxide is added, and the polyester is depolymerized at 260–270°C/0.15 mm
(127,128). The distillate—exaltolide—is purified by use of an alumina
column, mp 30.5–31°C.

U. $C_{16}H_{32}O_3$

1. JALAPINOLIC ACID (11-HYDROXYHEXADECANOIC ACID),
$CH_3(CH_2)_4CH(OH)(CH_2)_9COOH$

Ozonization of methyl 11,12-dodecenoate gives a 60% yield of methyl
10-aldehydododecanoate (126). When the aldo ester is treated with n-amyl

magnesium bromide, 33% of the hydroxy ester is obtained, mp 40–41°C (petroleum ether). Hydrolysis of the ester produces jalapinolic acid, mp 68–69°C (ethyl acetate).

2. JUNIPERIC ACID (16-HYDROXYHEXADECANOIC ACID), $HO(CH_2)_{15}COOH$

The first step in preparing juniperic acid is to heat 1,14-dichlorotetradecane (29 g) with sodium cyanide (4.5 g) in 2-ethoxyethanol at 105°C; after treatment and extraction with chloroform a crude product, bp 198–200°C is formed. After refluxing with a solution of hydrogen chloride in acetic acid, 15-chloropentadecanoic acid (8.1 g) mp 62–63°C is obtained. About 50% of the starting material is also recovered. The chloro acid is hydrolyzed with potassium hydroxide to yield 15-hydroxypentadecanoic acid, mp 82–83°C. When 15-hydroxy-7-oxopentadecanoic acid is heated to 195°C with nitrogen tetroxide–water in potassium hydroxide and diethylene glycol, 15-hydroxypentadecanoic acid is formed, mp 82–83°C (129).

Treatment of the monoethyl ester of hexadecanedicarboxylic acid with thionyl chloride in a sealed tube and subsequently with ethyl mercaptan in pyridine gives $EtOC(CH_2)_{14}CO_2Me$, bp 192–209°C/5 mm. Reduction with Raney nickel gives the methyl ester of juniperic acid, mp 51–54°C (130).

The hydroxy acid can be prepared by converting ω-bromopalmetic acid to 1,16-hexadecanolide (bp 124°C/0.5 mm, mp 18°C) which is saponified to the hydroxy acid in a 78% yield, mp 94–95°C (131). Aleuritic acid is converted via ethyl 16-bromo-9-hexadecenoate to ethyl 16-acetoxy-9-hexadecenoate. Alkaline hydrolysis gives 16-hydroxy-9-hexadecenoic acid, which is then converted to the polyester (127); this in turn is decomposed thermally to Δ^9-isoambrettolide, bp 135–136°C ν (liquid film) 1729 cm^{-1} (macrocyclic lactone). Hydrogenation and hydrolysis of ethyl 16-acetoxy-9-hexadecenoate directly produces 16-hydroxydecanoic acid, mp 95°C, which lactonizes to give dihydroambrettolide (cyclohexadecanolide), bp 118–119°C (136).

Cyclization of ω-bromo acids with potassium carbonate yielding the macrolactones of ω-hydroxy acids with 9-16 methylene groups is described (132) as a general procedure.

V. $C_{17}H_{34}O_3$ (also ω-Hydroxy Acids: $C_{18}H_{36}O_3$ to $C_{21}H_{42}O_3$)

Reduction of the dimethyl ester of 1,17-heptadioic acid gives the acetyl derivative (mp 67–68°C) in 11% yield according to the method of Chuit and Hausser (122). Saponification gives the free ω acid, mp 87–88°C.

Similarly, the C_{18} acid, mp 96–97°C; the C_{19} acid (10% yield), mp 91–92°C; the C_{20} acid (8% yield), mp 97–98°C; and the C_{21} acid (16% yield), mp 92–93°C, are formed.

IV. PROPERTIES OF MONOMERS

In Tables 1–7 the properties of the hydroxy acids and lactones are presented. The abbreviations used in the tables are listed in the following key:

Key to Tables 1–7

mp	Melting points °C.
Sub	Sublimes.
bp	Boiling point, °C, atmospheric pressure when no pressure given; otherwise, °C/mm.
dec	Decomposition.
d_4	Density, water at 4°C, temperature in parentheses.
η_D	Index of refraction, temperature in parentheses.
K	Ionization constant, temperature in parentheses.
Sol	Solubility; ben—(benzene); chl—(chloroform); δ—(partial solubility); Me₂CO—(acetone); h—(soluble hot).
$[\alpha]_D$	Specific rotation, temperature and solvent in parentheses.
dl	Racemic form.
l, d	Optically active forms.
pet eth	Petroleum ether.
(hygr)	Hygroscopic.

V. POLYMERIZATION, POLYMERS, AND APPLICATIONS

A. Condensation of Aliphatic Hydroxy Acids

The polymeric products of lactic acid were reported as long ago as 1833 by Gay-Lussac (139). It was also found that glycolic acid yielded polymeric materials. Polymerization of the naturally occurring ricinoleic acid was reported by Meyer in 1897. Ricinoleic acid was converted at room temperature to a mixture of polyesters, from which the original acid was obtained after saponification.

Carothers et al. (33) carried out the first thorough study on polymers derived from the aliphatic acids. The α-hydroxy acids, such as lactic acid, condense to give both the dimeric six-membered cyclic ester, called a lactide, and linear polymer (140). The dimers of the α-hydroxy acids have been found to undergo ready conversion to the linear polyester:

TABLE 1

Hydroxy Acids: $C_2H_4O_3$–$C_4H_8O_3$

Hydroxy acid	mp	bp	d_4	η_D	$[\alpha]_D$	K	Sol	Reference
Glycolic	α–63 β–79	dec	—	—	—	1.4×10^{-4} (25)	H$_2$O, Et$_2$O, EtOH	133,134
Lactic (α-Propanoic) *dl*	18	122/15	1.249 (20)	1.4414 (20)	—	3.1×10^{-4} (25)	H$_2$O, EtOH, δ-Et$_2$O	133,134
l	26	dec	1.2485	—	+3.82 (H$_2$O, 15)	—		133,134
Hydracrylic (β-Propanoic)	—	dec	—	—	—	3.11×10^{-3} (25)	Et$_2$O, H$_2$O, EtOH	133–135
α-Butyric (Butanoic)	43–44; 60–70 (Sub)	225–260 dec 140/14	1.125 (20)	—	—	7.5×10^{-5}	H$_2$O, EtOH, Et$_2$O	133–135
β-Butyric *dl*	48–50	130/12	—	—	—	5.1×10^{-5}	H$_2$O, EtOH, Et$_2$O	133–135
l	49–50 (hygr)	—	—	—	−24.5 (H$_2$O, 25)	—	H$_2$O, EtOH, insol. ben	134,135
γ-Butyric	−17	dec	—	—	—	1.9×10^{-5} (25)	—	134,135

349

TABLE 2
Hydroxy acids: $C_5H_{10}O_3$–$C_6H_{12}O_3$

Hydroxy acid	mp	bp	$[\alpha]_D$	K	Sol	Reference
α-Valeric (Pentanoic)	28–29	Sub	—	—	H₂O, Et₂O, EtOH	134,135
β-Valeric	−32	—	−10.0 (20)	—	Et₂O, EtOH, chl, ben	134
γ-Valeric	—	—	−9.3 (H₂O)	2.02 × 10⁻⁵ (25)	—	134
δ-Valeric	Oil	—	+10.5 (24)	—	—	134
α-Isovaleric dl	86	200 dec	—	—	H₂O, EtOH, Et₂O	134
β-Isovaleric	Syrup	—	—	—	H₂O, EtOH, Et₂O	134
Pivalic	125	86/16	—	1.39 × 10⁻⁵ (25)	Et₂O	134,16
α-Caproic (Hexanoic) dl	60–62 100 (Sub)	270	—	—	—	134,135
l	60	—	−4.68 (EtOH)	—	H₂O, EtOH, chl, Et₂O	134
β-Caproic	13	—	—	—	H₂O	53

TABLE 3

Hydroxy Acids: $C_7H_6O_3$–$C_8H_8O_3$

Hydroxy acid	mp	bp	d_4	$[\alpha]_D$	K	Sol	Reference
Salicylic (o-benzoic)	159	211/20	1.443 (20)	—	1.02×10^{-3}	EtOH, Et$_2$O, δ-H$_2$O	133,136,137
m-Benzoic	201	—	1.473	—	8.7×10^{-5}	δ-Et$_2$O, δ, H$_2$O, EtOH (h)	133,136,137
p-Benzoic	213 76 (Sub)	—	1.468	—	2.86×10^{-5}	δ-H$_2$O, EtOH, Et$_2$O	133,136,137
1-Cyclohexanecarboxylic Mandelic	108–119	90–100/0.5	—	—	—	—	16
dl	118–119	dec	1.300 (20)	—	4.17×10^{-4} (25)	Et$_2$O, EtOH	134,135
d	133	—	1.341 (25)	−159.73 (EtOH)(20) −187.44 (HOAc)	4.3×10^{-4} (25)	H$_2$O, Et$_2$O, EtOH	134,135
2-Phenylacetic	145–147	240.3 (Lactone)	—	—	—	Et$_2$O, δ-H$_2$O, Et$_2$O	134
3-Phenylacetic	129	190/11	—	—	—	H$_2$O, EtOH, Et$_2$O	134,135
4-Phenylacetic	148–150	Sub	—	—	—	EtOH, Et$_2$O, H$_2$O	134,135

351

TABLE 4

Hydroxy Acids: $C_8H_{16}O_3$–$C_{15}H_{30}O_3$

Hydroxy acid	mp	bp	d_4	$[\alpha]_D$	K	Sol	Reference
α-Caprylic (octanoic)	69.5	—	—	—	1.55×10^{-4}	Et₂O, EtOH, δ-H₂O	134
ε-Caprylic	58	dec	—	—	—	ben, EtOH, H₂O, δ-pet eth	134
Atrolactic							
dl	94	—	—	—	—	δ-H₂O	134
d	116–117	—	—	+37.7 (EtOH) (16.5)	—	δ-H₂O	134
ω-Pelargonic (nonanoic)	53–54	137–9/3	1.4438 (20)	—	—	EtOAC (h)	116,134
ω-Capric (decanoic)	75	145–7/3	1.4471 (20)	—	—	Et₂O, δ-pet eth	116,134
ω-Undecanoic	76					Et₂O, EtOH	134
Sabinic	78–79					EtOH	134
Convolvulinolic	63.5–64					EtOH	134
ω-Pentadecanoic	82–84					EtOH, EtOAc, Me₂CO, ben	134
Jalapinolic	68–69	—	—	+0.79 (chl)	—	EtOH, chl, Et₂O	134
Juniperic	95	—	—	—	—	EtOH, Et₂O (h)	134

TABLE 5
Lactones: $C_4H_4O_2-C_6H_{10}O_2$

Lactone	mp	bp	d_4	n_D	$[\alpha]_D$	Sol	Reference
Glycolide	84					H_2O (h), chl (h), δ-Et_2O, δ-EtOH	134
Lactide							
d	95	150/25			−298 (ben) (18)	Et_2O, ben, chl, δ-EtOH	134
dl	124.5	142/8	0.862 (10)	—	—	δ-Et_2O, EtOH	134,136
β-Propiolactone	−33	155 dec; 51/10	1.1460 (20)	1.4131 (20)	—	H_2O dec, EtOH dec, Et_2O, chl	135
γ-Butyrolactone	−42	203–204; 89/12	1.1441 (0)	1.4343 (26)	—	Et_2O, EtOH, H_2O	134,135
γ-Valerolactone	−31	206; 83/13 d, 86–90/14	1.0465 (25)	1.4305 (25)	d, +13.5 (20)	H_2O, Et_2O	135
δ-Valerolactone	−12.5	219–222	1.0794 (20)	1.4503 (25)	—	δ-H_2O, EtOH, Et_2O	135
β-Isovalerolactone	—	54–55/10	—	1.4126 (20)	—	Me_2CO	47
γ-Caprolactone	−18	215–216 107–109/17	—	1.4455 (20)	—	H_2O	134,135
δ-Caprolactone	17–19	230–231 107/14	1.0443 (20)	1.4451 (20)	—	H_2O, EtOH, Et_2O	71,34
ε-Caprolactone	—	98–99/2	1.0693 (20)	1.4611 (20)	—	Et_2O	134
γ-Isocaprolactone	7	207	—	—	—	Et_2O	16,77

TABLE 6

Lactones: $C_7H_{12}O_2$–$C_{16}H_{30}O_2$

Lactone	mp	bp	η_D	Sol	Reference
γ-Enantholactone	—	235	1.4385 (20) [d_4 (20), 0.9768]		100
ε-Enantholactone	—	112–113/14	1.4588 (20)	—	59
ζ-Enantholactone	—	94/5	1.4688 (20)	—	59
γ-Caprylolactone	142	70/5	—	EtOH	134,135
ε-Isopelargonolactone	—	132–133/20	1.4604 (20)	—	59
γ-Caprilactone	—	107/5	1.4470 (20)	δ-H_2O	119,134
		281			
ζ-Methyl-ε-pelargonolactone	—	109–110/2.5	1.4625	—	59
		113/4			
Exaltolide	31–32	176/15	1.4633 (41) [d_4 (20), 0.9383]	—	134

TABLE 7

Higher Lactones (132): $C_{10}H_{18}O_2$–$C_{17}H_{32}O_2$

$$O\text{—}(CH_2)_n\text{—}CO$$

n	9	10	11	12	13	14	15	16
mp	64	2.4	2	27.5	33	37	36	42–43
bp	114/14	103/5	111/3	122/3.8	135/3.5	137/2	128/1	155/2.8

Polymerization of the lactide is accelerated by small amounts of water (141). For example, with the lactide of lactic acid, the water hydrolyzes the lactide to lactyllactic acid, which, in turn, reacts with other lactic molecules by ester interchange:

$$CH_3HC \overset{\displaystyle \overset{O}{\underset{||}{C}}}{\underset{\underset{\displaystyle \overset{||}{O}}{\underset{\displaystyle C}{\overset{\displaystyle O}{\underset{||}{}}}}{}}} O \quad \underset{\overset{}{\longleftrightarrow}}{\overset{H_2O}{}} \quad CH_3CH_2\overset{O}{\overset{||}{C}} - O\overset{CH_3}{\underset{|}{CH}}\overset{O}{\overset{||}{C}} - OH \quad \underset{\overset{}{\longleftrightarrow}}{\overset{lactide}{}} \quad polymer$$

With β-hydroxy acids, dehydration to unsaturated acids occurs instead of lactone formation; that is, intramolecular dehydration proceeds exclusively to intermolecular dehydration. Thus hydracrylic acid, β-hydroxypropionic acid, yields acrylic acid on heating (192):

$$HOCH_2CH_2COOH \xrightarrow{-H_2O} CH_2{=}CHCOOH$$

However, when there are no hydrogens beta to the hydroxyl group as in hydroxypivalic acid, $HOCH_2C(CH_3)_2COOH$, dehydration leads to the formation of polyesters.

Recently, polyesters from β-hydroxy acids have been synthesized by enzymatic action of microorganisms (143). In particular, the polymer from β-hydroxybutyric acid has been studied extensively.

The γ-hydroxy acids give γ-lactones as their main product, because of the formation of the highly stable five-membered ring system:

$$HOCH_2CH_2CH_2COOH \xrightarrow{-H_2O} \begin{matrix} H_2C & \overset{O}{\diagup \diagdown} & CO \\ | & & | \\ H_2C & \!\!\!\!-\!\!\!\!- & CH_2 \end{matrix}$$

Six-membered cyclic esters can also be formed from δ-hydroxy acids. However, a reversible transformation between the cyclic ester and the corresponding open-chain polyester occurs with remarkable ease. These cyclic esters usually polymerize spontaneously on standing, even at room temperature. The polymers, on the other hand, can be converted to monomeric lactones by distillation of the more volatile cyclic ester out of the reaction mixture (as with α-hydroxy acids). Some seven-membered cyclic esters derived from ε-hydroxy acids behave like their six-ring homologs.

Cyclic lactones containing 8–15-membered rings are not easily formed because of steric hindrance arising from the crowding of hydrogen atoms within the rings (144). As the ring size is increased above 15 atoms, steric effects are minimized and cyclization is possible. However, in ordinary bifunctional condensations (no diluent), the primary product from mono-mers of 15 or more members is almost exclusively polymer. This results from the statistical improbability that the ends of a long chain of atoms, connected by valence bonds about which there is free rotation, will meet. The probability that two ends of a long chain will occupy positions adjacent to each other, according to Flory, varies approximately as the inverse 3/2 power of the number of chain atoms.

In order to obtain large-membered rings, special conditions are required. The condensation is conducted in highly diluted solutions to enhance the probability of intramolecular condensation; or the polymer is heated under vacuum in the presence of certain catalysts, and the cyclic compound distilled off as it is formed. The dehydration of ω-hydroxy acids in benzene in the presence of benzenesulfonic acid or with potassium carbonate in methyl ethyl ketone have been studied (145,132), and rings of 11–20 atoms have been isolated and characterized.

Thus we may say in summary that polyesters can be prepared by the condensation of most hydroxy acids. Polyesters, however, cannot be formed satisfactorily by the high-temperature polycondensation of glycolic or lactic acids, which yields the cyclic dimers. The polycondensations are more effectively carried out in the presence of benzene or other materials capable of forming azeotropes with water. The β-hydroxy acids dehydrate to un-saturated acids upon heating. Stable lactones are the main products in the condensations of γ-hydroxy acids, whereas δ-hydroxypentanoic acid and ε-hydroxycaproic acid yield significant amounts of lactones. Polyesters of all these acids are best obtained by way of polymerization of their lactones, as we discuss briefly below. Finally, linear aliphatic polyesters whose functional groups are separated by nine or more atoms can be polymerized without difficulty. Cyclic esters of ring sizes above 15 can be obtained only under special conditions.

B.   Polymerization of Lactones

The ease of polymerization of certain cyclic esters (e.g., δ-valerolactones), has been observed for many years and was characterized by Carothers et al. (33). Polymerization occurs by gentle heating or merely allowing the lactone to stand at room temperature. However, most lactones polymerize only under the influence of catalysts or initiators (146). In fact, until recently γ-butyro-lactone was thought to be unpolymerizable. Korte and Glet (147) succeeded in polymerizing γ-butyrolactone by heating the monomer at 160°C under a pressure of 20,000 atm. The degree of polymerization of the polymer was only in the range of 14–40, but the polymer did not show any tendency to depolymerize at 80°C. The tendency of poly(oxypentanoyl) to depolymerize can be blocked by acetylation of the end groups (148). Polymers such as poly(β-oxypropanoyl), which cannot be synthesized from the hydroxy acid because of olefin formation, can be prepared via the β-lactone. The β-lactones are prepared by special routes; e.g., β-propiolactone is prepared by the reaction of ketene with formaldehyde (36).

The polymerization of lactones is believed to occur via scission of the acyl–oxygen bond in the monomers (except in the cationic polymerization of some β-propiolactones). The polymerization sequences may be illustrated as follows:

cationic

$$X^+ + O\text{——}R \longrightarrow X\text{—}\overset{+}{O}\text{——}R \longrightarrow X O R \overset{+}{C} O \xrightarrow{\text{monomer}} X O R C O O R \overset{+}{C} O \xrightarrow{\text{etc.}}$$

anionic

$$Y^- + O\text{——}R \longrightarrow O\text{——}R \longrightarrow Y C O R O^- \xrightarrow{\text{monomer}} Y C O R O O C R O^- \xrightarrow{\text{etc.}}$$

Thermal or radical-initiated polymerization of β-propiolactone to polyester is also possible (149). Theoretical understanding of the reactivity differences in the polymerization of lactones is still incomplete. For example, the highly strained (four-membered ring) β-propiolactone would be expected to show exceptional reactivity in polymerization. In fact, however, it is one of the least reactive of the lactones. On the other hand, some studies (150) have led to the prediction that γ-butyrolactone should be more reactive than it is.

The formation of polyesters by the polymerization of lactones has considerable advantages over other methods. The relatively low temperatures involved permit the avoidance of cyclization, dehydration to olefins, and certain other side reactions normally encountered in the polycondensation of hydroxy acids. It is also possible to obtain polyesters from monomers having certain secondary reactive centers, such as vinyl or epoxy groups, which would not survive the conditions of conventional polycondensation. Moreover, as a polyaddition reaction, the polymerization of the lactones avoids the problem of removal of condensation products such as water, glycols, and alcohols.

Control over the nature of the end groups is another advantage. For example, production of hydroxy-terminated polyesterglycols (Union Carbide's "Niax Polyols") takes advantage of the glycol-initiated polymerization of ε-caprolactone. The lactones have also been copolymerized with other monomers, such as vinyl monomers (151), ethylene carbonate, oxiranes (152), lactones, amino acids, and nylon salts (153). An interesting case is the copolymerization of β-propiolactone and γ-butyrolactone catalyzed by a triethylaluminum-water complex. Copolymerization of β-propiolactone and γ-valerolactone with propylene oxide and epichlorohydrin gave high-molecular-weight, solid copolymers (154).

Clearly, the polymerization of lactones has provided a method of obtaining polyesters not possible by polycondensation of hydroxy acids. Even poly-glycolic acid (155) and polylactic acid are best prepared via their lactones. Lactide was also used in the preparation of the optically active poly-S-lactic acid (156). The formation of polyesters by polymerization of lactones, how-ever, is limited mainly by the availability of the monomers.

C. Polymerization of Aromatic Hydroxy Acids

Direct polymerization of hydroxybenzoic acid monomers is difficult, since the phenolic hydroxyl group is relatively inert to direct condensation with the carboxylic acid group. Some polymer residues may be obtained when such monomers are heated, but decarboxylation also occurs at an increasingly rapid rate at temperatures above 150°C. Polymerization of these monomers has been accomplished by changing the reactivity of either or both of the functional groups. For example, polymerization of the acetoxy derivatives of all three hydroxybenzoic acid monomers has been reported by an ester interchange reaction with the elimination of acetic acid:

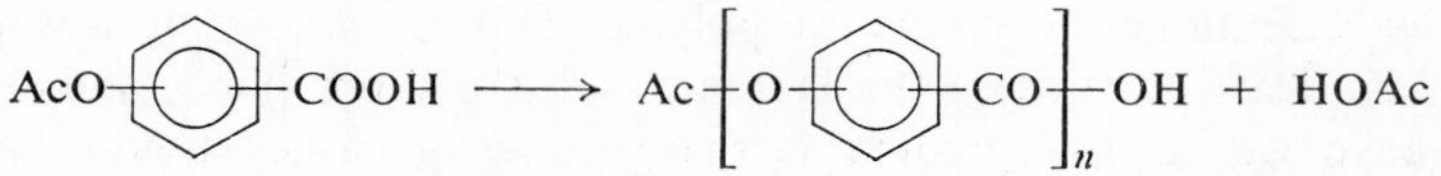

Another route involves the preparation of the acid chloride of the hydroxy-benzoic acid followed by condensation of hydrogen chloride (157):

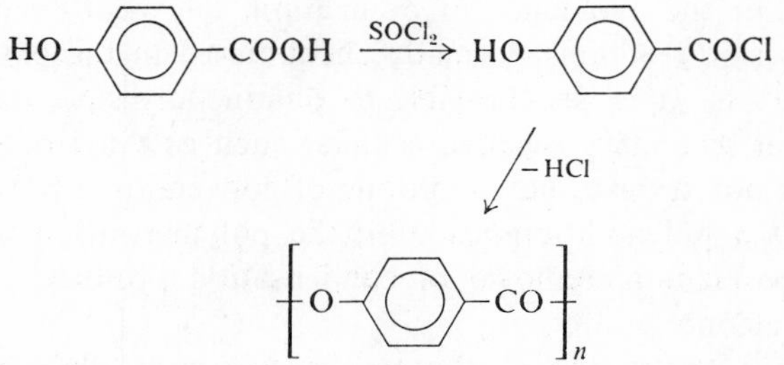

Very-low-molecular-weight polyesters are obtained by this method.

An unambiguous route to the polymer is only possible by blocking the carboxylic acid group to a reactive ester, which protects the monomer from decarboxylation and yet permits polymerization to proceed to high molecular weights. This approach has been used successfully for the preparation of the high-molecular-weight poly(p-oxybenzoyl), now commercially available as "EKONOL."

The earliest work on polymers derived from salicylic acid was carried out by Anschutz (158), who heated acetylsalicylic acid at 200°C under reduced pressure and obtained cyclic disalicylides and trisalicylides. In a later study (159), tetra and hexasalicylides were obtained by heating disalicylides under vacuum. Linear polymers were prepared by melt polymerization of the disalicylide or trisalicylide at 230°C under nitrogen, with lead stearate as a catalyst. At temperatures above 250°C, rapid decomposition of the polymer was observed. The polymer from the disalicylide had an inherent viscosity of 0.2 (m-cresol) and softened at temperatures above 150°C, whereas the polymer from the trisalicylide had an inherent viscosity of 0.39 (m-cresol) and softened at 190°C. Self-supporting films and fibers could be drawn from these polymers.

Dimers and octamers of m-hydroxybenzoic acid were reported by Schiff in 1882, but the syntheses of high-molecular-weight poly(m-oxybenzoyl) polymers were reported first by Gilkey and Caldwell (160). They obtained the polyester from the melt polymerization of m-acetoxybenzoic acid. The polyester had an inherent viscosity of 0.5 and a crystalline melting point at 176°C. Molecular weights of up to 25,000 were obtained by finishing the polymerization in the solid state under vacuum. Attempts to prepare copolymers of meta and paraacetoxybenzoic acid were not particularly successful because of differences in the reactivity of the two monomers. Block polymers containing up to 50% of the paraisomer melted at 205°C. When as much as 60% of the paraisomer was incorporated into the copolymer, the crystallinity of the polymer system increased markedly and the solubility decreased. Pressed films of this copolymer showed a transition point of 143°C and a flow point of greater than 295°C.

The most important polymers of commercial interest are derived from p-hydroxybenzoic acid. When ethylene carbonate is reacted with p-hydroxybenzoic acid, methyl ester, poly(p-oxyethoxybenzoyl), is formed. A surprising etherification also occurs when either p-hydroxybenzoic acid or its alkyl ester is heated with ethylene glycol in the presence of cobalt acetate (161). A commercial fiber, called Grilene, is manufactured by the copolymerization of p-hydroxybenzoic acid with glycol and dimethyl terephthalate (162). Attempts to polymerize p-hydroxybenzoic acid directly have not been very successful, since the monomer decomposes almost completely to phenol and carbon dioxide at 200–220°C (163). Interestingly enough, a residue can be obtained by heating the monomer at 300–350°C; but further heating under a carbon dioxide residue leads to the formation of phenyl p-phenoxybenzoate.

Fisher (164), on the other hand, reported a tetramer from *p*-hydroxybenzoic acid in 1910. Oligomeric materials were prepared by use of the acid chloride of the hydroxy acid (157). Polymerization of the acid was carried out to low molecular weights in the presence of triphenyl phosphite (165) and trifluoro-acetic anhydride (166). Caldwell and Gilkey were able to polymerize *p*-acetoxybenzoic acid directly in the melt, which led to an intractable product decomposing at 350°C (160).

By use of suitable blocking groups on *p*-hydroxybenzoic acid, thermally stable and fabricable poly(*p*-oxybenzoyl) polymers could be produced. The same polymer under its commercial name, "EKONOL," was mentioned previously. The polymer is highly crystalline and retains its crystallinity up to 340–360°C, where a crystalline transition occurs. However, the polymer does not melt below a temperature of 550°C, and at this point it decomposes rapidly. Isothermal thermogravimetric analyses in air show weight losses of less than 1%/hr at 400°C in the powder form of the polymer.

The fabrication of "EKONOL" homopolyester is of interest because it is formed below its melting point. Compression sintering of this material is possible at temperatures of 420°C and at pressures of greater than 5000 psi. Use of a high-energy rate forge provides a versatile technique for the rapid forming of "EKONOL" shapes. A preform is heated to 155–260°C and then subjected to forging. The energy generated during the forging is sufficient to produce flow and apparent fusion of the material. Plasma spraying has proved to be an excellent method for providing "EKONOL" coatings. Coatings from 0.5 to 200 mils thick can be produced by this technique.

The properties of compression-sintered "EKONOL" homopolyester compare favorably with those of other high-temperature polymers. The elastic modulus is the highest of any of the reported engineering plastics. In addition to its excellent thermal stability, "EKONOL" homopolyester has the highest thermal conductivity of any plastic material, most likely because of its high crystallinity. Most of the electrical properties of "EKONOL" homopolyester compare favorably with those of other engineering plastics. The coefficient of friction of the polyester is very low and is about twice the value of Teflon[®].

D. Polyester Applications

As stated earlier, the polymers obtained from aliphatic hydroxy acids have found little commercial utility because of their relatively low melting points, their tendency to cyclize, and their reactivity to moisture. There have been several attempts (138) to improve the thermal and hydrolytic stability of the aliphatic polyesters. Recently use has been made of aliphatic polyesters and copolyesters, alone or in combination with other materials, for varnishes, plasticizers, and epoxy-resin hardeners (167). Some aliphatic polyesters, such as poly(oxyhexanoyl), now serve as intermediates for polyurethanes (168), although most of the commercial polyester precursors are derived from AA/BB systems. Hydroxy-terminated poly(oxyhexanoyl) compounds, made by the polymerization of ε-caprolactone, have been introduced by Union Carbide Corporation under the trade name "Niax Polyols" as urethane precursors. They are claimed to give products with low brittle points and superior hydrolytic stability. Some interesting fibers, films, and thermoplastic molding resins have been prepared from high-melting aliphatic polyesters—e.g., poly(oxy-2,2-dimethylpropionyl) (169)—but none is yet produced commercially. Recently Union Carbide introduced the new polycaprolactone thermoplastic polymers "PCL-300" and "PCL-700," with molecular weights of about 15,000 and 40,000, respectively. Polycaprolactone polymers are used primarily for blends with other materials. In such applications it is reported that they improve the dyeability of polyolefin polymers; increase the impact resistance of selected polymers; increase stress-crack resistance, clarity, and antiblock characteristics of polyethylene; improve the gloss of extruded thermoplastics; and provide smooth surface properties to molded thermosets.

With aromatic hydroxy acids, only the fiber-forming polyester, poly-(oxyethoxybenzoyl) (170), and the homopolyester of p-hydroxybenzoic acid are being made commercially. The polyester fiber being developed in Japan is reported to have silklike properties.

The availability of rapid and versatile forming techniques, combined with unique properties, suggests numerous applications for the "EKONOL" (p-oxybenzoyl) polyester (171). Its resistance to creep or cold flow and its extraordinarily high elastic modulus mean that it provides an excellent material of construction for self-lubricating bearings. The bearing properties can be enhanced by the addition of fillers, such as polyfluorocarbons, graphite, or molybdenum disulfide. The high thermal conductivity of this polyester is extremely important in applications where localized heat buildup is a problem. Possible applications appear to be in the areas of self-lubricating bearings and bearing races.

Other areas of potential applications are in seals, rotors, or vanes of process pumps where wear resistance, dimensional stability, and corrosion resistance are highly critical. In the field of electrical insulation, the polyester is of particular interest for high-temperature electrical applications because of its high dielectric strength, its low dissipation factor, and its thermal stability. The immediate applications include high-temperature circuit boards, mechanical insulation components, and encapsulated diodes, transistors, and integrated circuits. The polyester may also be useful in handles and slip-free coatings, piston rings for automobile engines, and abradable seals.

References

1. E. Dane and J. Schmitt, *Ann. Chem.*, **537**:246 (1939).
2. A. Wurtz, *Compt. Rend.*, **76**:1167 (1873).
3. P. Friedlander and O. Low-Beer, *Chem. Zentr.*, **II**:471 (1906); German Patent 170,230.
4. K. Riemer and F. Tiemann, *Ber. Deut. Chem. Ges.*, **9**:1285 (1876); G. Hasse, *Ber. Deut. Chem. Ges.*, **10**:2186 (1877); German Patent 258,887.
5. J. Schmitt and K. Burkard, *Ber. Deut. Chem. Ges.*, **20**:2702 (1887).
6. R. R. Read, in "Organic Syntheses," Coll. Vol. I, H. Gilman, Ed.-in-Chief, Wiley, New York, pp. 314–315, 1932.
7. J. Wislicenus, *Justus Liebigs Ann. Chem.*, **128**:4 (1863); **167**:346 (1873).
8. V. V. Perekalin, A. K. Petrayareva, M. M. Zobacheva, and E. L. Metalkina, *Dokl. Akad. Nauk SSSR*, **166**(5):1129 (1966); *Chem. Abstr.*, **64**:15734 (1966).
9. L. Wolff, *Justus Liebigs Ann. Chem.*, **208**:104, 106 (1881).
10. A. H. Church, *J. Chem. Soc.*, **16**:302 (1863).
11. J. Dale, *J. Chem. Soc.*, **1965**:77.
12. G. C. Thomson, *Justus Liebigs Ann. Chem.*, **200**:76 (1880).
13. J. F. Thorpe and M. A. Whitely, "Thorpe's Dictionary of Applied Chemistry," 4th ed., Vol. VI, Longmans, Green and Co., London-New York, pp. 400–401, 1946.
14. On the Reformatsky reaction see: R. L. Shriner, in "Organic Reactions," Vol. I, R. Adams, Ed.-in-Chief, Wiley, New York, pp. 1–37, 1942.
15. For a discussion on hydroxy acids see: P. Karrer, "Organic Chemistry," Elsevier, London-New York, pp. 263–268, 1950.
16. See syntheses of hydroxy acids in "Richter's Organic Chemistry," 3rd ed., R. Anschutz and F. Reindel, Eds., Elsevier, London-New York, 1944, Chapter 7.
17. E. Erlenmeyer, *Justus Liebigs Ann. Chem.*, **191**:273 (1878).
18. A. Wolkow, *Z. Chem.*, **6**:326 (1870).
19. P. P. Schorigin and H. P. Kreshkov, *Chem. Abstr.*, **28**:6106 (1934); **29**:2147 (1935).
20. M. Guthzeit, *Justus Liebigs Ann. Chem.*, **209**:2346 (1881).
21. R. Marburg, *Justus Liebigs Ann. Chem.*, **294**:129 (1897).
22. A. T. Larson (to E. I. du Pont de Nemours Co., Inc.), U.S. Patent 2,153,064 (April 4, 1939); *Chem. Abstr.*, **33**:5006 (1939).
23. D. J. Loder (to E. I. du Pont de Nemours Co., Inc.), U.S. Patent 2,152,852 (April 4, 1939); French Patent 831,474 (Sept. 5, 1938), *Chem. Abstr.*, **33**:2539 (1939).
24. A. Sporzynski, W. Kocaz, and H. V. A. Briscoe, *Rec. Trav. Chim. Pays-Bas*, **68**:617 (1949), see also C. A. Bischoff and P. Walden, *Justus Liebigs Ann. Chem.*, **279**:45 (1894).
25. W. v. Heintz, *Pogg. Ann. Chem. Phys.*, **109**:489 (1860).
26. S. Chujo and S. Suzuki (to Dainippon Celluloid Co., Ltd.), Japanese Patent 3546('66) (March 1, 1963).
27. See also the excellent article on lactic acid by E. M. Filachione in "Encyclopedia of Chemical Technology," 2nd ed., Vol. 12, A. Standen, Ex. Ed., Interscience, New York, pp. 170–187, 1967.
28. D. B. Sabin, *Chemistry*, **40**(3):10–12 (1967).
29. J. Geoffrey and M. Thorne, *Chem. Process (London)*, **15**(1):8–9 (1969).
30. J. Biochard, B. Brossard, M. Gay, and R. Janin (to Société des Usines Chimiques Rhone-Poulenc), French Patent 1,465,640 (Jan. 13, 1967).
31. R. Platz et al. (to Badische Anilin- & Soda-Fabrik AG), German Patent 1,259,868 (Feb. 1, 1968).

32. T. E. Friedemann and J. B. Graeser, *J. Biol. Chem.*, **100**:291 (1933).
33. H. F. Mark and G. W. Whitby, Eds., "Collected Papers of Wallace Hume Carothers," Interscience, New York, 1940.
34. H. Byk (to Chem. Werke V. H. Pek), German Patent 267,826 (Feb. 20, 1912); French Patent 456,824 (April 18, 1913); *Chem. Abstr.*, **8**:559, 2034 (1914).
35. H. Moser (to Dynamit-Nobel AG). German Patent 1,083,275 (June 15, 1960).
36. F. E. Kung (to B. F. Goodrich Co.), U.S. Patent 2,356,459 (Aug. 22, 1944); *Chem. Abstr.*, **39**:88 (1945).
37. T. R. Steadman (to B. F. Goodrich Co.), U.S. Patent 2,424,589 (July 29, 1947); T. R. Steadman and P. L. Breyfogle (to B. F. Goodrich Co.), U.S. Patent 2,424,590 (July 29, 1947).
38. G. Machell, *Ind. Chem.*, **36**:13 (1960).
39. British Patent 909,341 (Aug. 29, 1962).
40. F. Dickens and H. E. H. Jones, *Brit. J. Cancer*, **15**:85 (1961); *Chem. Abstr.*, **55**:21347 (1961).
41. B. L. Van Duren, *Mycotoxins, Foodstuff, Proc. Symp. MIT*, **1964**:275; *Chem. Abstr.*, **64**:4064 (1966).
42. W. Markownikoff, *Justus Liebigs Ann. Chem.*, **153**:244 (1870).
43. H. Gilman and R. K. Abbott, Jr., *J. Org. Chem.*, **8**:224 (1943).
44. H. Terada, *Nippon Kagaku Zasshi*, **83**:1057 (1959).
45. J. Wislicenus, *Justus Liebigs Ann. Chem.*, **149**:205 (1869).
46. K. W. Tuerck and H. J. Lichtenstein (to Distillers Co., Ltd.), U.S. Patent 2,411,700 (Nov. 26, 1946).
47. Roger Adams, Ed., "Organic Reactions," Vol. VIII, Wiley, New York, p. 330, 1954.
48. J. Sixt (to Wacker Chemie GmbH), U.S. Patent 2,763,669 (Sept. 18, 1956).
49. H. F. McShane and W. W. Gilbert (to E. I. du Pont de Nemours Co., Inc.), U.S. Patent 2,772,291 (Nov. 27, 1956).
50. C. S. Marvel and E. R. Birkhimer, *J. Amer. Chem. Soc.*, **51**:260 (1929).
51. R. Fittig and W. Dannenberg, *Ann. Chem.*, **331**:132 (1904).
52. A. Menozzi, *Gazz. Chim. Ital.*, **14**:17 (1884).
53. F. A. Adickes and G. Andresen, *Ann. Chem.*, **555**:41 (1944); see also Ref. 14.
54. A. P. Dunlop and J. W. Maden (to Quaker Oats Co.), U.S. Patent 2,786,852 (March 26, 1957).
55. H. A. Schuette and R. W. Thomas, *J. Amer. Chem. Soc.*, **52**:3011 (1930).
56. A. M. Cloves, *Ann. Chem.*, **319**:357 (1902); see also Ref. 50.
57. S. K. Bhattacharyya and D. K. Nandi, *Ind. Eng. Chem.*, **51**:143 (1959).
58. Y. Y. Aliev and I. B. Romanova, *Neftekhim., Akad. Nauk, Turkm. SSR*, **1963**:204; *Chem. Abstr.*, **61**:6913 (1964).
59. P. S. Starcher and B. Phillips, *J. Amer. Chem. Soc.*, **80**:4079 (1958).
60. E. Fisher and G. Zemplen, *Ber. Deut. Chem. Ges.*, **42**:4891 (1909).
61. D. D. Coffman, R. Cramer, and W. E. Mochel, *J. Amer. Chem. Soc.*, **80**:2885 (1958).
62. Netherlands Appl. 6,506,637 (Dec. 6, 1965) (to Badische Anilin- & Soda-Fabrik AG); *Chem. Abstr.*, **64**:15746 (1966).
63. C. S. Marvel, D. W. MacCorquodole, F. E. Kendall, and W. A. Lazier, *J. Amer. Chem. Soc.*, **46**:2838 (1924).
64. S. Gronowitz, *Ark. Kemi*, **11**:519 (1951).
65. A. Adembri and G. Ghelardoni, *Gazz. Chim. Ital.*, **89**:700 (1959).
66. K. Fujiwara, *Nippon Kâgakû Zasshi*, **82**:627 (1961).
67. K. Fujiwara and K. Naruse, *Nippon Kâgakû Zasshi*, **82**:1400 (1961).
68. O. Riobe, *Compt. Rend.*, **247**:1016 (1958); *Chem. Abstr.*, **53**:7054c (1959).

69. K. W. Rosenmund and H. Bock (to Unilever Ltd.), U.S. Patent 3,048,599 (Aug. 7, 1962); see also British Patent 938,054 (Sept. 25, 1963).
70. R. Lukes, S. Dalizal, and K. Copek, *Collect. Czech. Chem. Commun.*, **27**:2408 (1962).
71. R. P. Linstead and H. N. Rydon, *J. Chem. Soc.*, **1934**:2000.
72. I. Ishimoto, H. Togawa, and T. Nachi (to Teijin Ltd.), Japanese Patent 656('66) (Jan. 22, 1966); *Chem. Abstr.*, **64**:12553 (1966).
73. A. Isard and F. Weiss (to Société d'Electro-Chimie, d'Electro-Metallurgie, et Aciéries Electriques d'Ugine), French Patent 1,411,213 (Sept. 17, 1965); *Chem. Abstr.*, **64**:3364 (1966).
74. F. Weiss (to Société d'Electro-Chimie d'Electro-Metallurgie et d'Aciéries Electriques d'Ugine), Belgian Patent 646,938 (Aug. 17, 1969); French Appl. (April 24, 1963, Dec. 3, 1963).
75. W. Goldsmith and D. F. Marples (to Union Carbide Corp.), French Patent 1,395,204 (April 9, 1965); *Chem. Abstr.*, **63**:11368 (1965).
76. W. A. Noyes, *J. Amer. Chem. Soc.*, **23**:393 (1961).
77. D. T. Jones and G. Tattersoll, *J. Chem. Soc.*, **85**:1962 (1904).
78. For a recent discussion on hydroxybenzoic acids, see R. T. Gottesman, in "Encyclopedia of Chemical Technology," Vol. 17, A. Standen, Ex. Ed., Interscience, New York, pp. 720–742, 1968.
79. J. E. Zajic, *Develop. Ind. Microbiol.*, **6**:16 (1964); *Chem. Abstr.*, **64**:1047 (1966).
80. W. W. Kaeding and E. J. Strojny (to Dow Chemical Co.), Belgian Patent 665,631 (Dec. 20, 1945).
81. R. Wakasa et al. (to Asaki Kogyo Kabushiki Kaisha), U.S. Patent 3,360,553 (Dec. 26, 1967).
82. C. Grabe and J. Kraft, *Ber. Deut. Chem. Ges.*, **39**:797 (1906); *Chem. Zentr.*, **II**:471 (1906).
83. A. Ree, *Justus Liebigs Ann. Chem.*, **233**:234 (1886).
84. L. E. Mills and W. W. Allen (to Dow Chemical Co.), U.S. Patent 1,937,477 (Nov. 28, 1933).
85. R. P. Berni (to Heyden Chemical Corp.), U.S. Patent 2,749,362 (June 5, 1956).
86. C. A. Buehler and W. E. Cote, in "Organic Syntheses," Coll. Vol. II, A. H. Blatt, Ed.-in-Chief, Wiley, New York, p. 341, 1943.
87. Y. Yasuhara and T. Nogi, *J. Org. Chem.*, **33**:4512 (1968).
88. Y. Yasuhara and T. Nogi, *Chem. Ind. (London)*, **1969**(3):77.
89. J. Boeseken and A. G. Lutgerhorst, *Rec. Trav. Chim. Pays-Bas*, **51**:164 (1932).
90. V. A. Zagorevshi, K. I. Loputina, and E. I. Fedin, *Zh. Organ. Khim.*, **2**(6):1035 (1966); *Chem. Abstr.*, **65**:15361 (1966).
91. D. L. Heywood, H. A. Stansbury, Jr., and B. Phillips (to Union Carbide Corp.), U.S. Patent 3,182,008 (May 4, 1945).
92. E. R. Marshall, J. A. Kuch, and R. C. Elderfield, *J. Org. Chem.*, **7**:454 (1942).
93. M. F. Clark and L. N. Owens, *J. Chem. Soc.*, **1950**:2108.
94. W. H. Perkin, *J. Chem. Soc.*, **55**:416 (1909).
95. L. N. Owens and R. A. Robins, *J. Chem. Soc.*, **1949**:332.
96. N. R. Campbell and J. H. Hunt, *J. Chem. Soc.*, **1950**:1381.
97. Ref. 96; see also R. Lukes, J. Trojanek, and K. Blaka, *Chem. Listy*, **49**:717 (1955); *Chem. Abstr.*, **49**:12369 (1955).
98. F. Balas and L. Srol, *Collect. Trav. Chim. Czech.*, **1**:658 (1929); see also Refs. 96 and 97.
99. D. Chanal, *Bull. Soc. Chim. Fr.*, **1950**:714.

100. N. I. Shuikin, V. V. An, V. M. Shostakovski, and I. F. Bel'skii, *Izv. Akad. Nauk SSSR, Ser. Khim.*, **1964**:111, 2102; *Chem. Abstr.*, **62**:9088 (1962).

101. Z. P. Golovena et al., *Chem. Abstr.*, **65**:8872f (1966).

102. A. E. Montagna, D. G. Kubler, and J. J. Brezinski (to Union Carbide Corp.), U.S. Patent 2,998,446 (Aug. 29, 1961).

103. R. Huisgen and H. Ott, *Angew. Chem.*, **70**:312 (1958).

104. R. Waskasa, K. Saotome, and Y. Okuni (to Asahi Chemical Industry Co., Ltd.), Japanese Patent 26,533 (Nov. 17, 1965).

105. S. A. Harris and J. S. Yeaw, in "Organic Syntheses," Coll. Vol. I, H. Gilman, Ed.-in-Chief, Wiley, New York, p. 329, 1932.

106. J. G. Aston, J. D. Newkirk, D. M. Jenkins, and J. Dorsky, in "Organic Syntheses," Coll. Vol. III, E. C. Horning, Ed.-in-Chief, Wiley, New York, p. 538, 1955.

107. S. Czaplicki, St. V. Kostanechi, and V. Lampe, *Ber. Deut. Chem. Ges.*, **42**:831 (1942)

108. J. H. Looker and D. N. Thatcher, *J. Org. Chem.*, **19**:784 (1954).

109. M. J. Boeseken, *Rec. Trav. Chim. Pays-Bas*, **37**:165 (1918).

110. E. E. Blaise and A. Koehler, *Bull. Soc. Chim. Fr.*, **7**:410 (1910).

111. E. L. Eliel and J. P. Freeman, in "Organic Syntheses," Coll. Vol. IV, N. Rabjohn, Ed.-in-Chief, Wiley, New York, p. 58, 1963.

112. A. McKenzie and G. W. Clough, *J. Chem. Soc.*, **97**:1022 (1910); Meilerig, *Justus Liebigs Ann. Chem.*, **209**:21 (1881).

113. G. du Pont, *Compt. Rend.*, **150**:1524 (1910).

114. R. Levine and J. R. Stephens, *J. Amer. Chem. Soc.*, **72**:1642 (1950).

115. E. Bowden and H. Adkins, *J. Amer. Chem. Soc.*, **62**:2422 (1940).

116. W. H. Lycan and R. Adams, *J. Amer. Chem. Soc.*, **51**:625 (1929).

117. M. D. Bogard, *Bull. Soc. Chim. Fr.*, **1**:350 (1907).

118. H. Thaler and G. Geist, *Biochem. Z.*, **302**:374 (1939).

119. T. M. Patrick, Jr., and F. B. Erickson in Ref. 111, p. 432.

120. F. L. Benton and A. H. Kiess, *J. Org. Chem.*, **25**:471 (1960).

121. G. H. Hargreaves and L. N. Owens, *J. Chem. Soc.*, **1947**:753.

122. For Syntheses of ω-hydroxy acids see, P. Chuit and J. Hausser, *Helv. Chim. Acta.*, **12**:436 (1929); *Chem. Abstr.*, **23**:3663 (1929).

123. K. Kimura, M. Takahashi, and A. Tanaka, *Chem. Pharm. Bull. (Tokyo)*, **8**:1059 (1960); *Chem. Abstr.*, **55**:24551b (1961).

124. T. Kariyone, H. Ageta, and K. Isoi, *Yakugaku Zasshi*, **79**:47, 54 (1954); *Chem. Abstr.*, **52**:18683i (1958); **53**:10031e (1959).

125. G. S. Hiers and R. Adams, *J. Amer. Chem. Soc.*, **48**:2385 (1926).

126. L. A. Davies and R. Adams, *J. Amer. Chem. Soc.*, **50**:1749 (1928).

127. H. H. Mathur and S. K. Bhattacharyya, *J. Chem. Soc.*, **1963**:3505.

128. M. Keschbaum, *Ber. Deut. Chem. Ges.*, **60**:902 (1927).

129. A. N. Nesmeyanov, L. I. Zakhaskin, T. A. Kort, and R. Kh. Freidlina, *Isv. Akad Nauk SSSR, Otd. Khim. Nauk.*, **1960**:211; *Chem. Abstr.*, **50**:7061f (1956); **54**:20864i (1960).

130. A. Tanaka, *Yakugaku Zasshi*, **79**:1327 (1959); *Chem. Abstr.*, **52**:16085h (1958); **54**:4381c (1960).

131. J. Plesch, *Chem. Listy*, **50**:561 (1956); *Chem. Abstr.*, **50**:13738a (1956).

132. H. Hunsdiecker and H. Erlbach, *Chem. Ber.*, **80**:129 (1947); see also R. Huisigen and H. Ott, *Tetrahedron*, **6**:253 (1959); *Chem. Abstr.*, **52**:15489b, 18351g (1958); **53**: 17896a (1959); **49**:10884d (1955); see also Ref. 103.

133. A. Standen, Ed., "Encyclopedia of Chemical Technology," Vol. 1, Interscience, New York, 1963, chapter on carboxylic acids, pp. 224–240.

134. I. M. Heilbron, "Dictionary of Organic Chemistry," Oxford University Press, New York, 1953.

135. R. C. Weast, Ed.-in-Chief, "Handbook of Chemistry and Physics," 45th ed., Chemical Rubber Co., Cleveland, Ohio, 1962.

136. N. A. Lange, Ed., "Lange's Handbook of Chemistry," rev. 10th ed., McGraw-Hill, New York, 1967.

137. B. Prager, P. Jacobson, P. Schmidt, and D. Stern. Eds., Beilsteins "Handbuch der Organische Chemie," 4th ed., Vol. X, Julius Springer, Berlin, 1927.

138. I. Goodman, in "Encyclopedia of Chemical Technology," 2nd Ed., Vol. 16, A. Standen, Ex. Ed., Interscience, New York, 1968, chapter on polyesters, pp. 159–189.

139. J. Gay-Lussac and J. Pelouze, *Justus Liebigs Ann. Chem.*, **7**:40 (1833).

140. W. H. Carothers, *Chem. Rev.*, **8**:353 (1931); R. Dietzel and R. Krug, *Ber. Deut. Chem. Ges.*, **58**:1307 (1925).

141. P. J. Flory, *J. Amer. Chem. Soc.*, **64**:2205 (1942).

142. E. E. Balise and L. Marcilly, *Bull. Soc. Chim. Fr.*, **31**(3):308 (1909).

143. G. Gottschalk, *Angew. Chem.*, **74**:342 (1962).

144. P. J. Flory, *Chem. Rev.*, **39**:137 (1946); see also E. W. Spanagel and W. H. Carothers, *J. Amer. Chem. Soc.*, **57**:935 (1935); **58**:654 (1936).

145. P. Ruggli, *Justus Liebigs Ann. Chem.*, **392**:92 (1912); M. Stoll and R. Rouve, *Helv. Chem. Acta*, **17**:1283 (1934).

146. H. Cherdron, H. Oshe, and F. Korte, *Makromol. Chem.*, **56**:178 (1962); S. Inoue, Y. Tomoi, T. Tsurutte, and J. Furukawe, *Makromol. Chem.*, **48**:229 (1961).

147. F. Korte and W. Glet, *J. Polym. Sci.*, **B4**:685 (1966).

148. K. Saotome and Y. Kodaira, *Makromol. Chem.*, **82**:41 (1965).

149. H. Oshe, H. Cherdron, and F. Korte, *Makromol. Chem.*, **86**:312 (1965).

150. Y. Yamashita, T. Tsuda, M. Okada, and S. Iwatsuki, *J. Polym. Sci.*, **A-14**:2121 (1966).

151. T. Tsuda, T. Shimizu, and Y. Yamashita, *Kôgyô Kâgakû Zasshi*, **67**:1661 (1964); J. L. Lary (to Dow Chemical Co.), German Patent 929,875 (1955); U.S. Patent 2,856,376 (1966).

152. H. Cherdron and H. Oshe, *Makromol. Chem.*, **92**:213 (1966).

153. R. Wakoda, K. Saotome, and N. Kavamoto (to Asahi Chemical Co., Ltd.), Japanese Patent 20522 (1964).

154. J. Falbe and F. Korte, *Angew. Chem.*, **74**:900 (1962).

155. G. E. Lowe, U.S. Patent 2,688,162 (1954); Y. Ishida, *J. Polym. Sci.*, **B3**:87 (1965).

156. M. Goodman and M. D'Alogni, *J. Polym. Sci.*, **B5**:515 (1962).

157. T. Kametani and K. Fukumoto, *Yakugaku Zasshi*, **80**:1188 (1960).

158. R. Anschutz, *Ber. Deut. Chem. Ges.*, **52B**:1815 (1919).

159. W. Baker, J. F. W. McOmie, W. D. Ollis, and T. S. Zealy, *J. Chem. Soc.*, **1951**:201.

160. R. Gilkey and J. R. Caldwell, *J. Polym. Sci.*, **2**:198 (1959); J. R. Caldwell (to Eastman Kodak Co.), U.S. Patent 2,600,376 (June 17, 1952).

161. Inventa A.G., British Patent 933,448 (1963).

162. W. Griehl, *Lenzinger Ber.*, **22**:55 (1966).

163. C. Graebe and A. Eichengrun, *Justus Liebigs Ann. Chem.*, **269**:325 (1892).

164. E. Fischer and F. Freundenberg, *Justus Liebigs Ann. Chem.*, **372**:32 (1910).

165. D. Aelony and M. M. Renfrew (to General Mills, Inc.), U.S. Patent 2,728,747 (Dec. 27, 1955).

166. E. J. Bourn, M. Stacy, J. C. Tatlow, and J. M. Tedder, *J. Chem. Soc.*, **1949**:2976.

167. British Patent 1,032,648 (1966) (to Asahi Kasei Kogyo Kabushiki Kaisha); D. A.

Lannon, E. T. Hoskins, and P. D. Ritche, Eds., "Physics of Plastics," Iliffe Books, Ltd., London, 1965.

168. "Modern Plastics Encyclopedia," 1967, McGraw-Hill, New York, 1966.

169. R. J. W. Reynolds and E. J. Vickers (to Imperial Chemical Industries, Ltd.), British Patent 766,347 (1957); T. A. Alderson (to E. I. du Pont de Nemours and Co., Inc.), U.S. Patent 2,658,055 (1953).

170. M. Korematsu, H. Masuda, and S. Kuriyama, *Kôgyô Kagaku Zasshi*, **63**:884 (1960).

171. J. Economy, B. E. Nowak, and S. G. Cottis, *SAMPE J.*, **6**(6):21 (1970); *Polym. Preprints*, **2**(1): February 1970.

5. DIISOCYANATES

A. A. R. SAYIGH, HENRI ULRICH, AND WILLIAM J. FARRISSEY, JR.

The Upjohn Company, Donald S. Gilmore Research Laboratories, North Haven, Connecticut 06473

Contents

I. INTRODUCTION

In 1848 Wurtz (1) reported the first synthesis of isocyanates from the reaction of diethyl sulfate and potassium cyanate. Shortly thereafter several prominent nineteenth-century scientists, such as Hofmann and Curtius, systematically investigated the chemistry of these "esters of isocyanic acid," and Hentschel (2) developed an alternate synthesis of isocyanates from the reaction of phosgene with the corresponding amines. Isocyanates, however, remained relatively unimportant as reactive chemical intermediates for some time, until the pioneering work of Staudinger and co-workers (3). Staudinger emphasized the structural similarities of isocyanates and ketenes, and he reported that like ketenes, isocyanates undergo reaction with certain unsaturated compounds. The remarkable utility of isocyanates in the preparation of synthetic plastics remained an untapped resource until the German scientists, intrigued by Carothers's synthesis of nylon, began looking for similar fibers. In 1937 Bayer (4) made known the first synthesis of a polyurethan from a diisocyanate. Unfortunately, the development of the polyurethan field was delayed by World War II, but after 1945 the German technical information was made available to the American industry.

Since 1946 the polyurethan market has enjoyed a rapid growth, 1 billion lbs of polyurethans (5) were used in the U.S. in 1970. Whereas the majority of polymers are based on a narrow range of chemical intermediates, the polyurethan polymers can be made from numerous types of raw materials, making possible the preparation of polymers with an extremely broad range of properties and applications. Aliphatic diisocyanates such as hexamethylene diisocyanate (HMDI), xylylene diisocyanate (XDI), and dicyclohexylmethane diisocyanate (H_{12}MDI), and aromatic diisocyanates such as tolylene diisocyanate (TDI), diphenylmethane diisocyanate (MDI), naphthalene 1,5-diisocyanate (NDI), tolidine diisocyanate (TODI), and dianisidine diisocyanate (DADI) are all commercially available materials.

Several articles that review the synthesis and chemistry of isocyanates include those by Siefken (6), Saunders and Slocombe (7), Arnold et al. (8), and Wilson (9). Both the syntheses and chemistry of sulfonyl isocyanates, and the cycloaddition reactions of isocyanates were reviewed recently by Ulrich (10,12). Petersen (11) has written an excellent review of the chemistry of aliphatic diisocyanates, and experimental details are included in his article. The chemistry and technology of polyurethans is discussed in recent books (13–18).

In view of the wealth of information already published, this chapter is primarily concerned with diisocyanates—their synthesis, reactions, properties, and analysis—and emphasis is placed on lately developed techniques and methods. Recently a variety of difunctional isocyanates in which the isocyanato group is attached to an atom other than carbon have emerged, and several novel polymers, based on diisocyanates, have been synthesized. Also, new analytical methods, utilizing modern instrumental techniques, have become available.

The diisocyanates, listed in the Tables 1–21, pages 372–427, are those for which reliable physical data have been reported. In some cases the evidence for the formation of the diisocyanate was the isolation of addition products, which could be identified readily. The numerous diisocyanates claimed in some of the patent literature without evidence of their syntheses have been omitted. The tables are to serve as a reference guide for research chemists. The one or two references provided for each diisocyanate represent those references which best describe its physical constants and synthesis. Similarly, the enormous number of patents that are related to minor improvements in the manufacture of the basic commercial diisocyanates are not included.

The number of new diisocyanates being offered in developmental quantities, together with the continuous search for a greater chemical understanding and a wider range of applications, should serve to ensure the continuation of growth of knowledge in this field, which is barely a quarter-century old.

II. SYNTHESIS OF DIISOCYANATES

Although a variety of methods can be used to synthesize isocyanates, only one method, the phosgenation of amines, has become of commercial importance. Of these other methods the Curtius, Lossen, and Hofmann rearrangements are used to prepare isocyanates in the laboratory, especially when the corresponding amines either are not readily available or are sensitive to phosgene itself. The thermolysis of "masked" isocyanates is described in the literature, as are "interchange" reactions, which are used in the synthesis of low-boiling monoisocyanates. The reaction of a halide or sulfate with a cyanate salt, the classical synthesis of an isocyanate, is applied

TABLE 1
Aliphatic diisocyanates, OCN–R–NCO

R	bp/mm, °C	Method of preparation[a]	Yield %	References
-CO-	103.5/705	C	72.7	99
-CH$_2$-	37–38/14	B	42	190
-(CH$_2$)$_2$-	65–68/7.5	A	69	41
	75/14	B	47	191
	125/100	D	—	105
-(CH$_2$)$_3$-	85–86/14	B	62	191
	103/30	B	73	187
-CH(CH$_3$)CH$_2$-	70/12.5	A	69	41
	83.5/25	B	42	191
-(CF$_2$)$_3$-	84–85	B	—	192
-(CH$_2$)$_4$-	102–104/14	A		6
	112–113/19	B	51	187
-(CF$_2$)$_4$-	105–106	B	—	192
-(CH$_2$)$_5$-	123–125/15	B	—	6
	116/11	B	—	193
-CH(CH$_3$)CH$_2$(CH$_2$)$_2$-	98–99/9	A	—	6
-(CH$_2$)$_6$-	130–132/14	A	—	6
	130/13	A	86	194
-(CH$_2$)$_7$-	140–142/14	B	—	6
	112–114/2.0	B	—	68
-(CH$_2$)$_3$C(CH$_3$)$_2$CH$_2$-	120–122/12	A	—	6
-(CH$_2$)$_8$-	146–148/11	B	—	6
	97–105/0.1	A	—	195
-(CF$_2$)$_8$-	105/220	B	—	192
-CH$_2$C(CH$_3$)$_2$CH$_2$CH(CH$_3$)CH$_2$-	106/5	A	—	6
-CH(C$_2$H$_5$)CH$_2$(CH$_2$)$_4$-	85–88/0.1	A	80	195
-CH(C$_2$H$_5$)CH$_2$CH$_2$CH(C$_2$H$_5$)-	66–70/0.5	A	93.5	195
-(CH$_2$)$_9$-	121/0.5	B	—	6
	128/2.5	B	33	196
-(CH$_2$)$_{10}$-	128/0.8	B	—	6
	134/0.54	B	—	196
-CH(C$_2$H$_5$)CH$_2$(CH$_2$)$_6$-	110–112/0.1	A	94.6	195
-CH$_2$CH(C$_2$H$_5$)CH$_2$CH$_2$CH(C$_2$H$_5$)CH$_2$-	95–96/2.0	A	55.0	195
-(CH$_2$)$_{11}$-	124/0.06	A	—	6
	151–152/3	B	—	197
-(CH$_2$)$_5$CO(CH$_2$)$_5$-	205–207/6	A	82.5	198
-(CH$_2$)$_{12}$-	135/0.08	A	—	6
	168–169/3	B	—	199
-(CH$_2$)$_{13}$-	155–160/0.04	B	50	197
-(CH$_2$)$_{14}$-	175/0.59	B	50	199
-(CH$_2$)$_{15}$-	157–160/0.2	B	56	197
-(CH$_2$)$_{16}$-	185/0.33	B	31	199

[a] In Tables 1–21 the letters under "Method of preparation" refer to the subsection of Section II in which the procedure is covered.

in the synthesis of organometallic isocyanates. The availability of isocyanic acid from the pyrolysis of either cyanuric acid or urethans has made available a direct synthesis of isocyanates from activated olefins. Also, nitro compounds and amines have been converted to the isocyanates by reaction with carbon monoxide in the presence of suitable catalysts.

Since the phosgenation procedures are those used commercially, the major portion of this part is concerned with the syntheses of isocyanates which employ phosgene. "Working examples" are described in Section II.G.

A. Phosgenation of Amines, Sulfonamides, Ureas, and Carbamates

The first synthesis of an isocyanate by the phosgenation of an amine or amine salt was reported by Hentschel in 1884 (2). In this procedure the diamine is dissolved in an inert solvent and added to an excess of phosgene in the same solvent at a temperature below 20°C (base phosgenation method). The slurry that is produced contains dicarbamoyl chlorides (1), and diamine dihydrochlorides (2). On heating at 50–70°C, 1 dehydrochlorinates to the diisocyanate 3. Subsequent heating above 100°C in the presence of excess phosgene slowly converts compounds 2 to 3 (Fig. 1).

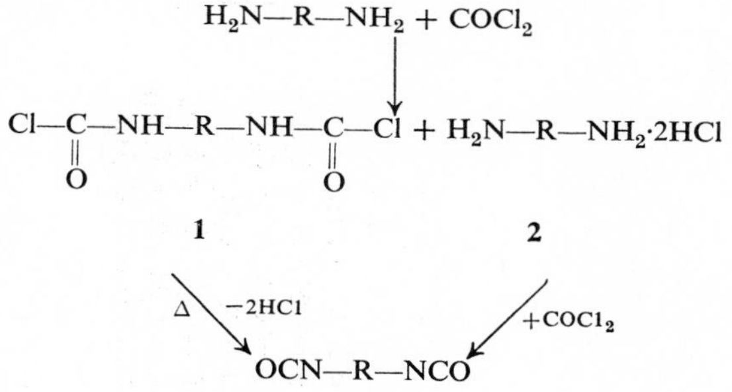

$$H_2N-R-NH_2 + COCl_2$$

$$Cl-\underset{\underset{O}{\|}}{C}-NH-R-NH-\underset{\underset{O}{\|}}{C}-Cl + H_2N-R-NH_2 \cdot 2HCl$$

$$\text{1} \qquad\qquad\qquad\qquad \text{2}$$

$$\Delta \searrow -2HCl \qquad\qquad \nearrow +COCl_2$$

$$OCN-R-NCO$$

$$\text{3}$$

Fig. 1. Phosgenation of diamines.

In an alternate procedure, the diamine is treated with hydrogen chloride or carbon dioxide to afford the dihydrochloride or dicarbamic acid, respectively, followed by phosgenation at temperatures above 100°C. The disadvantage of this method is that the gaseous phosgene reacts very slowly with the solid diamine salts. A relatively fast conversion is achieved with the less basic aromatic diamine dihydrochlorides, but the aliphatic diamine dihydrochlorides require exceedingly long reaction times. ω-Chloroalkylene isocyanates (4) are by-products, especially when higher boiling solvents are used. According to Siefken (6) the formation of 4 proceeds via the carbon-

$$OCNRNCO \rightarrow OCNRNCCl_2 \rightarrow OCNRCl + ClCN \qquad (1)$$

$$\text{4}$$

(Text continues on p. 382)

TABLE 2
Cycloaliphatic Diisocyanates

Structure	bp/mm, °C	Method of preparation	Yield, %	References
OCN $\diagup$ NCO (cyclopropane)	46/4.0	B	—	200
OCNOC $\diagup$ CONCO (cyclopropane)	90/5.0	B	—	200
cyclobutane with CH$_2$NCO, CH$_2$NCO	128–131/13–14	A	75	31
R = R′ = CH$_3$	91–93/5	A	56	201
R = CH$_3$; R′ = C$_2$H$_5$	84–88/1.0	A	—	201
R = CH$_3$; R′ = C$_3$H$_7$	100–102/1.0	A	—	201
R = R′ = C$_2$H$_5$	105–106/1.0	A	—	201
R = C$_2$H$_5$; R′ = C$_4$H$_9$	134–138/1.0	A	—	201
R = C$_2$H$_5$; R′ = C$_8$H$_{17}$	100/0.01	A	—	201
R = H; R′ = C$_{10}$H$_{21}$	85–90/0.006	A	—	201
cyclohexane NCO, NCO	119–120/15	B	52	191

374

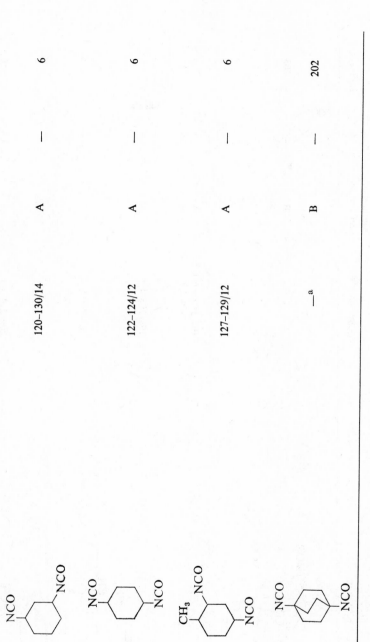

120–130/14	A	—	6
122–124/12	A	—	6
127–129/12	A	—	6
—[a]	B	—	202

TABLE 2 (continued)

Structure	bp/mm, °C	Method of preparation	Yield, %	References
(hexachloro bicyclic structure with two NCO, Cl, Cl–Cl, Cl)	—[a]	B	—	203
(OCN–cyclohexyl–CH₂–cyclohexyl–NCO)	165–180/0.6–0.5 179/0.9	A A	— 95	6 11
(OCN–cyclohexyl–CH(CH₃)–cyclohexyl–NCO)	198–208/4–5	A	—	6
(OCN–cyclohexyl–C(CH₃)₂–cyclohexyl–NCO)	208–230/8–10	A	—	6
(OCN, CH₃–cyclohexyl–CH₂–cyclohexyl–NCO, CH₃)	200–212/5.0	A	—	6

376

Structure	b.p. (°C/mm)			
(structure: OCN–cyclohexyl–CH₂–cyclohexyl–NCO with CH₃ groups)	198–210/3.0	A	—	6
(structure with CH₃ groups, OCN/NCO)	205–212/7.0	A	—	6
(C₂H₅, NCO, NCO cyclohexane)	140–144/19	A	—	6
CH₃ NCO cyclohexyl (CH₃)₂C–NCO	93–96/0.5	A	70	35
NCO, CH₂NCO, CH₃, (CH₃)₂ cyclohexane	158–159/15	A	83.5	32

377

TABLE 2 (*continued*)

Structure	bp/mm, °C	Method of preparation	Yield, %	References
CH_2NCO, CH_2NCO (on cyclohexane)	165–168/23–24	A	—	6
CH_2NCO ... CH_2NCO (on cyclohexane)	154–156/11	A	—	6
CH_2NCO, $(CH_2)_3$—NCO, CH_3, CH_3 (on cyclohexane)	165–172/4	A	—	6
H NCO H NCO ... H NCO H NCO (bicyclic)	130–134/0.5	A	90	—[b]

a The diisocyanate was not isolated.
b E. F. Cox, O. T. Manning, and H. A. Stansbury, U.S. Pat. 3,148,202 (1964); *Chem. Abstr.*, **61**, 11911 (1964).

TABLE 3
Unsaturated Aliphatic Diisocyanates

Structure	bp/mm, °C	mp, °C	Method of synthesis	Yield, %	References
OCN—CH=CH—NCO	150–155/750	67–69	B	—	205
OCN(Me$_2$N)C=C(NMe$_2$)NCO	125–128/1.2	—	F	—	206
OCNCH$_2$—C≡C—CH$_2$NCO	76–77/6	—	C	—	98
OCNCH$_2$CH=CHCH$_2$NCO	106–111/13–19	—	C	65	97
OCN(CH=CH)$_2$NCO	60–65/0.15[a]	84	B	73	207
OCNCH$_2$CH$_2$OOC—CH=CH—COOCH$_2$CH$_2$NCO	147/0.05	—	A	—	208
OCN(CH$_2$)$_4$C≡C(CH$_2$)$_4$NCO	181–182/0.1	—	A	—	209

[a] Sublimation.

TABLE 4
Aliphatic Ether Diisocyanates

Structure	bp/mm, °C	Method of preparation	Yield, %	References
OCNCH$_2$OCH$_2$NCO	90–91/28	B	34	191
OCNCH$_2$O(CH$_2$)$_2$OCH$_2$NCO	80–81/0.1	D	68	210
OCN(CH$_2$)$_3$O(CH$_2$)$_3$NCO	137/14	A	—	6
OCNCH(CH$_3$)O(CH$_2$)$_2$OCH(CH$_3$)NCO	69–70/1.0	D	74	210
	65/0.6	E	80	116
OCN(CH$_2$)$_2$CH(OCH$_3$)CH$_2$(CH$_2$)$_2$NCO	142–145/12	A	—	6
OCNCH(CH$_3$)O(CH$_2$)$_2$O(CH$_2$)$_2$OCH(CH$_3$)NCO	160–170/0.3	E	—	116
OCN[CH(OCH$_3$)]$_4$NCO	—	B	>90	211
OCN(CH$_2$)$_2$CH(OC$_4$H$_9$)CH$_2$(CH$_2$)$_2$NCO	142–155/2.0 (dec.)	A	—	6
OCNCH(CH$_3$)O(CH$_2$)$_4$OCH(CH$_3$)NCO	80–81/0.275	A	—	118
OCN(CH$_2$)$_3$O(CH$_2$)$_4$O(CH$_2$)$_3$NCO	130–135/0.1	A	—	6
OCN(CH$_2$)$_3$OCH$_2$C(CH$_3$)$_2$CH$_2$O(CH$_2$)$_3$NCO	109–115/0.11–0.16	A	—	212
OCN(CH$_2$)$_3$OCH$_2$C(C$_2$H$_5$)$_2$CH$_2$O(CH$_2$)$_3$NCO	152–158/2	A	81	212
OCN(CH$_2$)$_3$OCH$_2$C(C$_3$H$_7$)CH$_2$O(CH$_2$)$_3$NCO	160–165/1.5	A	70	212
OCN(CH$_2$)$_3$OCH$_2$—[cyclohexane ring with CH$_3$]—CH$_2$O(CH$_2$)$_3$NCO	175–185/1.0	A	75	212
[pyran ring structure with NCO, O, CH$_3$]	—ᵃ	B	—	213

ᵃ The diisocyanate itself was not isolated.

TABLE 5
Aliphatic Ester Diisocyanates

Structure	bp/mm, °C	Method of preparation	Yield, %	References
$OCN(CH_2)_2O$—CO—$O(CH_2)_2NCO$	120/1.0	A	76.3	38
$OCN(CH_2)_2O$—$CO(CH_2)_2CO$—$O(CH_2)_2NCO$	166/1.2	A	—	214
$OCN(CH_2)_2O$—$CO(CH_2)_4CO$—$O(CH_2)_2NCO$	287/1.2	A	—	214
$OCN(CH_2)_3CH(COOCH_3)NCO$	108–113/0.6–0.7	A	—	37
$OCN(CH_2)_4CH(COOCH_3)NCO$	123/0.45	A	—	37
$OCN(CH_2)_4CH(COOC_2H_5)NCO$	108.5–110/0.05	A	—	37
$OCN(CH_2)_4CH(COOC_3H_7)NCO$	122.5–123/0.275	A	—	37
$OCN(CH_2)_4CH(COOC_4H_9)NCO$	125–130/0.25	A	—	37
$OCN(CH_2)_4CH(COOC_8H_{17})NCO$	141/0.04	A	—	37
	146–151/0.12	D	—	39
	150–151/0.08	D	—	39
	161/0.07	D	—	39
	146–151/0.08	D	—	39
	101.5–103[a]	D	—[b]	39

[a] mp, °C.
[b] The yield is low.

381

imidoyl dichloride as shown (Eq. 1). Since the chloroisocyanates are potential chain terminators, they must be scrupulously removed, quite often by fractionation procedures. Hence the gain in phosgenation rate obtained at higher temperatures is offset to a large extent by the production of difficultly separable impurities.

The solvents most commonly used in phosgenation reactions include benzene, toluene, xylenes, chlorobenzene, o-dichlorobenzene, and 1,2,4-trichlorobenzene. Occasionally, more polar solvents, such as ethyl acetate, dimethyl sulfone, dioxane, and nitrobenzene are used. The highly polar solvents, such as dimethylsulfoxide (DMSO) and N,N-dimethylformamide (DMF) are not used because those solvents react with phosgene.

The reaction of amine hydrochlorides with phosgene is reported to be catalyzed by tertiary amines (20), metal halides (21), and boron trifluoride (22). The value of these catalysts is limited because the same substances can act as catalysts for the trimerization of isocyanates. Tetramethylurea is said to catalyze the reaction of aromatic amines with phosgene (23).

Many patents related to the phosgenation of diamines are issued, and they include processing details, such as continuous processes and operation under pressure. Vapor-phase reactions at high temperature are fast (24); however, undesired side reactions may also occur under these rigorous conditions.

A suitable phosgene precursor, such as trichloromethyl chloroformate, can be used instead of phosgene to convert primary amines into the corresponding isocyanates (25).

For economic reasons, only those diisocyanates derived from readily available, inexpensive diamines have assumed prominence on the market. For example, of the two aromatic diisocyanates which dominate the polyurethan market, one is derived from diaminotoluene (nitration of toluene and subsequent reduction) and the other from diaminodiphenylmethane (obtained by the condensation of aniline and formaldehyde). The dinitration of toluene affords mixtures of the 2,4 and 2,6-isomers, **5, 6,** and the commercial TDI (tolylene diisocyanate) is sold as distilled 80/20 or 65/35 mixtures of the two isomers. The pure 2,4-isomer, and crude undistilled variations are also offered. The synthesis of TDI (**7, 8**) is shown in Fig. 2 (26).

The dual reactivity of the two isocyanato groups in 7 renders this diisocyanate particularly useful for the preparation of prepolymers. However, this property acts as a disadvantage in the manufacture of polyurethan fibers (spandex-type materials).

The other important aromatic diisocyanate is diphenylmethane diisocyanate (MDI). The diamine precursor is obtained by the controlled condensation of aniline and formaldehyde in the presence of acid. In addition to the diamines, oligomeric polyamines are formed (Fig. 3). The commercial

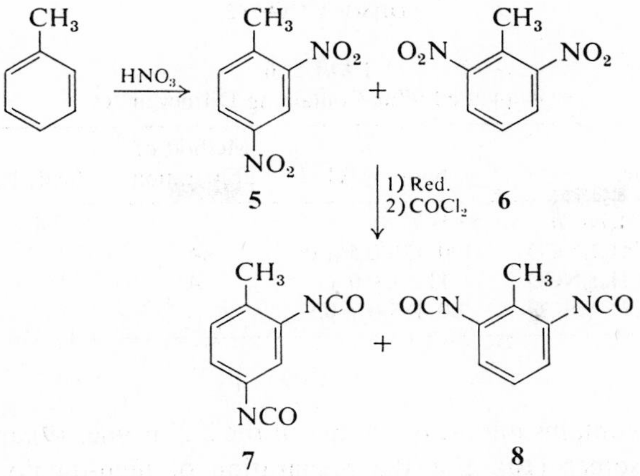

Fig. 2. Synthesis of TDI (tolylene diisocyanates).

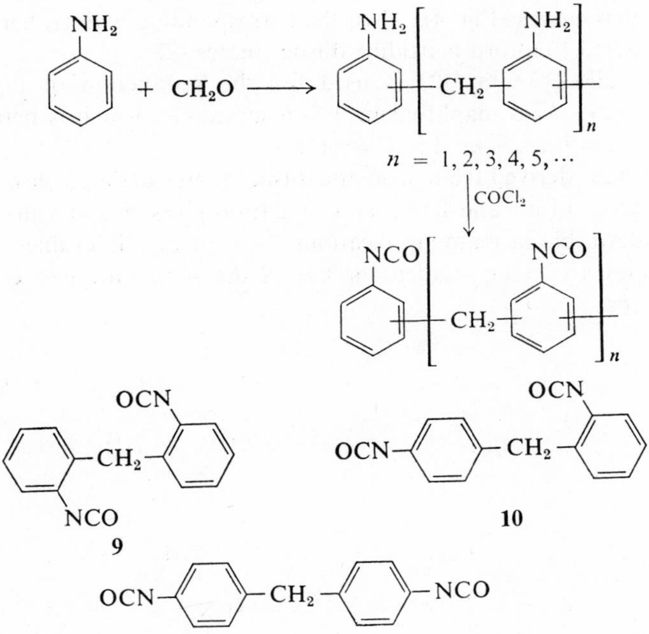

Fig. 3. Synthesis of MDI (diphenylmethane diisocyanate).

TABLE 6
Aliphatic Sulfur-Containing Diisocyanates

Structure	bp/mm, °C	Method of preparation	Yield, %	References
OCNCH$_2$SO$_2$CH$_2$NCO	53.5–54.5[a]	B	56	215
OCN(CH$_2$)$_2$S(CH$_2$)$_2$NCO	100–120/0.5–1.0	A	—	6
OCN(CH$_2$)$_3$S(CH$_2$)$_3$NCO	118/0.3–0.4	A	—	6
OCN(CH$_2$)$_6$S(CH$_2$)$_6$NCO	180–185/0.7–0.8	A	—	6

[a] mp, °C.

diisocyanate contains minute quantities of the 2,2′-isomer (9) and up to 10% of the 2,4′-isomer (10). For the preparation of high-quality fibers, pure MDI, which consists of 98% of the 4,4′-isomer (11) is required. The purification of the isomeric diamines is described in the patent literature (27). The undistilled polyisocyanates have found application in the production of rigid polyurethan foams.

A third type of commercially available diisocyanate is derived from the diaminobiphenyls (benzidines). For example, dianisidine and tolidine diisocyanate are produced by the phosgenation of the corresponding benzidine derivatives (Fig. 4). Also, the corresponding hydrazobenzenes can be phosgenated to afford benzidine diisocyanates (28).

Tolidine diisocyanate (12) is used for the production of high-quality elastomers (29). Also, naphthalene 1,5-diisocyanate, which is being offered in limited quantities, is used in elastomers.

Polyurethans derived from aromatic diisocyanates undergo slow oxidation in the presence of air and light. This oxidation gives rise to a discoloration that is undesirable in many applications. In view of this problem, aliphatic diisocyanates are being studied for use in the preparation of color-stable polyurethans.

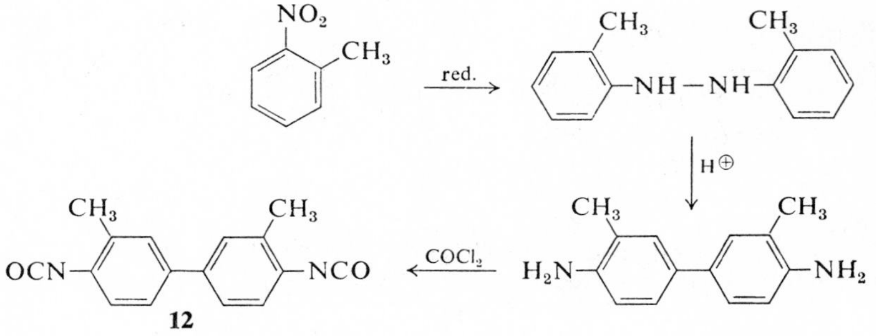

Fig. 4. Synthesis of TODI (tolidine diisocyanate).

Hexamethylene diisocyanate, prepared from the readily available hexamethylenediamine, has been studied extensively, but its low boiling point has created handling problems. Hexamethylene diisocyanate is a toxic chemical, and care should be exercised in its handling (see Section VI).

Other synthetic approaches to aliphatic diisocyanates evolve from the basic problem of the preparation of an inexpensive primary aliphatic diamine. Most approaches include reduction of nitrile to amine. For example, ammoxidation of xylene affords the phthalonitriles which, on hydrogenation and phosgenation (30) yield xylylene diisocyanate (XDI) (13) (Eq. 2). Dimerization of acrylonitrile followed by reduction and phosgenation is

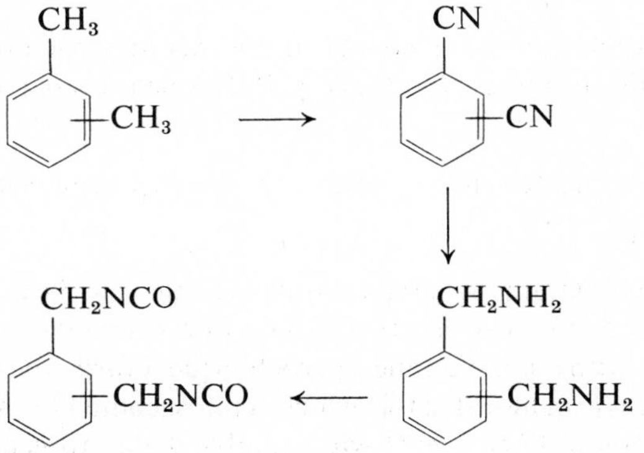

$$CH_3 \quad\quad\quad CN$$

(2)

13

another approach, but the product suffers from the same high vapor pressure and handling problems as HDI (31).

Two aliphatic diamines based on isophorone were converted to the corresponding diisocyanates, but again, the synthetic route requires reduction of nitrile to amine (32,33) (Fig. 5).

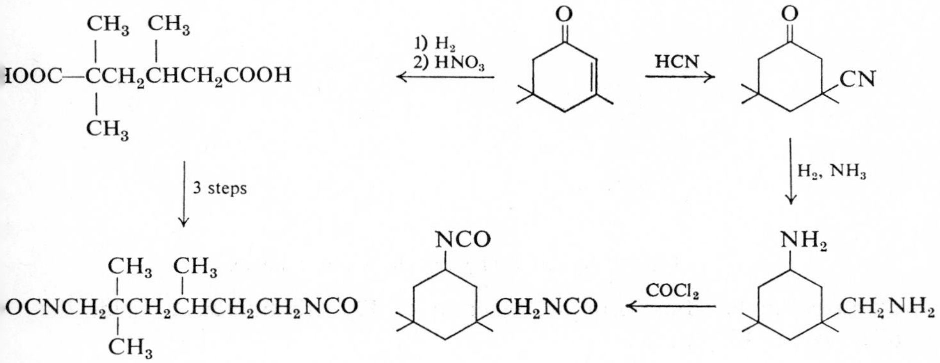

Fig. 5. Synthesis of diisocyanates based on isophorone.

Recently the direct conversion of nitriles into isocyanates by reduction in the presence of phosgene was reported (34) (Eq. 3).

$$RCN + H_2 + COCl_2 \longrightarrow RCH{=}NCOCl \xrightarrow[-HCl]{H_2} RCH_2NCO \qquad (3)$$

The Ritter reaction has been employed to directly introduce amines, or amine precursors, to unsaturated systems. Thus Bortnick (35) has prepared *p*-menthane 1,8-diisocyanate by phosgenating the diamine obtained from the reaction of the diolefin with hydrogen cyanide (Fig. 6). The *p*-menthane 1,8-diisocyanate can also be obtained directly from the bis-*N*-chloro amide (36).

The phosgenation of lysine esters (**14**) is another approach to the synthesis of the aliphatic diisocyanates, **15** (37) (Eq. 4). Lysine is prepared by fermentation methods.

$$H_2N(CH_2)_4CH(COOR)NH_2 + COCl_2 \rightarrow OCN(CH_2)_4CH(COOR)NCO \qquad (4)$$

14   **15**

The use of ethanolamine to prepare diisocyanates has been demonstrated by Brotherton and his co-workers (38,39). For example, esterification of maleic anhydride with ethanolamine hydrochloride affords the corresponding diesterdiamine dihydrochloride (**16**), which is phosgenated to the diisocyanate (**17**). The diisocyanate **17** can react with suitable dienes to generate a variety

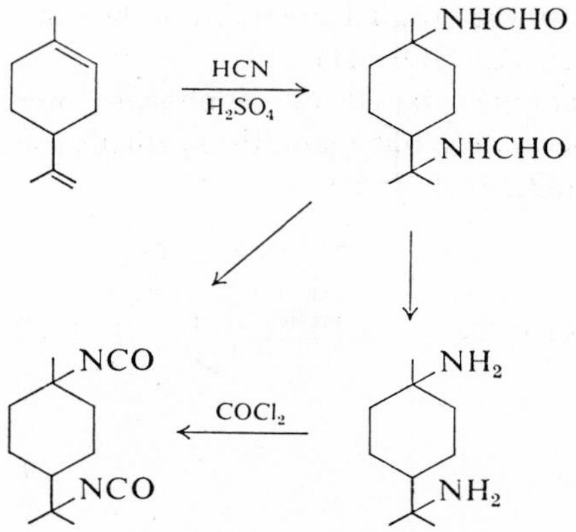

Fig. 6. Synthesis of *p*-menthane diisocyanate.

of diisocyanates. For example, reaction of **17** with butadiene yields the diisocyanate **18** (Eq. 5).

(*Text continues on p. 393*)

TABLE 7
Aliphatic Nitrogen-Containing Diisocyanates

Structure	bp/mm, °C	mp, °C	Method of preparation	Yield, %	References
OCNCH$_2$N(NO$_2$)CH$_2$CH$_2$NCO	—	1–2	B	70–75	216
OCN(CH$_2$)$_2$C(NO$_2$)$_2$CH$_2$CH$_2$NCO	—	39.5	A	80–85	217
OCN(CH$_2$)$_2$N(NO$_2$)CH$_2$CH$_2$NCO	—	35.5	B	72.5	216
OCNCH$_2$N(NO$_2$)CH$_2$CH$_2$N(NO$_2$)CH$_2$NCO	—	90–100	B	—	216
OCN(CH$_2$)$_2$N(NO$_2$)CH$_2$CH$_2$N(NO$_2$)CH$_2$CH$_2$NCO	—	91.5–93	B	76.5	216
OCN(CH$_2$)$_2$N(NO$_2$)CH$_2$(CH$_2$)$_2$N(NO$_2$)CH$_2$CH$_2$NCO	—	35	B	—	216
OCN(CH$_2$)$_2$N(NO$_2$)CH$_2$(CH$_2$)$_5$N(NO$_2$)CH$_2$CH$_2$NCO	—	75–76	B	—	216
OCN(CH$_2$)$_2$N(NO$_2$)CH$_2$CH$_2$N(NO$_2$)CH$_2$CH$_2$N(NO$_2$)CH$_2$CH$_2$NCO	—	90–95	B	—	216
OCN(CH$_2$)$_2$N(NO$_2$)CH$_2$CH$_2$C(NO$_2$)$_2$CH$_2$CH$_2$N(NO$_2$)CH$_2$CH$_2$CH$_2$NCO	—	88–90	B	—	218
OCN(CH$_2$)$_2$C(NO$_2$)$_2$CH$_2$CH$_2$N(NO$_2$)CH$_2$CH$_2$C(NO$_2$)$_2$CH$_2$CH$_2$NCO	—	92–95	B	—	219
OCN(CH$_2$)$_2$C(NO$_2$)$_2$CH$_2$CH$_2$C(NO$_2$)$_2$CH$_2$CH$_2$CH$_2$NCO	—	126–128	B	—	220
OCN(CH$_2$)$_3$N(COOCH$_3$)CH$_2$(CH$_2$)$_2$NCO	158–162/0.12	—	A	—	221
OCN(CH$_2$)$_3$N(COOC$_2$H$_5$)CH$_2$(CH$_2$)$_2$NCO	145–148/0.3	—	A	—	221
OCN(CH$_2$)$_3$N(CH$_3$)COO(CH$_2$)$_2$OCON(CH$_3$)CH$_2$(CH$_2$)$_2$NCO	—	120–122	A	—	221

387

TABLE 8
Aliphatic Diisocyanates Containing Aromatic Rings

Structure	bp/mm, °C	mp, °C	Method of preparation	Yield, %	References
OCNCH(C_6H_5)CH$_2$CH(C_6H_5)NCO	150–155/0.05	—	A	—	222
3-OCNCH$_2$C$_6$H$_4$CH$_2$NCO	159–162/12	—	A	—	6
4-OCNCH$_2$C$_6$H$_4$CH$_2$NCO	172/16	45–46	A	—	6
	144–145/46	46	A	58	223
3-OCNCH$_2$-4,6-(CH$_3$)$_2$C$_6$H$_2$CH$_2$NCO	165–167/3.0	—	A	—	224
4-OCN(CH$_2$)$_2$C$_6$H$_4$(CH$_2$)$_2$NCO	142–145/0.1–0.2	—	A	—	6
4-OCN(CH$_2$)$_3$C$_6$H$_4$(CH$_2$)$_3$NCO	180/1.0	—	A	—	225
3-OCN(CH$_3$)$_2$CC$_6$H$_4$C(CH$_3$)$_2$NCO	100–106/0.9	—	E	35	117
4-OCN(CH$_3$)$_2$CC$_6$H$_4$C(CH$_3$)$_2$NCO	110/0.55	78	E	16	117
1,4-(OCNCH$_2$)$_2$C$_{10}$H$_6$	183–184/1.0	—	A	—	6
1,5-(OCNCH$_2$)$_2$C$_{10}$H$_6$		88–89	A	—	6
4-OCNCH$_2$C$_6$H$_4$OC$_6$H$_4$CH$_2$NCO	210–220/1.0	54–55	A	56	223
4-OCN(CH$_2$)$_3$C$_6$H$_4$C$_6$H$_4$(CH$_2$)$_3$NCO	230–250/2–3	29	A	—	6
OCN(CH$_2$)$_3$ [fluorene] (CH$_2$)$_3$NCO	225–230/3	77–80	A	89	226
[anthracene] CH$_2$NCO / CH$_2$NCO	—	214–225	A	95	227
[bridged anthracene] NCO / NCO	188–192/0.8	83–85	A	—	228

388

TABLE 9
Aromatic-Aliphatic Diisocyanates

Structure	bp/mm, °C	Method of preparation	Yield, %	References
4-OCNC$_6$H$_4$CH$_2$NCO	152/19	A	—	6
4-OCNC$_6$H$_4$(CH$_2$)$_2$NCO	156/15	A	—	6
4-OCNC$_6$H$_4$CH$_2$C(CH$_3$)$_2$NCO	145–146/8	A	—	229
3-OCNC$_6$H$_4$CH(CH$_3$)NCO	134/11	A	—	6
3-OCNC$_6$H$_4$(CH$_2$)$_3$NCO	118–120/0.5	A	—	6
4-OCNC$_6$H$_4$(CH$_2$)$_4$NCO	100–115/0.5	A	—	6
3-OCNC$_6$H$_4$(CH$_2$)$_2$CH(CH$_3$)NCO	146–149/2	A	—	6
4-OCNC$_6$H$_4$(CH$_2$)$_2$CH(CH$_3$)NCO	142–145/3.5	A	—	6
	152–156/0.2–0.3	A	—	229
	168–169/3	A	—	6
	155–160/0.7	A	—	6
	117–118/0.04	A	—	6

389

TABLE 10

Aromatic Diisocyanates (Benzene Derivatives)

Structure	bp/mm, °C	mp, °C	Method of preparation	Yield, %	References
A: o-Diisocyanates (3,4-position)					
1-CF$_3$C$_6$H$_3$(NCO)$_2$	111–112/11.0	—	A	—	230
B: m-Diisocyanates (2,4-position)					
C$_6$H$_4$(NCO)$_2$	104–106/12	51	A	—	6
1-ClC$_6$H$_3$(NCO)$_2$	122–124/11	—	A	—	6
1,3-Cl$_2$C$_6$H$_2$(NCO)$_2$	120–126/1.5	64–65	A	—	6
1,5-Cl$_2$C$_6$H$_2$(NCO)$_2$	140–146/12	—	A	—	6
C$_6$CCl$_3$H(NCO)$_2$[a]	115–128/0.24	—	F	—	231
C$_6$CCl$_4$(NCO)$_2$	115–120/0.04	77–78	F	92	176,177
1-O$_2$NC$_6$H$_3$(NCO)$_2$	160–180/14	—	A	—	6
1-CH$_3$OC$_6$H$_3$(NCO)$_2$	—	75	A	—	6
1-C$_2$H$_5$OC$_6$H$_3$(NCO)$_2$	162–164/16	56	A	—	6
1,5-(CH$_3$O)C$_6$H$_2$(NCO)$_2$	—	125	A	—	6
1-C$_3$H$_7$OC$_6$H$_3$(NCO)$_2$	164–165/15	—	A	—	6
1-i-C$_4$H$_9$OC$_6$H$_3$(NCO)$_2$	180–181/14	—	A	—	6
1-CH$_3$C$_6$H$_3$(NCO)$_2$	124–126/18	33	A	—	6
1-CH$_3$(5-Br)C$_6$H$_2$(NCO)$_2$	116–126/1.0	59–60	F	—	232
1-CH$_3$C$_6$Cl$_3$(NCO)$_2$	154–158/0.6	101–104	F	85	178
1-CH$_2$ClC$_6$H$_3$(NCO)$_2$	152–155/11.0	—	F	89	179
1-CH$_2$BrC$_6$H$_3$(NCO)$_2$	112–115/0.06	—	F	84	179
	125–130/0.5	—	F	—	233
1-CF$_3$C$_6$H$_3$(NCO)$_2$	93–96/1.0	68	F	54	234
	104–105/5.0	36–37	B	58	230
1-CCl$_3$C$_6$H$_3$(NCO)$_2$	154–158/0.04	—	F	—	234
	134–140/0.3	—	F	—	179
	130–134/0.2	—	F	—	233

390

Compound	bp/mm	mp	Method		Ref.
$3\text{-}CH_3\text{-}(6\text{-}t\text{-}C_4H_9)C_6H_2(NCO)_2$	105–107/0.5	—	A	75–79	235
$3\text{-}CH_3C_6H_3(NCO)_2$	129–133/18	—	A	—	6
$3\text{-}CCl_3C_6H_3(NCO)_2$	130–135/0.4	—	F	—	233
$6\text{-}CH_3C_6H_3(NCO)_2$	133.5–134.5/17	—	A	—	6
$1,3\text{-}(CH_3)_2C_6H_2(NCO)_2$	140–144/17	—	A	—	6
$1,5\text{-}(CH_3)_2C_6H_2(NCO)_2$	139/15	70–71	A	—	6
$1\text{-}CH_3\text{-}3,5(C_2H_5)_2C_6H_1(NCO)_2$	109–113/0.6	71	A	94	236
$3\text{-}CH_3\text{-}1,5(C_2H_5)_2C_6H_1(NCO)_2$	108/0.8	—	A	—	135
$1\text{-}C_2H_5C_6H_3(NCO)_2$	124–126/12	—	A	—	135
$1,3,5\text{-}(C_2H_5)_3C_6H_1(NCO)_2$	98–101/4.0	—	A	—	237
$1\text{-}i\text{-}C_3H_7C_6H_3(NCO)_2$	161–171/14	—	A	—	135
$1,5\text{-}(i\text{-}C_3H_7)_2C_6H_2(NCO)_2$	136–140/12	—	A	—	6
$1,3,5\text{-}(i\text{-}C_3H_7)_3C_6H_1(NCO)_2$	152–156/12	59–61	F	66	175
$1\text{-}t\text{-}C_4H_9C_6H_3(NCO)_2$	144/0.4–0.45	—	A	—	135
$1\text{-}t\text{-}C_5H_{11}C_6H_3(NCO)_2$	138–140/8	—	A	—	238
$1\text{-}C_6H_{11}C_6H_3(NCO)_2$	160–174/0.3	—	A	—	239
$1\text{-}C_8H_{17}C_6H_3(NCO)_2$[a]	122–123/0.25	—	A	—	239
$1\text{-}C_9H_{19}C_6H_3(NCO)_2$[a]	236–240/2.6	—	A	—	238
$1\text{-}C_{12}H_{25}C_6H_3(NCO)_2$[a]	162–169/3.5	—	A	—	238
$(C_2H_5)_2C_6H_2(NCO)_2$[a]	138–140/8.0	—	A	—	238
$(i\text{-}C_3H_7)_2C_6H_2(NCO)_2$[a]	138–140/11	—	A	—	6
	148–160/12	—	A	—	6
C: p-Diisocyanates (1,4-position)					
$C_6H_4(NCO)_2$	110–112/12	94–96	A	—	6
$2\text{-}O_2NC_6H_3(NCO)_2$	165–173/4	59–61	A	—	6
$2,5\text{-}Cl_2C_6H_2(NCO)_2$	—	134–137	A	—	6
$2,6\text{-}Cl_2C_6H_2(NCO)_2$	154–158/16	64–67	A	—	240
$2\text{-}CH_3O(5\text{-}Cl)C_6H_2(NCO)_2$	—	134	A	—	6
$2\text{-}CH_3OC_6H_3(NCO)_2$	—	89	A	—	6
$2,5\text{-}(CH_3O)_2C_6H_2(NCO)_2$	—	180–181	A	—	6
$2,5\text{-}(C_2H_5O)_2C_6H_2(NCO)_2$	—	128	A	—	6

TABLE 10 (*continued*)

Structure	bp/mm, °C	mp, °C	Method of preparation	Yield, %	References
$C_6Cl_4(NCO)_2$	—	117–118	F	94.3	176
2-$CH_3C_6H_3(NCO)_2$	138–139/15	39	A	—	6
2-$CF_3C_6H_3(NCO)_2$	103–104/5.0	—	B	32	230
2,5-$(CH_3)_2C_6H_2(NCO)_2$	138–143/14	—	A	—	6
2,6-$(CH_3)_2C_6H_2(NCO)_2$	155–160/21	53–55	A	—	241
2-$CH_3(6$-$Br)C_6H_2(NCO)_2$	128–134/1.0	46–53	A	—	241
2,6-$(C_2H_5)_2C_6H_2(NCO)_2$	167–170/27	—	A	—	241
$(CH_3)_4C_6H_2(NCO)_2$	—	113.5–114	A	94	242

[a] Mixture of isomers.

392

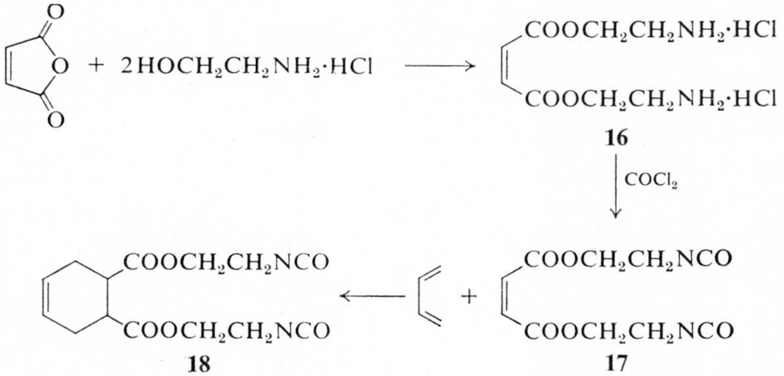

$$\text{maleic anhydride} + 2\,HOCH_2CH_2NH_2 \cdot HCl \longrightarrow \begin{array}{l} COOCH_2CH_2NH_2 \cdot HCl \\ \\ COOCH_2CH_2NH_2 \cdot HCl \end{array}$$

16

$$\downarrow COCl_2 \qquad\qquad (5)$$

$$\begin{array}{l} COOCH_2CH_2NCO \\ \\ COOCH_2CH_2NCO \end{array} \longleftarrow \quad + \quad \begin{array}{l} COOCH_2CH_2NCO \\ \\ COOCH_2CH_2NCO \end{array}$$

18   **17**

Ethanolamine hydrochloride reacts with phosgene to form 2,2′-diiso-cyanatodiethyl carbonate (**19**) (38) (Eq. 6).

$$2HOCH_2CH_2NH_2 \cdot HCl + COCl_2 \longrightarrow H_2NCH_2CH_2O-\underset{\underset{O}{\|}}{C}-OCH_2CH_2NH_2 \cdot 2HCl$$

$$\downarrow COCl_2$$

$$OCNCH_2CH_2O-\underset{\underset{O}{\|}}{C}-OCH_2CH_2NCO \qquad (6)$$

19

Isocyanates can also be synthesized by the phosgenation of N,N'-di-substituted ureas. This method is of importance for the synthesis of ethylene and trimethylene diisocyanate, which cannot be obtained by the direct phosgenation of the corresponding alkylene diamines.

Dialkylureas can react with phosgene either on nitrogen or oxygen. We have found that attack on nitrogen occurs preferentially if the substituents are primary alkyl groups (40,41). For example, ethylene urea (**20**) reacts with phosgene to form the N-carbonyl chloride (**21**). This readily dehydrochlorinates, preferably in the presence of a hydrogen chloride acceptor, to give ethylene diisocyanate (**22**) in good yield (40,41) (Eq. 7).

Several isocyanatobenzenesulfonyl isocyanates can be prepared from the phosgenation of the corresponding bisalkylureas. Similarly, the TDI homolog **24** is obtained by phosgenating the bisalkylurea **23** (42) (Eq. 8). A

$$\text{(7)}$$

$$OCNCH_2CH_2NCO$$
22

$$+ 2BuNCO$$

TABLE 11

Aromatic Diisocyanates (Naphthalene Derivatives)

Structure	bp/mm, °C	mp, °C	Method of preparation	Yield, %	References
1,4-$(OCN)_2C_{10}H_6$	—	67–70	A	—	6
1,5-$(OCN)_2C_{10}H_6$	—	130–132	A	—	6
2-CH_3-1,5$(OCN)_2C_{10}H_5$	165–170/1.0	59–60	A	—	241
2,6-$(OCN)_2C_{10}H_6$	—	152–154	A	—	6
2,7-$(OCN)_2C_{10}H_6$	—	152–153	A	—	6
	190–195/12	151–152	A	85	243
	—	132–133	A	—	6

review article has been published, relating to the synthesis of isocyanates and diisocyanates from sulfonamides and ureas (43).

A number of other substrates have been converted to isocyanates with phosgene. Although the examples referred to below are mainly mono-functional, the same reaction can be applied, in principle, to difunctional compounds.

The phosgenation of aziridines, using triethylamine as the hydrogen chloride acceptor, gives fair to good yields of 2-chloroalkyl isocyanates. For example, ethylene imine, in the presence of triethylamine, affords 2-chloro-ethyl isocyanate (**25**) in 53% yield (44) (Eq. 9).

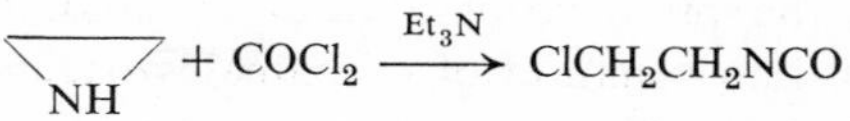

$$\text{(ethylene imine)} + COCl_2 \xrightarrow{Et_3N} ClCH_2CH_2NCO \qquad (9)$$

25

The phosgenation of amides and sulfonamides to acyl and sulfonyl isocyanates has been reported in the literature (45) (Eq. 10). However, since amides react with phosgene generally on oxygen to form imidoyl chlorides (46), this synthesis of acyl isocyanates must be approached with caution. Acyl isocyanates are best prepared by reacting primary amides with oxalyl chloride (47) (Eq. 11), or by reacting carboxylic acid chlorides with isocyanic acid in the presence of pyridine (48) (Eq. 12).

$$RCONH_2 + COCl_2 \rightarrow RCONCO + 2HCl \qquad (10)$$

$$RCONH_2 + (COCl)_2 \rightarrow RCONCO + CO + 2HCl \qquad (11)$$

$$RCOCl + HNCO \rightarrow RCONCO + HCl \qquad (12)$$

In contrast, sulfonamides can be phosgenated to sulfonyl isocyanates, although the reaction is sluggish and requires high temperatures (49). It has been demonstrated, however, that if an alkyl or aryl isocyanate is used as catalyst (42), sulfonamides phosgenate readily at relatively low temperatures

$$RSO_2NH_2 + COCl_2 \xrightarrow{R'NCO} RSO_2NCO + 2HCl \qquad (13)$$

(Eq. 13). Difunctional sulfonyl isocyanates can be prepared similarly.

Sulfanilamide (**26**) can be phosgenated to *p*-isocyanatobenzenesulfonyl isocyanate (**27**) in the absence of the catalyst, provided nitrobenzene is used as solvent (50) (Eq. 14). The diisocyanate **27** is useful in the synthesis of

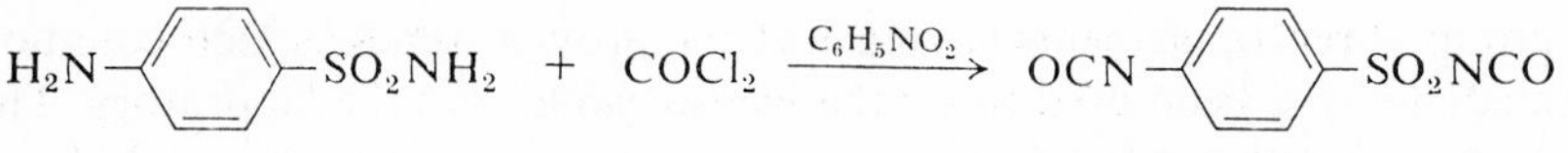

$$H_2N-\text{⟨C$_6$H$_4$⟩}-SO_2NH_2 + COCl_2 \xrightarrow{C_6H_5NO_2} OCN-\text{⟨C$_6$H$_4$⟩}-SO_2NCO \qquad (14)$$

26   **27**

color-stable polyurethans (51) because the regular oxidation pattern, observed in aromatic polyurethans, cannot occur. Furthermore, sulfonylarly-polyurethans are soluble in dilute base because the acidic hydrogen in the —SO₂NHCO—configuration allows salt formation with ammonia, amines, and alkali hydroxides (52).

The high-temperature phosgenation of carbamates and ureas has also been reported (53,54). Since it is well known that carbamates and ureas dissociate at high temperatures, these reactions may be those of the dissociated species only. However, in the presence of a catalytic amount of N,N-dimethyl-formamide (DMF), phosgenation of carbamates occurs at lower temperatures, indicating that a different mechanism may be operative (55). The chloroformamidinium chloride, generated from DMF and phosgene, reacts with the enol form of the carbamate to form the chloroformimidate **28**. The thermolysis of **28** to isocyanates and alkyl or aryl halides is known (56) (Eq. 15). DMF also catalyzes the phosgenation of isatoic anhydride (**29**) to

$$\text{RNHCOOR}' \rightleftarrows \underset{\underset{\text{OH}}{|}}{\text{RN}}{=}\text{C}{-}\text{OR}' + [\text{ClCH}{=}\overset{+}{\text{N}}(\text{CH}_3)_2]\text{Cl}^{\ominus} \longrightarrow$$

$$\underset{\underset{\text{Cl}}{|}}{\text{RN}}{=}\text{C}{-}\text{OR}' \overset{\Delta}{\longrightarrow} \text{RNCO} + \text{R}'\text{Cl} \quad (15)$$

28

2-isocyanatobenzoyl chloride (**30**) (Eq. 16) (57) and arylaminosulfonic acids to isocyanatobenzenesulfonyl chlorides (58).

$$(16)$$

29   30

B. Curtius, Hofmann, and Lossen Rearrangements

A group of rearrangements having acyl and aroyl nitrenes as their common intermediate have been used to synthesize isocyanates in the laboratory. The intermediate nitrene, RCON, can be generated by either thermolysis or photolysis. The reaction is often conducted in alcohol, and instead of the reactive isocyanates, the stable carbamates are isolated. The general reaction scheme is shown in Fig. 7.

The Curtius rearrangement in particular has been used extensively (6,59,60). A review article dealing with this reaction was published in 1946 (61) (see also Ref. 19).

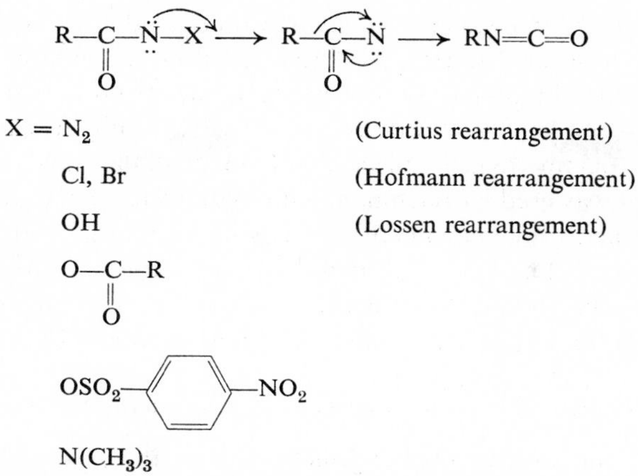

$$R-\underset{\underset{O}{\parallel}}{C}-\overset{..}{\underset{..}{N}}-X \longrightarrow R-\underset{\underset{O}{\parallel}}{C}-\overset{..}{\underset{..}{N}} \longrightarrow RN{=}C{=}O$$

$X = N_2$	(Curtius rearrangement)
Cl, Br	(Hofmann rearrangement)
OH	(Lossen rearrangement)

$$O-\underset{\underset{O}{\parallel}}{C}-R$$

$$OSO_2-\text{⟨benzene ring⟩}-NO_2$$

$$N(CH_3)_3$$

Fig. 7. Synthesis of isocyanates by rearrangement of acyl or aroyl nitrenes.

The Curtius rearrangement itself requires the synthesis of a carboxylic acid azide (31), which is generated either by the reaction of a carboxylic acid chloride with sodium azide or by the reaction of a carboxylic acid hydrazide with nitrous acid. The carboxylic acid azides are usually generated *in situ* in an inert organic solvent, such as benzene. By gently heating the reaction mixture, nitrogen is evolved concomitant with a rearrangement, thus affording the isocyanate (Eq. 17). Whereas this synthesis is of considerable

$$RCOCl + NaN_3 \longrightarrow RCON_3 \longleftarrow HNO_2 + H_2NNHCOR \qquad (17)$$

$$\mathbf{31}$$

$$\Big\downarrow -N_2$$

$$RNCO$$

use in the laboratory, it is very unattractive from a commercial point of view because of the cost of the starting materials and the inherent danger of handling large quantities of the thermally sensitive diazides.

Until recently, the Hofmann rearrangement was also very limited in its utility, because the isocyanate was generated in an aqueous phase (62). However, by using *t*-butyl hypochlorite instead of sodium hypobromite, this problem is solved. Lead tetraacetate was used recently to generate the nitrene intermediate (63,64), and potassium fluoride has also served to convert *N*-chlorobenzamide to phenyl isocyanate (65).

The Lossen rearrangement of hydroxamic acids to isocyanates has received very little use. A review of this reaction is available (66), as are patents related to the preparation of diisocyanates from dihydroxamic acids (67,68). For example, the disodium salt of sebacic dihydroxamic acid was converted to octamethylene diisocyanate in 50–60% yields. The Lossen rearrangement was used by Bachman and Goldmacher (69) in the conversion of carboxylic acids and nitromethane in polyphosphoric acid to the corresponding amines. Dimethyl sulfoxide was also used as solvent for this reaction (70). The alkali salts of acetyl (71) and benzoyl hydroxamates (72) react similarly to afford isocyanates as intermediates (Eq. 18). The dis-

$$\text{RCONHOCOR}' \xrightarrow{\text{NaOH}} \text{RCO}\overset{\ominus}{\text{N}}\text{OCOR}' \longrightarrow \text{RNCO} + \text{R}'\text{COO}^{\ominus} \qquad (18)$$

advantage of the Lossen rearrangement is that the isocyanate is usually generated in an aqueous system.

Several new types of reactions leading to nitrene intermediates are described in the recent literature (73–77). For example, when trimethylamine benzimide (32) prepared from the alkylation of benzhydrazide is heated, it yields phenyl isocyanate, which trimerizes in the presence of the generated trimethylamine (74–77) (Eq. 19). However, 1,8-octane diisocyanate was obtained in 60%

$$\text{C}_6\text{H}_5\text{CO}\overset{\ominus}{\overset{\cdot\cdot}{\text{N}}}\overset{\oplus}{\text{N}}(\text{CH}_3)_3 \xrightarrow{\Delta} (\text{C}_6\text{H}_5\text{NCO})_3 + \text{N}(\text{CH}_3)_3 \qquad (19)$$

32

yield from the thermolysis of bistrimethylamine sebacimide (78), and methyl isocyanate from 1-methyl-1-acetylimide-2-phenylpyrollidine (33) (74) (Eq. 20).

Several articles on the photolytic generation of acyl and aroyl nitrenes provide further evidence for both the existence and the rearrangement of these intermediate nitrenes (79–83).

C. Syntheses of Isocyanates by Thermolysis

"Masked" isocyanates are those compounds which regenerate isocyanates on heating. However, since these compounds are usually obtained from the reaction of the isocyanate itself with the corresponding masking agent, they are of little use synthetically.

Carbamates, the 1:1 addition products of isocyanates and alcohols, regenerate their starting materials at temperatures above 150°C. The carbamates from the more acidic alcohols and phenols dissociate at lower temperatures. This method has been used in the synthesis of low boiling monoisocyanates from carbonates and primary alkylamines. For example, the reaction of the cyclic carbonate **35** with ethylamine affords the ethyl carbamate **36,** which, on gentle heating to 250°C, yields ethyl isocyanate (6,84) (Eq. 21).

$$\tag{21}$$

Basic catalysts were used in the thermal dissociation of N-alkyl carbamates (85); however, these catalysts may be used only in the generation of secondary and tertiary isocyanates. Primary isocyanates undergo a rapid trimerization under basic conditions (85). The heating of carbamates with phosphorous pentoxide also produces isocyanates (86,87).

Transesterification processes are also used to generate a desired isocyanate as the lower boiling component. For example, heating tolylene diisocyanate with N-alkyl carbamates affords the lower boiling alkyl isocyanate in good yields (88). The dissociation of thiocarbamates proceeds in a similar manner, and a variety of mono isocyanates have been prepared by this method (89,90).

Likewise, alkylallophanates and alkylthioallophanates (the 2:1 adducts of isocyanates and alcohols, mercaptans, phenols, and thiophenols) undergo a rapid dissociation into 2 moles of isocyanate and the corresponding substrate (91). The allophanates and thioallophanates (**38**) (X = oxygen or sulfur), can be prepared from the corresponding allophanoyl chlorides (**37**), the latter being obtained from N,N'-dialkylureas and phosgene (40) (Eq. 22).

$$\text{RNHCONHR} + \text{COCl}_2 \longrightarrow \underset{\textbf{37}}{\text{RNHCON(COCl)R}} + \text{HCl}$$

$$\downarrow \text{R'XH} \qquad\qquad\qquad\tag{22}$$

$$2\text{RNCO} + \text{R'XH} \xleftarrow{\;\;\Delta\;\;} \underset{\textbf{38}}{\text{RNHCON(COXR')R}}$$

TABLE 12

Aromatic Diisocyanates (Biphenyl Derivatives)

Structure	bp/mm, °C	mp, °C	Method of preparation	Yield, %	References
2,4'-(OCN)$_2$C$_{12}$H$_8$	140–150/0.5	—	A	—	6
4,4'-(OCN)$_2$C$_{12}$H$_8$	150–160/0.5	—	A	—	6
2-O$_2$N-4,4'-(OCN)$_2$C$_{12}$H$_7$	—	123–125	A	81	182
3,3'-(CH$_3$)$_2$-4,4'-(OCN)$_2$C$_{12}$H$_6$	160–170/0.5	119–121	A	—	6
		68–69	A	—	6
		71–72	A	—	182
3,3'-(CH$_2$Cl)-4,4'-(OCN)$_2$C$_{12}$H$_6$	—	198–200	F	62	179
3,3'-(CH$_3$O)-4,4'-(OCN)$_2$C$_{12}$H$_6$	200–210/0.5	121–122	A	—	6,182

The pyrolysis of monosubstituted ureas is also a method for preparing isocyanates. For example, Sowa (92) synthesized isobutyl isocyanate (**40**) by heating isobutylurea (**39**) in the presence of boron trifluoride (Eq. 23).

$$(CH_3)_2CHCH_2NHCONH_2 \xrightarrow{BF_3} (CH_3)_2CHCH_2NCO + H_3NBF_3 \qquad (23)$$
$$\mathbf{39} \qquad\qquad\qquad \mathbf{40}$$

Stamm (93) has used urea in the synthesis of a variety of organometallic isocyanates of groups IV and V. For example, fusion of bis(tri-*n*-butyltin) oxide (**41**) with urea at 130–140°C for 1 hr affords tri-*n*-butyltin isocyanate (**42**) in 82% yield (Eq. 24). Trimethylsilicon isocyanate was similarly prepared

$$(Bu_3Sn)_2O + H_2NCONH_2 \rightarrow 2R_3SnNCO + 2NH_3 + H_2O \qquad (24)$$
$$\mathbf{41} \qquad\qquad\qquad \mathbf{42}$$

from trimethyl chlorosilane and urea (94). In general, the synthesis of isocyanates from monosubstituted ureas has found very little application in industry.

In contrast, substituted ureas, which undergo dissociation much more readily, have enjoyed a greater commercial utility. Thus methyl isocyanate can be prepared from the reaction of diphenylcarbamoyl chloride with methylamine (6) (Eq. 25). Heating *N,N'*-disubstituted ureas with diphenyl-

$$(C_6H_5)_2NCOCl + CH_3NH_2 \longrightarrow (C_6H_5)_2N-\underset{\underset{O}{\|}}{C}-NHCH_3$$

$$\downarrow \Delta$$

$$(C_6H_5)_2NH + CH_3NCO \qquad (25)$$

carbonate at 200–250°C is also another process for the synthesis of low boiling aliphatic monoisocyanates (95).

The reaction of 1,1'-carbonyldiimidazole **43** with primary amines affords the imidazolyl-substituted ureas **44**. Compounds **44** dissociate very readily into the corresponding isocyanate and imidazole (96) (Eq. 26). The urea

$$\mathbf{43} \qquad\qquad\qquad\qquad \mathbf{44}$$

44 (R = C_6H_5) in chloroform is dissociated to the extent of 16.1% at 20°C. This reaction forms the basis of a method for preparing isocyanates under mild conditions. For example, 1 mole of a primary amine is added to a solution of 1 mole of 1,1'-carbonyldiimidazole. Distillation of this mixture

(*Text continues on p. 409*)

TABLE 13

Aromatic Diisocyanates (Di- and Triphenylalkane Derivatives)

Structure	bp/mm, °C	mp, °C	Method of preparation	Yield, %	References
$2,2'\text{-}(OCN)_2C_{13}H_{10}$	142/0.5	46	A	>90	244
$2,4'\text{-}(OCN)_2C_{13}H_{10}$	152/0.5	35	A	>90	244
$4,4'\text{-}(OCN)_2C_{13}H_{10}$	160.5/0.5	40	A	>90	244
$3,3'\text{-}(Cl)_2\text{-}4,4'\text{-}(OCN)_2C_{13}H_8$	—	—	A	—	182
$3,3'\text{-}(Br)_2\text{-}4,4'\text{-}(OCN)_2C_{13}H_8$	195–205/1.0	—	F	—	245
$2,2',3,3',5,5',6,6'\text{-}(Cl)_8\text{-}4,4'\text{-}(OCN)_2C_{13}H_2$	Oil	—	F	—	176
$3\text{-}(CH_3)\text{-}4,4'\text{-}(OCN)_2C_{13}H_9$	190/1.0	—	F	—	174
$3\text{-}(CH_3O)\text{-}4,4'\text{-}(OCN)_2C_{13}H_9$	192/0.8	—	A	86	182
$3,5\text{-}(CH_3)_2\text{-}4,4'\text{-}(OCN)_2C_{13}H_8$	223–230/1.0	—	A	—	241
$2,2'\text{-}(CH_3)_2\text{-}4,4'\text{-}(OCN)_2C_{13}H_8$	208–211/5.0	65–67	A	—	6
$3,3'\text{-}(CH_3)_2\text{-}4,4'\text{-}(OCN)_2C_{13}H_8$	187–189/1.0	—	A	87.8	182
$2,4,6\text{-}(CH_3)_3\text{-}3,4'\text{-}(OCN)_2C_{13}H_7$	235–252/1.0	78–83	A	—	241
$2,2',5,5'\text{-}(CH_3)_4\text{-}4,4'\text{-}(OCN)_2C_{13}H_6$	200–240/1.2	127.6	A	—	6
$3,3'\text{-}(CH_3)_2\text{-}5,5'\text{-}(t\text{-}C_4H_9)_2\text{-}4,4'\text{-}(OCN)_2C_{13}H_6$	200–205/0.5	150–152	A	—	252
$3,5\text{-}(C_2H_5)_2\text{-}4,4'\text{-}(OCN)_2C_{13}H_8$	222–230/9.0	—	A	—	241
$3,3'\text{-}(CH_3O)_2\text{-}4,4'\text{-}(OCN)_2C_{13}H_8$	—	97–98	A	—	6
$4,4'\text{-}(CH_3O)_2\text{-}3,3'\text{-}(OCN)_2C_{13}H_8$	202/0.8	104	A	67	182
$3,3'\text{-}(C_2H_5O)_2\text{-}4,4'\text{-}(OCN)_2C_{13}H_8$	—	90–93	A	—	6
$4,4'\text{-}(C_2H_5O)_2\text{-}3,3'\text{-}(OCN)_2C_{13}H_8$	—	121–123.5	A	—	246
$6,6'\text{-}(C_2H_5O)_2\text{-}3,3'\text{-}(OCN)_2C_{13}H_8$	—	74	A	—	6
$2,2'\text{-}(CH_3)_2\text{-}5,5'\text{-}(CH_3O)_2\text{-}4,4'\text{-}(OCN)_2C_{13}H_6$	—	125–130	A	—	246
$3,3',5,5'\text{-}(C_2H_5)_4\text{-}4,4'\text{-}(OCN)_2C_6H_6$	185–190/0.2	128–129	A	—	247
$3,3',5,5'\text{-}(i\text{-}C_3H_7)_4\text{-}4,4'\text{-}(OCN)_2C_6H_6$	187–191/0.1	63–67	A	—	135

Compound	Bp (°C/mm)	Mp (°C)	Method	Yield (%)	Ref.
OCN—C$_6$H$_4$—CH$_2$—(naphthalene)—NCO	—	86–86.5	F	—	174
(4-OCNC$_6$H$_4$)$_2$CHCl	183–193/0.9	—	F	—	179
(4-OCNC$_6$H$_4$)$_2$C(CH$_3$)$_2$	212/5.0–6.0	91.5	A	—	6
[4-OCN(3-Cl)C$_6$H$_3$]$_2$C(CH$_3$)$_2$	—	100–101	A	—	6
(4-OCNC$_6$H$_4$)$_2$C(CF$_3$)$_2$	—[a]	—	—	—	248
(4-OCNC$_6$H$_4$)$_2$CHC$_6$H$_{11}$	195–200/0.1	—	A	—	6
(4-OCNC$_6$H$_4$)$_2$CH(2-O$_2$NC$_6$H$_4$)	—[a]	—	A	—	6
(4-OCNC$_6$H$_4$)$_2$CH(4-O$_2$NC$_6$H$_4$)	—[a]	—	A	—	6
[4-OCN-2,5-(CH$_3$)$_2$C$_6$H$_2$]$_2$CHC$_6$H$_5$	—	116–118	A	—	6
OCN—C$_6$H$_4$—CH$_2$CH$_2$—C$_6$H$_4$—NCO	223–227/20	—	A	—	6
OCN—C$_6$H$_4$—CH=CH—C$_6$H$_4$—NCO	—	89	A	—	249
	—	92–92.5	A	—	247
OCN—C$_6$H$_4$—CH=CH—C$_6$H$_4$—NCO[b]	—	149	A	82	250
OCN—C$_6$H$_4$—CF=CF—C$_6$H$_4$—NCO	—	148–150	B	—	251

[a] No physical constants are reported.
[b] Trans isomer.

TABLE 14
Aromatic Diisocyanates (Polybenzene Derivatives)

Structure	bp/mm, °C	mp, °C	Method of preparation	Yield, %	References
	—	133	A	—	6
	—	250–252	A	—	6
	—	234	A	—	6
	—	274	A	—	6

—	90–93	A	—	253
168/0.2	—	A	—	253

TABLE 15

Aromatic Diisocyanates (Miscellaneous Systems)

Structure	bp/mm, °C	mp, C°	Method of preparation	Yield, %	References
2-OCNC$_6$H$_4$OC$_6$H$_4$NCO-(4)	—	60–62	A	—	6
4-OCNC$_6$H$_4$OC$_6$H$_4$NCO-(4)	196/5.0	66–68	A	—	6
C$_6$Cl$_5$OC$_6$H$_3$(NCO)$_2$-(2,4)	175–190/0.1[a]	106–110	A	—	254
2,4,6-Br$_3$C$_6$H$_2$C$_6$H$_3$(NCO)$_2$-(2,4)	—	96–98	A	—	254
4-OCNC$_6$H$_4$O(CH$_2$)$_2$OC$_6$H$_4$NCO-(4)	—[b]	—	A	—	6
4-OCNC$_6$H$_4$O(CH$_2$)$_3$OC$_6$H$_4$NCO-(4)	—	103–105	A	70	255
4-OCNC$_6$H$_4$O(CH$_2$)$_2$O(CH$_2$)$_2$OC$_6$H$_4$NCO-(4)	—[b]	—	A	70	6
3-OCNC$_6$H$_4$-CO-C$_6$H$_4$NCO-(3)	—	118–120	A	—	6
3-OCN(4-CH$_3$)C$_6$H$_3$N=C=NC$_6$H$_3$(4-CH$_3$)NCO-(3)	—	113–115	E	—	256
3-OCN(4-CH$_3$)C$_6$H$_3$NHCONHC$_6$H$_3$(4-CH$_3$)NCO-(3)	—	241–245	F	—	257
(structure: bis(3,5-diethylphenyl)carbodiimide bearing two 3,5-diethyl-4-isocyanatobenzyl substituents)	—	88–90	E	—	258
4-OCNC$_6$H$_4$-N(CH$_3$)COO(CH$_2$)$_3$NCO	166/0.8	—	A	—	221
4-OCNC$_6$H$_4$-N(CH$_3$)COOC$_6$H$_4$NCO-(4)	—	85	A	—	221
[4-OCNC$_6$H$_4$-N(CH$_3$)COOCH]$_2$	—	97	A	—	221
4-OCNC$_6$H$_4$N=NC$_6$H$_4$NCO-(4)	—	158–161	A	—	6
	—	163	A	80	259
4-OCNC$_6$H$_4$N=NC$_6$H$_3$(3-CH$_3$)NCO-(4)	—	123–125	A	—	6

Compound		mp	Method	Yield	Ref.
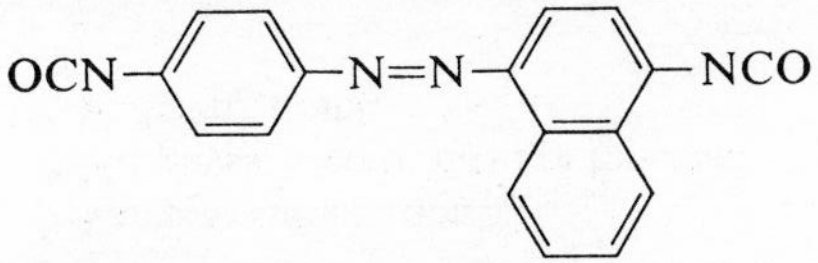 (structure above)	—	177–178	A	—	6
4-OCNC$_6$H$_4$O-CO-OC$_6$H$_4$NCO-(4)	—	110–118	A	52	260
2-OCNC$_6$H$_4$SC$_6$H$_4$NCO-(4)	—	73–74	A	—	6
4-OCNC$_6$H$_4$SC$_6$H$_4$NCO-(4)	175/1.0	—	F	—	173
4-OCN(3-CH$_3$O)C$_6$H$_3$CH$_2$SCH$_2$C$_6$H$_3$(3-CH$_3$O)NCO-(4)	—[b]	—	A	—	6
[3-OCN(4-CH$_3$O)C$_6$H$_3$SCH$_2$]$_2$	—	118	A	—	6
4-OCNC$_6$H$_4$-S-S-C$_6$H$_4$NCO-(4)	—	58–60	A	—	6
[3-OCN(6-CH$_3$)C$_6$H$_3$S]$_2$	—	77–78	A	—	6
[2-OCN(5-CH$_3$)C$_6$H$_3$S]$_2$	—	74–76	A	—	6
[3-OCN(4-CH$_3$)C$_6$H$_3$S]$_2$	—	88	A	—	6
[3-OCN(4-CH$_3$O)C$_6$H$_3$S]$_2$	—	66–67	A	—	6
[4-OCN(3-CH$_3$O)C$_6$H$_3$S]$_2$	—	101	A	—	6
4-OCNC$_6$H$_4$SO$_2$C$_6$H$_4$NCO-(4)	—	154	A	40–50	261
3-OCNC$_6$H$_4$SO$_2$C$_6$H$_4$NCO-(3)	—	126–129	A	68	261
3-OCN(4-Cl)C$_6$H$_3$SO$_2$C$_6$H$_3$(4-Cl)NCO-(3)	—	149–150	A	77	261
3-OCN(4-CH$_3$)C$_6$H$_3$SO$_2$C$_6$H$_3$(4-CH$_3$)NCO-(3)	—	132–134	A	—	6
3-OCN(4-CH$_3$O)C$_6$H$_3$SO$_2$C$_6$H$_3$(4-CH$_3$O)NCO-(3)	—	165	A	73	261
3-OCN(4-t-C$_4$H$_9$)C$_6$H$_3$SO$_2$C$_6$H$_3$(4-t-C$_4$H$_9$)NCO-(3)	—	120	A	—	6
4-CH$_3$C$_6$H$_4$SO$_2$C$_6$H$_3$(NO$_2$)-(2,4)	—	180^c	A	—	6
3-OCN(4-CH$_3$)C$_6$H$_5$SO$_2$OC$_6$H$_4$NCO-(4)	—	67.5	A	—	6
3-OCN(4-CH$_3$O)C$_6$H$_3$SO$_2$OC$_6$H$_4$NCO-(4)	—	104–106	A	—	6
4-OCN(3-CH$_3$O)C$_6$H$_5$CH$_2$SO$_2$CH$_2$C$_6$H$_5$(3-CH$_3$O)NCO-(4)	—[b]	—	A	—	6
[3-OCN(4-CH$_3$O)C$_6$H$_5$SO$_2$CH$_2$]$_2$	—	265–267	A	—	6
3-OCN(4-CH$_3$)C$_6$H$_3$SO$_2$NHC$_6$H$_3$(4-CH$_3$)NCO-(3)	—	191–196	A	—	6
(4-OCNC$_6$H$_4$SO$_2$NHCH$_2$)$_2$	—	178^c	A	—	6
[3-OCN(4-CH$_3$O)C$_6$H$_3$SO$_2$NHCH$_2$]$_2$	—	192	A	—	6
ClPO[OC$_6$H$_4$NCO-(4)]$_2$	—[b]	—	A	—	262

407

TABLE 15 (continued)

Structure	bp/mm, °C	mp, °C	Method of preparation	Yield, %	References
$C_6H_5CH=CHPO[OC_6H_3(4-CH_3)NCO-(3)]_2$	—[b]	—	A	72	262,263
$C_6H_5OPO[OC_6H_4NCO-(4)]_2$	—[b]	—	A	97	262
$CH_3PS[O-C_6H_3(4-CH_3)NCO-(3)]_2$	—[b]	—	A	—	262
$(CH_3)_2Si[O-C_6H_3(2-CH_3)NCO-(4)]_2$	—[b]	—	A	93	262
$4-OCNC_6H_4Si(CH_3)_2OSi(CH_3)_2C_6H_4NCO-(4)$	149–151/0.09	—	A	70	264
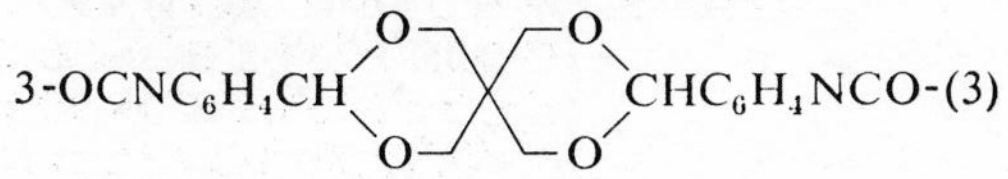 $3-OCNC_6H_4CH$ spiro-bis(1,3-dioxane) $CHC_6H_4NCO-(3)$	—	89–90	A	—	265
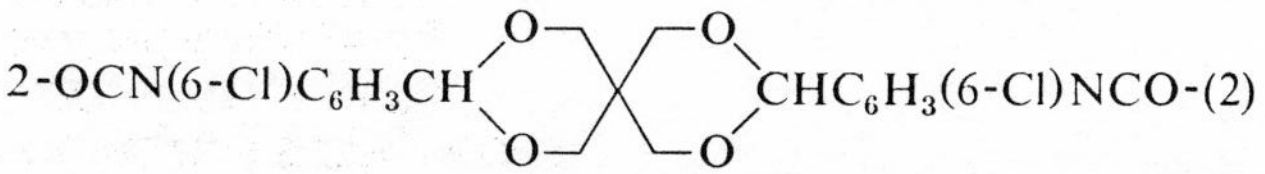 $2-OCN(6-Cl)C_6H_3CH$ spiro-bis(1,3-dioxane) $CHC_6H_3(6-Cl)NCO-(2)$	—	240–242	A	—	265
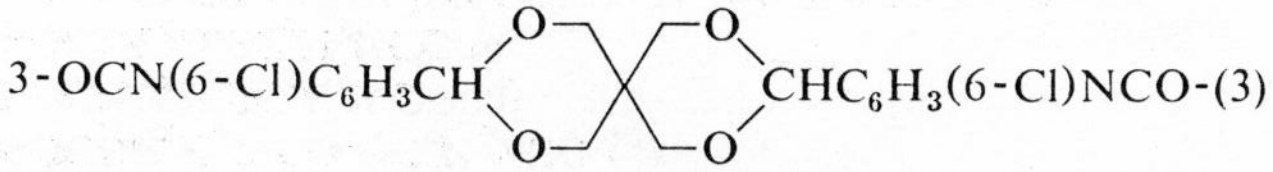 $3-OCN(6-Cl)C_6H_3CH$ spiro-bis(1,3-dioxane) $CHC_6H_3(6-Cl)NCO-(3)$	—	141	A	—	265

[a] Vacuum sublimation.
[b] No physical constants are reported.
[c] Softening point.

408

affords isocyanate and imidazole. The corresponding benzimidazole derivatives dissociate into benzimidazole and isocyanate.

Compounds **45** (R = *p*-tolyl) can be obtained in quantitative yield by reacting tosyl isocyanate with the corresponding amine (43). Since the arylsulfonylureas are stable compounds, they can be stored indefinitely, and the isocyanates can be generated by distillation (43) (Eq. 27). The recombination of the dissociation products is unlikely, because sulfonamides react with isocyanates at an exceedingly slow rate.

$$RSO_2NHCONHR \xrightarrow{\Delta} RSO_2NH_2 + R'NCO \qquad (27)$$
$$\mathbf{45}$$

The pyrolysis of secondary carbamoyl chlorides to isocyanates is also possible, provided a *t*-alkyl group is attached to the nitrogen atom. For example, 2-butene-1,4-diisocyanate **46** can be prepared by heating the biscarbamoyl chloride **47** in the presence of iron chloride (97) (Eq. 28).

$$ClCH_2CH{=}CHCH_2Cl + 2(CH_3)_3CNH_2 \longrightarrow$$

$$(CH_3)_3CNHCH_2CH{=}CHCH_2NHC(CH_3)$$

$$\Big\downarrow COCl_2$$

$$\xleftarrow[FeCl_3]{\Delta, -HCl} \; (CH_3)_3CNCH_2CH{=}CHCH_2NC(CH_3)_2$$
$$\underset{COCl}{|} \qquad\qquad \underset{COCl}{|}$$

$$OCNCH_2CH{=}CHCH_2NCO + 2(CH_3)_2C{=}CH_2 \qquad \mathbf{47}$$
$$\mathbf{46}$$
$$(28)$$

This method also was used to synthesize 2-butyne-1,4-diisocyanate (98).

Other pyrolysis procedures involve the decomposition of isocyanate dimers and trimers. Again, these compounds are best prepared from isocyanates, and therefore their synthetic value is *very* limited. However, in certain instances, isocyanates are prepared which cannot be obtained readily by other methods. For example, *N,N',N''*-trichlorocyanuric acid (**48**) gives chloroisocyanate (**49**) and carbonyl diisocyanate (**50**) on pyrolysis, depending on the reaction conditions (99) (Eq. 29). The carbonyl diisocyanate **50** is unstable, and polymerizes via a 1,4-cycloaddition sequence.

$$\xrightarrow{\Delta} 3ClNCO \longrightarrow CO(NCO)_2 + NCl_3 \qquad (29)$$
$$\mathbf{49} \qquad\qquad \mathbf{50}$$

TABLE 16
Heterocyclic Diisocyanates

Structure	bp/mm, °C	mp, °C	Method of preparation	Yield, %	References
(hydantoin with two 4-chloro-3-isocyanatophenyl groups)	—	172–174	E	—	8
(hydantoin with two 4-methyl-3-isocyanatophenyl groups)	—	165–166	E	—	266
(hydantoin with two 4-methoxy-3-isocyanatophenyl groups)	—	197	E	—	8
(hydantoin with two 4-phenoxy-3-isocyanatophenyl groups)	—	179–181	E	—	8
(hydantoin with two p-(2-isocyanato-2-methylpropyl)phenyl groups)	—	168–169	E	—	8

410

Structure	b.p./m.p.	m.p.			Ref.
OCNCH₂–(tetrahydrofuran ring)–CH₂NCO	76.2–77.5/0.01	—	A	—	267
dibenzofuran-diisocyanate (OCN···O···NCO)	—	164	A	—	268
OCN···NH···NCO (carbazole)	—	201–202	A	—	269
OCN···N–C₂H₅···NCO	—	122–123	A	—	6
OCN···NH···NCO (dichloro, Cl)	—	260–264	A	—	269
OCN···SO₂···NCO	—	288–290	A	—	6
OCN···CH₂/SO₂···NCO	—	166–168	A	—	6
	—	173–175	A	58.5	182

411

TABLE 17
Boron Diisocyanates

Structure	bp/mm, °C	mp, °C	Method of preparation	Yield, %	References
$C_6H_5B(NCO)_2$	55/0.4	—	D	—	270
$n\text{-}C_4H_9OB(NCO)_2$	45/0.35	—	D	—	270
$(CH_3)_2NB(NCO)_2$	—[a]	—	—	—	271
$(C_2H_5)_2NB(NCO)_2$	—[a]	—	—	—	271
$B_{10}H_{10}[C(NCO)]_2$	Oil	—	F	—	272

[a] Only spectral data are reported.

In contrast, sulfonyl diisocyanate (52) is considerably more stable. Compound 51, the major product obtained from the reaction of cyanogen chloride and sulfur trioxide, yields sulfonyldiisocyanate 52, sulfur dioxide, and chlorine when heated (100,101) (Eqs. 30, 31).

$$ClCN + SO_3 \longrightarrow ClSO_2NCO \tag{30}$$
$$\textbf{51}$$

$$2ClSO_2NCO \xrightarrow{\Delta} SO_2(NCO)_2 + SO_2 + Cl_2 \tag{31}$$
$$\textbf{52}$$

A variety of heterocyclic compounds give isocyanates on heating, but these reactions have very limited commercial importance. For example, when the thiazoline-4,5-diones (53) obtained from thioamides and oxalyl chloride are heated thioacyl isocyanates 54 and carbon monoxide result (102). The thioacyl isocyanates undergo rapid dimerization to form thiadiazindiones 55 via a 1,4-cycloaddition sequence (102) (Eq. 32). Also, 3,5-diphenyl-1,2,4-

$$\textbf{53} \qquad \textbf{54} \qquad \textbf{55} \tag{32}$$

oxadiazole (56) and 2,5-diphenyl-1,3,4-oxadiazole (57) undergo fragmentation to phenyl isocyanate and benzonitrile (103) (Eq. 33).

$$\textbf{56} \qquad\qquad\qquad\qquad \textbf{57} \tag{33}$$

D. Reaction of Organic Halides or Sulfates with Cyanates

The reaction of organic halides or sulfates with the salts of cyanic acid is of historic significance: Wurtz in 1848 (1) synthesized the first isocyanate by this procedure (Eq. 34). Ethyl isocyanate was prepared in 95% of theory

$$R_2SO_4 + KNCO \rightarrow RNCO + K_2SO_4 \qquad (34)$$

from the reaction of diethyl sulfate with potassium cyanate (104). A similarly high yield was obtained from ethyl bromide on reaction with potassium cyanate in dimethyl sulfone (105). The reaction of alkyl phosphates with potassium cyanate also affords isocyanates (106).

The use of such highly polar solvents as N,N-dimethylformamide (DMF) for this reaction results in extensive trimerization (107). Unless the generated isocyanate is intercepted by an active hydrogen compound, little isocyanate is obtained. However, carbamates can be had in good yield by reacting alkyl halides with potassium cyanate in DMF, in the presence of alcohols (108). This method is of general utility for the synthesis of isocyanato compounds of sulfur (109), phosphorus (110), silicon (111,112), germanium (113), and boron (114). Also, acyl isocyanates (115) can be obtained from the corresponding carboxylic acid chlorides and silver cyanate.

E. Addition and Interchange Reactions

The addition of isocyanic acid and iodine isocyanate to olefins affords isocyanate in low to good yield, depending on the reactivity of the olefin. Since isocyanic acid can be obtained from the pyrolysis of either cyanuric acid or urethans, this reaction will be of importance provided high yields are obtainable (116–118).

Hoover and Rothrock have demonstrated that isocyanic acid adds to activated olefins. For example, heating a mixture of isocyanic acid and p-diisopropenylbenzene (58) in toluene in the presence of hydroquinone and ammonium tosylate gave a mixture of products. Work-up of this mixture afforded both the mono and the diisocyanates, 59 and 60 (117) (Eq. 35).

TABLE 18
Silicon Diisocyanates

Structure	bp/mm, °C	mp, °C	Method of preparation	Yield, %	References
$(CH_3)_2Si(NCO)_2$	139.2	−31.2	D	84	112
	130–132	—	E	58	48
	72/70	—	D	75	273–275
	134–136	—	D	70.2	276,94
$(n\text{-}C_4H_9)_2Si(NCO)_2$	96/3.0	—	E	—	276
	100–105/3.0–5.0	—	D	79	277
$(C_6H_5)_2Si(NCO)_2$	319	22.9	D	—	111
	171/3.0	—	E	—	276
$(C_2H_5O)_2Si(NCO)_2$	175	—	D	—	111
	171	—	E	—	276
$(C_6H_5O)_2Si(NCO)_2$	172.5/10	—	D	—	273,274
$OCN(CH_3)_2Si\text{-}Si(CH_3)_2NCO$	105.5–106.5/30	—	D	54	278
$OCN(C_6H_5)_2Si\text{-}Si(C_6H_5)_2NCO$	—	123–124	D	84	278

TABLE 19

Tin Diisocyanates

Structure	bp/mm, °C	mp, °C	Method of preparation	Yield, %	References
$(C_4H_9)_2Sn(NCO)_2$	—	48–51	D	—	279
$[(C_4H_9)_2SnO \cdot (C_4H_9)_2Sn(NCO)_2]_n$	—	190–210	C	—	93
$(C_6H_5)_2Sn(NCO)_2{}^a$	—	—	D	—	279,280
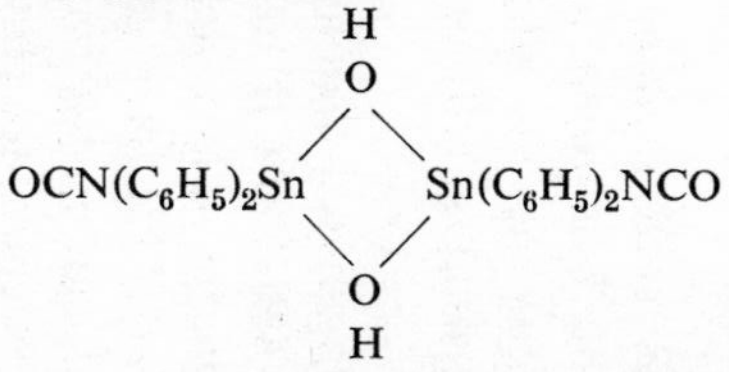 $OCN(C_6H_5)_2Sn \overset{H}{\underset{H}{\overset{O}{\underset{O}{\Large\diamond}}}} Sn(C_6H_5)_2NCO$	—	99–100	D	—	279

^{a} Isolated as 2,2'-bipyridine complex.

415

TABLE 20
Phosphorus, Arsenic, and Antimony Diisocyanates

Structure	bp/mm, °C	mp, °C	Method of preparation	Yield, %	References
$C_6H_5P(NCO)_2$	119–122/3.0	—	E	—	281
	76–78/0.2–0.25	—	E	61	48
$CH_3OP(NCO)_2$	30–35/2.0	—	E	41	281
	56/18.0	—	E	64.3	48
$C_2H_5OP(NCO)_2$	70/27	—	E	51.5	48
$CH_3PO(NCO)_2$	85–87/7.0	—	D	53	282
	71/2.0	—	D	30	283
$CH_2ClPO(NCO)_2$	80–82/0.9	—	D	—	110
$C_2H_5PO(NCO)_2$	80–81/7.0	—	D	39	282
	58–59/0.7	—	D	—	110
$i\text{-}C_3H_7PO(NCO)_2$	60–61/1.0	—	D	—	110
$C_{16}H_{33}PO(NCO)_2$	—[a]	—	D	—	110
$C_6H_5CH_2PO(NCO)_2$	145/1.0–2.0	—	D	—	110
$C_6H_5PO(NCO)_2$	118–122/3.0	—	D	—	110
	113–114/0.75	—	D	—	284
	95–100/0.1	—	D	—	285
	123/2.0	—	D	—	283
$C_4H_9OPO(NCO)_2$	80/3.0	—	D	—	286
$C_6H_5OPO(NCO)_2$	104/2.0	—	D	55	286,287
$(C_6H_5)_3As(NCO)_2$	—	40	C	96	93
$(C_4H_9)_3Sb(NCO)_2$	122/0.2	—	C	—	93
$(i\text{-}C_4H_9)_3Sb(NCO)_2$	122–124/0.15	—	C	75	93

[a] Crude liquid product.

The olefinic substrates used include vinyl ethers (116,118), isobutylene, isoprene, styrene, α-methylstyrene and m- and p-diisopropenylbenzenes (117). Isocyanic acid can also be added to certain aldehydes and ketones to form α-hydroxy isocyanates (119) and to imines to give α-amino isocyanates (120).

The reaction of halogenated aromatic hydrocarbons with isocyanic acid in the vapor phase (450–550°C), in the presence of cupric chloride on pumice, affords isocyanates in low yield (121).

Saunders and Bennet (122) have reacted primary amines with isocyanic acid and hydrogen chloride in the vapor phase, obtaining mixtures of isocyanates and carbamoyl chlorides (Eqs. 36, 37).

$$RNH_2 + HNCO + HCl \rightarrow RNCO + NH_4Cl \tag{36}$$

$$RNCO + HCl \rightarrow RNHCOCl \tag{37}$$

Iodine isocyanate, which can be generated readily from silver cyanate and iodine (123), adds to a wide variety of olefins to afford α-iodo isocyanates (124). A recent article by Hassner and co-workers (125) describes the scope and synthetic utility of this reaction. The progress of the reaction can be followed by noting the disappearance of the iodine color. The reaction time depends on the olefin used, but it is usually several hours. Stereospecific addition occurs; the iodine and isocyanate functions are introduced trans to each other. For example, from *cis* and *trans*-2-butene (126) there is obtained the *threo* and *erythro* 3-iodo-2-butyl isocyanates (61) and (62), respectively, in 85% yields (Eqs. 38, 39).

$$\underset{\substack{\text{H} \quad\quad \text{H}}}{\overset{\substack{\text{CH}_3 \quad\quad \text{CH}_3}}{\text{C}=\text{C}}} + \text{INCO} \longrightarrow \quad \mathbf{61} \tag{38}$$

$$\underset{\substack{\text{H} \quad\quad \text{CH}_3}}{\overset{\substack{\text{CH}_3 \quad\quad \text{H}}}{\text{C}=\text{C}}} + \text{INCO} \longrightarrow \quad \mathbf{62} \tag{39}$$

Elimination reactions, involving a four-center-type transition state, also serve to generate isocyanates. For example, Staudinger and Hauser (3) have demonstrated that addition of carbon dioxide to phosphine imides affords mixtures of isocyanates and carbodiimides. The formation of carbodiimide results from the consecutive cycloaddition of the generated isocyanate to the phosphine imide (Eqs. 40, 41).

$$R_3P{=}NR' + CO_2 \longrightarrow \left[\begin{array}{c} R_3P\text{---}NR' \\ | \quad | \\ O\text{---}C \\ \quad\quad O \end{array} \right] \longrightarrow R_3PO + R'NCO \tag{40}$$

$$R_3P{=}NR' + R'NCO \longrightarrow \left[\begin{array}{c} R_3P\text{---}NR' \\ | \quad | \\ O\text{---}C \\ \quad\quad NR' \end{array} \right] \longrightarrow R_3PO + R'N{=}C{=}NR' \tag{41}$$

Wadsworth and Emmons (127,128) obtained the carbamates **64** from the reaction of the phosphoramidate anion **63** with carbon dioxide. Compound **64,** when heated to 80°C, generates the corresponding isocyanate in good yield, provided R is a secondary or tertiary alkyl group (Eq. 42). In the

$$(C_2H_5O)_2\overset{\underset{O}{\uparrow}}{P}\overset{\ominus}{N}R + CO_2 \longrightarrow (C_2H_5O)_2\overset{\underset{\underset{\underset{\overset{\ominus}{O}}{O}}{C=O}}{|}}{\overset{\uparrow}{P}}{-}NR \longrightarrow RNCO + (C_2H_5O)_2\overset{\underset{\overset{O}{\diagdown}}{\underset{\ominus}{O}}}{\overset{O}{\diagup}}{P} \qquad (42)$$

63   **64**

reaction of **63** (R = C$_6$H$_5$) with carbon dioxide, however, no phenyl isocyanate was obtained, the basic reaction medium effecting a rapid trimerization (128).

The phosphorimidates **65,** on heating to 200°C, generate alkyl isocyanate and trialkyl phosphate (129) (Eq. 43). Trichlorophosphazenes react analogously with carbon dioxide to afford isocyanates (130). The drawback of

$$ROOC{-}N{=}P(OR')_3 \longrightarrow [ROOC{-}\overset{\overset{\displaystyle R'}{|}}{N}{-}\overset{\overset{\displaystyle O}{\uparrow}}{P}(OR')_2]$$

65

$$(43)$$

$$R'NCO + RO{-}\overset{\overset{\displaystyle O}{\uparrow}}{P}(OR')_2$$

the synthesis of isocyanates by addition of carbon dioxide to phosphorus–nitrogen double-bonded compounds is the subsequent reaction of the generated isocyanate with the substrate to form the corresponding carbodiimide.

1,2-Cycloaddition reactions of isocyanates across carbon–nitrogen double bonds are also reported. For example, heating benzaldehyde methylimine **(66)** with phenyl isocyanate affords methyl isocyanate in 90% yield (131) (Eq. 44). This method can be used to generate a wide variety of mono-isocyanates (131). Likewise, the exocyclic imino group in 3-cyclohexyl-2-cyclohexylimino-4,5-diphenyl-4-oxazoline **(67)** undergoes exchange reaction with phenyl isocyanate to yield cyclohexyl isocyanate (132) (Eq. 45).

$$C_6H_5CH{=}NCH_3 + C_6H_5NCO \longrightarrow \begin{bmatrix} C_6H_5CH{-}\!\!-\!\!NCH_3 \\ \\ C_6H_5N\!\!-\!\!\diagdown \\ \qquad\qquad O \end{bmatrix} \qquad (44)$$

66

$$CH_3NCO + C_6H_5CH{=}NC_6H_5$$

$$\underset{\substack{\\ \\ NC_6H_{11} \\ \\ \textbf{67}}}{\overset{C_6H_5 \;\rule[0.5ex]{2em}{0.4pt}\; C_6H_5}{C_6H_{11}N\diagdown\!\!\diagup O}} + C_6H_5NCO \longrightarrow \underset{NC_6H_5}{\overset{C_6H_5 \;\rule[0.5ex]{2em}{0.4pt}\; C_6H_5}{C_6H_{11}N\diagdown\!\!\diagup O}} + C_6H_{11}NCO$$

$$(45)$$

The "interchange" reactions of heterocumulenes form another useful method for synthesizing isocyanates. In 1959 Case (133) heated phenyl isothiocyanate with α-naphthyl isocyanate and obtained phenyl isocyanate. Apparently this reaction occurs via the equilibria indicated in Eq. 46. The

$$RNCO + R'NCS \; \underset{\longleftarrow}{\overset{\longrightarrow}{\rule{0pt}{1.5ex}}} \; \begin{bmatrix} & O \\ & \| \\ RN & NR' \\ & \diagdown\!\!\diagup \\ & \| \\ & S \end{bmatrix} \; \underset{\longleftarrow}{\overset{\longrightarrow}{\rule{0pt}{1.5ex}}} \; RNCS + R'NCO{\uparrow} (46)$$

continued removal of the lowest boiling component shifts the equilibria in the desired direction. Tolylene diisocyanate can be used as the high boiling isocyanate source. The heating of tolylene diisocyanate with allyl isothiocyanate yields allyl isocyanate (134). Neumann and Fischer used the interchange reaction of carbodiimides and isocyanates to synthesize low boiling isocyanates (135), whereas Clemens et al. (136) showed that sulfurdiimides undergo the same reaction on being heated with isocyanates. Similarly, isocyanates can be obtained from the reaction of tosyl isocyanate with carbodiimides (137).

F. Miscellaneous Syntheses of Isocyanates

Isocyanides (isonitriles) can serve as intermediates for isocyanate synthesis. They can be oxidized to isocyanates with mercuric oxide (138), dimethylsulfoxide (139,140), pyridine N-oxide (141), nitrile oxides (142), and ozone (143). Isocyanates can be prepared also by the oxidation of isothiocyanates with mercuric oxide (144) (Fig. 8).

$$RNC + [X] \rightarrow RNCO$$

$$RNCS + HgO \rightarrow RNCO + HgS$$

$$X = HgO, (CH_3)_2SO, C_5H_4NO, RC{\equiv}N \rightarrow O, O_3$$

Fig. 8.   Preparation of isocyanates from isonitriles and isothiocyanates.

Isocyanides also can be converted to isocyanates via the carbonimidoyl dichlorides (68) (139,145–149) (Eq. 47). Thus addition of trace amounts of

$$RN{=}CCl_2 + H_2O \rightarrow RNCO + 2HCl \tag{47}$$

$$\textbf{68} \qquad\qquad \textbf{69}$$

water to 68 (R = SF$_5$) yields 60% of 69 (R = SF$_5$) (145). Formic acid (146,147) and acid anhydrides (148) can effect a similar conversion. In a related reaction, chloroformamidates, readily obtained from the dichloride and sodium alcoholate, decompose to isocyanate and alkyl halide (149) (Eq. 48). Trifluoromethylcarbonimidoyl difluoride has been converted with

$$RN{=}\underset{\underset{Cl}{|}}{C}{-}OR' \longrightarrow RNCO + R'Cl \tag{48}$$

water to trifluoromethylisocyanate (150).

Two classes of compounds, nitrile oxides (Eq. 49) and cyanates (Eq. 50), isomeric with isocyanates, may rearrange to isocyanates under certain conditions. The thermal rearrangement of nitrile oxides to isocyanates is restricted in general utility because nitrile oxide dimerization competes

$$RC{\equiv}N \longrightarrow O \underset{\Delta}{\longrightarrow} R{-}N{=}C{=}O \tag{49}$$

effectively with isocyanate formation. Where dimerization is prevented by steric hindrance, rearrangement to isocyanates is observed (151).

Cyanates, the other isocyanate isomers, can be obtained readily by a variety of methods (152). In contrast to nitrile oxides, only primary aliphatic cyanates rearrange readily to the corresponding isocyanates, and the tendency toward isomerization decreases with increasing alkyl chain length (153–156). The bridgehead dicyanate, 71, is prepared from 1,4-dihydroxybicyclo[2,2,2]-octane (70) by treatment with sodium hydroxide and cyanogen chloride. On heating with boron trifluoride trietherate, 71 rearranges to the diisocyanate 72 (156) (Eq. 50).

$$(50)$$

The direct conversion of amino and nitro compounds to isocyanates has received considerable attention of late. For example, Hardy and Benett (157) obtained phenyl isocyanate in 35% yield by reacting nitrobenzene with carbon monoxide (500 atm) in the presence of rhodium-on-carbon and anhydrous ferric chloride (Eq. 51). In other cases, the isocyanates obtained

$$C_6H_5NO_2 + 3CO \rightarrow C_6H_5NCO + 2CO_2 \qquad (51)$$

from nitrobenzenes and carbon monoxide are scavenged with alcohols. The following catalysts have been described: rhodium trichloride/ferric chloride (158), $IrCl(CO)(PPh_3)_2$ (159), palladium/alumina, and ferric chloride (160). Similarly, the carbonylation of amines to isocyanates, using palladium dichloride as the catalyst, was accomplished (161,162). Moderate yields of monoisocyanates and a 9.6% yield of tolylene diisocyanate were observed (162). Tolylene diisocyanate was also obtained from the corresponding diamine, carbon monoxide, and oxygen in the presence of nickel tetracarbonyl (163). Again, the isocyanates generated were scavenged by conducting the reaction in the presence of an alcohol (164).

Elimination reactions have been used to prepare isocyanates. For example, N-chloroformamides eliminate hydrogen chloride to yield isocyanates. Hurwitz and Auten have used this reaction to synthesize p-menthane 1,8-diisocyanate 74 from the corresponding bis-N-chloroformamide 73 (165) (Eq. 52). The use of quinoline as the hydrogen bromide scavenger in the

$$(52)$$

generation of isocyanates from N-bromoformamides was described in 1966 (166). However, Kühle (167) has shown that the reaction can be conducted without the base, especially when bromine or sulfuryl chloride is used as the oxidizing agent. For example, N-benzylformamide (75) is converted by sulfuryl chloride in thionyl chloride to benzyl isocyanate (76) in 71% yield (167) (Eq. 53). Chloroformamidine N-carbonyl chlorides (77), readily

$$C_6H_5CH_2NHCHO + SO_2Cl_2 \longrightarrow C_6H_5CH_2\underset{\underset{Cl}{|}}{N}CHO \qquad (53)$$

$$\mathbf{75}$$

$$\downarrow -HCl$$

$$C_6H_5CH_2NCO$$

$$\mathbf{76}$$

available from carbodiimide and phosgene, are in equilibrium with isocyanates and carbonimidoyl dichlorides at elevated temperatures (168) (Eq. 54).

$$RN{=}C{-}\underset{\underset{Cl}{|}}{N}(COCl)R \; \rightleftarrows \; RN{=}CCl_2 + RNCO \qquad (54)$$

$$\mathbf{77}$$

The reaction of certain ureas and carbamates with phosphorus pentachloride affords isocyanates (169). For example, from 1-p-toluenesulfonyl-3-n-butylurea (78) and phosphorus pentachloride, a mixture of n-butyl isocyanate and p-toluenesulfonyltrichlorophosphazene (79) is obtained (Eq. 55).

$$CH_3{-}\!\!\left\langle\!\!\bigcirc\!\!\right\rangle\!\!{-}SO_2NHCONHBu \; + \; PCl_5$$

$$\mathbf{78}$$

$$BuNCO \; + \; CH_3{-}\!\!\left\langle\!\!\bigcirc\!\!\right\rangle\!\!{-}SO_2N{=}PCl_3 \quad (55)$$

$$\mathbf{79}$$

Although the synthesis of isocyanates from carbamates and phosphorus pentachloride (170) and pyrocatechylphosphorus trichloride (171) has been demonstrated, these reactions have found no utility because of the problem of separating the phosphoryl chlorides from the generated isocyanates. An interesting method of synthesis for phosphorus-containing isocyanates was reported by Kirsanov (172): ethyl carbamate, when heated with phosphorus pentachloride, affords dichlorophosphoryl isocyanate (80) in good yield (Eq. 56).

$$H_2NCOOC_2H_5 + PCl_5 \longrightarrow \overset{\overset{\textstyle O}{\uparrow}}{Cl_2}PNCO + C_2H_5Cl + 2HCl \qquad (56)$$
$$\mathbf{80}$$

Diisocyanates can also be obtained by Friedel-Crafts-type reactions. For example, phenyl isocyanate reacts with sulfur dichloride to give the diisocyanate **81** (173) (Eq. 57). Likewise, chloromethylaryl isocyanates can be

$$(57)$$
$$\mathbf{81}$$

reacted with aromatic substrates to the mono and diisocyanates. For example, reaction of **82** with phenyl isocyanate in the presence of ferric chloride affords 4,4'-diphenylmethane diisocyanate (**83**) (174) (Eq. 58). The alkylation of aromatic isocyanates with olefins can be conducted using hydrofluoric acid as both the solvent and the catalyst (175).

$$(58)$$
$$\mathbf{83}$$

The halogenation of aromatic isocyanates is feasible, and reaction in the nucleus as well on the side chain has been reported (176–179). Radical addition to unsaturated isocyanates is known (180). The halogenated isocyanates are useful in the synthesis of flame-retardant polymers.

G. Examples

1. PREPARATION OF 80:20 TOLYLENE DIISOCYANATE FROM CRUDE TOLYLENEDIAMINE (181)

a. Phosgenation

The apparatus used for the atmospheric-pressure hot phosgenations was a fully baffled stirred reactor, consisting of a 4000-ml resin reaction kettle (Pyrex glass, 5 in. diameter, 14 in. length), fitted with a turbine-type agitator, a thermometer, a reflux condenser, a phosgene inlet, and an inlet for the diamine feed solution. Both inlets dip below the stirrer.

Dry o-dichlorobenzene (1000 ml) was placed in the reactor and saturated with phosgene at the reaction temperature (120–160°C). A preheated solution of 255 g (∼2.0 moles) of crude tolylene diamine in 1500 ml of o-dichlorobenzene was pumped into the reactor through a metering pump at a uniform rate of about 0.5–0.6 mole tolylenediamine (TDA) per hour. Simultaneously, an excess of gaseous phosgene was fed into the reaction mixture at a phosgene: TDA mole ratio of approximately 8:1. The reaction mixture was stirred at about 1500 rpm and maintained at reaction temperature during the TDA and phosgene feeding operation. Hydrogen chloride and excess phosgene were withdrawn through the water-cooled condenser and separated by means of an acetone–dry-ice-cooled condenser. The condensed phosgene may be evaporated and recycled or discarded. All gas streams are metered.

b. Work-Up Procedure

The crude reaction mixture was heated to and kept at 170°C, while a moderate stream of dry nitrogen was passed through the solution for about 30 min in order to remove dissolved hydrogen chloride and excess phosgene. The purged crude reaction mixture was then fed into a long tube externally heated by steam at atmospheric pressure. This long-tube flash-distillation apparatus is attached to a cyclone separator. During this operation, the distillation apparatus was kept under a vacuum of between 5–10 mm Hg. By this procedure a substantial part of the solvent, containing approximately only 1–3% of pure tolylene diisocyanate, was removed from the crude reaction mixture (which then consists of ca. 20–40% of crude TDI in o-dichlorobenzene).

This concentrated solution of crude TDI was then subjected to a second flash distillation by adding it slowly through a dropping funnel into a round-bottom flask, kept at 180–190°C and 1–2 mm Hg pressure, to distill the TDI and the solvent, leaving the resinous by-product as a residue in the flask.

The TDI content of the distillates is determined by vapor-phase chromatography, by titration with dibutylamine, or by a combination of these methods. Total TDI yields, ranging from 88–96% (based on 96% metaisomers in the crude TDA) are obtained.

The concentrated pure TDI solution from the second flash distillation and optionally, the dilute TDI solution from the first long-tube flash distillation, were distilled at reduced pressure through a fractionating column. Tolylene diisocyanate, boiling at 120°C (10 mm Hg) was recovered almost quantitatively from the fractional distillation; 2, 4/2, 6 isomer ratio 80:20 ± 2% η_D^{25} 1.5665–1.5667, mp 13.6°C, assay 99.5–99.9, hydrolyzable chloride < 0.005%.

2. 4,4'-Diphenylmethane Diisocyanate (6)

In 2000 ml of o-dichlorobenzene there was dissolved 800 g (8 moles) of phosgene, and the resulting solution was cooled with an ice-salt bath. To this was added slowly with stirring, through a heated dropping funnel, a hot solution of 200 g (1.01 moles) of 4,4'-diaminodiphenylmethane in 1000 ml of o-dichlorobenzene. The rate of addition was regulated so that the temperature of the phosgene solution did not rise substantially above 0°C. The fine suspension that resulted was slowly heated and additional phosgene (700 g, 7.1 moles) was added at 130°C until a clear solution appeared. After purgation with carbon dioxide, the solvent was removed under vacuum and the product purified by vacuum distillation. There was obtained 215 g (0.85 mole, 84% yield) of 4,4'-diphenylmethane diisocyanate at 156–158°C (0.1 mm).

3. 3,3'-Dimethoxy-4,4'-diphenyl Diisocyanate (6)

In 2500 ml of chlorobenzene there was suspended 600 g (1.9 moles) of dianisidine dihydrochloride. The suspension was heated to 150–160°C and treated with about 2.5 kg (25.2 moles) of phosgene over a period of 16 hr. Hot filtration on a steam-heated filter yielded only a minimal residue. The filtrate was cooled with stirring and 562 g (1.6 moles, 85% yield) of a complex containing 2 moles of diisocyanate and 1 mole of chlorobenzene, mp 125–126°C was obtained. Heating the complex under vacuum at 150°C generated the diisocyanate, mp 121–122°C. The filtrate could be used without further purification for a subsequent phosgenation.

4. p-PHENYLENE DIISOCYANATE—HYDROCHLORIDE METHOD (182)

A 12-liter reactor was charged with 625 g (5.75 moles) of p-phenylene-diamine in 8500 g (6500 ml) of dry o-dichlorobenzene, and the solution was heated to 100°C. Gaseous hydrogen chloride, mixed with nitrogen to prevent plugging, was added over 1.75 hr until 495 g (13.5 moles) had been added. The reactor temperature was allowed to rise to 145°C without external heating. Phosgene (1700 g, 16.8 moles) was then admitted over a 3-hr period at 140–150°C. After a 1.5-hr nitrogen purge at 150°C, the solution was cooled to 25°C and filtered to remove unconverted amine hydrochloride and ureas (ca. 20%). The filtrate was evaporated under reduced pressure in a long-tube evaporator, and the residue was fractionated to afford 508 g (55%) of p-phenylene diisocyanate (assay: 99.5%). Considerable sublimation accompanied the distillation. The solvent contained an additional 84 g (9.1%) of the diisocyanate.

5. m-XYLYLENE DIISOCYANATE—HYDROCHLORIDE METHOD (183)

In a 5-liter round-bottom flask equipped with a stirrer, an efficient con-denser, a thermometer, and a gas-inlet tube were placed 136 g (1 mole) of commercial m-xylylene diamine and 3100 ml of chlorobenzene (solvent/diamine weight ratio = 24.6). While the mixture was agitated, dry hydrogen chloride was passed through at a rate of 1690 cc/min, allowing the tempera-ture to rise to 60°C. After 1 hr complete conversion of the diamine to its dihydrochloride occurred. The resulting rather thin and easily agitated slurry was heated to reflux and 70 ml of chlorobenzene-water azeotrope was removed (solvent/diamine weight ratio now = 24). Phosgene was then passed through a flowmeter to the dry reaction mixture at a rate of 221 cc/min while the temperature was maintained between 120 and 125°C. After 10–12 hr, a clear yellow solution resulted. The solution contained a very small amount of white solid (very fluffy). It was purged at reflux with dry nitrogen for 1–2 hr and cooled to room temperature (25–30°C). The product was filtered with suction and a 50-g portion of the filtrate analyzed; from this it was found that the total yield of m-xylylene diisocyanate was 183 g or 98.0% yield. The chlorobenzene was removed under vacuum and the residue was transferred to a 300-ml flask and distilled under vacuum. The fraction of m-xylylene diisocyanate boiling at 126°C/1 mm was collected and found to weigh 178 g (91.4% yield). Analysis revealed that it was 99.3% pure and contained 0.08% total chloride.

TABLE 21

Sulfonyl Diisocyanates

Structure	bp/mm, °C	mp, °C	Method of preparation	Yield, %	References
$SO_2(NCO)_2$	139	—	D	80	188
$S_2O_5(NCO)_2$	40–41/10	—	C	40	288,101
	46/0.2	26.8	C	17	288
	—	26	D	—	289
$OCNSO_2(CH_2)_4SO_2NCO$	200/0.1	60–63	A	57	50
$OCNSO_2(CH_2)_5SO_2NCO$	95–110/0.5	—	A	60.2	50
$OCNSO_2(CH_2)_6SO_2NCO$	135–136/1.1	—	C	67	290
$3\text{-}OCNC_6H_4SO_2NCO$	120/0.4	—	A	64.3	42
$4\text{-}OCNC_6H_4SO_2NCO$	108–109/0.15	40–44	A	45.7	42
$3\text{-}OCNSO_2C_6H_4SO_2NCO$	—	42.5	A	86.7	50
$3\text{-}OCNSO_2(6\text{-}CH_3)C_6H_3SO_2NCO$	171–175/1.8	70–73	C	29	290
		—	A	38.8	42
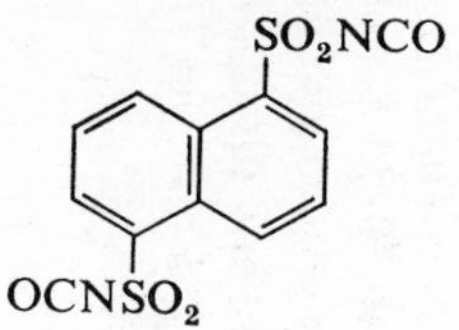 (naphthalene bearing SO_2NCO and $OCNSO_2$ groups)	—	185–190	A	54	50

6. p-XYLYLENE DIISOCYANATE (182)

A 12-liter reactor was charged at 25°C with 7000 g (6.350 ml) of chlorobenzene and 1000 g (10.1 moles) of gaseous phosgene dissolved in the chlorobenzene. The phosgene solution was cooled to 10°C and a hot (80°C) solution of 448 g (3.25 moles) of p-xylylenediamine in 1500 g (1350 ml) of chlorobenzene was added over 45 min. The phosgene solution was held at 10–25°C throughout the addition. The mixture was heated to 120°C over a period of 3 hr, and the hot phosgenation was begun. While a reflux temperature of 120–130°C was maintained, a total of 750 g of phosgene was added over a 21-hr period, a total phosgene excess of 170%. The solution was then purged with nitrogen at 130°C for 2 hr to remove residual hydrogen chloride and excess phosgene; then it was cooled to room temperature. Filtration removed 62 g (10%) of dried solids. The filtrate was stripped of solvent at reduced pressure, and the product was distilled at 129–138°C (1.5–2.7 mm). There was obtained 426 g (69%) of p-xylylene diisocyanate, which was analyzed and found to be 98.4% pure.

7. 1,2-DIISOCYANATOETHANE (41)

To a suspension of 78 g (0.525 mole) of 2-imidazolidinone-N-carbonyl chloride in 1500 ml of benzene, a solution of 50.5 g (0.50 mole) of triethylamine in 250 ml of benzene was added at 30–35°C over a period of 40 min. After the solution had been allowed to stand for 30 min at 45–50°C, 66.1 g (96%) of triethylamine hydrochloride was removed by filtration. The filtrate was heated to 70–75°C and approximately 10 g of phosgene was added. The solution was cooled and filtered, producing 1.3 g of solids. After evaporation of the solvent and distillation in vacuo, 38.6 g (69%) of ethylene diisocyanate, bp 65–68°C/7.5 mm, bp 77–80°C/16 mm, lit. bp 75–76°C/14 mm, η_D^{23} 1.4472, was obtained.

8. 1,2-DIISOCYANATOPROPANE (41)

To a stirred suspension of 16.3 g (0.1 mole) of methyl-2-imidazolidinone-N-carbonyl chloride in 350 ml of benzene there was added dropwise during 10–20 min 10.0 g (0.099 mole) of triethylamine. The temperature rose from 31.5 to 35°C from the heat of reaction. The mixture was stirred for another hour at 35–45°C and worked up in the same way as in the ethylene diisocyanate preparation. The resulting 13.3 g of pale yellow, crude concentrate was distilled, yielding 8.7 g (69%) of 1,2-diisocyanatopropane, bp 81°C/ 19 mm, η_D^{24} 1.4407.

9. PENTAMETHYLENE DIISOCYANATE (183)

To prepare pentamethylene diisocyanate, 51 g of pentamethylenediamine (0.5 mole) in 1500 g of anhydrous chlorobenzene was converted to the desired product by following the procedure outlined in Section II.G.5 (solvent/diamine weight ratio = 29). The yield was 69 g or 90% of theory, bp 105°C/4.0 mm (assay: 99%).

10. HEXAMETHYLENE DIISOCYANATE— HYDROCHLORIDE METHOD (183,184)

a. Hexamethylenediammonium Chloride

To a solution of 116 g (1.0 mole) of hexamethylenediamine in 145 ml of methanol in a 1-liter beaker there was added slowly from a dropping funnel 175 ml of concentrated hydrochloric acid (specific gravity 1.19). The mixture was well stirred during the addition and cooled externally to keep the contents below 30°C. The hexamethylenediammonium chloride was then precipitated by adding the solution slowly with stirring to approximately 2 liters of acetone. The precipitate was collected on a Büchner funnel, washed with 100 ml of cold acetone, and dried in a vacuum oven at 75°C for 12–18 hr. The yield of dry product was 170–187 g (90–99%), mp 243–246°C.

b. Hexamethylene Diisocyanate

A suspension of 94.5 g (0.50 mole) of finely powdered hexamethylenediammonium chloride in 500 ml of anhydrous redistilled amylbenzene (or tetralin) was prepared in a 1-liter, three-necked flask fitted with an efficient mechanical stirrer, a water-cooled reflux condenser, a thermometer, and a phosgene inlet tube extending well below the surface of the suspension. Stirring was started, the mixture was heated to 180–185°C, and gaseous chlorine–free phosgene was delivered to the mixture at a rate of 33 g (0.33 mole) per hour. Hydrogen chloride and excess phosgene were allowed to escape through the condenser. The temperature was carefully maintained between 180 and 185°C. After 8–15 hr the formation of the hexamethylenediisocyanate was essentially complete and hydrogen chloride no longer evolved. The reaction mixture was filtered and the filtrate distilled at reduced pressure through a fractionating column to give amylbenzene—bp 65–75°C/ 10 mm, and 70–80 g (84–95%) of hexamethylene diisocyanate—120– 125°C/10 mm, 92–96°C/1 mm, 108–111°C/5 mm, η_D^{20} 1.4585, d_4^{20} 1.0528.

11. NONAMETHYLENE DIISOCYANATE (183)

According to the procedure outlined in Section II.G.5, 158 g (1 mole) of nonamethylenediamine was dissolved in 3950 g of anhydrous chlorobenzene and converted to nonamethylene diisocyanate (solvent/diamine weight ratio = 25). The yield was 191 g (91%), bp 134°C/0.6 mm (assay: 99%).

12. DECAMETHYLENE DIISOCYANATE (182,183)

Decamethylenediamine, 345 g (2 moles), was dissolved in 5400 ml of dry chlorobenzene at 50°C, and dry hydrogen chloride was passed into the solution with agitation until complete conversion occurred (in 1–2 hr). It was observed that the reaction mixture became quite thick at about 50–85% conversion to the dihydrochloride, and it became quite thin and easily agitatable toward the end. The hydrogen chloride was replaced with phosgene and the reaction mixture was heated to reflux temperature, 125–130°C. After phosgenation for 5.5 hr at a phosgene rate of 540 ml/min, a clear, amber-colored reaction mixture was obtained which contained a very small amount of white solids. The hot solution was purged with nitrogen for 1–2 hr, filtered, and distilled. There was obtained 406 g (90%) of decamethylene diisocyanate, dp 135–143°C/1–1.5 mm, η_D^{25} 1.4541 (assay: 99.1%).

13. *p*-1,8-MENTHANE DIISOCYANATE (183)

A solution of 85 g (0.5 mole) of 1,8-diamino-*p*-menthane in 1615 ml of chlorobenzene was converted to the dihydrochloride. The slurry was then heated to reflux and trace amounts of water were removed azeotropically with approximately 100 ml of solvent (solvent/diamine weight ratio = 20). Phosgene was added at a rate of 220 cc/min and the unreacted phosgene was collected in a flask cooled with dry ice and acetone; the temperature of the reaction was 125–126°C. When 120% of the theoretical amount of phosgene had been added, the collected phosgene plus hydrogen chloride was recycled through the reaction mixture. This operation was repeated three or four times over a period of 18 hr. At the end of this time, the reaction mixture was clear amber in color. The solvent was evaporated, and vacuum distillation of the residue afforded 98.5 g (88.7%), of *p*-1,8-menthane diisocyanate, bp 94–110°C/0.4–0.5 mm (assay: 99.1%).

14. 4,4'-DICYCLOHEXYLMETHANE DIISOCYANATE— CARBAMIC ACID METHOD (19)

4,4'-Diaminodicyclohexylmethane (430 g, 2.05 moles) was dissolved in 4500 ml of *o*-dichlorobenzene and heated on a water bath to 90–95°C. At this temperature a stream of dry carbon dioxide was introduced with stirring until no more absorption occurred. The resulting fine suspension of the carbamic acid was stirred an additional 4 hr at this temperature. After cooling to 0°C, 400 g (4.04 moles) of phosgene was admitted, which reduced the viscosity of the suspension. The temperature was then raised gradually with continuous addition of phosgene (1000 g, 10.2 moles over a period of 7–8 hr) until, at 150°C, a clear solution resulted. The solvent was removed and vacuum distillation of the residue afforded 455 g (84%) of 4,4'-dicyclohexylmethane diisocyanate, bp 165–180°C/0.6–0.5 mm. The diisocyanate gradually crystallized to a waxy mass.

15. Bis(2-isocyanatoethyl)fumarate (39)

A stream of phosgene (at 0.02 mole/hr) was passed into a stirred mixture of 989 g (16 moles) ethanolamine and 490 g (5 moles) of maleic anhydride dissolved in 2096 g of o-dichlorobenzene over a period of 10 hr at 75–78°C, and the precipitated bis(2-aminoethyl)fumarate dihydrochloride (1259 g; 91.5%) was collected by filtration. To a suspension of 1067 g (3.86 moles) of the dihydrochloride in 9962 g of o-dichlorobenzene was added phosgene at a rate of 1 mole/hr over a period of 6 hr at 140–160°C. After purgation with nitrogen, removal of the solvent, and vacuum distillation of the residue, 800 g (81.3%) of bis(2-isocyanatoethyl)fumarate, bp 147°C/0.05 mm was obtained.

16. 4-Isocyanatobenzenesulfonyl Isocyanate (50,185)

A suspension of 172 g (1 mole) of 4-aminobenzenesulfonamide (sulfanilamide) in 1000 g of nitrobenzene was added to a solution of 396 g (4 moles) of phosgene in 548 g of nitrobenzene at a temperature between −10 and 0°C. The temperature of the reaction mixture was then raised to 157°C and phosgene was added for 5 hr. After purgation with nitrogen, evaporation of the solvent, and vacuum distillation, 194 g (86.7%) of 4-isocyanatobenzenesulfonyl isocyanate, bp 108–109°C/0.15 mm, mp 42.5°C was obtained.

17. Terephthaloyl Diisocyanate (186)

To a suspension of 32.8 g (0.2 mole) of terephthalic acid diamide in 200 ml of anhydrous carbon tetrachloride, a total of 152.4 g (1.2 mole) of oxalyl chloride was added and the reaction mixture was refluxed for approximately 10 days until a clear solution has been obtained. Evaporation of the solvent afforded 43.2 g (98%) of terephthaloyl diisocyanate, mp 100–102°C.

18. 2-Butene-1,4-diisocyanate (97)

A solution of 10 g (0.03 mole) of 1,4-bis-t-butylaminobutene-2-N,N'-dicarbonyl chloride and 30 mg of ferric chloride (hexahydrate) in 90 ml of ethylene dichloride was refluxed for 25 hr with a stream of dry nitrogen passing through the reaction mixture in order to facilitate removal of gaseous products. Evaporation of the solvent and vacuum distillation of the residue yielded 2.8 g (65%) of 2-butene-1,4-diisocyanate, bp 106–111°C/13–19 mm, mp −2 to 0°C, η_D^{23} 1.4728

19. TETRAMETHYLENE DIISOCYANATE (187)

To a solution of 211 g (1.21 mole) of adipoyl dihydrazide in 1.4 liters of aqueous hydrochloric acid containing 2.42 moles of the acid at 0–5°C a solution of 173 g (2.5 moles) of sodium nitrite in 200 ml of water was added with stirring and cooling. After addition of approximately half of the nitrite solution, 400 ml of benzene was added. The benzene layer was separated when the addition had been completed, and the aqueous layer was extracted with two 100-ml portions of benzene. The combined benzene solution was dried for 2 hr over anhydrous calcium chloride and heated slowly with good stirring until the evolution of nitrogen ceased. Evaporation of the solvent under vacuum and fractional distillation of the residue afforded 86.8 g (51%) of tetramethylene diisocyanate, bp 112–113°C/19 mm.

20. SULFONYL DIISOCYANATE (188)

A suspension of 80 g of dry silver cyanate in 70 g of chlorosulfonyl isocyanate was refluxed for 45 hr at 150–160°C. During the heating period, the reaction mixture has to be protected from atmospheric moisture, because of the exceedingly fast reaction of sulfonyl isocyanates with water. Vacuum sublimation (2–5 mm) into two traps cooled with dry-ice–methanol, afforded 62 g of a product containing 0.5% of chlorine. A repeated treatment with an additional amount of 15 g of silver cyanate gave 58.3 g (80%) of sulfonyl diisocyanate, bp 139°C/760 mm; d_4^{22} 1.588.

21. TRIISOBUTYLANTIMONY DIISOCYANATE (93)

In 200 ml of acetonitrile was dissolved 36.4 g (0.1 mole) of triisobutyl-antimony dichloride. To this solution was added 26 g (0.4 mole) of powdered sodium cyanate. The agitated mixture was refluxed for 3 hr. The warm solution was quickly filtered, and the solids were washed with fresh acetonitrile. After flash evaporation at 0.2 mm, 38 g (100%) of a clear, oily liquid was recovered. This could be purified by two quick high-vacuum distillations, during which the bath temperature was kept below 170°C. Then 25 g (75%) of colorless tri-isobutylantimony diisocyanate was recovered, bp 122–124°C/ 0.15 mm, η_D^{20} 1.5128, strong infrared band at 4.6 μ (NCO).

22. TRIPHENYLARSENIC DIISOCYANATE (93)

Triphenylarsenic oxide (32.2 g, 0.1 mole) was mixed with powdered urea (12 g, 0.2 mole) and heated to fusion at 135°C. After 0.5 hr at 140–150°C, reduced pressure was applied in order to remove all water and ammonia from the reaction. Upon cooling to room temperature, the reaction product triphenylarsenic diisocyanate (37.5 g, 96%) solidified to an amorphous, glassy mass, mp about 40°C. The product showed a strong NCO absorption at 4.68 μ. The analysis calculated for $C_{20}H_{15}N_2O_2$ was as follows: arsenic, 19.22; nitrogen, 7.19; the amounts found were: arsenic, 19.96; nitrogen, 7.2.

23. 1,1'-(ETHYLENEDIOXY)DIETHYL DIISOCYANATE (116)

To a mixture of 36 g (0.84 mole) of isocyanic acid and 0.1 g of p-toluenesulfonic acid in 100 ml of benzene, 22.8 g (0.2 mole) of the divinyl ether of ethylene glycol was added over a period of 20 min. The temperature rose to 50°C, and after stirring for an additional 20 min, benzene and excess isocyanic acid were removed under vacuum. The residue was distilled through a short pass at 0.3 mm, and redistillation through a short Vigreux column yielded 31.8 g (80%) of 1,1'-(ethylenedioxy)diethyl diisocyanate, bp 65°C/0.6 mm; η_D^{25} 1.4360.

24. BIS(2-ISOCYANATOETHYL)-4-CYCLOHEXENE-trans-1,2,-DICARBOXYLATE (189)

A mixture of 140 g (0.55 mole) of bis(2-isocyanatoethyl)fumarate and 94 g (1.74 moles) of butadiene in 200 ml of xylene was heated for 5 hr at 150°C in a steel rocking autoclave. After removal of the solvent and vacuum rectification of the residue, 122 g (72%) of bis(2-isocyanatoethyl)-4-cyclohexene-trans-1,2-dicarboxylate, bp 146–151°C/0.12 mm was obtained.

III. REACTIONS OF DIISOCYANATES

Isocyanates, the esters of isocyanic acid, are characterized by the heterocumulene double-bond system —N═C═O. Heterocumulenes as a class are quite reactive substances, and isocyanates are certainly not exceptions. A general survey of the chemistry of isocyanates is presented in this part. Although many of the reactions are described for monofunctional materials, difunctional derivatives react analogously.

A. Nucleophilic Reactions

Isocyanates react with a broad spectrum of compounds containing "active hydrogen." The reaction can be described as the attack of a nucleophile on the electrophilic carbon of the heterocumulene system (Eq. 59). Because of the reactivity of the isocyanate function, the number of compounds encompassed by X in Eq. 59 is very large. For example, nucleophiles containing

$$R—N═C═O + X—H \rightleftarrows R—NH—\overset{\displaystyle O}{\underset{\displaystyle X}{C}} \qquad (59)$$

hydrogen–oxygen bonds (water, alcohols, phenols, acids); hydrogen–sulfur bonds (hydrogen sulfide, mercaptans, thiophenols); hydrogen–nitrogen bonds (amines, amides, ureas, hydrazines, etc.); hydrogen–carbon bonds (nitroalkanes, malonates, hydrogen cyanide, etc.); hydrogen–phosphorus bonds (phosphines, hydrogen phosphites), and hydrogen–halogen bonds have been added to aliphatic and aromatic isocyanates. A summary of these reactions is presented in Table 22. The literature references regarding the indicated reactions can be found in the various monographs related to the chemistry of isocyanates (7–10,13).

The reactivity of a particular attacking species is proportional to its nucleophilicity. A rough order of reactivity is primary aliphatic amines > primary aromatic amines > alcohols > phenols > thiophenols. The reactivity of the slower reacting phenols and thiophenols can be increased enormously by forming the corresponding salts using strong bases as catalysts. Likewise, sulfonamides and hydrogen phosphites are best reacted in the form of their sodium salts. These sodium salts react sufficiently fast to permit the use of aqueous acetone as the reaction medium.

The products obtained from the reaction of nucleophiles with isocyanates are not always the simple addition compounds. To be sure, primary and secondary alcohols react rapidly and quantitatively to yield carbamates. In fact, this reaction forms the backbone of the polyurethan industry. However, the carbamate produced from a tertiary alcohol, like other tertiary esters, may decompose to olefin and acid. The acid, a carbamic acid, loses carbon dioxide spontaneously, and the resultant amine consumes another mole of isocyanate with formation of urea (291,292) as in Eq. 60. Analogously,

$$2C_6H_5NCO + (CH_3)_3COH \rightarrow C_6H_5NHCONHC_6H_5 + CO_2 + (CH_3)_2C{=}CH_2 \quad (60)$$

triphenylmethanol reacts to form the trisubstituted urea, since the trityl carbonium ion cannot eliminate a proton to form olefin (293) (Eqs. 61, 62).

$$RNCO + (C_6H_5)_3COH \longrightarrow RNHC(C_6H_5)_3 + CO_2 \quad (61)$$

$$RNHC(C_6H_5)_3 + RNCO \longrightarrow RNHCONC(C_6H_5)_3 \quad (62)$$
$$\underset{R}{|}$$

Other nucleophiles add to isocyanates to give unstable intermediates which decompose to the ultimate products. These are indicated in Table 22 also. For example, water forms unstable carbamic acids that readily lose carbon dioxide.

As described previously, the generated amine appears as urea. Acids form mixed anhydrides that are of isolatable stability in some cases, although they decompose to substituted urea and symmetrical an-hydride. Further reaction of these initial products yields the amide (Eqs. 62–65) (294).

$$RCOOH + R'NCO \rightarrow RCOOCONHR' \quad (62)$$

$$RCOOCONHR' + RCOOH \rightarrow RCOOCOR + R'NH_2 + CO_2 \quad (63)$$

$$R'NH_2 + R'NCO \rightleftarrows R'NHCONHR' \quad (64)$$

$$R'NH_2 + RCOOCOR \rightarrow R'NHCOR + RCOOH \quad (65)$$

The mixed anhydrides of sulfuric and boronic acids similarly are unstable and decompose to amides and carbon dioxide.

The reaction of isocyanates with the slower reacting phenols illustrates some of the problems of isocyanate chemistry. First, the reaction is readily reversible and heating favors the dissociated species (Eq. 66). Second, the

$$RNCO + C_6H_5OH \rightleftarrows RNHCOOC_6H_5 \qquad (66)$$

$$RNHCOOC_6H_5 + RNCO \rightleftarrows RNHCON COOC_6H_5 \qquad (67)$$
$$\underset{R}{|}$$

(84)

reaction product is itself an "active hydrogen" compound and can undergo further reaction to form allophanate (84) (Eq. 67). Thus if heat is applied to speed up the phenol-isocyanate reaction, a mixture of products ensues. To combat these problems, phenol reactions are catalyzed by bases of varying strength. Very strong bases, such as potassium t-butoxide, promote the stepwise reaction of phenols at low temperatures to carbamate or to allophanate (91). Depending on the stoichiometry employed, a good yield of either product is obtained. Mercaptans and thiophenols respond similarly to strong base catalysis (91).

Primary and secondary amines combine with isocyanates rapidly and quantitatively. This reaction forms the basis of most wet methods for isocyanate assay (see Section V). A host of linear, cyclic, and heterocyclic amines have been reacted with isocyanates. References to these reactions are readily available in the several review articles already mentioned.

In compounds containing two "active-hydrogen" types, distinction can be made sometimes on the basis of nucleophilicity. For example, the hydroxyl group of an aliphatic hydroxy acid is more reactive than the carboxyl group (295), whereas hydroxybenzoic acids afford the corresponding amide exclusively (296) (i.e., attack occurs on the carboxyl group in preference to the phenolic hydroxyl). Amino alcohols can be reacted on the amino group (297). By contrast, however, S,N-disubstitution occurs with certain mercapto-amines (298). In particularly structured substrates, the initial isocyanate reaction product can undergo further reaction. For example, acetylenic alcohols react with isocyanates to afford the corresponding carbamates, which cyclize in the presence of base to 4-methylene-2-oxazolidinones (85) (Eq. 68) (299–301).

A variety of heterocyclic systems have been obtained from o-difunctional aromatic compounds such as salicylaldehyde (Eq. 69) (302,303), o-amino-acetophenone (Eq. 70) (304,305), o-aminobenzonitrile (Eq. 71) (306), o-aminobenzoic acids (Eq. 72) (307,308), o-hydroxybenzoic acid (Eq. 73) (309), and phthalic anhydride (Eq. 74) (310).

B. Catalysis of Nucleophilic Reactions

The ability of isocyanates to undergo the reactions described varies considerably with the structure of the isocyanate. Generally, the order of

TABLE 22

Nucleophilic Reactions of Isocyanates, $RN{=}C{=}O + HX \rightarrow RNHCOX$

Active hydrogen compound	Formula	Intermediates	Product
Water	HOH	$[RNHCOOH, RNH_2]$	$RNHCONHR + CO_2$
Alcohols (phenols)	$R'OH$		$RNHCOOR'$
Hydroxysilanes	$R_3'SiOH$		$RNHCOOSiR_3'$
Hydroperoxides	$R'OOH$	$[RNHCOOCOR']$	$RNHCOOOR'$
Boronic acids	$R'B(OH)_2$	$[R'B(OCONHR)_2]$	$R'B(NHR)_2 + 2CO_2$
Sulfuric acid	$HOSO_3H$	$[RNHCOOSO_3H]$	$RNHSO_3H + CO_2$
Phosphinic acids	$R'P(OH)H \;\downarrow\; O$		$RNHCOP(OH)R' \;\downarrow\; O$
Phosphonic acids	$R'P(OH)_2 \;\downarrow\; O$		$RNHCOOP(OH)R' \;\downarrow\; O$
Hydroxamic acids	$R'CONHOH$		$RNHCOONHCOR'$
Oximes	$R_2'C{=}NOH$		$RNHCOON{=}CR_2'$
Hydrogen sulfide	HSH	$[RNHCOSH, RNH_2]$	$RNHCONHR + COS$
Mercaptans (thiophenols)	$R'SH$		$RNHCOSR'$
Sodium bisulfite	$NaHSO_3$		$RNHCOSO^{\ominus}Na^{\oplus}$
Ammonia	NH_3		$RNHCONH_2$
Amines	$R_2'NH$		$RNHCONR_2'$
Cyanamide	$NCNH_2$		$RNHCONHCN$
Amidines	$R'(NH_2)C{=}NH$		$RNHCONHC(R'){=}NH$
Guanidines	$(R'NH)_2C{=}NH$		$RNHCON{=}C(NHR')_2$
Nitramines	$R'NHNO_2$		$RNHCON(R')NO_2$
Amides	$R'CONH_2$		$RNHCONHCOR'$
Ureas	$R'NHCONHR'$		$R'NHCON(R')CONHR'$
Urethanes	$R'NHCOOR''$		$RNHCON(R')COOR''$
Sulfonamides	$R'SO_2NH_2$		$RNHCONHSO_2R'$
Hydrazine	H_2NNH_2		$RNHCONHNH_2$
Hydroxylamine	$HONH_2$		$RNHCONHOH$
Hydrogen cyanide	HCN		$RNHCOCN$
Hydrogen fluoride	HF		$RNHCOF$
Hydrogen chloride	HCl		$RNHCOCl$
Hydrogen bromide	HBr		$RNHCOBr$
Hydrogen iodide	HI		$RNHCOI$
Nitroalkanes	$R'CH_2NO_2$		$RNHCOC(R')HNO_2$
Acetylacetone	$CH_3COCH_2COCH_3$		$RNHCOC(COCH_3)HCOCH$
Acetoacetic esters	CH_3COCH_2COOR'		$RNHCOC(COCH_3)COOR'$
Malonates	$CH_2(COOR')$		$RNHCOCH(COOR')_2$
Phosphine	PH_3		$P(CONHR)_3$
Hydrogen phosphites	$(R'O)_2HPO$		$RNHCOP(OR')_2 \;\downarrow\; O$

effect of the *o*-methyl group (311). For reaction with 2-ethylhexanol, 2-4- and 2,6-toluenediisocyanates exhibit activation energies of 8.1 and 7.9 kcal/mole, respectively (312).

$$OCN—R—NCO + R'OH \xrightarrow{k_1} OCN—R—NHCOOR \qquad (75)$$

$$OCN—R—NHCOOR + R'OH \xrightarrow{k_2} R'OOCNH—R—NHCOOR' \qquad (76)$$

Benzylisocyanates, such as *m*-xylylene diisocyanate, react approximately 20 times faster than HDI (first step) and 40 times faster in the second step (311).

Kaplan (313) has measured the influence of substituents on the rates of reactions of substituted phenyl isocyanate with 2-ethylhexanol. He finds a correlation with the Hammett equation with relative reactivities as follows: *p*-NO$_2$ ≳ *m*-NO$_2$ ≫ *m*-CF$_3$ ≳ *m*-Cl ≳ *m*-Br > *m*-NCO > *p*-NCO ≳ *p*-Cl > *m*-CH$_3$O > phenyl isocyanate > *m*-CH$_3$ ≳ *p*-CH$_3$ > *p*-CH$_3$O. Of course, other strong electron-withdrawing groups, such as sulfur dioxide and R$_2$P(O) are comparable to the nitro group in their ability to enhance the reactivity. If the isocyanato group is bonded to an acyl, aroyl, or sulfonyl group, its reactivity is vastly enhanced, and disulfonyl isocyanates react extremely rapidly with active hydrogen compounds.

In addition to the electronic effects, steric considerations are of importance, and primary alkyl isocyanates react faster than secondary and tertiary isocyanates. Likewise, orthosubstitution in aryl isocyanates lowers their relative reactivity. For example, the methyl group in 2,4-TDI lowers the reactivity of the orthoisocyanate function relative to the unhindered para-isocyanato group (304). The result of the combined steric and electronic effects of the reaction rate for several diisocyanates has been reported (315).

Most studies of isocyanate reactivity have been concerned with alcohols as nucleophiles. Generalizations derived from these studies may not be valid for all the varied substrates reactive toward isocyanates.

Although the effect of basic and acidic catalysts on the rate of the reaction of isocyanates with active hydrogen compounds has been known since the turn of the last century, quantitative data were not available until 1942 (316). Tarbell and co-workers found that triethylamine is a better catalyst than pyridine, and that strong acids HCl > CCl$_3$COOH > CH$_3$COOH and Lewis acids, such as boron trifluoride, are good catalysts as well. However, triethyl-amine was the most effective catalyst in this study. Baker and co-workers (317) studied the catalysis of the reaction of aryl isocyanates with alcohols. Summaries of the various catalysts and other investigations related to the catalysis of this reaction can be found in Refs. 13 and 311.

Since most of the rate studies have been conducted on monofunctional model systems, this work is not discussed here, but important developments in the catalysis of isocyanates, especially recent results, are included.

The need of polyurethan producers for very fast reaction rates led to the development of highly active catalysts, such as triethylenediamine (1,4-diaza[2,2,2]bicyclooctane, DABCO), approximately 10 times more efficient

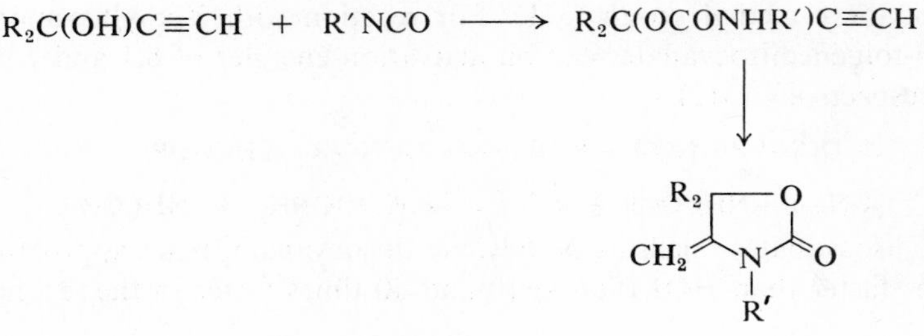

$$R_2C(OH)C\equiv CH \ + \ R'NCO \ \longrightarrow \ R_2C(OCONHR')C\equiv CH$$

(68)

85

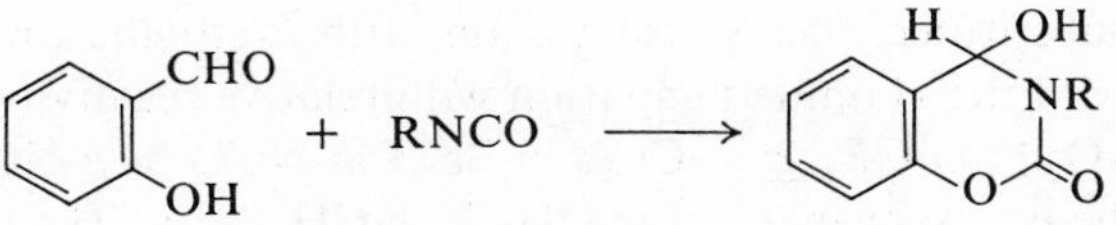

(69)

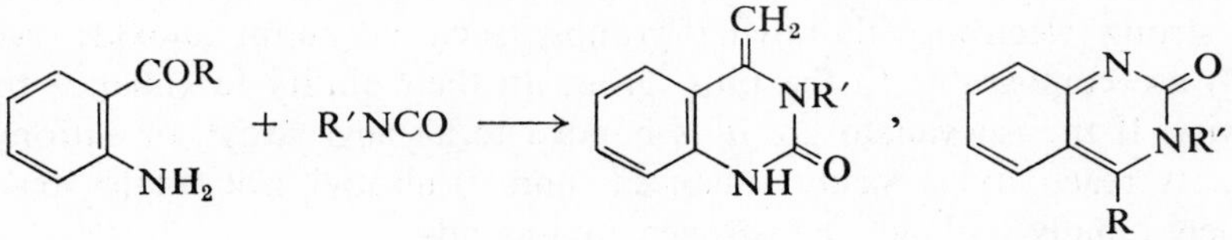

(70)

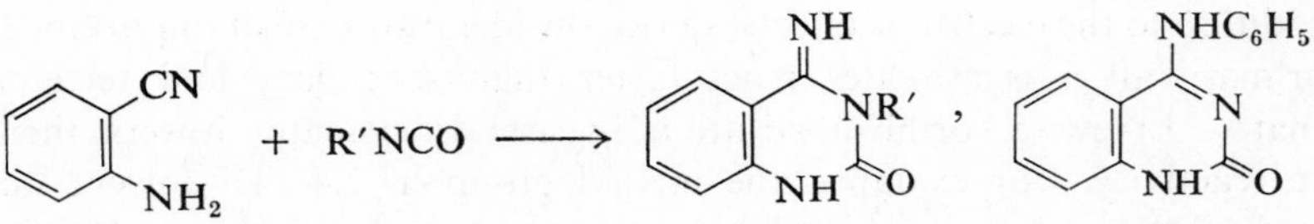

(71)

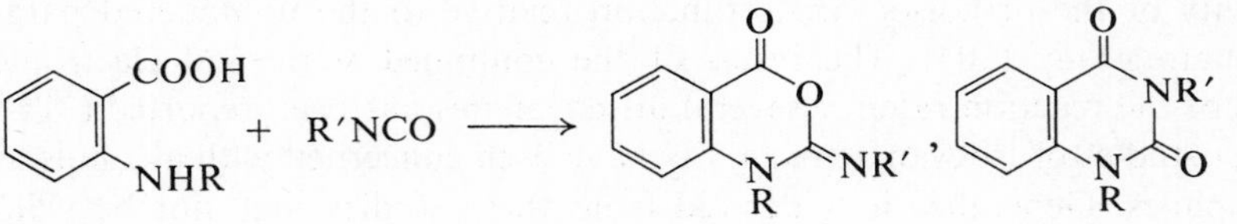

(72)

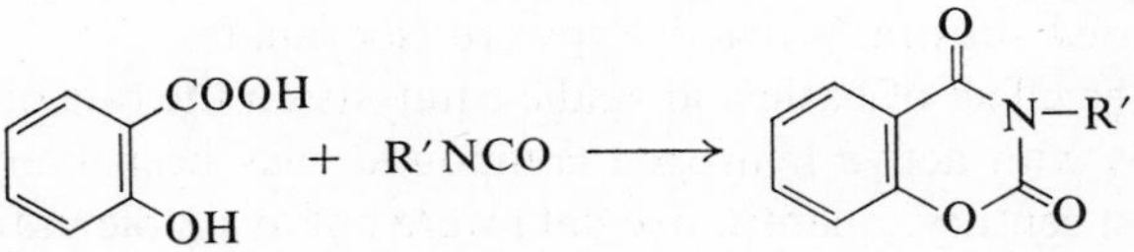

(73)

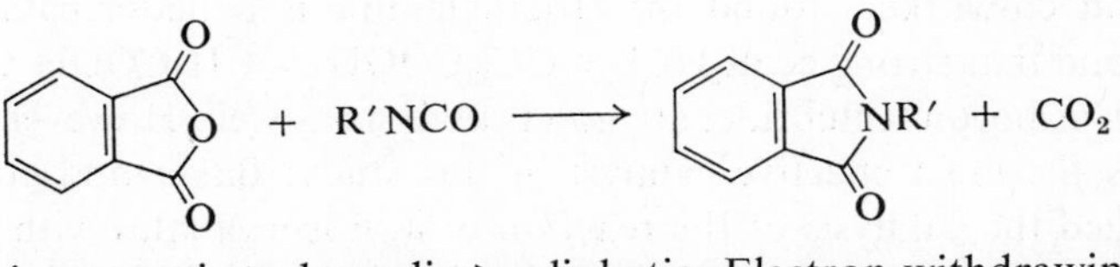

(74)

reactivity is aromatic > benzylic > aliphatic. Electron-withdrawing substituents increase and electron-donating groups decrease the reactivity of isocyanates toward nucleophilic attack. For example, 2,4-TDI reacts approximately 350 times faster than hexamethylene diisocyanate in the initial formation of the monocarbamate (Eq. 75). In the second step (Eq. 76), this difference is decreased: TDI reacts only 60 times faster than HDI because of the change in the substituent from isocyanate to the more electron-donating carbamate function and, perhaps even more important, because of the steric than triethylamine (311) and di-*n*-butyltin diacetate, 2400 times more effective than triethylamine (318). A synergistic effect was also found when a combination of tertiary amine and tin catalysts was used (318). Although most model rate studies were conducted using the infrared method (314), Britain and Gemeinhardt (319) measured the gelation time of the reaction of diisocyanate with a polyether triol of molecular weight 3000, at 70°C. The order of catalytic efficiency of various metal organic compounds was as follows: Bi > Pb > Sn (319). This technique was recently mechanized (182) and found to be an exceedingly useful method to screen a wide variety of catalysts under conditions that approximate actual polymerization conditions.

The kinetics and catalysis of urethan foam formation, using stannous-octoate–tertiary-amine combination catalysts, have been studied. It was demonstrated that tertiary amines containing aliphatic ether linkages in their molecular structure are equivalent to DABCO in the alcohol-isocyanate reaction. In contrast, for the water-isocyanate reaction, the ether amines are superior to DABCO (320). Furthermore, Flynn and Nenortas (321) have demonstrated that heptamethylisobiguanide has the highest catalytic activity of any amine catalyst. Burkus (322) has shown that pyridine *N*-oxides are effective catalysts for the isocyanate-alcohol reaction.

The mechanism of the reaction of isocyanates with hydroxyl compounds, in the presence of a tertiary amine catalyst, is not too well established. Recent results indicate that the tertiary amine forms a complex with the isocyanate (321), whereas in the phenol-isocyanate reaction, the catalyst generates the phenolate ion (323). For the noncatalyzed reaction, hydrogen bonding involving several molecules of the alcohol (rather than formation of an isocyanate-alcohol complex) has been discussed recently (324,325).

Smith (326) has studied the mechanism of the catalysis of TDI and phenyl isocyanates with tin and cobalt catalysts. In the phenyl isocyanate reaction, cobalt stearate > stannous 2-ethylhexoate > DABCO; whereas in the TDI reaction, the tin compound is more effective than the cobalt compound. The author postulates the intermediacy of a ternary complex of the three reagents.

Frisch and his co-workers (327,328) and Entelis et al. (329–331) have discussed the intermediacy of a 1:1 complex of the alcohol and dibutyltin dilaurate, lead naphthate and triethylamine, respectively. Likewise, tetraalkyl-distannoxanes are catalysts for urethan formation (332,333), and tin phenolates have been used effectively to produce polyurethans (334). The kinetics of the reaction of ureas and phenyl isocyanate in the presence of tin catalysts has been investigated also (335).

The rapid reaction of alkyl and aryl isocyanates with tin oxides and tin alkoxides (336) may have some relation to the catalytic effect of the tin compounds.

A series of metal acetylacetonates has also been evaluated for catalytic activity in the reaction of diisocyanates and polyester, and the following order of activity was noted: $Mn > V > Fe > Cu > Co > Cr$ (337). Whereas Weisfeld (337) assumes that more than one metal site per isocyanate group is required in the transition state, Bruenner and Oberth (338) postulate the intermediacy of an alkoxy metal complex in the metal chelate catalysis of the isocyanate-alcohol reaction. Iron acetylacetonate is an excellent catalyst for the reaction of nitrodiisocyanates with nitrodiols (339).

In the reaction of phenyl isocyanate and n-butanol in the presence of cobalt caprylate, the formation of the active complex occurs during the initial mixing (340). Robins (341) classified the metal ion catalysts in two groups: catalysts that maintain a fair portion of their activity even in systems containing hindered isocyanates, such as lead, bismuth, and tin, and catalysts that are very sensitive to changes in steric factors at the isocyanate group, such as cobalt, zinc, and cadmium. Robins indicates that the function of the metal salts is twofold: to increase the electrophilicity of the center carbon atom in the isocyanate group, and to bring the alcohol and the isocyanate in closer proximity (i.e., he leans toward the ternary complex transition state).

Organolead compounds are supposed to catalyze both the alcohol and the water reaction effectively (342,343). For example, phenyllead triacetate was found to be comparable to stannous dioctoate in the gel test, but considerably superior in the generation of carbon dioxide from the water reaction (343).

In summary, it can be concluded that the catalysis of the isocyanate-alcohol reaction is extremely effective, but despite the tremendous efforts of numerous investigators, the mechanism is incompletely understood. Perhaps each of the various groups of catalysts operates by a different mechanism, and generalizations cannot be made at the present state of the art.

C. Insertion Reactions

A related nucleophilic reaction of isocyanates involves the apparent "insertion" of the carbon-nitrogen function of the isocyanate between, say, the metal–oxygen bond of a trialkyltin oxide (Eq. 77). This reaction may be

$$R'_3Sn-OR'' + RNCO \longrightarrow \begin{bmatrix} R'_3Sn\text{-----}OR'' \\ \vdots \quad \vdots \\ N\text{-----}C \\ R \quad\quad O \end{bmatrix} \longrightarrow R'_3Sn-\underset{R}{\underset{|}{N}}-\overset{O}{\overset{\|}{C}}OR'' \quad (77)$$

viewed as involving the four-center transition state shown or as a simple extension of the reactions of "active hydrogen" compounds, with trialkyltin in place of hydrogen. The complexing ability of the metals in question must certainly influence the reaction and foster a more concerted process, such as in Eq. 77. Numerous examples of "insertion" are described in Table 23. For a review of this type of reaction and literature references, see Ref. 12.

TABLE 23
Insertion Reactions of Isocyanates, $RN\!=\!C\!=\!O + X\!-\!Y \rightarrow XN(R)COY$

Substrate	Formula	Products
Boron trichloride	BCl_3	$ClB[N(R)COCl]_2$
Phenyldichloroborane	$C_6H_5BCl_2$	$ClB\begin{smallmatrix}\nearrow N(R)COCl\\ \searrow N(R)COC_6H_5\end{smallmatrix}$
Triphenylborane	$(C_6H_5)_3B$	$C_6H_5B[N(R)COC_6H_5]_2$
Carbonyl fluoride	COF_2	$RN(COF)_2$
Phosphorus pentachloride[a]	PCl_5	$RN\!=\!PCl_3 + COCl_2$
Sulfur tetrafluoride[b]	SF_4	$RN\!=\!SF_2 + COF_2$
Trialkyltin alkoxides	$R_3'SnOR''$	$R_3'SnN(R)COOR''$
Trialkyllead alkoxides	$R_3'PbOR''$	$R_3'PbN(R)COOR''$
Mercury alkoxides	$R'HgOR''$	$R'HgN(R)COOR''$
Acetals	$CH_2(OR')_2$	$CH_2[N(R)COOR']_2$
Aminals	$CH_2(NR_3)_2$	$R_2'NCH_2N(R)CONR_2'$
Orthoesters	$CH(OR')_3$	$(R'O)_2CHN(R)COOR'$
Aminoboranes[c]	$B(NR_2')_3$	$B[N(R)CONR_2^1]_3$
Silylamines	$R_3'SiNR_2'$	$R_3'SiN(R)CONR_2''$
Silazanes	$(R_3'Si)_2NH$	$R_3'SiN(R)CONHSiR_3'$
Trialkyltin amides	$R_3'SnNR_2''$	$R_3'SnN(R)CCNR_2''$
Trialkyllead amides	$R_3'PbNR_2''$	$R_3PbN(R)CONR_2''$
Arsenic triamides[c]	$As(NR_2')_3$	$As[N(R)CONR_2']_3$
Sulfenamides	$S(NR_2')_2$	$R_2'NSN(R)CONR_2'$
Bis(trialkyltin) oxides	$R_3'SnOSnR_3'$	$R_3'SnN(R)COOSnR_3'$

[a] This reaction occurs with aromatic isocyanates only; aliphatic isocyanates react with phosphorus pentachloride to afford carbonimidoyl dichlorides.

[b] Inorganic isocyanates react with sulfur tetrafluoride with formation of N-fluoroformyl-iminosulfur difluoride ($F_2S\!=\!N\!-\!COF$).

[c] 1:1, 2:1, and 3:1 adducts are formed, depending on stoichiometry.

Recent evidence for the four-centered transition state in the reaction of organosilyl amines and phenyl isocyanate has been reported by Kaufmann and co-workers (408). Further work related to the insertion of isocyanates into tin–oxygen bonds (409,410), tin–phosphorus bonds (411), and lead–oxygen bonds (412) has been published since 1967. The formation of a 1:1 adduct of bis(dimethylamino)phenylphosphine and tosyl isocyanate has also been observed (413).

D. Cycloaddition Reactions

In addition to the nucleophilic (or electrophilic) reactions of isocyanates, several types of cycloaddition reactions have been reported in the literature. The best known example of 1,2-cycloaddition is perhaps the dimerization of aryl isocyanates (Eq. 78). Upon heating, the cyclodimers dissociate into

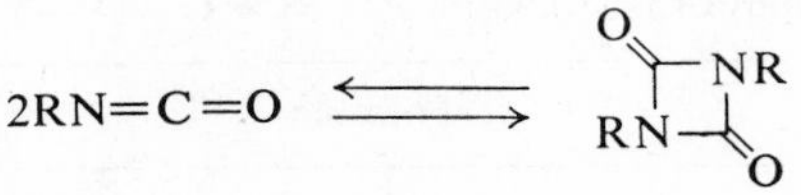

$$2RN{=}C{=}O \rightleftharpoons \begin{array}{c} O \\ \diagdown \\ RN{-} \\ \diagup \\ O \end{array}\!\!\!\begin{array}{c} {-}NR \\ \\ \end{array} \qquad (78)$$

the starting isocyanates. Cycloaddition reactions of isocyanates with a wide variety of double bonds have been encountered (12) and a summary appears in Table 24. Often the four-membered ring 1:1 adducts **86** are isolated; however, isomerization to the linear 1:1 adducts **87** occurs if a hydrogen is attached to the β position of the substrate (Fig. 9).

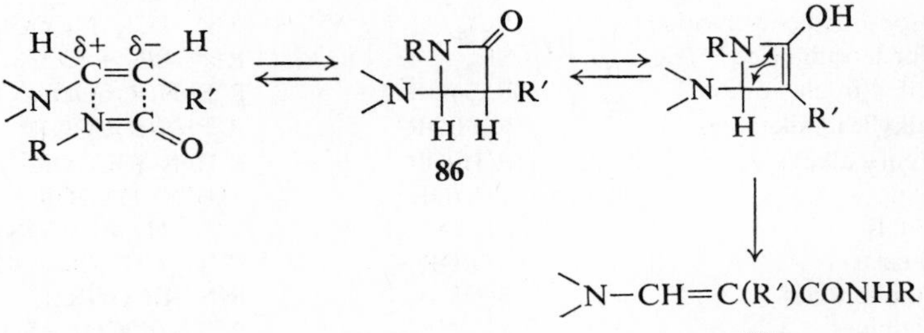

$$>N{-}CH{=}C(R')CONHR$$
87

Fig. 9.   Cycloaddition reactions of isocyanates.

The four-membered 1:1 cycloadducts are thermally unstable, and they equilibrate upon heating. If the lowest boiling component is constantly removed from the equilibrium mixture, a clean exchange reaction occurs. For example, heating of naphthyl isocyanate with dicyclohexylcarbodiimide affords the mixed α-naphthylcyclohexylcarbodiimide, provided 1 eq of cyclohexyl isocyanate is removed by distillation (Eq. 79) (135).

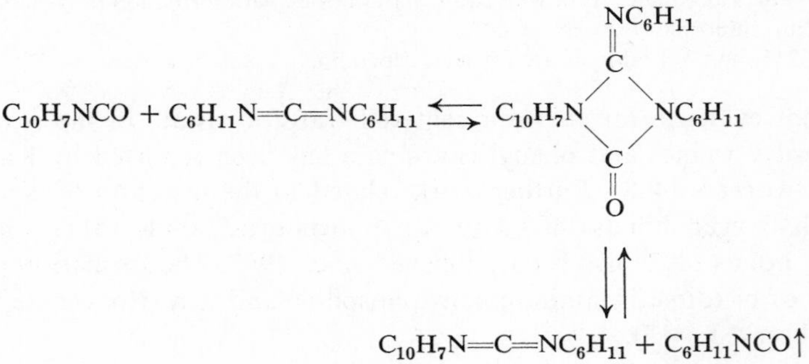

$$C_{10}H_7N{=}C{=}NC_6H_{11} + C_6H_{11}NCO{\uparrow} \qquad (79)$$

The use of a second mole of α-naphthyl isocyanate and complete removal of cyclohexyl isocyanate affords di-α-naphthylcarbodiimide with simultaneous conversion of 1 mole of dicyclohexylcarbodiimide to 2 moles of cyclohexyl isocyanate. The "exchange" reaction can be used to synthesize a wide variety of heterocumulenes (see Table 24). The kinetic and thermodynamic properties of the isocyanate-carbodiimide reaction (344) and the isocyanate dimerization (345) have been examined in some detail.

The cycloaddition reaction may proceed by a stepwise or concerted mechanism, depending on the substrates used and the presence or absence of catalyst. For example, the catalyzed dimerization of phenyl isocyanate involves the steps outlined in Eqs. 80–82 (345).

$$C_6H_5NCO + B: \rightleftarrows C_6H_5\overset{|}{\underset{B^+}{N}}-C=O \longleftrightarrow C_6H_5N=\overset{|}{\underset{B^+}{C}}-O^- \tag{80}$$

$$A$$

$$A + C_6H_5NCO \rightleftarrows C_6H_5N\overset{O}{\overset{||}{C}}-\overset{|}{\underset{C_6H_5}{N}}-\overset{O}{\overset{||}{C}}-B^+ \tag{81}$$

88

$$88 \rightleftarrows C_6H_5N \underset{\overset{||}{O}}{\overset{\overset{O^-\ \ B^+}{\diagup C \diagdown}}{\diagdown C \diagup}} NC_6H_5 \rightleftarrows C_6H_5N \underset{\overset{||}{O}}{\overset{\overset{O}{\overset{||}{\diagup C \diagdown}}}{\diagdown C \diagup}} N-C_6H_5 + B \tag{82}$$

We may apply a similar mechanism to the uncatalyzed dimerization and to other cycloadditions wherein another double-bond system functions as B in Eq. 80. The 1:1 adduct may react with a further mole of isocyanate or the substrate to yield the 2:1 adducts, as in Eqs. 83–85. Thus both the 1:1 and

$$R-NCO + A=B \rightleftarrows R-\overset{-}{\underset{+A-B}{N}}-C=O \rightleftarrows R-\overset{|}{\underset{A-B}{N}}-\overset{|}{C}=O \tag{83}$$

$$R-\overset{-}{\underset{^+A-B}{N}}-C=O + R-NCO \longrightarrow \begin{array}{c} \overset{O}{\overset{||}{C}} \\ R-N \diagup \ \diagdown B \\ | \qquad | \\ O=C \diagdown \ \diagup A \\ N \\ | \\ R \end{array} \tag{84}$$

$$R-\overset{-}{\underset{^+A-B}{N}}-C=O + A=B \longrightarrow \begin{array}{c} A \\ R-N \diagup \ \diagdown B \\ | \qquad | \\ O=C \diagdown \ \diagup A \\ B \end{array} \tag{85}$$

TABLE 24
1,2-Cycloaddition Reactions of Isocyanates,

$$A = B \ + \ RN = C = O \ \longrightarrow \ \begin{array}{c} A \!-\!\!\!-\!\!\!\overset{\displaystyle O}{\overset{\|}{C}} \\ | \quad \ \ \backslash \\ B \!-\! NR \end{array}$$

Double-bond system	Formula	Products
Olefins (enamines, ketenactals)		
Ketenes	$R_2'C = C = O$	
Aldehydes, ketones (amides, ureas)	$R_2'C = O$	$\longrightarrow R_2'C = NR + CO_2$
Thiocarbonyl compounds	$R_2'C = S$	$\longrightarrow R_2'C = NR + COS$
Azomethines	$R_2'C = NR''$	
Carbodiimides	$R'N = C = NR'$	
Nitroso compounds	$R'N = O$	$\longrightarrow R'N = NR + CO_2$
Sulfoxides	$R_2'S = O$	$RN = SR_2' + CO_2$
Sulfurdiimides	$R'N = S = NR'$	
Alkylidene phosphoranes	$R_3'P = CR_2''$	$R_2'C = C = NR + R_3'PO$
Phosphine imides	$R_3'P = NR''$	$R''N = C = NR + R_3'PO$
Phosphine oxides	$R_3'P = O$	$R_3'P = NR + CO_2$

2:1 adducts of sulfonylisocyanates with carbodiimide have been isolated (137).

The trimerization of isocyanates to the isocyanurates, which occurs readily in the presence of strong bases, probably follows the same sequence outlined in Eqs. 80 and 81. Intermediate **88** adds another mole of isocyanate before cyclizing, yielding the trimer (Eq. 86) (91). Trimer formation is also

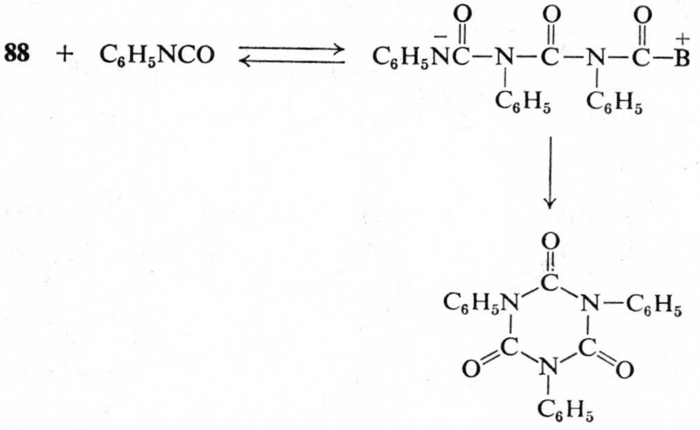

(86)

promoted by acids, as outlined below for hydrogen chloride catalyst (Eqs. 87–89) (40).

$$RNCO + HCl \longrightarrow RNHCOCl \qquad (87)$$

$$RNHCOCl + RNCO \longrightarrow RNHCON(R)COCl \qquad (88)$$

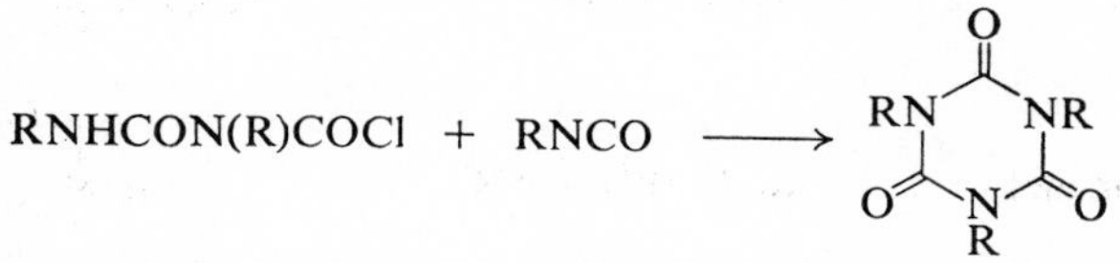

(89)

Isocyanates can also participate as dipolarophiles in typical 1,3-cyclo-addition reactions (346), and the hitherto reported 1,3-cycloaddition reactions are summarized in Table 25 (12). Generally, the carbon–nitrogen double bond becomes attached to the 1,3 system, but in certain special cases the carbon-oxygen double bond has been added as well.

Isocyanates having a carbonyl, thiocarbonyl, or imidoyl group adjacent to the cumulative double-bond system can react as dienes as well as dieno-philes in 1,4-cycloaddition reactions. A typical example of this reaction is the dimerization of thioacyl isocyanates (Eq. 90) (347). However, more

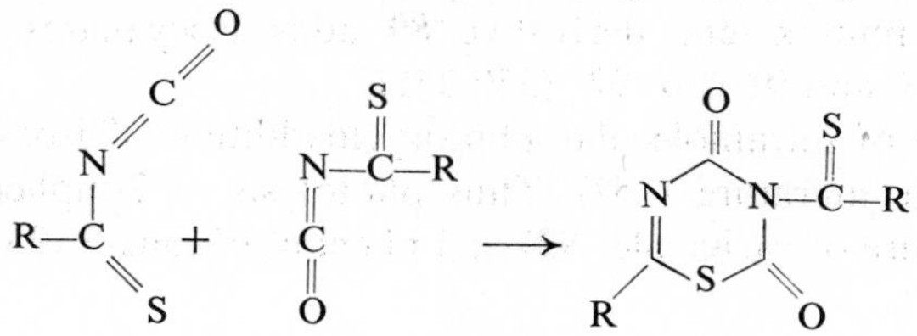

(90)

TABLE 25

1,3-Cycloaddition Reactions of Isocyanates,

$$RN{=}C{=}O \ + \ ^{\oplus}A{-}B{-}C^{\ominus} \ \longrightarrow$$

1,3 System	Formula	Products
Epoxides (ethylene carbonate)		(CO_2)
Cyclic anhydrides		$+ CO_2$
Ketocarbenes		
Nitrones (*N*-oxides)	$R'CH{=}N(R') \rightarrow O$	
Nitrile imines	$RC{\equiv}N \rightarrow NR'$	
Azomethine imines	$R_2C{=}(R') \rightarrow NR'$	
Hydrazoic acid	HN_3	

reactive systems, such as acetylenes (348) and benzyne (349), can be added in 1,4 fashion, even to phenyl isocyanate (Eqs. 90, 91).

Reactive 1,4-dipoles, which can be generated by thermolysis of the appropriate precursors, can add isocyanates across their 1,4 system. For example, the iminoketene derivative **89** adds isocyanates to afford the cycloadducts **90** and **91** (Eq. 92) (350,351).

An example of intramolecular photocycloaddition of isocyanates is also reported in the literature (352). Thus photolysis of 2-biphenyl isocyanate affords a mixture of carbazole (**92**) and phenanthridone-3 (**93**) (Eq. 93).

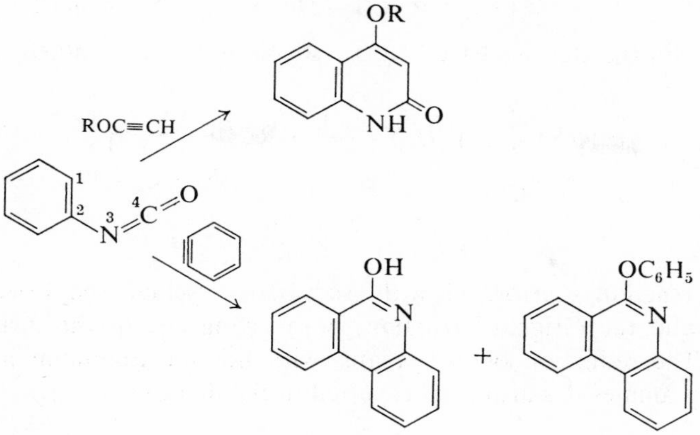

OR

ROC≡CH

$ROC\equiv CH$

NH   O

$_1$
$_2$ $_3$N $_4$C=O

OH

+

OC$_6$H$_5$

N

N

(91)

When properly catalyzed, isocyanates will react to form carbodiimides. Cyclic phosphorus compounds are unusually effective for this reaction (258,353). Metal carbonyls also function as catalysts by the following sequence of reactions (Eqs. 94, 95), involving an isonitrile complex as the intermediate (354).

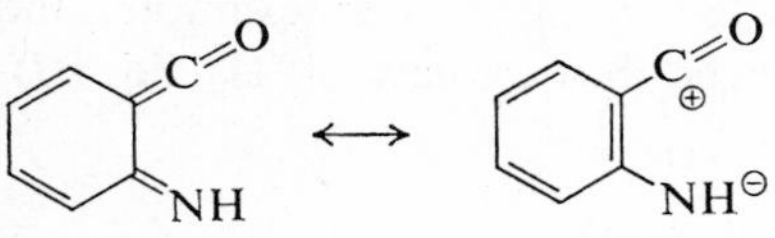

C=O

⟷

C=O
⊕

NH

NH⊖

89

RNCO

O

O

+

O
NR

NH NR

NH O

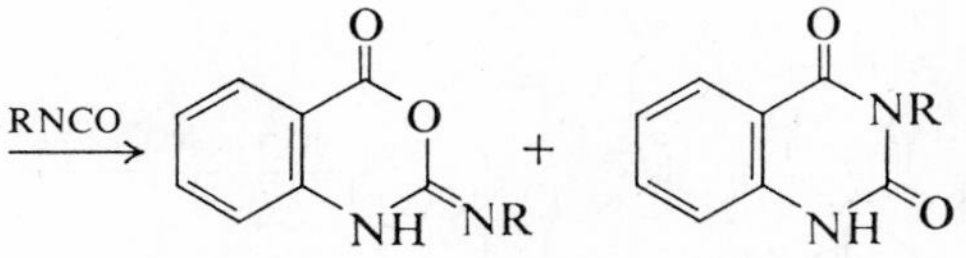

(92)

90   **91**

Recently Ramirez and co-workers (355) obtained 5-acylhydantoins (**94**) from a reaction of diketones with isocyanates in the presence of trialkyl phosphites (Eq. 96). The 5-acylhydantoins are precursors of β-keto-α-amino acids.

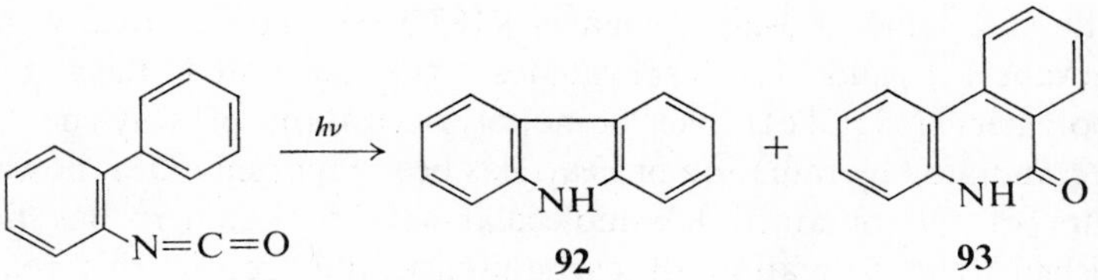

hν

+

N=C=O

NH

NH O

(93)

92   **93**

$$M(CO)_n + RNCO \rightarrow M(CO)_{n-1}(C{=}N{-}R) + CO_2 \qquad (94)$$

$$M(CO)_{n-1}(C{=}N{-}R) + RNCO \rightarrow R{-}N{=}C{=}N{=}R + M(CO)_n \qquad (95)$$

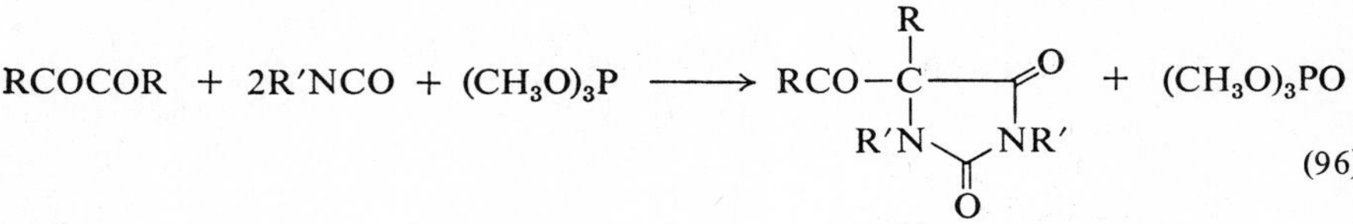

$$RCOCOR + 2R'NCO + (CH_3O)_3P \longrightarrow RCO{-}\underset{\underset{O}{\underset{\displaystyle R'N \diagdown \diagup NR'}{|}}}{\overset{\displaystyle R}{\underset{}{C}}}{\diagup}^{O} + (CH_3O)_3PO \qquad (96)$$

94

Other reactions performed with isocyanates include the Friedel-Crafts reaction and the Grignard reaction, both giving rise to the formation of amides. The reduction of isocyanates with lithium aluminum hydride to secondary amines has also been reported in the literature.

IV. POLYMERIZABILITY AND APPLICATIONS

The polymerizations of diisocyanates may be characterized by three general modes of reaction: homopolymerization, polyaddition, and poly-condensation.

Homopolymerization is limited to the short-chain, aliphatic diisocyanates. These diisocyanates were cyclopolymerized by an alternating, intermolecular-intramolecular propagation mechanism (187,191,356). For example, the cyclopolymers **95** are obtained from ethylene diisocyanate (191) (Eq. 97).

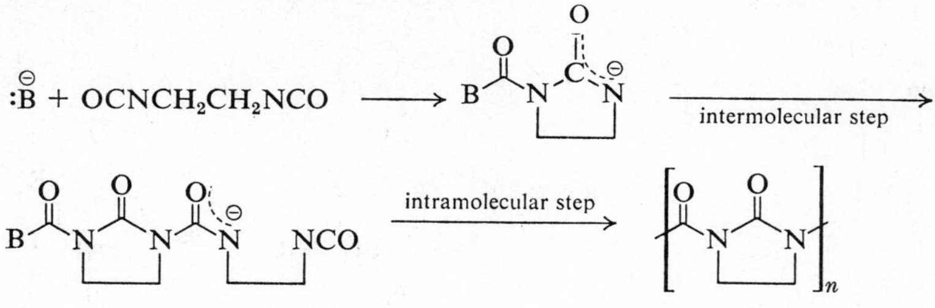

95

The most effective catalyst so far for the cyclopolymerization is sodium cyanide in *N,N*-dimethylformamide, a system used by Shashoua in the homopolymerization of monoisocyanates (357).

The cyclopolymerization of ethylene diisocyanate has been initiated also by irradiation with X-rays (187). The scope of the reaction is limited to aliphatic 1,2- and 1,3-diisocyanates (187,191). The sterically hindered cyclohexane-1,2 and 1,3-diisocyanates also generated their respective cyclopolymers (191,356). The homopolymerization of tolylene-2,4-diiso-cyanate, initiated by radiation or heat, has been reported (358); however, the structure of the obtained low-molecular-weight oligomers has not been established. The formation of carbodiimide linkages in this reaction is indicative of a different reaction mechanism (see Eq. 94, 95). Copolymerization of aldehydes and aromatic diisocyanates in the presence of anionic catalysts has been reported (359).

The *addition polymerization* of diisocyanates with bifunctional hydroxy compounds to afford polyurethans is well described and documented in three books (13,14,18), and duplication of these efforts would be presumptuous. However, a general discussion on the applications of polyurethans is included at the end of Section IV (see pp. 451–454).

The highly fluorinated polyurethans **96** have been prepared from the reaction of perfluorinated diisocyanates with fluorinated hydroxyl terminated polyesters and polyethers (360) (Eq. 98). Diisocyanates react readily with

$$OCN-\!\!\left\langle\!\!\begin{array}{c}F\end{array}\!\!\right\rangle\!\!-NCO \ + \ HO-R-OH \ \longrightarrow \ \left[CONH-\!\!\left\langle\!\!\begin{array}{c}F\end{array}\!\!\right\rangle\!\!-NHCOO-R-O\right]_n$$

$$\textbf{96} \qquad (98)$$

active hydrogen compounds other than those of difunctional alcohols. Whereas the reactions of diisocyanates with water and hydrogen sulfide yield complex reaction mixtures (11), reactions with dicarboxylic acids afford mixed anhydrides that can dissociate either to polyamides or to polyureas and polycarboxylic acid anhydrides. For example, decamethylene diisocyanate, on reaction with sebacic acid, yields the corresponding polyamide **97** (361) (Eq. 99).

$$OCN(CH_2)_{10}NCO + HOOC(CH_2)_8COOH \ \xrightarrow{-CO_2} \ -\!\!\left[NH(CH_2)_{10}NHCO(CH_2)_8CO\right]_n$$

$$\textbf{97} \qquad (99)$$

This reaction is not useful for aromatic difunctional reagents because the undesirable side reactions are more pronounced. For example, MDI reacts with benzoic acid in dimethyl sulfoxide to give poly[methylene bis(4-phenylurea)] (362). Although the initial products are those of polyaddition, the net reaction is a polycondensation.

The reaction of diisocyanates with primary and secondary amines is the simplest preparation of polyureas. The solution polymerization of numerous diamines and diisocyanates is reported in the patent literature, and these reactions are summarized in a book (363).

The 1:1 adducts of diisocyanates and ethyleneimine have been polymerized to form clear and insoluble resins (364).

Diisocyanates can react with hydrazine and carboxylic acid dihydrazides to yield polysemicarbazides. For example, the polymerization of MDI with isophthalic dihydrazide in dimethylsulfoxide produces the polysemicarbazide

98 (365–367) (Eq. 100). Dioximes and diisocyanates react in a similar manner to give polyaddition products (365,368).

$$OCN\text{—}\bigcirc\text{—}CH_2\text{—}\bigcirc\text{—}NCO \ + \ H_2NNHCO\text{—}\bigcirc\text{—}CONHNH_2 \longrightarrow$$

$$\left[\text{—}COHN\text{—}\bigcirc\text{—}CH_2\text{—}\bigcirc\text{—}NHCONHNHCO\text{—}\bigcirc\text{—}CONHNH\text{—}\right]_n$$

98   (100)

Several interesting combinations of addition/condensation-type polymerizations have been reported recently. The interfacial polymerization of isocyanatobenzenesulfonyl chlorides with diamines (Eq. 101) generated the polysulfonamideureas **99** (58,369). The amido (NH) group adjacent to the

$$OCNC_6H_4SO_2Cl + H_2N\text{—}R\text{—}NH_2 \rightarrow \text{—}NHCONHC_6H_4SO_2NH\text{—}R\text{—}_n \quad (101)$$

99

sulfonyl moiety acts as an acidic site for subsequent salt formation (52). Similarly, glycols and isocyanatocarboxylic acid chlorides give polyesterurethans (370).

A two-step addition/condensation sequence employing diisocyanates as monomers has served in the preparation of thermally stable heterocyclic polymers. Thus reaction of 4,4′-diaminobiphenyl-3,3′-dicarboxylic acid with p-phenylene diisocyanate yielded the polyurea **100**, which, on dehydration at 160–180°C, gave the condensation polymer **101**. Compound **101**, on further heating to 230–250°C, rearranged to poly-[3,3′-(p-phenylene)-6,6′-bis-quinazolinedione] (**102**), a polymer exhibiting outstanding thermal stability (371) (Eq. 102).

Another type of addition polymerization involves a 1,2-, 1,3-, or 1,4-cycloaddition sequence. Enamines, such as **103** on reaction with MDI, undergo a 1,2-cycloaddition reaction before rearranging to the polyamides **104** (Eq. 103). With formic acid in DMF, compounds **104** hydrolyze to the keto polymers **105** (372).

The 1,3-cycloaddition polymers prepared from diisocyanates with suitable substrates are important because a variety of monomers is available. Linear poly-2-oxazolidinones (**107**) are obtained from the diglycidyl ether of 2,2-bis(4-hydroxyphenyl)propane (**106**) and TDI (373) (Eq. 104).

Various carbonyl, imidoyl, and thioacyl isocyanates readily undergo 1,4-cycloaddition reactions. For difunctional compounds, the corresponding polymers are obtained: carbonyl diisocyanate, on standing, polymerizes to

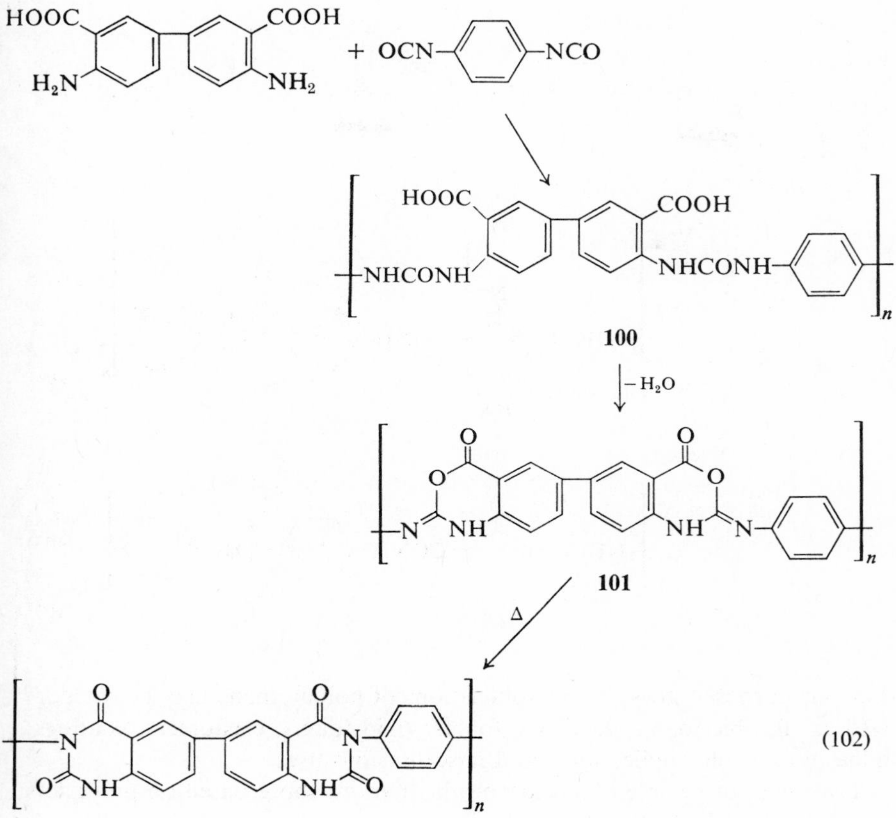

100

101

102

(102)

polyoxadiazine-2,4-dione **108** (99) (Eq. 105). Vinyl isocyanate would be expected to polymerize in a similar manner.

Heating of diisocyanates in the presence of a phosphorus catalyst yields polycarbodiimides **109** (353) (Eq. 106). This reaction can be used in the synthesis of diisocyanates containing carbodiimide linkages (258,353).

The several examples cited serve to demonstrate the versatility of diiso-cyanates in the synthesis of polymers. However, only the addition polymers prepared from diisocyanates with hydroxy-terminated polyesters or polyethers have received extensive commercial development.

The extraordinary development of this market during the twenty years since Bayer reported the first synthesis of polyurethans from diisocyanates (4,14) is remarkable. The number of patents related to polyurethans well exceeds one thousand, and the scope of these applications is indeed broad.

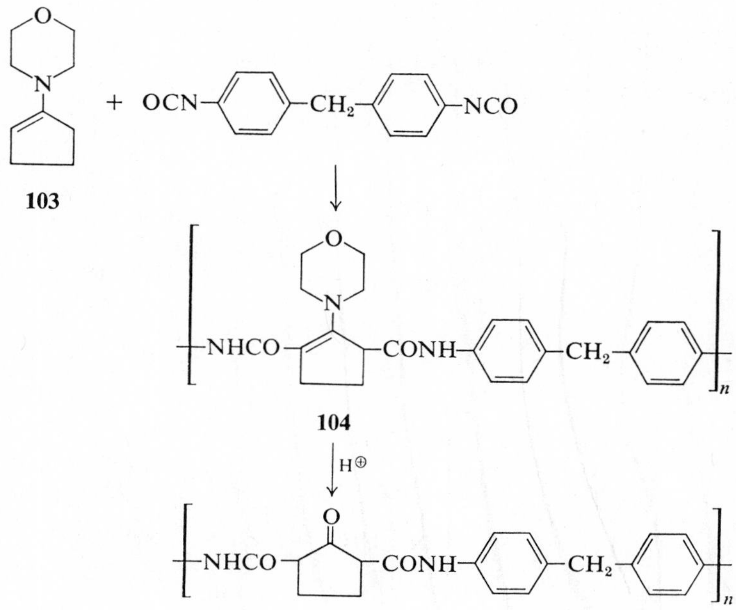

103

104

$$\downarrow H^{\oplus}$$

105   (103)

For our purposes, these many applications of polyurethans may be categorized as: flexible foams, semirigid foams, rigid foams, elastomers, coatings, adhesives, textile applications, and miscellaneous uses.

Two types of flexible foams are on the market: those based on polyesters and those based on polyethers. The former are used in clothing interlinings, rug underlays, and packaging, and the latter in general upholstery. Tolylene

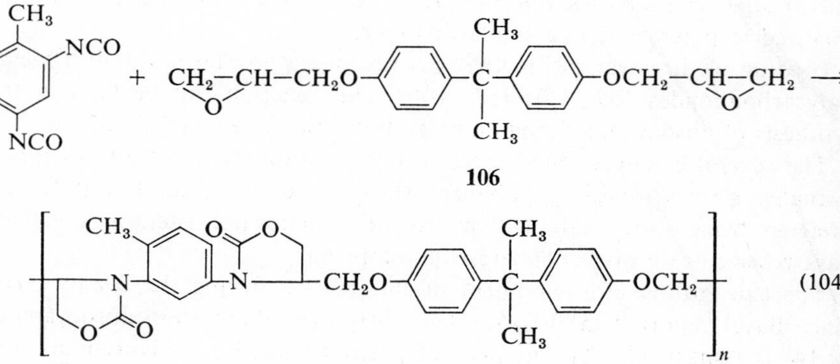

106

107   (104)

$$\text{OCN} \diagdown \overset{\oplus}{N} = \overset{C=O}{\underset{O^{\ominus}}{C}} \quad + \quad \overset{N-CONCO}{\underset{\underset{O}{\parallel}}{\underset{C}{\parallel}}} \quad \longrightarrow \quad \text{OCN} \left[\overset{O}{\underset{\underset{O}{\parallel}}{N}} \overset{}{\underset{O}{\diagup}} N \right]_n - \text{CONCO} \tag{105}$$

108

$$\text{OCN—R—NCO} \xrightarrow{\; -CO_2 \;} [\, N{=}C{=}N{-}R \,]_n \tag{106}$$

109

diisocyanate (TDI)* and 4,4′-diphenylmethane diisocyanate (MDI)† are predominantly used in this application. Molded parts or slabs are prepared in continuous or discontinuous one-shot or prepolymer operations featuring mechanical metering, mixing, and dispensing systems. Additives, such as silicon surfactants, blowing agents (freons or water), and catalysts [usually a mixture of stannous octoate and triethylenediamine (DABCO)] are added to the major components. Pigments may be added to produce colored foams.

Branching can be controlled by the functionality of the two major components, and a wide range of polyurethans can be prepared. For example, semirigid foams produced from branched polyethers and diisocyanates are used in the automotive industry for the preparation of crash pads, head rests, arm rests, and A-posts for safety and comfort. The obvious advantage of semirigid foams is their capacity for "impact absorption."

Rigid foams are prepared from polyfunctional, crude MDI-type‡ isocyanates and polyfunctional hydroxy components. Polyethers, rather than polyesters, are the more economic hydroxy components. Since these foams exhibit very low thermal conductivity, they are used primarily in insulation. Houses, storage tanks, pipe lines, ships, railroad cars, trailer trucks, and refrigerators are insulated with rigid polyurethan foam. Furthermore, laminated roofing and construction materials made from rigid polyurethan

* The various isomer mixtures are produced by National Aniline Division of Allied Chemical Corporation as Nacconate 65, 80, 100, and 4040; by E. I. du Pont de Nemours and Company as Hylene T, TM, TM-65, and TIC; by Mobay Chemical Company as Mondur TD, TD-80, TDS, and MT-40; by Union Carbide Corporation as Niax TDI-P and TDR; by Olin Chemical Corporation as TDI-80; and by Rubicon as Vibrathane 4080.

† MDI is produced and sold by The Upjohn Company, Polymer Chemicals Division, as Isonate 125M, 125MF, and Isonate 143L; by Mobay Chemical Company as Multrathane M; by National Aniline Division of Allied Chemical and Dye Corporation as Nacconate 300, by E. I. du Pont de Nemours and Company as Hylene M and Hylene M-50; and by Kaiser Aluminum and Chemicals Corporation as 2002 and 2006.

‡ Products of this type are produced by The Upjohn Company, Polymer Chemicals Division, as PAPI® and Isonate 390P, by Mobay Chemical Company as Mondur MR, by Kaiser Aluminum and Chemicals Corporation as NCO-10, -20, and -120, by Rubicon as Suprasec DN, and by Union Carbide as AFPI.

foam serve in the building industry. High-density rigid foams, of light weight and excellent structural properties, are being used in the manufacture of furniture.

One of the most promising applications of polyurethans is in elastomeric materials. For example, spandex-type fibers are suitable for replacing rubber thread, and they are, in fact, being used extensively, especially in foundation garments. Elastomers of cast polyurethans are used in the manufacture of machinery parts, shock absorbers, industrial tires, etc. Other applications of these materials range from industrial truck tires to seamless floors, parts for shoes, and components for space vehicles. Recently cast polyurethans based on color-stable aliphatic diisocyanates have been used successfully in coatings and optical applications. The numerous advantages of these elastomers include high abrasion resistance, high tear strength, excellent shock absorption, hardness, a wide range of elasticity, solvent resistance, and resistance to oxygen and ozone aging. The several prepolymer systems and extrudable pellets on the market are Vulcollan, Chemigum SL, Vulcaprene A, Adiprene, Estane, Texin, and Pellethane.

Polyurethan elastomers are prepared either by one-shot or prepolymer methods. Extenders such as diols, diamines, and hydrazines are used in obtaining the elastomeric properties desired for a particular application. Since thermoplastic granules are on the market, injection-molding techniques can be used to prepare myriad types and shapes of products.

Color-stable coatings prepared from aliphatic isocyanates are used with paper, wood, textiles, leather, metals, and other substrates to produce films that exhibit extraordinary adhesion; high gloss; low permeability; resistance to solvents, abrasion, and weather; and excellent electrical properties. Thus leather and textile finishes, wire enamels, and paints all take advantage of polyurethan coatings.

In adhesives, polyurethans offer the property of strong and flexible bonding. Although triphenylmethane triisocyanate was used originally, polyaryl-polyisocyanate-type materials, which are economically far superior, have replaced the triisocyanate in adhesives. The bonding of elastomers to metals, wood, leather, and other substrates is quite easily accomplished.

Polyurethans and polyureas are used in the treatment of textiles and leather in order to improve their dyeability and waterproofing. Artificial leathers recently developed by several companies use a polyurethan in the upper layer for scuff resistance.

Thus diisocyanates are used in an *enormous* variety of applications. For actual procedures, technical bulletins are available from Upjohn, Union Carbide, Allied, du Pont, Mobay, Olin, Wyandotte, and other leaders in the field of urethans. The books (13,14,18) previously recommended should also be considered.

V. ANALYTICAL PROCEDURES

The identification and quantitative analysis of isocyanates can be achieved by both spectroscopic and chemical methods.

Isocyanates exhibit a characteristic strong absorption in the infrared region at approximately 2300–2200 cm^{-1}, arising from the asymmetric stretching mode of the cumulative double-bond system. Since this absorption band occurs in a region in which few other functional groups exhibit a strong absorption, it is well suited for the detection of isocyanates in complex reaction mixtures. The position of the absorption band is influenced somewhat by conjugation and by neighboring polar groups: acyl isocyanates absorb at 2275–2225 cm^{-1} (42), thioacyl isocyanates at 2235–2218 cm^{-1} (374), and sulfonyl isocyanates at 2240–2220 cm^{-1} (10).

Infrared analysis has been used successfully in the determination of the amount of unreacted isocyanate in rigid urethan foams (375,376) and in kinetic studies on the reactions of isocyanates (314). The absorption of aliphatic and aromatic isocyanates was also measured in the near-infrared region, and a correlation between chain length and branching was found (377).

NMR spectroscopy is also used in the identification of isocyanates. The aliphatic isocyanates having α-hydrogen atoms are particularly well suited to NMR analysis (12). Polyurethans in arsenic trichloride at 100°C were analyzed with NMR, and the diisocyanate component as well as the polyether or polyester component could be identified (378). Dimethylsulfoxide and N,N-dimethylacetamide are often used as solvents for NMR studies (379) of isocyanates and derivatives.

The mass spectra of aliphatic (380) and aromatic (381) isocyanates have been examined and the fragmentation patterns analyzed.

Although spectral methods are valuable in the identification and characterization of isocyanates, the most widely used analytical method for the determination of the isocyanato group is a quantitative titration with aniline or di-n-butylamine. This method was developed by Spielberger (382), and several modifications are available (383,384). In fact the titration of isocyanates with di-n-butylamine has become a standard procedure of the American Society for Testing and Materials (385). A micromethod titration using n-butylamine with methyl red as the indicator has also been developed (386).

Titration with di-n-butylamine can be used in the end-group analysis of prepolymers. For the determination of small amounts of free isocyanate in polymeric compositions, the colorimetric method of Kubitz (387) is the method of choice: the polymer is titrated with n-butylamine in tetrahydrofuran and the unreacted n-butylamine is determined with malachite green. An alternate way of determining the free diisocyanate in prepolymer systems takes advantage of the difference in solubility between diisocyanates and their derivatives. For example, by extracting prepolymers with an inert solvent, such as hexane, it is possible to selectively remove the diisocyanate, which can then be determined accurately by titration with di-n-butylamine (388).

Tolylene diisocyanate produces a characteristic color on treatment with nitrite ion in an alcoholic solution (389,390). However, the determination of trace amounts of tolylene diisocyanate can be done more efficiently by electron-capture gas chromatography (391).

An alternate method for the detection of trace amounts of diisocyanates is based on the hydrolysis of diisocyanates to the corresponding diamines, followed either by diazotization and coupling with N-1-naphthylethylene-diamine (392) or by reaction of the diamine with thiotrithiazyl chloride (393).

The diisocyanate component in polyurethans can also be detected by a degradative analysis in which 50% sulfuric acid serves for the hydrolysis of the polymer. The diamine that is obtained can be separated and identified by standard procedures (394).

The ratio of isomers in tolylene diisocyanate has been determined both by infrared methods (395) and by measurement of dielectric constants (396). Gas chromatography (397) and NMR spectrometry (379) can be used here also. The bismethylcarbamates, stable, easily prepared derivatives of diiso-cyanates, can be used in these determinations (379).

Determination of Isocyanate Equivalent Weights: Modification of ASTM Method D1638-59T

Apparatus:

1. 250-ml Erlenmeyer flasks equipped with ground glass necks and stoppers.
2. Two graduated cylinders, one 25 ml, one 100 ml.
3. One 20-ml pipet.
4. Water condensers and hot plates.
5. One 50-ml buret.
6. Magnetic stirrer and bars.

Reagents:

1. Toluene (dry) A.R.
2. di-n-Butylamine (ca. 40N).
3. Methyl alcohol (absolute) A.R.
4. Hydrochloric acid (1N).
5. Bromo-phenol blue indicator.

Preparation of solutions:

1. di-n-Butylamine—weigh 260 g of di-n-butylamine into a 1-liter volumetric flask. Dilute to mark with dry toluene A.R. and shake well.
2. Bromo-phenol blue—dissolve 0.1 g of bromo-phenol blue per 100 ml of methyl alcohol.
3. All other reagents are purchased in the needed grades.

Procedure:

1. Weigh 2–3 g sample into flask.
2. Dissolve sample in 25 ml of toluene.
3. Pipet 20 ml of di-*n*-butylamine into flask; stopper.
4. Prepare blank of 25 ml of toluene, 20 ml of di-*n*-butylamine in a second flask; stopper.
5. Reflux sample and blank, using water condensers at approximately 250°C (see note 4).
6. Wash down flasks with 15 ml of toluene, cool, and add 100 ml of absolute methyl alcohol.
7. Add indicator and titrate with $1N$ HCl to a yellow endpoint, agitating solution during titration.

Calculations:

$$\frac{\text{weight of sample}}{(\text{blank titer-sample titer}) \; N \text{ of HCl}} \times 1000 = \text{equivalent weight}$$

$$\frac{\text{calculated equivalent weight}}{\text{theoretical equivalent weight}} = 1000 = \% \text{ assay}$$

Notes:

1. A blank should be run daily. The concentration of the di-*n*-butylamine varies slightly from day to day. This is particularly important when the stock solution is low.
2. Always use dry toluene, not allowing the sample to sit in toluene without adding di-*n*-butylamine for any length of time. Moisture can be absorbed by the sample and toluene, effecting the final results.
3. Use water condensers, *not* air condensers. Samples refluxed on air condensers do not give valid results.
4. Aliphatic isocyanates should be refluxed approximately 15 min and cooled 15 min. Aromatic isocyanates should be refluxed approximately 30 min and cooled 15 min. Polymers should be refluxed approximately 45 min and cooled 15–20 min.
5. A precipitate may form in the flasks upon refluxing. This should dissolve when the methyl alcohol is added.
6. Samples should be run in duplicate and agree within 1%.

VI. STORAGE AND HANDLING

Since isocyanates in general exhibit a high reactivity, precautions must be used in their handling. Water is the most ubiquitous and persistent contaminant of isocyanates. The formation of insoluble ureas is the common signal of improper protection against either moisture in the air or moisture from techniques employed in transferring. Water has a low equivalent weight, and a *large* amount of isocyanate can react with a small quantity of water or moisture. A concomitant complication is the pressure buildup that can occur in closed systems from the carbon dioxide gas generated in the reaction of water with isocyanates (Eq. 107).

$$2RNCO + H_2O \rightarrow RNHCONHR + CO_2 \qquad (107)$$

Another problem encountered in the storage of isocyanates is represented by the facility with which these substances dimerize or trimerize (Eqs. 108,

109). Whereas the formation of trimer can be controlled by the exclusion of

$$2RNCO \rightleftarrows \quad \text{(structure)} \tag{108}$$

$$3RNCO \longrightarrow \quad \text{(structure)} \tag{109}$$

catalysts (such as tertiary amines, alkoxides, and phosphines, which are known to enhance trimerization), the formation of dimer cannot be easily controlled. In fact, there is evidence that the dimerization is a noncatalyzed, thermal process and that it can be slowed down only by storage at lower temperatures (182,398).

An interesting aspect of this problem of dimerization is that the rate of formation of dimer is not directly proportional to the reaction temperature—a break in the curve occurs near the melting point of the crystalline isocyanate. At comparable temperatures, the rate of formation of dimer in the crystalline solid is much faster than in the liquid, an observation suggesting that the alignment of the molecules in the solid state favors the formation of dimer.

Of the two commercially important diisocyanates, TDI and MDI, the former dimerizes very slowly at ambient conditions, whereas in the solid state the latter is kept in the cold for long-term storage (398). In contrast, o-tolidine diisocyanate is exceptionally stable in the solid state at room temperature (182). This is probably because both isocyanato groups are protected with the o-methyl group substituent. No dimers of aliphatic isocyanates have been reported. Specific storage instructions for individual commercial isocyanates are available from the manufacturer and should be consulted.

Encapsulation, another method for protecting isocyanates from atmospheric contamination, can be used for long-term storage of isocyanates blended with other active ingredients (399). No reaction can occur before the encapsulating envelope is destroyed, so the danger of premature polymerization is minimal.

Isocyanates in general are *toxic* compounds, and great care must be exercised in handling them. Oral ingestion of substantial quantities can be tolerated by the human body, but *acute* symptoms may develop from the inhalation of *much smaller* amounts. This toxicity in inhalation of isocyanates

TABLE 26
Toxicity of Diisocyanates

Substance[a]	LD_{50}, g/kg[b]	LC_{50},[c] ppm	MAC,[d]	Reference
HDI	0.35–1	44–370[a]		402,404,405
XDI	0.84			405
PPDI	0.9			402
TDI	1.95–5.8	4–13.5	0.02	402,404,406
MDI	>31.6		0.02	402
DDI	34.6			407

[a] HDI, hexamethylene diisocyanate; XDI, xylylene diisocyanate; PPDI, p-phenylene diisocyanate; TDI, tolylene diisocyanate; MDI, p,p'-diphenylmethane diisocyanate; DDI, long-chain aliphatic diisocyanate (General Mills).

[b] Oral lethal dose for 50% of test animals, expressed in grams per kilogram of body weight.

[c] Inhalation lethal concentration for 50% of test animals.

[d] Maximum allowable concentration for continuous long-term exposure.

presents a great danger to people who work with these chemicals. Thus the less volatile isocyanates are the less toxic ones (see Table 26). For TDI and MDI, the maximum allowable concentration (MAC) for continuous exposure is 0.02 ppm. This dose is below the limit that can be detected by smell alone (400).

The more volatile a diisocyanate, the more apparent are its lachrymatory properties. In spray applications, however, regardless of the vapor pressure of the isocyanate, atomization of the isocyanate may raise to intolerable levels, the concentration of isocyanate that is inhaled (400,401). Appropriate breathing equipment should be employed in such cases. Sufficient ventilation is exceedingly important in the handling of isocyanates under *any* circumstances.

Most isocyanates act as irritants to both the skin and eyes (26,402), and MDI has been found to stain skin. In particular, discoloration has been observed on the hands of foundrymen using certain no-bake mold compositions (403). Allergic reactions have appeared in people who work with diisocyanates. Repeated exposure can cause a high sensitivity to even minute concentrations of isocyanates (400,402).

APPENDIX

Synthesis of Diisocyanates

A series of aromatic diisocyanates containing fluorinated side chains were synthesized by Malichenko and co-workers (415–418). These authors also prepared tetrafluoro-m-phenylene diisocyanate (414), and the method used in all cases involved nitration, reduction, and phosgenation. Trifluoromethyl-phenylene diisocyanates were also obtained by Inukai and Maki (419). An MDI derivative, 3-methoxy-4,4'-diisocyanato-diphenylmethane, in which the two isocyanato groups have different reactivities, was synthesized recently (420). Also, aromatic o-diisocyanates were prepared for the first time; however, these compounds undergo rapid homopolymerization (see second part of this appendix). N-Methyldiphenylamine 4,4'-diisocyanate was synthesized by phosgenation of the corresponding diamine precursor (421). Apparently the basicity of the tertiary amine function is rather weak, and the hydrochloride dissociates under the phosgenation conditions, giving rise to the formation of the free diisocyanate. Heterocyclic diisocyanates are best prepared by the reaction of the corresponding heterocyclic diamine and oxalyl chloride (422). Gizycki (422) has used this method to synthesize several 1,3,5-triazinyl diisocyanates. Pyridyl 2,6-diisocyanate was synthesized by the thermolysis of the corresponding diazide (423). The azide method was also used to prepare several substituted m-phenylene diisocyanates (424). However, an explosion occurred in the preparation of ethylene diisocyanate by the azide method, rendering this procedure *hazardous* for small molecules (425).

An interesting thermolysis procedure for the preparation of diisocyanates was reported recently (426,427). In this reaction dioxathiazole derivatives, such as **1**, are heated to produce the corresponding diisocyanate **2** and sulfur dioxide (424).

The dioxathiazole derivative can be prepared from the bis-hydroxamic acid and thionyl chloride. In a similar manner, heterocyclic dioxazolinones, obtained from hydroxamic acids and phosgene, are thermolyzed to yield the corresponding isocyanate and carbon dioxide (427).

Aromatic diisocyanates were also prepared by coupling of monoisocyanates, provided another functional group is present in the isocyanate. For example, reaction of the monoisocyanate **3** in a two-phase system (methylene chloride/water) with base produces the diisocyanate **4** (428). Likewise, reaction of **3** with bisphenol A yields the diisocyanate **5** (429).

Caprolactam was also used as a raw material to produce aliphatic diisocyanates. For example, phosgenation of 6-aminocaproic acid gives 6-isocyanatocaproyl chloride (**6**), which upon coupling with bisphenol A

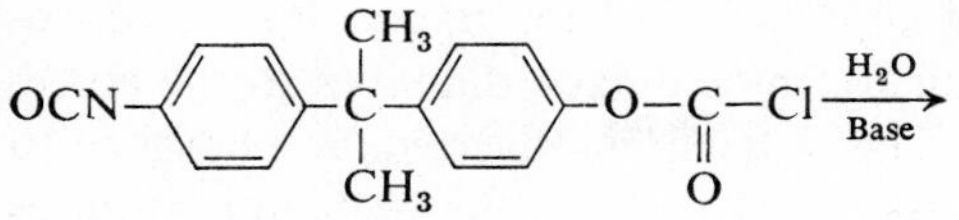

$$3$$

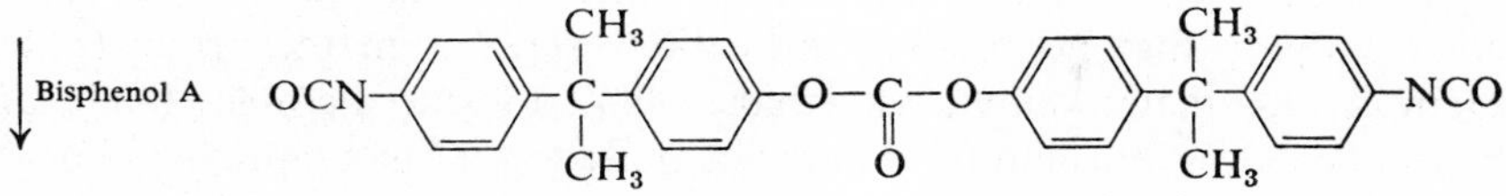

Bisphenol A

$$4$$

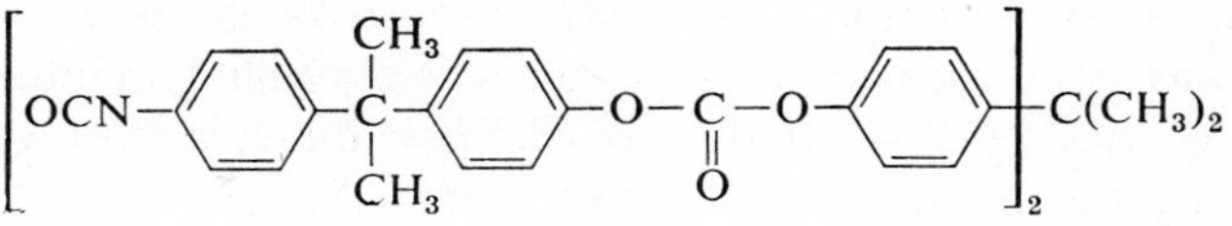

$$5$$

afforded the diisocyanate **7** (430). Reaction of **6** with ethanolamine hydrochloride, followed by phosgenation, produces the diisocyanate **8** (431).

$$OCN(CH_2)_5COCl \xrightarrow[\text{COCl}_2]{\text{HOCH}_2\text{CH}_2\text{NH}_2 \cdot \text{HCl}} OCN(CH_2)_5COOCH_2CH_2NCO$$

$$\text{\textbf{6}} \qquad\qquad\qquad\qquad\qquad\qquad\qquad \text{\textbf{8}}$$

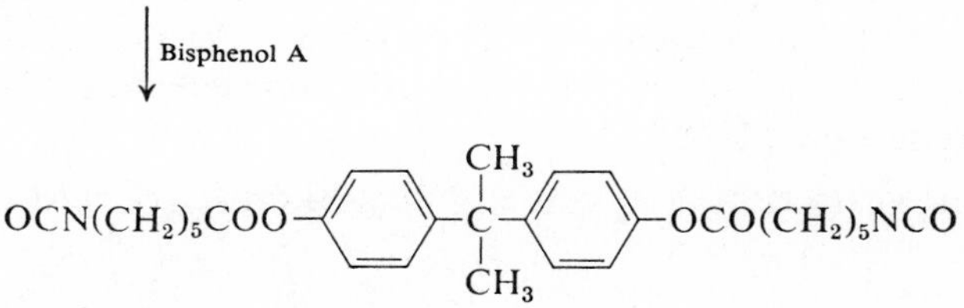

Bisphenol A

$$7$$

The reaction of isocyanatocarboxylic acid chlorides with ethanolamine hydrochloride was used to prepare a variety of ester diisocyanates, similar in structure to **8** (432). Pentamethylene diisocyanate was also prepared, starting with caprolactam (433). Ring opening of caprolactam gives the hydroxamic acid derivative **9**, which yields pentamethylene diisocyanate (**10**) upon

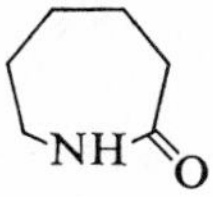

$+ \; HONH_2 \cdot HCl \longrightarrow HONHCO(CH_2)_5NH_2 \cdot HCl$

$$\text{\textbf{9}}$$

$COCl_2$

$$OCN(CH_2)_5NCO$$

$$\text{\textbf{10}}$$

phosgenation (433). A commercial approach to color stable aliphatic diisocyanates involves the catalytic reduction of aromatic diamines to the corresponding cycloaliphatic diamines. The latter are converted by phosgene to the corresponding aliphatic diisocyanates. However, the reduction requires high pressure and noble metal catalysts; also, it is absolutely essential for color stability that no residual aromaticity remains in the products. For example, 4,4'-diaminodiphenylmethane (434), toluene diamines (435), and benzidine (436) were converted in this manner to the corresponding aliphatic diisocyanates.

The reaction of fumaro-, adipo-, and sebacodihydroxamic acid with thionyl chloride was also used to prepare the corresponding aliphatic diisocyanates in high yields (437). Unsaturated aliphatic diisocyanates were prepared in good yields using the Curtius degradation (438).

The reaction of xylylene dichlorides with sodium cyanate in a mixed solvent (dimethylsulfone/o-dichlorobenzene) yields xylylene diisocyanates (439,440). In a similar manner 3,6-bis(isocyanatomethyl)durene was prepared from the corresponding dichloride and potassium cyanate, using a N,N-dimethylacetamide/chlorobenzene solvent mixture (441). The reaction of reactive halogen compounds with silver cyanate was used to prepare a series of aliphatic diisocyanates, having the isocyanato group on the same carbon atom that is bonded to oxygen (442). For example, reaction of the chloroethers 11 ($n = 2$–4) with silver cyanate gave the corresponding diisocyanates in yields ranging from 51–68%.

$$\text{ClCH}_2\text{O(CH}_2)_n\text{OCH}_2\text{Cl} + \text{AgOCN} \rightarrow \text{OCNCH}_2\text{O(CH}_2)_n\text{OCH}_2\text{NCO}$$
11

Polymerizations

The homopolymerization of isocyanates to form 1,1-polyamides has received considerable attention. Although most information concerns the polymerization (443–447) and structure (448–450) of the monoisocyanate polymers, including optically active ones (451), some diisocyanates have been examined (452,453). Of interest is the cyclopolymerization of the *cis*- and *trans*-1,3 diisocyanatocyclohexanes (453). Homopolymers of the 1,1-polyamide type are produced spontaneously from aromatic o-diisocyanates (454).

Trimerization of polyisocyanates has received considerable investigation since the discovery that cellular materials of excellent fire resistance can be prepared utilizing polyisocyanurates (455–458).

Polyurethanes and polyamides have been prepared from hydroxyl- and carboxyl-terminated polyisobutylenes (459). Polycarbodiimides are obtained from diisocyanates by the pholene oxide catalyzed condensation-elimination sequence described earlier (460).

A wide variety of heterocyclic polymers have been prepared from isocyanates. Dihydroxyaryldicarboxylic esters react in separate addition-condensation stages to form polyoxazindiones. These polymers exhibit interesting high temperature properties (461). Similarly, polyhydantoins can be prepared from diisocyanates and bis-glycine derivatives (462,463), and polyquinazolinediones from diisocyanates and diamino dicarboxylic acids (464), or diester-diisocyanates and diamines (465).

An interesting series of polyiminoimidazolidinediones and polyparabanic acids has been prepared from diisocyanates and hydrogen cyanide (466). Again, good high temperature properties for these materials are claimed. High molecular weight polyoxazolidinones from diisocyanates and diepoxides have been achieved (467,468). From appropriately substituted amines and diisocyanates were prepared substituted polyureas that could be cyclized to polyhydrouracils and polyiminoimidazolidinones (469). Polymers from Schiff bases and diisocyanates were shown to be complicated and of non-regular structure (470).

Polyisocyanates and dianhydrides have been used to prepare polyimide foams of outstanding thermal stability and flame resistance (471,472). Polyamide-imide foams and solid polymers have been prepared from polyisocyanates and trimellitic anhydride (473–475). Linear polyimides have been synthesized from diisocyanates and dianhydrides (472,476–478). The complex nature of this reaction has been examined in some detail (471,478–480). A review of the synthesis of heterocyclic polymers from polyisocyanates has been published recently (481).

References

1. A. Wurtz, *Compt. Rend.*, **27**:242 (1848).
2. W. Hentschel, *Ber.*, **17**:1284 (1884).
3. H. Staudinger and E. Hauser, *Helv. Chim. Acta*, **4**:861 (1921).
4. O. Bayer, *Angew. Chem.*, **59**:257 (1947).
5. *Chem. Eng. News*, Jan. 25, 1971.
6. W. Siefken, *Ann.*, **562**:75 (1949).
7. J. H. Saunders and R. J. Slocombe, *Chem. Rev.*, **43**:203 (1948).
8. R. G. Arnold, J. A. Nelson, and J. J. Verbanc, *Chem. Rev.*, **57**:47 (1957).
9. C. V. Wilson, "Organic Chemical Bulletin" (published by the Research Laboratories of the Eastman Kodak Company), **35**:2, 3 (1963).
10. H. Ulrich, *Chem. Rev.*, **65**:369 (1965).
11. S. Petersen, *Ann.*, **562**:205 (1949).
12. H. Ulrich, "Cycloaddition Reactions of Heterocumulenes," Academic Press, New York, 1967, pp. 177–191.
13. J. H. Saunders and K. C. Frisch, "Polyurethane Chemistry and Technology," Parts I and II, Interscience, New York, 1962.
14. O. Bayer, "Das Diisocyanat-Polyadditionsverfahren," Carl Hauser, Münich, 1963.
15. T. H. Ferrigno, "Rigid Plastic Foam," Reinhold, New York, 1963.

16. T. T. Healy, "Polyurethane Foams," Iliffe Books, London, 1964.
17. L. N. Phillips and D. B. V. Parker, "Polyurethanes," Iliffe Books, London, 1964.
18. B. A. Dombrow, "Polyurethanes" 2nd ed., Reinhold, New York, 1965.
19. S. Petersen et al., in Houben-Weyl, "Methoden der Organischen Chemie," 4th ed., Vol. 8, p. 119.
20. J. G. Lichty and N. V. Seeger, U.S. Patent 2,362,648 (1944); Chem. Abstr., 39:2764 (1945).
21. E. Burgoine and R. G. A. A. New, British Patent 574,222 (1945); E. S. Gutsell, Jr., U.S. Patent 3,262,960 (1966).
22. T. C. Allen and D. H. Chadwick, U.S. Patent 2,733,254 (1956); Chem. Abstr., 50:13093 (1956).
23. W. W. Thompson, U.S. Patent 2,689,861 (1954); Chem. Abstr., 49:11712 (1955).
24. R. J. Slocombe, E. E. Hardy, J. H. Saunders, and R. L. Jenkins, J. Amer. Chem. Soc., 72:1888 (1950).
25. W. Hentrich and H. J. Engelbrecht, U.S. Patent 2,261,156 (1941); Chem. Abstr., 36:1045 (1942).
26. R. E. Kirk and D. F. Othmer, Eds., "Encyclopedia of Chemical Technology," Vol. 12, Interscience, New York, 1967, p. 45.
27. F. X. Demers and T. R. Fink, U.S. Patent 2,938,054 (1960); Chem. Abstr., 54:19592 (1960).
28. E. Klauke and O. Bayer, U.S. Patent 3,253,010 (1966); see German Patent 1,154,091 (1963); Chem. Abstr., 60:457 (1964).
29. K. W. Rausch, R. F. Martel, and A. A. R. Sayigh, I & EC Prod. R & D, 3:125 (1964).
30. Chem. Eng. News, Oct. 24, 1966, p. 38.
31. O. Bayer, R. Schröter, W. Siefken, and K. Wagner, U.S. Patent 3,232,973 (1966); see German Patent 1,110,859 (1961); Chem. Abstr., 55:27970 (1961).
32. K. Schmitt, F. Gude, K. Rindtorff, Jr., and J. Disteldorf, German Patent 1,102,785 (1965); Chem. Abstr., 63:17933 (1965).
33. K. Schmitt, Angew. Chem. Int. Ed., 6:375 (1967).
34. N. B. Rainer, U.S. Patent 3,218,345 (1966).
35. N. M. Bortnick, U.S. Patent 2,692,275 (1954); Chem. Abstr., 49:11699 (1955).
36. M. D. Hurwitz and R. W. Auten, U.S. Patent 2,728,787 (1955); Chem. Abstr., 50:10770 (1956).
37. J. D. Garber, R. A. Gasser, and D. Wasserman, French Patent 1,351,368 (1964); Chem. Abstr., 60:15740 (1964); see also Canadian Patent 727,644 (1966).
38. T. K. Brotherton and J. W. Lynn, French Patent 1,353,461 (1964); Chem. Abstr., 61:574 (1964); see also U.S. Patent 3,162,664 (1964); U.S. Patent 3,256,220 (1966).
39. T. K. Brotherton, J. W. Lynn, and R. J. Knopf, Belgian Patent 635,304 (1963); Chem. Abstr., 61:11909 (1964).
40. H. Ulrich, J. N. Tilley, and A. A. R. Sayigh, J. Org. Chem., 29:2401 (1964).
41. A. A. R. Sayigh, J. N. Tilley, and H. Ulrich, J. Org. Chem., 29:3344 (1964); U.S. Patent 3,275,618 (1966).
42. H. Ulrich, B. Tucker, and A. A. R. Sayigh, J. Org. Chem., 31:2658 (1966); see also Ref. 51.
43. H. Ulrich and A. A. R. Sayigh, Angew. Chem. Int. Ed., 5:704 (1966).
44. C. K. Johnson, J. Org. Chem., 32:1508 (1967).
45. E. Waltmann and E. Wolf, U.S. Patent 2,346,202; Chem. Abstr., 38:3990, 4957 (1944).
46. H. Eilingsfeld, M. Seefelder, and H. Weidinger, Angew. Chem., 72;836 (1960).
47. A. J. Speziale and L. R. Smith, J. Org. Chem., 27:3742 (1962); A. J. Speziale, L. R. Smith, and J. E. Fedder, J. Org. Chem., 30:4306 (1965).

48. P. R. Steyermark, *J. Org. Chem.*, **28**:586 (1963).
49. H. Krzikalla, U.S. Patent 2,666,787 (1954); *Chem. Abstr.*, **48**:3995 (1954).
50. J. Smith, Jr., T. K. Brotherton, and J. W. Lynn, *J. Org. Chem.*, **30**:1260 (1965).
51. H. Ulrich, U.S. Patent 3,330,848 (1967).
52. L. D. Taylor, M. Plukar, and L. E. Rubin, *Polym. Lett.*, **5**:73 (1967).
53. H. A. Piggott and F. S. Statham, British Patent 485,761 (1938); *Chem. Abstr.*, **32**:7926 (1938).
54. P. Chabrier, *Compt. Rend.*, **214**:362 (1942).
55. H. Ulrich, B. Tucker, and A. A. R. Sayigh, *Angew. Chem. Int. Ed.*, **6**:636 (1967).
56. T. Mukaiyama, T. Fumisawa, and O. Mitsunobo, *Bull. Chem. Soc. (Jap.)*, **35**:1104 (1962); *Chem. Abstr.*, **57**:9733 (1962).
57. H. Ulrich, B. Tucker, and A. A. R. Sayigh, *J. Org. Chem.*, **32**:4052 (1967).
58. L. M. Alberino, H. Ulrich, and A. A. R. Sayigh, *J. Polym. Sci. A*, **5**:3212 (1967).
59. G. Schroeter, *Ber.* **42**:2336, 3356 (1909).
60. T. Curtius and W. Hechtenberg, *J. Prakt. Chem.*, **105**:289 (1923); T. Curtius, G. v. Brüning, and H. Derlon, *J. Prakt. Chem.*, **125**:63 (1930).
61. P. A. S. Smith, "Organic Reactions," *Coll. Vol. III*, Wiley, New York, 1946, p. 337.
62. F. L. Pyman, *J. Chem. Soc.*, **103**:852 (1913).
63. B. Acott, A. L. J. Beckwith, A. Hassanali, and J. W. Redmond, *Tetrahedron Lett.*, **1965**:4039; B. Acott and A. L. J. Beckwith, *Chem. Commun.*, **1965**: 151.
64. H. E. Baumgarten and A. Staklis, *J. Amer. Chem. Soc.*, **87**:1141 (1965).
65. L. Rand and M. J. Albinak, *J. Org. Chem.*, **25**:1837 (1960); L. Rand and R. J. Dolinski, *J. Org. Chem.*, **30**:48 (1965).
66. H. L. Yale, *Chem. Rev.*, **33**:209 (1943).
67. M. E. Cupery, U.S. Patent 2,346,665 (1944); *Chem. Abstr.*, **38**:5845 (1944).
68. J. B. Dickey, J. M. Straley, and T. E. Stanin, U.S. Patent 2,394,597 (1946); *Chem. Abstr.*, **40**:2848 (1946).
69. G. B. Bachman and J. E. Goldmacher, *J. Org. Chem.*, **29**:2576 (1964).
70. D. C. Berndt and W. J. Adams, *J. Org. Chem.*, **31**:976 (1966).
71. L. W. Jones, *Amer. Chem. J.*, **42**:515 (1909); **48**:5 (1912); *J. Amer. Chem. Soc.*, **38**:413 (1917).
72. D. C. Berndt and H. Schechter, *J. Org. Chem.*, **29**:916 (1964).
73. W. Lwowski, T. J. Maricich, and T. Mattingly, *J. Amer. Chem. Soc.*, **85**:1200 (1963).
74. S. Wawzonek and R. C. Gueldner, *J. Org. Chem.*, **30**:3031 (1965).
75. R. F. Smith and P. C. Briggs, *Chem. Commun.*, **1965**:120.
76. M. S. Gibson and A. W. Murray, *J. Chem. Soc.*, **1965**:880.
77. W. J. McKillip, L. M. Clemens, and R. Haugland, *Can. J. Chem.*, **45**:2613 (1967); W. J. McKillip and R. C. Slogel, *ibid.*, **45**:2619 (1967); R. C. Slogel and A. E. Blomquist, *ibid.*, **45**:2625 (1967).
78. Archer-Daniels Midland Co., Belgian Patent 686,027 (1967).
79. R. Puttner and K. Hafner, *Tetrahedron Lett.*, **1964**:3119.
80. R. Kreher and G. H. Berger, *Tetrahedron Lett.*, **1965**:369.
81. R. Huisgen and J. P. Anselme, *Chem. Ber.*, **98**:2998 (1965).
82. J. P. Anselme, *Chem. Ind. (London)*, **1966**:1794.
83. J. S. McConaghy and W. Lwowski, *J. Amer. Chem. Soc.*, **89**:2357 (1967) and previous articles by W. Lwowski and his co-workers.
84. R. Barclay, Jr., and R. P. Kurkjy, U.S. Patent 3,076,007 (1963).
85. N. M. Bortnick, L. S. Luskin, M. D. Hurwitz, and Q. W. Rytina, *J. Amer. Chem. Soc.*, **78**:4358 (1956).
86. A. W. Hofmann, *Ber.*, **3**:653 (1870).

87. H. J. Wenker, *J. Amer. Chem. Soc.*, **58**:2608 (1936).
88. G. Müller and R. Merten, German Patent 1,207,378 (1965); see also French Patent 1,381,845 (1964); *Chem. Abstr.*, **62**:11687 (1965).
89. H. Pelster, E. Muehlbauer, and O. Delfs, Belgian Patent 631,961 (1963); *Chem. Abstr.*, **61**:6919 (1964).
90. W. Will and O. Bielschowski, *Ber.*, **15**:1309 (1882).
91. H. Ulrich, B. Tucker, and A. A. R. Sayigh, *J. Org. Chem.*, **32**:3938 (1967).
92. F. J. Sowa, U.S. Patent 3,013,045 (1961); *Chem. Abstr.*, **56**:8632 (1962).
93. W. Stamm, *J. Org. Chem.*, **30**:693 (1965).
94. J. Goubeau and D. Paulin, *Chem. Ber.*, **93**:1111 (1960).
95. H. Morschel and C. Skopalic, U.S. Patent 3,190,905 (1965); see also German Patent 1,154,090 (1963); *Chem. Abstr.*, **60**:2758 (1964).
96. H. A. Staab, *Angew. Chem. Int. Ed.*, **1**:351 (1962).
97. J. N. Tilley and A. A. R. Sayigh, *J. Org. Chem.*, **28**:2076 (1963).
98. R. Harada and M. Hayashi, Japanese Patent 1740 (1967); *Chem. Abstr.*, **66**:85457 (1967).
99. E. Nachbaur, *Monatsh. Chem.*, **97**:361 (1966); *Chem. Abstr.*, **65**:10482 (1966).
100. R. Graf, *Chem. Ber.*, **89**:1071 (1956).
101. K. Matterstock and R. Graf, German Patent 1,152,093 (1963).
102. J. Goerdeler and K. Jonas, *Chem. Ber.*, **99**:3572 (1966) and earlier publications by Goerdeler and his students.
103. J. L. Cotter and G. J. Knight, *Chem. Commun.*, **1966**:336.
104. K. H. Slotta and L. Lorenz, *Ber.*, **58B**:1320 (1925).
105. W. D. Schaeffer, U.S. Patent 3,017,420 (1962); *Chem. Abstr.*, **57**:664 (1962).
106. T. I. Bieber, *J. Amer. Chem. Soc.*, **74**:4700 (1952).
107. K. Fukui and H. Kitano, British Patent 858,810 (1961); *Chem. Abstr.*, **55**:14491 (1961).
108. P. A. Argabright, H. D. Rider, and R. Sieck, *J. Org. Chem.*, **30**:3317 (1965).
109. O. C. Billeter, *Ber.*, **36**:3213 (1903).
110. A. C. Haven, Jr., *J. Amer. Chem. Soc.*, **78**:842 (1956).
111. G. S. Forbes and H. H. Anderson, *J. Amer. Chem. Soc.*, **70**:1043 (1948), and earlier articles by these authors.
112. G. S. Forbes and H. H. Anderson, *J. Amer. Chem. Soc.*, **70**:1222 (1948).
113. H. H. Anderson, *J. Amer. Chem. Soc.*, **72**:193 (1950).
114. J. Goubau and H. Gräber, *Chem. Ber.*, **93**:1379 (1960).
115. A. J. Hill and W. M. Degnan, U.S. Patent 2,379,486 (1949).
116. F. W. Hoover and H. S. Rothrock, *J. Org. Chem.*, **28**:2082 (1963).
117. F. W. Hoover and H. S. Rothrock, *J. Org. Chem.*, **29**:143 (1964).
118. J. L. Harper, U.S. Patent 3,168,545 (1965).
119. F. W. Hoover, H. B. Stevenson, and H. S. Rothrock, *J. Org. Chem.*, **28**:1825 (1963).
120. W. J. Middleton and C. G. Krespan, *J. Org. Chem.*: **30**:1398 (1965).
121. M. Manes and J. S. Mackay, U.S. Patent 3,201,433 (1965).
122. J. H. Saunders and W. B. Bennet, U.S. Patent 2,732,392 (1956); *Chem. Abstr.*, **50**:12103 (1956).
123. S. Rosen and D. Swern, *Anal. Chem.*, **38**:1392 (1966).
124. L. Birkenbach and M. Linhard, *Ber.*, **64B**:961, 1076 (1931).
125. A. Hassner, M. E. Lorber, and C. Heathcock, *J. Org. Chem.*, **32**:540 (1967).
126. A. Hassner and C. C. Heathcock, *Tetrahedron Lett.*, **1964**:1125.
127. W. S. Wadsworth and W. D. Emmons, *J. Amer. Chem. Soc.*, **84**:1316 (1962).
128. W. S. Wadsworth and W. D. Emmons, *J. Org. Chem.*, **29**:2816 (1964).

129. I. T. Kay and B. K. Snell, *Tetrahedron Lett.*, **1967**:2251.
130. H. Ulrich and A. A. R. Sayigh, *Angew. Chem. Int. Ed.*, **1**:595 (1962).
131. W. Merz, German Patent 1,222,042 (1966); *Chem. Abstr.*, **65**:13604 (1966).
132. T. Kumamoto, T. Nagaoka, and T. Mukaiyama, *Bull. Chem. Soc.* (*Jap.*), **39**:1765 (1966); *Chem. Abstr.*, **65**:15361 (1966).
133. L. C. Case, *Nature*, **183**:675 (1959).
134. W. E. Erner, *J. Org. Chem.*, **29**:2091 (1964).
135. W. Neumann and P. Fischer, *Angew. Chem. Int. Ed.*, **1**:621 (1962).
136. D. H. Clemens, A. J. Bell, and J. L. O'Brien, *Tetrahedron Lett.*, **1965**:1491.
137. H. Ulrich, B. Tucker, and A. A. R. Sayigh, *J. Amer. Chem. Soc.*, **90**:528 (1968).
138. A. Gautier, *Ann.*, **149**:311 (1869).
139. H. W. Johnson, Jr., and P. H. Daughhetee, Jr., *J. Org. Chem.*, **29**:246 (1964).
140. D. Martin and A. Weise, *Angew. Chem. Int. Ed. Eng.*, **6**:168 (1967).
141. H. W. Johnson, Jr., and H. Krutzsch, *J. Org. Chem.*, **32**:1939, **1967**.
142. P. V. Finzi and M. Arbasino, *Tetrahedron Lett.*, **1965**, 4645.
143. H. Feuer and H. Rubinstein, U.S. Patent 3,002,013 (1961).
144. B. Kühn and M. Lieber, *Ber.*, **23**:1536 (1890).
145. C. W. Tullock, D. D. Coffman, and E. L. Mutterties, *J. Amer. Chem. Soc.*, **86**:357 (1964).
146. A. V. Kirsanov, G. I. Derkach, and N. I. Liptuga, *Zh. Obshch. Khim.*, **34**:2812 (1964); *Chem. Abstr.*, **61**:14514 (1964).
147. G. I. Derkach and N. I. Liptuga, *Zh. Obshch. Khim.*, **36**:461 (1966); *Chem. Abstr.*, **65**:634 (1966).
148. E. Kühle, *Angew. Chem. Int. Ed.*, **1**:647 (1962).
149. Y. Kodama and T. Sekiba, Japanese Patent 11,389 (1965); *Chem. Abstr.*, **63**:13293 (1965).
150. J. A. Young, W. S. Durell, and R. D. Dresdner, *J. Amer. Chem. Soc.*, **81**:1587 (1959).
151. C. Grundmann and J. M. Dean, *Angew. Chem. Int. Ed.*, **3**:585 (1964).
152. E. Grigat and R. Pütter, *Angew. Chem. Int. Ed.*, **6**:206 (1967).
153. K. A. Jensen and A. Holm, *Acta Chem. Scand.*, **18**:826 (1964).
154. K. A. Jensen, M. Due, and A. Holm, *Acta Chem. Scand.*, **19**:438 (1965).
155. D. Martin, *Tetrahedron Lett.*, **1964**:2829.
156. J. C. Kauer and W. W. Henderson, *J. Amer. Chem. Soc.*, **86**:4732 (1964).
157. W. B. Hardy and R. P. Benett, *Tetrahedron Lett.*, **1967**:961.
158. Imperial Chemical Industries, Ltd., Netherlands Application 6,609,601 (1967).
159. Imperial Chemical Industries, Ltd., Netherlands Application 6,603,612 (1966); *Chem. Abstr.*, **66**:28511 (1967).
160. Imperial Chemical Industries, Ltd., Netherlands Application 6,609,480 (1967); *Chem. Abstr.*, **66**:32450 (1967).
161. J. Tsuji and N. Iwamoto, *Chem. Commun.*, **1966**:828.
162. E. W. Stern and M. L. Spector, *J. Org. Chem.*, **31**:596 (1966).
163. F. E. Drummond, U.S. Patent 3,070,618 (1962); *Chem. Abstr.*, **59**:9886 (1963).
164. D. M. Fenton, U.S. Patent 3,277,061 (1966); *Chem. Abstr.*, **66**:10851 (1967).
165. M. D. Hurwitz and R. W. Auten, U.S. Patent 2,640,846 (1953); *Chem. Abstr.*, **48**:7633 (1954).
166. J. J. Donovan and K. F. Gosselin, U.S. Patent 3,277,140 (1966).
167. E. Kühle, German Patent 1,090,197 (1960); *Chem. Abstr.*, **55**:19799 (1961).
168. H. Ulrich and A. A. R. Sayigh, *J. Org. Chem.*, **28**:1427 (1963).
169. H. Ulrich and A. A. R. Sayigh, *J. Org. Chem.*, **30**:2779 (1965).
170. O. Folin, *Amer. Chem. J.*, **19**:323 (1897).

171. H. Gross and J. Gloede, *Chem. Ber.*, **96**:1387 (1963).
172. A. V. Kirsanov, *J. Gen. Chem. (USSR)*, **24**:1033 (1954); *Chem. Abstr.*, **49**:8787 (1955).
173. H. F. McShane, Jr., and J. J. Verbanc, U.S. Patent 3,041,364 (1962).
174. H. F. McShane, Jr., French Patent 1,377,888 (1964); *Chem. Abstr.*, **62**:9082 (1965).
175. S. Hartung, E. Klauke, and H. Schwarz, U.S. Patent 3,118,922 (1964); see German Patent 1,092,462 (1960); *Chem. Abstr.*, **55**:22238 (1961).
176. H. Holtschmidt, O. Bayer, and E. Degener, German Patent 1,157,601 (1963); *Chem. Abstr.*, **60**:5394 (1964).
177. J. F. Start, U.S. Patent 3,281,488 (1966); *Chem. Abstr.*, **66**:2327 (1967).
178. W. Zecher and H. Holtschmidt, French Patent 1,375,462 (1964); *Chem. Abstr.*, **62**:9059 (1965).
179. Farbenfabriken Bayer, AG, British Patent 752,931 (1956); *Chem. Abstr.*, **51**:7420 (1957).
180. W. J. Farrissey, Jr., F. P. Recchia, and A. A. R. Sayigh, *Angew. Chem. Int. Ed.*, **5**:607 (1966).
181. W. J. Schnabel, Olin Chem. Corp., private communication.
182. The Upjohn Company, Donald S. Gilmore Research Laboratories, unpublished results.
183. A. A. R. Sayigh, British Patent 1,086,782 (1967).
184. M. W. Farlow, *Org. Syn.*, **Coll. Vol. IV**, p. 521.
185. For a different synthesis of this compound, see also Ref. 51.
186. R. Neidlein and R. Bottler, *Ber.*, **100**:698 (1967).
187. Y. Iwakura, K. Uno, and K. Ichikawa, *J. Polym. Sci.*, **A2**:3387 (1964).
188. R. Appel and H. Gerber, *Chem. Ber.*, **91**:1200 (1958).
189. R. J. Knopf and T. K. Brotherton, *J. Chem. Eng. Data*, **12**:421 (1967).
190. R. Roesch and M. H. Gold, *J. Amer. Chem. Soc.*, **73**:2959 (1951).
191. C. King, *J. Amer. Chem. Soc.*, **86**:437 (1964).
192. I. L. Knunyants, M. P. Krasuskaya, and D. P. Del'tsova, *Izv. Akad. Nauk SSSR, Ser. Khim.*, **1966**:1110; *Chem. Abstr.*, **65**:10482 (1966).
193. Y. Iwakura, Y. Matsumoto, and K. Fujikawa, *Chem. High Polym. (Jap.)*, **8**:524 (1951); *Chem. Abstr.*, **48**:11345 (1954).
194. E. Burgoine, B. Collie, and R. G. A. New, U.S. Patent 2,379,948 (1945); *Chem. Abstr.*, **39**:5254 (1945).
195. J. F. Nobis and H. Greenberg, U.S. Patent 2,865,940 (1958); *Chem. Abstr.*, **53**:9069 (1959).
196. Y. Iwakura, K. Uno, K. Suzuki, and K. Fujii, *Nippon Kâgakû Zasshi*, **78**:1504 (1957); *Chem. Abstr.*, **54**:1560 (1960).
197. Y. Iwakura, K. Uno, K. Suzuki, and K. Fujii, *Nippon Kâgakû Zasshi*, **78**:1511 (1957); *Chem. Abstr.*, **54**:1541 (1960).
198. K. Ueda, I. Kimura, and T. Ogawa, Japanese Patent 215 (1966); *Chem. Abstr.*, **64**:11085 (1966).
199. Y. Iwakura, K. Uno, K. Suzuki, and K. Fujii, *Nippon Kâgakû Zasshi*, **78**:1507 (1957); *Chem. Abstr.*, **54**:1540 (1960).
200. T. Shono, T. Morkawa, R. Okayama, and R. Oda, *Makromol. Chem.*, **81**:142 (1965).
201. J. C. Martin, U.S. Patent 3,161,665 (1964); *Chem. Abstr.*, **62**:7655 (1965).
202. E. L. Martin, British Patent 1,026,506 (1966); *Chem. Abstr.*, **65**:3768 (1966).
203. P. E. Hoch, U.S. Patent 3,151,143 (1964); *Chem. Abstr.*, **61**:15991 (1964).

204. E. F. Cox, D. T. Mannig, and H. A. Stansbury, Jr., U.S. Patents 3,069,468 (1962); 3,148,202 (1964). See also British Patent 876,013 (1961); *Chem. Abstr.*, **57**:2098 (1962).

205. M. H. Gold, U.S. Patent 2,680,131 (1954); *Chem. Abstr.*, **49**:4710 (1955).

206. M. Brown, U.S. Patent 3,214,412 (1965).

207. L. R. Moffett, Jr., and W. E. Hill, Jr., *J. Org. Chem.*, **27**:1454 (1962).

208. T. K. Brotherton and J. W. Lynn, Canadian Patent 730,160 (1966).

209. F. Huba, J. H. Wotiz, and F. B. Slezak, U.S. Patent 3,178,465 (1965); *Chem. Abstr.*, **63**:6860 (1965).

210. K. F. Zenner, G. Oertel, and H. Holtschmidt, German Patent 1,205,087; *Chem. Abstr.*, **64**:19413 (1966).

211. V. Prey and A. Aszalos, Austrian Patent 226,673 (1963); *Chem. Abstr.*, **59**:9790 (1963).

212. W. Lehmann and H. Ziemann, German Patent 1,154,092 (1963); *Chem. Abstr.*, **60**:1587 (1964).

213. A. Zamojski and K. Jankowski, *Rocz. Chem.*, **38**:707 (1966); *Chem. Abstr.*, **62**:1560 (1965).

214. T. K. Brotherton and J. W. Lynn, Canadian Patent 723,173 (1966).

215. J. Strating, H. O. van Oven, and A. M. van Leusen, *Rec. Trav. Chem. Pays-Bas*, **85**:631 (1966); *Chem. Abstr.*, **65**:8748 (1966).

216. K. Klager and C. R. Vanneman, U.S. Patent 2,967,193 (1961); *Chem. Abstr.*, **55**:8299 (1961).

217. L. Herzog, M. H. Gold, and R. D. Geckler, *J. Amer. Chem. Soc.*, **73**:749 (1951).

218. K. Klager and C. R. Vanneman, U.S. Patent 2,978,476 (1961); *Chem. Abstr.*, **55**:19799 (1961).

219. K. Klager and M. B. Frankel, U.S. Patent 2,978,474 (1961); *Chem. Abstr.*, **55**:19799 (1961).

220. K. Klager, U.S. Patent 2,978,475 (1961); *Chem. Abstr.*, **55**:19799 (1961).

221. G. Oertel and H. Holtschmidt, Belgian Patent 672,551 (1966); *Chem. Abstr.*, **65**:13558 (1966).

222. R. Merten and G. Mueller, *Angew. Chem.*, **74**:866 (1962).

223. E. Cocea, L. Stoicescu-Crivat, A. Petrus, L. Mandanescu, and I. Matei, *Acad. Rep. Pop. Rom., Fil. Iasi, Studii Cercet. Stiint. Chim.*, **14**:213 (1963); *Chem. Abstr.*, **62**:1584 (1965).

224. C. L. Parris, U.S. Patent 3,062,882 (1962); *Chem. Abstr.*, **58**:8986 (1963).

225. M. Imoto and M. Kumada, Japanese Patent 14,222 (1964); *Chem. Abstr.*, **61**:16014 (1964).

226. J. O'Brochta and S. C. Temin, U.S. Patent 3,160,648 (1964).

227. M. W. Miller, R. W. Amidon, and P. O. Tawney, *J. Amer. Chem. Soc.*, **77**:2845 (1955).

228. T. W. Campbell, V. E. McCoy, J. C. Kauer, and V. S. Foldi, *J. Org. Chem.*, **26**:1422 (1961).

229. O. Stallmann, U.S. Patent 2,729,666 (1956); *Chem. Abstr.*, **50**:12107 (1956).

230. K. Inukai and Y. Maki, *Kôgyô Kagaku Zasshi*, **68**:315 (1965); *Chem. Abstr.*, **63**:2913 (1965).

231. J. J. Tazuma, U.S. Patent 2,915,545 (1959); *Chem. Abstr.*, **54**:4494 (1960).

232. F. W. Berk and Co., French Patent 1,315,191 (1963); *Chem. Abstr.*, **59**:2713 (1963).

233. Farbenfabriken Bayer, AG, British Patent 1,024,803 (1966); see also French Patent 1,375,462 (1964); *Chem. Abstr.*, **62**:9059 (1965).

234. E. Klauke, H. Schwarz, and H. Holtschmidt, German Patent 1,138,391 (1962); *Chem. Abstr.*, **58**:8969 (1963).
235. G. H. Swart, U.S. Patent 3,322,809 (1967).
236. D. Delfs and F. Münz, U.S. Patent 2,901,497 (1959); *Chem. Abstr.*, **54**:9869 (1960).
237. T. Kimura and S. Yamada, *Yakugaku Zasshi*, **77**:888 (1957); *Chem. Abstr.*, **52**:1095 (1958).
238. G. A. Bonetti and V. J. Keenan, U.S. Patent 2,986,576 (1961); see also British Patent 852,988 (1960); *Chem. Abstr.*, **55**:14381 (1961).
239. H. France, A. Lambert, and D. Lees, British Patent 907,559 (1962); *Chem. Abstr.*, **58**:8970 (1963).
240. L. C. Case, U.S. Patent 3,180,883 (1965).
241. L. C. Case, *J. Appl. Polym. Sci.*, **8**:533 (1964).
242. L. C. Fetterly, D. O. Collamer, and C. W. Smith, U.S. Patent 3,105,845 (1963); *Chem. Abstr.*, **60**:4059 (1964). See also U.S. Patent 3,089,862 (1963); *Chem. Abstr.*, **60**:453 (1964).
243. H. Krimm and H. Schnell, German Patent 1,233,854 (1967); *Chem. Abstr.*, **67**:21617 (1967).
244. W. J. Rabourn, The Upjohn Company, unpublished results.
245. F. W. Berk and Co., British Patent 971,168 (1964); *Chem. Abstr.*, **61**:14582 (1964).
246. F. Long and R. J. Roberts, British Patent 990,399 (1965); *Chem. Abstr.*, **63**:2928 (1965); see also U.S. Patent 3,274,226 (1966).
247. D. J. Lyman, J. Heller, and M. Barlow, *Makromol. Chem.*, **84**:64 (1965).
248. N.P. Gambaryan, E. M. Rokhlin, Y. V. Zeifman, C. Ching-Yun, and I. L. Knunyants, *Angew. Chem. Int. Ed.*, **5**:947 (1966).
249. M. Yoshida and T. Miki, Japanese Patent 6583 (1966); *Chem. Abstr.*, **65**:5421 (1966).
250. H. Bertsch, E. Ulsperger, and M. Bock, *J. Prakt. Chem.*, **11**:225 (1960).
251. L. M. Yagupol'skii and B. F. Malichenko, *Zh. Obshch. Khim.*, **35**:490 (1965); *Chem. Abstr.*, **63**:531 (1965).
252. J. F. Olin, French Patent 1,447,964 (1966); *Chem. Abstr.*, **66**:104805 (1967).
253. J. C. Petropoulos, U.S. Patent 2,855,420 (1958); *Chem. Abstr.*, **53**:11330 (1959).
254. R. J. Knopf and T. K. Brotherton, U.S. Patent 3,281,447 (1966).
255. F. H. McMillan, *J. Amer. Chem. Soc.*, **74**:5229 (1952).
256. K. C. Schmeltz, U.S. Patent 2,840,489 (1958); *Chem. Abstr.*, **52**:16290 (1958).
257. R. L. Pelley, U.S. Patent 2,757,184 (1956); *Chem. Abstr.*, **51**:470 (1957).
258. H. Ulrich, B. Tucker, and A. A. R. Sayigh, *J. Org. Chem.*, **32**:1360 (1967).
259. H. Fasold and F. Turba, *Biochem. Z.*, **337**:80 (1963); *Chem. Abstr.*, **59**:8863 (1963).
260. T. K. Brotherton and J. W. Lynn, U.S. Patent 3,162,664 (1964); see also French Patent 1,353,461 (1964); *Chem. Abstr.*, **61**:574 (1964).
261. W. F. Hart, M. E. McGreal, and P. E. Thurston, *J. Org. Chem.*, **27**:338 (1962).
262. H. Holtschmidt and G. Oertel, *Angew. Chem. Int. Ed.*, **1**:617 (1962).
263. G. Oertel and H. Holtschmidt, German Patent 1,129,149 (1962); *Chem. Abstr.*, **57**:11238 (1962).
264. Union Carbide Corporation, Netherlands Application 6,410,323 (1965); *Chem. Abstr.*, **63**:8403 (1965).
265. E. Müller, H. Wilms, H. Kritzler, and K. Wagner, U.S. Patent 3,426,011 (1966); *Chem. Abstr.*, **65**:2275 (1966).
266. J. H. Wild, British Patent 944,309 (1963); *Chem. Abstr.*, **60**:836 (1964).
267. J. D. Garber, U.S. Patent 3,049,552 (1962); *Chem. Abstr.*, **58**:1434 (1963).
268. J. J. Jaruzelski and W. E. Smith, U.S. Patent 2,980,705 (1961); *Chem. Abstr.*, **55**:21145 (1961).

269. H. M. Grotta, U.S. Patent 3,074,959 (1962); *Chem. Abstr.*, **59**:577 (1963).
270. M. F. Lappert and H. Pyszora, *Proc. Chem. Soc.*, **1960**:350.
271. M. F. Lappert and H. Pyszora, *J. Chem. Soc.*, **1965**:854.
272. T. L. Heying, J. A. Reid, and S. I. Trotz, U.S. Patent 3,291,820 (1966).
273. J. Prejznev, *Rocz. Chem.*, **39**:747 (1965); *Chem. Abstr.*, **63**:9846 (1965).
274. K. Matterstock, German Patent 1,205,099 (1965); *Chem. Abstr.*, **64**:8236 (1966).
275. R. G. Neville and J. J. McGee, *Inorg. Syn.*, **8**:23 (1966).
276. G. K. Weisse and R. M. Thomas, U.S. Patent 3,093,451 (1963); *Chem. Abstr.*, **60**:552 (1964).
277. K. Thinius, *Plaste Kautsch.*, **12**:389 (1965); *Chem. Abstr.*, **63**:10124 (1965).
278. R. M. Pike and E. B. Moynahan, *Inorg. Chem.*, **6**:168 (1967).
279. A. S. Mufti and R. C. Poller, *J. Chem. Soc.*, **1965**:5055.
280. A. S. Mufti and R. C. Poller, *J. Organometal. Chem.*, **3**:99 (1965).
281. G. K. Weisse and R. M. Thomas, German Patent 1,161,559 (1964); *Chem. Abstr.*, **60**:10719 (1964).
282. G. I. Derkach, E. I. Slyusarenka, B. Y. Lipman, and N. I. Liptuga, *Zh. Obshch. Khim.*, **35**:1881 (1965).
283. L. I. Bai, A. Y. Yakubovich, and L. I. Muler, *Zh. Obshch. Khim.*, **34**:3609 (1964); *Chem. Abstr.*, **62**:9168 (1965).
284. I. C. Popoff and J. P. King, *J. Polym. Sci.* **B1**:247 (1963).
285. K. Utvary, E. Freundlinger, and V. Gutmann, *Monatsh. Chem.*, **97**:680 (1966); *Chem. Abstr.*, **65**:13747 (1966).
286. H. C. Fielding, British Patent 968,886 (1964); *Chem. Abstr.*, **61**:13194 (1964).
287. H. C. Fielding, British Patent 891,861 (1962); *Chem. Abstr.*, **57**:7170 (1962).
288. R. Graf, German Patent 940,351 (1956); *Chem. Abstr.*, **52**:12344 (1958).
289. R. Appel and H. Gerber, *Angew. Chem.*, **70**:271 (1958).
290. H. R. Davis, U.S. Patents 3,185,677 (1965); 3,280,184 (1966).
291. G. Karmas, U.S. Patent 2,574,484 (1951); *Chem. Abstr.*, **46**:7123 (1952).
292. M. P. Thorne, *Can. J. Chem.*, **45**:2537 (1967).
293. J. W. McFarland, D. E. Lenz, and O. J. Grosse, *J. Org. Chem.*, **31**:3798 (1966).
294. W. van Pee and J. C. Jungers, *Bull. Soc. chim. Fr.*, **1967**:158; E. de Cooman and I. de Aguirre, *ibid.*, **1967**:165; W. D. Olleslager and I. de Aguirre, *ibid.*, **1967**:179.
295. E. E. Blaise and I. Hermann, *Ann. Chim.* (8), **17**:393 (1909).
296. W. Humnicki, *Rocz. Chem.*, **11**:674 (1931); *Chem. Abstr.*, **26**:5556 (1932).
297. L. Knorr and P. Rossler, *Ber.*, **36**:1280 (1903).
298. A. F. Ferris and B. A. Schutz, *J. Org. Chem.*, **28**:3140 (1963).
299. K. Sisido, K. Hukuoka, M. Tuda, and H. Nozaki, *J. Org. Chem.*, **27**:2663 (1962).
300. N. R. Easton, D. R. Cassady, and R. D. Dillard, *J. Org. Chem.*, **27**:2927 (1962).
301. N. Shachat and J. J. Bagnell, Jr., *J. Org. Chem.*, **28**:991 (1963).
302. R. E. Strube and F. A. McKellar, *Rec. Trav. Chim. Pays-Bas*, **83**:1191 (1964).
303. G. Bobowski and J. Shavel, Jr., *J. Org. Chem.*, **32**:953 (1967).
304. W. Metlesics, G. Silverman, V. Toome, and L. Sternbach, *J. Org. Chem.*, **31**:1007 (1966).
305. R. F. Smith, *J. Heterocycl. Chem.*, **3**:535 (1966).
306. E. C. Taylor and R. V. Ravindranathan, *J. Org. Chem.*, **27**:2622 (1962).
307. J. C. Sheehan and G. D. Daves, Jr., *J. Org. Chem.*, **29**:3599 (1964).
308. M. Kurihara and N. Yoda, *Tetrahedron Lett.*, **1965**:2597.
309. Farbenfabriken Bayer, Belgian Patent 691,900 (1967).
310. C. D. Hurd and A. G. Prapas, *J. Org. Chem.*, **24**:388 (1959).

311. A. Farkas and G. A. Mills, "Advances in Catalysis," Vol. 13, Academic Press, New York, 1962, p. 393.
312. N. Onodera, *Kôgyô Kagaku Zasshi*, **65**:1249 (1962); *Chem. Abstr.*, **61**:6886 (1964).
313. M. Kaplan, *J. Chem. Eng. Data*, **6**:272 (1961).
314. M. E. Baily, V. Kirss, and R. G. Spaunburgh, *Ind. Eng. Chem.*, **48**:794 (1956).
315. Carwin Chemical Company Bulletin "Carwin Diisocyanates," March 15, 1955.
316. D. S. Tarbell, E. C. Mallatt, and J. W. Wilson, *J. Amer. Chem. Soc.*, **64**:2229 (1942).
317. J. W. Baker et al., *J. Chem. Soc.*, **1947**:713; **1949**:9, 19, 24, 27.
318. E. F. Fox and F. Hostettler, "Abstracts of Papers," 135th Meeting of the American Chemical Society, Boston, 1959, p. 112-O.
319. J. W. Britain and P. G. Gemeinhard, *J. Appl. Polym. Sci.*, **4**:207 (1960).
320. F. G. Willboordse, F. E. Critchfield, and R. L. Meeker, *J. Cell. Plastics*, **1**:3 (1965).
321. K. G. Flynn and D. R. Nenortas, *J. Org. Chem.*, **28**:3527 (1963).
322. J. Burkus, *J. Org. Chem.*, **27**:474 (1962).
323. A. Farkas and P. F. Strohm, *Ind. Eng. Chem.*, **4**:32 (1965).
324. O. V. Nesterov and S. G. Entelis, *Kinetika i Kataliz*, **6**:178 (1965); *Chem. Abstr.*, **62**:14442 (1965).
325. I. A. Pronina, Y. L. Spirin, A. A. Blagonravova, S. M. Arefeva, and A. R. Gantmakher, *Kinetika i Kataliz*, **7**:439 (1966); *Chem. Abstr.*, **65**:10448 (1966).
326. H. A. Smith, *J. Appl. Polym. Sci.*, **7**:85 (1963).
327. L. Rand, B. Thir, S. L. Reegen, and K. C. Frisch, *J. Appl. Polym. Sci.*, **9**:1787 (1965).
328. K. C. Frisch, S. L. Reegen, W. V. Floutz, and J. P. Oliver, *J. Polym. Sci.*, *A*, **5**:35 (1967).
329. S. G. Entelis and O. V. Nesterov, *Kinetika i Kataliz*, **7**:464 (1966); *Chem. Abstr.*, **65**:10462 (1966).
330. S. G. Entelis, O. V. Nesterov, and V. B. Zabrodin, *Kinetika i Kataliz*, **7**:627 (1966); *Chem. Abstr.*, **65**:18478 (1966).
331. O. V. Nesterov, V. B. Zabrodin, Y. N. Chrkov, and S. G. Entelis, *Kinetika i Kataliz*, **7**:805 (1966); *Chem. Abstr.*, **66**:37000 (1967).
332. M. Yokoo, J. Ogura, and T. Kanzawa, *Polym. Lett.*, **5**:57 (1967).
333. H. L. Heiss, French Patent 1,438,111 (1966); *Chem. Abstr.*, **66**:38050 (1967).
334. R. Hulse and H. J. Twitchett, British Patent 957,841 (1964); *Chem. Abstr.*, **61**:9526 (1964).
335. E. Dyer and R. B. Pinkerton, *J. Appl. Polym. Sci.*, **9**:1713 (1965).
336. A. J. Bloodworth and A. G. Davies, *Proc. Chem. Soc.*, **1963**:264, 315; *J. Chem. Soc.*, **1965**:5238, 6245.
337 L. B. Weisfeld, *J. Appl. Polym. Sci.*, **5**:424 (1961).
338. R. S. Bruenner and A. E. Oberth, *J. Org. Chem.*, **31**:887 (1966).
339. J. R. Fischer, *Tetrahedron*, **19**: Suppl. 1, 97 (1963).
340. I. A. Pronina, Y. L. Spirin, A. A. Blagonravova, S. M. Arefeva, A. R. Gantmakher, and S. S. Medvedev, *Dokl. Akad. Nauk SSSR*, **161**:362 (1965); *Chem. Abstr.*, **63**:451 (1965).
341. J. Robins, *J. Appl. Polym. Sci.*, **9**:821 (1965).
342. G. J. M. Van der Kerk, *Ind. Eng. Chem*, **58**:29 (1966).
343. H. G. J. Overmars and G. M. van der Want, *Chimia*, **19**:126 (1965).
344. W. J. Farrissey, Jr., R. J. Ricciardi, and A. A. R. Sayigh, 153rd Meeting of the American Chemical Society, April 9, 1967, Abstracts of Papers, p. Q-60; *J. Org. Chem.*, **33**:1913 (1968).
345. R. E. Buckles and L. A. McGrew, *J. Amer. Chem. Soc.*, **88**:3582 (1966).
346. R. Huisgen, *Angew. Chem. Int. Ed.*, **2**:565 (1963).

347. J. Goerdeler and H. Schenk, *Angew. Chem. Int. Ed. Eng.*, **2**:552 (1963).
348. J. Nieuwenhuis and J. F. Arens, *Rec. Trav. Chim. Pays-Bas*, **76**:999 (1957).
349. J. C. Sheehan and G. D. Daves, Jr., *J. Org. Chem.*, **30**:3247 (1965).
350. H. Herlinger, *Angew. Chem. Int. Ed. Eng.*, **3**:378 (1964).
351. R. Staiger, C. L. Moyer, and G. R. Pitcher, *J. Chem. Eng. Data.*, **8**:454 (1963).
352. J. S. Swenton, *Tetrahedron Lett.*, **1967**:2855.
353. T. W. Campbell and K. C. Smeltz, *J. Org. Chem.*, **28**:2069 (1963).
354. H. Ulrich, B. Tucker, and A. A. R. Sayigh, *Tetrahedron Lett.*, **1967**:1731.
355. F. Ramirez, S. B. Bhatia, and C. P. Smith, *J. Amer. Chem. Soc.*, **89**:3030 (1967).
356. G. C. Corfield and A. Crawshaw, *Chem. Commun.*, **1966**:85.
357. V. E. Shashoua, *J. Amer. Chem. Soc.*, **81**:3156 (1959).
358. H. C. Beachell and C. P. Ngoc Son, *Polym. Lett.*, **1**:25 (1963).
359. H. Takida and K. Noro, *Kobunshi Kagaku*, **22**:463 (1965); *Chem. Abstr.*, **64**:2174 (1966).
360. J. Hollander, F. D. Trischler, E. S. Harrison, Polymer Preprints, 154th Meeting of the American Chemical Society, Chicago, 1967, pp. 1, 149.
361. British Patent 543,297 (1942); the experimental procedure is outlined in W. R. Sorenson and T. W. Campbell, "Preparative Methods of Polymer Chemistry," Interscience, New York, 1961, p. 88.
362. W. R. Sorenson, *J. Org. Chem.*, **24**:978 (1959).
363. P. W. Morgan, "Condensation Polymers," Interscience, New York, 1965, pp. 217–223.
364. P. Esselman and J. Düsing, German Patent 745,472 (1943).
365. T. W. Campbell, V. S. Foldi, and J. Farago, *J. Appl. Polym. Sci.*, **2**:155 (1959).
366. T. W. Campbell and E. A. Tomic, *J. Polym. Sci.*, **62**:379 (1962).
367. E. A. Tomic, T. W. Campbell, and V. S. Foldi, *J. Polym. Sci.*, **62**:387 (1962).
368. T. W. Campbell, V. S. Foldi, and R. G. Parrish, *J. Appl. Polym. Sci.*, **2**:81 (1959).
369. Y. Iwakura, K. Hayashi, and K. Inagaki, *Makromol. Chem.*, **100**:22 (1967).
370. Y. Iwakura, K. Hayashi, S. Kang, and K. Inagaki, *Makromol. Chem.*, **95**:205 (1966).
371. N. Yoda, R. Nakanishi, M. Kurihara, Y. Bamba, S. Tohyama, and K. Ikeda, *Polym. Lett.*, **4**:11 (1966).
372. W. H. Daly, Polymer Preprints of the 152nd Meeting of the American Chemical Society, New York, 1966, p. 569.
373. S. R. Sandler, F. Berg, and G. Kitazawa, *J. Appl. Polym. Sci.*, **9**:1994 (1965).
374. J. Goerdeler and H. Schenk, *Ber.*, **98**:2954 (1965).
375. E. B. Murphy and W. A. O'Neil, *SPE J.*, **18**:191 (1962); *Chem. Abstr.*, **57**:4871 (1962).
376. G. G. Greth, R. G. Smith, and G. O. Rudkin, Jr., *J. Cell. Plastics*, **1**:159 (1965).
377. D. J. David, *Anal. Chem.*, **35**:1647 (1963).
378. E. G. Brame, Jr., R. C. Ferguson, and G. J. Thomas, Jr., *Anal. Chem.*, **39**:517 (1967).
379. M. Sumi, Y. Chokki, Y. Nakai, M. Nakabayashi, and T. Kanzawa, *Makromol. Chem.*, **78**:146 (1964).
380. J. M. Ruth and R. J. Philippe, *Anal. Chem.*, **38**:720 (1966).
381. J. L. Cotter, *J. Chem. Soc.*, **1964**:5491.
382. G. Spielberger; see W. Siefken, *Ann.*, **562**:99 (1948).
383. H. E. Stagg, *Analyst*, **71**:557 (1946).
384. S. Siggia and J. G. Hanna, *Anal. Chem.*, **20**:1084 (1948).
385. American Society for Testing and Materials, Philadelphia, Pa., Method D1638-59T.
386. B. S. Karten and T. S. Ma, *Microchem. J.*, **3**:507 (1959).
387. K. A. Kubitz, *Anal. Chem.*, **29**:814 (1957).
388. W. R. McElroy, U.S. Patent 2,969,386 (1964).

389. H. Bank, *Kunststoffe*, **37**:102 (1947).
390. J. A. Zapp, *AMA Arch. Ind. Health*, **15**:324 (1957).
391. B. B. Wheals and J. Thomson, *Chem. Ind. (London)*, 1967:753.
392. K. Marcali, *Anal. Chem.*, **29**:552 (1957).
393. V. Levin, B. W. Nippoldt, and R. L. Rebertus, *Anal. Chem.* **39**:581 (1967).
394. E. Schroeder, *Plaste Kautsch.*, **9**:121 (1962); *Chem. Abstr.*, **57**:15337 (1962).
395. S. S. Lord, Jr., *Anal. Chem.*, **29**:497 (1957).
396. S. Steingiser, W. C. Darr, and E. E. Hardy, *Anal. Chem.*, **31**:1261 (1959).
397. N. R. Neubauer, G. R. Skrekoski, R. G. White, and A. J. Kane, *Anal. Chem.*, **35**:1647 (1963).
398. Mobay Chemical Company Data Sheet, February, 1962.
399. National Cash Register Co., British Patent 1,038,739.
400. *Lancet*, January 1, 1966, p. 32; A. Munn, *Chem. Ind. (London)*, 1968:172.
401. K. E. Grim and R. E. Knox, *Amer. Ind. Hyg. Assoc. J.*, **27**:62 (1966).
402. "Toxicity and Safe Handling of Isocyanates," Mobay Chemical Co., Pittsburgh, Pa., reported as "single oral lethal dose."
403. "Occupational Health," Michigan Department of Public Health, **11**(4): 1966.
404. International Research and Development Corp., studies for The Upjohn Company.
405. T. Kanzawa and K. Naito, Japan Chemical Quarterly, **III–IV**:38 (1967); Takeda Chemical Ind., Ltd., 400 Park Avenue, New York, N.Y., 10022.
406. Manufacturing Chemists Assoc., Chemical Safety Data Sheet SD-73, "Tolylene Diisocyanate."
407. General Mills Chemicals, Kankakee, Ill., Data Sheet, DDI®.
408. K. D. Kaufmann, H. Bormann, K. Rühlmann, G. Engelhardt, and H. Kriegsmann, *Chem. Ber.*, **101**:984 (1968).
409. A. J. Bloodworth, A. G. Davies, and S. C. Vashishtha, *J. Chem. Soc. (C)*, 1967:1309.
410. A. G. Davies and P. G. Harrison, *J. Chem. Soc. (C)*, 1967:1313.
411. H. Schumann and R. J. Jutzi, *Chem. Ber.*, **101**:24 (1968).
412. A. G. Davies and R. J. Puddephatt, *J. Chem. Soc. (C)*, 1967:2663; 1968:1479.
413. R. H. Cragg, *Chem. Ind. (London)*, 1967:1751.
414. B. F. Malichenko and V. V. Penchuk, *Zh. Obshch. Khim.*, **38**:2497 (1968); *Chem. Abstr.*, **70**:46986 (1969).
415. B. F. Malichenko and A. V. Yazlovitskii, *Zh. Obshch. Khim.*, **39**:299 (1969); *Chem. Abstr.*, **70**:3089 (1969).
416. B. F. Malichenko and A. V. Yazlovitskii, *Zh. Obshch. Khim.*, **39**:2323 (1969); *Chem. Abstr.*, **72**:43022 (1970).
417. B. F. Malichenko and O. N. Tsypina, *Zh. Obshch. Khim.*, **39**:2515 (1969); *Chem. Abstr.*, **72**:78563 (1970).
418. B. F. Malichenko and O. N. Tsypina, *Zh. Org. Khim.*, **6**:2293 (1970); *Chem. Abstr.*, **74**:42109 (1971).
419. K. Inukai and Y. Maki, Japanese Patent 67,24,890; *Chem. Abstr.*, **69**:43609 (1968).
420. A. A. R. Sayigh, J. N. Tilley, and H. Ulrich, U.S. Patent 3,375,264 (1968); *Chem. Abstr.*, **68**:104705 (1968).
421. K. A. Kornev, A. G. Panteleimonov, and V. G. Ostroverkhov, *Ukr. Khim. Zh.*, **34**:1046 (1968); *Chem. Abstr.*, **70**:67779 (1969).
422. U. v. Gizycki, *Angew. Chem. Int. Ed. Engl.*, **10**:403 (1971).
423. S. Hyden and G. Wilbert, *Chem. Ind. (London)*, 1967:1406.
424. E. W. Crandall and L. Harris, *Org. Prep. Proc.* **1**:147 (1969).
425. J. A. Maclaren, *Chem. Ind. (London)*, 1971:395.
426. E. H. Burk and D. D. Carlos, *J. Heterocycl. Chem.*, **7**:177 (1970).

427. J. Sauer and K. K. Mayer, *Tetrahedron Lett.*, **1968**:319.
428. H. Krimm and H. Schnell, German Patent 1,232,133 (1967); *Chem. Abstr.*, **68**:95497 (1968).
429. H. Krimm and H. Schnell, German Patent 1,244,769 (1967); *Chem. Abstr.*, **68**:68686 (1968).
430. H. Krimm, G. Malamet, and H. Schnell, German Patent 1,231,688 (1967).
431. Rohm and Haas Co., British Patent 1,170,777 (1969).
432. W. D. Emmons and J. F. Levy, French Patent 1,507,036 (1967); *Chem. Abstr.*, **70**:19556 (1969).
433. R. Merten and C. Weber, *Synthesis*, **1970**:589.
434. W. Mesch, *Chem. Ztg.*, *Chem. App.*, **95**:554 (1971); *Chem. Abstr.*, **75**:63203 (1971).
435. M. Cenker and T. Y. P. Kahn, German Offen, 2,005,297 (1970); *Chem. Abstr.*, **73**:98475 (1970).
436. R. J. Freure and M. Moyle, German Offen, 1,936,430 (1970); *Chem. Abstr.*, **72**:89886 (1970).
437. E. H. Burk, Jr. and D. D. Carlos, U.S. Patent 3,423,449 (1969); *Chem. Abstr.*, **70**:67581 (1969).
438. A. V. Fokin, M. A. Rasksha, B. V. Bocharov, T. M. Potarina, and G. A. Osipova, *Zh. Org. Khim.*, **3**:1746 (1967); *Chem. Abstr.* **68**:38974 (1968).
439. A. L. McMaster, U.S. Patent 3,440,269 (1969); *Chem. Abstr.*, **70**:3123 (1969).
440. A. L. McMaster and K. E. Davis, U.S. Patent 3,440,270 (1969); *Chem. Abstr.*, **70**:3122 (1969).
441. D. C. Eaton and W. G. Healey, German Offen. 2,031,289 (1971); *Chem. Abstr.*, **74**:87591 (1971).
442. H. Böhme and W. Pasche, *Arch. Pharm.*, **302**:617 (1969).
443. C. G. Overberger and J. A. Moore, "Encyclopaedia of Polymer Science and Technology," Vol. 7, Interscience, New York, 1969, p. 743.
444. Y. Iwakura, K. Uno, and H. Kobayashi, *J. Polym. Sci.*, A-1, **6**:793 (1968).
445. Y. Kitahama, H. Ohama and H. Kobayashi, *J. Polym. Sci.*, A-1, **7**:935 (1969).
446. G. Odian and L. S. Hiraoka, *Polym. Preprints, Vol. II*, **1**:82 (1970).
447. K. Harada, A. Deguchi, J. Furukawa, and S. Yamashita, *Makromol. Chem.*, **132**:281 (1970).
448. J. B. Milstein and E. Charney, *Macromolecules*, **2**:678 (1969).
449. A. J. Bur and D. E. Roberts, *J. Chem. Phys.*, **51**:406 (1969).
450. H. Plummer and B. R. Jennings, *Eur. Polym. J.*, **6**:171 (1970).
451. M. Goodman and S. C. Chen, *Macromolecules*, **3**:398 (1970).
452. H. Matsui, K. Yasuda, and J. Goto, *Japan 71*, **15**:298 (1971).
453. G. C. Corfield and A. Crawshaw, *J. Macromol. Sci. Chem.*, **A5**:3 (1971).
454. W. J. Schnabel and E. Kober, *J. Org. Chem.* **34**:1162 (1969).
455. H. G. Nadeau, R. A. Kolakowski, H. E. Reymore, Jr., R. L. Grieve, and A. A. R. Sayigh, to be presented at the Americal Chemical Society Meeting, April 9–14, 1972, Boston, Mass.; G. V. Comunale, H. G. Nadeau, German Offen. 1,904,575 (1969); *Chem. Abstr.*, **71**:102680e (1969).
456. G. W. Ball, G. A. Haggis, R. Hurd, and J. F. Wood, *J. Cell. Plast.*, **4**:248 (1968); G. A. Haggis, U.S. Patent 3,516,950 (1970); *Chem. Abstr.*, **73**:56803p (1970); G. W. Ball and P. J. Briggs, British Patent 1,229,983 (1971); *Chem. Abstr.*, **75**:50065u (1971).
457. H. J. Diehr, R. Merten, and H. Piechota, U.S. Patent 3,580,868 (1971); *Chem. Abstr.*, **75**:64856q (1971).
458. K. Ashida and T. Yagi, British Patent 1,155,768 (1969).
459. R. L. Zapp, G. E. Seriuk, and L. S. Minkler, *Rubber Chem. Technol.*, **43**:1154 (1970).

460. T. W. Campbell and V. S. Foldi, *Macromol. Syn*, **3**:109 (1968).

461. L. Bottenbruch, *Angew. Makromol. Chem.*, **13**:109 (1970).

462. M. Russo, *Mater. Plast. Elastomeri*, **36**:645 (1970); *Chem. Abstr.*, **73**:8828 (1970).

463. R. Merten and G. D. Wolf, German Offen., 1,906,492; *Chem. Abstr.*, **73**:131619z (1970); R. Merten and W. Zecher, U.S. Patent 3,448,170 (1969); *Chem. Abstr.*, **71**:39659n (1969).

464. E. I. Khofbauer and E. I. Nesterova, Russian Patent 219,784 (1969); *Chem. Abstr.*, **70**:12136g (1969).

465. Stammicarbon N.V., Netherlands Patent 68,03527 (1969); *Chem. Abstr.*, **72**:33367q (1970).

466. T. L. Patton, *Polym. Preprints*, 12 (1), 162 (1971); German Offen. 2,003,938 (1970); *Chem. Abstr.*, **73**:99411b (1970).

467. R. R. Dileone, *J. Polym. Sci.*, A-1, **8**:609 (1970).

468. J. E. Herweh, W. Y. Whitmore, *J. Polym. Sci.*, A-1, **8**:2759 (1970).

469. E. Dyer and J. Hartzler, *J. Polym. Sci.*, A-1, **7**:833 (1969).

470. K. Harada, Y. Mizoe, J. Furukawa, and S. Yamashiita, *Makromol. Chem.*, **132**:295 (1970).

471. W. J. Fassissey, Jr., J. S. Rose, and P. S. Carleton, *J. Appl. Polym. Sci.*, **14**:1093 (1970).

472. W. J. Farrissey, Jr., A. McLaughlin and J. S. Rose, U.S. Patent 3,562,189 (1971).

473. S. T. Kus and F. W. Koenig, U.S. Patent 3,479,305 (1969); *Chem. Abstr.*, **72**:22309w (1970).

474. W. Zecher and R. Merten, German Offen., 1,956,512 (1971); *Chem. Abstr.*, **75**:65475u (1971).

475. S. Terney, J. Keating, J. Zielinski, J. Hakala, and H. Sheffer, *J. Polym. Sci.*, A-1, **8**:683 (1970).

476. P. S. Carleton, W. J. Farrissey, Jr., and J. S. Rose, German Offen., 2,001,914 (1970); *Chem. Abstr.*, **73**:89264s (1970).

477. Societé Rhodiaceta, *Fr. Addn.*, 94,881 (1970); *Chem. Abstr.*, **73**:99542r (1970).

478. R. A. Meyers, *J. Polym. Sci.*, A-1, **7**:2757 (1969).

479. P. S. Carleton and W. J. Farrissey, Jr., *Tetrahedron Lett.*, **40**:3485 (1969).

480. P. S. Carleton, W. J. Farrissey, Jr., and J. S. Rose, *J. Appl. Polym. Sci.*, in press.

481. R. Merten, *Angew. Chem. Int. Ed. Engl.*, **10**:294 (1971).

6. AROMATIC DIACIDS, ESTERS, AND ACID CHLORIDES

WILLIAM F. BRILL, *Halcon International Inc., Little Ferry, New Jersey* AND
JOSEFINA T. BAKER, *Princeton Chemical Research Inc., Princeton,
New Jersey*

Contents

I. INTRODUCTION

Aromatic diacids and their esters, acid chlorides, and anhydrides are used in the preparation of polyesters and polyamides. The polyfunctional and acidic character of diacids makes them useful and theoretically interesting chemicals. The most important aromatic acids are the three isomeric benzene dicarboxylic acids, which account for nearly all the aromatic acids produced commercially. The orthoisomer, phthalic acid, is the only one to form an intramolecular anhydride. The paraisomer is known as terephthalic acid, and the less important metaisomer is called isophthalic acid. Other diacids, particularly those described in this chapter, are receiving increasing attention as the raw materials for their preparation become more available. Diesters and diacid chlorides are interesting because their physical and chemical properties allow greater versatility in the preparation of condensation polymers and in their purification.

The single most important method of preparing aromatic polycarboxylic acids is the oxidation of alkyl aromatic hydrocarbons. For almost all carboxylic acids, the choice of a synthetic method becomes one of finding the most appropriate and convenient oxidation procedure. Oxidation with molecular oxygen is usually the method of choice, if economics and product purity are of great importance; but the use of such oxidizing agents as permanganate, chromate, or nitric acid may become the preferred method if convenience is a factor. When aromatics are orthosubstituted and the direct production of a sufficiently volatile anhydride is possible, air oxidation may be conducted in the vapor phase over a heterogeneous catalyst of vanadium oxide. Other alkylsubstituted aromatics must be oxidized in the liquid phase. The foregoing observations are clearly generalizations and, in fact, a multitude of synthetic methods have been reported. Specific synthetic procedures usually reflect, in various degrees, individual differences in the chemical reactivity and physical properties of the starting materials and of the product acids.

The production of esters and acid chlorides can be accomplished in most cases by well-known reactions of the carboxylic acids. However, in a few cases, oxidation or chlorination reactions have been reported which lead directly to these derivatives from alkyl aromatics or aldehydes, without the formation or isolation of acid.

The enormous number of recent publications on the preparation and purification of aromatic dicarboxylic acids makes it difficult and unwise to give a comprehensive account of work in this area. More than 1200 references from *Chemical Abstracts* covering the period from 1960 to the middle of 1967 on phthalic anhydride and the benzenedicarboxylic acids, esters, and acid chlorides were examined. Most of these references consist of patent literature and seem to the authors to be of uncertain value. As perhaps may be expected, they often appear to be of even more uncertain novelty. Nevertheless, an attempt has been made to bring the discussion up to date with a brief description of enough recent work to illustrate the current trends in research.

II. PHTHALIC ANHYDRIDE AND PHTHALIC ACID

A. History

Phthalic acid was first prepared by Laurent in 1836 by the oxidation of 1,2,3,4-tetrachloronaphthalene with chromic acid; phthalic anhydride was produced by sublimation of the acid (1). Growing demand for the acid as a dye intermediate at the end of the nineteenth century led to the development of a large-scale process for oxidizing naphthalene in sulfuric acid in the presence of mercury salts (2). Numerous other liquid-phase oxidation methods were also developed, but none became important. For example, naphthalene was oxidized to phthalic acid by potassium permanganate in boiling water (2). Alkali chlorate solutions activated with osmium tetroxide were also used as oxidizing agents. Naphthols and other naphthalene derivatives were oxidized by heating alkaline solution with various metallic oxides and peroxides. Electrolytic oxidation of naphthalene in acid solutions and air oxidation using copper or nickel oxides in the liquid phase were reported.

Phthalic acid was also prepared by the liquid-phase oxidation of *o*-xylene and various orthosubstituted benzene derivatives, but this reaction never acquired great synthetic interest. In 1885 Nölting oxidized *o*-xylene with potassium permanganate while demonstrating the structure of the three isomeric xylenes (3). Other oxidizing agents that have been used include calcium permanganate, chromic acid, and fuming nitric acid. Early preparations of phthalic acid involved the oxidation of *o*-toluic acid, *o*-xylylene glycol, *o*-phthalaldehyde, and trichloro-*o*-tolunitrile (4).

Phthalic anhydride became available as an inexpensive chemical intermediate only after the discovery that it could be produced by the direct vapor-phase oxidation of naphthalene over a catalyst based on vanadium or molybdenum oxides. The interesting history and the commercial development of this process have been reviewed (2,5). Essentially the same process was discovered independently by Gibbs and Conover (6,7) in the United States and by Wohl at I. G. Farben in Germany. In the United States, a number of companies joined in the commercial development.

In 1946 production of phthalic anhydride by the catalytic oxidation of *o*-xylene was initiated. A review of the earliest efforts to oxidize *o*-xylene in the vapor phase over vanadia catalysts (8) indicates that it had proved more difficult to obtain good yields than had been the case working with naphthalene. Considerable amounts of methyl benzaldehyde and phthalaldehyde were produced. Parks and Allard studied a variety of supported catalysts and obtained the best yield of 18% phthalic anhydride over vanadium pentoxide on Alfrax at 530°C.

B. Commercial Utility and Production Volume

Early predictions that the availability of phthalic anhydride by catalytic air oxidation would lead to expanding uses and production on a very large scale have proven correct. The dyestuffs industry, which stimulated the early synthetic work and took the largest part of the phthalic anhydride output in 1918, now consumes less than 5% of the total production (9). The major use of phthalic anhydride, which accounts for 48% of its consumption, is in the preparation of vinyl plasticizers (10). Alkyd resins, which now take 30% of the anhydrides' end use, were once the largest outlet, but they have been losing a portion of their market to other surface coatings. The most rapidly growing market for phthalic anhydride, consuming 14% in 1968, is polyester resins. Miscellaneous uses include pharmaceuticals, dielectric materials, and decorative laminates.

In 1968 more than 800 million lb of phthalic anhydride was produced in the United States. As a result of recent expansion and the construction of very large plants (10), phthalic anhydride production capacity expanded from 928 million lb to well over 1 billion lb/year in 1970.

C. Vapor-Phase Catalytic Oxidation of Aromatic Hydrocarbons

1. REACTION CONDITIONS

The oxidations of naphthalene and of o-xylene are strongly exothermic reactions, and reactors must be designed to provide good heat removal. Although the heat of reaction for the formation of anhydride is 3033 cal/g

$$\text{o-xylene} + 3O_2 \longrightarrow \text{phthalic anhydride} + 3H_2O \quad (-2528 \text{ cal/g})$$

$$\text{naphthalene} + \tfrac{9}{2}O_2 \longrightarrow \text{phthalic anhydride} + 2H_2O + 2CO_2 \quad (-3033 \text{ cal/g})$$

for naphthalene and 2528 cal/g for o-xylene, a considerable amount of complete combustion occurs (11), and the heat liberated in practice may be up to 5500 cal/g—almost twice that expected for a selective oxidation. The problem is solved by passing the vapor mixture of hydrocarbon and air over the granular-supported catalyst packed in small-diameter tubes. The heat is partially removed by a fluid heat-exchange medium, usually molten salt and sometimes boiling mercury, which contacts the exterior of the catalyst tubes. An appreciable part of the generated heat is carried out of the reactor as sensible heat of reaction (12). In recently constructed commercial reactors, the catalyst is packed in more than 8000 tubes, each about 25 mm in diameter and 3 m long, all contained in a single molten salt medium. Another solution to the thermal problems involved in catalytic oxidations involves the use of a fluidized catalyst bed. The use of fluidized catalyst beds was initiated with a 3-million-lb Sherwin-Williams Company pilot plant in 1945. Fluid beds allow heat to be removed by means of cooling coils placed directly in the bed. Other advantages over fixed beds are easy replacement of catalyst and ability to accept a liquid feed directly, thereby avoiding explosive mixtures (9). Recent large plants producing 75 million lb/year operate so efficiently that it has been estimated that raw material accounts for one-half of the production cost.

Phthalic anhydride is removed from the reactor effluent most simply by reducing the temperature of the hot gases to just above the dew point (125–130°C) and passing these cooled vapors into large air-cooled condensers. The crude phthalic anhydride crystallizes on the walls of the condensers, known as "hay barns." In modern plants, tubular exchangers are used in cyclical operation in which the solid condensed product is periodically melted and drained out of the condenser (5). Cyclones or water scrubbers are also used.

Modern manufacturing processes differ in the choice of naphthalene or o-xylene as feed stock as well as in the use of fixed-bed or fluidized-bed reactors. It is also possible to categorize processes on the basis of the different temperatures and reaction times employed. Each process requires a specific vanadium catalyst, which does not function well under all reaction conditions. Naphthalene can be oxidized at lower temperatures than o-xylene. Fixed-bed naphthalene plants generally operate at 350°C at a contact time of 3–5 sec with an air feed containing 0.5 mole % hydrocarbon. Fluid beds require contact times as great as 15 sec, but they allow the use of hydrocarbon concentrations of 1 mole %.

o-Xylene is oxidized by both low-temperature and high-temperature processes (13). The leading low-temperature fixed-bed process, developed by Chemische Fabrik von Heyden, operates at 350–360°C at reaction times of about 1 sec and is believed to require the addition of sulfur dioxide to the feed. Other fixed-bed processes use temperatures between 420 and 450°C and reaction times of about 0.1 sec. Yields as high as 76% have been claimed. The conditions employed in commercial fluidized o-xylene plants have not been revealed. Bromine may be used as a promoter. Laboratory studies with fused vanadium pentoxide indicated optimum conditions—490°C, 3–4 sec contact time with about 1% o-xylene—allow a yield of 68% phthalic anhydride (14). Details on many industrial reactor designs and processing conditions, as described in the patent literature, have been reviewed by Sittig (15).

The increasing attention given to o-xylene as a raw material for the production of phthalic anhydride arises partly from the higher weight yields possible in theory. If quantitative yields could be achieved, 100 g of hydrocarbon would yield 140 g of phthalic anhydride from o-xylene, whereas only 116 g of product could be produced from naphthalene. In practice, it is difficult to obtain yields above 60 mole % from o-xylene, although yields of from 85–90% are commonly obtained from naphthalene. Economic factors are increasingly favoring the use of petroleum o-xylene, since the supply of naphthalene from coal tar is limited and the processing of petroleum naphthalene is expensive (16). For laboratory oxidations, o-xylene is preferred because it is an easily handled liquid at room temperature and naphthalene is a solid that melts at 80°C.

Phthalic anhydride may also be obtained by the oxidation of phenanthrene, anthracene, and other polynuclear aromatics. The use of various commercially available coal tar fractions has been suggested (17), and in Russia, there has been extensive work directed to the utilization of anthracene oil (18). Various crude naphthalenes containing quantities of methyl naphthalene have also received attention. Less work has been reported on the oxidation of pure polynuclear hydrocarbons and substituted naphthalenes, and these investigations have been of interest largely in demonstrating structural effects (19). Interestingly, monoalkyl benzenes, if the number of carbon atoms in the side chain is three or more, also yield phthalic anhydride over a vanadium catalyst (20). Best results are obtained with *tert*-alkyl benzene, the yield from *tert*-butyl benzene being 36%.

It is seldom possible to approach the reported commercial yields of phthalic anhydride from either *o*-xylene or naphthalene in the laboratory. The availability of a laboratory reactor with good heat-transfer properties is an obvious factor. However, the need for detailed procedures for the preparation, conditioning, and use of proprietary catalysts is of even greater importance; and, of course, such data are not accessible. Well-described studies (21) indicate that although maximum theoretical yields of 60% are obtained from the oxidation of *o*-xylene over a fixed-bed, without optimizing conditions and catalyst considerably poorer results may be obtained. In one otherwise comprehensive report on the kinetics of *o*-xylene oxidations, in which selectivities to phthalic anhydride appear to rise above 70%, the catalyst is described only as the French Synoxy catalyst, a doped vanadium pentoxide (22).

2. HETEROGENEOUS CATALYSTS

All catalysts for the oxidation of aromatic hydrocarbons to phthalic anhydride are based on vanadia. The commercial catalyst for naphthalene oxidations is modified with potassium sulfate and supported on silica gel. This type of catalyst has been well described (23–25) and extensively studied (26). Commerical catalysts may be purchased for laboratory use (Davison) in forms suitable for fixed or fluidized beds.

Titanium dioxide and stannic oxide have also been used as modifiers, and these compounds are believed to stabilize the structure of the reduced form of the catalyst (27). Many more modifiers, including oxides of uranium, molybdenum, cobalt, chromium, silver, boron, rubidium, and cesium, have been incorporated in *o*-xylene oxidation catalysts; but there is some skepticism about the superiority of such catalysts over a properly supported, unmodified vanadium catalyst. Vanadia catalysts modified with potassium sulfate, although preferred for naphthalene oxidations, are poor for the oxidation of *o*-xylene. The effects of crystal structure and the mechanisms of adsorption and oxidation on vanadium oxide surfaces are receiving increasing attention, but these subjects must still be considered poorly understood.

D. Production of Phthalic Acid

Phthalic acid is most readily obtained by hydrolysis of phthalic anhydride. Hydrolysis is most simply accomplished by boiling the anhydride in water and allowing the acid to crystallize from the solution on cooling. The reaction is not catalyzed by acid and, in fact, it has been observed (28) that perchloric acid slows the rate of hydrolysis in water and aqueous dioxane by an assumed salt effect at concentrations around $1.0M$.

The liquid-phase air oxidation of o-xylene has also been investigated as a synthetic route to phthalic acid. A three-step procedure in which toluic acid is prepared and esterified and the methyl ester oxidized in high yield has been described (29,30). The halogen-activated direct oxidation of o-xylene is also claimed to give high yields in a continuous commercial process (31). Cobalt acetate and metal bromide catalysts are used (32) in an acetic acid solvent.

E. Laboratory Preparation of Phthalic Anhydride

1. CATALYTIC REACTOR

Numerous variations in equipment are possible in the catalytic reactor itself, in the associated feed system, and in the product collection system. For laboratory synthetic purposes, the simple system (Fig. 1) used in the first example below can prove adequate. More elaborate and expensive equipment (Fig. 2) will allow optimum yields and simplify the operating procedure. Modifications recommended depend on the facilities available and the skill of the experimenter.

The simplest reactor consists of a 3-ft 1-in stainless steel tube heated by a Hevi-Duty tubular furnace. Reactors with better heat-transfer behavior may be constructed by placing the tube in a close-fitting brass block heated electrically with strip heaters. Carefully constructed tubular mercury (33) or salt bath (22) reactors may be used in the laboratory. Various U-shaped reactors (21) are convenient to load and heat, but generally they cannot hold large amounts of catalysts. Fluidized-bed reactors are also easily assembled and generally consist simply of a Pyrex or quartz tube fitted with an axial thermowell and a porous disk on which the catalyst is supported (14). Feed lines and a thermocouple may enter the reactor tube through a silicone rubber stopper, the unheated top portion of the tube being sufficiently cool if it is at least 6 in. long. More flexibility is provided by using a separate reactor head of Vycor or stainless steel, which is attached with ground ball joints.

Cylinder or laboratory air may be used and should be dried and filtered. A flow regulator (Moore) is desirable, and air may be metered through calibrated flow meters or rotameters. Hydrocarbon may be aspirated or

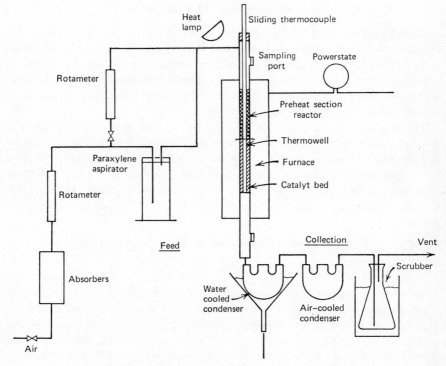

Fig. 1.   Simplified reactor for *o*-xylene oxidation.

pumped. If naphthalene is used as a feed, the aspirator or pump and the feed lines must be heated.

2.   PROCEDURES

a.   Simplified Reactor, Aspirated o-Xylene

The reactor (Fig. 1) is charged through the reactor head with 70 ml of prepared catalyst (described in Section II.E.3.a) by removing the silicone rubber stopper and clamping the thermocouple well with its catalyst support screen attached, or by resting the well on a rod temporarily placed through the bottom of the reactor. The section of the reactor tube between the top of the catalyst bed and the point where it enters the furnace is then packed with 8–10 mesh Crystolon silicon carbide chips, α-alumina, or any available inert refractory material.

The aspirator, which may be an ordinary gas scrubbing bottle, is filled about half full with *p*-xylene (Phillips Petroleum Company, 98% pure grade) and weighed. The reactor is preheated to between 350 and 400°C. The air and xylene are then introduced at the desired flow rates. Air is metered at a rate of 625 l/hr with part of this by-passed through the

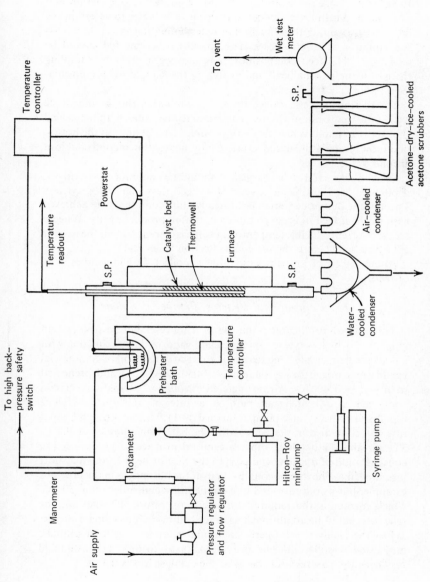

Fig. 2. Schematic of reactor for *o*-xylene oxidation: S.P. = Sampling point.

487

aspirator to produce a feed stream containing about 0.2% xylene. The wanted air flow rate through the aspirator may be calculated to be about 190 l/hr; but since the aspirator may not saturate the side stream and since variations in room temperature occur, it is better to ascertain the flow rate by venting the effluent and determining the xylene loss from the aspirator at several flows. The amount of xylene fed should be 6.04 ml/hr. The feed composition may also be learned by sampling the gas at the reactor head and analyzing the sample by gas chromatography.

The thermocouple is moved up and down within the thermocouple well until the location of the maximum temperature within the bed (hot spot) is found. While the thermocouple is maintained at the hottest point, the temperature input is varied until the reaction is controllable at around 475°C.

When a new catalyst is charged, it should be conditioned overnight, if possible, by introducing a very dilute feed (less than 0.2% xylene). The effluent should be vented until steady-state conditions are achieved with the final feed in order to obtain optimum product purity. When a reasonably reproducible temperature profile is demonstrated, the reactor effluent is switched to the product collection assembly.

About 4 g/hr of phthalic anhydride is collected. The yield is 54% of theory.

b.   Conventional Reactor, Pumped o-Xylene

This procedure illustrates the use of a more elaborate feed system (Fig. 2), in which o-xylene is pumped and vaporized before mixing with a preheated air supply. The reactor is charged with 70 ml of commercial vanadium oxide catalyst (Harshaw V-050 1-S-$\frac{1}{8}$ in.) and preheated to about 300–400°C. Air is metered through a stainless steel coil contained in a heated salt bath at a rate of 160 l/hr. o-Xylene is pumped with a syringe pump (Harvard) at 11.9 ml/hr through a separate coil contained in the same bath. The bath is maintained at 300°C. The air and the o-xylene vapor are mixed in a tee to produce a 1% concentration of o-xylene and pass to the reactor head through a short insulated line. The thermocouple is kept at the hot spot as in the previous example, but its output is fed to a temperature controller set at 500°C, which maintains the required furnace temperature. If practical, the reaction should be conditioned as described in the preceding example. When the catalyst is properly conditioned, up to 7 g/hr of phthalic anhydride is collected. The yield is 47% of theory. The estimated residence time in the void spaces of the catalyst bed is 0.3 sec.

To prepare 100-g quantities of phthalic anhydride, the reaction may be operated unattended for long periods of time. It is recommended that a back-pressure safety switch, placed in the air line, be set to turn off the pump and air supply if the pressure drop through the system rises. Obtaining kilogram quantities of product necessitates a salt or mercury bath reactor having a 3-m long reaction tube and a diameter of 3.3 cm. With a good commercial catalyst, it is possible to prepare 4 kg of product a day.

To optimize reaction conditions for catalysts not described previously or any catalyst used in a scaled-up reactor, it is useful to sample the feed at the reactor head and the hot effluent continually. By using a jacketed thermostated syringe, constant volume samples can be analyzed immediately by gas chromatography, and conditions can be adjusted to give the highest concentration of phthalic anhydride in the effluent. Initial air flow and xylene feed rates should produce a xylene concentration of from 0.1 to 0.2%.

3.   Preparation of Vanadium Oxide Catalyst

A solution of vanadium oxalate is first prepared by dissolving 500 g of oxalic acid (reagent grade) in 600 ml of distilled water by warming at 60–80°C with stirring. Vanadium pentoxide, 260 g (Vanadium Corp. of America, 99.5%) is added gradually, and heating is continued until a clear solution is obtained. The concentrated solution is diluted to 1 liter with distilled water.

a.   Conventional Method

To 227 g of ⅛ × ⅛ in. cylinder-sintered α-alumina carrier (Carborundum Company SAHT-96) in a 1-liter round-bottom flask is added 100 ml of the previously prepared vanadium oxalate solution. The flask is fixed with a reflux condenser and the mixture refluxed for 1 hr. The flask is transferred to a Rinco rotary evaporator and evacuated and heated with tumbling until the pellets appear dry. The pellets are placed in a suitable container, such as a quartz boat, and heated in an oven at 200°C for 1 hr and then at 500°C for 1 hr.

b.   Improved Method

To 500 g of ⅛ × ⅛ in. cylinder-sintered α-alumina carrier in a 1-liter round-bottom flask is added 350 ml of the previously prepared vanadium oxalate solution. The flask is fixed with a reflux condenser and the mixture refluxed for 1 hr. The solution is allowed to cool to 40°C, and 245 ml of solution is drained off. The flask is attached to a Rinco rotary evaporator and the pellets tumbled and heated at 60°C under vacuum until dry. The pellets are dried at 200°C for 1 hr and finally at 500°C for 1 hr. This procedure avoids the formation of fines and "cement" and produces an easily handled, free-flowing catalyst.

The carrier selected should be a low-surface-area (~1 m²/g) high-porosity material. Silicon carbide may also be used, but it is difficult to coat evenly.

F. Purification

The purity of phthalic anhydride produced by catalytic air oxidation depends on the hydrocarbon oxidized and reaction conditions. The most important by-products from *o*-xylene are *o*-tolualdehyde, phthalide, and maleic anhydride, as well as carbon monoxide, carbon dioxide, and water (34). Traces of citraconic acid, benzoic and toluic acids, and benzaldehyde may be present. Typical impurities produced during the oxidation of naphthalene are described in the analytical section.

It was thought very early that phthalic anhydride separates with ease on sublimation or recrystallization from all the other substances formed by catalytic oxidation (7). In the process of fractional sublimation, water and hydrocarbon pass off at lower temperatures than the anhydride and separate quite sharply from it. The anhydride in turn is sublimed away from the colored materials present and condensed in a high state of purity. Crystallization from organic solvents works well. For instance, the crude product may be dissolved in warm carbon tetrachloride and treated with decolorizing charcoal to remove the color.

In fact, however, neither of these processes completely frees phthalic anhydride of all impurities, and other treatments are used (2). Impurities may be caused to condense or polymerize by heating with or without special agents. Subsequent sublimation results in purer product. The usual heat treatment is at 180–280°C for from 2–10 hr. Agents used are zinc chloride, manganese dioxide, silica, and zeolites. Important purification of phthalic anhydride occurs also during the condensation of the vapors from the catalytic reaction, and much attention has been given to the process of crystallization from the vapor phase.

Phthalic anhydride is purified in current commercial processes by distillation, generally after some type of simple treatment of the crude product. A column of about 5 theoretical plates is generally adequate. In a typical operation, a center cut consisting of from 85–90% of the total crude is collected at about 80 mm with a 3:1 reflux ratio, the still pot being maintained at about 220°C. The forecut is removed at much higher reflux ratios, typically 15:1. Stills may be staged and the distillation conducted in a continuous (35) or stepwise manner (36). Product may be recovered from still residues by extraction, using solvents such as methyl ethyl ketone (37), or further distillation under more drastic conditions may take place (38).

Pretreatment of phthalic anhydride involves heating the crude anhydride at temperatures up to 270°C in the presence of numerous additives. Recently claimed purification procedures describe the additions of alkali earth metals, zinc hydrosulfites, hydrosulfites, formaldehyde sulfoxylates (39), sodium carbonate (40), zinc or tin chloride (41), mixtures of hydrogen peroxide and sodium carbonate (42), and sodium borohydride (43).

Zone refining of phthalic anhydride is used to obtain very pure samples and gives an odorless product (44).

III. TEREPHTHALIC ACID

Terephthalic acid has been prepared by the oxidation of many para-substituted benzene derivatives. A detailed laboratory procedure has been described for the successive oxidation of p-methylacetophenone with nitric acid and potassium permanganate, giving terephthalic acid yields of 88% (45). Older procedures (46) involve the permanganate oxidation of p-toluic acid or dihydro-p-tolualdehyde and the chromic acid oxidation of p-cymene. Other preparative routes are illustrated by the reaction of dibromobenzene or p-chloro or p-bromobenzoic acid with potassium and cuprous cyanides and by the reaction of p-dibromobenzene or p-iodobenzoic acid with butyl lithium and carbon dioxide.

A. Commercial Production

The commercial production of terephthalic acid and its methyl ester in the United States today is based entirely on the liquid-phase air oxidation of p-xylene (47). du Pont initiated the production of terephthalic acid using a nitric acid oxidation process that gives yields of about 85%. About 2 lb of nitric acid per pound of terephthalic acid are required. Hercules Powder Company produces dimethyl terephthalate by a four-step process in which terephthalic acid is not an intermediate. Toluic acid produced by the air oxidation of p-xylene is esterified, and the product—methyl toluate—is air oxidized to the monomethyl ester of terephthalic acid. From the monoester, the dimethyl ester is prepared in about 75–80% yield on the p-xylene consumed. This process is frequently called the Witten or Imhausen process.

Mobil Chemical uses a process in which oxygen serves to oxidize p-xylene in an acetic acid solvent. Cobalt catalyst and methyl ethyl ketone activator are used, and yields of terephthalic acid are 95%. Tennessee Eastman appears to use a catalytic process featuring acetaldehyde as an activator. In another well-publicized process used by Amoco, bromine is the activator at relatively high temperatures and pressures.

In Japan several processes were developed which are based on the disproportionation of a potassium salt. Mitsubishi Chemical Industries starts with the salt of benzoic acid from toluene. Teijin rearranges the dipotassium o-phthalate from phthalic anhydride to produce an unusually pure product. Another process under development by Ube Industries, Ltd., treats potassium benzoate with carbon monoxide and potassium carbonate. Yields based on the starting toluene are about 70%. A process used by Toyo Hoatsu also starts with toluene. Chloromethylation yields isomeric chloromethyl toluenes, which are oxidized to a mixture of terephthalic and phthalic acids.

Nearly all the terephthalic acid produced is consumed in the preparation of polyethylene terephthalate. Production data are not available because large quantities are used captively. About 3 million lb is used for such miscellaneous applications as an intermediate in herbicide production; formulating adhesives, printing inks, coatings, and paints; and preparing animal-grade feed supplements.

B.   Autoxidation of *p*-Xylene

Good yields of terphthalic acid are produced directly by oxidation of *p*-xylene with molecular oxygen only in a carboxylic acid solvent in the presence of an organic activator and concentrations of metal catalyst higher than are normally useful in hydrocarbon autoxidations. By far the most efficient system for oxidizing alkyl aromatic groups to carboxylic acid consists of a cobalt acetate catalyst in acetic acid. Although in principle a variety of initiators or activators capable of readily forming peroxide or radical intermediates may be used, practical synthetic procedures have been developed using only aldehydes, methylene ketones and bromine, or their precursors. Recently a combination of cobalt and lanthanide salts, particularly mixtures of cobalt and zirconium acetate, was claimed to be sufficiently effective catalytically to allow the production of terephthalic acid in good yields without the use of an organic activator (48). Unless reaction conditions are carefully defined, the autoxidation of dialkyl benzene tends to stop when the monocarboxylic acid is produced. The increased resistance of toluic acids to oxidation results primarily from the electrophilic character of autoxidations and the electron-withdrawing nature of the carboxyl group (49). Interestingly, the long-known difficulty in finding conditions under which toluic acid would oxidize at a practical rate was not at first recognized as an indication that polar effects are important in free radical reactions.

1.   ALDEHYDE ACTIVATION

The value of acetaldehyde and other aldehydes in activating cobalt-catalyzed oxidations stems from the efficiency with which cobaltic ion is produced from cobalt salts in the presence of oxidizing aldehyde (50). In early work this system proved effective for autoxidation of toluene to benzoic acid and *o*-xylene to *o*-toluic acid (51). The production of terephthalic and isophthalic acid can now be accomplished in high yields, but large quantities of acetaldehyde are needed (49), and the addition of this compound should be continued during the course of the reaction. The exact quantities consumed depend on the reaction temperature and the mode and rate of addition. A commercial process in which reactants are continuously fed to an oxidation tower (52) requires 0.5 kg of acetaldehyde and 0.7 kg of xylene for every kilogram of acid produced.

A detailed study of experimental factors in the acetaldehyde-activated oxidation in a stirred-tank reactor at atmospheric pressure has been reported (53). Although the reaction was diffusion controlled because agitation was limited by the character of the equipment, yields of up to 99% terephthalic acid were obtained. In a typical experiment, 50 g of p-xylene was oxidized at 90°C in 47.2 g of acetic acid containing 24 g of cobalt acetate and 6.6 g of acetaldehyde. Additional acetaldehyde was added at 12 ml/hr over 13 hr. Paraldehyde may be used as an activator in place of acetaldehyde, and procedures have been described in which the best yields are obtained by continuous addition to the reaction (54). Aromatic aldehydes have been used as activators, and these may find practical applications (especially if they form terephthalic acid as an end-product when they are consumed, as is the case for tolualdehyde.

To perform the acetaldehyde-activated oxidation of p-xylene (55), a 10-liter stainless steel autoclave is charged with 1.5 kg of glacial acetic acid, 0.5 kg of p-xylene, 90 g of cobaltous acetate tetrahydrate, and 110 g of acetaldehyde. The autoclave is equipped with an efficient agitator, a pressure gauge, an internal cooling coil, a thermocouple, a heating jacket, a gas dispersion inlet, and an outlet purge system. The solution is continuously stirred and heated to 95°C. Air is introduced to produce a pressure of 5 atm while an exit gas, regulated during the course of the reaction to contain between 4 and 5% oxygen, is maintained. The strongly exothermic reaction commences immediately, and cooling water is circulated through the internal cooling coil to maintain the reaction temperature between 110 and 115°C. While this temperature is maintained a solution of 0.9 kg of p-xylene and 0.28 kg of acetaldehyde in 3.0 kg of acetic acid is gradually added, over a 90-min period, to the reaction vessel. This may be accomplished by pumping or, with more difficulty, from a small pressure vessel through a metering valve. Air is passed through the reaction mixture for an additional hour.

The autoclave is cooled to 95°C and emptied. The precipitated terephthalic acid is separated by filtration and washed, first with acetic acid and then with distilled water. Additional washing with methanol may be required if much toluic acid is formed. After air drying, the white precipitate weighs 2.15 kg, corresponding to a 97.4% yield.

2. BROMIDE ACTIVATION

Soluble bromide salts used with cobalt and manganese salts in acetic acid are another effective system for the oxidation of alkyl aromatics. Catalysis depends on the formation of hydrogen bromide in the reaction solution and its oxidation to atomic bromine. The kinetics and mechanism of the oxidation of p-toluic acid (56) have been studied at 130°C, and with sodium bromide and cobalt acetate the reactions obeys the rate law:

$$\frac{-d(O_2)}{dt} = K[Co(II)][NaBr]^{1/2}[O_2]^{1/2}$$

This reaction has been extensively promoted, initially as a batch process, for the commercial production of terephthalic acid from p-xylene. A continuous process, claimed to be capable of yielding product with improved color, was developed later (57,58). Reaction temperatures as high as 275°C are used, and pressures up to about 40 atm are required to maintain the liquid phase and the required partial pressure of oxygen. Heat is removed from the reaction by condensing and refluxing solvent vapors. Oxygen is supplied to the reaction zone in excess of the stoichiometric requirements, since oxygen-deficient conditions lead to the formation of by-products, many of them colored. In the application of the process, conversions of p-xylene are typically greater than 95% and the yield of terephthalic acid is 90%. The high-temperature acidic system containing bromine is corrosive, and under some modes of operation equipment constructed of Hastelloy C or titanium is required.

> In the autoxidation of p-xylene in the presence of manganese acetate and ammonium bromide (59), a 1-liter corrosion-resistant stirred autoclave is used. The autoclave is equipped with a gas inlet tube, a heating jacket, an internal cooling coil, a reflux condenser, and a vent for passing off noncondensable gases. In the reactor is placed 146.4 g of p-xylene (95% pure), 375 g of glacial acetic acid, 1.8 g of manganese acetate, and 1.5 g of ammonium bromide.

> Air is passed into the reaction mixture and purged to the atmosphere at a rate of about 1500 liters/hr, while the mixture is maintained at 195°C with vigorous agitation. The gauge pressure is maintained at 400 psi. After 2 hr the reactor is cooled and the contents removed. The crude solid terephthalic acid is separated by filtration and washed successively with three 400-ml portions of acetic acid and three 400-ml portions of distilled water. About 170 g of light-colored terephthalic acid is obtained (75% of theory).

3. KETONE ACTIVATION

The oxidation of p-xylene using a ketone containing an α-methylene group and a high concentration of cobalt acetate catalyst in acetic acid is probably the most convenient laboratory autoxidation procedure available (49). It may be conducted at atmospheric pressure at 90°C to give almost quantitative yields of terephthalic acid. As is the case when acetaldehyde is used at 1 atm, efficient agitation is absolutely essential and long reaction times are required. With methyl ethyl ketone, however, as little as 0.2 mole of activator for each mole of xylene oxidized need be used, and may be added at one time. Other ketones vary in efficiency, and their effect on the rate and induction period have been studied in detail (60).

Commercially practical reaction times of less than 1 hr are achieved at a temperature of 130°C using oxygen pressures of up to 500 psig (61). Generally, larger quantities of ketone are required, but not all of it is consumed.

In common with other metal-catalyzed processes using acetic acid, terephthalic acid precipitates from the reaction as the oxidation proceeds and is isolated readily by filtration. The filtered reaction solution is recycled after removal of the by-product water (which has an inhibiting effect on the oxidation) and the addition of makeup xylene and activator.

A study of the rate of oxidation of the xylenes and some substituted toluenes indicates several interesting theoretical aspects to the ketone-activated reaction (49). The oxidation is favored by a high electron density at the reaction site and has been shown to have a Hammett ρ value of approximately -1. At 1 atm the partial pressure of oxygen exerts not only a strong effect on the reaction rate but also an inverse effect on the induction period, which is difficult to explain in terms of conventional initiation processes. It has been postulated that direct oxidation of hydrocarbon by cobaltic ion plays a key role in the process.

To begin the methyl–ethyl-ketone-activated oxidation of p-xylene (49), a 2-liter glass reaction flask is fitted with a gas inlet tube, a 0.2°C aniline point thermometer, and a reflux condenser. The lower third portion of the flask is heated with a Glas-col mantle, which may be connected to a capacitance-actuated control relay to regulate the reaction temperature. Stirring is provided by a Teflon-covered magnetic stirring bar capable of being driven by a strong external magnet at speeds above 800 rpm. A conventional high-speed stirrer of 316 stainless steel with an airtight bearing may also be used. Good agitation is achieved when a vortex is produced in the solution exposing the rotating bar or blades.

To the flask is added 106 g (1.0 mole) of p-xylene (98% purity), 24.9 g (0.1 mole) of cobaltous acetate tetrahydrate (reagent grade, air dried at room temperature), 21.6 g (0.3 mole) of methyl ethyl ketone, and 900 g (15 moles) of glacial acetic acid (99.5%). The solution is heated to 90°C with rapid stirring while passing extra dry grade oxygen into the flask just above the solution surface. Oxygen is fed through a sensitive flow regulator (Moore), which is adjusted to maintain a flow from the condenser exit of about 40 ml/min. The solution gradually turns from purple to dark green as cobaltic ion is formed and the rate of oxygen uptake strongly increases. Within 6 hr it becomes necessary to pass a stream of cooling air around the reactor to help dissipate the heat of reaction and maintain the reaction temperature at 90°C. Adjustment of the flow regulator is some-times required at this time to maintain the off gas rate. After a total reaction time of 24 hr, the flow of oxygen is discontinued and the green solution cooled.

The precipitate of terephthalic acid is removed by filtration and washed with 500 ml of water and with 50–100 ml of ethanol. After the precipitate has dried at 100°C, the yield of pure white terephthalic acid is 154 g (93%). The neutralization equivalent is 83 (theoretical, 83), and ignition residue is 0.04%. The dimethyl ester after one recrystallization melts at 140.3–140.6°C.

C. Disproportionation or Rearrangement of Potassium Carboxylates

The preparation of dipotassium terephthalate in high yield by heating dipotassium phthalate at 400–450°C with a cadmium or zinc salt catalyst and the similar preparation of terephthalate from potassium benzoate have been described by Raecke (62). Older reactions, which are similar but were not of synthetic importance because of poor yields, involve the fusion of dipotassium p-sulfobenzoate with sodium formate or the heating of sodium benzoate. Considerable activity by many companies, some previously mentioned, resulted in the development of commercial processes for the production of terephthalic acid based on the catalyzed reactions, the basic patents for which are held by Henkel & Cie.

In the process developed by Mitsubishi (63), benzoic acid is reacted with electrolytic-grade caustic and the potassium salt separated from the accompanying mother-liquor in a centrifuge. When it has been blended with a sludge of recycled catalyst, the mixture is dried and pelletized. The reaction, producing 1 mole of terephthalate and 1 mole of benzene from 2 moles of benzoate, takes place continuously in the solid phase. The reactor is designed with mechanical features that avoid caking out of material and allow the reaction to take place under an inert blanket of carbon dioxide. The product is dissolved in water and reacted with sulfuric acid. Both the by-product potassium sulfate and a significant quantity of impurities are separated from the acid in a centrifuge. Nothing further is done to the crude terephthalic acid if it is to be used for preparing dimethyl terephthalate.

The quality of terephthalic acid produced by the disproportionation of potassium benzoate is claimed to be high, with only traces of isophthalic and other polycarboxylic acids and no p-carboxybenzaldehyde present. The light yellow tint is removed from the product in the purification process.

Ogata (64) has made an extensive investigation of the kinetics and mechanism of the thermal reaction of various potassium benzene carboxylates. Studies were conducted at 365°C in melted potassium cyanate (because the salts were readily soluble in this medium), using cadmium iodide catalyst. The rates of reaction of potassium isophthalate, benzoate, and phthalate were reported to follow the kinetic equations $v = k$ (CdI_2) (isophthalate), $v = k$ (CdI_2) (benzoate), and $v = k_1$ (CdI_2) [phthalate $+ k_2$ (CdI_2)] (phthalate)^2, respectively. It was proposed that both the rearrangement and the disproportionation reactions proceed through a sandwich-type activated complex between Cd^{2+} and two carboxylate compounds. Some doubt has been cast on the kinetic results by Ratusky (65), who reported that under the reaction conditions existing in the potassium cyanate melt, the same uniform reaction mixture is formed from any of the various benzenecarboxylate salts. The salts investigated included phthalate, isophthalate, terephthalate, benzoate, and mixtures of benzoate with trimesoate, trimellitate, and hemimellitate. In each case the composition of the mixture obtained from potassium cyanate at 425°C, after transformation to the mixed esters, was: 18–20% methyl benzoate, 18–20% dimethyl terephthalate, 31–33% dimethyl isophthalate, 5–6% dimethyl phthalate, 9–10% trimethyl trimesoate, 12% trimethyl trimellitate, and 1–2% trimethyl hemimellitate. These results confirm an earlier assumption that enhanced formation of potassium terephthalate in the Hinkle reaction results from a shift in equilibrium as the terephthalate recrystallizes from the reaction mixture.

To rearrange dipotassium phthalate (62), a carefully dried mixture of 25 g of dipotassium phthalate and 1 g of cadmium oxide is first placed in a large test tube. The tube is heated for 1 hr in a metal block or an air bath at 410°C, while a slow stream of dry carbon dioxide is passed in. After the tube has been cooled, the glass is crushed and the contents dissolved in hot water. The solution is filtered and acidified with concentrated hydrochloric acid. A yield of about 11–12 g of terephthalic acid is obtained.

D. Other Synthetic Methods

In a reaction that appears to be related to the Hinkle process, it is claimed that benzoic acid may be carboxylated to give yields in excess of 80% of a mixture of phthalic and terephthalic acids. Alkali metal carbonates, prepared by the reaction of carbon dioxide with metal alkoxide, are used as carboxylating agents (66).

Benzonitrile can also be used to prepare terephthalic acid by heating a mixture of the nitrile with potassium carbonate in a sealed tube at 360–500°C. In the presence of cadmium iodide, the yield is 33% or less, whereas in the absence of a catalyst, the yield is only 8%. Considerable isophthalic acid is formed as a by-product (67).

Alkyl-substituted aromatic compounds are oxidized to the corresponding carboxylic acids by sulfur dioxide under pressure in the absence of catalysts or solvents (68). The overall reaction is represented by the equation:

$$RCH_3 + 3/2SO_2 \rightarrow RCOOH + 3/2S + H_2O$$

where sulfur and water are by-products. In the presence of excess sulfur dioxide, yields of 90% terephthalic acid are obtained from *p*-xylene in 6 hr at 280°C and 300 atm. The reaction appears to be homogeneous, with the rate being proportional to the square of the hydrocarbon concentration. It is catalyzed by bromine and iodine, inhibited by silver and copper metals, and unaffected by free radical inhibitors or initiators. Reaction conditions, including such factors as reactor size, have been studied and optimized for the synthesis of commercially important acids.

Oxidations effected by sulfur compounds under other conditions have been summarized (68). Xylene heated with sulfur compounds at 300°C in the presence of a large excess of water around 170 atm yields phthalic acids. In aqueous systems, sulfur dioxide has been used alone or in conjunction with hydrogen sulfide, with an alkali or ammonium sulfite, or with an alkali sulfide-sulfate mixture (69). Elemental sulfur has been used in water either alone or with a base. It has been suggested that the oxidant in these reactions is a combination of water-soluble inorganic polysulfides.

Dilute aqueous hydrogen bromide (5%) catalyzes the oxidation of water-soluble methyl-aromatic compounds, such as substituted benzoic acids, to the corresponding carboxylic acid (70). The reaction requires times of from 2–4 hr at temperatures of 200°C and high oxygen pressures. Yields of terephthalic acid exceeding 90% are obtained from toluic acid. The presence of a vanadium compound improves the oxidation of isopropyl aromatic compounds and of water-insoluble compounds.

Terephthalic acid has been prepared from many dialkyl benzenes other than *p*-xylene and from various benzene derivatives. Alkyl groups whose oxidations have been recently of interest include ethyl, isopropyl, isobutyl and *sec*-butyl (70–80). Other derivatives that yield terephthalic acid are *p*-acetylcumene (81), chloromethyltoluene, chloromethylcumene (82), diacetyl benzene (83–85), *p*-cyanotoluene, *p*-cyanoacetophenone (86), acetyl toluic acid, α,α'-dihydroxy-*p*-diisopropylbenzene (85), and *p*-tolyl carbinol (87).

Benzene carboxylic acids can be obtained by the oxidation of coal, but thus far this process has been more important as a method of studying the structure of coal than as a synthetic method. In the first systematic work, air was used to oxidize suspensions of coal in aqueous sodium carbonate. Most of the possible benzene carboxylic acids were isolated and identified from the products of the oxidation of coal in alkaline permanganate solution. By using oxygen in aqueous caustic, total yields of aromatic acids of 43% may be obtained (88). However, the yield of the benzenedicarboxylic acids amounts to only 7.1%, the individual yields being phthalic acid 5.9%, isophthalic acid 1.1%, and terephthalic acid 0.1%. The remainder consists of equal amounts of tricarboxylic and tetracarboxylic acids.

A number of oxidizing agents have been reexamined and oxidation conditions have been described which seem to increase their promise as synthetic routes. By passing nitrogen dioxide through p-xylene in acetic acid containing a cobalt acetate catalyst at 150°C under autogenous pressure, an 80% yield of terephthalic acid is obtained (89). Trichlorobenzene has also been used as a solvent for oxidizing various dialkyl benzenes with nitrogen dioxide (90).

Aqueous sodium dichromate is an excellent reagent which allows yields of 78% terephthalic acid to be obtained from p-xylene by oxidation under pressure at 280–290°C (91). Friedman (92) obtained a 92% yield of terephthalic acid from p-cymene by using a 50% excess of dichromate. The reaction represents a good general method for the oxidation of alkylarenes. The detailed experimental procedure is given in Sec. VII-C-2 dealing with the preparation of 2,3-naphthalene dicarboxylic acid.

Nitric acid oxidations of dialkyl benzenes are continuing to receive attention (79,80). Yields of around 90% terephthalic acid are obtained from p-xylene when careful attention is given to the method of adding the nitric acid, the reactor design, and the reaction temperatures and pressures (93).

With the development of methods for the production of aromatic nitriles by the catalytic oxidation of alkyl benzenes in the presence of ammonia, there has been increasing interest in the hydrolysis of terephthalonitrile. Ammonium salts of terephthalic acid are obtained directly by hydrolysis of the nitrile with aqueous ammonia under pressure (94).

The interchange of nitrile and carboxyl groups has been utilized in an interesting preparation of terephthalic acid. When p-toluic acid and p-cyanobenzoic acid are heated at 250°C, terephthalic acid and p-tolunitrile are produced. The continuous removal of the tolunitrile vapor shifts the equilibrium and allows quantitative yields of dicarboxylic acid to be obtained (92). The recovered tolunitrile is autoxidized in another step to provide p-cyanobenzoic acid for the process.

E. Purification

The impurities present in terephthalic acid are strongly dependent on the method used in its preparation. For example, nitric acid oxidation may be expected to produce an impure product, since nitration of the ring cannot be completely avoided. In fact, a typical analysis of crude product produced by nitric acid oxidation at about 270°C is: terephthalic acid—93.1%, p-nitrobenzoic acid—3.6%, 3-nitrotoluic acid—1.8%, p-toluic acid—1.1%, and cyanobenzoic acid—0.23%. Modification of reaction conditions may reduce the amount of analyzable impurities (93).

Air oxidation is capable of producing relatively purer product. However, in processes using high catalyst concentrations, careful washing of the crude product is required in order to minimize metal ion contamination. When bromine is used as an activator, additional impurity producing side reactions are possible. Corrosion is a problem in commercial practice, too, and contributes to contamination of the product. Identified reaction intermediates include p-methyl benzaldehyde, p-toluic acid, p-hydroxymethyl benzaldehyde, p-hydroxymethyl benzoic acid, and traces of hydroperoxide; by-products that have been isolated include formaldehyde, p-cresol, and resinous matter (96). By extraction of esterified crude acid with chloroform, followed by column chromatography on alumina, it has been shown that 4,4'-dicarboxy-benzophenone and 3,6-dicarboxy-9-fluorenone are also major impurities (97).

As a result of two factors—its limited solubility at moderate temperatures in nearly all solvents and its relatively low vapor pressure—terephthalic acid is traditionally purified by esterification with methanol, recrystallization or distillation of the methyl ester, and hydrolysis of the purified ester to the pure acid. Most often it is the methyl ester itself which is wanted for the preparation of polyesters by transesterification, so that hydrolysis is not usually required. However, since 1960, the preparation of polyester directly from a very pure grade of terephthalic acid has been developed and purification of crude terephthalic acid, without conversion to an ester, has received increasing attention. Advantages of the preparation of polyester fiber by processes not involving dimethyl terephthalate are claimed to lie in improved production costs, equal or lower installation costs, and weight yields as much as 15% higher.

Methods available for the production of polymerization-grade terephthalic acid include fractional and size-gradation crystallization, selective solvent extraction, sublimation, and conversion to more easily purified derivatives. Only recrystallization and solvent extraction appear to be inexpensive enough to be commercially attractive. It has been reported that methyl or ethyl cellulose, diethylene glycol, diethyl ether, and dioxane have been used to purify the product from the bromine and ketone activated autoxidation processes (98).

The major problem in purifying terephthalic acid produced by air oxidation of p-xylene lies in removing the p-formylbenzoic acid by-product. Although p-formylbenzoic acid is very readily oxidized and is quite soluble in the reaction solvents, it is present in the crude dicarboxylic acid in amounts in the range of 0.2–20% by weight, depending on the synthesizing conditions. Fujita has shown that a definite and irreducible amount of p-formylbenzoic acid becomes incorporated in the crystal lattice when terephthalic acid crystallizes from a solvent (99). When crude terephthalic acid is recrystallized at high temperatures and pressure from water, acetonitrile, acetic acid, or dimethyl sulfoxide, the amount of the formylbenzoic acid present in the product is never reduced to less than one-tenth of its original concentration. Recrystallization of 5 g of terephthalic acid containing 10% of the aldehyde at 260°C from 50 g of acetic acid reduces the aldehyde concentration to 1.1%. Extraction with solvents such as water, ether, acetone, acetonitrile, or acetic acid is also ineffective and produces little change in the formylbenzoic acid content in terephthalic acid.

A review of the recent patent literature indicates a variety of procedures effecting purification by oxidation, hydrogenation, or acid treatment. Temperatures up to 300°C are required to produce solution of crude acid, but suspensions can be treated. Oxidation is carried out usually with air in the presence of a catalyst (100–103). Crude acid dissolved in alkali sulfites can be treated with sulfur trioxide (104). Alkaline permanganate is often used (105) as the oxidizing agent. Hydrogenation may be carried out in water using the usual supported palladium or nickel catalysts (106). In acid treatment, crude terephthalic acid is heated with several percent phosphorous acid (107) or sulfuric acid (108,109). Recrystallization has been effected from concentrated sulfuric acid as well as from molten benzoic acid at 150°C (110).

Purification of product prepared by nitric acid oxidation is directed primarily at the removal of nitrogen-containing impurities. Solvents used for recrystallization include tetrahydrofuran-water (111), cyclooctane diol or ketol (112), and carboxylic anhydrides such as benzoic anhydride (113). Improved results are obtained on recrystallization from water by decompressing and cooling in stages (114,115). Heating with aqueous solutions of sodium pyrosulfite or sodium sulfite also helps remove nitrogen containing impurities (116). Salt formation with pyrrolidone forms the basis of another process (117).

Treatment of aqueous solutions of alkali terephthalate is most conveniently applied directly to the product from the Hinkle process. Reagents used include sodium hypochlorite (118), potassium permanganate (119), active carbon or coal (120), and air in the presence of a noble metal.

IV. ISOPHTHALIC ACID

A. Preparation

Most of the preparative methods that yield terephthalic acid when applied to *p*-xylene or other paradisubstituted benzene compounds yield isophthalic acid when applied to *m*-xylene and metadisubstituted benzene compounds. Usually no or few modifications in procedure are needed. Before pure meta and paraxylene isomers were available commercially by low-temperature crystallization techniques, many oxidation processes were applied to mixed xylenes to obtain mixtures of the meta and paradicarboxylic acids. Methods were developed to obtain isophthalic acid by separation from such mixtures by utilizing minor solubility differences or the large difference in esterification rates. These practices are of less preparative utility today, but they serve in the purification of isophthalic acid, since the commercial product contains up to 2% terephthalic acid.

Isophthalic acid was first produced commercially in 1956 by the oxidation of *m*-xylene with sulfur in aqueous ammonia reaction medium (121). At 260–315°C and 1000–2000 psig, amides are produced which are hydrolyzed with sulfuric acid (122). Small amounts of benzoic and toluic acid are formed as by-products. Today isophthalic acid is also produced commercially by the autoxidation of *m*-xylene in acetic acid in the presence of a cobalt salt catalyst and a bromide salt. This type of oxidation is described in Section III. Yields of isophthalic acid are about 75%

B. Production Data and Commercial Utility

In 1965 it is estimated that 65 million lb of isophthalic acid was produced (123). Of this, 25 million lb was consumed in the manufacture of unsaturated polyester resins. Isophthalic acid polyester has proved competitive with the less expensive phthalic-based polyesters because of the better performance and fabricating properties of the product. Chemical-resistant applications represent a growing outlet. Another 20 million lb of isophthalic is used for alkyd resins, which are finding growing acceptance in paints and industrial coatings. Miscellaneous applications, accounting for the consumption of 10 million lb of isophthalic acid, are in vinyl plasticizers, polyester fibers and films, polyamide fibers, and high-temperature-resistant polymers, as well as in the preparation of isophthaloyl chloride. Exports account for 15% of domestic production (121).

C. Purification

Many of the procedures useful for the purification of terephthalic acid can be applied to isophthalic acid to remove the same types of impurities. For example, alkaline solutions are treated with oxidizing agents such as sodium hypochlorite (124) or with adsorbants like activated carbon (120), primarily to remove carboxyaldehydes. Many methods have been developed to separate and remove terephthalic acid, which is present in isophthalic acid in quantities dependent on the freedom from the paraisomer of the *m*-xylene used in its preparation.

Most purifications of product containing terephthalic acid are based on the relatively greater solubility of isophthalic acid in water and organic solvents. Mixtures of acids are separated by leaching with water at about 150°C (125, 126) or by extraction with hot methanol (127) or dimethylformamide (128), Slow cooling of aqueous solutions of mixed acids yields a mixture in which the crystal sizes differ for each isomer, allowing separation by sieving (129). Mixed crystals can also be separated by flotation on a toluene–carbon tetrachloride mixture; isophthalic acid tends to form a surface layer, and terephthalic acid sinks to the bottom (130).

Other processes are based on differences in solubilities of salts formed from the mixed acids. For example, ammonium (131), pyridinium (132), and thallous salts have been used. The thallous salt of terephthalic acid is almost insoluble in water, but the isophthalic acid salt is readily soluble (133). The process is carried out by adding thallous acetate to a cold, slightly ammoniacal solution of the two acids.

Purification is also effected by using differences in esterification rates of the isomeric acids. A continuous process (134) based on the 12–14 times faster esterification of isophthalic acid with methanol at 180°C in the absence of a catalyst has been reported.

Much has been published on the separation of mixtures of the three isomeric benzene dicarboxylic acids. Separation may be carried out on the basis of the different solubilities of the ammonium, calcium, and barium salts in water. Terephthalic acid is removed as the insoluble ammonium salt, whereas phthalic acid is partially separated from isophthalic as the insoluble calcium salt (80). A complete commercial procedure has been described (135).

V. ESTERS

A. General Preparative Methods

Nearly all the workable methods for the preparation of carboxylic acid esters have been used to prepare the esters of aromatic dicarboxylic acids. As may be expected, the most important method is the direct reaction of the aromatic dicarboxylic acid with an alcohol in the presence of a strong mineral acid. Other commonly used procedures involved the reaction of alcohols with acid chlorides, anhydrides, or other esters that are more easily prepared.

Carboxylate salts can be reacted with alkyl halides or with thionyl chloride and alcohols. Alcoholysis of aromatic nitriles works well and may prove to be particularly valuable when the nitrile is available by the ammoxidation reaction. For analytical purposes, methyl esters may be prepared using diazomethane or, more conveniently, trimethylsilyl esters may be prepared using bis(trimethylsilyl) acetamide.

Newer methods of esterifying aromatic dicarboxylic acids include heating the acid with trimethyl phosphate in the presence of an acid catalyst (136) or treating the anhydride with dimethyl sulfite in the presence of potassium fluoride (137). The reaction of alkyl halides with carboxylate salts is catalyzed by tertiary amines (138) or by quaternary ammonium salts (139). Solvents used for this reaction are dimethylformamide and dimethylsulfoxide (140).

A large number of catalysts, are known for the preparation of esters of aromatic carboxylic acids with aliphatic alcohols or glycols, and many of these appear to have advantages over mineral acids. Recently described examples are methanedisulfonic acid and methanetrisulfonic acid (141), N,N'-carbonyldiimidazol (142), and sulfuric acid together with ammonium formate (143). Metal catalysts include antimony trioxide (144,145), antimony oxalate (146), stannous salts of carboxylic acids (147), stannous oxide (148), tetra-bis(2-ethylhexyl) titanate (149), and other tetraalkyl titanates (150), and hydrated titanium dioxide (151).

Heterogeneous catalysts are receiving increasing attention for esterification reactions in the vapor phase as well as the liquid phase. For example, silica gel or alumina is mixed with terephthalic acid and methanol is passed through the mixture (152) at elevated temperatures, preferably in a fluidized bed (153). Other solid catalysts are phosphoric acid on silica (154), sodium hydroxide on activated carbon (155), various salts of cation ion-exchange resin (156), and finely divided titanic acid (157).

B. Phthalic Esters

The esterification of phthalic anhydride is primarily of interest in the direct preparation of alkyd-type polyesters. However, esterification to mono and dialkyl esters has received much attention because of the importance of phthalate esters as plasticizers, and it would be useful to briefly examine some typical syntheses.

1. MONOESTERS

a. Monoesters of Normal Alcohols (158)

Esters from methyl through butyl are prepared by heating 4 g of phthalic anhydride with 1 g of the alcohol at the reflux temperature for 30 min. From amyl to eicosyl,, the alcohol is heated with twice the required amount of anhydride at 105–110°C for periods varying from 30 min up to 2 hr, as the molecular weight of the alcohol increases.

The reaction mass from reactions of methyl through hexyl alcohols is shaken with 25 ml of benzene, filtered from excess anhydride, and neutralized with dilute sodium bicarbonate, leaving the mixture slightly acidic. The aqueous layer is extracted three times with 50-ml portions of benzene to remove any alcohol or diester that may be present. The esters are precipitated as colorless liquids or white crystalline solids from the water solution by the addition of dilute hydrochloric acid. The product is recrystallized from a mixture of 90% petroleum ether and 10% benzene. The high-molecular-weight esters, which usually form an emulsion when neutralized by a dilute base, were extracted with ether. The recovered solid is treated with water for 45 min at 60°C and the dried residue is extracted with chloroform.

b. sec-Octyl Hydrogen Phthalate (159)

The preparation of sec-octyl hydrogen phthalate is a step in the resolution of d- and l-octanol-2. A mixture of 130 g (1 mole) of sec-octyl alcohol and 148 g (1 mole) of phthalic anhydride is heated and stirred for 12–15 hr in a flask surrounded by an oil bath at 110–115°C. The cooled reaction mixture is added to about 8 liters of water containing 150 g (1.4 moles) of anhydrous sodium carbonate. The solid material gradually goes into solution. If the solution is not clear, it is first extracted with ether. The sec-octyl phthalate is precipitated as an oil, which rapidly turns to a solid when dilute hydrochloric acid is added in slight excess. The product is filtered, washed with water, ground thoroughly with water in a mortar, and finally filtered and dried. It is crystallized from petroleum ether (bp 60–70°C) or glacial acetic acid, from which it separates as needles melting at 55°C. The yield is nearly quantitative if the sec-octyl alcohol is pure.

2. DIESTERS

The monoesters are converted to normal diesters by heating with an excess of alcohol and a catalyst. One manufacturing process, described in detail (160), produces esters for use as plasticizers by reacting olefins having more than five carbons with alkyl acid phthalate in the presence of an acid catalyst.

Diesters are more often formed directly from the anhydride (161). Concentrated sulfuric acid and its acid salts, aryl sulfonic acids, and phosphoric acid, are useful as catalysts and are used in relatively high concentrations to achieve rapid esterification rates. The formation of by-product olefins and ethers is avoided by using catalysts of amphoteric oxides and oxide hydrates of the elements in groups I–IV and VII of the periodic table (162). Aluminum hydroxide is particularly effective, but amphoteric compounds of copper, magnesium, zinc, aluminum, gallium, indium, thorium, tin, lead, arsenic, and antimony are also active. Esterification is catalyzed homogeneously and may be conducted at the boiling temperatures of the alcohols without any entrainers. Amphoteric compounds are most interesting as transesterification catalysts.

C. Esters from Terephthalic Acid

1. DIMETHYL TEREPHTHALATE

Polymer-grade dimethyl terephthalate is produced commercially from terephthalic acid in a methanol reaction medium both with and without catalysts. In the earliest processes, high yields were obtained by reaction at 105°C and 50 psig in the presence of sulfuric acid, but excessive conversion of methanol to dimethyl ether occurred. At higher pressures and temperatures above 200°C, the esterification rate is relatively fast even without a catalyst. As the insoluble terephthalic acid reacts with methanol, a homogeneous solution is produced because even the monomethyl ester has good solubility. Conversion is governed by the equilibrium of the esterification, and excess methanol usually serves to counteract the effect of the water formed. Overall yields exceeding 95% are obtained commercially (121). The product is purified by crystallization, distillation, or a combination of both techniques. Dimethyl terephthalate is also obtained by the esterification of monomethyl terephthalate produced by the oxidation of methyl toluate as previously described (Imhausen process).

High yields of aromatic methyl esters are achieved by using 2 eq of commercial boron trifluoride–methanol complex in a considerable excess of methanol. The procedure described (163) is claimed to constitute a simple, convenient method for the preparation of many aromatic esters.

a.   Acid-Catalyzed Esterification of Terephthalic Acid in Excess Methanol

In a 10-gallon Pfaudler kettle, 1.5 kg of terephthalic acid and 22.5 kg of methanol are charged. While agitating the mixture, 1.5 kg of sulfuric acid (94–96%) is slowly added. Continuing the agitation, the mixture is heated rapidly to 64°C and allowed to reflux for 5 hr. Heating and agitation are stopped and the reaction solution is cooled to 10°C. After 1 hr at this temperature, the solid product is filtered out in a Büchner funnel. The ester is washed with 1.0 kg of cold methanol.

The filtered cake is returned to the kettle with 14.5 kg of fresh methanol. The contents is agitated and heated at reflux for 1 hr. The solution is discharged from the kettle through a filter that has been warmed to 60°C and the dimethyl terephthalate is allowed to crystallize from solution at 10°C for 1 hr. The ester is separated on a Büchner funnel, washed with 1 kg of cold methanol, and dried in a vacuum oven at 70°C.

b.   Esterification of Terephthalic Acid in Excess Methanol at Elevated Temperatures (163)

In a pressure vessel are placed 1.5 kg of terephthalic acid and 22.5 kg of methanol. The vessel is heated to 270–275°C, while the contents is agitated. The pressure rises to 125–140 atm. The esterification is 92–95% complete in 45 min. The heating and agitation is terminated and the reaction mixture cooled to 10°C and maintained at this temperature for 1 hr. The dimethyl terephthalate is separated by filtration.

c. Esterification of Terephthalic Acid with Boron Trifluoride–Methanol (164)

In a 2-liter round-bottom flask equipped with a reflux condenser are placed 166 g (1.0 mole) of terephthalic acid, 220 ml (2 eq) of boron trifluoride–methanol reagent (51 % BF$_3$), and 1100 ml of dry methanol. The mixture is refluxed for 6 hr. After cooling, the mixture is poured into a saturated solution of sodium bicarbonate and the product isolated. The yield of dimethyl terephthalate is 95 %.

2. BIS(HYDROXYALKYL) TEREPHTHALATE

Hydroxyalkyl esters of aromatic dicarboxylic acids are usually prepared by reaction of either glycols or epoxides with the acids or their dialkyl esters. The most important hydroxyalkyl ester is bis(2-hydroxyethyl) terephthalate, which is an intermediate for the production of poly(ethylene terephthalate). It is formed by transesterification of dimethyl terephthalate in excess ethylene glycol. While the reaction is being conducted in a melt at 197°C in the presence of a catalyst, such as calcium acetate and antimony trioxide, the methanol formed is distilled off. The hydroxyethyl ester is not usually isolated but heated at 283°C to form the polymer, while the ethylene glycol formed by ester interchange is removed.

Many processes, most of them continuous, have been described for preparing bis(2-hydroxyethyl) terephthalate (165–178). The numerous catalysts that may be used for the transesterification reaction include titanium oxide, zinc acetate, and ion-exchange resins. The relative efficiencies of sodium alkoxides, phenoxides, and carboxylates have been studied, and the electronegativities of metals have been correlated with catalytic activity (169). Catalysts for direct esterification of terephthalic acid with ethylene glycol are similar or identical to those already described for other esterification reactions and include titanium oxide (179) and many other metal oxides (180).

Terephthalic acid has been reacted with ethylene oxide in a variety of catalysts and solvents. Triphenyl phosphine in methyl ethyl ketone (181), tributyl phosphine in isopropanol (182), anhydrous triethyl amine in absolute methanol (183,184), tripropyl amine in methyl ethyl ketone (185), dimethyl aniline in propanol (186), and N,N'-dimethylcyclohexylamine in butanol (187) have been used. However, aqueous sodium hydroxide appears to be adequate (188–191). Other solvents include methyl cyanide (192), dimethylformamide (193,194), and various halogenated hydrocarbons (195).

To prepare bis(2-hydroxyethyl) terephthalate from terephthalic acid and ethylene oxide (196), a slurry of 50 parts of terephthalic acid, 159 parts of methyl ethyl ketone containing 26.7 parts of dissolved ethylene oxide, and 2.0 parts of tri-*n*-propyl amine is placed in an autoclave equipped with a stirrer, a gas inlet tube, a pressure gauge, a thermo-couple well, and a heating jacket. The air is displaced with nitrogen and the stirred reaction mixture is heated to 110°C. This temperature and a nitrogen pressure of 225 psig are maintained for 2 hr. The hot reaction mixture is filtered from 2.5 parts of unreacted terephthalic acid. After the filtrate has cooled to 10°C, the precipitated bis(hydroxyethyl) terephthalate is separated by filtration. The yield of diester is 94.5%. The product, recrystallized from hot water, melts at 110.4°C. An additional 12 parts of product is obtained from the reaction solution by distilling off the solvent.

D. Other Methods for Preparing Aromatic Esters

Of the methods that may be used to prepare aromatic esters from intermediates other than the carboxylic acid, the esterification of nitriles appears to have the most potential usefulness. Particularly interesting is the solvolysis of terephthalonitrile or isophthalonitrile under neutral or basic conditions. Reaction with glycols or epoxides produces hydroxyalkyl esters (197–199) directly with the liberation of ammonia, which may be recovered. Solvolysis in glycols requires temperatures in the neighborhood of 200°C and the presence of water is necessary. Reaction with ethylene oxide proceeds more readily at lower temperatures, as shown in the example following. Best yields are obtained with metal salts as catalyst (200,201). A wide variety of catalysts are known; for example, the salts of copper, zinc, cadmium, mercury, nickel, manganese, and cobalt.

Various processes are known for preparing terephthalate esters directly by the oxidation of *p*-xylene or *p*-cymene in the presence of methanol and thereby avoiding a separate esterification of the terephthalic acid. High yields are obtained with a cobalt bromide catalyst by feeding methanol and isophthalic or terephthalic acid into a solution initially containing 85–95% alcohol at 200°C and 60 atm (202). Methyl esters of aliphatic acids such as methyl acetate have been used as solvents and esterification agents in place of alcohol (203).

Depolymerization of poly(ethylene terephthalate) in methanol and glycols at temperatures up to 300°C in the presence of a catalyst has been used to prepare dimethyl terephthalate or bis(hydroxyethyl) terephthalate.

Bis(hydroxyethyl) terephthalate may be obtained from terephthalonitrile and ethylene oxide (204) in the following manner. A 1-liter, three-necked flask is equipped with a reflux condenser, an agitator, a thermometer and an inlet tube for feeding ethylene oxide to the bottom of the flask. Then 50 parts of terephthalonitrile, 30 mole% magnesium oxide, and 400 parts of water are added. The suspension is heated and agitated at 85–90°C while ethylene oxide is introduced at the rate of 6 parts per hour. After 10 hr, the reaction mixture is cooled. The precipitated crystals are separated by filtration. A yield of 79.5 parts (80.2%) of bis(hydroxyethyl) terephthalate, melting at 108–110 °C, is obtained.

E. Purification

Dialkyl esters of benzene dicarboxylic acids are often purified by distillation under vacuum as well as by crystallization, already illustrated in the preparative examples. The esterification catalyst is first removed from the reaction mixture. Treatment with dilute base is customary when acid catalysts have been used; for example, washing with aqueous sodium carbonate to remove benzenesulfonic acid. Catalysts such as zinc, tin, or titanium compounds may be removed from the esterification mixture by treating the solution with activated carbon (205).

Many procedures have been described for pretreating the esterification mixture to prevent the formation of degradation products during distillation and to improve the purity and color of the distillate. For example, alkyl phthalates have been distilled in the presence of activated alumina (206), and dimethyl terephthalate has been distilled with the addition of adipic acid (207) or triphenylphosphite (208).

VI. ACID CHLORIDES

A. Use as a Monomer and Preparation

Aromatic diacid chlorides are interesting monomers because they react rapidly at low temperatures, in the presence of a proton acceptor, with the active hydrogen of amines and alcohols to form condensation polymers. The common method of carrying out the reaction, called interfacial condensation, it is related to the classical Schotten-Bauman reaction for acylating amines and alcohols and the Hinsberg method for separating amines. Reaction occurs at the interface between a solution of a diacid chloride in a water-immiscible solvent and a water solution of the diamine or alcohol. The active hydrogen compounds may be aliphatic primary or secondary diamines or diphenols as their sodium salts. Aromatic diamines and aliphatic diols react too slowly to yield high polymer.

Good laboratory procedures are available for the polycondensation of phythaloyl chloride with a secondary cyclic amine, piperazine (209), and of terephthaloyl chloride with a primary aliphatic diamine, ethylene diamine, and of isophthaloyl chloride with the sodium salt of a diphenol, diphenylolpropane (210). Polyesters from terephthaloyl chloride and isophthaloyl chloride and aliphatic glycols are prepared in an anhydrous melt system.

Aromatic acid chlorides are prepared by all the methods generally used for converting carboxylic acids to acid chlorides, particularly by reaction with thionyl chloride or phosphorus pentachloride. If phosphorus pentachloride is used, approximately equimolecular portions of the chloride and the acid or its salt are mixed. After the initial reaction has subsided, the formation of the acid chloride is completed by warming. To moderate the reaction, inert solvents such as benzene or chloroform can be used.

Thionyl chloride has the advantage of giving entirely gaseous by-products and allowing easier purification of the acid chloride. However, it cannot be used to prepare phthaloyl chloride unless the reaction is carried out at elevated temperatures in the presence of anhydrous zinc chloride (211). With more moderate conditions, the anhydride forms from phthalic acid and no further reaction occurs. A small amount of pyridine is often added when aromatic acids are treated with thionyl chloride, but the use of amides, especially dimethylformamide, is receiving increasing attention (212,213).

Many other reagents, including acyl chlorides, may be used to prepare acyl chlorides. Benzotrichloride gives good yields of phthaloyl chloride from phthalic anhydride at 110–120°C in the presence of zinc chloride (211). An interesting application of this reaction is the preparation of terephthaloyl chloride by the reaction of hexachloro-p-xylene with terephthalic acid (214). Triphenyl phosphite also has been used as a catalyst for this reaction (215). The use of phosgene as the chlorinating agent has received much attention recently. Reaction with intramolecular anhydrides is conducted at temperatures between 40 and 150°C in the presence of dimethylformamide (216) or ferric chloride (217).

Acyl chlorides are also prepared directly by chlorination of the aromatic dialdehyde (218). Terephthalaldehyde is converted to terephthaloyl chloride with a 94% selectivity by passing a mixture of the aldehyde and chlorine diluted with nitrogen through a tube at 400°C, using a residence time of around 1 sec. Terephthaloyl chloride has also been prepared by photochlorination of dimethyl terephthalate at about 180°C (219–221).

Aromatic sulfonyl chlorides are obtained by treatment of sulfonic acids or their salts. If neither is available, the sulfonyl chloride can be prepared from an arene by reaction with chlorosulfonic acid. In fact, the direct preparation of sulfonyl chlorides using chlorosulfonic acid, followed by hydrolysis of the product, represents an alternate route to sulfonic acids which may have an advantage over sulfonation. The sulfonyl chloride is usually soluble in organic solvents, and it may be easier both to separate from the reaction mixture and to purify than the sulfonic acid.

B. Preparative Procedures

1. Terephthaloyl Chloride from Terephthalic Acid and Phosphorus Pentachloride (209)

Into a dry beaker, 63 g (0.3 mole) of phosphorus pentachloride is rapidly weighed out. The phosphorus pentachloride is poured into a dry, 2-liter round-bottom flask equipped with a thermometer and a dry condenser connected to a hydrogen chloride trap. A quantity of 25 g (0.15 mole) of terephthalic acid is added to the phosphorus pentachloride. After the solids have been mixed, they are heated to 100°C. The solids turn to a liquid and the reaction begins. The mixture is stirred or swirled slowly to get the solids into the liquid. When most of the solids have reacted (ca. 10 min), the mixture is heated to a gentle reflux for 1 min. There should be little or no solid remaining. The reaction mixture is cooled, and 250 ml of low boiling petroleum ether (40–60°C) is added. To dissolve the terephthaloyl chloride, the solution is heated to reflux. The hot solution is filtered, the filtrate is allowed to cool to room temperature, and then it is placed in an ice bath.

The solids are separated by filtration and washed with 50–100 ml of ice-cold low boiling ligroine. If the terephthaloyl chloride is to be used for the preparation of a polyamide by interfacial condensation, it should be recrystallized a second time from low boilling ligroine. After air drying for 5–10 min, 15 g of acid chloride is obtained, mp 83–84°C.

2. Terephthaloyl Chloride from Terephthalic Acid and Thionyl Chloride (210,222)

A mixture of 100 g of terephthalic acid, 500 g of thionyl chloride and 2 ml of pyridine is refluxed for 12 hr. The excess thionyl chloride is distilled at water-aspirator pressure on a steam bath. The residual acid chloride is distilled under vacuum through a short Vigreux column. The boiling point is 115–116°C at 3.0 mm. The product can be recrystallized from dry hexane (100 g/700 ml), mp 81–82°C.

The reaction may also be conducted in carbon tetrachloride (222) as follows. Into a solution of 16 g of terephthalic acid and 1.5 ml of thionyl chloride in 300 ml of carbon tetrachloride at 80°C, 20 g of pyridine is added dropwise. The solution is kept at 80°C for 10–12 hr. The yield of terephthaloyl chloride is 85%.

Isophthaloyl chloride can be prepared from isophthalic acid by the procedures given. It can be purified by vacuum distillation at 110°C/8 mm or recrystallized from dry hexane using 50 ml of hexane for 100 g of acid chloride. The crystals are allowed to form at 21–24°C; ice cooling should not be used. The melting point is 42–43°C.

3. Phthaloyl Chloride from Phthalic Anhydride and Phosphorus Pentachloride (223)

A mixture of 148 g (1 mole) of phthalic anhydride and 220 g (1.06 mole) of phosphorus pentachloride is placed in a 500-ml Claisen flask. The flask is equipped with a reflux condenser whose upper end is provided with a calcium chloride tube; the side arm of the flask is closed with a cork. The flask is inclined slightly to allow any phosphorus oxychloride that collects from the reaction to run back into the flask. After the mixture has been heated in an oil bath at 150°C for 12 hr, the reflux condenser and the stopper in the end of the side arm are removed and a water-cooled condenser and a distillation receiver are connected to the flask. The temperature is gradually raised to 250°C, and during this time most of the phosphorus oxychloride produced in the reaction distills over. The liquid residue is distilled under reduced pressure. After a small quantity of phosphorus oxychloride has been distilled, the o-phthaloyl chloride is collected at 131–133°C/9–10 mm. The product contains a small amount of phthalic anhydride. It solidifies on cooling in an ice salt mixture and melts at 11–12°C. The yield is 187 g (92%).

4. 4,4'-Biphenyldicarboxylic Acid Chloride (224)

The acid chloride is prepared by heating 12.5 g of powdered 4,4'-biphenyldicarboxylic acid with 50 g of biphenyl to 150°C under a reflux condenser while 21 g of phosphorus pentachloride is slowly added. After removal of the phosphorus oxychloride, the diacid chloride is purified by distilling off most of the biphenyl and pouring the residue into hot benzene. The 4,4'-biphenyldicarboxylic acid chloride, crystallized in long white needles, is sparingly soluble in cold benzene; mp 184°C (yield, 80%).

5. 4,4'-Sulfonyldibenzoyl Chloride (225)

To 424.2 g 4,4'-sulfonyldibenzoic acid (1.38 mole) suspended in 690 ml of phosphorus oxychloride is added in small portions with stirring and heating 738.4 g of phosphorus pentachloride (3.55 mole). The solution becomes clear shortly after the last phosphorus pentachloride has been added. The mixture is heated under reflux for 6 hr. The product, which separates on cooling, is removed by filtration and washed liberally with hexane. Additional product, which separates in the filtrate, is also collected by filtration. The total product is suspended in 2.5 liters of hexane and heated under reflux for 30 min. The product is removed by filtration and dried in a vacuum oven at 60°C; yield 377.1 g (79%). The entire yield is recrystallized once from pure, dry trichloroethylene and then distilled through a 16-in. Widmer column to give the following cuts: (a) 78.3 g, bp up to 290°C/0.9 mm; (b) 47.8 g, bp up to 300°C/1.0 mm; (c) 41.8 g, bp up to 300°C/1.0 mm;

and (d) 81.6 g, bp 282–284°C/0.275 mm. The Widmer column is removed and replaced by a 4-in. Vigreux column before distillation of the fourth cut. The inherent viscosities of the polyamides from *trans*-2,5-dimethylpiperazine prepared by interfacial condensation from the foregoing fractions are as follows: (a) 0.24, (b) 0.85, (c) 1.15, and (d) 3.13.

6. *m*-Benzenedisulfonyl Chloride from *m*-Benzenedisulfonic Acid and Phosphorus Pentachloride (226,227)

In a 3-liter three-necked flask fitted with a condenser, a thermometer, and a stirrer are placed 1360 g (6.55 mole) of ground phosphorus pentachloride and 727 g of phosphorus oxychloride. To the stirred mixture is added, during 30 min, 770 g (2.94 mole) of *m*-benzenedisulfonic acid (90%). The temperature is not allowed to exceed 70°C. The mixture is refluxed for 3 hr; then phosphorus oxychloride is distilled, first at atmospheric pressure and later under vacuum, until approximately 700 ml has been collected. The dark-colored liquid reaction mixture is poured with stirring into a 5-liter beaker two-thirds full of cracked ice. The cold mixture is stirred about 20 min and filtered. The solid is dissolved in 1 liter of benzene and washed three times with 250-ml portions of 5% sodium bicarbonate solution and once with 250 ml of water. After it has dried over anhydrous calcium sulfate and received treatment with decolorizing carbon, the filtrate is passed with suction through a $1\frac{1}{4}$-in. diameter column packed with 15 in. of activated alumina to remove the last traces of charcoal and to thoroughly dry the solution. The column is washed with 200 ml of benzene. To the solution is added 2500 ml of olefin-free *n*-hexane. The oil that separates is cooled to 20°C and scratched to cause crystallization. The solid is filtered on a large Büchner funnel and washed twice with 500-ml portions of olefin-free *n*-hexane to remove a slight yellow color. After it has dried in an evacuated desiccator containing calcium chloride for 3 hr, the solid weighs 553 g (68%). It melts at 62.0–62.5°C. The filtrate is concentrated to 350 ml and 500 ml of *n*-hexane is added. An oil is produced, and following seeding and subsequent drying, there is 90 g of crystalline material, melting at 61.0–61.5°C. The total yield is 643 g (80%).

7. *m*-Benzenedisulfonyl Chloride from Benzene and Chlorosulfonic Acid (226)

One mole of benzene is added with cooling to 2 moles of sulfur trioxide, in the form of 65% oleum, and 8 moles of chlorosulfonic acid. The mixture is then brought to reflux. No hydrogen chloride is liberated during the reaction. The cooled, dark brown liquid is poured onto ice and the mixture of diacid chlorides of *m*-benzenedisulfonic acid and *m, m'*-diphenyl sulfone disulfonic acid is sucked off. The *m*-benzene-disulfonyl chloride is separated by treatment with carbon tetrachloride, in which it dissolves very readily. *m,m'*-Diphenyl sulfone disulfonyl chloride is very sparingly soluble in carbon tetrachloride. The total yield of the chlorides is about 70%. The *m*-benzenedisulfonyl chloride may be purified by recrystallization from ether or by distillation, mp 62–63°C.

8. 4,4'-Biphenyldisulfonyl Chloride from Calcium 4,4'-Biphenyldisulfonate and Phosphorus Pentachloride (228)

Calcium 4,4'-biphenyldisulfonate (see Section VII) is dried for 10 hr at 150°C. A mixture of 78 g (0.2 mole) of the finely ground salt and 84 g (theory 83 g) of phosphorus pentachloride is warmed in a round-bottom flask in phosphorus oxychloride as diluent at 70°C for 8 hr. After cooling, the reaction mixture is poured into ice water. The acid chloride is removed by filtration and washed with water until the washed filtrate no longer gives an acid reaction. The product is dried over calcium chloride and phosphorus oxychloride. The 4,4'-biphenyldisulfonyl chloride is soluble in acetic acid, and it may be purified by treatment of the solution with activated charcoal. The product melts at 203°C.

VII. OTHER DICARBOXYLIC ACIDS

A. Biphenyldicarboxylic Acids

Biphenyldicarboxylic acids are useful for the preparation of polymers and plasticizers. The most readily available biphenyldicarboxylic acid is the 2,2'-isomer, diphenic acid. All the isomers may be prepared by oxidation of the corresponding substituted or alkylated biphenyls, but diphenic acid is unique in that it is produced readily from a polycyclic aromatic hydrocarbon, phenanthrene. With the exception of anthracene from which it is somewhat difficult to separate, phenanthrene is the most abundant single component in high-temperature coal tars. Phenanthrene can also serve as a potential source of the more interesting monomer 4,4'-biphenyldicarboxylic acid, inasmuch as the 4,4'-isomer may be prepared from diphenic acid by thermal rearrangement of its dipotassium salts (62).

1. Diphenic Acid

Phenanthrene is oxidized to diphenic acid with chromium trioxide in acetic acid, potassium dichromate in sulfuric acid, or hydrogen peroxide in acetic acid (229,230). It is possible also to conduct the oxidation of phenanthrene so that phenanthraquinone is obtained as the major product. Vapor-phase catalytic oxidation over vanadia yields either phthalic anhydride or the lactone of 2-hydroxydiphenyl-2'-carboxylic acid as the main product. Conditions for the peracetic oxidation have been carefully worked out by many researchers (231–233), and yields as high as 70–75% of diphenic acid have been obtained by this method from coal tar fractions containing up to 30% anthracene and carbazole impurities (229,234).

Ozonization of phenanthrene has been of theoretical interest, since ozone attacks the bond possessing the lowest bond-localization energy (235). Oxidation of the purified ozonide with potassium permanganate yields diphenic acid. There appears to be some recent commercial interest in ozonization as a synthetic method. Although the direct formation of diphenic acid using 2–3 mole of ozone has been described (236,237), other processes use sodium hypochlorite (238), potassium permanganate (239), or nitric acid (240) to oxidize the ozonide or its decomposition products in a second stage. The one-stage ozonization can be catalyzed by cobalt salts to give yields as high as 56% diphenic acid (241).

A large number of esters and amides have been prepared from diphenic acids. These derivatives are useful as plasticizers (242) and high boiling solvents and have been considered for use as insecticides and herbicides (243). Alkyd resins prepared from diphenic anhydride possess properties comparable to those of phthalic anhydride. Cast films prepared with pentaerythritol show good flexibility, hardness, and resistance to acids (244). Attempts to prepare high-molecular-weight condensation polymers with ethylenediamine and ethylene glycol have been less successful because of the tendency of the 2,2'-dicarboxylic acid to form cyclic oligomers (245).

Diphenic acid may be obtained by the oxidation of phenanthrene with hydrogen peroxide in acetic acid (229). Phenanthrene (1 mole) is dissolved in glacial acetic acid (2 liters) in a 5-liter three-necked flask fitted with a stirrer, a thermometer, and a reflux condenser, and warmed to 85°C on the water bath. Hydrogen peroxide (5.5 moles) as a 30% aqueous solution is added during 40 min; the temperature drops to 80°C and some phenanthrene comes out of solution. When the addition is complete, the mixture is heated on the water bath for 3–4 hr more.

The volume of the solution is then reduced to about half by distillation under reduced pressure. A substantial amount of the diphenic acid crystallizes out on cooling; this is filtered off, and the filtrate is then evaporated to dryness under reduced pressure. The residue is extracted with 750 ml of 10% aqueous sodium carbonate by warming on the water bath. The extract is boiled with a small amount of animal charcoal, filtered, and then brought to pH 4.5 by the addition of dilute hydrochloric acid. The tarry material precipitated is filtered off after the solution has been stirred with a further small amount of active charcoal. The clarified filtrate is cooled to 0°C and acidified with dilute hydrochloric acid. The acid that precipitates is filtered off, washed with water, and dried at 110°C. The yield of diphenic acid, mp 228°C, is 160–170 g (65–70%). The acid may be purified by recrystallizing from acetic acid until the melting point is 233.5–234°C.

This method can be applied to crude phenanthrene samples, but with such starting materials it is preferable to recover as much of the diphenic acid as possible by fractional crystallization, accomplished by successive reduction of the volume of the acetic acid solution. The crystals so obtained are often brown and contain some anthraquinone. They may be purified by solution in aqueous sodium carbonate, removal of tarry matter by partial neutralization, and precipitation with hydrochloric acid.

2. 4,4'-BIPHENYLDICARBOXYLIC ACID

4,4'-Biphenyldicarboxylic acid has been used in the preparation of a large number of fiber-forming polyesters and polyamides. It has been synthesized by dimerization of sodium benzoate and from the products of the diazotization of benzidine and of the chloromethylation or acylation of biphenyl. 3,3'-Biphenyldicarboxylic acid has received little attention. It has been made by the oxidation of 3,3'-ditolyl with potassium permanganate (246).

When sodium benzoate is ground in a ball mill with sodium for 14 hr at a maximum temperature of 242°C and the product dissolved in water, filtered, and precipitated with acid, a 37% yield of 4,4'-biphenyldicarboxylic acid is obtained (247). Hypochlorite oxidation of 4,4'-diacetylbiphenyl, produced in 50% yield by acylation of biphenyl with acetic anhydride, gives the 4,4'-diacid in 87% yield (248). Chloromethylation of diphenyl with paraformaldehyde and hydrogen chloride using stannic chloride yields 4,4'-bis-(chloromethyl) diphenyl in 59% yield (249). Oxidation of the chloromethyl groups with sodium dichromate gives the diacid in 88% yield, whereas poor results are obtained with potassium permanganate.

From 2 moles of biphenyl, 5 moles of dimethylcarbamyl chloride, and 5 moles of aluminum chloride, a 59% yield of N,N,N',N'-tetramethyl-4,4'-biphenyl dicarboxamide (mp 187–205°C) can be obtained (250). Hydrolysis of the amide with sodium hydroxide in boiling alcohol, followed by acidification of the reaction mixture, gives the diacid.

4,4'-Biphenyldicarboxylic acid may be prepared from benzidine (224). Benzidine (37 g) is dissolved in concentrated hydrochloric acid (100 ml), and ice (400 g) is added. Sodium nitrite (about 90 ml of 30%) is added slowly until some free nitrous acid is present; the diazonium chloride solution is then neutralized by addition of saturated aqueous sodium carbonate. This mixture is added slowly with stirring to a solution of cuprous cyanide at 90°C from 171.6 g of potassium cyanide and

159.8 g of copper sulfate (in 1060 ml of water). After the mixture has been allowed to cool and stand overnight, the solid product is collected. dried, and extracted with alcohol (Soxhlet). The alcohol-soluble material (2.7 g), mp 230–233°C, is biphenyldinitrile. The alcohol-insoluble material (32.7 g) appears to be a copper complex, which is decomposed by warming on a water bath for 30 min with excess concentrated hydrochloric acid. The acid is diluted and the solid material filtered off, washed free from acid, and extracted with acetone. The acetone-soluble extract crystallizes readily, mp 230–233°C, and is identical with the previous crop of biphenyl nitrile. The total yield is 45%. The dinitrile (10 g) is mixed with 70% sulfuric acid (90 ml) and refluxed for 45 min. The product is cooled and poured into water. The biphenyldicarboxylic acid is removed by filtration, washed, and dried. The yield is 95%.

B. 4,4′-Sulfonyldibenzoic Acid

The patent literature contains many examples of polyesters based on 4,4′-sulfonyldibenzoic acid. Various polyamides, suitable for the formation of fibers, have also been prepared by interfacial condensation. Poorer polymers result from conventional melt methods (225).

4,4′-Sulfonyldibenzoic acid [bis(p-carboxyphenyl)sulfone] is prepared by the oxidation of di-p-tolylsulfone. Air oxidation in the presence of an activator in acetic acid gives good yields and appears to be the method of choice (251,252). Similar aldehyde and ketone-activated autoxidations have already been discussed. Of the conventional oxidizing reagents, chromic oxide in acetic acid works well as does potassium dichromate (253,254). Poorer results are obtained with potassium permanganate (253,255). Ditolyl sulfone, which smells like onions, may be prepared by the Friedel-Crafts reaction of p-methylbenzenesulfonyl chloride (254).

4,4′-Sulfonyldibenzoic acid has also been prepared from diphenyl sulfide by condensation with acetyl chloride followed by the sodium dichromate oxidation of the product produced (255).

1.   Oxidation of di-p-Tolylsulfone with Chromic Acid (253)

A solution of 50 g of di-p-tolylsulfone in 400 ml of glacial acetic acid is warmed to 40–50°C and added during 30 min to 150 g of chromium trioxide in a previously warmed solution prepared from 250 ml of water, 400 ml of glacial acetic acid, and 250 ml of concentrated sulfuric acid (70°C). The reaction temperature is maintained at 100–106°C during the addition and then at 100°C for 15 min longer. The mixture is diluted with 1600 ml of water and treated with sulfur dioxide for 10 min to reduce the excess chromium trioxide. The p,p′-sulfonyldibenzoic acid is separated by filtration and washed with hot water. The yield of dicarboxylic acid is 57 g (92%).

The dimethyl ester, prepared by reaction with phosphorus pentachloride and methanol, melted at 193°C.

2. AUTOXIDATION OF DI-*p*-TOLYLSULFONE (252)

The quantities of 100 ml of acetic acid and 7 g of cobalt acetate are mixed and heated to 90–95°C. Oxygen is bubbled into the solution through a dispersion plate, and 25 g of acetaldehyde is pumped in during a period of 1 hr. The color of the solution changes from pink to dark green (cobaltic ion), showing that the catalyst has been activated. A solution of 30 g of bis(*p*-methylphenyl)sulfone in acetic acid is then added dropwise over a period of 12–14 hr. During this time, oxygen is continuously bubbled into the solution and a total of 70–80 g of acetaldehyde is added. The aldehyde feed is then stopped and the reaction mixture filtered. A cake of bis(*p*-carboxyphenyl)sulfone is obtained. It is washed with water followed by dilute hydrochloric acid and then more water. The yield is 36 g of bis(*p*-carboxyphenyl)sulfone which is 96% of theoretical. It titrates to give an equivalent weight of 154; the calculated weight is 153. The product can be further purified by dissolving it in aqueous alkali and precipitating with dilute hydrochloric or sulfuric acid.

The alkyl diesters can be prepared by refluxing the free acid with a large excess of an alkyl alcohol. Standard esterification catalysts and procedures can be employed.

C.   Naphthalene Dicarboxylic Acids

Naphthalene dicarboxylic acids and anhydrides have utility in the preparation of resins, plasticizers, polyesters, and polyamides. The condensation polymers of the 2,6-diacid have been claimed to have premium characteristics (256). Nevertheless, only the anhydride of 1,8-naphthalene dicarboxylic acid is presently produced in commercial quantities. Worldwide production is probably under 2 million lb/year (257), and nearly the entire amount is used for the manufacture of dyestuffs.

1,8-Naphthalic anhydride is manufactured by the vapor-phase oxidation of acenaphthene. A weight yield of 100% is obtained. The catalytic reaction produces also phthalic anhydride and maleic anhydride, which are not collected with the product. The only impurity found in practice is hemimellitic acid, and this is readily removed (257). Acenaphthene may also be oxidized in the liquid phase in a lower aliphatic acid solvent with a cobalt catalyst and an aldehyde activator. The quantity of crude naphthalic anhydride obtained is almost equal to the weight of acenaphthene oxidized, but more than three times that amount of butyraldehyde is consumed in the process (258).

Naphthalic acid may serve as the source of 2,6-naphthalenedicarboxylic acid by the thermal rearrangement of its dipotassium salt (259). The procedure is described in detail below. 2,6-Naphthalenedicarboxylic acid has also been prepared by hydrolysis of the corresponding nitrile (available by the fusion of dipotassium 2,6-naphthalenedisulfonate with potassium cyanide). Other methods for its preparation include the thermal disproportionation of the potassium salt of 1- or 2-naphthoic acid and the oxidation of 2-methyl-6-acetylnaphthalene with dilute nitric acid at 200°C.

All the dimethylnaphthalenes are potential starting materials for the production of naphthalene carboxylic acids, and all are present in varying concentrations in catalytic gas oils. The 2,6-dimethyl isomer can be isolated in high purity and has been used to prepare the 2,6-dicarboxylic acid on a pilot-plant scale by oxidation with nitrogen dioxide and selenium dioxide (90). In general the methods used for the oxidation of dialkyl benzenes will serve to oxidize the dialkyl naphthalenes. These processes include oxidation in acetic acid with air in the presence of cobalt and bromide salts (260,261), oxidation in trichlorobenzene with nitrogen dioxide (90), oxidation with sulfur dioxide or with sulfur in the presence of ammonia and water (68), and oxidation with aqueous alkaline dichromate (92). Complete experimental details for the sodium dichromate oxidation of 2,3-dimethylnaphthalene are described (262).

1. 2,6-NAPHTHALENEDICARBOXYLIC ACID FROM NAPHTHALIC ANHYDRIDE (259)

A solution of 66.5 g (1.01 moles) of 85% potassium hydroxide in 300 ml of water is heated to 60–70°C, and 100 g (0.505 mole) of 1,8-naphthalic anhydride is stirred in. The pH value is adjusted to 7 with $6N$ hydrochloric acid and $3N$ potassium hydroxide, and the solution treated twice with 10 g of decolorizing charcoal. The filtered solution is concentrated in a 1.5-liter beaker on a steam bath to about 180 ml. The concentrate is cooled to room temperature, 800 ml of methanol is added with vigorous stirring, and the mixture cooled to 0–5°C. The precipitated dipotassium naphthalate is separated by filtration, washed with 150 ml of methanol, and dried in a vacuum oven at 150°C/150 mm. The dried salt weighs 130–135 g (88–92% yield).

A mixture of 100 g of dipotassium naphthalate and 4 g of anhydrous cadmium chloride is ground in a ball mill for 4 hr. The mixture is placed in a 0.5-liter autoclave that can be rocked, rolled, or shaken. The line to the pressure gauge should be so placed that clogging is prevented. The autoclave is evacuated as oxygen lowers the yield of product. The autoclave is then filled with 30 atm of carbon dioxide. It is heated with agitation to 400–430°C during 1–2 hr, causing a pressure rise to about 90 atm.

After an additional 1.5 hr at 400–430°C, the autoclave is cooled and the carbon dioxide is bled off. The solid product is removed, pulverized, and dissolved in 1 liter of warm water. The solution is well stirred with 10 g of decolorizing charcoal and filtered to remove cadmium salts and carbon. The filtrate is heated to 80–90°C and carefully acidified with concentrated hydrochloric acid to pH 1 with stirring. The precipitated 2,6-naphthalenedicarboxylic acid is recovered by filtration and washed successively with 300 ml of water, 300 ml of 50% ethanol, and 300 ml of 90% ethanol. After the filtrate has been dried at 100–150°C/150 mm, the yield of acid with a decomposition point of 310–313°C is 42–45 g (57–61%).

2. Oxidation of 2,3-Dimethylnaphthalene with Sodium Dichromate (262)

A stirred autoclave is charged with 200 g (1.28 moles) of 2,3-dimethylnaphthalene, 940 g (3.14 moles, 23% excess) of sodium dichromate dihydrate, and 1.8 liters of water. The autoclave is heated to 250°C with agitation. After 18 hr at this temperature, the autoclave is cooled with continued agitation and the pressure is released. The autoclave is opened and the still-warm contents transferred to a 10-quart polyethylene pail or other container. To effect complete transfer, the autoclave is rinsed with several 500-ml portions of hot water. Green hydrated chromium oxide in the reaction mixture is separated on a large Büchner funnel and washed with warm water until the filtrate is colorless. The combined filtrates (7–8 liters) are acidified with 1.3 liters of 6N hydrochloric acid. The acidified mixture is allowed to cool to room temperature overnight. The precipitated 2,3-naphthalenedicarboxylic acid is separated by filtration and washed with water on the filter until the filtrate is colorless. After drying to constant weight at 50°C/20 mm, 240–256 g of white 2,3-naphthalenedicarboxylic acid is obtained, mp 239–241°C.

D. 2,6-Pyridinedicarboxylic Acid

One of the less readily available isomeric pyridinedicarboxylic acids, 2,6-pyridinedicarboxylic acid has been used extensively in the preparation of polyesters and polyamides. In addition to condensation polymers prepared from hexamethylene diamine (263) and ethylene glycol (264), a wide variety of polymers with other aliphatic and aromatic diols and diamines have been reported. Polymers of 2,6-pyridinedicarboxylic acid were found to have poorer thermal stability than the corresponding polymers of isophthalic acid (264). Major biochemical interest in 2,6-pyridinedicarboxylic acid is due to its isolation, as its calcium salt, from bacterial spores. The contribution of the dicarboxylic acid to spore structure is regarded as a fundamental problem in spore research. 2,6-Pyridinedicarboxylic acid, commonly called dipicolinic acid, may be prepared by fermentation or by the oxidation of 2,6-dimethylpyridine.

A fermentation process for the manufacture of 2,6-pyridinedicarboxylic acid developed in Japan cultivates a penicillium on a glucose medium (265). Many other bacteria are effective for the production of dipicolinic acid, and diverse materials—including hydrocarbons, the polymer of 3-hydroxybutyric acid, and 2,6-dioxohexandioic acid—have been used as culture media. Yields of 2,6-pyridinedicarboxylic acid from media containing hydrocarbons as the sole carbon source range from 0.7 mg/ml for n-decane to 8.1 mg/ml for n-dodecane (266), compared with 13.4 mg/ml for a glucose medium. Other studies have shown that optimum yields are obtained under conditions in which less than optimum growth is achieved (267).

Like other alkyl pyridines, the oxidation of 2,6-lutidine, 2,6-dimethylpyridine is facilitated by the stability of the pyridine ring. Although 2,6-lutidine is available in relatively good purity via isolation from coal tar, it does not now appear to provide an adequate basis for the economic production of 2,6-pyridinedicarboxylic acid in volume. Conditions for the oxidation of 2,6-lutidine with potassium permanganate have been carefully worked out (268). Direct oxidation to 2,6-pyridinedicarboxylic acid may also be accomplished by refluxing 2,6-lutidine with about three times its weight of selenium dioxide in dilute xylene (269). Only a little aldehyde is formed. If ester or acid chloride derivatives are wanted, the chlorination of 2,6-lutidine with thionyl chloride and hydrolysis of the hexachloromethylpyridine obtained, appears to have merit as a synthetic procedure (270).

Gas-phase oxidation at 380–400°C of 2,6-lutidine over a catalyst of silica gel impregnated with 15–20% vanadium and molybdenum oxides produces 2,6-pyridinedicarboxaldehyde (271). The dialdehyde oxidizes readily to diacid, even when it is standing in air at room temperature (272). Good yields of pure diacid may be prepared by oxidation with hydrogen peroxide. In one example (271), 40 g of dialdehyde is slowly added to 100 g of 30% hydrogen peroxide heated at 60°C. After 1 hr at 70°C, the solution is cooled to produce crystals of 2,6-pyridinedicarboxylic acid.

1. 2,6-PYRIDINEDICARBOXYLIC ACID BY FERMENTATION (265)

Penicillium citreoviride, No. 406, is cultured in 1 liter of a medium composed of 10% glucose, 2% sodium nitrate, and 0.5% monobasic potassium phosphate. The culture is kept at 30°C for ten days, and then it is filtered. The 660-ml filtrate is concentrated *in vacuo* to 400 ml. The concentrate is treated with 100 ml of concentrated hydrochloric acid, heated, and allowed to cool slowly to room temperature. It is kept at 2–5°C overnight and filtered to give 14.1 g of pyridinedicarboxylic acid.

2. Oxidation of 2,6-Lutidine with Potassium Permanganate (268)

A 5-liter four-necked flask is equipped with a small water-cooled condenser wide enough to accommodate the addition in portions of permanganate, a thermometer, a mercury seal stirrer, and a large capacity reflux condenser. To 107 g (1.0 mole) of 2,6-lutidine in 2500 ml of water is added 838 g (5.3 moles) of potassium permanganate in ten portions during 17 hr. The first five additions are made with the solution at 70°C, the last five at 85–90°C. Each addition is made only after the preceding amount has been consumed and washed down the condenser with water (total 400 ml). After the last addition has been decolorized, the hot reaction mixture is filtered by suction. The manganese dioxide is washed on the filter with 1500 ml of hot water in four portions, allowing each portion to soak into the cake without application of vacuum; finally, the cake is sucked dry before the fresh wash water is added.

The combined filtrate-wash is concentrated to 2 liters and made 1.5 normal by the addition of about 500 ml of concentrated hydrochloric acid. The mixture is heated to dissolve the precipitated solid, allowed to cool slowly to room temperature, kept overnight at 5°C, and filtered. The 2,6-pyridinedicarboxylic acid is washed with 50 ml of cold water and air dried. The filtrate is concentrated to 1200 ml, cooled at 5°C overnight, and filtered. The first crystal crop weighed 130 g (78% yield). The product may be recrystallized from 5N hydrochloric acid.

3. 2,6-Pyridinedicarboxylic Acid from 2,6-Bis(trichloromethyl)pyridine (270)

A solution of 100 g of 2,6-lutidine chlorohydrate in 1000 ml of thionyl chloride is heated in a pressure vessel at 180°C for 20 hr. When the excess thionyl chloride has been distilled off, the reddish yellow residue is taken up in ether. The unreacted lutidine chlorohydrate is removed by filtration and the ether filtrate shaken with a solution of sodium bicarbonate. The ether is removed and the fibrous crystalline residue vacuum distilled. A light yellow oil, which soon crystallizes, is obtained between 165–170°C/12 mm. The yield of 120 g of 2,6-bis(trichloromethyl)pyridine is purified by recrystallization from methanol and sublimation. The melting point is 86–87°C.

The hexachlorolutidine is hydrolyzed by heating in 800 ml of 80% sulfuric acid until the evolution of hydrogen chloride is complete. On dilution, colorless crystals of 2,6-pyridinedicarboxylic acid are obtained. The product is removed by filtration, recrystallized from dilute hydrochloric acid, and washed with cold water. It melts at 236–237°C and is found by analysis to be completely free of chlorine-containing by-products.

The methyl ester of 2,6-pyridinedicarboxylic acid is obtained by heating 100 g of hexachlorolutidine, 150 ml of concentrated sulfuric acid, and 10 ml of methanol for 2 hr at 220°C. From the dark-colored reaction solution, 50 g of ester melting at 120–122°C is isolated.

VIII. SULFONIC ACIDS

A. *m*-Benzenedisulfonic Acid

m-Benzenedisulfonic acid is of interest commercially for the production of resorcinol by fusion with sodium hydroxide. The preparation of a large number of its esters and amide derivatives have been reported (226). The acid chloride derivative, which may be prepared directly from benzene, is used to make polyamides by interfacial condensation (273).

Aromatic disulfonic acids are best produced by sulfonation (274) with the same reagents used for monosulfonation; chlorosulfonic acid, sulfuric acid, oleum, or sulfur trioxide. More drastic conditions are required for the introduction of the second sulfonic acid group. The metaisomer is practically the sole product when benzene is treated for a short time at a moderate temperature (8 hr at 85°C). However, the equilibrium mixture of acids, obtained by prolonged heating at 235°C with 87% acid, consists of 66.3% meta and 33.7% parabenzenedisulfonic acid.

In the commercial production of *m*-benzenedisulfonic acid, as an intermediate in the production of resorcinol, the major problems are the consumption of sulfuric acid and the formation of by-product sulfone. If sulfuric acid is used for the monosulfonation step and 65% oleum is then added to introduce the second group, the process yields spent sulfuric acid by lime neutralization; the spent acid is equivalent to 6.5 tons of gypsum for each ton of resorcinol produced. A theoretically perfect process using sulfur trioxide for both steps would require only 1.45 tons of reagent and would produce no gypsum. The use of stronger reagents, however, leads to an increase in the production of diphenyl sulfone and sulfonated diphenyl sulfones. An increase in strength of the oleum used to introduce the second group from 17 to 70% results in an increase in the sulfone content from 3 to 31%.

In one procedure used commercially, benzene is monosulfonated with low-strength oleum, and liquid sulfur trioxide is added to the reaction mixture to produce the disulfonic acid. This procedure represents a compromise between high sulfone formation and high spent acid formation. A similar method, in which the monosulfonic acid is first prepared by distillation under partial pressure and reacted with 65% oleum, gives 8% sulfone. In a procedure that simplifies product isolation, the second stage of sulfonation is carried out under a vacuum of at least 60 mm at a temperature not exceeding 275°C. The water is distilled off first. The temperature is kept at

210°C for 3 hr, whereupon all the sulfuric acid except about 10% of the total mass is distilled over.

The most effective way of reducing sulfone formation appears to be the addition of sodium sulfate to the sulfonation mixture. When 100% sulfuric acid is used in the first step and 65% oleum in the second step, the addition of 0.5 mole of sodium sulfate per mole of benzene reduces sulfone from 24.3 to 1.7% (227). One current commercial approach involves concurrently adding sulfur trioxide, benzene, and sodium sulfate to the reaction mixture at 140–160°C, to give a 90% yield of a product containing 90% m-benzenedisulfonic acid.

The mechanism of sulfonation involves the stepwise attachment of sulfur trioxide and H^+ to the substrate, followed by loss of a ring proton. It has been postulated that the sulfonated sulfones do not arise from m-benzenedisulfonic acid but are formed by sulfonation of diphenyl sulfone from the monosulfonic acid (227).

In preparing m-benzenedisulfonic acid by two-stage sulfonation (227,275), first 60 g of benzene is added to 151 g of 100% sulfuric acid over 2 hr at 50°C. The temperature is raised over 3 hr to 100°C and maintained for an additional 3 hr. To this sulfonation mixture, 234 g of 64.5% oleum is then introduced over 3 hr while the temperature is uniformly raised from 30 to 80°C. The mixture is held at 80°C for 2 more hours. Only 1.71% of diphenylsulfone-3,3'-disulfonic acid is found to be present in the sulfonation mixture.

The disulfonic acid may be isolated after neutralization of the sulfonation mixture with lime or by treatment with lead carbonate. The syrupy sulfonation mixture is slowly poured with stirring into ice contained in a large crock. The solution is neutralized by adding lead carbonate in portions with occasional stirring, waiting until the evolution of gas slackens before each new addition. The pasty mass is filtered by suction and the lead sulfate is washed with water. The combined filtrates are concentrated by evaporation and the free m-benzenedisulfonic acid is liberated with sulfuric acid. The lead sulfate is filtered off and the acid is isolated from the filtrate by removal of the water.

The acid may be purified by recrystallization of its potassium salt, formed by neutralization of the disulfonic acid with potassium carbonate. The free acid forms deliquescent crystals containing 2.5 moles of water of hydration. The hemihydrate is obtained after drying at 135°C (276).

B. Biphenyldisulfonic Acid

4,4'-Biphenyldisulfonic acid is best prepared by direct sulfonation of molten biphenyl (228,227) using an excess of reagent. Quantitative yields have been reported with sulfuric acid. Sulfonation with less than 5 moles of sulfuric acid gives a mixture of 4-monosulfonic and 4,4′-disulfonic acids. The monosulfonic acid is readily freed from the disulfonic acid through its sparingly soluble copper salt; the disulfonic acid remains in solution and can be recrystallized as the potassium salt (278). Potassium 4-biphenyl-sulfonate is converted by heating into biphenyl and the 4,4′-disulfonate (279). The disproportionation of arylmonosulfonic acids is catalyzed by such heavy metal catalysts as mercury, chromium, vanadium, or silver (278). The reaction is carried out at 300°C in an inert atmosphere.

2,2′-Biphenyldisulfonic acid is not as readily synthesized as the 4,4′isomer. It has been prepared by an involved procedure: sulfonation of nitrobenzene to *m*-nitrobenzenesulfonic acid, reduction to azobenzene disulfonic acid by means of zinc dust and alkali, and further reduction and rearrangement by means of stannous chloride and hydrochloric acid to 4,4′-diaminobiphenyl-2,2′-disulfonic acid (280). Removal of the two amino groups by diazotization and treatment with copper powder in ethyl alcohol produces 2,2′-biphenyl-disulfonic acid. The same acid has also been prepared by a coupling reaction from *o*-aminobenzenesulfonic acid (281). The amino acid is diazotized and the product reduced with cupro-ammonia ion in dilute ammonium hydroxide to produce a 20% yield of 2,2′-disulfonic acid. The meta and paraamino sulfonic acids do not give this reaction.

1. 4,4′-BIPHENYLDISULFONIC ACID BY SULFONATION OF BIPHENYL (228)

In a round-bottom flask equipped with a stirrer and a reflux condenser is placed 154 g (1.0 mole) of biphenyl. The flask is warmed to melt the biphenyl, and 1500 g of sulfuric acid monohydrate is added with vigorous stirring. The sulfonation is completed by heating with continued stirring at 140°C for 1 hr. The flask is cooled and the sulfonation mixture poured with stirring onto ice in a large beaker. The calcium disulfonate is obtained by treatment with 20% calcium chloride (calculated on the volume of liquid). The yield is quantitative. By recrystallizing the calcium salt twice from water, colorless crystals are obtained. If the disulfonyl chloride is wanted, the free acid is not liberated but the procedure described in Section VI.B.8 is followed.

2. 2,2′-BIPHENYLDISULFONIC ACID (280)

Dilute sodium hydroxide is added to a mixture of 100 g of 4,4′-diaminobiphenyl-2,2′-disulfonic acid in 1.5 liters of ice water until solution is effected. Then 41 g of sodium nitrite is added and, after solution is complete, a cold solution of 120 ml of concentrated sulfuric acid in 500 ml of water is slowly added. The clear solution is allowed to remain at 0°C for three days, while long yellow needles crystallize out. These are removed, washed with 100 ml of cold water, and air dried for

15 min. The weight of the partially dried product is 100 g. The product is placed in an 800-ml beaker, covered with 200 ml of 95% ethyl alcohol, and copper bronze is added slowly with stirring. The mixture warms up and nitrogen is evolved. When the reaction is ended, the mixture is heated on the steam cone for 15 min and filtered from the unreacted copper bronze. The filtrate is concentrated until it is almost free of alcohol, and then it is taken up in 200 ml of water. A solution of 80 g of barium hydroxide in 600 ml of water is then added, and the 800 ml of solution is filtered hot from a small amount of black insoluble material. On cooling a 50-g crop of white crystalline barium salt of 2,2'-biphenyldisulfonic acid comes out. This precipitate is removed and warmed with 12 g of concentrated sulfuric acid in 200 ml of water and filtered from the precipitated barium sulfate. The precipitated barium sulfate is then removed by filtration and the filtrate evaporated to a thick syrup, which solidifies when it is allowed to stand over phosphorus pentoxide for two days. 2,2'-Biphenyldisulfonic acid is very soluble in water and alcohol, soluble in acetone, and insoluble in chloroform, benzene, and petroleum ether. It is very deliquescent.

IX. ANALYTICAL METHODS

Almost all the analytical procedures used for carboxylic acids, anhydrides, esters, and chlorides may be applied to the analysis of aromatic dicarboxylic acids and their derivatives. These methods have been summarized, and in most cases detailed laboratory procedures are described (282).

A. Acidimetry

Titration with base is the most convenient and frequently reported method for analyzing acids. For dicarboxylic acids that are poorly soluble in water, nonaqueous systems are used or the sample of acid is dissolved in excess aqueous caustic and the excess caustic back-titrated. Titrimetric results are seldom adequate for the characterization of dicarboxylic acids to be used for the preparation of polymers, since the presence of impurities such as aldehydes, quinones, nitrocompounds, and poorly defined colorbody precursors is of primary interest. However, where the presence of other carboxylic acids is the principal concern, potentiometric titrations are often useful, since large differences in pK_a values can be utilized in the proper system. For example, the isomeric phthalic acids in mixtures with maleic or oxalic acid or in three-component systems with maleic and fumaric acids, may be titrated potentiometrically with benzene-methanol solutions of $0.1N$ tetraethyl ammonium hydroxide (283). A similar titration has been used with diphenic and o-phthalic acids in mixtures obtained by the oxidation of phenanthrene (284). Numerous other bases and solvents have been reported which permit selective titration of aromatic dicarboxylic acids, including alcoholic potassium hydroxide in aliphatic ketones, dimethylformamide, or pyridine-benzene (284–286).

B. Spectroscopy

Spectrophotometric methods have often been used to analyze mixtures of aromatic acid. The content of isophthalic acid and terephthalic acid in a mixture can be determined by measurements in methanol (287) using absorption of λ_{max} 279.5 and 284 nm, respectively. A method for the quantitative analysis of all three benzene dicarboxylic acids gives the best results using a hydrogen chloride–aqueous methanol medium (288). The maximum absorptions for the ortho, meta, and paraisomers occur at 275.5, 289, and 298 nm, respectively. Allowances are made for the deviations from linearity of the absorbance versus concentration curves.

C. Gas Chromatography

Analysis of phthalic anhydride for trace impurities has been conducted by conventional chemical, polarographic, and ultraviolet spectrophotometric procedures. However, gas chromatography provides a rapid and convenient method for all the substances of interest. Best results are obtained with ⅛-in. diameter columns packed with silicone absorbants such as SE-52 and SE-30, on silanized supports. The following procedures, developed for particular products, can probably find general application to the analysis of crude aromatic phthalic anhydride from various sources and to the analysis of other aromatic anhydrides. It should be pointed out that free o-dicarboxylic acid present in the sample will complicate the analysis, since it will give rise to anhydride in the heated injection port and on the column during chromatography.

1. PHTHALIC ANHYDRIDE FROM THE OXIDATION OF NAPHTHALENE (289)

A 6-ft length of ¼-in. stainless steel tubing is packed with 30% Silicone SF 96 on 60–80 mesh acid-washed Chromosorb W. The column is conditioned at 220°C and operated at that temperature isothermally with a helium flow of 60 ml/min (30 psig at inlet). The chromatograph is operated with the injection port at 250°C and the detector cell at 200°C.

A well-ground sample of phthalic anhydride is transferred to a 50-ml volumetric flask and dissolved in o-dichlorobenzene. If the solution is hazy, about 5 ml of acetone is used to replace part of the solvent. A 10-μl sample is injected into the chromatograph. Attenuations are adjusted to keep all peaks at least 15% and not more than 85% of full scale on the recorder. The weight in milligrams of each component present for each integrated and normalized peak is obtained directly from calibration curves prepared using pure samples. The retention times in minutes (measured from inject) and relative to the solvent peak are: acetone (1.1) 0.24; maleic anhydride (2.4) 0.52; o-dichlorobenzene (4.6) 1.00; benzoic acid (6.2) 1.35; naphthalene (8.4) 1.83; phthalic anhydride (11.4) 2.48; 1,4-naphthoquinone (16.3) 3.54.

2. PHTHALIC ANHYDRIDE FROM THE OXIDATION OF o-XYLENE (290)

Samples are dissolved in acetone and analyzed using a 4-ft column packed with 5% SE-52 on 60–80 mesh Gas-Chrom Q. The temperature is programmed at 10°/min from 60 to 175°C. Retention times in minutes are: acetone 0.2, maleic anhydride 3.3, o-xylene 3.7, citraconic anhydride 4.7, benzyl alcohol 6.2, p-tolualdehyde 6.9, naphthalene 8.4, toluic acid 9.6, phthalic anhydride 10.3. Separation of maleic anhydride and o-xylene may not be complete if they are present in large quantities. Naphthalene is used as an internal standard; a weighed amount being added to the acetone solution. Other solvents that do not interfere with the product peaks are acetonitrile, carbon tetrachloride, ethylene glycol, and toluene.

3. SILYLATION OF CARBOXYLIC ACIDS (291)

Aromatic dicarboxylic acids are not sufficiently volatile to allow analysis directly by gas chromatography. A volatile derivative must first be prepared. Usually the dimethyl ester is obtained by esterification with methanol or diazomethane. A more conveniently prepared derivative is a silyl carboxylate and several silylating agents are available.

A slurry of 10 mmoles of the acid in 5 ml of ether is vigorously swirled as 2 ml of bis(trimethylsilyl) acetamide (Applied Science Laboratories) is slowly added. Samples are chromatographed as soon as solution is complete, usually within several minutes. Good resolution of trimethylsilyl carboxylates is obtained using a silicone gum, SE-30 or SE-52 column.

D. Liquid–Solid Chromatography

Column chromatography and thin-layer chromatography (292–300) have often been used to separate mixtures of aromatic acids, particularly the side products obtained in the oxidation of p-xylene to terephthalic acid or of mixed xylenes. Partition chromatography using mineral acid as the stationary phase appears to be the preferred method for the separation and quantitative estimation of mixtures when unknown acids may be present. Thin-layer chromatography using silica gel or mixed sorbents and a variety of solvent mixtures is far less time consuming if it is applicable. In both methods care must be exercised to effect good separation of isophthalic and terephthalic acids. Anion-exchange resins may also be used to separate aromatic and aliphatic acids by elution chromatography. The elution order of each acid with formic acid is a function of the molarity of the formic acid and depends on both the structure and the acid strength of the particular acid (301).

The partition chromatography of benzenecarboxylic acids (302) requires a glass tube (1.8-cm diam.) fitted with a tap and a sintered glass disk at the bottom. The height of the tube above the disk is at least 40 cm. Solvent mixtures are prepared by shaking appropriate volumes of n-butanol and chloroform (analytical quality) with a little distilled water and separating the organic layer.

Silica gel (20 g) and 0.5N sulfuric acid (10 ml) are mixed and then slurried with chloroform (80 ml). The column is packed at atmospheric pressure, so that the rate of flow is about 3 ml/min.

The acid mixture (up to about 0.1 g) is dissolved in methanol (0.4 ml) and diluted with chloroform (0.6 ml). The solution is added to the top of the column, from a capillary held just above the surface, and allowed to drain into the column. The procedure is repeated twice with washings from the flask, which consist successively of 10% methanol-chloroform (1 ml) and chloroform (1 ml).

Chloroform (50 ml) is added to the column, and elution is commenced, the eluant being collected in 10-ml fractions. More solvent is added in portions (each 50 ml) before the last 10 ml of the preceding solvent has drained into the column. The complete range of solvents is chloroform (50 ml), followed by these n-butanol–chloroform mixtures: 2–5%, 150 ml; 10%, 100 ml; 12.5%, 100 ml; 17.5%, 100 ml; 25%, 100 ml; 30%, 100 ml; 40%, 100 ml; 50%, 100 ml; 70%, 200 ml.

With concentrations of alcohol up to about 40%, a flow rate of 2–3 ml/min can be maintained at atmospheric pressure. At higher concentrations it is necessary to apply slight positive pressure to maintain the rate.

The acid in each fraction is estimated, after addition of water (50 ml) and a little phenolphthalein, by titration with 0.02N sodium hydroxide. The following values (in milliliters) are obtained, with each value being the mean of several determinations of peak effluent volumes: benzoic acid, 20; isophthalic acid, 83; phthalic acid, 108; trimesic acid, 225; trimellitic acid, 254; hemimellitic acid, 359; benzene-1,2,3,5 tetracarboxylic acid, 452; pyromellitic acid, 508; benzene-1,2,3,4 tetracarboxylic acid, 631; benzenepentacarboxylic acid, 808; and mellitic acid, 1040.

E. Polymer-Grade Terephthalic Acid

Probably one of the most interesting and important analytical problems involving monomers is the determination of the suitability of terephthalic acid for direct polymerization to polyethylene terephthalate. The carboxyl function as such is not reduced at the dropping mercury electrode, but polarography appears to be commonly used to analyze aldehydic impurities in commercial terephthalic acid. No completely satisfactory procedure for the determination of p-formylbenzoic acid has been published, although it must be assumed that the problems posed by the insolubility of terephthalic acid and the poor extractability of the aldehyde have been resolved. The reduction half-wave potential of p-formylbenzoic acid occurs at -1.12 V in a supporting electrolyte of $0.1 M$ boric acid–$0.01 M$ potassium hydroxide in 1:1 water-methanol. The proprietary nature of analytical procedures for polymer-grade terephthalic acid and perhaps other monomers is illustrated by a patent (303) entitled "Method of Testing Terephthalic Acid to be Converted into Polyethylene Terephthalate." In the method described, 7.5 g of terephthalic acid is dissolved in 50 ml of $2N$ potassium hydroxide and the light transmittance of the solution measured at 340 nm. Polymers of good color can be obtained from acid having light transmittances of 93–97%. Previous methods depending on the Hazen number, the optical density, or the acid value are claimed to be less sensitive to colorless impurities, such as isophthalic, benzoic, and p-toluic acid, which reduce the quality of the polymer obtained commercially.

X. PHYSICAL PROPERTIES

Aromatic dicarboxylic acids are white crystalline solids whose properties depend on the positions of the two carboxyl groups in relation to the aromatic ring and to each other. Physical properties are summarized in Tables 1–16. Although very high melting points have been reported for most of the dicarboxylic acids, they are often indefinite (304) and not useful for characterization of the compound or indicating purity. The limited utility of the melting point is illustrated by the thermal behavior of 2,6-pyridinedicarboxylic acid, for which melting points of from 225° to 236°C have been reported. From the rate of thermal dicarboxylation, the true melting point is estimated to be higher than 243°C (305). Unlike aromatic tribasic and polybasic acids and disulfonic acids, hydrate formation has not been reported for the dicarboxylic acids.

A. Acidity

The most important physical characteristic of aromatic diacids is due to the acidic properties of the carboxyl and sulfonyl groups. The inductive effect of one acidic group is expected to enhance the acidity of the other and from Tables 1, 5, and 9–11, it can be seen that the acid strength of dicarboxylic acids as measured by the first acid-dissociation constant, K_1, is higher than that of benzoic acid, 6.5×10^{-5} (H_2O at 25°C), or 1- or 2-naphthoic acid,

2.0 × 10^{-4} and 6.9 × 10^{-5}, respectively. As is the case with aliphatic di-carboxylic acids, the second acid-dissociation constant is much smaller because it is more difficult for a proton to separate under the electrostatic attraction of the nearby carboxylate anion. Of the three isomeric benzene dicarboxylic acids, phthalic acid has the strongest first ionization and weakest second ionization constant. This phenomenon is explained (306) by hydrogen bonding, which helps to form a seven-membered ring in the orthoisomer but not in the meta and paraisomers. The hydrogen bonding enhances the first ionization and hinders the second ionization.

B. Spectra

1. INFRARED

The infrared spectra of the three pure benzene dicarboxylic acids obtained as Nujol mulls have been published, along with the spectra of the remaining nine benzene polycarboxylic acids (307). All spectra are highly specific, showing a great deal of fine structure in the range 1700–650 cm^{-1}, and may be used for ready characterization of any acid. No free hydroxyl absorption is exhibited, indicating a high degree of inter and intramolecular association. Dimeric COOH gives rise to five bands between 5000 and 665 cm^{-1}.

The infrared spectra of the aromatic diesters have also been examined in detail. Apparent extinction coefficients have been reported, and assignments to specific molecular modes (308) have been made. The nature of the alkyl group was found to have little effect on the band characteristics of the acyloxy group.

2. ULTRAVIOLET

Ultraviolet absorption studies have been largely directed to the development of analytical procedures. One interesting exception is the correlations found between resonance structures for the diphenyl dicarboxylic acids or esters and the values for their extinction coefficients and wavelengths of maximum absorption (309). The absorption of 2,2′-biphenyldicarboxylic acid (2800 Å) differs little from that of benzoic acid. The expected resonance effect does not occur, since the o-carboxyl groups make coplanarity impossible. On the other hand, the 4,4′ isomer can assume exact coplanarity and does show greatly enhanced absorption. Interestingly, the 3,3′ isomer shows less absorption and a shift in the maximum toward shorter wavelengths. Since no resonance structure involving a simultaneous double bond between the rings and the carboxyl carbon at the same time, can be written for this isomer, decreased conjugation results. The spectra of 2,6-pyridine-dicarboxylic acid have also received increased attention, primarily because of its usefulness in spore research (310). In ethanol the major peak at 268–270 nm has two distinct shoulders at 275 and 262 nm, which disappear in aqueous solutions.

3. Nuclear Magnetic Resonance

The NMR spectra of dicarboxylic acids possess the features expected for aromatic acids. The benzene protons show a negative shift relative to that of benzene at 2.73τ and the acidic proton signal occurs between 0 and -3τ. The displacement for ring protons is greater for anhydrides and esters. The differences in the chemical shifts for phthalic anhydride ($\tau 1.84$) and phthalic acid ($\tau 2.31$, 2.28) have been used to study hydrolysis rates (311).

TABLE 1
Physical Constants of Phthalic Anhydride and Phthalic Acid (121)

Properties	Anhyride	Acid
Liquid		
Normal boiling point, °C	284.5	
Freezing point (dry air), °C	131.11 ± 0.01	191 (approx)
Triple point, °C	131.100 ± 0.001	
Heat of vaporization, at 131°C, kcal/g-mole	15.6	
Solid		
Specific gravity, at 4°C	1.527	
Specific heat, cal/g		
at 300°K	0.2627	0.2722
at 200°K	0.1824	0.1912
at 90°K	0.1010	0.1035
Entropy, at 25°C ($S°$ 298), cal/(g-mole)(°C)	42.9	49.7
Heat of combustion, at 25°C, kcal/g-mole	779.02	770.49
Heat of formation, at 25°C, kcal/g-mole	110.03	186.88
Heat of sublimation, at 131°C, kcal/g-mole	21.2	
Heat of fusion, at 131°C, kcal/g-mole	5.6	
Heat of solution in water, at 25°C, kcal/g-mole	10.5	4.9
Ionization constants in water		
first		1.1×10^{-3}
second		5.5×10^{-6}

TABLE 2
Approximate Solubilities of Phthalic Anhydride and Phthalic Acid g/100 g of Solvent (121)

Solvent	Anhydride	Phthalic Acid
Water		
at 14°C		0.54
at 25°C	0.62	
at 99°C		18.0
at 135°C	95	
Ethyl alcohol, at 18°C		11.7
Ethyl ether, at 15°C		0.68
Carbon disulfide, at 20°C	0.7	
Formic acid (95%), at 20°C	4.7	
Pyridine, at 20–25°C	80	
Glacial acetic acid, at 100°C		12.0

TABLE 3
Properties of Molten Phthalic Anhydride (121)

Temperature, °C	Surface tension, dynes/cm	Density, g/ml	Viscosity, mP	Vapor pressure, mm Hg
132			11.9	6
135		1.208		7
140		1.202		8.7
155	35.49			
160		1.181		20.5
180	32.70	1.161		41
197			6.4	75
200		1.142		80.5
220		1.124	5.5	
240		1.105		
284.5				760

TABLE 4
Properties of Esters of Phthalic Acid (121)

Ester	mp, °C	bp, °C/mm Hg	Specific gravity at 20°C	at 25°C
Dimethyl	0	282/760	1.192	
Diethyl	−40(−4)	296/760	1.118	
Diallyl	65	156-175/4	1.120	
Dibutyl	−40	340/760		1.042
Butyl cyclohexyl		189-222/5		1.076
Diamyl	< −55	342/760		1.022
Butyl benzyl		370/760		1.111
Dicyclohexyl	58–65	212-218/5	1.148	
Butyl octyl	< −50	225/5		0.993
Butyl decyl	−50	220/5		0.991
Di-n-octyl	−25	220-248/4	0.978	
Diisooctyl	< −50	228-239/5	0.986	
Bis-2-ethylhexyl	−46	231/5	0.986	
n-Octyl n-decyl	−28	250/5		0.970
Isooctyl isodecyl	−48	235-248/4		0.967
Diisodecyl	−48	255/5		0.961

TABLE 5

Physical Constants of Terephthalic and Isophthalic Acids (121)

Properties	Terephthalic Acid	Isophthalic Acid
Triple point, °C	427	
Molar magnetic susceptibility, $-\chi \times 10^6$	83.51	
Heat of combustion, kcal/mole	770.9	
Heat of sublimation, kcal/mole	23.5	25.5 (250°)
Sublimation point, °C	402	
Specific heat, cal/(g)(°C)	0.2873	
Density, g/ml	1.510	1.507
Ionization constants		
in water, first	3.1×10^{-4}	3.3×10^{-4}
in water, second	1.5×10^{-5}	3.2×10^{-5}
in 50% aqueous methanol, first	4.1×10^{-5}	
second	8.3×10^{-7}	
Vapor pressures, mm Hg		
100°C		0.07
120°C	0.5	
125°C		0.60
130°C	1.0	
140°C	2.0	
150°C	4.7	
Solubilities, 25°C, g/100 g		
Sulfuric acid (95%)	2	
Glacial acetic acid	0.035	0.078
Water	0.0019	0.013
Methanol	0.1	2.1
Benzene	Insol.	Insol.
Dimethylformamide	6.7	
Dimethyl sulfoxide	20	
Solubilities, higher temperatures	(200°C)	(190°C)
Methanol	15	
Water	1.8	16.5
Glacial acetic acid	1.8	11.0
Acid chloride		
mp, °C	78	
bp, °C	259	

TABLE 6
Properties of Esters of Terephthalic and Isophthalic Acids (121)

	mp, °C	bp, °C/mm Hg	Density
Terephthalate			
Diethyl	43–44	140–142/3	1.1022^{45}_{22}
Di-n-propyl	25	164–166/4	1.0715^{28}_{25}
Di-n-butyl	16.6–18.1	181–189/2	1.0452^{25}_{25}
Di-2-ethylhexyl	−63.5	186–192/0.1	0.9823^{25}_{25}
Diisooctyl	Amorphous solid	189–190/0.05	0.9809^{25}_{25}
Diallyl	−10	140–143/1	1.1380^{45}_{25}
Isophthalate			
Dimethyl	67–68	124/12	
Diethyl	11.5	302/760	1.1389^{20}_{20}
Diisooctyl		127–138/0.4	0.987^{20}_{20}
Di(2-ethylhexyl)		223–225/1	0.980^{20}_{20}
Diallyl		150–152/0.9	1.1143^{20}_{20}
Diphenyl	134–138		

TABLE 7
Properties of Dimethyl Terephthalate (121)

Property	Value
Heat of combustion, kcal/mole	1113.2
Specific heat, cal/(g)(°C)	6.326
Liquid	
Freezing point, °C	
in N_2	140.655
in air	140.648
Cryoscopic constant, mole %/°C	1.4
Normal boiling point, °C	288
Heat of fusion, cal/g	38.0
Heat of vaporization, cal/g	70.5
Specific heat, cal/(g)(°C)	0.464
Density, g/ml at 150°C	1.068
Viscosity, cP, at 150°C	0.965
Vapor pressures, mm Hg	
141°C	10
210°C	100
232°C	200
260°C	400
288°C	760
Reduction potential, mercury cathode (312)	
$E_{0.5}$, pH 1.2–4.4	−1.19–1.39
4–12.2	−1.54
9.5–12.2	−1.83

TABLE 8
Solubility of Dimethyl Terephthalate at Various Temperatures, g/100 g Solvent (121)

		Solubility		
Solvent	(bp of solvent, °C)	25°C	60°C	bp of solvent, °C
Ethylene glycol		0.8		
Diethylene glycol	(245)			400
Methanol		1.0	5.7	
Carbon tetrachloride	(77)	1.5	3.6	25
Ethyl ether		1.6		
Acetone	(56)			25
Methyl ethyl ketone		1.6	12.5	
Benzene		2.0	14.0	
Ethyl acetate		3.5	16.0	
Butyl acetate	(120)			400
Toluene	(111)	4.3	10.4	>100
Ethylenediamine	(117)			25
Ethylene dichloride	(84)	6.4	18.8	100–400
Dioxane		7.5	28.5	
Chloroform		10.0	23.0	

TABLE 9
Properties of Diphenyldicarboxylic Acid (313,229)

Properties	2,2'-	3,3'-	4,4'-
mp, °C			
H_2O	228–229		
AcOH	233.5		
EtOH (309)	231.5	356–357	
Solvent			
H_2O	mod. hot	0.0052N (25°C)	
	sol.		
Et_2O	sol.	sol.	
λ_{max}, nm (309)	280	243	
Dissociation constants (314)			
H_2O, pK_1	3.17		
pK_2	5.35		
50% EtOH–H_2O, pK_1	3.98		
pK_2	7.90		
Dimethyl ester			
mp, °C MeOH	74	104	214 (217)
bp/mm	204–6/14		
λ_{max}, nm (309)			280
Diethyl ester, mp, °C	42	68	112
dipole moment, D			2.2
Anhydride, mp, °C	217		
Solvent			
Et_2O	sparingly		
H_2O	insol.		
Dichloride (benzene) (224)	93.4		184

TABLE 10
Diesters of Diphenic Acid (315)

R	mp, °C	bp, °C/mm	D_4^{20}	n_D^{25}
Methyl	72–73	—	—	—
Ethyl	41.5–42	—	—	—
n-Propyl	—	201–202/4	1.118	1.5441
Isopropyl	74.0–74.5	—	—	—
n-Butyl	—	192–193/1.5	1.082	1.5339
Isobutyl	—	191–192/2	1.085	1.535
sec-Butyl	81–82	—	—	—
tert-Butyl	132–133	—	—	—
n-Pentyl	—	201.5–202.5/1	1.064	1.529
Isopentyl	—	229.5–230.5/9	1.063	1.529
sec-Pentyl	74.5–75.5	—	—	—
n-Hexyl	44.5–45.5	—	—	—
n-Heptyl	—	248–249/5	1.018	1.5184
n-Octyl	—	246–247.5/1	1.015	1.5143
n-Decyl	—	272–273/1.5	1.013	1.5093
Cyclopentyl	121–121.5	—	—	—
Cyclohexyl	51–52	—	—	—

TABLE 11
4,4′-Sulfonyldibenzoic Acid

Properties	Value
mp, °C	
(benzene) (316)	371
pK_a	4.73
σ (Hammett sigma)	0.62
Solubility	
H_2O	insol.
Alcohol	insol.
Dimethyl ester, mp, °C	194
Diethyl ester, mp, °C	156–156.5
Acid chloride, mp, °C (228)	203

TABLE 12

Melting and Boiling Points, °C, of Naphthalenedicarboxylic Acids (313)

Isomer	1,2-	1,3-	1,4-	1,5-	1,6-	1,7-	1,8-	2,3-	2,6-	2,7-
mp										
H$_2$O	175									
EtOH		267–268								
AcOH			309			294–296d	260	239–241	>300d	>300d
Sublimed					310			246		
bp				315–320d						
mp										
Dimethyl ester, MeOH	85		67	119	98					
AcOH, bp/mm			195–197/12							
EtOH						90	104		186	141
Et$_2$O–petroleum ether								47		
mp										
Diethyl ester, EtOH							59–60			
bp/mm							238–239/19			
mp										
Anhydride, EtOH	168–169			123–124						
AcOH							274	246		238
mp										
Dichloride, petroleum ether			80	155–156						
CHCl$_3$										
CS$_2$							84–86			
bp/mm							195–200/0.2			

TABLE 13
Solubilities of Naphthalenedicarboxylic Acids[a]

Isomer	H_2O	EtOH	Et_2O	AcOH	C_6H_6	Ligroine	$CHCl_3$	CS_2
1,2-		s	s	s	ss	ss	ss	ss
1,3-								
1,4-	i	s						
1,5-		(insoluble ordinary solvents)						
1,6-		sh		sh				
1,7-		(sol. ordinary solvents)						
1,8-	i	sh	ss					
2,3-	ss	sh	ss	ssh	ss	ss		ss
2,6-				ih	ih			
2,7-				ssh	ssh			

[a] s = soluble, ss = sparingly sol., sh = sol. hot, i = insoluble.

TABLE 14
2,6-Pyridinedicarboxylic Acid

mp, °C	
(dilute HCl) (270)	252, 236–237
decomposes (305)	228, 243
λ_{max}, nm (317)	273
Dimethyl ester	
mp, °C	121
bp at 0.5 mm, °C	155–160
Diethyl ester	
mp, °C	28, 41–42
bp at 1 mm, °C	158
Dichloride	
mp, °C	61
bp, °C	284
Solubility	
H_2O	s
EtOH	s

TABLE 15
m-Benzenedisulfonic Acid (313)

Property	Value
mp, °C	
2.5 H$_2$O hydrate	100
0.5 H$_2$O hydrate	135
Solubility	
H$_2$O	s
Dimethyl ester, mp, °C	52–54
Diethyl ester	liquid
Dichloride	
mp (ether), °C	63
bp at 20 mm, °C	211

TABLE 16
4,4'-Biphenyldisulfonic Acids (313)

Properties	Value
mp, °C	
Monohydrate	72.5
Dihydrate	137.5–138.5
Deliquescent (280)	
Solubility, H$_2$O	s
Dimethyl ester, mp, °C	193
Dichloride, mp, °C (228)	203 d.
Solubility	
alcohol, ether	sol.
CS$_2$	insol.

References

1. A. Laurent, *Ann.*, **19**:38 (1836).
2. L. F. Marek and D. A. Hahn, "The Catalytic Oxidation of Organic Compounds in the Vapor Phase," Chemical Catalog Co., New York, 1932, Chapter 8.
3. E. Nolting, *Ber.*, **18**:175 (1885).
4. F. Beilstein's *Handbuch der organischen Chemie*, 3rd ed., **9**:791.
5. R. Landau and H. Harper, *Chem. Ind. (London)*, **1961**:1143.
6. J. Gibbs, *Ind. Eng. Chem.*, **11**:1031 (1919).
7. C. Conover and H. D. Gibbs, *Ind. Eng. Chem.*, **14**:120 (1922).
8. W. G. Parks and C. E. Allard, *Ind. Eng. Chem.*, **31**:1162 (1939).
9. D. Geldhart, *Chem. Ind. (London)*, **1967**(35):74.
10. *Oil, Paint, Drug Rep.*, Oct. 21, 42 (1968).
11. I. E. Levine, *Chem. Eng. Progr.*, **43**:168 (1947).
12. R. I. Stirton, in "Encyclopedia of Chemical Technology," R. E. Kirk and D. F. Othmer, Eds., Vol. 10, Interscience, New York, 1953, p. 584.

13. *Eur. Chem. News*, Sept. 22, 40 (1967).
14. S. K. Bhattacharyya and R. Krishnamurthy, *J. Appl. Chem.*, **13**:547 (1963).
15. M. Sittig, "Dibasic Acids and Anhydrides," Noyes Development Corp., Park Ridge, N.J., 1966.
16. R. B. Stabaugh, *Hydrocarbon Process.*, **45**:149 (1966).
17. C. R. Kinney and I. Pincus, *Ind. Eng. Chem.*, **43**:2880 (1951).
18. N. D. Rusyanova, A. S. Kostromin, and A. A. Belyaeva, *Koks. i Khim*, **1964**:25.
19. J. K. Dixon and J. E. Longfield, in "Catalysis," Vol. 7, P. Emmett, Ed., Reinhold, New York, p. 183.
20. S. Morita, *Bull. Chem. Soc., Jap.*, **33**:309 (1960).
21. S. K. Bhattacharyya and I. B. Gulati, *Ind. Eng. Chem.*, **50**:1719 (1958).
22. J. Herten and G. F. Froment, *Ind. Eng. Chem., Process Design Develop.*, **7**:516 (1968).
23. B.I.O.S., Final Rep. No. 753; item No. 22.
24. F.I.A.T., Final Rep. No. 649, (1947).
25. N. Chomitz and W. R. Rathjins, U.S. Patent 2,973,371 (1961).
26. V. Yu Vol'fson and L. N. Ganyuk, *Kinetika i Kataliz*, **6**:306 (1965).
27. R. J. Sampson and D. Shooter, "The Heterogeneous Selective Oxidation of Hydrocarbons," in "Oxidation and Combustion Reviews," Vol. 1, C. F. H. Tipper, Ed., Elsevier, New York, 1965, p. 223.
28. C. A. Bunton et al., *J. Chem. Soc.*, **1965**:6174.
29. K. E. Khcheyan et al., *Khim. Prom.*, **1962**:392.
30. A. I. Kamneva et al., *Neftekhimiya*, **2**:524 (1962).
31. *Chem. Eng. News*, Nov. 14, p. 29 (1966).
32. Progil, French Patent 1,441,453 (1964).
33. I. E. Levine, U.S. Patent 2,521,466 (1945).
34. F. Bernardini, M. Ramacci, and A. Paolocci, *Chim. Ind. (Milan)*, **47**:485 (1965).
35. E. P. Jaeger, German Patent 1,181,691 (1964).
36. H. Suter and G. Poehler, German Patent 1,097,427 (1961).
37. M. D. Kulik, U.S. Patent 3,040,060 (1962).
38. V. N. Kochetkov, V. V. Smirnova, and R. I. Zaitseva, *Vestn. Tekhn. i Ekon. Inform., Nauchn.-Issled. Inst. Tekhn.-Ekon. Issled. Gos. Kom. po Khim. pri Gosplane SSSR*, **3**:23 (1963).
39. S. Meinstein, U.S. Patent 3,303,203 (1965).
40. V. Kh. Katunin and M. V. Kuznetsova, *Khim. Prom.*, **42(9)**:670 (1966).
41. V. Kh. Katunin et al., USSR Patent 160,176 (1962).
42. S. Meinstein, U.S. Patent 3,338,924 (1967).
43. S. Meinstein and C. Fuchs, U.S. Patent 3,328,429 (1967).
44. British Patent 920,627 (1963).
45. C. F. Koelsch, *Org. Syn.*, **Coll. Vol. III**:791 (1965).
46. F. Beilstein's *Handbuch der organischen Chemie*, 4th Ed., **14**: p. 841; 1st Suppl. **9**: p. 373.
47. P. F. Lewis, "Chemical Economics Handbook," 695.4020 A, Stanford Research Institute, Menlo Park, Calif., April, 1966.
48. Y. Ichikawa, U.S. Patent 3,299,125 (1967).
49. W. F. Brill, *Ind. Eng. Chem.*, **52**:837 (1960).
50. W. O. Walker, U.S. Patent 1,976,757 (1934).
51. D. C. Hull, U.S. Patent 2,673,217 (1954).
52. B. Thompson and S. D. Neely, U.S. Patent 3,240,803 (1966).
53. M. Yamamoto et al., *J. Appl. Chem.*, **17**:293 (1967).
54. Toyo Rayon, French Patent 1,384,187 (1963).

55. H. S. Rudzki, Belgian Patent 648,155 (1964).
56. D. A. S. Ravens, *Trans. Faraday Soc.*, **55**:1768 (1959).
57. P. H. Towle and R. H. Baldwin, *Hydrocarbon Process.*, **43**:149 (1964).
58. R. H. Baldwin, U.S. Patent 3,170,768 (1965).
59. A. Saffer, U.S. Patent 3,089,906 (1963).
60. Y. Junzo et al., *J. Chem. Soc., Japan*, **67**:1153 (1964).
61. A. E. Ardis, F. L. Nasti, and A. A. Vaitekunas, U.S. Patent 3,036,122 (1962).
62. B. Raecke, *Angew. Chem.*, **70**:1 (1958).
63. N. P. C., *Chem. Eng.*, April 26, 1965, p. 71.
64. Y. Ogata and K. Sakamoto, *Chem. Ind. (London)*, **1964**:2012.
65. J. Ratusky, *Chem. Ind. (London)*, **1967**:1093.
66. R. F. Ruthruff, U.S. Patent 3,038,006 (1962).
67. Y. Ogata et al., *Bull. Chem. Soc., Jap.*, **37**:1648 (1964).
68. A. J. Shipman, *Adv. Chem.*, **51**:52 (1965).
69. W. G. Toland, *J. Amer. Chem. Soc.*, **82**:1911 (1960).
70. J. E. McIntyre and R. S. Ravens, *J. Chem. Soc.*, **1961**:4082.
71. J. P. Fortuin et al., Netherlands Patent 100,907 (1962).
72. British Patent 843,180 (1960).
73. W. A. O'Neill and J. S. M. Robertson, British Patent 798,342 (1958).
74. N. Ota, *Tokyo Kogyo Shikensho Hokoku*, **54**:337 (1959).
75. S. Tsutsumi, Japanese Patent 2974 (1959).
76. M. M. Movsumzade and A. P. Petrov, *Izv. Vyssh. Ucheb. Zaved., Neft i Gaz*, **10**:65 (1961).
77. V. Kudlacek and Z. Pokorny, *Sb. Ved. Praci Vys. Skola Chem. Technol. Pardubice*, **1961**:Pt. 1, 95.
78. J. P. Fortuin et al., Netherlands Patent 108,519 (1964).
79. I. V. Butina and V. G. Plyusnin, *Tr. Vses. Soveshch. po Khim. Pererabot. Neft. Uglevodorodov v Poluprod. Ilya Sint. Volokon i Plast. Mass, Baku*, **1957**:131.
80. I. V. Butina and V. G. Plyusnin, *Tr. Inst. Khim., Akad. Nauk SSSR, Ural. Filial*, **1960**(4):73.
81. G. Midorikawa and H. Mochida, Japanese Patent 9082 (1962).
82. A. E. Kretov et al., *Zh. Prikl. Khim.* **33**:2329 (1960).
83. W. Schemuth, British Patent 843,180 (1960).
84. A. I. Kamneva et al., *Neftekhimiya*, **2**:536 (1962).
85. A. J. Harding, German Patent 1,147,934 (1963).
86. Y. Fujita, Japanese Patent 23,495 (1965).
87. R. Motoyama et al., Japanese Patent 470 (1964).
88. Y. Kamija, *Bull. Chem. Soc., Jap.*, **33**:1656–1660 (1960).
89. W. A. O'Neill, British Patent 823,437 (1959).
90. J. J. Melchiore, H. R. Moyer, and L. J. Christman, *Adv. Chem.*, **51**:89 (1965).
91. Z. Eckstein and E. Grochowski, Polish Patent 48,260 (1962).
92. L. Friedman, D. L. Fisher and H. Shecter, *J. Org. Chem.*, **30**:1453 (1965).
93. E. Bartholeme, U.S. Patent 3,271,445 (1966).
94. A. Moor and E. P. Manasse, Netherlands Patent 101,362 (1962).
95. W. G. Toland, Jr., U.S. Patent 2,795,599 (1957).
96. M. Shigeyasu, *Kôgyô Kagaku Zasshi*, **67**:(9), 1396 (1964).
97. M. Shigeyasu and T. Ozaki, *Kôgyô Kagaku Zasshi*, **68**:304 (1965).
98. *Chem. Week*, March 9, 1963, p. 58.
99. Y. Fujita, A. Takeda, and T. Tonaka, Symposia Preprints, "Foreign Developments in Petrochemicals," Division of Petroleum Chemistry, Inc., ACS **13**(4):A85 (1968).

100. Teijin Ltd., Belgian Patent 660,335 (1965).
101. Teijin Ltd., Netherlands Patent 6,411,770 (1965).
102. Mitsui Petrochemical, French Patent 1,355,273 (1964).
103. D. G. R. Grundy, British Patent 1,047,433 (1966).
104. R. H. Baldwin and P. H. Towle, U.S. Patent 3,095,445 (1963).
105. Institut Français du Petrole, des Carburants et Lubrifiants, Netherlands Patent 6,505,950 (1965).
106. Standard Oil, British Patent 994,769 (1965).
107. D. H. Meyer, U.S. Patent 3,288,849 (1966).
108. H. Walz et al., German Patent 1,245,356 (1967).
109. E. L. Ringwald, German Patent 1,144,255 (1963).
110. P. Nesbitt and J. S. M. Robertson, British Patent 816,892 (1959).
111. H. Moell and H. Soenksen, German Patent 1,112,059 (1960).
112. H. Moell, A. Kreyer, and H. Soenksen, German Patent 1,126,855 (1962).
113. H. Moell, A. Kreyer, and H. Soenksen, German Patent 1,136,999 (1962).
114. B. Popp, O. Sherer, H. Wolfram, and A. Steinmetz, German Patent 1,047,192 (1958).
115. D. W. Williston, Belgian Patent 619,788 (1962).
116. O. Scherer and W. Wetzel, German Patent 1,175,661 (1964).
117. G. E. Ham, U.S. Patent 2,949,483 (1960).
118. R. H. Wise and D. H. Meyer, U.S. Patent 3,102,137 (1963).
119. C. W. Tate, U.S. Patent 3,047,621 (1962).
120. A. L. Hensley, Jr., and P. H. Towle, U.S. Patent 3,344,177 (1967).
121. P. H. Towle, R. H. Baldwin, and D. H. Meyer, in "Encyclopedia of Chemical Technology," 2nd ed. Vol. 15, R. E. Kirk and D. F. Othmer, Eds., Interscience, New York, 1968, p. 444.
122. W. G. Toland, Jr., et al., *J. Amer. Chem. Soc.*, **80**:5423 (1958).
123. C. Dean, "Chemical Economics Handbook," 667, 5020A (1966).
124. California Research Corp., British Patent 951,279 (1964).
125. P. R. Hines and R. P. Taylor, U.S. Patent 3,036,123 (1962).
126. Standard Oil Co., British Patent 970,781 (1964).
127. G. N. Friedlin and V. N. Davydov, *Zh. Prikl. Khim.*, **35**:2520–6 (1962).
128. J. P. O'Brien, U.S. Patent 2,897,232 (1959).
129. C. A. Spiller, Jr., and R. V. Malo, U.S. Patent 3,029,278 (1962).
130. W. H. Bowman, U.S. Patent 3,244,744 (1966).
131. S. J. Hetzel, U.S. Patent 3,043,869 (1962).
132. J. O. Knobloch and D. H. Meyer, U.S. Patent 3,059,025 (1962).
133. D. Smith, *Chem. Ind. (London)*, **1953**:244.
134. G. N. Friedlin and V. N. Davydov, *Zh. Prikl. Khim.*, **35**:1150–1153 (1962).
135. R. H. Baldwin and C. A. Spiller, U.S. Patent 3,082,250 (1963).
136. R. D. Gorsick, U.S. Patent 3,109,017 (1963).
137. H. J. Hagemeyer, U.S. Patent 2,921,089 (1960).
138. R. H. Mills and O. J. Weinkauff, U.S. Patent 3,148,200 (1964).
139. H. P. Crocker and R. W. Kay, British Patent 917,568 (1963).
140. S. Yoneda, Z. Yoshida, and K. Fukin, *Kôgyô Kagaku Zasshi*, **69**(4):641 (1966).
141. G. P. Toney and R. H. Goins, U.S. Patent 3,053,884 (1959).
142. H. A. Staab, German Patent 1,111,186 (1959).
143. V. L. Hughes et al., U.S. Patent 2,903,477 (1959).
144. Standard Oil, British Patent 879,799 (1961).
145. C. D. Kaldafelis and D. H. Meyer, U.S. Patent 3,022,333 (1962).
146. M & T Chemicals, British Patent, 1,011,660 (1965).

147. Imperial Chemicals, Belgian Patent 636,545 (1964).
148. E. Katzchmann, German Patent 1,091,556 (1960).
149. W. P. Barie, Jr., N. W. Franke, and A. C. Whitaker, U.S. Patent 3,332,983 (1967).
150. F. X. Werber, German Patent 1,083,265 (1956).
151. F. X. Werber and S. J. Averill, U.S. Patent 3,056,817 (1962).
152. Chemische Werke Huels A.-G., French Patent 1,367,278 (1964).
153. R. A. Novotny and A. Benning, German Patent 1,103,334 (1961).
154. J. Stresinka et al., Czechoslovak Patent 93,689 (1960).
155. J. Stresinka and M. Ciha, Czechoslovak Patent 104,664 (1962).
156. M. A. Baker, R. L. Friedman, and W. J. Raab, French Patent 1,365,733 (1964).
157. Beck, Koller and Co., British Patent 1,058,242 (1967).
158. J. R. Goggans and J. E. Copenhauer, J. Amer. Chem. Soc., 61:2909 (1939).
159. J. Kenyon in "Organic Synthesis," 2nd ed., Coll. Vol. I, H. Gilman, Ed., Wiley, New York, 1932, p. 419.
160. L. O. Raether and H. R. Gamrath, Adv. Chem., 48:66 (1965).
161. E. W. Eckey, in "Encyclopedia of Chemical Technology," Vol. 5, R. E. Kirk and D. F. Othmer, Eds., Interscience, New York, 1950, p. 797.
162. A. Colnen, Adv. Chem., 48:76 (1965).
163. A. A. Artemev et al., Khim Prom. 1960:627.
164. G. Hallar, J. Chem. Soc., 1965:5770.
165. Farbwerke Hoechst, A.-G., Belgian Patent 634,031 (1963).
166. E. Katzschmann, U.S. Patent 2,937,197 (1960).
167. Farbwerke Hoechst A.-G., Belgian Patent 633,878 (1963).
168. A. Girantet et al., French Patent 1,449,727 (1966).
169. K. Yoda et al., Kôgyô Kagaku Zasshi, 67(6):909 (1964).
170. G. Torraca and R. Turriziani, Chim. Ind. (Milan), 44(5):483 (1962).
171. Farbwerke Hoechst A.-G., Netherlands Patent 6,504,627 (1965).
172. Société Rhodiaceta, Netherlands Patent 6,513,513 (1966).
173. Metallgesellschaft A.-G., Netherlands Patent 6,513,493 (1966).
174. R. Inone et al., Japanese Patent 18,353 (1967).
175. E. Katzschmann, German Patent 1,117,560 (1967).
176. British Patent 903,099 (1962).
177. R. Vilencie, German Patent 1,168,888 (1964).
178. F. J. Sebelist and R. H. Weir, U.S. Patent 3,057,909 (1962).
179. O. York, Jr., U.S. Patent 2,906,737 (1959).
180. G. P. Roeser, French Addn. Patent 83,866 (1964).
181. Mitsui Petrochemical, Netherlands Patent 6,601,077 (1966).
182. P. Lafont and H. Menand, French Patent 1,408,874 (1965).
183. British Patent 851,029 (1960).
184. Japan Soda Co., Belgian Patent 660,257 (1965).
185. Olin Mathiesen, British Patent 915,891 (1963).
186. A. A. Vaitekunas, U.S. Patent 3,037,049 (1962).
187. F. Mares et al., German Patent 1,261,497 (1968).
188. H. W. Burns and R. D. Slockett, U.S. Patent 3,062,862 (1962).
189. R. C. Glogan and R. H. Weir, U.S. Patent 3,052,711 (1960).
190. A. A. Vaitekunas and H. C. Weinreb, U.S. Patent 3,101,366 (1963).
191. Socony Mobil Oil, French Patent 1,355,290 (1964).
192. Mitsui Petrochemical, Netherlands Patent 6,413,334 (1965).
193. K. E. Kolb, U.S. Patent 2,901,505 (1959).
194. Belgian Patent 616,238 (1962).

195. Mitsui Petrochemical, Netherlands Patent 6,506,220 (1965).
196. A. A. Vaitekunas, U.S. Patent 3,037,049 (1962).
197. B. Robinson, *J. Chem. Soc.*, **1963**:2417–2419.
198. F. C. Schaefer and G. A. Peters, *J. Org. Chem.*, **26**:412 (1961).
199. E. J. Gossen, British Patent 800,875 (1958).
200. Toyo Rayon, French Patent 1,365,841 (1964).
201. T. Tsutsumi, U.S. Patent 3,378,577 (1968).
202. Badische Anilin- &-Sodafabrik, British Patent 878,269 (1959).
203. M. Fenske, U.S. Patent 3,036,111 (1962).
204. Toyo Rayon, French Patent 1,401,427 (1964).
205. Imperial Chemicals, Netherlands Patent 6,602,422 (1966).
206. Esso, British Patent 844,033 (1960).
207. K. Scherf, German Patent 1,119,250 (1959).
208. D. H. Meyer and M. M. Garvey, U.S. Patent 3,076,018 (1963).
209. N. C. Rose, *J. Chem. Educ.*, **44**:283 (1967).
210. W. R. Sorenson and T. W. Campbell, "Preparative Methods of Polymer Chemistry," Interscience, New York, 1961.
211. L. P. Kyrides, *J. Amer. Chem. Soc.*, **59**:207 (1937).
212. Farbenfabriken Bayer A.-G., British Patent 942,621 (1963).
213. R. Mory, E. Stocklin, and M. Schmid, German Patent 1,026,750 (1958).
214. A. Hlynsky, British Patent 949,574 (1964).
215. E. Zinn et al., British Patent 946,491 (1964).
216. F. J. Christopher et al., U.S. Patent 3,318,950 (1967).
217. T. K. Brotherton, U.S. Patent 3,337,622 (1967).
218. R. W. Etherington, Jr., and W. F. Brill, U.S. Patent 3,274,242 (1966).
219. Chemische Werke Witten GmbH., Belgian Patent 630,675 (1963).
220. B. F. Malichenko, *Zh. Prikl. Khim.*, **40(6)**:1385 (1967).
221. E. Katzschmann, German Patent 1,130,432 (1962).
222. I. Reichel and E. Schonberger, *Acad. Rep. Pop. Rom.*, *Baza Cercet. Stünt. Timisoara, Studii Cercet. Stünt. Chim.* **6**:81 (1959).
223. E. Ott, in "Organic Synthesis," Coll. Vol. II, A. H. Blatt, Ed., Wiley, New York, 1943, p. 528.
224. T. S. Work, *J. Chem. Soc.*, **1940**:1315.
225. C. W. Stephens, *J. Polym. Sci.*, **40**:359 (1959).
226. A. V. Kirsanov and N. H. Kirsanova, *J. Gen. Chem. USSR*, **29(6)**:1774 (1959).
227. A. P. Shestov and N. A. Asipova, *J. Gen. Chem. USSR*, **29**:59 (1959).
228. J. Feldman, *Helv. Chim. Acta*, **14**:751 (1931).
229. R. E. Dean, E. N. White, and D. McNeil, *J. Appl. Chem.*, **3**:473 (1953).
230. W. F. O'Connor and E. J. Mariconi, *J. Amer. Chem. Soc.*, **73**:4044 (1951).
231. S. Kato and H. Shioda, *Yûki Gôsei Kagaku Kyokai Shi*, **15**:637 (1957).
232. L. D. Gluzman et al., *Sb. Nauchn. Tr. Ukr. Nauchn.-Issled. Uglekhim. Inst.*, No. 13, 144 (1962).
233. Rütgerswerke A.-G., British Patent 853,369 (1960).
234. N. D. Rus'yanova, Z. K. Gordeeva, and G. F. Belyaeva, *Plast. Massy*, **1960**:No. 5, 43–46.
235. P. G. Copeland, R. E. Dean, and D. McNeil, *J. Chem. Soc.*, **1961**:1232.
236. R. H. Callighan and J. O. Hawthorne, U.S. Patent 3,118,934 (1964).
237. R. H. Callighan, U.S. Patent 3,007,963.
238. N. P. Greco, U.S. Patent 3,291,825 (1966).
239. P. G. Copeland, British Patent 1,007,012 (1965).

240. A. K. Lebedev and V. P. Shabrov, USSR Patent 189,829 (1966).
241. N. D. Rusyanova and V. G. Koksharov, *Katalit. Reakts. v Zhidk. Faze, Akad. Nauk Koz. SSR, Kazakhsk. Gos. Univ., Kazakhsk. Resp. Pravl. Mendeleevskogo Obshch., Tr. Vses. Kof., Alma-Ata* **1962**:433–437.
242. L. P. Kulev et al., *Izv. Tomsk. Politekhn., Inst.*, **111**:26 (1961).
243. L. P. Kulev, R. N. Gireva, and G. M. Stepnova, *Zh. Obshch. Khim.*, **32**:2812 (1962).
244. V. Kabaivanov, N. Natov, and M. Georgieva, *God. Khim. Tekhnol. Inst.*, **5**:147 (1958).
245. H. Shioda and S. Kato, *Yuki Gosei Kagaku Kyokai Shi*, **18**:263 (1960).
246. H. Hauptman, W. Walter, and C. Marino, *J. Amer. Chem. Soc.*, **80**:5832 (1958).
247. D. O. DePree, U.S. Patent 3,108,135 (1963).
248. M. M. Dashevskii and E. M. Shamis, *Ukr. Khim. Zh.*, **30**(9):938 (1964).
249. G. I. Golivetz, *Nauk Zap. Odes'k. Politekhn. Inst.*, **50**:88 (1963).
250. R. J. Slocombe and E. E. Hardy, *J. Amer. Chem. Soc.*, **72**:3312 (1950).
251. J. R. Caldwell, U.S. Patent 2,673,218 (1954).
252. J. R. Caldwell, U.S. Patent 2,614,120 (1952).
253. M. Adamek and J. Novak, *Sb. ved Proci, Vys. Skola Chem—Technol. Pardubice*, **1960**(2):87; *Chem. Abstr.*, **55**:22205i.
254. C. H. Buehler and J. E. Masters, *J. Org. Chem.*, **4**:262 (1939).
255. H. H. Szmant and F. P. Palopali, *J. Amer. Chem. Soc.*, **72**:1757 (1950).
256. G. Suld, *Chem. Eng. News*, **40**(39):68 (1962).
257. V. G. Obing, *Erdol Kohle Erdgas Petrochem.*, **21**(2):81 (1968).
258. J. Straley and C. Wayman, U.S. Patent 2,578,759 (1951).
259. B. Raecke and H. Schirp, in "Organic Synthesis," Vol. 40, M. S. Newman, Ed., Wiley, New York, 1960, p. 71.
260. R. S. Baker and A. Saffer, U.S. Patents 2,963,508; 2,963,509 (1958).
261. A. Saffer and R. S. Baker, U.S. Patent 2,833,816 (1955).
262. L. Friedman, in "Organic Synthesis," Vol. 43, B. C. McKusick, Ed., Wiley, New York, p. 80, 1960.
263. H. Hopff and A. Krieger, *Makromol Chem.*, **47**:93 (1961).
264. M. Hasegarva and F. Suzuki, *Kôgyô Kogaku Zasshi*, **66**:1230 (1963).
265. J. Oyoma, *Hakko Kyokaishi*, **19**:340 (1961); Japanese Patent 14,394 (1963).
266. P. H. Hodson and W. H. Darlington, *J. Bacteriol.*, **88**(3):803 (1964).
267. A. E. Humphrey, A. Kitai, and C. L. Cooney, *Hakko Kogaku Zasshi*, **46**(6):283 (1966).
268. G. Blacke, E. Depp, and B. B. Corson, *J. Org. Chem.*, **14**:14 (1949).
269. M. Henze, *Ber.*, **67**:750 (1934).
270. R. Graf and F. Zettl, *J. Prakt. Chem.*, **147**:188 (1936).
271. W. Mathes, W. Sanermilch, and T. Klein, *Chem. Ber.*, **86**:584 (1953).
272. J. Klosa, *Arch. Pharm.*, **288**:426 (1955).
273. S. A. Sundet, W. A. Murphy, and S. B. Speck, *J. Polym. Sci.*, **40**:389 (1959).
274. E. E. Gilbert, "Sulfonation and Related Reactions," Interscience, New York, 1965, p. 69.
275. A. F. Holleman and J. J. Polak, *Rec. Trav. Chim., Pays-Bas*, **29**:416 (1910).
276. L. Barth and C. Senhofer, *Ber.*, **8**:1477 (1875).
277. J. Rahm and F. Juracka, *Chem. Listy*, **50**:837 (1956).
278. C. M. Suter in "Organic Reactions," Vol. III, Roger Adams, Ed., Wiley, London, 1946, p. 155.
279. H. Engelhardt and P. Latschinov, *Z. Chem.*, **1871**:259.
280. W. M. Stanley and R. Adams, *J. Amer. Chem. Soc.*, **52**:4474 (1930).
281. E. Atkinson et al., *J. Amer. Chem. Soc.*, **67**:1513 (1945).

282. S. Veibel, "Carboxyl and Derived Functions," in "Treatise on Analytical Chemistry," Part II, Vol. 13, I. M. Kolthoff and P. J. Elving, Eds., Interscience, New York-London, 1966, p. 223.

283. A. P. Kreshkov, L. N. Bykova, and N. T. Smolova, *J. Anal. Chem.*, *USSR*, **19**:144 (1964).

284. V. K. Kondratov et al., *J. Anal. Chem.*, *USSR*, **20**:1297 (1965).

285. A. P. Kreshkov, *Wiss. Z. Tech. Hochsch. Chem. Leuna-Merseburg* **6**(314):255 (1964).

286. N. Van Meurs and E. A. M. F. Dahmen, *Anal. Chim. Acta*, **19**:64 (1958).

287. J. Ratusky, *Chem. Ind. (London)*, **1962**:1093.

288. I. V. Butina, V. G. Plusnin, and N. A. Shevchenko, *J. Anal. Chem.*, *USSR*, **18**:1203 (1963).

289. H. Trackman and R. Zucher, *Anal. Chem.*, **36**:269 (1964).

290. J. Magder and W. F. Brill, unpublished results.

291. J. F. Klebe, *J. Amer. Chem. Soc.*, **88**:3390 (1966).

292. J. W. Frankenfeld, *J. Chromatogr.*, **18**(1):179 (1965).

293. J. Kolesinska, T. Urbanski, and A. Wielopolski, *Chem. Anal.*, **10**(6):1107 (1965).

294. J. Kolesinska, T. Urbanski, and A. Wielopolski, *Chem. Anal.*, **11**(3):473 (1966).

295. J. Kulicka et al, *Chem. Anal.*, **12**(1):171 (1967).

296. A. I. Kamneva and E. S. Panfilova, *Zavodsk. Lab.*, **29**(6):666 (1963).

297. D. Salbut et al., *Chem. Anal.*, **10**(6):1099 (1965).

298. N. C. Saha, G. D. Mitra, and A. N. Basu, *Indian J. Technol.*, **2**(11):385 (1964).

299. D. Salbut et al., *Chem. Anal.*, **11**(4):673 (1966).

300. B. Calmanovici, *Rev. Chim.*, **17**(6):374 (1966).

301. C. Davies, R. D. Hartley, and G. Lawson, *J. Chromatogr.*, **18**:47 (1965).

302. B. Fishiwick, *J. Chem. Soc.*, **1957**:1196.

303. Toyo Rayon, British Patent 1,000,045 (1965).

304. F. Gonzalez-Sanchez, *Tetrahedron*, **1**:231 (1957).

305. A. Bylicki, *Bull. Acad. Polon. Sci.*, **7**:111 (1959).

306. L. Hunter, *Chem. Ind. (London)*, **1953**:155.

307. F. Gonzales-Sanchez, *Spectrochim. Acta*, **12**:17 (1958).

308. R. Katrilzky, J. M. Lagowski, and J. A. Beard, *Spectrochim. Acta*, **16**:964 (1960).

309. B. Williamson and W. H. Rodebush, *J. Amer. Chem. Soc.*, **63**:3018 (1941).

310. J. H. Holsinger, L. C. Blankenship, and M. J. Pollansch, *Arch. Biochem. Biol. Phys.*, **1119**:282 (1967).

311. J. A. Krenz, R. J. Angelo, and W. E. Barth, *J. Polym. Sci.*, **5**:2961 (1967).

312. S. Ono and J. Nokaya, *J. Chem. Soc.*, *Jap.*, **74**:907–911 (1953).

313. J. Pollock and R. Stevens, Eds., "Dictionary of Organic Compounds," 4th Ed., Oxford University Press, New York, 1965.

314. V. Mazzucato and A. Foffani, *Ric. Sci.*, **26**:2409 (1956).

315. R. Hill, R. Sublett, and H. Oshburn, *J. Chem. Eng. Data*, **8**:233 (1963).

316. J. D. Loudon, *J. Chem. Soc.*, **1936**:221.

317. H. H. Martin and J. W. Foster, *Arch. Mikrobiol.*, **31**:171 (1958).

7. AROMATIC DIAMINES

J. PRESTON, H. C. BACH, AND J. B. CLEMENTS, *Chemstrand Research Center, Inc., Durham, North Carolina*

Contents

I. INTRODUCTION

Aromatic diamines are a very distinct class of organic compounds, their history dating back to the middle of the nineteenth century.* These widely used and numerous compounds are produced and employed commercially on a very large scale, particularly in the manufacture of azodyes, in the fabrication of leather, and in the synthesis of commercially important intermediates, for example, diisocyanates. Table 1 lists annual production figures, trade names, and prices of certain of the more important aromatic diamines.

In recent years the use of aromatic diamines as building blocks of linear aromatic polymers such as polyamides and polyimides has shown a great increase. This growing interest in aromatic diamines as monomers for linear polymers is due to the very desirable combination of properties possessed by polymers composed solely of aromatic units. Some of these properties are thermal stability in inert as well as oxidative atmospheres far greater than can be realized in aliphatic polymers, high melting or softening points, and high tensile moduli due to the inherently greater stiffness of aromatic polymer chains (when compared with the rather flexible aliphatic polymer chains).

An almost endless variety of aromatic diamines is known, but very few have yet attained importance as monomers for commercial, or potentially commercial aromatic polymers, and only these are treated extensively in this chapter. They are: *o*-, *m*, *p*-phenylene diamines (1), †benzidine (2), 4,4′-methylenedianiline (3), bis(4-aminophenyl)ether, and some of the so-called ordered aromatic diamines. Diamines that are not used directly as monomers

---

　　* Of the more than 500 references considered during the preparation of this chapter, only some 100 plus were found to be suitable for inclusion in a work that would be consistent with the intent of this volume.

　　† References 1–3 are to some reviews on the technology of the various diamines.

TABLE 1

Production Figures and Prices of Aromatic Diamines

Diamine	Trade names	Annual production (or sales), 1000 lb	Price, $/lb
o-Phenylenediamine	CI oxidation base 16, Orthamine	120[a]	1.44[a] (1.35)[b]
m-Phenylenediamine		836[a]	(1.08)[b]
p-Phenylenediamine	P.P.D.⁻, Ursol®D	446[a]	1.50[a]
Benzidine (hydrochloride and sulfate)		1610[c]	1.02[c]
4,4'-Methylenedianiline	Tonox	1086[a]	(0.56)[b]
Bis(4-aminophenyl)ether	Oxydianiline	data not available	(4.50)[b,d]

[a] U.S. Tariff Commission Reports, Synthetic Organic Chemicals, TC Publication 206, 1965.
[b] Prices quoted by suppliers.
[c] U.S. Tariff Commission Reports, Synthetic Organic Chemicals, TC Publication 34, 1960.
[d] This diamine was once offered by Ott Chemical Co. for about $3.25/lb.

for polymers but serve only as intermediates for monomers (e.g., diiso-cyanates) have been omitted. Selected physical data for the phenylenediamines and for benzidine and the bridged biphenylenediamines are summarized in Tables 2 and 3 respectively.

Some of these diamines have been known for a long time and are used widely; however, in certain cases, such as bis(4-aminophenyl)ether, very little information is available in the literature on processes used in their manufacture. This is probably because of their mostly captive use by in-dustrial companies.

II. GENERAL CHEMISTRY AND SYNTHESIS OF AROMATIC DIAMINES

The chemistry of the aromatic diamines covered in this chapter is for the most part straightforward along the lines of the chemistry of aromatic amines in general. The most important reaction of these diamines is their reaction (4,5) with nitrous acid. The tetrazonium salts formed couple easily with phenols and amines to yield a wide variety of industrially important azodyes. There is a great deal of published information on diazotization products of other diamines, and it is easily found in the chemical and dyestuff literature. For example, the azodyes derived from methylene dianiline are discussed in detail by Kouris (3). In the case of o-phenylene diamine, diazotization is not applicable to the synthesis of azodyes because of the sole formation of benzotriazole upon treatment of o-phenylene diamine with nitrous acid. This reaction can be used as a preparative method for the following heterocyclic compound (6):

$$\text{(1)}$$

Another important class of reactions of the diamines discussed in this chapter is comprised of condensation reactions with diacid chlorides, dianhydrides, and diisocyanates. These reactions, leading to the formation of linear polymers, are discussed in greater detail at the end of this chapter.

Other typical reactions of the diamines described are: Schiff base formation with benzaldehyde; reaction with phosgene to yield industrially important diisocyanates (7); alkylation; acylation with formic acid, benzoyl chloride, acetic acid, or acetic anhydride; and reactions involving the aromatic ring system, such as halogenation, nitration, and sulfonation.

o-Phenylenediamine differs in its chemistry from the other diamines discussed because of the proximity of the two amino groups. In most cases,

TABLE 2

Some Physical Properties of Phenylenediamines

Properties	Phenylenediamine		
	ortho	*meta*	*para*
mp, °C	99–101 (103–104)	62.8	139.7 (139–141)
bp, °C	256–258 dec/760 mm 142–143/28 mm	285.5/760 mm 147.0/10 mm 131.2/5 mm	267/760 mm
pK_a, 21°C (H$_2$O)	4.47	4.48	6.08
Solubility[a]			
Water	4.2 (35°C) 733 (81°C)	35.1 (25°C)	3.8 (24°C)
Alcohol	v.s.	v.s.	s.
Diethyl ether	v.s.	v.s.	s.
Dipole moment in benzene at 25°C	1.44D	1.79D	1.56D
Molecular diamagnetic susceptibility, $-\chi_M \times 10^6$	71.98	70.53	70.28
Ionization potential, eV	7.45	—	7.15 (1st) 8.0 (2nd)
Heat of combustion, kcal/mole	—	—	843

[a] v.s. = very soluble; s. = soluble.

553

TABLE 3

Some Physical Properties of Benzidine and Bridged Biphenylenediamines

	mp, °C	bp, °C	pK_a at 21°C, H_2O	Acetone	Benzene	Diethyl ether	Alcohol	Water	Dipole moment	Molecular diamagnetic susceptibility, $-\chi_M \times 10^6$
4,4'-Benzidine	127.5–128.7	401.7/760 mm 400/740 mm				2.2 (20°C)	8 (20°C)	0.1 (12°C) 1.0 (100°C)		117.8
4,4'-Methylene-dianiline	92–93 (93–94)	263/25 mm 249/15 mm 221/5 mm	4.81	273.0	9.0	9.5	143.0 (methanol)	0.1–0.14	1.94D	
Bis(4-amino-phenyl)ether	193–195	subl.		15	insoluble			insoluble		

instead of the formation of linear derivatives, ring closure occurs to heterocyclic compounds such as benzotriazole (6), benzimidazoles, quinoxalines (8–10), 2-mercaptobenzimidazoles (11), or benzodiazepines (12).

(2)

The syntheses of aromatic diamines to be detailed later usually involve the reduction of a nitro group at some point in their preparation. Catalytic hydrogenation of nitro groups is currently in favor and is slowly replacing the older iron-aqueous acid Bechamp procedures in industrial processes. For o- and p-phenylenediamine, ammonolysis (13,14) of the corresponding dihalobenzenes has also been reported.

III. SPECIFIC AROMATIC DIAMINES

A. o-Phenylenediamine

1. SYNTHESIS

o-Phenylene diamine, a colorless, crystalline solid that rapidly becomes colored in air, has been commercially produced in a way similar to the laboratory procedure (15) by reduction of o-nitroaniline with zinc dust and alkali, or by sodium sulfide (17,18) and by treating o-dichlorobenzene or o-chloraniline with aqueous ammonia at 150°C under pressure (19). Martin (15) provides detailed directions for the zinc and sodium hydroxide reduction of o-nitroaniline to give 75–85% of purified o-phenylenediamine. This convenient laboratory procedure has been checked by Hartman and Fierke (15).*

* Another small-scale laboratory procedure of merit is found in Ref. 16, p. 640.

In a 1-liter, three-necked, round-bottom flask, fitted with a liquid-sealed mechanical stirrer and a reflux condenser, are placed 69 g (0.5 mole) of *o*-nitroaniline (*Org. Syn.* **coll. vol. I**, 1941, 388), 40 cc of a 20% solution of sodium hydroxide, and 200 cc of 95% ethanol. The mixture is stirred vigorously and heated on a steam bath until the solution boils gently. The steam is turned off, and 10-g portions of 130 g (2 g-atoms) of zinc dust [The zinc dust should be at least 80% pure, and the amount used should be equivalent to 130 g of 100% material. A large excess of zinc dust has been used without changing the yield]* are added frequently enough to keep the solution boiling. [Great care must be taken not to add too much zinc dust at first, since the reaction becomes very vigorous. It is well to have a bath of ice and wet towels at hand in order to control the reaction if it should become too violent. Occasionally the reaction suddenly stops and it is necessary to add an additional 10 cc of 20% sodium hydroxide solution, which causes the reaction to proceed.] After the addition of zinc dust has been completed the mixture is refluxed with continued stirring for 1 hr, the color of the solution changes from a deep red to nearly colorless. The hot mixture is filtered by suction, and the zinc residue is returned to the flask and extracted with two 150-cc portions of hot alcohol. To the combined filtrates is added 2–3 g of sodium hydrosulfite, and the solution is concentrated under reduced pressure (using a water pump), on a steam bath, to a volume of 125–150 cc. After thoroughly cooling in an ice-salt bath, the faintly yellow crystals are collected, washed once with a small amount of ice water, and dried in a vacuum desic-cator. The yield of crude *o*-phenylenediamine melting at 97–100°C is 46–50 g (85–93% of the theoretical amount). If a purer product is desired, the material is dissolved in 150–175 cc of hot water containing 1–2 g of sodium hydrosulfite and treated with decolorizing charcoal. After thoroughly cooling in an ice-salt mixture, the colorless crystals are filtered by suction and washed with 1–15 cc of ice water. The purified *o*-phenylenediamine weighs 40–46 g (74–85% of the theoretical amount) and melts at 99–100°C. [The product can also be purified by distillation under reduced pressure in an inert atmosphere; but unless the material is very nearly pure, considerable decomposition occurs and the distilled product darkens rapidly in contact with air.]

The dihydrochloride of the free diamine may be obtained and purified as follows (15):

The crude *o*-phenylenediamine is dissolved in a mixture of 90–100 cc of concentrated hydrochloric acid (sp gr 1.19) and 50–60 cc of water containing 2–3 g of stannous chloride, and the hot solution is treated with decolorizing charcoal. To the hot colorless filtrate is added 150 cc

---

* In this and succeeding Preparation sections, the material in brackets has been added by the chapter authors to replace NOTES in the text of the original quotation.

of concentrated hydrochloric acid, and the mixture is cooled thoroughly in an ice-salt bath. The colorless crystals are filtered by suction, washed with a small amount of cold concentrated hydrochloric acid, and dried in a vacuum oven over solid sodium hydroxide. The yield of o-phenylenediamine dihydrochloride is 77–81 g (85–90% of the theoretical amount based on the weight of o-nitroaniline used).

A patent assigned to Universal Oil Products (20) describes the catalytic reduction of o-nitroaniline using palladium on charcoal at 70–90°C under 60-psig pressure to give o-phenylenediamine in approximately the same amounts.

One truly large-scale preparation has been made public (21).

4700 kg of a 24% sodium sulfide solution is diluted in an iron vessel by addition of 1930 liters of water to give a 17% solution. Then at temperatures below 50°C, 581 kg of o-nitroaniline is slowly added as a wet paste. In the closed reactor the reaction mixture is heated in 3 hr to a temperature of 105°C and is stirred at this temperature for 5 hr. After cooling to 10°C and stirring for 12 hr, the crystalline o-phenylenediamine is filtered off. Yield is 83.5% of theory.

2. ANALYSIS

Gas chromatographic, potentiometric, and polarographic methods have been used to analyze o-phenylenediamine. Bryan (22) reported a rapid gas chromatographic method for determining all the phenylenediamines, and deviations less than 1% are claimed. Kreshkov (23,24) developed potentiometric methods for the quantitative determination of diamines in nonaqueous solvents and claims that o-phenylenediamine can be determined by this method. Mark (25) used a polarographic method for determining trace amounts of o-phenylenediamine in the presence of up to a tenfold excess of either or both the *meta* and *para*-isomers. He also suggested that the method might be useful for quantitatively determining o-phenylenediamine in the presence of the other isomers.

3. TOXICOLOGY

o-Phenylenediamine is a toxic substance but somewhat less so than p-phenylenediamine. It is also a skin irritant. The minimum lethal dose (MLD) subcutaneous in rats for o-phenylenediamine is 600 mg/kg.

In handling o-phenylenediamine, care should be exercised to avoid bodily contact. Special care should be taken to avoid breathing the dust or vapor of o-phenylenediamine.

B. *m*-Phenylenediamine

1. SYNTHESIS

m-Phenylenediamine, a colorless crystalline solid that quickly becomes colored in air, has been prepared commercially by the reduction of *m*-dinitrobenzene with iron and hydrochloric acid (13,26), but current methods favor catalytic continuous liquid-phase hydrogenation of *m*-dinitrobenzene at moderate temperature (27,28). In these catalytic processes for reducing polynitroaromatic (29) compounds it is important that only small amounts of unreduced polynitroaromatic compounds be present in the reactor at any time and that vigorous agitation be used. These are safety measures, and they serve to prevent explosions. The favored catalyst is palladium on carbon and, because of its pyrophoric nature, it is best stored and used as an aqueous slurry. Raney nickel has also been mentioned (30) as a catalyst for this hydrogenation. One specific medium-scale preparation using iron and hydrochloric acid as a reducing agent is given as an illustration of the Bechamp procedure (21).*

> 1.5 liters of water, 400 g of fine iron filings, and 20 ml of concentrated hydrochloric acid are stirred together in a flask provided with an efficient stirrer. The mixture is heated to boiling for at least 5 min. During 40 min 168 g of *m*-dinitrobenzene is added in portions of 2 g each to the boiling mixture in such a manner that each portion has been completely reduced before a new addition is made.
>
> Upon complete reduction, a test sample will give a colorless filtrate.
>
> After addition of the *m*-dinitrobenzene, the reaction mixture is boiled for a few minutes, with the volume being held constant by addition of water. In order to remove any soluble iron salts, the reaction mixture is made alkaline (litmus) by addition of about 10 g of calcined sodium carbonate, boiled for a few minutes and filtered. The pure *m*-phenylenediamine is isolated either by vacuum distillation or by crystallization at 0°C.

Numerous other methods have been used to obtain *m*-phenylenediamine, but none has the commercial significance of the continuous liquid-phase catalytic hydrogenation process. For example, *m*-dinitrobenzene has been reduced electrolytically, and substituted dinitrobenzenes have served as starting materials for *m*-phenylenediamine. Thus 4-chloro-1,3-dinitrobenzene, 4-bromo-1,3-dinitrobenzene, 2,4,6-tribromo-1,3-dinitrobenzene, 2-4-dinitrobenzoic acid, 2,6-dinitrobenzoic acid, and 2,4-dinitrobenzenesulfonic acid have all been reduced to *m*-phenylenediamine by various metal and acid combinations. Resorcinol (1,3-dihydroxybenzene) has also been converted to

---

* A small-scale laboratory procedure using tin for reduction of *m*-dinitrobenzene is found in Vogel (16), p. 641.

m-phenylenediamine by heating with ammonium sulfate and aqueous ammonia at 125–150°C or with calcium chloride and ammonia at 280–300°C or in the vapor phase at 420°C with ammonia over aluminum oxide. Good yields of *m*-phenylenediamine have been obtained by reducing *m*-dinitrobenzene with hydrazine using palladium on charcoal as a catalyst (30,31). Partially reduced aromatic compounds can also serve as a source of hydrogen, and cyclohexene has been successfully used to reduce *m*-dinitrobenzene to *m*-phenylenediamine (32). Sodium borohydride in conjunction with palladium on carbon has also been used as a source of hydrogen to reduce *m*-dinitrobenzene to *m*-phenylenediamine (33).

2. PURIFICATION

The treatment of commercial *m*-phenylenediamine in aqueous solution with a heavy metal salt such as nickel chloride or with an alkali metal chromate such as sodium dichromate at 80°C, followed by removal of the water and vacuum distillation of the residue, is claimed to give pure *m*-phenylenediamine stable to light, heat, and air (34). Another purification procedure recommends that molten technical *m*-phenylenediamine be passed at the rate of about 100 g/hr through a 2-in. silica gel bed at 120°C to give a product of considerably increased storage properties and light stability (35). More commonly, the diamine is purified by vacuum distillation from zinc dust.

3. ANALYSIS

Gas chromatographic and potentiometric methods of analysis have been used with *m*-phenylenediamine much in the same fashion as with *o*-phenylenediamine. Thus the gas chromatographic method of Bryan (22) and the potentiometric methods of Kreshkov (23,24) will serve to analyze *m*-phenylenediamine. The polarographic method of Mark (25) for *o*-phenylenediamine apparently is unsuitable with *m*-phenylenediamine. A titration method using standardized benzenediazonium chloride for making a diamine assay of *m*-phenylenediamine is available from Allied Chemical Corporation (36).

4. TOXICOLOGY

m-Phenylenediamine is reported to be highly toxic, producing serious liver damage if ingested. The oral MLD for rabbit and cat is reported to be 300 mg/kg. Unlike the *para*-isomer, *m*-phenylenediamine does not seem to be a skin sensitizer or an asthmatic allergen. Great care should be taken to avoid any contact with the dust or vapors of this diamine.

C. *p*-Phenylenediamine

1. SYNTHESIS

p-Phenylenediamine is also a colorless crystalline solid that darkens in air. It is manufactured commercially by reduction of *p*-nitroaniline (obtained from *p*-chloronitrobenzene by ammonolysis) with iron and aqueous acid. One large-scale process using iron with hydrochloric acid in large wooden vats has been described (13); another features sulfuric acid, superheated steam, and a special flow apparatus, yielding a product of high purity (37). Catalytic reduction with hydrogen has not been described, but presumably the process given for polynitroaromatic compounds (29) could be used for *p*-phenylenediamine.

A small-scale preparation of *p*-phenylenediamine from *p*-nitroaniline can be carried out as follows (13):

> *p*-Nitroaniline, 300 parts, is stirred with 100 parts of 20% ammonia, 10 parts of sulfur, and 10 parts of reduced iron for 5 hr at 180°C under a pressure of 120 atm of hydrogen. The hydrogen consumed is replaced by degrees to maintain the pressure. The product is *p*-phenylenediamine, with traces of unconverted *p*-nitroaniline.

As an example of a large scale preparation (13) of *p*-phenylenediamine, the following procedure is given:

> I. *Reduction.* A large wood vat is used for this operation. The batch is stirred by means of a steel shaft which carries two or more sets of plows. A heavy-duty motor is provided to take care of the heavy initial reduction load and also the starting up of tubs that have been left quiescent overnight. Two tubs are required for plant operations. One is used for preparing a new batch while the other is delivering the reduced charge to the filter.
>
> The operation is started by pumping up sufficient wash water from previous charges to cover the paddles of the agitator. The stirrers, which operate at 40 rpm, are then started, and 1000 lb of iron borings is slowly put into the tub. Then 100 lb of 20° Bé hydrochloric acid is added. The mixture is agitated and heated until the iron is etched and a good paste of ferrous chloride is made, which should react immediately and distinctly when spotted with a weak sodium sulfide solution. Either dry or moist *p*-nitroaniline is then introduced, about 50 lb at a time. After each addition, sufficient time must elapse to ensure the presence of ferrous chloride, for it will be noticed that the spot test for soluble iron fails immediately after the addition of nitro compound. The charge must be kept sufficiently warm that a foaming reaction prevails. It

has been found that a ratio of 1.20 lb of iron to 1 lb of p-nitroaniline (molar ratio 3:1) gives the best results, although it is possible (as in the production of aniline) to lower this ratio slightly. Toward the close of the feeding operation, the balance of the iron necessary for reduction is introduced. Thus when 750 lb out of a total of 1200 lb of nitro compound has been fed into the vat, the balance of approximately 500 lb of fine iron is added. It takes about 12 hr to feed the nitro compound into the reducer.

Toward the end of the run, the reaction slows down, and it is necessary to introduce steam to carry on and complete the reduction. The test for soluble iron also becomes less distinct. As long as there is any p-nitroaniline present, a yellow spot test will be obtained on filter paper. p-Phenylenediamine yields a purple spot with a perfectly clear ring around the sludge spot. It is always advisable to test for soluble iron with sodium sulfide to ensure completeness of reduction. If the reduction is not carried on at the boiling temperature, intermediate azo and hydrazo products are sure to be formed. These are less easily reduced and cause a lowering of the yield.

II. *Filtration of Reduction Batch.* In order to prevent any oxidation of the p-phenylenediamine liquor before filtration, it is advisable not to neutralize the charge until this phase of the process begins. An excess of soda ash is used, and a test for alkalinity with phenolphthalein is required. Then 2 lb of sodium bisulfite and 3 lb of sodium sulfide (30 % crystals) are added to precipitate soluble iron salts and to ensure against subsequent oxidation.

A large plate-and-frame filter is customarily used to separate the residual iron sludge from the p-phenylenediamine liquor. The press is fitted with a pump for delivering the charge from the tubs, air and water lines, and an ejector for delivering hot water during the washing period. The press is first warmed by passing in live steam, and the delivery pump is then started. The filtrate is tested for completeness of reduction and for clarity. A spot on filter paper should be very light purple with no traces of yellow. The presence of a blue tint indicates the formation of indulines.

As soon as the batch is on the filter, it is washed with wash water from previous batches. This liquid follows the rest of the batch into one of two storage tanks, which are placed below the press level so that the filtrate flows into them by gravity. Hot water is then introduced into the press, and this filtrate runs into a large tank. Sufficient hot water is used to maintain a water balance; that is, to have enough hot water to provide a heel for the reducer tub and to replace the original mother liquor on the filter. The filter is then blown with air until the cake is dry. The cake is dropped into a pan underneath the press and removed. It is analyzed regularly for amino content.

III. *Dehydration of Diamine Liquor.* The filtrate from the press is a dilute solution of p-phenylenediamine containing a small amount of iron oxide in suspension. Since only 936 lb of *p*-phenylenediamine can theoretically be obtained from 1200 lb of *p*-nitroaniline, and since 9000–10,000 lb of water is used during the reduction, it is manifest that a 10% solution of phenylenediamine is delivered to the liquor-storage tank. Although it is feasible and practicable to deliver a solution of such strength directly to the dryer, it is more economical from the standpoints of both yield and steam consumption first to concentrate this liquor. Two-stage vertical evaporators and film evaporators have been used for effecting the final dehydration. On top of the dryer is an upright steel shell 2 ft in diameter and 4 ft high. This is packed with suitable material to prevent any entrainment, and it leads to a tubular condenser. A sight box at the base of the condenser permits the operator to inspect the condensate, which should not contain more than a trace of color—otherwise a loss of *p*-phenylenediamine is indicated. When the sight box indicates that most of the water has been distilled off, the jacket steam pressure is reduced from 15 to 5 lb and after 1 hr is turned off completely. If the batch in the dryer is now further agitated for 1 hr, it will be ready for dumping; but a sample must always be taken first to ensure absolute dryness. The crude product must be black with a purplish tinge; a gray tinge indicates moisture. The presence of 1% of water in the crude product is detrimental to the production of a good *p*-phenylenediamine.

The material in the dryer is discharged into large steel cans and weighed. From this weight, the yield of crude *p*-phenylenediamine is obtained, and the charges for the vacuum still are made up. It is advisable to analyze the crude *p*-phenylenediamine for salt and iron at regular intervals, as these figures give valuable information regarding the operations. About 82 lb of crude product is obtained for every 100 lb of *p*-nitroaniline reduced.

IV. *Vacuum Distillation of Crude p-Phenylenediamine.* A cast iron still capable of holding 1000 lb of crude *p*-phenylenediamine is placed on a masonry setting, with a solid arch built underneath it running back within 10 in of the rear wall. Thus the heat, furnished either by gas or by fuel oil, reaches the still mainly by radiation. When optimum conditions prevail, the temperature of the batch in the still will be 230–250°C, and the temperature of the vapors in the line leaving the still will be 180–190°C. It is essential that an efficient vacuum pump be installed on this system in order to obtain a very attractive crystalline product. The crystallizing pans that receive the condensed vapors from the still are placed into water-cooled castings. At the close of each run, the pans are allowed to stand and cool for several hours and then pulled out. It is necessary to cool the distilled product about 36 hr before breaking it up, because it remains molten on the inside for a long time. Furthermore, its color deteriorates if the fused cake is broken prematurely.

If the crude product is moist, the final product will be very poor and sometimes sloppy. A distinct red coloration of the finished product indicates either moisture or acidity in the crude product.

Yields. The yields obtained by the process outlined should be about 90% of theory, or slightly more than 70 lb of finished *p*-phenylenediamine per 100 lb of *p*-nitroaniline used. The sources of loss in the system are as follows:

Process	Loss, %
Reduction	3
Filtration	1.5
Dehydration	1.5
Distillation	3
Total	9

A number of other methods have been used to prepare *p*-phenylenediamine, none of any current commercial significance. A procedure for reducing 4-aminoazobenzene on a scale of 6.7 kg using zinc and hydrochloric acid is described in the older German literature (38). In that method the 4-aminoazobenzene was separately prepared; but a more recent procedure is based on the preparation of benzenediazonium chloride by diazotizing aniline, then coupling with excess aniline present, and immediately thereafter reducing with iron and sodium hydroxide to give crude *p*-phenylenediamine. The crude material is then purified by vacuum distillation. Electrolytic reduction of *p*-nitroaniline has been described (39). Sodium borohydride in conjunction with palladized charcoal has also been used to reduce *p*-nitroaniline to *p*-phenylenediamine (32).

2. ANALYSIS

The same comments made about the analysis of *m*-phenylenediamine are applicable to *p*-phenylenediamine. Bryan's gas chromatographic method (22) and Kreshkov's potentiometric titration (23,24,40) can be used to analyze *p*-phenylenediamine.

3. TOXICOLOGY

Of the diamines treated in this chapter, *p*-phenylenediamine is by far the most toxic. It is a powerful skin irritant and is also responsible for asthmatic and other respiratory problems of workers in the fur dye industry. *p*-Phenylenediamine can cause kerato-conjunctivitis and eczema. Liver damage, fatal in at least one case, has been reported.

Especially with this diamine, extreme care should be taken to avoid any bodily contact with the fumes, dust, or crystals.

D. Benzidine*

1. SYNTHESIS

Benzidine (4,4'-diaminobiphenyl), a colorless crystalline solid that darkens in air, has been known in the chemical literature for more than 120 years (2).

* The reader is referred to Ref. 2 for the numerous articles that could be cited in connection with benzidine, particularly the reports made by the British and American investigating teams after World War II of the wartime I. G. Farbenindustrie manufacturing processes.

Indeed, for all practical purposes, the original preparative method—the well-known benzidine rearrangement—is still the only process used for manufacturing benzidine. Since nitrobenzene, the starting material for this rearrangement, is relatively inexpensive, benzidine can be produced on a large industrial scale, most commonly as the hydrochloric acid or sulfuric acid salt. The free base is obtained from these salts by means of a weak inorganic base such as sodium carbonate.

Benzidine has also been prepared by reaction of 4,4'-dichlorobiphenyl with ammonia under pressure, but the method (41) is limited by the scarcity of the starting material.

The overall benzidine rearrangement is a two-step reaction sequence in which nitrobenzene is first reduced to hydrazobenzene and then a second step is required to rearrange the hydrazobenzene to benzidine.* The first step generally is done under basic conditions; the second step (the true benzidine rearrangement) requires acid conditions, which accounts for the isolation

nitrobenzene   azobenzene   hydrazobenzene

$$\xrightarrow{\text{acid}} \quad NH_2-\!\!\bigcirc\!\!-\!\!\bigcirc\!\!-NH_2\cdot\text{acid salt} \qquad\qquad (3)$$

benzidine

of the acid salt of benzidine in the process. The reduction of nitrobenzene to hydrazobenzene actually goes through an intermediate stage of reduction to give azobenzene, and a process used at one time in the Leverkusen plant of I. G. Farben actually carried out the reduction of nitrobenzene to hydrazobenzene in two steps (2). Current practice seems to favor zinc and alkali (2) or iron and water (2,42) as the reducing agents to obtain hydrazobenzene. Electrolytic (2) and noble metal catalysts with aqueous alkali (2) are other reducing conditions used to produce hydrazobenzene. The actual rearrangement of hydrazobenzene to benzidine is done in a straightforward manner in aqueous hydrochloric acid (or sulfuric acid) to obtain directly the dihydrochloride (or sulfate).

* A small-scale laboratory procedure based on this rearrangement is found in Ref. 16, p. 633.

Current practice in the United States favors use of zinc for the reduction of nitrobenzene in the synthesis of benzidine (2); the reduction with iron should be more economical, however, but the process is quite troublesome on a commercial scale (2). The process given here overcomes the usual difficulties by use of an especially prepared iron powder (42).

100 parts of nitrobenzene is stirred with 50 parts of very finely divided iron obtained by reducing iron oxide with hydrogen or by thermal decomposition of iron carbonyl, in a vessel containing 1 atm of nitrogen. The mixture is heated to about 90°C, when 80 parts of 60% caustic soda is gradually added, the temperature being maintained at about 100°C. The mixture is well stirred, and when the formation of azoxybenzene is complete (as shown by the setting point of the mixture) a paste of 16 parts of iron and 40 parts of 60% caustic soda is added. The mixture is stirred at the same temperature until formation of azobenzene is complete. Again 16 parts of iron and 40 parts of caustic soda are added, the mass being maintained in a liquid state by raising the temperature toward the close of the reaction to about 130°C. When reduction is complete, the crude hydrazobenzene is separated from the iron oxide and caustic alkali. To convert it into benzidine it is stirred for several hours in the cold with hydrochloric acid of 24°Bé and the mass then heated to 80°C, the benzidine being soluble in the hot liquor. The hot solution is then filtered and the benzidine precipitated as the sulfate by the addition of sulfuric acid. The benzidine sulfate is finally filtered off and washed, and the base is obtained in good yield by treatment with ammonia.

2. PURIFICATION

Purification of benzidine is generally accomplished by way of the sparingly soluble sulfate salt with subsequent regeneration of the free base. The free base can also be vacuum distilled.

3. ANALYSIS

The Kreshkov potentiometric titration of diamines, previously mentioned in connection with the phenylenediamines can be used in assaying benzidine (40). As was the case with the other diamines, a mixture of chloroform and acetonitrile (4:1) is the best nonaqueous solvent, and the two amino groups of benzidine can be individually titrated in this medium.

4. TOXICOLOGY

Unlike the phenylenediamines, benzidine is only a weak methemoglobin former and consequently has a lower order of toxicity, although it is likely to cause dermatitis. The lethal dosage in dogs is 400 mg/kg. However, benzidine is strongly suspected as a carcinogen responsible for bladder tumors.

Great care should be exercised to avoid any contact with benzidine powder or vapor. Protective clothing and respiratory masks should be worn when handling this material.

E. 4,4'-Methylenedianiline

1. SYNTHESIS

4,4'-Methylenedianiline, also called *p,p'*-methylenedianiline or *p',p'*-diaminodiphenyl methane, is commercially produced by the condensation of aniline with formaldehyde in the presence of catalytic amounts of strong acid.

A German patent (43) describes the preparation on a scale of slightly more than 1 kg of 4,4'-methylenedianiline in 98% yield, using aqueous hydrochloric acid as a catalyst. Many modifications of this method are available (44–50), but none gives the yield claimed by the German patent procedure. The commercial preparation is exemplified by the following method (translated from the original) (43):

> To 2280 parts of aniline are added 32.5 parts of 28% hydrochloric acid. Then 700 parts of 30% aqueous formaldehyde is added with stirring. During addition, the temperature rises from 20 to 28°C. Subsequently, the mixture is heated to reflux, and 616 parts of water together with 101 parts of aniline are distilled off, with the temperature rising to 130°C. This temperature is maintained for 8 hr under stirring. Then excess aniline (1008 parts) and traces [*sic*] of water (26 parts) are removed by distillation *in vacuo*, and at the end of the distillation the pot temperature is 170°C at 0.8 torr. Yield: 1222 parts of 4,4'-methylenedianiline (98.2% of theory).

A few other methods have been used to prepare 4,4'-methylenedianiline, but none has the significance of the formaldehyde-aniline condensation procedure. 4,4'-Dinitrodiphenylmethane has been reduced in steps using phenylhydrazine, then hydrazine, to give 4,4'-methylenedianiline (51). The dinitro compound has also been reduced with zinc and hydrochloric acid in one step. Aniline also condenses with 4-aminobenzyl alcohol or bis(4-aminobenzyl)sulfide under acid conditions to give 4,4'-methylenedianiline. The diamine can be obtained as well by decarboxylation of 4,4'-diaminodiphenyl acetic acid with hydrogen chloride at 180–220°C.

2. PURIFICATION

According to Demers and Fink (7), crude 4,4'-methylenedianiline can be purified by fractional distillation, followed by precipitation as the hydrochloride. The fraction boiling at 235–250°C/8 mm is converted to the salt by treatment with aqueous hydrochloric acid at 75°C for 16 hr; the insoluble salt is filtered off at 30°C and washed. The free diamine is then regenerated with 50% sodium hydroxide, filtered, washed, and dried. In this way it is claimed (7) that the melting point is raised from 90 to above 92°C.

3. ANALYSIS

Both infrared spectroscopy and potentiometric titrations in nonaqueous solvents have been used to analyze 4,4'-methylenedianiline. Scheddel (52) has given an infrared procedure suitable for the analysis of mixtures of aniline and 4,4'-methylenedianiline. Kreshkov (40) has developed a potentiometric method in which a standard solution of perchloric acid in methylethylketone is titrated against the diamine, dissolved in a mixture of chloroform and acetonitrile. There is also available from the Dow Chemical Company another nonaqueous titration in which 4,4'-methylenedianiline is titrated potentiometrically with perchloric acid, using glacial acetic acid as the solvent (53).

4. TOXICOLOGY

4,4'-Methylenedianiline has a moderately high oral toxicity. It can cause appreciable skin irritation in some human subjects (54). Also, temporary liver damage and methemoglobinemia (cyanosis) have been reported in workers exposed to excessive amounts of the dust (3). Contact with the skin and inhalation of the dust should be avoided.

F. Bis(4-aminophenyl)ether

Bis(4-aminophenyl)ether finds wide application in polymers because of the added flexibility imparted by the ether bridge to polymeric chains. Thus improvements in impact strength are noted in epoxy resins, and lower brittleness is observed in polyimides, with little penalty in thermal stability.

Commercial production figures for bis(4-aminophenyl)ether are not available, but production undoubtedly is high and increasing because of captive use of the diamine. du Pont's Kapton film, for example, is based on the polypyromellitimide of this diamine, and as more of this high-performance polymer finds use, more diamine surely will be consumed.*

1. SYNTHESIS

Although several syntheses have been reported for bis(4-aminophenyl)-ether, no large-scale method of preparation has been published. Two methods favored are ammonolysis of the corresponding dibromo compound and

---

* Several domestic producers have marketed bis(4-aminophenyl)ether in the past (e.g., the Dow Chemical Company and the Ott Chemical Company), but at present the only domestic source of the diamine is Wallach-Gracer Export Corporation, who import the material from Japan.

reduction of the corresponding dinitro compounds. The dinitro compounds can be synthesized in purest form via reaction of a *p*-halonitrobenzene with a metal *p*-nitrophenolate:

$$NO_2-\!\!\left\langle\bigcirc\right\rangle\!\!-OK + Br-\!\!\left\langle\bigcirc\right\rangle\!\!-NO_2 \longrightarrow NO_2-\!\!\left\langle\bigcirc\right\rangle\!\!-O-\!\!\left\langle\bigcirc\right\rangle\!\!-NO_2 \quad (4)$$

The nitration of diphenyl ether followed by reduction yields the desired diamine, but mixed isomers are obtained (55).

The most probable commercial synthesis involves the catalytic reduction of 4,4′-dinitrodiphenyl ether, thus paralleling the reported laboratory procedure (56).

> 4,4′-Dinitrodiphenyl ether (25 g, 0.01 mole) was dissolved in 200 ml of absolute ethanol. About 5 g (wet weight) of Raney nickel was added, and the mixture was reduced in the Parr apparatus at 50 psi at 75°C. The theoretical drop in pressure took place in approximately 30 min. The mixture was filtered hot and the filtrate allowed to cool. The product was collected on a Büchner funnel and air dried (mp 186–187°C). The yield was 18 g (90%).

Another laboratory procedure employs platinum dioxide as catalyst, and the reduction is carried out at 100 psi (57). Atmospheric-pressure reductions may be performed using hydrazine hydrate and Raney nickel (58).

2.　PURIFICATION

Sublimation of bis(4-aminophenyl)ether under reduced pressure (200°C/0.05 min) affords a product suitable for polymerization (59).

3.　TOXICOLOGY

No data on the toxicity of bis(4-aminophenyl)ether are available, but we might reasonably presume that the compound is at least as toxic as 4,4′-methylenedianiline.

G.　4,4′-Diaminoazobenzene

Although 4,4′-diaminoazobenzene (4,4′-azodianiline) is produced in limited commercial quantities for use in dyestuffs, it is included here because of the potential use of the azolinkage in polymers (60). Methods of preparation of the diamine are reviewed by Santurri et al. (61), whose published procedure has been checked and found to be operable.

> In a 1-liter three-necked round-bottom flask equipped with an efficient stirrer, a reflux condenser, and a thermometer are placed 500 ml of glacial acetic acid, 29.0 g (0.19 mole) of *p*-aminoacetanilide (mp 158–160°C) 40 g (0.26 mole) of sodium perborate tetrahydrate,

and 10 g (0.16 mole) of boric acid. The mixture is heated with stirring to 50–60°C and held at this temperature for 6 hr. Initially, the solids dissolve, but, after heating for approximately 40 min, the product begins to separate. At the end of the reaction period, the mixture is cooled to room temperature and the yellow product is collected on a Büchner funnel. It is washed with water until the washings are neutral to pH paper and then dried in an oven at 110°C [If the product is not washed well, the dried material will turn violet, indicating unreacted p-aminoacetanilide.] The yield of 4,4'-bis(acetamido)azobenzene, mp 288–293°C (dec.), is 16.5 g (57.7%). It is used as such for the hydrolysis step. [This compound may be recrystallized from glacial acetic acid or ethanol.]

In a 500-ml round-bottom flask equipped with a reflux condenser and a magnetic stirrer* are placed 150 ml of methanol, 150 ml of 6N hydrochloric acid, and the total yield of 4,4'-bis(acetamido)azobenzene. The mixture is heated under reflux for 1.5 hr. The reaction mixture is cooled and the violet solid collected on a Büchner funnel. The damp product is suspended in 500 ml of water in a 1-liter beaker equipped with a stirrer, and the mixture is slowly neutralized by the addition of 2.5N sodium hydroxide. In the course of the neutralization, the salt dissolves and the free base separates. The 4,4'-diaminoazobenzene is collected on a Büchner funnel, washed with water, and dried under reduced pressure. The yield of yellow product, mp 238–241°C (dec.), is 11–12 g. The overall yield from p-aminoacetanilide is 52–56%.

H. "Ordered Diamines"

Recently several diamines have been reported which are unique in that they contain within their molecules preformed amide linkages or heterocyclic rings. These diamines easily react with appropriate difunctional monomers, giving rise to a wide variety of condensation polymers that have been termed "ordered copolymers." Examples of these polymers are those which contain aromatic rings and amide groups (62,64); aromatic rings, heterocyclic rings, and amide groups (66); aromatic rings and at least two types of heterocyclic rings, one being preferably an imide ring system (67,68); and aromatic rings, imide groups, and amide groups (69). The "ordered diamines" used to form these ordered copolymers can be either symmetrical (i.e., there exists an element of symmetry in the molecule) or unsymmetrical. In general, the symmetrical ordered diamines have been more thoroughly investigated and their properties are discussed below; a few unsymmetrical diamines, such as the diaminobenzanilides†, appear to be useful in preparing polyamides (70), polyamide-imides (69), and polyheterocyclo-imides (71,73).

---

* [It has been found that bumping occurs during the reflux period if stirring is omitted].

† One such diamine, 3,4'-diaminobenzanilide, has recently become commercially available at approximately $10/lb from American Aniline Products, Inc.

TABLE 4
Selected Diamino-Diamides

Diamine	mp,[a] °C
N,N'-m-Phenylenebis(m-aminobenzamide)[b]	213–214 (269–271)
N,N'-m-Phenylenebis(p-aminobenzamide)[b]	227–228 (279–281)
N,N'-p-Phenylenebis(m-aminobenzamide)[b]	289–291 (222–335)
N,N'-4,4'-Biphenylenebis(m-aminobenzamide)[c]	332–334 (357–362)
N,N'-Bis(4-aminophenyl)isophthalamide[d]	239–240 (278–280)[e]

[a] The values given in parentheses are the melting points of the corresponding dinitro compounds.
[b] Ref. 63.
[c] Ref. 64.
[d] Ref. 62.
[e] Ref. 76.

TABLE 5
Heterocyclic Diamines

Diamine	mp,[a] °C
2,5-Bis(m-aminophenyl)-1,3,4-oxadiazole[b]	257–258 (234–236)[c]
2,5-Bis(p-aminophenyl)-1,3,4-oxadiazole[b]	260–262 (319–321)[c]
4,4'-Bis(p-aminophenyl)-2,2'-bithiazole[b]	284–286 (327–329)[c]
m-Bis(4-p-aminophenyl-thiazol-2-yl)benzene[b]	230–232 (271–272)[c]
2,2'-Bis(m-aminophenyl)-5,5'-bibenzinidazole[b]	297–299
2,2'-Bis(p-aminophenyl)-5,5'-bibenzimidazole[d]	321–322[e]
2,2'-Bis(m-aminophenyl)-5,5'-bibenzoxazole[f]	308–309
2,2'-Bis(p-aminophenyl)-5,5'-bibenzoxazole[f]	346–349
2,2'-p-Phenylenebis(5-aminobenzoxazole)[f]	382–384 (358–360)[c]
2,2'-m-Phenylenebis(5-aminobenzoxazole)[f]	300–302 (279–280)[c]
2,2'-Bis(5-aminobenzoxazole)-4,4'-diphenyl[f]	360–363
2,2'-Bis(5-aminobenzoxazole)-4,4'-diphenylether[f]	314–316
2,2'-Bis(m-aminophenyl)-5,5'-bibenzothiazole[f]	292–293
2,2'-Bis(p-aminophenyl)-5,5'-bibenzothiazole[f]	373–374
2,2'-m-Phenylenebis(5-aminobenzothiazole)[f]	290–292[g]

[a] Uncorrected.
[b] Ref. 78.
[c] Melting point of the corresponding dinitro.
[d] Ref. 78.
[e] mp. 334 °C; Ref. 82.
[f] Ref. 84.
[g] mp. 270°C; Ref. 85.

Obviously, not all such diamines will achieve commercial importance; three that are potentially important are described in some detail; several others are merely listed in Tables 4 and 5. A fairly general method of preparing benzheterocyclic-ordered diamines is also given, and certain of these diamines are presented in Table 5.

1. N,N'-m-Phenylenebis(m-aminobenzamide)

The diamine N,N'-m-phenylenebis(m-aminobenzamide) was first reported without properties or method of synthesis as a monomer for the preparation of polyamide-imides (69). Later the synthesis (Eq. 5) of this diamine was reported in connection with the preparation of ordered copolyamides (63).

$$(5)$$

1

Similar diamines (Table 4) may be prepared by the reaction of aromatic diamines with nitroaroylchlorides, followed by reduction of the dinitro intermediate. Reduction has been effected by stannous chloride and hydrochloric acid and by catalytic reduction (63,64). A large-scale preparation of the diamine **1** may be made in which the dinitro precursor is not isolated from solution but is reduced *in situ* (74). Typical preparations of this diamine and its precursor are carried out as follows.

a. Dinitro Precursor (65)

A solution of 38 g of m-nitrobenzoyl chloride in 40 ml of dry tetrahydrofuran was poured all at once into a Blendor jar containing 10.8 g of m-phenylenediamine, 8 g of sodium hydroxide, and 200 ml of ice-cold water. The reaction mixture was agitated rapidly; the product was filtered off, washed with warm water, and dried; 34.7 g of N,N'-m-phenylenebis(m-nitrobenzamide) was obtained (85% yield); mp 269–270°C.

The dinitro intermediate may also be prepared in solution as follows (75):

To a solution of 1.08 g (0.01 mole) of *m*-phenylenediamine in 15 ml of dry dimethylacetamide, 3.75 g (0.02 mole) of *m*-nitrobenzoyl chloride was added all at once. A clear solution was obtained; after 15 min a solid began to crystallize; product of somewhat better purity is obtained and the yield is improved if the contents of the flask is heated to about 100°C. The product may be collected on a filter or poured into 150 ml of rapidly stirred water in a Blendor jar. Yield: 3.5 g (86%).

Pyridine may be substituted for dimethylacetamide in the foregoing reaction. Yield: 3.1 g (76%). Although dimethylformamide may not be used as a reaction medium because of its reaction with the nitro acid chloride, it is suitable for recrystallization of the dinitro intermediate.

b. Diamine

(*1*) *By Reduction with Stannous Chloride* (65). A 5-g portion of the dinitro precursor was placed in a 250-ml flask with 50 ml of absolute ethanol. A solution of 25 g of stannous chloride hydrate in 30 ml of concentrated hydrochloric acid and 50 ml of absolute ethanol was added and the mixture was refluxed about 30 min. From the cooled solution, crystalline diamine dihydrochloride was filtered off, washed with ethanol, and dried. The diamine was isolated from the dihydrochloride by neutralization with sodium carbonate solution. The product was obtained in 70% yield and a melting point of 212°C.

(*2*) *By Catalytic Reduction.* The dinitro intermediate may be reduced in dimethylacetamide solution using relatively low pressures (48 psi) and Raney nickel (64) or palladium on charcoal (65) as catalysts. The diamine is isolated by filtering off the catalyst and pouring the filtrate into water. Yields obtainable are 87–100%.

c. Synthesis of Dinitro Intermediate and in situ Reduction (74)

During a period of 7 min 373 g of molten *m*-nitrobenzoyl chloride (2.01 mole) was added under nitrogen to a stirred solution of 108 g (1.0 mole) of *m*-phenylenediamine in 1381 g of dimethylacetamide (DMAc)

The reaction temperature was allowed to increase from 20 to 52°C and was held there for 18 min. The reaction mixture was then neutralized at 50°C with 357 g of sodium bicarbonate over a 28-min period. The solution was filtered at 50°C and the sodium chloride/sodium bicarbonate cake was washed with 200 g of DMAc. The filtrate had a pH of 8.9.

The filtrate was charged to a 1-gal stirred autoclave along with 35 g of Raney nickel catalyst. Hydrogenation was carried out at 100°C and

25 psig. The rate of hydrogen consumption—3.7 standard liters/min kg charged—was essentially constant for 19 min, decreasing rapidly to zero at this point. The charge was held an additional 10 min at 100°C and 25 psig with continued stirring. The charge was drained from the autoclave through a filter into a stirred vessel under nitrogen and heated to 85°C. Then 4911 g of water at 85°C was added over a 30-min period. The precipitated N,N'-m-phenylenebis(m-aminobenzamide) was recovered by filtration, reslurried in 4911 g of water at 85°C, refiltered, dried in a vacuum oven at 100–110°C. The dried product weighed 306.5 g (88.6%) and contained less that 50 ppm nitro groups. The melting point of the light amber-colored product was 212.0–213.5°C.

d. Toxicology (75)

Both the dinitro precursor and the diamine N,N'-m-phenylenebis(m-aminobenzamide) appear to show low toxicity, although the long-term accumulative effects are not known and reasonable precautions against repeated contact, inhalation, or ingestion should be taken.

Oral ingestion of N,N'-m-phenylenebis(m-aminobenzamide) was non-lethal at the highest dose of 15,800 mg/kg on male and female rats and may be classified as practically nontoxic. The compound may be classified as practically nontoxic by skin absorption in male and female rabbits at the highest application of 5010 mg/kg of weight. Application of the diamine as a finely ground powder or as a 20% aqueous suspension to the clipped intact skin of male and female rabbits indicated that the material was nonirritating. The compound was classed as a slight eye irritant in male and female rabbits, but the irritation appeared to be of a mechanical nature, since rapid recovery occurred 24 hr after contact and following irrigation.

2. N,N'-Bis(3-aminophenyl)isophthalamide

Three syntheses for N,N'-bis(3-aminophenyl)isophthalamide (2) have been reported (62,77). One type of synthesis (Eq. 6) depends on the reaction of isophthaloyl chloride with a large excess of m-phenylenediamine in methylene chloride or dimethylacetamide (DMAc) (62). Another synthesis (Eq. 7)

(6)

2

depends on the reaction of a fivefold excess of *m*-phenylenediamine with isophthalic acid at an elevated temperature (62). The former reaction yields

$$NH_2 \quad \text{---} \quad NH_2 \;+\; HO\!-\!\overset{\overset{O}{\|}}{C}\!\quad\!\overset{\overset{O}{\|}}{C}\!-\!OH \quad \xrightarrow[\text{NaH}_2\text{PO}_2\cdot\text{H}_2\text{O}]{250°C} \quad 2 \qquad (7)$$

(excess)

much polymer (~13%), and the latter employs a very large quantity of excess diamine.

A more convenient synthesis would involve the reaction (77) (Eq. 8) of *m*-nitroaniline with isophthaloyl chloride, followed by reduction. No details of the preparation via this method are available; however, the methods given

$$NO_2 \quad\text{---}\quad NH_2 \qquad \overset{\displaystyle Cl-\overset{O}{\overset{\|}{C}}\quad\overset{O}{\overset{\|}{C}}-Cl}{\underset{}{\xrightarrow{\hspace{3cm}}}}$$

$$NO_2\;\text{---}\;NH\!-\!\overset{\overset{O}{\|}}{C}\quad\overset{\overset{O}{\|}}{C}\!-\!NH\;\text{---}\;NO_2 \quad \xrightarrow{\text{reduction}} \quad 2 \qquad (8)$$

for *N,N'-m*-phenylene-bis(*m*-aminobenzamide) should be readily adaptable.

a. Preparation via High-Temperature Reaction (62)

Into 3.23 parts of isophthalic acid which was previously ball-milled to less than 100-mesh size, and 0.50 part of sodium hypophosphite monohydrate is distilled at atmospheric pressure about 25 parts of *m*-phenylene diamine, which represents a fivefold excess. The mixture is heated to 250°C with stirring for 10 min. The reaction mixture is dumped into 800 parts of water and acidified with dilute hydrochloric acid. The mixture is then heated to a boil and filtered hot. Upon cooling to room temperature, a small amount of cloudiness develops and the mixture is filtered again. The clear solution is treated with an excess of saturated aqueous sodium bicarbonate solution. After removal by filtration, the product is washed thoroughly with water and dried overnight in a vacuum oven at 70°C. Yield: 78% (based on isophthalic acid).

b. Preparation via Low-Temperature Method (62)

A solution of 5.06 parts of isophthaloyl chloride in 66.8 parts of methylene chloride is added to a solution of 21.63 parts of *m*-phenylene diamine in 400 parts of methylene chloride in an "Osterizer" over a 1-min period. The mixture is stirred as rapidly as possible for 2 min after the addition is completed. The precipitated reaction product is removed by filtration, and the filtrate is treated with 8.66 parts of m-phenylene diamine and made up to the original volume with methylene chloride. Isophthaloyl chloride in an amount of 4.06 parts, in 66.8 parts of methylene chloride, is added over a period of about 1 min with continued rapid stirring. The product is filtered off and combined with the first precipitate. The filtrate is used in the same manner for another 18 times and the combined filter cakes are treated with 2500 parts of water and heated to 80°C to drive off methylene chloride. After cooling to room temperature, the mixture is diluted to about 7000 parts and acidified with concentrated hydrochloric acid. The insoluble portion, which is a low [molecular weight] polymer of inherent viscosity 0.15 (0.5 g/100 cc of solvent) in concentrated sulfuric acid, is removed by filtration and the filtrate is treated with excess concentrated aqueous ammonia to precipitate the product, which again is removed by filtration. The low [molecular weight] polymer precipitate mentioned previously is extracted with 7000 parts dilute hydrochloric acid to dissolve residual N,N'-bis(3-aminophenyl)isophthalamide, the latter precipitated with aqueous ammonia and combined with the above-mentioned ammonia insoluble portion. The combined products are dried overnight in a vacuum oven at 70°C, and they represent 85.5 parts, which corresponds to a yield of 62%. After it has been re-crystallized twice from 80% aqueous acetone, the product is suitable for the preparation of high-molecular-weight polymer.

3.   2,5-BIS(p-AMINOPHENYL)-1,3,4-OXADIAZOLE

Several convenient syntheses are available for the preparation of 2,5-bis(p-aminophenyl)-1,3,4-oxadiazole (78) from the corresponding dinitro compound (78,79). The latter is readily prepared from a dinitro dibenzoyl hydrazide (3) made via a Schotten-Bauman type reaction or in solution. Compound 3 is converted to 2,5-bis(p-nitrophenyl)-1,3,4-oxadiazole (4) via

$$(9)$$

$$3$$

$$(10)$$

cyclodehydration reactions effected by heat or by reaction with thionyl chloride, phosphorus oxychloride, or fuming sulfuric acid.

$$3 \xrightarrow{-H_2O} \quad NO_2\text{-}\underset{N-N}{\overset{O}{\bigcirc\!\!\bigcirc}}\text{-}NO_2 \tag{11}$$

4

Fuming sulfuric acid may be used (79) to obtain the 2,5-bis(p-nitrophenyl)-1,3,4-oxadiazole directly from p-nitrobenzoic acid (Eq. 12).

$$NO_2\text{-}\bigcirc\text{-}\overset{O}{\overset{\|}{C}}\text{-}OH \xrightarrow[\text{fuming } H_2SO_4]{NH_2NH_2 \cdot H_2SO_4} \quad 4 \tag{12}$$

Reduction of the dinitro oxadiazole intermediate may be effected by means of zinc or stannous chloride and hydrogen chloride, iron, and acids, or by means of catalytic hydrogenation.

The synthesis of the dinitro precursor and the preparation of 2,5-bis(p-aminophenyl)-1,3,4-oxadiazole via reduction are described below.

a. Preparation of 2,5-Bis(p-nitrophenyl)-1,3,4-oxadiazole in Oleum

The following process for the preparation of 2,5-bis(p-nitrophenyl)-1,3,4-oxadizaole using oleum for condensation and dehydration reagent has been described (79):

> 70 parts of 4-nitrobenzoic acid are introduced while cooling into a solution of 26 parts hydrazine sulfate in 480 parts of oleum (24 % SO$_3$), and the mixture is then heated at about 60–80°C until the quantity of carboxylic acid used in excess no longer diminishes [sic]. The mixture is then poured on to ice, filtered with suction, and the filter residue is washed and dried.
>
> There is obtained 61–92 parts of 2,5-bis(p-nitrophenyl)-1,3,4-oxadiazole in the form of an almost colorless powder melting at 301–302°C.

The dinitro dihydrazide precursor (**3**) may be prepared and converted via cyclodehydration of dihydrazide to (**4**) as follows (80):

> To a glass-lined, 30-gal Pfaudler fitted with a crow-foot agitator was added 25 liters of water and 15 kg of ice. The mixture was stirred at 175 rpm while 7300 g of hydrazine sulfate and 5100 g of sodium bicarbonate was added. A solution of 3750 g of pure p-nitrobenzoyl chloride in 8 liters of dry tetrahydrofuran (THF) was added gradually over a period of about 10–15 min (the temperature of the reaction mixture after addition of all the THF solution was 0°C). The mixture was stirred for 2 hr, then filtered.

The yellow product was washed with hot water, then methanol. The product was reslurried in hot water, filtered, and washed on the filter with hot water, then methanol. The yield of product after drying at 110°C under vacuum for 60 hr was 3160 g (95.8% of theory).

The foregoing dihydrazide was converted to 2,5-bis(p-nitrophenyl)-1,3,4-oxadiazole as follows (80):

To a glass-lined 5-gal Pfaudler fitted with stirrer and containing 8 kg of phosphorous oxychloride 2739 g of the above-described bis(p-nitrobenzoyl) hydrazide, C, was added. The mixture was stirred under reflux (106°C) for 5 hr. A 10-ml portion of dimethylformamide (DMF) initially was added to catalyze the reaction; 20 ml of DMF was added after 5 hr and the mixture was refluxed an additional 2 hr.

The mixture was cooled to 38°C and filtered through a coarse glass filter funnel. The yellow filter cake was washed with 4 liters of toluene to remove excess phosphorous oxychloride, then it was slurried in 24 liters of water to complete the destruction of residual phosphorous oxychloride. Next, the crude product was washed with hot water, then methanol (the latter extracted an orange-colored impurity). The yield of 2,5-bis(p-nitrophenyl)-1,3,4-oxadiazole (mp 312–315°C) after drying for 60 hr in a forced air oven was 2061 g (78%).

b. Preparation of 2,5-Bis(p-aminophenyl)-1,3,4-oxadiazole

Small-scale preparations of 2,5-bis(p-aminophenyl)-1,3,4-oxadiazole may employ catalytic reduction or reduction using acid and a reducing agent. For large-scale preparations, the following catalytic reduction is convenient (80):

The product (4) and 360 g of Raney nickel, type #28, washed with dimethylacetamide (DMAc), were placed in a 10-gal stirred autoclave with 8 liters of DMAc plus an additional 2.5 liters used to transfer the product and catalyst to the reactor. After the autoclave was purged twice, with nitrogen, it was pressured to 55 psi with hydrogen and the autoclave stirrer was started. Hydrogenation was carried out over a period of 4 hr, the maximum temperature never exceeding 125°C.

The contents of the reactor were filtered under 10-psi pressure through a cartridge filter (10-μ pore size) followed by a Sparkler L-1000 filter. To the clean filtrate was added 10 ml of hydrazine hydrate (90%) to scavenge any unreduced dinitro intermediate. Approximately 25 gal of hot water (90°C) was added to the filtrate in a 30-gal Pfaudler, blanketed with nitrogen. The precipitated product was filtered and dried at 110°C for 44 hr (the oven was allowed to cool to room temperature prior to exposure to air). The yield of pure 2,5-bis(p-aminophenyl)-1,3,4-oxadiazole, mp 260–262°C, was 980 g (55%). Higher yields of diamine may be obtained by recovery of diamine from the filtrates and the washings.

4. 2,2′-p-PHENYLENEBIS(5-AMINOBENZOXAZOLE)

The preparation of 2,2′-p-phenylenebis(5-aminobenzoxazole) is included less for its importance than because the synthesis of the diamine may be applied to a number of benzheterocyclic diamines (Table 5). The synthesis is based on the method of preparation of benzoxazoles and benzimidazoles reported by Hine (81). Others have reported the preparation of diamines such as benzimidazoles (78,82,83), benzoxazoles (84), and benzothiazoles (84).

a. Preparation (84)

(1) Crude Materials. A three-necked, round-bottom 500-ml flask, equipped with an all-glass Trubore stirrer, a nitrogen inlet, and a drying tube containing Drierite was charged with 20 g (slightly more than 0.1 mole) of 2,4-diaminophenol dihydrochloride, 8.3 g (0.05 mole) of terephthalic acid, and 70 ml of 116% polyphosphoric acid (PPA). The flask was placed in an oil bath and heated slowly to control excessive foaming, to 110°C; the reactants were stirred continuously under 1 atm of nitrogen. The flask was kept at 110°C for 1 hr, then the temperature was increased slowly to 210°C and maintained at the latter temperature for 3 hr. The contents of the flask were allowed to cool and were then poured into 600 ml of water, forming a brownish precipitate. The crude product was soaked overnight in 600 ml of an aqueous 10% sodium bicarbonate solution, filtered, washed with water, and dried under vacuum at 60°C; yield, 96%.

(2) Purification. The crude material was extracted in a Soxhlet apparatus with acetone (or alcohol) to remove unreacted starting materials; it was next dissolved in a large quantity of hot dimethylacetamide. The solution of the diamine was treated with charcoal and filtered; addition of water to the filtrate precipitated the purified product. The diamine could also be sublimed under high vacuum at 340–350°C; diamines purified via the sublimation method are of higher purity than those produced by the extraction and precipitation method.

IV. STORAGE

The diamines of this chapter vary greatly in their abilities to resist degradation by such agents as air and light. Certain of the "ordered" diamines may be left in direct sunlight in air, but diamines such as p-phenylenediamine discolor very rapidly in air.

In view of the corrosive and toxic nature of a good many of the diamines, it is a good precaution to store the various diamines in dry glass bottles (preferably dark) and to seal around the caps with friction tape or masking tape. For the more sensitive diamines, such as the phenylenediamines and benzidine, an added precaution is to blanket the diamine with an inert gas (e.g., nitrogen or argon) prior to sealing the container.

Another safeguard that is worthwhile observing to preserve the diamines in a state suitable for polymerization is to store the materials in a cool, dry area in the absence of light.

V. POLYMERIZATION

A. Polymer-Forming Reactions

Aromatic diamines of high purity may be used for the preparation* of linear, high-molecular-weight polyamides, polyureas, polyurethanes, and polyimides. However, o-phenylenediamine might be mentioned as an exception because of its tendency to undergo ring-forming rather than chain-extending reactions. Sometimes the molecular weights of certain polyamides and ureas (e.g., those from benzidine) are limited because of precipitation due to polymer insolubility in the polymerization media or because of limited solubility of the diamine.

At least two polymers based on aromatic diamines have achieved some commercial significance. One is the polyamide obtained by polycondensation of m-phenylenediamine with isophthaloyl chloride (87,88); this polymer may be formed in high molecular weight and in high percentage polymer solids in dimethylacetamide, a solvent convenient for the spinning of fibers. It is thought that poly-m-phenylene isophthalamide is the polymer from which the heat-resistant fiber and paper Nomex® is produced at the annual rate of 10–15 million lb/yr. (Note: The production figures for m-phenylene-diamine given in Table 1 do not include the captive production of this diamine.)

The other polymer based on an aromatic diamine and being marketed is an aromatic polyimide, the polypyromellitimide of bis(4-aminophenyl)ether (Kapton® film).† Polyimides can be prepared from all the diamines discussed in this chapter (except for o-phenylenediamine) because of the excellent solubility of the polyamic-acid precursor in polar solvents. The diamine of choice for commercial preparation of films is bis(4-aminophenyl)ether because the films from this polyimide, unlike those derived from most other diamines, are nonbrittle at all points of cure from the polyamide-acid to polyimide. For polyimide varnishes for wire coating and adhesives for composites, the less expensive diamines, m-phenylenediamine and 4,4'-methylenedianiline, are suitable for condensation with pyromellitic dian-hydride, benzophenone dianhydride, or both (89–92).

---

* The references to the preparation of the several types of polymers that may be prepared from aromatic diamines are too numerous to be included here. An excellent review is found in Morgan's book (86).

† Also, precursors of the polyimides in solution are available from du Pont Company and Monsanto Company for use as laminating resins (Pyro-ML®, Skybond® 700–703, etc.)

Large quantities of aromatic diamines, especially *m*-phenylenediamine and 4,4'-methylenedianiline, are also used as hardening agents to crosslink epoxy resins.

B. Condensation of Diamines to Determine their Purity

It would be desirable to have a general test of purity for the diamines discussed here—for example, the ability to polymerize them to polyamides having an inherent viscosity of a specified value. However, relatively few truly high-molecular-weight polyamides can be made from aromatic diamines by simple, reproducible methods. As an example of a reproducible method, the preparation of poly[2,5-bis(*p*-phenylene-1,3,4-oxadiazole)] isophthalamide, (93,94) is given.

It is also possible to prepare high-molecular-weight polyimides (95) from all the diamines discussed here (with the exception already noted of *o*-phenylenediamine), and a method of preparation reported by Sroog is reproduced here (96). The inherent viscosity of the high-molecular-weight, polyamic-acid precursor to the polyimide can be determined by dilution techniques (96) or by dissolving films of polyamic-acid in a suitable solvent.*

Polyureas having high inherent viscosities can be prepared from selected diamines and diisocyanates, but the number of soluble compositions are limited.

Polyurethanes can be prepared either directly from the aromatic diamines by reaction with bischloroformates (although this reaction is of little importance) or by the reaction of diisocyanates, derived from the diamines, with diols. Of the diamines included in this chapter, only the phenylene diamines, 4,4'-methylenedianiline and benzidine, are of importance in producing the phenylenediisocyanates, 4,4'-diphenylmethanediisocyanate, and 4,4'-biphenyldiisocyanate ("xenylene diisocyanate"). (Note: The captive use of diamines for the preparation of diisocyanates is not included in the production figures for the diamines given in Table 1.)

In Table 6 the inherent viscosities and methods for polymerization for a number of types of polymers are summarized.

1. POLYAMIDE FORMATION

Aromatic polyamides may be formed by low-temperature polymerizations employing either solution polycondensation or interfacial polymerization (86). Two typical procedures are described below.

---

* Inherent viscosities of the polyimides are difficult to determine because of the insolubility of the polymer in organic solvents. Sulfuric acid is commonly used as a solvent, but viscosity values obtained are probably questionable because polymer degradation occurs. Arsenic chloride and antimony chloride (and mixtures of the two) have been mentioned as nondegrading solvents for polyimides (97–99).

TABLE 6
Polymers from Aromatic Diamines Prepared by Low-Temperature Polymerization

Diamine	Comonomer	Polymerization method[a]	η_{inn}[b]	Reference
Polyamides				
m-Phenylenediamine	Isophthaloyl chloride	S(DMAc)[c]	>1.8	87
	Isophthaloyl chloride	I	2.47	88
	Terephthaloyl chloride	S(DMAc)[c]	1.53	87
p-Phenylenediamine	Terephthaloyl chloride	S(HPT)[d]	1.9	87
4,4'-Methylenedianiline	Isophthaloyl chloride	S(DMTMS)[e]	1.4	87
	Isophthaloyl chloride	I	1.86	100
Bis(4-aminophenyl)ether	Terephthaloyl chloride	S(NMP)[f]	1.70	101
Polyureas				
m-Phenylenediamine	4,4'-Methylenediphenyl-diisocyanate	S$\left(\begin{array}{c}\text{dioxane-}\\\text{benzene}\end{array}\right)$	0.24	102
p-Phenylenediamine	4,4'-Methylenediphenyl-diisocyanate	S(DMF)[g]	0.36	102
4,4'-Methylenedianiline	2-Methyl-1,3-phenylene-diisocyanate	S(DMF)[g]	0.66	102
Bis(4-aminophenyl)ether	4-Methyl-1,3,-phenylene-diisocyanate	S(DMF)[g]	0.40	59
Polyurethanes				
m-Phenylenediamine	2,2-Dimethyltrimethylene bischloroformate	I	0.85	103
p-Phenylenediamine	2,2-Dimethyltrimethylene bischloroformate	I	1.20	104
4,4'-Methylenedianiline	2,2-Dimethyltrimethylene bischloroformate	I	2.3	103,104
	Hexamethylene bischloroformate	I	0.60	105

[a] S = solution polycondensation, I = interfacial polymerization.

[b] Determined at 30°C on a solution of 0.5 g of polymer/100 ml of solvent.

[c] DMAc = dimethylacetamide.

[d] HPT = hexamethylphosphorictriamide.

[e] DMTS = dimethyltetramethylenesulfone.

[f] NMP = N-methylpyrrolidone.

[g] DMF = dimethylformamide.

a. Solution Method

As an example of the polycondensation of aromatic diamines with diacid chlorides in solution, the preparation of poly[2,5-bis(p-phenylene-1,3,4-oxadiazole)] isophthalamide is given. The procedure described (94), with suitable modifications can be used to make several aromatic polyamides,* ordered copolyamides, and ordered heterocyclecopolyamides.

* Cf. the procedure of Ref. 87.

To a three-necked resin pot equipped with a heavy-duty stirrer with an efficient stirrer blade and a nitrogen bleed are added 7.56 g (0.03 mole) of 2,5-bis(p-aminophenyl)-1,3,4-oxadiazole [mp 260–262°C; this monomer should be dried thoroughly before use] and 60 ml of N,N-dimethylacetamide containing 3% dissolved lithium chloride. [The solvent should be of high purity and free of water, which can be reduced to a suitably low level by storing the solvent over Linde molecular series (type 5A, $\frac{1}{16}$-in. pellets. The lithium chloride should be dried in an oven at 150°C before addition to the solvent. As a precaution, the freshly prepared solution containing dissolved lithium chloride is stored for several hours over molecular sieves before use in order to remove any water that might have been picked up by the hydroscopic lithium chloride. Another solvent that can be used in this polymerization is N-methylpyrrolidone.] The solution [the mixture first can be warmed if necessary to effect solution of the diamine] is cooled to −30°C and 6.09 g (0.03 mole) of isophthaloyl chloride [mp 42–43°C (from hexane)] is added all at once. The reaction mixture is stirred rapidly for 30 min, the cooling bath is removed, and the temperature of the solution is maintained at 0°C for 30 min. Then the temperature of the solution is allowed to rise to room temperature, and the solution is stirred for at least 1 hr.

The ordered oxadiazole-amide copolymer is isolated [if desired, the hydrochloric acid in the polymer solution can be neutralized · · · by the addition of 2.52 g (0.06 mole) of lithium hydroxide monohydrate] by pouring the clear, viscous solution into rapidly stirred water in a Blendor jar. [Care should be taken to add the polymer solution very slowly to the water; otherwise the tough polymer lumps up and stalls the Blendor. Alternatively, the very viscous solution may be thinned with additional solvent before precipitation.] The tough, cream-colored polymer is washed with water (two 100-ml washes) and is then dried in a vacuum oven at 60°C. The yield of polymer is 96–100%; the inherent viscosity of the polymer is 1.65 determined at 30°C on a 0.5% solution in N,N-dimethylacetamide (containing 5% dissolved lithium chloride). The polymer does not melt below 400°C.

b. Interfacial Method

The interfacial polymerization technique may be used to prepare certain aromatic polyamides, provided the diamine used is soluble or is swollen by the aqueous phase and the polymer is sufficiently soluble or swollen in the organic phase. The ordered copolymers, in general, are not obtained in high molecular weight by this technique, nor are polymers derived from diamines such as benzidine.

The method may be illustrated by the procedure reported by Hill et al.

for the polyisophthalamide of 4,4′-methylenedianiline (100):

A solution of 5.95 parts of bis(4-aminophenyl)methane in 150 parts of water and 111 parts of tetrahydrofuran with 6.36 parts of sodium carbonate is prepared. This solution is rapidly agitated in a Waring Blendor and a solution of 6.06 parts of isophthaloyl chloride in 44.4 parts of tetrahydrofuran added. Agitation is continued for 10 min and at the end of that time polymer is separated and found to have an inherent viscosity of 1.86 and at melting point of 350°C.

2. POLYIMIDE FORMATION

The preparation of aromatic polyimides has been described by several workers, and a method that has been checked is reported by Sroog (96). The procedure also is applicable for the polycondensation of dianhydrides with most of the diamines described in this chapter.

A 500-ml flask fitted with mercury seal stirrer, nitrogen inlet drying tube, and stopper is carefully flamed to remove traces of water on the walls and is allowed to cool under a stream of dry nitrogen in a dry box. In the dry box 10.0 g (0.05 mole) of bis(4-aminophenyl)ether [obtained from Du Pont Organic Chemicals Department, . . . purified by simple vacuum drying at 50°C at ≤ 1 mm Hg for 8 hr, mp 193–194°C] is added to the flask through a dried powder funnel, and residual traces are flushed in with 160 g of dry dimethylacetamide [du Pont technical grade · · · distilled from phosphorus pentoxide at 300 mm Hg]. Pyromellitic dianhydride (10.90 g, 0.05 mole) technical-grade material of the du Pont Explosives Department was sublimed through silica-gel at 220–260°C and 0.25–1 mm Hg is added to the flask through a second dried powder funnel over a period of 2–3 min with vigorous agitation. A red-orange color may be observed at the solid-liquid interface which rapidly lightens to a lemon-yellow as the pyromellitic dianhydride dissolves and reacts with the diamine. Residual dianhydride is washed in with 28 g of dry dimethylacetamide. The powder funnel is replaced by the stopper, and the mixture is stirred for 1 hr. A small surge in temperature to 40°C occurs as the dianhydride is first added, but the mixture rapidly returns to room temperature. Cooling may be necessary to prevent an initial temperature surge. Best results are obtained at temperatures of 15–75°C; above 75°C a decrease in the molecular weight of the polyamic-acid becomes marked.

This procedure yields a 10% solution whicn can be stirred without difficulty. Sometimes the mixture becomes extremely viscous, and dilution to 5–7% solids may be required for efficient stirring. Polyamic acids so prepared exhibit η_{inh} 1.5–3.0 at 0.5% in dimethylacetamide at 30°C.

The polymer solutions may be stored in dry, sealed bottles at −15°C.

The polyamic acid may be isolated from solution by casting the solutions to film as follows (96):

> Thin layers (10–25 mils) of polyamic-acid solutions are doctored onto dry glass plates and dried for 20 min in a forced draft oven (with nitrogen bleed) at 80°C. The resulting colorless-to-pale-yellow films are only partly dry and, after cooling, can be peeled from the plates, clamped to frames, and dried further under reduced pressure at room temperature.

3. POLYUREA FORMATION

Polyureas from the aromatic diamines can be prepared by reaction with phosgene or with diisocyanates; when the diisocyanate is based on a diamine other than the one with which it is reacted, the polyurea formed is of the alternating type. The alternating copolymer illustrated here is more soluble

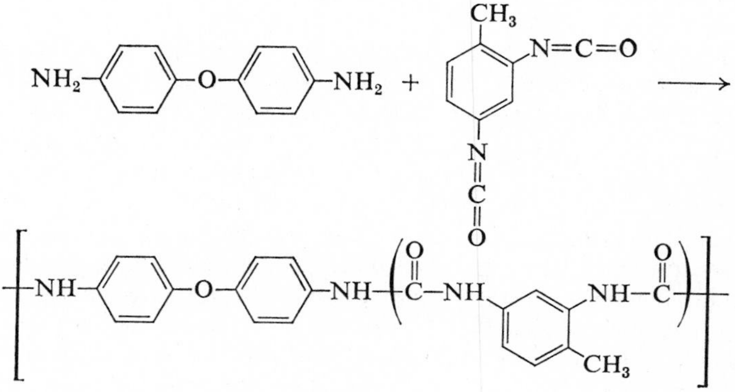

(13)

than most aromatic polyureas (a) because the ether bridge of the diamine contributes to solubility and (b) due to the copolymeric effect produced because the 2,4-tolylenediisocyanate moiety can enter the polymer in a head-to-head or head-to-tail fashion. The following preparation is taken from Sorenson and Campbell (59).

> In a 100-ml three-necked flask equipped with a stirrer is placed 5.00 g (0.025 mole) of 4,4′-diaminophenyl ether [purified by recrystallization from ethanol or by vacuum sublimation (200°C/0.05 mm)] in 25 ml of dimethylformamide [fractionally distilled (bp 154°C) from 2,4-toluenediisocyanate, which is added to the pot in the amount of 8 g/liter]. To it is added, all at one time, with stirring a solution of 4.35 g (0.025 mole) 2,4-toluenediisocyanate [purified by distillation (bp 75–78°C/1 mm)] in 25 ml of dimethylformamide. The diisocyanate solution should be prepared last and added as soon as prepared.
>
> The reaction is stirred for 0.5 hr during which time some heat is generated and the solution becomes somewhat viscous. The solution

may be cast directly onto glass plates and dried in a vacuum oven at 70–80°C for about 3 hr to give clear, flexible films. The polymer may be isolated by pouring into 400 ml rapidly stirred water (as in a home blender), washing further with water, and drying in vacuum at 70°C. The inherent viscosity in sulfuric acid is 0.4 (0.5% concentration, 25°C); the polymer melt temperature is about 320°C with decomposition.

References

1. "Encyclopedia of Chemical Technology," 1st ed., Vol. 10, R. E. Kirk and O. F. Othmer, Eds., Wiley, New York, 1953, p. 378; see also, *op. cit*, 2nd ed., Vol. 15, 1968, p. 216.
2. R. E. Kirk and D. F. Othmer, Eds., 2nd ed., Vol. 3, 1964, p. 408.
3. C. S. Kouris, *Dyestuffs*, **44**:3 (1963).
4. C. L. Butler, Jr., and R. Adams, *J. Amer. Chem. Soc.*, **47**:2610 (1925).
5. Ref. 2, pp. 413–414.
6. R. E. Damschroder and W. D. Peterson, *Org. Syn.*, **Coll. Vol. 3**:106 (1955).
7. F. X. Demers, Jr., and T. R. Fink (to Allied Chemical Corp.), U.S. Patent 2,938,054 (1960); *Chem. Abstr.*, **54**:19592d (1960).
8. R. G. Jones and K. C. McLaughlin, *Org. Syn.*, **Coll. Vol. 4**:824 (1963).
9. O. Hinsberg, *Ann.*, **237**:340 (1887).
10. D. C. Morrison, *J. Amer. Chem. Soc.*, **76**:4483 (1954).
11. J. A. Van Allan and B. D. Deacon, *Org. Syn.*, **Coll. Vol. 4**:569 (1963).
12. J. Davoll, *J. Chem. Soc.*, **1960**:308.
13. "Encyclopedia of Chemical Technology," 2nd ed., Vol. 2, R. E. Kirk and D. F. Othmer, Eds., New York, 1963, p. 76, see also, *op. cit.*, p. 332.
14. A. J. Quick, *J. Amer. Chem. Soc.*, **42**:1033 (1920).
15. E. L. Martin, *Org. Syn.*, **Coll. Vol. 2**:501 (1947).
16. A. I. Vogel, "A Text-book of Practical Organic Chemistry," Longmans Green, London, 1956.
17. D. A. W. Adams, H. Greaves, T. Harrington, P. C. Holmes, and A. Y. Livingston, "I. G. Farbenindustrie, The Manufacture of Miscellaneous Dyestuff Intermediates (Excluding Naphthalene Derivatives)," Off. Tech. Serv. Ret. PB 85687 (1947); BIOS Final Rep. No. 1153, p. 261.
18. D. B. Andrews et al., "German Dyestuff Intermediates," Field Information Agency, Tech. Rep. No. 1313, Vol. **I**, pp. 29, 175, 229, 231 (1948).
19. French Patent 788,348 (1935) (Société pour l'Industrie Chimique à Bâle).
20. J. Levy (to Univ. Oil Products), U.S. Patent 3,230,259 (1966).
21. "Ullmanns Encyklopädie der technischen Chemie," Vol. 13, Urban & Schwarzenberg, Münich, 1962, p. 488.
22. W. H. Bryan, *Anal. Chem.*, **36**(10):2025 (1964).
23. A. P. Kreskov, L. N. Bykova, and I. D. Pevzner, *Zh. Analit. Khim.*, **19**(7):890 (1964); see also, *Dokl. Akad. Nauk*, *USSR*, **150**:99 (1963).
24. A. P. Kreshkov and N. Sh. Aldarova, *Izv. Vyssikh Uchebn. Zaved., Khim. i Khim. Tekhnol.*, **8**(2):316 (1965).
25. H. B. Mark, Jr., *Anal. Chem.*, **36**(4):940 (1964).
26. P. H. Groggins, "Unit Processes in Organic Synthesis," 4th ed., McGraw-Hill, New York, 1953.
27. R. G. Benner and A. C. Stevenson (to du Pont Co.), U.S. Patent 2,619,503 (1952).

28. D. E. Graham (to General Aniline & Film Corp.), U.S. Patent 2,894,036 (1959); *Chem. Abstr.*, **53**:16060A (1959).
29. L. F. Albright, F. H. Van Munster, and J. C. Forman, *Chem. Eng.*, Nov. 6, 1967, p. 251.
30. S. Pietra, *Ann. Chim. (Rome)*, **45**:850–853 (1955).
31. L. P. Kuhn (to Ringwood Chem. Corp.), U.S. Patent 2,768,209 (1956).
32. British Patent 705,919 (1954) (to National Research & Development Corp.).
33. T. Neilson, H. C. S. Wood, and A. G. Wylie, *J. Chem. Soc.*, **1962**:371.
34. British Patent 794,639 (1958) (to General Aniline Film Corp.).
35. L. M. Scheneck and A. Bloom (to General Aniline & Film Corp.), U.S. Patent 2,950,319 (1960).
36. Technical Data Sheet 761-1 "*m*-Phenylenediamine," 1961 Allied Chemical Corp., National Aniline Division.
37. H. V. Bramer, J. E. Magoffin, and M. Clemens (to Eastman Kodak Co.), U.S. Patent 2,578,328 (1951).
38. L. Paul, *Z. angew. Chem.*, **10**:145 (1897).
39. J. F. Norris and E. O. Cummings, *Ind. Eng. Chem.*, **17**:305 (1925).
40. A. P. Kreshkov, L. N. Bykova, and I. D. Pevzner, *Dokl. Akad. Nauk USSR*, **150**:99 (1963).
41. C. F. Booth (to Swann Research), U.S. Patent 1,954,469 (1934).
42. H. Dreyfus, U.S. Patent 2,010,067 (1935).
43. German Patent 1,205,975 (1965) (to Badische Anilin- & Soda-Fabrik AG).
44. F. B. Erickson (to Monsanto Co.), U.S. Patent 2,818,433 (1957).
45. V. N. Kuchinkii and V. E. Griz, *Neftekhimiya*, **2**:624 (1962).
46. French Patent 1,335,124 (1963) (to Société Toulousaine de Produits Chimiques Tolochimie).
47. German Patent 1,179,945 (1964) (to Badische Anilin- & Soda-Fabrik AG).
48. French Patent 1,376,288 (1964) (to Jefferson Chemical Co., Inc.).
49. German Patent 1,210,872 (1966) (to Badische Anilin- & Soda-Fabrik AG).
50. Fr. Pat. 1,398,418 (1965) (to Mobay Chem. Co.).
51. L. M. Litvinenko and N. F. Levchenko, *Zh. Obshch. Khim.*, **29**:3079 (1959).
52. R. T. Scheddel, *Anal. Chem.*, **30**:1303 (1958).
53. Analytical Method No. 55616, The Dow Chemical Company, Midland, Mich., Jan. 8, 1967.
54. "*p,p'*-Methylenedianiline," Technical Service and Development, The Dow Chemical Company, Midland, Mich.
55. J. L. Towle (to Harshaw Chemical Co.), U.S. Patent 3,140,316 (1964).
56. G. A. Reynolds, *J. Amer. Chem. Soc.*, **73**:4996 (1951).
57. K. V. Martin, *J. Amer. Chem. Soc.*, **80**:233 (1958).
58. D. Balcom and A. Furst, *J. Amer. Chem. Soc.*, **75**:4334 (1953).
59. W. R. Sorenson and T. W. Campbell, "Preparative Methods of Polymer Chemistry," Interscience, New York, 1961, p. 94; see also, 2nd ed., 1968.
60. H. C. Bach, ACS Polymer Preprints, **7**(1):576 (1966); *ibid.*, **8**(1):610 (1967); see also, H. C. Bach and W. B. Black, *J. Polym. Sci.*, **C22** (2):799 (1969).
61. P. Santurri, F. Robbins, and R. Stubbings, *Org. Syn.*, **Coll. Vol. 40**: Wiley, New York, 1960, p. 18.
62. C. W. Stephens (to du Pont), U.S. Patent 3,049,518 (1962).
63. J. Preston, *J. Polym. Sci.*, *A-1*, **4**:529 (1966); J. Preston (to Monsanto Co.), U.S. Patent 3,232,910 (1966).
64. F. Dobinson and J. Preston, *J. Polym. Sci.*, *A-1*, **4**:2093 (1966).

65. J. Preston and F. Dobinson (to Monsanto Co.), U.S. Patents 3,240,760 (1966) and 3,242,213 (1966).

66. J. Preston and W. B. Black, *J. Polym. Sci.*, *B*, **4**:267 (1966).

67. J. Preston and W. B. Black, *J. Polym. Sci. A-1*, **5**:2429 (1967).

68. J. Preston, W. DeWinter, and W. B. Black, *J. Polym. Sci. A-1*, **7**:283 (1969).

69. G. M. Bower and L. W. Frost, *J. Polym. Sci. A-1*, **1**:3135 (1963); L. W. Frost and G. M. Bower (to Westinghouse), U.S. Patent 3,179,635 (1965).

70. J. Preston and R. W. Smith, *J. Polym. Sci.*, *B*, **4**:1033 (1966); see also, U.S. Patent 3,354,125 (1967) (to Monsanto Co.).

71. J. S. Rodia (to 3M Co.), U.S. Patent 3,247,165 (1966).

72. L. W. Frost, G. M. Bower, J. H. Freeman, H. A. Burgman, E. J. Traynor, and C. R. Ruffing, *J. Polym. Sci. A-1*, **6**:215 (1968).

73. J. Preston, W. DeWinter, and W. L. Hofferbert, *J. Heterocycl. Chem.*, **5**:269 (1968).

74. R. W. Smith, C. R. Campbell, M. J. Mathews, III (to Monsanto Co.), U.S. Patent 3,499,031 (1970).

75. W. H. Hunt, unpublished work, Monsanto Co. No responsibility for the accuracy of the data given is claimed or implied by Monsanto Co.

76. J. Murphy, unpublished work, Monsanto Co.

77. M. A. Dahlen (to du Pont), U.S. Patent 2,001,526 (1935).

78. J. Preston, *J. Heterocycl. Chem.*, **2**:441 (1965).

79. A. E. Siegrist, E. Moergeli, and K. Hoelzle (to Ciba Ltd.), U.S. Patent 2,765,304 (1956).

80. R. W. Smith, unpublished work, Monsanto Co.

81. D. W. Hine, R. J. Alheim, and J. J. Leavitt, *J. Amer. Chem. Soc.*, **79**:427 (1957).

82. Y. Iwakura, K. Uno, Y. Imai, and M. Fukui, *Makromol. Chem.*, **72**:41 (1964).

83. T. Kurosaki and P. R. Young, *J. Polym. Sci.*, *C*, **23**:57 (1968).

84. J. Preston, W. DeWinter, and W. L. Hofferbert, Jr., *J. Heterocycl. Chem.*, **5**:269 (1968).

85. P. Petitcolas, R. Sureau, J. Frenkiel, and R. Goupil, *Bull Soc. Chim. Fr.*, **1949**:103.

86. P. W. Morgan, "Condensation Polymers: By Interfacial and Solution Methods," Interscience, New York, 1965.

87. S. L. Kwolek, P. W. Morgan, and W. R. Sorenson (assigned to du Pont Co.), U.S. Patent 3,063,966 (1962).

88. H. W. Hill, Jr., S. L. Kwolek, and W. Sweeny (assigned to du Pont Co.), U.S. Patent 3,094,511 (1963).

89. W. M. Edwards (to du Pont Co.), U.S. Patent 3,179,614 (1965).

90. W. M. Edwards (to du Pont Co.), U.S. Patent 3,179,634 (1965).

91. E. Lavan, A. H. Markhart, and R. E. Kass (to Shawinigan Resins Corp.), U.S. Patent 3,190,856 (1965).

92. E. Lavan, A. H. Markhart, and R. F. Kass (to Monsanto Co.), U.S. Patent 3,347,808 (1967).

93. J. Preston and W. B. Black, *J. Polym. Sci. B*, **4**:267 (1966); see also, *ibid.*, *C*, **19**:17 (1967).

94. J. Preston and W. B. Black, "Macromolecular Syntheses," Vol. 3, Interscience, New York, 1969.

95. C. E. Sroog, A. L. Endrey, S. V. Abramo, C. E. Berr, W. M. Edwards, and K. L. Oliver, *J. Polym. Sci.*, *A*, **3**:1373 (1965).

96. C. E. Sroog, "Macromolecular Syntheses," Vol. 3, Interscience, New York, 1969.

97. H. A. Szymanski and A. Bluemle, *J. Polym. Sci.*, *A*, **3**:63 (1965).

98. H. A. Szymanski, W. Collins, and A. Bluemle, *J. Polym. Sci.*, *B*, **3**:81 (1965).

99. R. S. Irwin and W. Sweeney, *J. Polym. Sci.*, *C*, **19**:41 (1967).

100. H. W. Hill, Jr., S. L. Kwolek, and P. W. Morgan (to du Pont Co.), U.S. Patent 3,006,899 (1961).

101. J. R. Holsten, J. Preston, and M. R. Lilyquist, *Appl. Polym. Symp.*, **9**:63 (1969).

102. M. Katz (to du Pont Co.), U.S. Patent 2,888,438 (1959).

103. M. Katz and E. L. Wittbecker (to du Pont Co.), U.S. Patent 2,973,333 (1961); see also (to du Pont Co.), U.S. Patent 3,089,864 (1963).

104. E. L. Wittbecker and M. Katz, *J. Polym. Sci.*, **40**:367 (1959).

105. E. Dyer and R. J. Hammond, *J. Polym. Sci.*, *A*, **2**:1 (1964).

Suggested Readings

Sax, N. I., "Dangerous Properties of Industrial Materials," Reinhold, New York, 1957.

8. BIPSHENOLS AND THEIR BIS(CHLOROFORMATES)

ROBERT BARCLAY, JR., *Thiokol Chemical Corporation, Chemical Division, Trenton, New Jersey*, THEODORE SULZBERG, *Sun Chemical Corporation, Corporate Reseach Laboratory, Carlstadt, New Jersey*

Contents

Bisphenols are organic compounds containing two hydroxyl groups which are bonded directly to an aromatic ring. They have been studied extensively for the following reasons: many of them are easy to prepare from inexpensive starting materials; the compounds are highly reactive, and they readily form high-molecular-weight linear and crosslinked polymers. The bisphenols discussed in this chapter were chosen because of their actual or potential utility as monomers in commercial applications. By necessity, therefore, a degree of arbitrariness was required, and only eleven compounds are included for review. In general, these bisphenols are the ones whose polymers are detailed in Morgan's well-known treatise (1). The literature covered includes material appearing in *Chemical Abstracts* through Volume 69.

I. BISPHENOL A

A. Introduction and Survey of Chemistry

The bisphenol that plays the largest role in condensation polymers is 2,2-bis(4-hydroxyphenyl)propane. In Europe this compound is generally referred to as dian, whereas the trivial name used in the United States is bisphenol A. The latter designation reflects the most common method of preparation, i.e., from phenol and acetone. (The trivial names bisphenol C, bisphenol F, and bisphenol S are similarly applied to the bisphenols derived from cyclohexanone, formaldehyde, and sulfuric acid, respectively.)

Bisphenol A (2) was first prepared by Dianin in 1891 by the hydrochloric acid-catalyzed reaction of phenol with acetone. Its history has paralleled that of the condensation polymers industry, although some other uses for it, particularly as an antioxidant, have been reported. Its first commercial development followed the discovery by Beatty (3) that thermoset castings prepared from bisphenol A–formaldehyde resins are clear, tough, and light in color. Production of bisphenol A by Chemische Werke Albert in Germany began in 1923 (4), and it was used as an intermediate in the manufacture of modified phenolic resins called Albertols; similar resins were introduced in the United States under the name Amberol. Although the first Chemische

Werke Albert bisphenol A plant employed hydrochloric acid as the condensing agent, sulfuric acid was used in most installations during this period because it does not present the problems of volatility and corrosiveness associated with hydrochloric acid. The product is difficult to purify when sulfuric acid is used, but bisphenol A of only moderate purity was satisfactory for use in bisphenol A–formaldehyde resins. [Some of the patented procedures using sulfuric acid were reviewed by Leibnitz and Naumann (5).]

Epoxy resins, which saw commercial production in the years following World War II, require the use of somewhat better quality bisphenol A. It was the introduction of linear polymers, such as the polycarbonates, however, which made it essential to develop practical routes to high-purity (99% or better) bisphenol A. Major advances toward this goal were made in a process (6) that involved treating acetone with an excess (7.5 moles) of phenol and anhydrous hydrogen chloride at 35–40°C, recovering a crystalline molecular compound of phenol and bisphenol A, and then recycling the filtrate containing the by-products. High-purity bisphenol A, fp 157.3°C, was obtained in a 74% yield by distilling phenol from the filter cake, and more bisphenol A was obtained from the recycled by-products.

The mechanism by which bisphenol A is formed is believed (7) to involve electrophilic attack of a protonated acetone molecule on the aromatic ring, followed by acid-catalyzed elimination of water to give the cation **1**. This

$$HO-\hspace{-2mm}\underset{}{\bigcirc}\hspace{-2mm}-\underset{\underset{CH_3}{|}}{\overset{\overset{CH_3}{|}}{C}}\oplus$$

1

intermediate is probably involved in the formation of not only bisphenol A **(2)** but also two of the major by-products: the 2,4′ isomer **3** and the trisphenol **4**. Dianin's compound (codimer) **5**, the third major by-product, may be formed by the reaction of phenol and mesityl oxide. It would not be surprising if most or all of the reactions by which these compounds are interrelated prove to be reversible. It has been shown that **2** and **3** can be equilibrated in a phenol solution containing hydrogen chloride, with the proportion of **2** decreasing as the temperature is increased (8). Under preparative conditions, the concentration of **3** is initially higher than the equilibrium value, but later part of it is converted to **2** (8,9). On the other hand, some tendency for **5** to accumulate on recycle has been reported, although **3** and a compound identified as the 2,2′-isomer were smoothly converted to bisphenol A (10). Most modern bisphenol A processes involve a crystallization step.

Another important contribution to bisphenol A technology was the discovery (11) that many compounds of bivalent sulfur markedly accelerate the acid-catalyzed reaction of a ketone with a phenol. The most widely

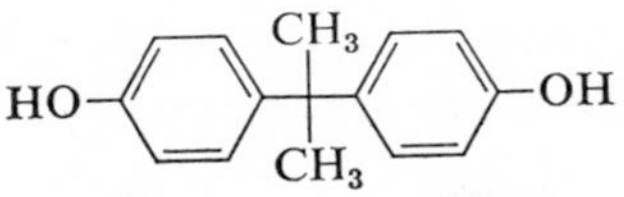

2

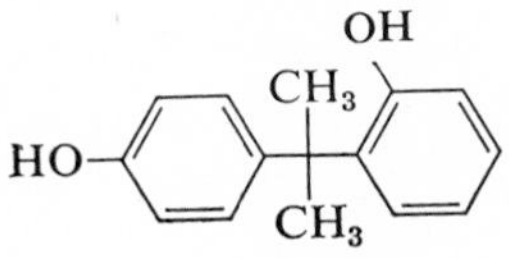

3

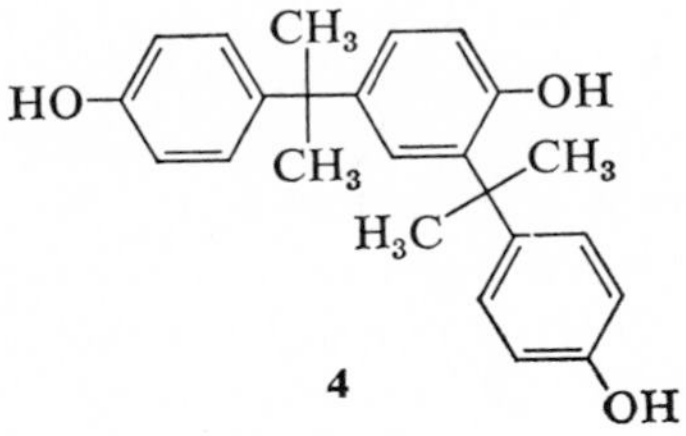

4

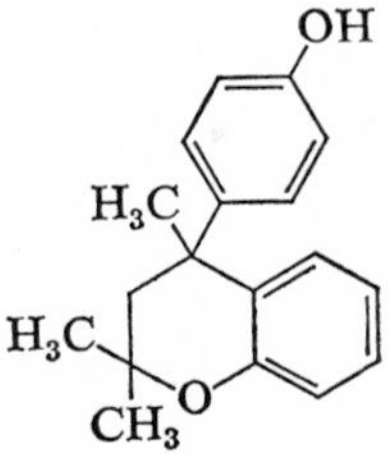

5

studied promoters are mercaptans (11) and mercaptoalkanoic acids (12). The effectiveness of these compounds has been attributed (13) to the formation of intermediate mercaptols or hemimercaptols, which are supposed to be more reactive than the ketones themselves. Their value consists mainly in permitting satisfactory reaction rates at relatively low temperatures, thus reducing the formation of **3** and other by-products.

Continuous processes based on catalysis by hydrogen chloride or hydrogen chloride plus a mercaptan promoter have been devised (14,15). These processes necessarily involve such operations as neutralization or stripping of the catalyst, or both. A catalyst particularly suited to continuous operation is an insoluble, strongly acidic cation-exchange resin (16) such as a sulfonated styrene-divinylbenzene copolymer. Reaction streams can simply be pumped through beds of this catalyst, thus largely eliminating problems of corrosion and catalyst recovery. A promoted cation-exchange resin catalyst, which can be prepared by partial esterification of the sulfonic acid groups by a lower alkyl mercapto alcohol (17), has been used to prepare bisphenol A continuously (18,19).

Bisphenol A can also be prepared by a base-catalyzed reaction of phenol with acetone (20), but prolonged reaction at high temperatures is necessary; the method has no obvious advantages.

The method required to purify bisphenol A depends, of course, on the method of preparation. Examples of the more important purification methods are given in Section I.B.

Although the reaction of acetone with phenol is the most important preparative method for bisphenol A and the only one known to have been operated on a commercial scale, several others have been reported in the literature. One of these (21,22) is a modification of the well-known manufacture of phenol and acetone by the acid-catalyzed rearrangement of cumene hydroperoxide (Eq. 1). After decomposition of the hydroperoxide is

$$
\text{[benzene ring]}-\underset{\underset{CH_3}{|}}{\overset{\overset{CH_3}{|}}{C}}-OOH \xrightarrow{\text{H}^+} \text{[benzene ring]}-OH + CH_3COCH_3 \qquad (1)
$$

complete, the reaction is continued under more forcing conditions, preferably after addition of a mercaptan and either addition of phenol (23) or removal of acetone (24) to adjust the stoichiometry. Yields as high as 96% have been reported (23).

Since p-isopropenylphenol can be converted to the cation 1 by addition of a proton, it is not surprising that it reacts with excess phenol under acidic conditions to give bisphenol A in yields exceeding 80% (25). The linear dimer of p-isopropenylphenol, a mixture consisting largely of 2,4-bis(4-hydroxyphenyl)-4-methyl-1-pentene, gives similar results (26,27).

Many compounds that can be regarded as simple functional derivatives of acetone have been converted to bisphenol A by reaction with phenol and a strong acid; examples include 2,2-bis(alkylthio)propanes (28), isopropenyl acetate (29), methyl isopropenyl ether (30), and 2-chloropropene (31,32). Of greater potential importance than these, however, are allene, methylacetylene, and mixtures of the two. These materials are available from petrochemical feedstocks by the cracking of isobutylene or propylene, and they react with phenol to give bisphenol A in good yields. Boron trifluoride catalysts have been most widely studied (33), but strongly acidic cation-exchange resins can also be used (34). Soviet investigators have suggested (35,36) that this route might be economically competitive with the conventional ones based on acetone.

The demand for bisphenol A in the United States in 1968 was 142 million lb, and production capacity in 1969 was 188 million lb. About 75% of the bisphenol A produced was used in the manufacture of epoxy resins and 18% for polycarbonates (37).

Bisphenol A apparently presents no significant toxicity problems. Occasional exposure to its vapors does not appear to be harmful. It may cause mild skin irritations and should be removed by washing with soap and water. Since eye damage may result from contact with either solid or dissolved bisphenol A, thorough flushing of an eye after contact and prompt medical attention are recommended (38).

B.   Synthesis

High-purity bisphenol A is commercially available. Further purification, if necessary, can be accomplished by recrystallization from toluene (39) or chlorobenzene (40).

The four examples that follow illustrate the most important modifications of the general acetone-phenol method. In the first two, only a slight excess of phenol is used, and bisphenol A is isolated directly either by extraction of impurities with chloroform or by recrystallization from toluene. In the latter two, which employ excess phenol, bisphenol A is crystallized in the form of the phenol adduct. The first three are suitable for a laboratory preparation; the fourth is probably similar to actual manufacturing operations.

Freshly distilled phenol should always be used for the preparation of bisphenols, and, whenever feasible, the reactions should be carried out under a blanket of an inert gas, such as argon or nitrogen.

1. METHOD OF DUNCAN AND WIDIGER (41)

In a 3-liter glass flask equipped with a stirrer, a condenser, and a gas inlet dip tube are placed 775 g of 36.5% hydrochloric acid, 790 g (8.4 moles) of phenol, 232 g (4.0 moles) of acetone, 137 g of o-dichlorobenzene, and 8.4 g of n-octylmercaptan. Anhydrous hydrogen chloride is passed into the mixture throughout the run. The flask is immersed in a water bath, preheated to 40°C. The bath temperature is raised to 50°C over 1 hr, maintained there for 4 hr, and lowered to 40°C over 1 hr. The slurry is diluted with 200 ml of chloroform and the resulting mixture filtered. The filter cake is washed with five 100-ml portions of chloroform and then with 1 liter of water, and dried. The yield of bisphenol A, fp 157.0°C, is 716 g (78.5%).

Distillation of the organic layer from the combined filtrate and washings gives an additional 128 g (14%) of a mixture of bisphenols.

2. METHOD OF DE JONG (42)

A solution of 395 g (4.2 moles) of phenol in 736 g of toluene is cooled to 20°C. To this is added with stirring, and while maintaining the temperature at 20°C, 522 g of 77.5% sulfuric acid and 1.5 ml of thioglycolic acid. Acetone (116 g, 2.0 moles) is added to the stirred mixture over 3 hr at 25°C. The mixture is stirred for an additional 7 hr at 35°C, then diluted with 1060 g of water and heated to 82°C to effect complete solution. The aqueous layer, which contains most of the sulfuric acid, is removed. One liter of water is added to the organic layer; then solid sodium bicarbonate is added in small portions until the pH rises to 4. The stirred mixture is allowed to cool to 25°C over about 3 hr, and the crystallized bisphenol A is recovered by centrifugation and dried. The yield is 425 g (93%), mp 156.4–158°C.

Analysis of the toluene mother liquor showed that it contained 14 g of phenol and 17.4 g of bisphenols. The aqueous layers contained 10.2 g of phenol and 7.6 g of phenolsulfonic acid.

3. METHOD OF STOESSER AND SOMMERFIELD (6)

In a glass flask equipped with a stirrer, a condenser, a thermometer, and a gas inlet tube is placed 530 g (5.6 moles) of phenol and 43.5 g (0.75 mole) of acetone. This mixture is heated to 40°C and held at that temperature with stirring while 13 g of hydrogen chloride gas is being added. Stirring continues for 19 hr at 35–40°C; during this time crystallization occurs. The resulting slurry is filtered, and the filter cake is washed with 150 g of molten phenol at 42–45°C. (The temperature is critical, since the solubility of the crystals increases rapidly with a rise in temperature. At substantially lower temperatures the phenol would freeze.) The filter cake is then stripped by gradually increasing the pot temperature to 210°C and decreasing the pressure to 3 mm. The yield of residue product bisphenol A, fp 157.3°C, is 127.3 g (74%).

The combined filtrate and washings is freed of hydrogen chloride and water by stripping to a pot temperature of 60–65°C/20 mm. It is recycled, whereupon it gives additional bisphenol A of high purity.

4. METHOD OF GROVER AND RICHARDSON (19)

The continuous process of Grover and Richardson represents an improvement over that of Apel, et al. (18). The catalyst is a sulfonated styrene-divinylbenzene copolymer, such as Dowex 50-X-4, in which about 20% of the sulfonic acid groups have been esterified with mercaptoethanol. It may be prepared in the laboratory by drying the ion-exchange resin (in the H^+ form) by azeotropic distillation with n-heptane, adding the mercaptoethanol, and heating under reflux until the calculated amount of water is evolved (17).

The apparatus consists essentially of a reaction zone, a concentrator, a crystallization zone, a solid-liquid separation zone, a final evaporation zone, and a rearrangement reactor. The feed to the reaction zone contains fresh phenol and acetone along with recycled phenol, acetone, bisphenol A, and by-products. Typical analysis is 84.7% phenol, 4.5% acetone, 0.1% water, 6.4% bisphenol A, and 4.3% by-products. The reaction zone is an elongated chamber containing a fixed bed of catalyst; with a temperature of 75°C and a residence time of 1 hr, about half of the acetone is converted to bisphenol A. The effluent from the reaction zone passes through the concentrator, in which the pressure is about 200 mm and the temperature high enough to effect overhead removal of all acetone and water plus some phenol. The concentrator bottoms pass into the crystallization zone and are cooled to about 40°C to bring about crystallization of the bisphenol A-phenol adduct. The resulting slurry passes into the solid-liquid separation zone, in which the crystals are recovered by centrifugation, washed with phenol at about 40°C, and then passed to the final evaporation zone. Here high-purity bisphenol A is obtained as a residue product by heating to a temperature not exceeding 200°C and at a pressure low enough to effect complete removal of phenol. The combined filtrate and washing from the solid-liquid separation zone—a mixture containing about 86.4% phenol, 7.0% bisphenol A, and 6.6% by-products—is passed into the rearrangement reactor (containing the catalyst) and maintained at about 75°C. The effluent, about 86.1% phenol, 8.3% bisphenol A, and 5.6% by-products, is recycled to a feed tank and thence to the reaction zone. Although this change in composition is small, it permits a substantial reduction in the steady-state concentration of by-products recycled to the reaction zone.

Auxiliary equipment includes distillation columns, whose main function is to remove water from the concentrator overhead product before recycle to the reaction zone.

II. BISPHENOL C

A. Introduction and Survey of Chemistry

The bisphenol of cyclohexanone, 1,1-bis(4-hydroxyphenyl)cyclohexane, was first prepared by Schmidlin and Lang in 1910 by the sulfuric acid-catalyzed reaction of phenol with cyclohexanone (43). The chemistry of its preparation appears to be entirely analogous to that of bisphenol A, and the synthesis has been described as being particularly easy (44). Bisphenol C polycarbonate has attracted considerable attention in the Soviet Union, where it is called Ilon (45). Nevertheless, fully detailed directions for the preparation of the monomer do not seem to be available. The two examples given illustrate catalysis by both acid and base, as well as different purification methods.

B. Synthesis

1. METHOD OF MEYER AND SCHNELL (46)

A solution containing 400 kg (4250 moles) of phenol and 78 kg (790 moles) of cyclohexanone is cooled to 10°C and mixed with 80 kg of 38% hydrochloric acid. An exothermic reaction takes place with the temperature rising spontaneously to about 65°C. The mixture, which solidified to a mass of crystals after 15–20 hr, is allowed to stand for 6 days. It is then heated by passing in steam while 60 kg of a 50% sodium hydroxide solution is being added with stirring. When the mixture has fully liquefied, the lower aqueous layer is removed. The organic phase is then stirred with 500 kg of water at 90°C, and the phases are again allowed to separate while hot. The organic layer, which is now the lower one, is filtered hot and allowed to cool. When crystallization is complete, the mixture is centrifuged to give 252 kg of crystals and 250 kg of aqueous phenol. Distillation of the crystals gives

50.4 kg of phenol overhead and, as residue, 176 kg(82%) of bisphenol C, mp 186°C. Distillation of the mother liquor yields 175.8 kg of phenol and 16.4 kg of isomeric bisphenols. Extraction of the combined aqueous layers gives 35.8 kg of phenol.

2. METHOD OF BENDER, CONTE, AND APEL (20)

In a 3-liter, three-necked flask equipped with a stirrer, a thermometer, and a reflux condenser is placed 1410 parts (14.98 moles) of phenol, which was melted by heating to 50–60°C. To this is slowly added, with stirring, 726 parts of potassium hydroxide pellets (85% pure); the temperature is maintained at about 100°C by external cooling. Then 490 parts (4.99 moles) of cyclohexanone is added, and the stirred mixture is heated to the reflux temperature (pot temperature 145°C, vapor 105°C). Heating under reflux continues for 24 hr (final pot temperature 140°C, vapor 101°C). The reaction mixture is cooled and then acidified by the cautious addition of $6N$ hydrochloric acid to a final pH of 2. The precipitate is collected by filtration and washed with distilled water until tests for hydrogen ion and chloride ion are negative. The dried solid is stripped under reduced pressure to a final pot temperature of 260°C/20 mm. The residue is dissolved in an equal amount of boiling toluene, and the solution is allowed to cool to room temperature. The yield of light tan crystals of bisphenol C, mp 187°C, is 85%.

III. BISPHENOL F

A. Introduction and Survey of Chemistry

The reaction of phenol with formaldehyde forms the basis of the phenolic resin industry. For the purposes of this Chapter, however, we are concerned with it mainly as a preparative method for bis(4-hydroxyphenyl)methane, commonly known as bisphenol F (47). Formaldehyde is much more reactive toward the ortho positions of the phenol nucleus than either the higher aldehydes or ketones are because of its smaller steric requirements. Therefore, although bisphenol F was isolated from the products of phenol-formaldehyde reactions as early as 1892, the development of a practical synthetic method has been very slow.

Because the phenol molecule is di or trifunctional in its reaction with formaldehyde, polymeric products normally predominate. Typical phenolic resins are solid condensation products of relatively low molecular weight which, after blending with fillers and (sometimes) crosslinking agents such as hexamethylenetetramine, are cured to three-dimensional networks in a mold. In practice, it is desirable to effect the curing reaction as rapidly as possible. Phenolic resins, such as 6, having a high ortho content are therefore preferred, since the para positions are most reactive in crosslinking. [For references to the work of Bender and others in this field, see the papers of

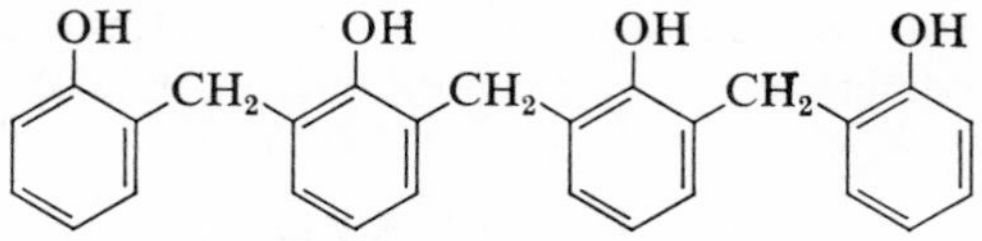

6

Peer (48) and Partansky (49).] The formation of high-ortho structures is favored by the presence of bivalent cations such as magnesium and zinc. They are believed to form chelate rings with the phenolic hydroxyl and an *ortho*-hydroxymethyl group. In 1967 it was reported (49) that parasubstitution is favored by the use of strongly acidic catalysts at low temperatures and in the presence of alcoholic diluents.

In view of the foregoing considerations, it is not surprising that published procedures for preparing bisphenol F from phenol and formaldehyde have generally given poor yields (e.g., 22%) (50). The crude products, however, consist largely of the isomeric dihydroxydiphenylmethanes, and mixtures suitable for polymer preparation have been isolated by distillation (51). The proportion of the 4,4′ isomer was reportedly (52) improved by using dimethylolurea as the source of formaldehyde. The mechanism (53) of this reaction appears to be complex, and somewhat similar results can be obtained by starting with a mixture of urea and formaldehyde (54).

Various indirect routes to bisphenol F have also been proposed, such as the following sequence shown in Eq. 2 as reported by Zigeuner et al. (55).

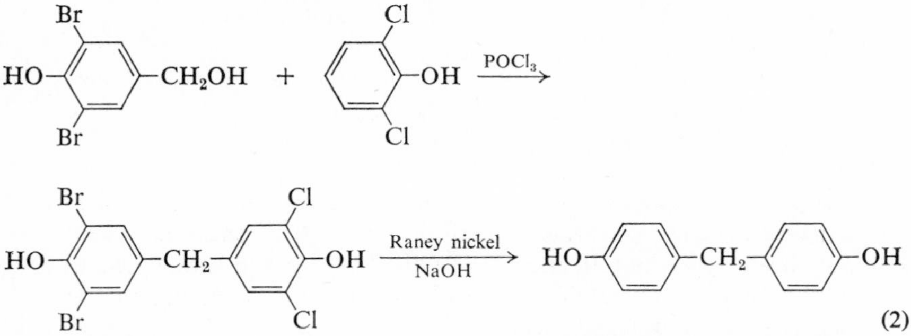

(2)

Levine and Temin (56) prepared bisphenol F by the catalytic hydrogenation of 4,4′-dihydroxybenzophenone (see Section IX). Bisphenol F has also been prepared (57) by the sequence of Bender et al. in Eq. 3. Pure bisphenol F,

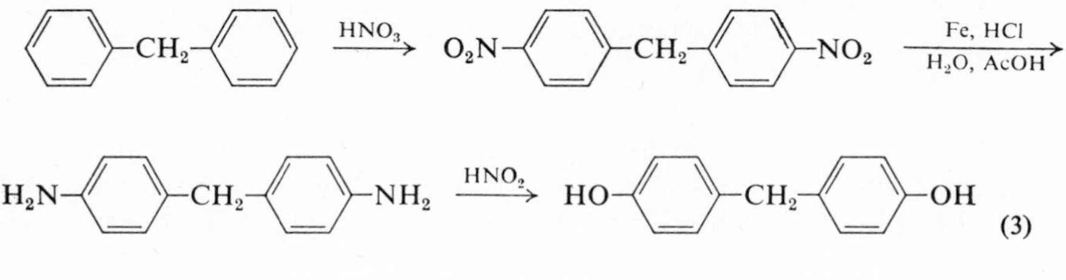

(3)

however, is not commercially available, even though phenolic resins rich in dihydroxydiphenylmethanes play a significant role in epoxy resins.

B. Synthesis

According to the older literature, the melting point of pure bisphenol F is about 158°C (55). Analysis of solubility data, however, shows that this type of product is only about 95% pure (57). Pure bis(4-hydroxyphenyl)-methane, mp 162–163°C, was obtained by two recrystallizations of the impure product, first from a mixture of *n*-butanol and toluene at room temperature and then from pure toluene. It was concluded that bisphenol F must be appreciably labile under the recrystallization conditions previously used.

1.   PHENOL-FORMALDEHYDE REACTIONS

a.   Method of Partansky (49)

A mixture of 100 parts (1.06 mole) of phenol, 7 parts (0.21 mole) of 91% paraformaldehyde, 14 parts of 37% hydrochloric acid, and 100–150 parts of 50% ethanol is allowed to react for about 25 days at room temperature. The yield of 4,4'-methylenediphenol is 85–90%. (No further details are given.)

b.   Method of Farnham and Klosek (50)

To a stirred mixture of 2000 g (21.25 moles) of phenol, 2700 g of 37% hydrochloric acid, and 1300 g of water, is added dropwise 292 g of 37% formalin solution (containing 108 g, or 3.60 moles, of formaldehyde). Addition time is 1 hr, and the temperature is maintained at 25–30°C. Stirring is stopped, and the mixture is allowed to stand for 1.5 hr. The upper organic layer is separated, washed once with water, and stripped under reduced pressure to a final pot temperature of 180°C (ca. 50 mm). The yield of residue product is 633 g.

(When a portion of this residue was analyzed by fractional distillation, the diphenol fraction, bp 170–210°C/0.5 mm, amounted to 65.7%. From the melting range it was estimated to contain about 53% of 4,4'-methylenediphenol.)

Finally 200 g of the original residue is dissolved in 150 ml of hot 40% acetic acid. On cooling there is obtained 49.5 g (22%) of bisphenol F, mp 156–158°C.

2.   PHENOL-DIMETHYLOLUREA REACTION (52)

Dimethylolurea is recrystallized twice from alcohol; mp 125–126°C.

A mixture of 3000 g (31.88 moles) of phenol and 10 ml of concentrated hydrochloric acid is heated to 45°C. A slurry of 300 g (2.50 moles) of dimethylolurea in 600 ml of water is added over 1 hr with rapid stirring; the temperature is held at 45–60°C during addition. Stirring is continued for about 45 min with the temperature falling to 40°C; the mixture is then heated for 30 min at 60°C. Concentrated hydrochloric acid (200 ml) is added, and the solution is heated under reflux for 1 hr. The reaction mixture is distilled under reduced pressure to a pot temperature of 130°C/20–25 mm; the residue is extracted twice with boiling water and then repeatedly with boiling toluene. The cooled extracts deposit 898.5 g of crude crystalline product. Recrystallization from water gives 479 g (48%) of snow-white bisphenol F, mp 154–157°C. Distillation of a portion of the product gives purer material, bp 237–243°C/12–13 mm, mp 159–161.5°C.

3. PURIFICATION BY METHOD OF HELLER, BOTTENBRUCH, AND SCHNELL

According to the patent held by Heller et al. (58), bisphenols can be purified by recrystallization from aqueous solutions containing only minor amounts of base (i.e., much less than would be needed to form the monosodium salts). Examples were given for seven bisphenols.

First 100 g (0.50 mole) of partially purified bisphenol F, fp 157.4°C, is suspended in 300 ml of water, and the slurry is heated to 100°C. A 45% solution of sodium hydroxide is then added in small portions until the solution becomes clear; 5.5 ml (containing about 0.093 mole of sodium hydroxide) is required. The solution is allowed to cool; the crystals are recovered by filtration, washed with two 300-ml portions of water, and dried *in vacuo* at 80°C. The yield of purified bisphenol F, fp 161.1°C, is 85 g (85%).

IV. TETRACHLOROBISPHENOL A

A. Introduction and Survey of Chemistry

The synthesis of 4,4′-isopropylidene-bis(2,6-dichlorophenol), usually known as tetrachlorobisphenol A, was first reported by Moss (59) in 1938. Although this compound and its derivatives have been studied in many biological applications, tetrachlorobisphenol A is practically nontoxic, whether ingested orally or encountered topologically. Like other chlorine-containing phenols, it should be handled with care to prevent dermatitis in hypersensitive people. The only known significant use of tetrachlorobisphenol A is as a flame-resistant component in polymers.

Tetrachlorobisphenol A is prepared by chlorination of bisphenol A as shown in Eq. 4. The chlorinating agent is generally chlorine (60–63), but

$$\text{(4)}$$

reference has been made to the use of sulfuryl chloride (64). Several solvents have been used, however, including tetrachloroethylene (61), aqueous tetra-chloroethylene (60), 1,2-dichloroethane (63), aqueous benzene (62), and glacial acetic acid (64,65). Some methods employ Lewis acid catalysts such as ferric chloride (66) and zinc chloride (67).

B. Synthesis

The method of synthesizing tetrachlorobisphenol A described is that of Bryner and Dietzler (60). Water is used to improve the color and purity of the product. The function of the tetrachloroethylene added with the chlorine is to prevent plugging of the chlorine inlet, and a phenolic stabilizer is added to prevent its reacting with chlorine.

> Bisphenol A (228 g; 1 mole) is mixed with 600 ml (978 g) of tetrachloroethylene containing 9.8 g of water. Chlorine (292 g, 4.12 moles) is added as rapidly as it can be absorbed in the reaction mixture through a dip tube over a 3.5-hr period at 55–60°C, after being bubbled through a tetrachloroethylene solution containing 0.1% tetrachloro-bisphenol A. While the reaction mixture is still hot, nitrogen is swept through to remove excess chlorine and dissolved hydrogen chloride. The batch is cooled to 25°C, and the crystals are recovered by filtration, then washed with two 50-ml portions of tetrachloroethylene and four 100-ml portions of water. On drying there is obtained 275 g (75% yield) of tetrachlorobisphenol A, mp 133–134°C.

The utility of this procedure was demonstrated in a series of eleven experiments in which the mother liquor from previous runs was used in each succeeding run. The average yield of tetrachlorobisphenol A in these twelve runs was 91%, and the melting point was always 133–134°C or better.

V. PHENOLPHTHALEIN

A. Introduction and Survey of Chemistry

Phenolphthalein (7) (68) has usually been prepared by one method; namely, the reaction of phenol with phthalic anhydride (Eq. 5). Some

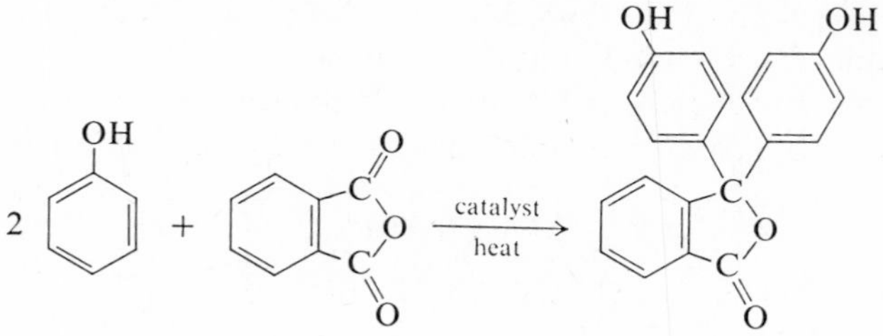

(5)

7

investigators have used phthaloyl chloride in place of the anhydride, but this method has little significance.

In the first recorded synthesis, by Baeyer in 1871, the catalyst was sulfuric acid; this method was further studied by Molle (69). Some of the other agents used to effect this condensation are: zinc chloride (70), zinc chloride plus inorganic (71) and organic (72) sulfonyl chlorides, aluminum chloride plus ferric chloride (73), and zinc chloride plus sulfuric acid (74).

Phenolphthalein seems to be made in the United States by only one manufacturer, the Monsanto Chemical Company. Therefore, no production figures are available. The volume is probably relatively small, however, since the known uses are only as a pH indicator (75) and as a laxative (76).

B. Synthesis

The preparative method described here is on a laboratory scale, yielding about 300 g of crude phenolphthalein. The process was readily amenable to scale-up to the level of almost 1.4 tons (72). The two purification procedures given are those that appear to give material of the highest purity.

1. METHOD OF GAMRATH (72)

A mixture of 148 g (1 mole) of phthalic anhydride, 188 g (2 moles) of phenol, 68 g (0.5 mole) of zinc chloride, and 47.6 g (0.25 mole) of p-toluenesulfonyl chloride is charged into a 1-liter, three-necked flask equipped with a paddle stirrer, a thermometer, and a vent line. The stirred vessel is closed, heated in an oil bath to 120°C, and maintained at that temperature for 18 hr. To the vessel is added 100 ml of warm water; the mixture is stirred until a homogeneous mass is obtained. The contents of the flask is emptied into 500 ml of cold water, the mixture is stirred for 10 min, and the slurry is filtered to collect the solid. It is washed with warm water until the filtrate is no longer acid to methyl orange test paper. After drying there is obtained 303 g (95%) of crude phenolphthalein.

2. PURIFICATION FROM ETHANOL (74)

In the purification of phenolphthalein from ethanol, which gives U.S.P. grade material, the crude phenolphthalein (150 g) is slurried with 110 g of ethanol; the solution is heated to 60°C, then cooled to 20°C and held there for 12 hr. After centrifuging, the phenolphthalein is washed again with ethanol to give a recovery of 140 g. The latter is added to 500 g of ethanol; the mixture is heated to effect solution, 6 g of decolorizing charcoal is added, and the mixture is refluxed for 30 min. After filtering, 350 g of alcohol is distilled off and the residue cooled. The solid is collected, recrystallized again from ethanol, and dried in vacuo. The phenolphthalein, which is collected in >90% recovery, melts at 261–262°C.

3. Purification from Tetrahydrofuran (77)

The phenolphthalein (150 g) is mixed with 400 ml of tetrahydrofuran and heated to effect solution. After slurrying with 10 g of decolorizing charcoal, the mixture is filtered and the filtrate cooled to 5°C for 48 hr. The resulting solid is collected to give 106 g of the 1:1 complex of phenolphthalein and tetrahydrofuran. From the filtrate, which is cooled again under the same conditions, an additional 63 g of complex is obtained. The combined crystal crops are dried at 30°C *in vacuo* to give 138 g (92% recovery) of phenolphthalein, mp 260.5–262.5°C.

VI. 4,4′-DIHYDROXYDIPHENYLSULFONE

A. Introduction and Survey of Chemistry

The preparation of 4,4′-dihydroxydiphenylsulfone (usually referred to as bisphenol S) was first reported by Glutz (78) in 1868. Bisphenol S (**8**) is available by a variety of routes; the most attractive seem to be the reaction of phenol with sulfuric acid and the hydrolysis of 4,4′-dichlorodiphenylsulfone. Because of its potentially low price, bisphenol S has been studied extensively in polymer and in other applications. In the latter category, its use in U.S. Steel's Ferrostan tin-plating process may be mentioned (79,80).

Bisphenol S resembles other phenolic compounds in its toxicological behavior; that is, it is a mild irritant to the skin and should be removed immediately with water. By acute oral ingestion, it is practically nontoxic, since the approximate lethal dose is 10 g/kg of body weight for rats and 7 g/kg for rabbits (81).

In theory, the direct condensation of phenol with concentrated sulfuric acid, shown in Eqs. 6 and 7, should be the most inexpensive route to 4,4′-dihydroxydiphenylsulfone (**8**). However, this reaction gives a mixture of isomers containing 60–70% of the desired compound (82,83).

$$HO\!-\!\langle\ \rangle + H_2SO_4 \longrightarrow HO\!-\!\langle\ \rangle\!-\!SO_3H + H_2O \tag{6}$$

$$HO\!-\!\langle\ \rangle\!-\!SO_3H + HO\!-\!\langle\ \rangle \longrightarrow HO\!-\!\langle\ \rangle\!-\!SO_2\!-\!\langle\ \rangle\!-\!OH + H_2O \tag{7}$$

By the use of an azeotropic agent to remove the by-product water, higher yields of a cleaner product were obtained (83). Several solution techniques have been reported for the isolation of the less soluble 4,4′-dihydroxydiphenylsulfone (82,84). Previous work using sulfur trioxide as sulfonating agent gave a dihydroxydiphenylsulfone mixture of unstated composition (85).

A method that should be receiving increased attention and is based on the alkaline hydrolysis of 4,4′-dichlorodiphenylsulfone is presented in Eq. 8 (86). Methods have been developed which give high yields of pure 4,4′-dichloro-

$$Cl\!-\!\langle\bigcirc\rangle\!-\!SO_2\!-\!\langle\bigcirc\rangle\!-\!Cl + 4NaOH \longrightarrow NaO\!-\!\langle\bigcirc\rangle\!-\!SO_2\!-\!\langle\bigcirc\rangle\!-\!ONa$$

$$+ \; 2NaCl \; + \; 2H_2O$$

(8)

diphenylsulfone, a monomer required for the commercial Polysulfone resins (87). The sulfonylation of chlorobenzene is effected by reaction with sulfur trioxide and excess diethyl sulfate, with the reaction probably involving diethyl pyrosulfate as an intermediate (Eqs. 9 and 10). The sulfonylation of

$$SO_3 + (C_2H_5)_2SO_4 \xrightarrow{<10°C} C_2H_5O\!-\!SO_2\!-\!O\!-\!SO_2\!-\!OC_2H_5 \qquad (9)$$

$$C_2H_5O\!-\!SO_2\!-\!O\!-\!SO_2\!-\!OC_2H_5 \; + \; 2\langle\bigcirc\rangle\!-\!Cl \; + \; SO_3 \xrightarrow{<15°C}$$

(10)

$$Cl\!-\!\langle\bigcirc\rangle\!-\!SO_2\!-\!\langle\bigcirc\rangle\!-\!Cl \; + \; 2C_2H_5O\!-\!SO_2\!-\!OH$$

chlorobenzene can also be effected by reaction with sulfur trioxide and phosphorus pentoxide (88).

A third approach involves the synthesis of bis(hydroxyphenyl) compounds that contain sulfur in a lower oxidation state, followed by oxidation to bisphenol S (8). Both the sulfide 9 and the sulfoxide 10 can readily be prepared in high purity from appropriate sulfur chlorides. Uncatalyzed reactions of phenol with either sulfur dichloride (89,90) or thionyl chloride (91,92) give the sulfide 9, as shown in Eqs. 11 and 12, respectively. Purification of 9 is readily accomplished from dilute sodium bicarbonate solution (58). If a Lewis acid catalyst is used, the product of the reaction between phenol and thionyl chloride is the sulfoxide 10 (Eq. 13) (93).

$$2\ \underset{OH}{\langle\bigcirc\rangle} \; + \; SCl_2 \longrightarrow HO\!-\!\langle\bigcirc\rangle\!-\!S\!-\!\langle\bigcirc\rangle\!-\!OH \qquad (11)$$

9

$$2\ \underset{OH}{\langle\bigcirc\rangle} \; + \; SOCl_2 \quad \begin{array}{c} \xrightarrow{CHCl_3}\ \textbf{9} \\[4pt] \xrightarrow[C_2H_2Cl_4]{AlCl_3}\ HO\!-\!\langle\bigcirc\rangle\!-\!SO\!-\!\langle\bigcirc\rangle\!-\!OH \end{array}$$

(12)

(13)

10

The sulfide **9** was oxidized to the sulfone **8** by Tassinari (78) in 1890. Oae and co-workers converted **9** to the bis(*p*-nitrobenzoate), oxidized it with perbenzoic acid in chloroform, and hydrolyzed the diester-sulfone to **8** with alcoholic potassium hydroxide (90). The oxidation of **10** with hydrogen peroxide in acetic acid also gave **8** (93).

B.   Synthesis

Three detailed preparations are given: 4,4′-dichlorodiphenylsulfone, bisphenol S from 4,4′-dichlorodiphenylsulfone, and bisphenol S from phenol and sulfuric acid.

1.   4,4′-DICHLORODIPHENYLSULFONE (87)

To 462 g (3 moles) of diethyl sulfate in a 2-liter vessel equipped with a thermometer and a good agitator, 80 g (1 mole) of stabilized sulfur trioxide is added with stirring. The mixture is maintained at a temperature below 10°C for 3.5 hr, when 450 g (4 moles) of chlorobenzene is added to the reaction mass. A 1°C rise in temperature is noted. Sulfur trioxide (80 g, 1 mole) is then added dropwise over a 1.5-hr period while the temperature is kept below 15°C. An equal volume of methylene chloride is added, and the reaction mixture is washed with water and then with a 10% aqueous sodium hydroxide solution. The organic layer is separated, dried over anhydrous magnesium sulfate, and filtered. Methylene chloride is removed by evaporation. The excess diethyl sulfate is then recovered by distillation under reduced pressure, leaving a colorless residue that solidifies upon cooling. When the solid, 222.4 g (77.5%), is analyzed by vapor-phase chromatography, 98.9% pure 4,4′-dichlorodiphenylsulfone is noted.

2.   4,4′-DIHYDROXYDIPHENYLSULFONE FROM 4,4′-DICHLORODIPHENYLSULFONE

The procedure for obtaining 4,4′-dihydroxydiphenylsulfone from 4,4′-dichlorodiphenylsulfone is based on an elaboration of a published synthesis (86) as developed by F. N. Apel and L. B. Conte, Jr, of Union Carbide Corporation.

A mixture of 254 g (0.88 mole) of 4,4′-dichlorodiphenylsulfone, 214 g (5.35 moles) of sodium hydroxide in 725 ml of water, and 5 g of cuprous oxide is placed in the copper liner for a 2.8-liter rocking bomb. The liner is placed in the bomb, heated to 280–290°C, and maintained at that temperature for 6 hr.

The bomb is cooled, the liner removed, the contents filtered, and the clear filtrate acidified with concentrated hydrochloric acid to a pH of 1. The resulting pasty mass is filtered and washed with water until the filtrate gives a negative chloride ion test. The solid is dried at 80°C under reduced pressure to give 202 g (92%) of crude 4,4′-dihydroxydiphenyl-sulfone, mp 244–246°C.

Purification is effected by dissolving 1816 g of crude 4,4'-dihydroxy-diphenylsulfone in 2500 ml of hot ethanol (94). Charcoal (25 g) is added, the solution is filtered hot, and 4280 ml of hot water is added to the filtrate. The mixture is heated to effect complete solution and then cooled slowly to room temperature (ca. 18 hr). The resulting solid is filtered and dried at 80°C *in vacuo* to give 1730 g (95% recovery) of white needles of 4,4'-dihydroxydiphenylsulfone, mp 245–246°C.

3. 4,4'-DIHYDROXYDIPHENYLSULFONE FROM PHENOL AND SULFURIC ACID (83)

Phenol (376 g, 4 moles) is charged into a 1-liter flask equipped with a mechanical stirrer, a Dean-Stark trap, and a condenser; it is heated to 150°C. Concentrated sulfuric acid (196 g, 2 moles) is slowly added to the stirred reaction mixture, the temperature rising to 160°C. After 2 hr, the trap is filled with benzene, and 20 ml more is added to the reaction mixture while the temperature is maintained. At the end of 6 hr, benzene addition is made in 2-ml portions, to maintain reflux. Water is removed intermittently in 2-ml increments until a total reaction time of 13.25 hr has elapsed and 33 ml of benzene has been used. Isolation of the dihydroxydiphenylsulfone is accomplished simply by adding 300 ml of hot water continuously to the reaction mixture and removing the benzene by distillation. After all the benzene has been removed, the resulting solid is collected by filtration and washed with water to remove excess phenol and phenolsulfonic acid.

Purification is carried out as follows (94): crude dihydroxydiphenyl-sulfone (500 g) is dissolved in 850 ml of acetone, and 1700 ml of warm benzene is added. The solution is chilled, some seed crystals of 4,4'-dihydroxydiphenylsulfone are added, and the resulting solid is collected and dried. It is purified further by repeating this procedure and using 500 ml of acetone and 1000 ml of warm benzene to give 163 g of solid after drying at 70°C *in vacuo*.

The solid is dissolved in 210 ml of hot ethanol to give an amber solution. After treatment with 10 g of charcoal, the mixture is filtered, and 670 ml of hot water is added. After the liquid has been cooled slowly and chilled overnight, the white solid is collected and dried at 80°C under reduced pressure. The product is 150 g of pure 4,4'-dihydroxydiphenylsulfone, mp 245.5–247°C.

Material made by this procedure is readily converted quantitatively to a high polymer. Further scale-up to the 5-lb level is easily achieved (94).

VII. RESORCINOL

A. Introduction and Survey of Chemistry

Resorcinol (*m*-dihydroxybenzene) (95) was first prepared by Hlasiwetz and Barth in 1864 by the alkali fusion of galbanum and asafetida resins. In

1866 it was synthesized by Körner, who fused *m*-iodophenol with potassium hydroxide. Its value as an intermediate for the manufacture of azo, triphenylmethane, and other dyestuffs was soon recognized, and a commercial process for making it by alkali fusion of *m*-benzenedisulfonic acid, introduced in 1875, was quickly devised. This process is still the only one known to be used to manufacture resorcinol.

Dyestuffs, notably eosin (**11**), still constitute a significant end use for

11

resorcinol, but larger amounts are consumed by the tire and rubber industry. Resorcinol-formaldehyde condensation products, which can be cured under quite mild conditions, are used as adhesives to bond the tire cord to the rubber stock; resorcinol itself is used as an antioxidant for rubber products. Resorcinol-formaldehyde resins are also used in laminated products such as plywood. A pharmaceutically important derivative of resorcinol is 4-*n*-hexylresorcinol, which is used in antiseptics (96,97). The manufacturing capacity for resorcinol in the United States in 1967 was estimated (97) as 24 million lb/year.

The two-step process by which resorcinol is manufactured is shown in Eqs. 14 and 15. In the classical process (98,99), described in Section VII.B,

$$\text{(14)}$$

$$\text{(15)}$$

the sulfonic acid was first converted to the soluble calcium salt, which was separated from sparingly soluble calcium sulfate and then converted to the sodium salt by treatment with sodium carbonate. The alkali fusion step required twice the calculated amount of sodium hydroxide. It is certain that modern industrial practice involves improvements over these operations, but the details have been kept secret. Some attention has been given to improvements in the sulfonation step; thus stabilized γ-sulfur trioxide has been proposed (100) as the active reagent, and anhydrous sodium sulfate has been added to inhibit the formation of sulfones (101). Improvements in the fusion reaction have also been achieved by modifying the apparatus. By carrying out the fusion in batches in an Inconel kneader mixer, it is possible to obtain high yields of resorcinol while using only a slight excess (5–20%) of sodium hydroxide (102). The stoichiometric amount of sodium hydroxide has been used successfully in a fluidized-bed reaction (103). A rotating continuous reactor has been described in which good conversions were obtained using 35% excess sodium hydroxide with a residence time of only about 30 sec (104).

An attractive route to resorcinol involving a sequence (105) starting with m-diisopropylbenzene (**12**) is shown in Eq. 16. This route has been

investigated in detail in several industrial laboratories, notably those of Distillers Company Ltd. in the United Kingdom and Hercules Inc. in the United States. The opinion has been expressed (106) that this process would be the preferred one in new plants of very high capacity (substantially greater than 6 million lb/year). Since no directions suitable for use as laboratory procedures were located, the process is described in some detail here.

Either m-diisopropylbenzene (**12**) or the paraisomer can be obtained in high purity and efficiency by a process (107) based on the alkylation of cumene with propylene; high temperatures (near 250°C) favor the formation of **12**. After purification of the desired isomer by fractional distillation, the other isomers and higher alkylation products are fed to a transalkylation reactor, where treatment with benzene converts them to cumene and diisopropylbenzenes for recycle.

The dihydroperoxide **13** is obtained by treating **12**, preferably in the presence of an initiator such as the monohydroperoxide **14**, with air at temperatures near 90°C. Oxidation is carried out in a continuous reactor coupled with a continuous extractor, in which the oxidate is contacted with a dilute

14

15

(ca. 4%) solution of sodium hydroxide (108); the aqueous phase contains **13** and some **15**, and the organic phase, consisting mainly of **12** and **14**, is recycled to the oxidizer. Since bases apparently inhibit certain reactions by which hydroperoxides decompose (109), oxidation is normally performed at pH 6–10 (106). Conditions are controlled to give a predetermined total hydroperoxide content; analytical methods have been described by Mair and Graupner (110).

Procedures have been developed for the separation of **13** from **15**. This can be accomplished by fractional extraction of the alkaline solution with solvents such as methyl isobutyl ketone (111), or by exhaustive extraction followed by precipitation of the disodium salt of **13** by the addition of a concentrated sodium hydroxide solution (112). However, this step is apparently not essential to efficient production of resorcinol.

The decomposition of **13** is usually carried out in acetone or methyl isobutyl ketone. Suitable catalysts include sulfuric acid (113), acid-washed clays (114), and clays containing free mineral acids (115). The recovery of resorcinol is complicated by the formation of high boiling by-products, which are probably formed in part from the **15** normally present in the feed. Thermolysis of these by-products increases the yield of resorcinol, but it also forms *m*-isopropenylphenol (**16**) as a distillable by-product. In a continuous

16

process (116) for cracking and recovery, the resorcinol and **16** were finally separated by distribution between water and toluene.

Resorcinol and the other dihydroxybenzenes have also been prepared by the hydrolysis of dichlorobenzenes or chlorophenols. The reaction of *o*-chlorophenol with sodium hydroxide in iron vessels at high temperatures (265–280°C) gave mixtures of resorcinol and catechol, both in the melt (117) and in solution (118). It has been shown by tracer techniques (119) that benzyne is the probable intermediate in some of the reactions that occur when chlorobenzene is treated with aqueous sodium hydroxide at 395°C.

The reactions cited previously probably also proceed via a benzyne intermediate as shown in Eq. 17. Mixtures attributable to the formation of inter-

(17)

mediate benzyne derivatives have been obtained analogously from p-chlorophenol (117,118) and m-chlorophenol (118). By-products also included dihydroxydiphenyl ethers and trihydroxybiphenyls (118). According to another report, the products of the reaction of p-chlorophenol with sodium hydroxide were resorcinol, 2,4'-biphenol, 4,4'-biphenol, phenol, and formic acid (98).

The alkaline hydrolysis of aryl chlorides proceeds at lower temperatures in the presence of copper and copper compounds. This appears to have been demonstrated in the case of p-chlorophenol, although the reports of different investigators conflict to some extent (120–122). In 1968 it was reported (123) that the copper-catalyzed reaction of aryl chlorides is faster with the sodium salt of a carboxylic acid than with sodium hydroxide. Equation 18 describes the overall course of the reaction. Under these conditions, none of the

$$RCO_2Na + ArCl + H_2O \rightarrow ArOH + RCO_2H + NaCl \qquad (18)$$

isomeric products attributable to aryne intermediates are observed. This can be rationalized by noting that the carboxylate anions are too weakly basic to generate arynes. An example of the preparation of resorcinol from m-dichlorobenzene by this method is given in Section VII.B.

Resorcinol has also been prepared from a technical grade of m-dichlorobenzene by Kamlet (124). This method, which is also outlined in Section VII.B, depends on the reaction sequence as depicted in Eq. 19.

Resorcinol is probably more toxic than bisphenol A, but it does not present serious hazards under ordinary conditions of use. It should not, of course, be taken internally, and inhalation of its dust or vapor should be avoided. In particular, skin contact should be minimized, since resorcinol may be readily absorbed through the skin. Both localized dermatitis and more widespread effects may follow, particularly in sensitive individuals (125–128).

B. Synthesis

Resorcinol of analytical reagent quality is commercially available. It is hygroscopic, however, and small amounts of water should be removed by vacuum distillation or sublimation. Trace amounts of phenol, colored impurities, etc., are also removed in this way.

1. PREPARATION FROM BENZENE BY SULFONATION AND ALKALI FUSION (98,99)

To 78 g (1 mole) of benzene is added, over about 2 hr with stirring, 250 g of 20% oleum (containing 0.62 mole of sulfur trioxide); the temperature is not allowed to exceed 45°C. Then 200 g of 66% oleum (1.65 moles of sulfur trioxide) is added over 2 hr, the temperature rising to about 70°C. Finally the mixture is heated for 1 hr at 90°C. The mixture is cooled and poured slowly into 2 liters of water; the resulting solution is heated and neutralized by the gradual addition, with stirring, of about 400 g of calcium carbonate. The resulting precipitate of calcium sulfate is collected by filtration, washed thoroughly with water, and discarded. The combined filtrate and washings are heated, stirred, and treated with small portions of solid sodium carbonate until the solution turns phenolphthalein paper red; about 100 g is required. The precipated calcium carbonate is collected by filtration, washed with water, and discarded. The combined filtrate and washings are evaporated to dryness and then dried at 130–140°C. The yield of disodium m-benzenedisulfonate is about 250 g (89%).

The alkali fusion reaction is conducted in a "baking" apparatus (99) consisting of an iron vessel, heated by an oil bath, and equipped with two thermometers and a wide line connected to a vacuum pump. The reactants are charged to an iron tray that can be inserted into the oil-heated vessel. Because the mixture of reactants foams greatly during dehydration, the tray cannot be more than half-filled. In the tray are placed 32 g (0.80 mole) of sodium hydroxide and 20 g of water. To this mixture, maintained at 200°C, 28.0 g (0.0993 mole) of disodium m-benzenedisulfonate is added slowly with good mixing. The tray is inserted into the baking apparatus, and its contents is dehydrated by heating at 200°C under reduced pressure. (If the mixture is not fully dehydrated at this relatively low temperature, the yield of resorcinol is diminished.) The pressure is reduced to 12–15 mm, and the mixture is heated at 320°C for 6 hr. The residue is dissolved in the minimum amount of water, and the solution is acidified with hydrochloric acid to a pH of 3 or less, heated to boiling to expel sulfur dioxide, and cooled. (In technical procedures less water can be used, the bulk of the sodium sulfite being recovered by filtration.) The solution is filtered, when necessary, then extracted continuously with ether until the aqueous layer gives a negative test with ferric chloride solution. The ether extract is dried over anhydrous sodium sulfate, the ether is removed by evaporation, and the residue is fractionally distilled under reduced pressure. The result is 0.5 g (5%) of phenol, bp 80°C/12 mm, and 8.5 g (78%) of pure, white resorcinol, mp 109–110°C.

The relation between the vapor pressure of resorcinol and temperature (125) is as follows:

Temperature, °C	Vapor pressure, mm
108.4	1
130	3.0
150	8.5
152.3	10
170	23.5
190	53.0
210.0	100
276.7	760

2. METHOD OF BURSAK, MOLTZAN, AND JOHNSTON (123)

In a 1-liter copper-lined autoclave are placed 0.4 mole of m-dichlorobenzene, 1.2 moles of sodium acetate, 4 g of cuprous oxide, and 500 ml of water. The mixture is heated under autogenous pressure to 290°C. At intervals samples are removed for analysis. By the time the temperature reaches 290°C, the pH has fallen to 4.5; it does not change thereafter. A sample removed after 3.5 hr at 290°C contains 93.4% resorcinol, 3.3% m-chlorophenol, 3.2% phenol, and 0.3% m-dichlorobenzene. Catechol and hydroquinone are not detected.

3. METHOD OF KAMLET (124)

A mixture of the three isomeric dichlorobenzenes is available from the high-temperature vapor-phase chlorination of benzene. The orthoisomer can be separated by fractional distillation, but the fraction boiling between 170–175°C contains both the meta and paraisomers. A typical mother liquor remaining after crystallization of p-dichlorobenzene from this fraction was Kamlet's starting material; it contained 72% m-dichlorobenzene and 28% p-dichlorobenzene.

First 2575 parts (17.5 moles) of this meta-para fraction is cooled to 10–20°C. To it is added, with stirring, 800 parts (10.0 moles; i.e., enough to sulfonate 80% of the m-dichlorobenzene present) of sulfur trioxide. It is cooled to keep the temperature below 40°C. The mixture is stirred for 1 hr at 35–40°C, after which the temperature is increased slowly to 95–100°C and held there for 1 hr. (Under these conditions, p-dichlorobenzene is not sulfonated.) Then 10,000 parts (10.0 moles) of a 4% sodium hydroxide solution is added, the mixture is stirred for a few minutes at 95–100°C, and the liquid layers are separated. To the aqueous layer, which contains the sodium salt of 2,4-dichlorobenzene-sulfonic acid, 1760 parts (44.0 moles) of solid sodium hydroxide is added. The resulting solution is heated for 3 hr at 180–200°C in a copper-lined autoclave. (Agitation is optional.) The contents of the auto-clave is discharged, acidified with 3675 parts (30 moles) of an 80% sulfuric acid solution, and heated in a lead-lined autoclave at 180–200°C for 30 min. The autoclave is allowed to cool and is vented, whereupon its contents is exhaustively extracted with ether. Evaporation of the ether extract leaves essentially pure resorcinol, mp 109–110°C. The yield is 70–85%, based on 2,4-dichlorobenzenesulfonic acid.

The organic layer remaining after sulfonation is cooled to 10–15°C and centrifuged. Thus it is possible to recover 510 parts of *p*-dichloro-benzene and 550 parts of a liquid mixture containing 62% meta and 38% paradichlorobenzene.

VIII.   HYDROQUINONE

A.   Introduction and Survey of Chemistry

Hydroquinone (*p*-dihydroxybenzene) (129) was first prepared in 1820 by Pelletier and Caventou as a product of the destructive distillation of quinic acid. It was more fully characterized in 1844 by Wöhler, who also prepared it by the reduction of quinone (17) and named it accordingly. Its value as a photographic developer was recognized after 1880, and practical manu-facturing processes were then devised. These involved the oxidation of aniline (18) to 17 followed by reduction to hydroquinone. The preferred oxidizing agent is pyrolusite, a natural form of manganese dioxide (Eq. 20).

$$2 \overset{NH_2}{\underset{18}{\bigcirc}} + 4MnO_2 + 5H_2SO_4 \longrightarrow$$

$$2 \overset{O}{\underset{17}{\bigcirc}} + 4MnSO_4 + (NH_4)_2SO_4 + H_2O \qquad (20)$$

Reduction of 17 is normally accomplished with iron dust, with the by-product iron oxide apparently being essentially ferrous oxide (Eq. 21).

This method, the only one known to be used commercially, was described in some detail in 1952 (130). After oxidation of aniline in batches at tempera-tures not exceeding 8°C (131), the quinone was separated from inorganic

$$\text{17} \quad + \quad Fe \quad + \quad H_2O \quad \longrightarrow \quad \text{(hydroquinone)} \quad + \quad FeO \qquad (21)$$

substances by vacuum steam distillation and then condensed in and reduced by a slurry of iron dust in water (132). Hydroquinone was recovered by crystallization under controlled conditions. Although there is no indication that this process has been improved chemically, a more recent Japanese process (133,134), differing from it significantly only in apparatus design, has appeared. Although the oxidation can be run at a higher temperature (50–60°C) and under reduced pressure, thus removing the **17** by distillation as it is formed (135), the reduction operation apparently gives lower yields.

Hydroquinone can also be prepared from p-diisopropylbenzene via the dihydroperoxide (105). Apparently this route has not been studied as much as the analogous one leading to resorcinol. The separations are simpler, however, because fairly pure **19** can be recovered directly by crystallization from the mixture of oxidation products.

$$HOO-\underset{\underset{CH_3}{|}}{\overset{\overset{CH_3}{|}}{C}}-\text{⟨benzene⟩}-\underset{\underset{CH_3}{|}}{\overset{\overset{CH_3}{|}}{C}}-OOH$$

19

Two variations of this procedure have been reported. One of them involves formation of the hydroperoxide from an ester of p-isopropylphenol, followed by rearrangement (136,137). In the other, the oxidation of p-diisopropyl-benzene and rearrangement of the resulting hydroperoxides were carried out in one step by using acetic anhydride as the reaction medium and p-toluene-sulfonic acid as catalyst; hydroquinone diacetate was obtained in a 4% yield (138). It is impractical to convert p-isopropylphenol directly to the hydroperoxide, since it is an inhibitor for radical reactions.

An oxidation route to hydroquinone, from either benzene or phenol, has been studied by many workers and is treated in a review by London (139). Among the reagents that have been used to oxidize phenol are Fenton's reagent (hydrogen peroxide plus a ferrous salt), oxygen and water in the presence of X-rays, Fremy's salt $[ON(SO_3K)_2]$, and persulfate ion. The reactions, which differ widely in their product distributions and mechanisms, do not seem to be attractive preparative methods for hydroquinone. The use of hydrogen peroxide or peralkanoic acids as hydroxylating agents for phenol (140) or phenyl alkyl ethers (141,142) can be quite efficient at low conversions, but orthosubstitution usually predominates. Similarly, reaction of phenyl acetate with electrolytically generated acetoxyl radicals gave somewhat more catechol diacetate than hydroquinone diacetate (143).

The oxidation of benzene to quinone has been effected at the lead dioxide anode of an electrolytic cell by using dilute sulfuric acid as the supporting electrolyte. In the earlier studies (144,145), the benzene was suspended in the sulfuric acid; emulsification with the aid of nonionic or cationic surfactants reportedly (146) allowed the use of higher current densities and gave higher yields based on current. It is possible that problems of cell design have inhibited further progress in this area.

Another oxidative reaction used to prepare hydroquinone is the Dakin reaction (147) as given in Eq. 22.

$$HO-\langle\bigcirc\rangle-CH{=}O + H_2O_2 \longrightarrow HO-\langle\bigcirc\rangle-OH \qquad (22)$$

One other hydroquinone preparation, which is probably oxidative in nature, is the hydrolysis of nitrobenzene with 5% sulfuric acid (148). Yields of about 55% were obtained with heating times of about 9 hr at $230°C$. It was speculated that the first step was hydrolysis to give phenol and nitrogen oxides, which then oxidized the phenol to hydroquinone.

A unique route to hydroquinone was discovered by Reppe in the course of his pioneering research in the field of acetylene chemistry. When iron pentacarbonyl or iron dihydrotetracarbonyl is heated under pressure with acetylene, water (or an alcohol), and basic catalyst, hydroquinone is formed (149,150). As first practiced, this method required stoichiometric amounts of the organometallic compound and was formulated as in Eq. 23. The yields

$$Fe(CO)_5 + 4C_2H_2 + 2H_2O \rightarrow 2C_6H_6O_2 + FeCO_3 \qquad (23)$$

were poor, however, since not more than 22%, based on iron pentacarbonyl, resulted. Later a catalytic hydroquinone synthesis, such as depicted in Eq. 24

$$2C_2H_2 + 3CO + H_2O \rightarrow C_6H_6O_2 + CO_2 \qquad (24)$$

was developed, with such complex salts as hexammineiron(II) bis[tetracarbonylcobaltate(—I)] acting as catalysts. Since these substances could not be reused, however, the process was still unsatisfactory. Recently it was found that iron carbonyls could serve catalytically at very high pressures if incremental amounts of water were added to the system (152).

As might be expected, certain other group VIII elements are useful catalysts for hydroquinone synthesis, particularly ruthenium and rhodium (153). Factors influencing the reactions catalyzed by triruthenium dodecacarbonyl have been studied by Italian workers (154,155), who reported such reactions as Eq. 25. A Japanese group (156) has reported a detailed

$$2C_2H_2 + 2CO + H_2 \rightarrow C_6H_6O_2 \tag{25}$$

study of the rhodium-catalyzed reactions. However, the very high price of these catalysts is a barrier to their commercial acceptance.

Hydroquinone, unlike resorcinol, cannot be made by the alkali fusion of p-benzenedisulfonic acid (157). After one of the sulfonic acid groups has been replaced by the oxide anion, the ring apparently is too negative for a second displacement reaction.

Hydroquinone is readily oxidized to quinone, and most of its commercial uses are related to this property. Thus hydroquinone, usually in combination with p-methylaminophenol sulfate, is very widely used as a photographic developer. Silver halide grains that have been exposed to light are reduced by hydroquinone to metallic silver much more rapidly than unexposed grains. Hydroquinone is also used as an antioxidant for paints, fish and vegetable oils, etc., and as a polymerization inhibitor for vinyl monomers, because it rapidly reacts with and destroys free radicals. It also serves as a dyestuff intermediate. Manufacturing capacity for hydroquinone in the United States was estimated in 1965 as about 25 million lb/year (158).

Hydroquinone is said to be at least as toxic as phenol; its mode of action apparently involves oxidation to quinone. Nevertheless, it does not present serious hazards unless taken internally. Production workers have suffered eye injuries as a result of prolonged exposure to hydroquinone dust or vapor (126–128,130).

B. Synthesis

High-purity hydroquinone is commercially available, and the compound can be further purified by recrystallization from carefully deoxygenated water (159).

1. PREPARATION FROM ANILINE

A fully satisfactory laboratory procedure for preparing hydroquinone from aniline does not appear to be available, mainly because of the difficulties involved in handling the intermediate quinone, a reactive and volatile solid. In commercial practice (130,133) quinone is separated from inorganics by flash steam vacuum stripping and condensed in and immediately reduced by a slurry of iron powder in water. Therefore, the directions given for oxidation (131) and reduction (160) should be regarded as illustrative only.

a. Preparation of Quinone (131)

In a reaction vessel equipped with a stirrer, a thermometer, and the means for the separate addition of a liquid and a solid, are placed 240 lb of water, 237.5 lb (2.26 moles) of 93.2% sulfuric acid, and 75 lb (0.73 mole) of 85% manganese dioxide (ground to pass a 200-mesh screen). This mixture is cooled to 5–8°C. A solution containing 158.3

lb (1.700 moles) of aniline, 475 lb (4.52 moles) of 93.2% sulfuric acid, and 1600 lb of water is prepared separately and cooled to room temperature. This aniline sulfate solution and 308 lb (3.01 moles) of ground 85% manganese dioxide are added concurrently to the reaction vessel, whose contents is maintained below 8°C, over a period of 5–5.5 hr. The rates of addition are such that manganese dioxide is always present in a slight excess. After the end of the addition, the mixture is stirred for 4–24 hr at 5–8°C to complete the reaction. The yield of quinone is stated to be 90–92%, but the work-up is not described. Recovery by flash vacuum distillation would probably give a somewhat lower actual yield.

b.   Preparation of Hydroquinone (160)

The starting material for the reduction of quinone is a recycled mother liquor from an earlier run; if the crystallization temperature was 15°C, it should contain 58 g of hydroquinone per liter. Sodium bisulfite (25 g/liter) is added to inhibit the formation of highly colored oxidative by-products. The solution is placed in a reaction vessel equipped with an efficient agitator; nitrogen gas is bubbled into it throughout the run. (The quantities given are based on 1000 liters of mother liquor.) The solution is heated to 75–85°C, and 10 kg of 200-mesh gray cast iron powder is added, followed by 20 kg of moist quinone in portions. When reaction appears to be complete, iron powder and quinone are added in portions alternately. The total amounts used are 50 kg (463 moles) of quinone and 40 kg (716 moles) of iron. The reduction should always be complete before more quinone is added. One can test for this by removing a sample of the reaction mixture, extracting with ether, and allowing the extract to evaporate. The residue should be white crystals of hydroquinone; a green color is due to quinhydrone, the molecular complex formed from quinone and hydroquinone. When the reaction is complete, filter aid is added, the solution is filtered hot to remove iron and iron oxides, and the filtrate is cooled rapidly to 15°C. The crystals are recovered by filtration, washed with a little water, and dried in a vacuum oven below 40°C. The yield of hydroquinone, mp 169–171°C, is 49.8 kg (98%, based on quinone). The mother liquor is saved for recycle.

Hydroquinone made in this way is of technical grade. The means used to purify it to the photographic grade are described in general terms in the literature (127,130). Polymer-grade material was obtained by recrystallizing three to four times from deoxygenated water, which was prepared by boiling and cooling in a nitrogen stream.

2.   PREPARATION FROM p-DIISOPROPYLBENZENE

a.   Method of Kruzhalov and Fedorova (161)

p-Diisopropylbenzene containing 2.5 wt.% of its monohydroperoxide is treated with air for 8 hr at 110°C, and a mixture of the mono and dihydroperoxides is obtained. The reaction mixture is decomposed by treatment with sulfuric acid (0.1 wt. % based on the organic substrate) in acetone at 55°C. The reaction mixture is neutralized, the acetone is removed by distillation, and the residue is extracted with hot water. Steam distillation of the extract gives p-isopropylphenol as distillate and hydroquinone (68% yield) as residue. Vacuum distillation of the water-insoluble material gives additional p-isopropylphenol to a total yield of 21%.

b. Preparation of p-Diisopropylbenzene Dihydroperoxide (162)

Lorand and Reese (162) give slightly more detailed directions for the preparation of the dihydroperoxide **19**. In a glass reactor equipped with a condenser, an inlet tube for oxygen reaching to the bottom, and a high-speed stirrer are placed 1200 parts of technical diisopropylbenzene (composition not given), 400 parts of a 2% aqueous sodium hydroxide solution, 36 parts of p-diisopropylbenzene monohydroperoxide, and 4 parts of stearic acid. To this mixture, maintained at 90°C, oxygen is added at a rate of 12 liter/hr for 90 hr. The organic layer, on cooling, deposits 234 parts of crystals, containing 87% **19**. The yield of **19**, based on total diisopropylbenzene, is 12%.

c. Rearrangement of p-Diisopropylbenzene Dihydroperoxide to Hydroquinone (163)

Domencali and Canti (163) described the preparation of pure hydroquinone from crude **19**. A suspension of 250 g of crude **19** (containing 186.7 g of the pure substance) in 440 g of benzene is added over 70 min to a solution of 5 g of 96% sulfuric acid in a mixture of 158 g of acetone and 176 g of benzene, maintained at 25–35°C. The mixture is stirred 15 min longer, then neutralized by adding 11 g of sodium carbonate and heating to 60°C. The inorganic substances are removed by filtration, and the filtrate is fractionally distilled; 278 g of acetone is collected as distillate. The residue is cooled and filtered; the yield of hydroquinone, mp 173°C, 99.61% pure, is 92 g (quantitative). Removal of benzene from the filtrate gives 40 g of phenolic by-products.

Although these operations do not appear to be particularly hazardous, it is advisable to take suitable precautions in any experiments involving significant amounts of peroxides.

3. METHOD OF REPPE AND MAGIN (152)

The use of acetylene under high pressures, unless conducted by persons expert in the required techniques, is extremely hazardous. Therefore, the following method should never be attempted as a laboratory preparation and is given mainly to illustrate its relative simplicity as a manufacturing process.

A 50-liter stirred stainless steel autoclave is swept with nitrogen; to it is added, while excluding air, 20 kg of dioxane, 160 g of water, and 800 g (4.08 moles) of iron pentacarbonyl. The autoclave is twice filled with nitrogen to a pressure of 15 atm and vented. Then, with stirring, a gas mixture (11 vol.% acetylene and 89 vol.% carbon monoxide) is forced in until the pressure reaches 370 atm. The mixture is heated over 4–5 hr to 160–165°C with the pressure rising to 550 atm. More of the gas mixture is added to increase the pressure to 700 atm, and the pressure is then maintained at 680–700 atm for 10 hr by continuous addition of the gas mixture. The autoclave is cooled and vented, 160 g of water is added, and the procedure is repeated. The autoclave is cooled and vented again, 100 g of water and 800 g (4.08 moles) of iron penta-carbonyl are added, and the same procedure is repeated again. Finally, another 100 g of water is added, and a fourth reaction cycle is carried out. The autoclave is cooled, vented, emptied, and rinsed with 5 kg of dioxane. Dioxane is removed from the rinsings by distillation under reduced pressure, and the residue is added to the original reaction mixture. The resulting mixture is stirred vigorously to permit removal of a representative sample (1940 g of a total 23,500 g).

The sample is filtered with suction, and the solids are washed with cold dioxane. The filtrate and washings are combined. The solids are extracted with 400 ml of dioxane at 80°C; removal of dioxane from the extract by distillation under reduced pressure leaves 48.7 g of slightly impure hydroquinone, mp 168°C.

The combined filtrate and washings are distilled under reduced pressure; 119 g of crude hydroquinone is collected as a sublimate at oil bath temperature 130–220°C/20 mm. Extraction of the sublimate with 500 ml of toluene at 50°C leaves 93.4 g of practically pure hydro-quinone, mp 170–171°C. The combined yield from the sample, 142.1 g, is equivalent to a total yield of 1722 g (69% based on acetylene, or 96% based on iron pentacarbonyl).

In other runs on a smaller scale, hydroquinone yields substantially greater than stoichiometric (based on iron carbonyl) were obtained. When the procedure was repeated with all the water in the original charge, the yield of hydroquinone was lower (50%, based on acetylene).

IX.   4,4'-DIHYDROXYBENZOPHENONE

A.   Introduction and Survey of Chemistry

4,4'-Dihydroxybenzophenone (164) was reported in 1878 by four different groups: Staedel and Gail by oxidizing bisphenol F dibenzoate; Lieberman by degrading pararosaniline; Caro and Graebe by decomposing aurin; and Baeyer and Burkhardt (165) by alkaline fusion of phenolphthalein (7). Although the compound probably has not been manufactured on a large scale, it has attracted the attention of polymer chemists because certain polyethers (166,167), and other polymers derived or derivable from it exhibit ready crystallizability.

The literature suggests many routes to 4,4′-dihydroxybenzophenone, but only four appear worthy of mention. One of the oldest (165), alkali fusion of phenolphthalein (7), has been repeated several times (Eq. 26). However,

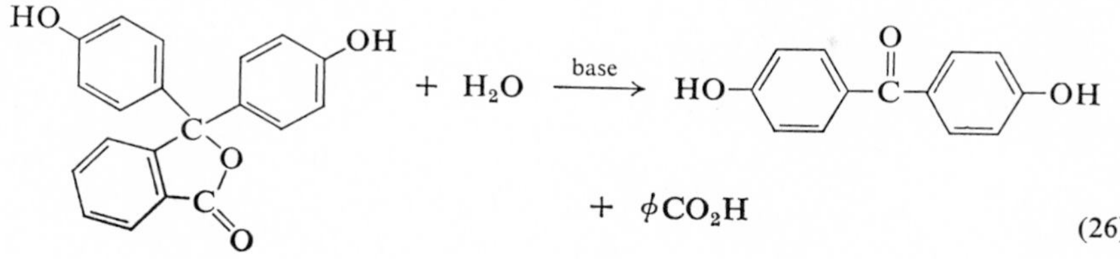

(26)

unless the conditions are carefully controlled, entirely different products may result (168).

4,4′-Dihydroxybenzophenone has also been prepared by the copper-catalyzed base hydrolysis of 4,4′-dichlorobenzophenone (169), as is described in Eq. 27. The latter is available from the Friedel-Crafts acylation of chloro-

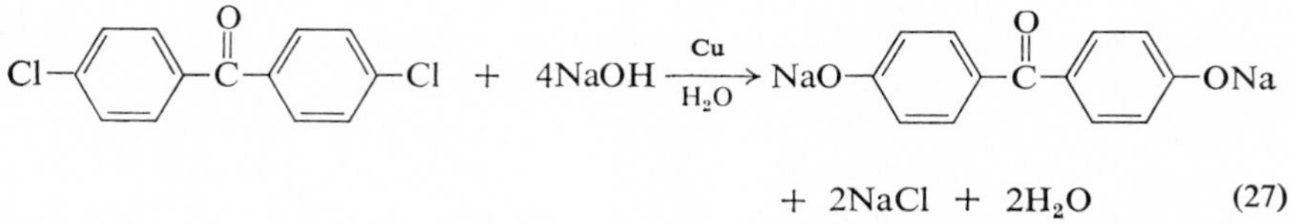

$$+ \ 2NaCl \ + \ 2H_2O \qquad (27)$$

benzene with phosgene or, better, with oxalyl chloride (170). The hydrolysis sometimes gave a polymerization-grade product, but other batches of dihydroxybenzophenone were contaminated with a difficult-to-remove impurity having a hydroxyl functionality greater than 2 (170).

The Fries rearrangement of phenyl p-hydroxybenzoate (20) also leads to 4,4′-dihydroxybenzophenone. A patented (171) procedure apparently involves the formation and rearrangement of 20 in one operation (Eq. 28).

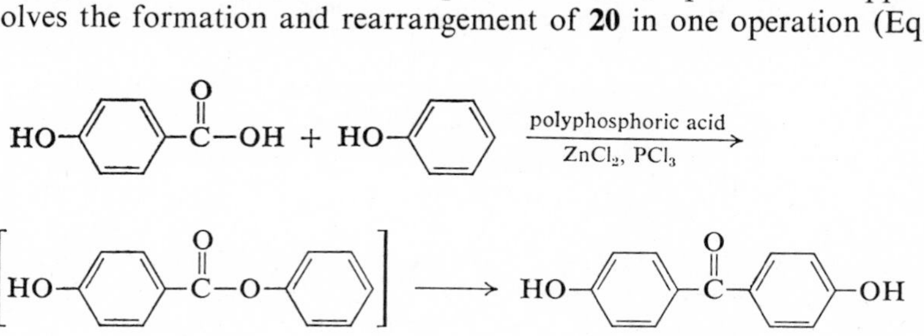

(28)

4,4′-Dihydroxybenzophenone is conveniently prepared in the laboratory (170) by the iodine-catalyzed reaction of anisoyl chloride with anisole (172) followed by the cleavage of the resulting 4,4′-dimethoxybenzophenone with pyridine hydrochloride using the method of Prey (173). This is shown in Eq. 29.

$$(29)$$

B. Synthesis

1. PREPARATION FROM ANISOYL CHLORIDE AND ANISOLE (170)

In a 500-ml flask equipped with a magnetic stirrer and an efficient reflux condenser with drying tube are placed 100.8 g (0.59 mole) of anisoyl chloride, 127 ml (1.17 moles) of anisole, and 4.5 g of iodine. The mixture is heated under reflux for 15 hr and allowed to cool. It then is dissolved in 2.3 liters of isopropanol, and the solution is allowed to stand at room temperature overnight. The crystals are collected by filtration, washed with 200 ml of isopropanol, and dried in a vacuum oven at 60–65°C. The yield of 4,4′-dimethoxybenzophenone, mp 141–143°C, is 114.7 g (80%).

In a 100-ml flask equipped with a reflux condenser with drying tube and heated by a silicone oil bath are placed 19.4 g (0.08 mole) of 4,4′-dimethoxybenzophenone, 60 g (0.52 mole) of pyridine hydrochloride (98% pure), and a few boiling chips. The oil bath is preheated to 214–216°C before the flask and contents are immersed in it, and heating at that temperature is continued for 4.75 hr after the reaction mixture has become homogeneous. While still warm, the reaction mixture is diluted with 750 ml of a 2.8% hydrochloric acid solution; after the mixture has been allowed to cool to room temperature, the solid is recovered by filtration, washed with water, and dried in a vacuum oven at 60°C. The yield of crude 4,4′-dihydroxybenzophenone, mp 214.3–216°C, is 15.9 g (93%). This material (14.8 g) is purified by recrystallization from a mixture of 200 ml of ethanol and 500 ml of water containing 1.0 g of activated charcoal; after hot filtration, the filtrate is kept overnight at 3°C. The resulting crystals are collected and dried in a vacuum oven at 90–95°C. Recovery is 13.28 g (90%), mp 214.3–217°C.

Scale-up of the pyridine hydrochloride cleavage appears to involve no major problems. For larger runs the proportion of pyridine hydrochloride is reduced from 6.5 to 4 moles/mole of dimethoxybenzophenone (100% excess), and a heating mantle and mechanical stirrer are used. On a larger scale it is more convenient to purify the product by recrystallization from 30 vol.% acetic acid as described below (170).

4,4'-Dihydroxybenzophenone (100 g) is suspended in a mixture of glacial acetic acid (160 ml) and water (375 ml); the mixture is heated almost to boiling, and activated charcoal (6–7 g) is added. The boiling solution is stirred for a few minutes and filtered hot; the carbon cake is washed with 95 ml of boiling 30% acetic acid. The crystals are collected after they have been allowed to stand overnight at room temperature; then they are washed with a little 30% acetic acid and dried in a vacuum oven at 100–105°C. Prolonged drying is required for complete removal of acetic acid. Recoveries rang from 82 to 86%.

2. PREPARATION FROM p-HYDROXYBENZOIC ACID (171)

A mixture of 30 g (0.217 mole) of p-hydroxybenzoic acid, 20.4 g (0.217 mole) of phenol, 67 g (0.49 mole) of zinc chloride, and 317 g of 103% phosphoric acid (prepared by blending commercial polyphosphoric acid with commercial phosphoric acid) is heated with stirring to 40°C. To it is added, with stirring over 1.5 hr, 38 g (0.28 mole) of phosphorus trichloride. The mixture is then heated slowly to 60°C, held at that temperature for 16 hr, and diluted with cold water. The precipitate is recovered by filtration, washed successively with water, a 2.5% sodium bicarbonate solution, water again, and then dried. The yield of dihydroxybenzophenone, mp 210.4–212°C, is 42 g (90%).

For this method the phosphoric acid concentration is critical and should be in the range 95 to 106%.

X. 4,4'-BIPHENOL

A. Introduction and Survey of Chemistry

The synthesis of 4,4'-biphenol (174) was first reported in 1876 by Döbner, who carried out an alkali fusion of biphenyl-4,4'-disulfonic acid. The compound has been prepared by a variety of methods and has been studied in some detail, but it is still an experimental material.

The method best suited to the laboratory preparation of 4,4'-biphenol (22) is the replacement of the amino groups of benzidine (21) by tetrazotization, followed by hydrolysis (Eq. 30). This procedure has been modified to

$$H_2N-\langle\rangle-\langle\rangle-NH_2 \longrightarrow \overset{(+)}{N_2}-\langle\rangle-\langle\rangle-\overset{(+)}{N_2} \xrightarrow[H_2O]{H^+}$$

21

$$HO-\langle\rangle-\langle\rangle-OH$$

22   (30)

produce 4,4′-biphenol continuously on a plant scale (175). Benzidine, which is readily prepared from nitrobenzene via hydrazobenzene as in Eq. 31, is commercially available.

$$\langle\rangle-NO_2 \xrightarrow{reduction} \langle\rangle-NH-NH-\langle\rangle \xrightarrow[H_2O]{H^+} \textbf{21}$$

(31)

Another procedure that has been studied in some detail is the sulfonation of biphenyl to biphenyl-4,4′-disulfonic acid (**23**) and subsequent conversion to **22** by treatment with alkali (Eq. 32) (176). Although the first step has not

$$\langle\rangle-\langle\rangle \xrightarrow{H_2SO_4} HO_3S-\langle\rangle-\langle\rangle-SO_3H \xrightarrow[heat]{OH^-} \textbf{22} \quad (32)$$

23

been optimized, the second has been and should be amenable to further scale-up.

Of great potential interest is the oxidative dimerization of phenol:

$$2\langle\rangle-OH \xrightarrow[-2[H]]{catalyst} \textbf{22} \tag{33}$$

The reported attempts using chemical and irradiative methods usually led to low yields of isomerically impure 4,4′-biphenol. However, vanadium tetrachloride has been reported to give yields as high as 46% (177).

An elegant improvement in phenol-coupling technique has been described by Hay (178,179) and is detailed in Eq. 34. Since **24** has a sterically hindered hydroxyl group with no free ortho positions, carbon–carbon coupling occurred exclusively to give **25**. Subsequent reaction with more **24** gave the bisphenol **26**, and dealkylation give 4,4′-biphenol. The latter two reactions can conveniently be combined into a single step (180). This method has been reported on a laboratory scale only, but it should be readily adaptable to scale-up, since the by-product isobutylene can be recycled. The overall

$$(34)$$

yield of highly pure 4,4'-biphenol, based on 2,6-di-*t*-butylphenol, was reported to be >80%.

Two other methods that resulted in reasonable yields of 4,4'-biphenol have been reported. Chlorination of biphenyl gave largely 4,4'-dichloro-biphenyl (181), and subsequent hydrolysis gave **22** (182), as in Eq. 35.

$$(35)$$

4,4'-Biphenol has also been prepared from phenanthraquinone (**27**) (183) (Eq. 36). Further study could lead to the increased practicality of this procedure.

$$(36)$$

B. Synthesis

Although 4,4'-biphenol is not commercially available in large quantities, it has stimulated much interest. Of the methods just discussed, we feel that two, based on present knowledge, have the best potential for future utilization. They are the syntheses involving benzidine tetrazotization and 2,6-di-*t*-butylphenol dimerization.

1. TETRAZOTIZATION OF BENZIDINE

The following method for the tetrazotization of benzidine is on a laboratory scale and is based on the procedure of Wilds (184). More recently, a detailed continuous process for the manufacture of phenol by hydrolysis of diazonium salts was claimed (175).

a. 4,4'-Diacetoxybiphenyl

Benzidine (200 g, 1.09 mole) is dissolved in 2.2 liters of hot water containing 210 ml of concentrated hydrochloric acid. The solution is cooled to 25°C, and an additional 235 ml of hydrochloric acid is added. The resulting suspension is cooled below 10°C and a solution of sodium nitrite (151 g, 2.19 mole) in 450 ml of water is added slowly. After the solution has been stirred for 20 min longer, 1 g of urea is added. The 4,4'-biphenylenebis(diazonium chloride) solution is added slowly to a solution of 215 ml of concentrated sulfuric acid in 4.4 liters of water at 85–90°C. After it has been heated 10 min longer, the reaction mixture is cooled; the resulting crude 4,4'-biphenol is filtered, collected, and dried. It is then added to 400 ml of acetic anhydride; the mixture is heated under reflux for 3 hr, then cooled. The crude product (295 g) is collected by filtration. Distillation of the solid at 0.1–0.5 mm in a flask-to-flask setup and subsequent recrystallization from benzene gives 172 g of 4,4'diacetoxybiphenyl, mp 160–162°C. The acetic anhydride is distilled from the initial filtrate to give a residue, which is likewise distilled and recrystallized to give another 50 g of product (mp 158–160°C). Total yield is 222 g (84%). Further purification raises the melting point to 163–164°C.

b. 4,4'-Biphenol

A solution of sodium hydroxide (22 g) in 50 ml of water is added to a slurry of 4,4'-diacetoxybiphenyl (75 g, 0.28 mole) in 200 ml of ethanol over a 5-min period. The mixture is heated for about 30 min under reflux, then diluted with 750 ml of water. The solution is strongly acidified with 20% hydrochloric acid, heated on a steam bath for 1 hr, and cooled. The resulting solid is filtered, washed with water, and dried at 80°C. The yield of 4,4'-biphenol, mp 279–282°C, is 51 g (97%). Recrystallization can be effected from ethanol to give shiny white leaflets, mp 279.5–281.5°C.

2. PREPARATION FROM 2,6-di-t-BUTYLPHENOL (180)

The simplicity of the synthesis of 4,4'-biphenol from 2,6-di-t-butylphenol via 3,3',5,5'-tetra-t-butyldiphenoquinone should make 4,4'-biphenol and its derivatives readily available for further study (185).

a. 3,3',5,5'-Tetra-t-butyldiphenoquinone

Oxygen is passed through a vigorously stirred solution containing 0.2 g of copper(I) chloride and 0.24 g of N,N,N',N'-tetramethylethylenediamine in 300 ml of t-butanol in a bath at 30°C. After all the copper salt has dissolved there is added 50 g of 2,6-di-t-butylphenol. Over a period of 18 min the temperature of the reaction rises to 43.5°C. After 23 min the reaction subsides, and a brownish solid has precipitated. The product is removed by filtration and washed twice with 50-ml portions of *t*-butanol containing about 1 ml of concentrated hydrochloric acid. After drying *in vacuo*, 48.7 g of 3,3′,5,5′-tetra-*t*-butyldiphenoquinone is obtained. Dilution of the filtrate with water gives an additional 0.6 g of product. When a 20-g portion of this product is recrystallized from 250 ml of acetic acid in a Soxhlet apparatus, there is obtained 19.1 g of red-brown crystals having a metallic sheen, mp 245°C (dec.). The total yield after recrystallization is 47.1 g (95 %).

b. 4,4′-Biphenol

To a 250-ml, three-necked flask equipped with a condenser, a nitrogen gas inlet, a thermometer, and a stirrer is added 50 g (0.12 mole) of 3,3′,5,5′-tetra-*t*-butyldiphenoquinone, 55 g (0.26 mole) of 2,6-di-*t*-butylphenol, and 5 ml of pyridine. Over a 30-min period the temperature is raised to 260°C, then the reaction mixture is cooled to 200°C, and there is added 1.0 g of aluminum isopropoxide. The reaction mixture is then heated to 250°C, and isobutylene is vigorously evolved. After 30 min the reaction mixture has solidified, so the temperature is raised to 280°C for 15 min. The reaction mixture is cooled and dissolved in 800 ml of hot ethanol containing 10 ml of concentrated hydrochloric acid, treated with charcoal, and filtered. The filtrate is evaporated down to 350 ml and cooled in ice. There is deposited 37.5 g (0.2 mole, 82.5 % yield) of 4,4′-biphenol, mp 279–81°C. Recrystallization from ethanol containing a small amount of aqueous hydrochloric acid gives 4,4′-biphenol, which melts at 285°C.

XI. ANALYTICAL PROCEDURES FOR BISPHENOLS

A. General

The problem of analyzing a given sample of a condensation monomer may be considered from several points of view. To simplify the discussion, we assume that the sample is known to consist mainly of the compound of interest and that its melting point is in reasonable agreement with the values reported in the sections of this chapter dealing with synthesis.

The first question that normally arises is whether the sample is pure enough to be used in the preparation of polymers having useful molecular weights. A test polymerization quite generally applicable to bisphenols involves reaction of the sample with bisphenol A bis (chloroformate) and pyridine in boiling 1,2-dichloroethane (186). This reaction generally goes to completion within a few minutes and, if both monomers are very pure, gives polymers having reduced viscosities (measured in chloroform at a concentration of

0.2 g/100 ml) of 1 or higher. Any doubts about the purity of the bisphenol A bis(chloroformate) can be resolved by treating it under the standard conditions with the readily available high-purity bisphenol A.

The polymerization test gives information about the functionality of the sample. It is useful in the detection of monofunctional impurities, which limit the polymer molecular weight, and of tri or tetrafunctional impurities, which may give insoluble, crosslinked polymers. However, the test may give misleading results unless supplemented by more fundamental analytical methods. In particular, the presence of isomers of the monomer is seldom detected.

Further information is sometimes provided by measurement of the freezing (setting) point, but this has distinct limitations. Resorcinol, for example, may commonly contain both water and phenol, which cause markedly different depressions of the freezing point (187). The presence of water, which can be determined by the Karl Fischer method, can often be tolerated in small amounts. Phenol, on the other hand, is an effective monofunctional terminator in many reaction systems.

The most useful analytical methods are the chromatographic ones, which permit the identification and determination of individual impurities, even when present at very low levels. Reported methods for the quantitative analysis of bisphenol A include thin-layer, paper, and gas chromatography.

As stated in Section I-A, the major impurities in commercial bisphenol A are the 2,4' isomer 3, the trisphenol 4, and Dianin's compound or codimer 5. Anderson et al. (188) were unable to effect complete resolution of these substances by one-dimensional paper-strip chromatography even after the chemicals had been concentrated in the mother liquor by recrystallization of the original sample from chlorobenzene. They solved this problem by making two chromatograms; 3 and 5 appeared as separate spots when the eluent was carbon tetrachloride, and 4 could be measured on a chromatogram eluted by water even though it was not fully resolved from 5. Like later workers, they developed the chromatograms by spraying with p-nitrobenzenediazonium fluoborate; spot areas were measured with a planimeter. Challa and Hermans (189) used vacuum sublimation to preconcentrate the impurities 3 and 5, which are more volatile than bisphenol A, and then achieved complete resolution by eluting the paper with a mixture of propanol and kerosene in an atmosphere of ammonia. Reinking and Barnabeo (190) analyzed bisphenol A semiquantitatively by reversed-phase circular-paper chromatography, using a 0.125M trisodium phosphate solution to elute the phenols from paper impregnated with tricresyl phosphate. They estimated the amount of each impurity by bracketing around the minimum detectable quantity. Concentration of impurities prior to analysis was not necessary. A quantitative thin-layer chromatographic method involving concentration of the impurities by recrystallization from chlorobenzene and use of a mixture of chloroform and acetone as eluent has also been reported (191).

Tominaga (192) reported a direct gas-chromatographic analysis of bisphenol A, but other workers (188,193,194) have encountered tailing or decomposition under ordinary operating conditions. For this reason, Gill (193) acetylated the phenols prior to analysis. More recently, Brydia (195) reported a method in which the phenols were converted to the trimethylsilyl ethers and then analyzed on a 1% OV-1 silicone rubber on Chromosorb G (HP) column. Best results were obtained by operating isothermally until **3, 5**, and **2** had been eluted, then rapidly programming the temperature to cause elution of **4**. By using a hydrogen-flame ionization detector it was possible to determine impurities at the 0.005% level without a concentration step. Water was determined by the Karl Fischer method and phenol by a separate gas-chromatographic analysis (196).

Gas chromatography has been applied with some success to the analysis of hydroquinone and resorcinol (156,197–202). Products of the reaction between phenol and formaldehyde, including the more important bisphenol F isomers, have been analyzed by thin layer- and paper chromatography (203).

B. Test Polymerization Method (186)

Bisphenol A bis(chloroformate) is prepared as described in Section XII·B·1 and recrystallized from heptane; mp 95–96°C. 1,2-Dichloro-ethane purchased from Matheson, Coleman, and Bell is dried over molecular sieves (Linde, type 5A). Pyridine is distilled and stored over molecular sieves.

In a 100-ml, two-necked flask equipped with a magnetic stirrer, an inlet for dry nitrogen gas, and a reflux condenser with a drying tube is placed 3.532 g (0.1000 mole) of bisphenol A bis(chloroformate), 0.1000 mole of the bisphenol (weighed to the nearest milligram), and 40 ml of 1,2-dichloroethane. The mixture is purged with nitrogen for 10 min and then heated to about 65–70°C. Pyridine (3.2 ml, 0.40 mole) is added in one portion through the condenser. The mixture begins to boil almost immediately, and a significant increase in viscosity is noted within 1 min. The solution is heated under reflux for 30 min, cooled, and poured into 250 ml of methanol in a high-speed Blendor. The precipitated polymer is collected by filtration, washed with a little methanol, washed in the Blendor with two portions of water, collected, and dried in a vacuum oven at 75°C for 24 hr. The reduced viscosity is measured in chloroform (0.2 g/100 ml) at 25°C.

The polymer yields are usually between 85 and 95%. Monomers of satisfactory purity should give polymers with reduced viscosity at least 0.5, preferably at least 0.75.

This method has also been applied successfully to 2,2'-(*m*-phenylenedioxy)-diethanol (204) and other nonphenolic diols.

XII. BIS(CHLOROFORMATES)

A. Introduction and Survey of Chemistry

Chloroformates are prepared from bisphenols by reaction with phosgene and a catalyst (Eq. 37). Preparation of bisphenol bis(chloroformates)

$$HOArOH + 2COCl_2 \xrightarrow{\text{catalyst}} Cl-\overset{\overset{\displaystyle O}{\|}}{C}-O-Ar-O-\overset{\overset{\displaystyle O}{\|}}{C}-Cl + 2HCl \qquad (37)$$

was first reported by Oesper et al. in 1925 (205). They synthesized both the hydroquinone and resorcinol derivatives, using dialkylanilines as catalysts. Catalysts reported more recently include stearyltrimethylammonium chloride (206,207), tertiary amides (208), anion-exchange resins (209), and alkylamines (210).

Of the 11 bisphenols discussed here, only the bis(chloroformate) of phenolphthalein has not been characterized, although it was prepared as a viscous oil (208). The bis(chloroformates) of resorcinol (205) and 4,4'-dihydroxydiphenylsulfone (211) were characterized only by melting point, however.

B. Synthesis

The procedures that follow, illustrative of those used with stearyltrimethylammonium chloride and *N,N*-dimethylaniline, respectively, as catalysts, are general enough to be applicable to the synthesis of most bis(chloroformates).

1. Bisphenol A bis(chloroformate) (206)

Stearyltrimethylammonium chloride is obtained as an aqueous solution ("Aliquat 7") from General Mills. It is isolated as a solid by vacuum stripping, followed by repeated azeotropic distillation with benzene and final drying *in vacuo* at 50°C, mp 185–221°C (dec.).

> The apparatus consists of a 500-ml flask equipped with a thermometer, a mechanical stirrer, a condenser topped with a dry-ice–acetone condenser, and a gas inlet tube reaching below the surface of the liquid. A mixture of 2,2-bis(4-hydroxyphenyl)propane (22.8 g, 0.1 mole), 3.45 g (0.01 mole) of stearyltrimethylammonium chloride, and 250 ml of chlorobenzene is charged and cooled to 0°C. Phosgene (22 g, 0.22 mole), which has been liquefied, is added all at once, and the reaction

mixture is slowly warmed. At 20°C, the phosgene refluxes, and at 40°C hydrogen chloride is evolved. The temperature rises rapidly to 90°C and then more slowly to 125°C. After 75 min at 125°C (3 hr total reaction time), the reaction mixture is cooled to room temperature, the catalyst is filtered off and washed with chlorobenzene, and the combined filtrates are treated on the steam bath under reduced pressure to remove the chlorobenzene. The residue is dissolved in 150 ml of toluene, and the solution is filtered to remove insolubles. The solution is then placed on a silica gel column (3 × 30 cm using Davison Grade 12, mesh 28–200) and further eluted with 300 ml of toluene. After the toluene has been distilled out, 33.2 g (94%) of bisphenol A bis-(chloroformate) is obtained. Recrystallization is effected from 150 ml of n-hexane to give polymerization grade material, mp 93–95°C.

2. 4,4'-DIHYDROXYBENZOPHENONE BIS(CHLOROFORMATE) (170)

In a 2-liter, three-necked flask equipped with a gas inlet tube, a mechanical stirrer, a thermometer, a dropping funnel, and a dry-ice-cooled condenser with drying tube are placed 53.6 g (0.250 mole) of 4,4'-dihydroxybenzophenone and 600 ml of methylene chloride. To this mixture, maintained at 7–9°C, 74 g (0.75 mole) of phosgene gas is added over 25 min. Then a solution of 60.6 g (0.500 mole) of N,N-dimethylaniline is added over a period of 90 min at 8–14°C; the solids gradually dissolve, giving a red-orange solution. The cooling bath is then removed, and the mixture is allowed to warm up to room temperature (26°C) over a 65-min period. The methylene chloride and excess phosgene are removed by distillation under reduced pressure, and the residue is extracted with two 500-ml portions of toluene. The red toluene extract is filtered through a column (1-⅝ in. diam. × 18 in. long) of Davison Chemical grade 12 silica gel, and the column is eluted with toluene until about 5.5 liters of colorless eluate has been collected. Toluene is removed by distillation under reduced pressure, and the residue is recrystallized from 700 ml of 50/50 (vol.) toluene-heptane, giving 57.2 g (67%) of 4,4'-dihydroxybenzophenone bis(chloroformate), mp 130.3–131.9°C. An additional 4.3 g (5%) of slightly less pure crystals is obtained from the chilled mother liquor.

Analysis. Calculated for $C_{15}H_8Cl_2O_5$: C, 53.13; H, 2.38; Cl, 20.91. Found: C, 53.14; H, 2.36; Cl. 21.31,

C. Physical Properties

The only physical data reported for the arylene bis(chloroformates) are melting points, recrystallization solvents, and (in a few instances) boiling points. The data are presented in Table 1. The carbonyl absorption in the infrared region is very characteristic and appears at 5.62–5.65 μ (212).

TABLE 1

Physical Properties of Bisphenol Bis(chloroformates)

Bis(chloroformate) of	mp, °C	bp, °C	Recrystallization solvent	Reference
Bisphenol A	93–95		n-Hexane	206,208,209,211
	96–98	183–187 at 0.2 torr	Distilled	213
Tetrachlorobisphenol A	164–166		n-Heptane	206
	163–165		Benzene	213
Hydroquinone	100–101		n-Hexane or n-heptane	206
	98–100		n-Hexane	208
4,4'-Dihydroxydiphenyl-methane	60–61		n-Hexane or n-heptane	206
4,4'-Biphenol	127		Residue product	210
4,4'-Dihydroxybenzophenone	130–132		Toluene-n-heptane (1:1)	170
1,1-Bis(4-hydroxyphenyl)-cyclohexane	50–51	228 at 0.7 torr	Distilled	213
4,4'-Dihydroxydiphenylsulfone	139–141		Distilled	211
Resorcinol	46	155 at 30 mm	Distilled	205

D. Analysis

The isomeric purity of a given bis(chloroformate) sample obviously depends, in part at least, on that of the bisphenol from which it was prepared. Since a bis(chloroformate) is usually purified by recrystallization, in most cases the isomeric purity is at least as good as that of the bisphenol. Among the tests applicable in further characterization are test polymerization with high-purity bisphenol A, as described in Section XI.B, and infrared analysis. Even quite small amounts of free hydroxyl or symmetrical carbonate groups can be detected through the infrared spectrum; the exact positions of these impurity bands can be established with model compounds.

XIII. POLYMERS DERIVED FROM BISPHENOLS

The number of polymers derived from bisphenols, which have been described in the last 20 years, is so large that only a very cursory treatment is possible in the space available. Further information about many of these polymers may be found in the text by Sorenson and Campbell (214) and in the "Modern Plastics Encyclopedia" (215).

A. Commercial Polymers

Although the first commercial use of bisphenol A was in phenolic resins (4)' other applications are now much more important. By far the largest amount of bisphenol A is used in epoxy resins, which are prepared according to the schematic Eq. 38. The value of n depends on the mole ratio of epichlorohydrin

$$HO-\underset{CH_3}{\overset{CH_3}{C}}-OH + CH_2-CH-CH_2Cl + NaOH \longrightarrow$$

$$CH_2-CH-CH_2-O\left[\overset{CH_3}{\underset{CH_3}{C}}-OCH_2-\underset{OH}{CH}-CH_2O\right]_n CH_2-CH-CH_2$$

$$+ NaCl + H_2O \tag{38}$$

to bisphenol A. If a substantial excess of epichlorohydrin is used, the value of n approaches zero, and the product is mostly bisphenol A diglycidyl ether. With smaller amounts of epichlorohydrin, the value of n rises, and still higher polymers can be made by treating a given resin with bisphenol A. Resins with n values up to about 15 are commercially available.

The true epoxy resins are used as chemical intermediates, since their applications involve operations such as crosslinking through reaction of the epoxide groups with amines or anhydrides, esterification of hydroxyl groups by fatty acids, etc. Major uses of epoxy resins are in coatings, adhesives, laminates and other reinforced plastics, and castings.

Closely related to the epoxies are the polyhydroxyethers or phenoxy resins (27a) (216,217). The phenoxies have no epoxy groups, however, and their molecular weights are high enough to impart good mechanical properties without further reaction. Thus they are true thermoplastics and can be fabricated by injection molding, extrusion, etc. They are also useful in solution coatings, which can, if desired, be crosslinked by reaction with diisocyanates, and other compounds.

$$\left[\overset{CH_3}{\underset{CH_3}{C}}-OCH_2-\underset{OH}{CH}-CH_2O\right]_n$$

27a

Next to the epoxy resins in production volume are the polycarbonates (218), which are usually made by the reaction of bisphenol A with either phosgene or diphenyl carbonate, as in Eqs. 39 and 40. The polycarbonates are char-

$$\text{HO}-\!\!\!\left\langle\bigcirc\right\rangle\!\!\!-\underset{\underset{\text{CH}_3}{|}}{\overset{\overset{\text{CH}_3}{|}}{\text{C}}}-\!\!\!\left\langle\bigcirc\right\rangle\!\!\!-\text{OH} \ + \ \text{COCl}_2$$

$$\left[-\text{O}-\!\!\!\left\langle\bigcirc\right\rangle\!\!\!-\underset{\underset{\text{CH}_3}{|}}{\overset{\overset{\text{CH}_3}{|}}{\text{C}}}-\!\!\!\left\langle\bigcirc\right\rangle\!\!\!-\text{O}-\overset{\overset{\text{O}}{\|}}{\text{C}}-\right]_n$$

$$+ \ 2\text{HCl} \tag{39}$$

$$\text{HO}-\!\!\!\left\langle\bigcirc\right\rangle\!\!\!-\underset{\underset{\text{CH}_3}{|}}{\overset{\overset{\text{CH}_3}{|}}{\text{C}}}-\!\!\!\left\langle\bigcirc\right\rangle\!\!\!-\text{OH} \ + \ \left\langle\bigcirc\right\rangle\!\!-\text{O}-\overset{\overset{\text{O}}{\|}}{\text{C}}-\text{O}-\!\!\left\langle\bigcirc\right\rangle \longrightarrow$$

$$\left[-\text{O}-\!\!\!\left\langle\bigcirc\right\rangle\!\!\!-\underset{\underset{\text{CH}_3}{|}}{\overset{\overset{\text{CH}_3}{|}}{\text{C}}}-\!\!\!\left\langle\bigcirc\right\rangle\!\!\!-\text{O}-\overset{\overset{\text{O}}{\|}}{\text{C}}-\right]_n \ + \ 2\left\langle\bigcirc\right\rangle\!\!-\text{OH}$$

$$\tag{40}$$

acterized by excellent impact strength and dimensional stability over a wide range of temperatures, good electrical properties, and transparency. The resin has a fair resistance to burning, but its flammability can be further reduced by chemical modification, possibly effected by replacing part of the bisphenol A with tetrachloro or tetrabromobisphenol A. Also commercially available is a blend or alloy of bisphenol A polycarbonate with an acrylonitrile-butadiene-styrene (ABS) copolymer called Cycoloy 800 (219). Bisphenol C polycarbonate has attracted considerable attention in the Soviet Union (45).

Interest has been centered on block copolymers of bisphenol A polycarbonate, i.e., copolymers in which relatively long sequences of units of polycarbonate and of another polymer are present in a single linear molecule. An example is a copolymer (220) of polycarbonate and a silicone (28).

$$\left[-\text{O}-\!\!\left\langle\bigcirc\right\rangle\!\!-\underset{\underset{\text{CH}_3}{|}}{\overset{\overset{\text{CH}_3}{|}}{\text{C}}}-\!\!\left\langle\bigcirc\right\rangle\!\!-\text{O}-\overset{\overset{\text{O}}{\|}}{\text{C}}-\right]_m\!\!\left[-\text{O}-\!\!\left\langle\bigcirc\right\rangle\!\!-\underset{\underset{\text{CH}_3}{|}}{\overset{\overset{\text{CH}_3}{|}}{\text{C}}}-\!\!\left\langle\bigcirc\right\rangle\!\!-\text{O}-\underset{\underset{\text{CH}_3}{|}}{\overset{\overset{\text{CH}_3}{|}}{\text{Si}}}-\right]_n$$

28

Recent additions to the family of commercial polymers derived from bisphenols are the poly(aryl ethers). These are prepared in either of two related systems as illustrated by Eqs. 41 and 42. In 4,4'-dichlorodiphenyl-

$$NaO-\underset{\underset{CH_3}{|}}{\overset{\overset{CH_3}{|}}{C}}-ONa \;+\; Cl-\!\!\!\!\!\!-SO_2-\!\!\!\!\!\!-Cl \quad\xrightarrow[C_6H_5Cl]{DMSO}$$

29

$$\left[O-\underset{\underset{CH_3}{|}}{\overset{\overset{CH_3}{|}}{C}}-O-\!\!\!\!\!\!-SO_2-\!\!\!\!\!\! \right]_n \;+\; 2\,NaCl \qquad (41)$$

30

$$NaO-\underset{\underset{CH_3}{|}}{\overset{\overset{CH_3}{|}}{C}}-ONa \;+\; Br-\!\!\!\!\!\!-\!\!\!\!\!\!-Br \quad\xrightarrow[\phi_2CO,\ \phi CH_3]{CuCl,\ C_5H_5N}$$

31

$$\left[O-\underset{\underset{CH_3}{|}}{\overset{\overset{CH_3}{|}}{C}}-O-\!\!\!\!\!\!-\!\!\!\!\!\! \right]_n \;+\; 2\,Na\,Br \qquad (42)$$

sulfone (29) the chlorine atoms are activated to nucleophilic displacement by the electron-withdrawing sulfonyl group, and reaction to form the high polymer proceeds rapidly in solutions containing the polar, aprotic solvent dimethyl sulfoxide (DMSO) at temperatures below 160°C (166). With nonactivated aryl halides like 31, it is necessary to use a cuprous salt as catalyst and to carry out the reaction under more forcing conditions (221). The polyether 30 has been made commercially under the name Polysulfone. It is a tough, rigid thermoplastic characterized by resistance to acids and bases, good electrical properties, excellent creep resistance, and low flammability. Its resistance to thermal oxidation is outstanding; it is recommended by the manufacturer for continuous use in air at 170°C.

Bisphenol A finds some use in unsaturated polyester resins. One commercial product (222), made by reaction of bisphenol A with propylene oxide followed by esterification with fumaric acid, has a structure schematically shown as

$$\left[\underset{O}{\overset{O}{\parallel}}CCH=CH\underset{O}{\overset{O}{\parallel}}CO-\underset{}{\overset{CH_3}{|}}CH-CH_2O-\!\!\!\!\!\!-\underset{\underset{CH_3}{|}}{\overset{\overset{CH_3}{|}}{C}}-\!\!\!\!\!\!-OCH_2\underset{}{\overset{CH_3}{|}}CH-\!\!\!\! \right]_n$$

The finished article is obtained by copolymerization of this polyester with an unsaturated monomer such as styrene, in the presence of glass or some other reinforcing filler.

B. Experimental Polymers

Among the first linear polymers derived from bisphenols to be seriously investigated where those obtained by reaction (223) with a bifunctional phosphorus halide such as phenylphosphonic dichloride (Eq. 43). For some

$$
\text{HOArOH} + \text{Cl}-\overset{\overset{\displaystyle O}{\|}}{\underset{\phi}{P}}-\text{Cl} \longrightarrow \left[\!\!-\text{OArO}-\overset{\overset{\displaystyle O}{\|}}{\underset{\phi}{P}}\!-\!\!\right]_n + 2\text{HCl} \qquad (43)
$$

years the polymers obtained all had low molecular weights and relatively low softening points (224,225). Progress has since been made toward increased molecular weights, but no commercial polymers of this type appear to have been produced. Polymers derived from trivalent phosphorus compounds have little or no value as primary plastics, but both monomeric (226) and polymeric (227,228) bisphenol A phosphites have been claimed as stabilizers for poly(vinyl chloride) and other plastics.

Another early reported application of bisphenols to polymer chemistry was the use of their bis(allyl carbonates) in thermoset castings (229,230). Today other materials, including diallylphthalate, diallyl isophthalate, and diethylene glycol bis(allyl carbonate), are used commercially. Linear polyesters from bisphenols and dibasic acids, also known for many years, have been studied extensively by Conix in Belgium, by Korshak and Vinogradova in the Soviet Union, and by several other groups. Some of the fully aromatic polyesters, especially bisphenol A terephthalate and its copolymers (231,232), have properties comparable to those of certain engineering thermoplastics, but we know of none that is manufactured commercially. These materials are covered in reviews by Goodman and Rhys (233) and by Korshak and Vinogradova (234).

The condensation polymers that have been dealt with thus far all have oxygen atoms in the backbone. Other polymers exist in which some or all of these oxygens are replaced by other groups, such as –S– or –NR–. Replacement of one chain oxygen atom of a polycarbonate by a nitrogen atom, for example, leads to a polyurethane. Polyurethanes, which may be prepared by the reaction of a disecondary amine with the bisphenol bis(chloroformate) (214) are described in Eq. 44.

$$
\text{Cl}-\overset{\overset{\displaystyle O}{\|}}{C}-\text{OArO}-\overset{\overset{\displaystyle O}{\|}}{C}-\text{Cl} + \text{H}-\overset{\overset{\displaystyle R}{|}}{N}-\text{R}'-\overset{\overset{\displaystyle R}{|}}{N}-\text{H} \longrightarrow
$$

$$
2\text{HCl} + \left[\!\!-\overset{\overset{\displaystyle O}{\|}}{C}-\text{OArO}-\overset{\overset{\displaystyle O}{\|}}{C}-\overset{\overset{\displaystyle R}{|}}{N}-\text{R}'-\overset{\overset{\displaystyle R}{|}}{N}\!-\!\!\right]_n \qquad (44)
$$

The reaction of the diglycidyl ether of a bisphenol with either a primary amine (235) or a disecondary amine (236) gives polymers analogous to the phenoxy resins (Eqs. 45 and 46).

$$CH_2-CH-CH_2-OArO-CH_2-CH-CH_2 + H-\overset{\overset{\displaystyle R}{\vert}}{N}-H \longrightarrow$$

$$\left[OCH_2-\underset{\underset{\displaystyle OH}{\vert}}{CH}-CH_2OArOCH_2-\underset{\underset{\displaystyle OH}{\vert}}{CH}-CH_2-\overset{\overset{\displaystyle R}{\vert}}{N}\right]_n \quad (45)$$

$$CH_2-CH-CH_2-OArO-CH_2-CH-CH_2 + H-\overset{\overset{\displaystyle R}{\vert}}{N}-R'-\overset{\overset{\displaystyle R}{\vert}}{N}-H \longrightarrow$$

$$\left[OCH_2-\underset{\underset{\displaystyle OH}{\vert}}{CH}-CH_2-OArO-CH_2-\underset{\underset{\displaystyle OH}{\vert}}{CH}-CH_2-\overset{\overset{\displaystyle R}{\vert}}{N}-R'-\overset{\overset{\displaystyle R}{\vert}}{N}\right]_n \quad (46)$$

Treatment of a diglycidyl ether with ammonium thiocyanate gives an epithio compound such as **32**. These materials react with various curing agents to give thermoset products analogous to those obtained from epoxy resins (237).

$$CH_2-CH-CH_2O-\langle\ \rangle-\underset{\underset{\displaystyle CH_3}{\vert}}{\overset{\overset{\displaystyle CH_3}{\vert}}{C}}-\langle\ \rangle-OCH_2-CH-CH_2$$

32

The reaction of a bisphenol (usually as the disodium salt) with a disulfonyl chloride gives a polysulfonate (214,238), as pictured in Eq. 47. Polymers of

$$NaOArONa + Cl-SO_2-Ar'-SO_2-Cl \rightarrow 2NaCl + (OArO-SO_2-Ar'-SO_2)_n \quad (47)$$

this general type were made on pilot-plant scale in the United States (239) but are not available commercially.

The following tabulation presents some of the other polymer types that can be prepared from bisphenols and various dihalides.

Dihalide	Polymer type	Reference
$CSCl_2$	Polythiocarbonate	240
$(\phi O)_2CCl_2$	Polyorthocarbonate	241
CH_2X_2	Polyformal	242,243

Copolymers of mixed types can be made by the reaction of a bisphenol (or bisphenol derivative) with a mixture of reagents of different types. Examples include polycarbonate-esters (244), polyester-amides (245), polycarbonate-urethanes (246), polycarbonate-urethane-ureas (247), poly-ester-anhydrides (248), and polyester-sulfonates (249). Polycarbonate-urethanes have been prepared in large experimental or semi-commercial quantities.

Also to be classified as experimental polymers are modifications of commercial bisphenol A polycarbonate. Photochemical chlorination (250) gave substitution products predominantly in the methyl groups, rather than in the aromatic rings. The polymers had improved resistance to flammability, but their mechanical properties were adversely affected. Soluble graft copolymers can be prepared (251) by allowing styrene to polymerize in the presence of a copolymer containing units of bisphenol A polycarbonate and of **33**. Conversely, graft copolymers can be made (252) by phosgenating

33

mixtures of bisphenol A with a condensation polymer having pendant hydroxyl or carboxyl groups.

A wide variety of other polymers have been prepared from bisphenols in less familiar ways. A few examples follow.

1. POLYESTER-IMIDES (253)

2. Polymers with Triazines (254)

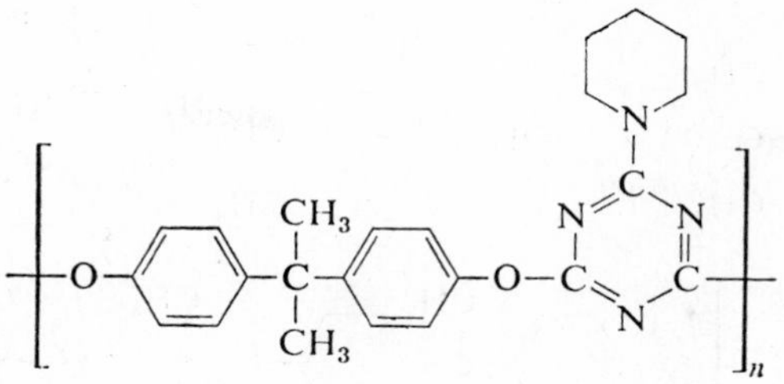

3. Phosphonitrilic Polymers (255)

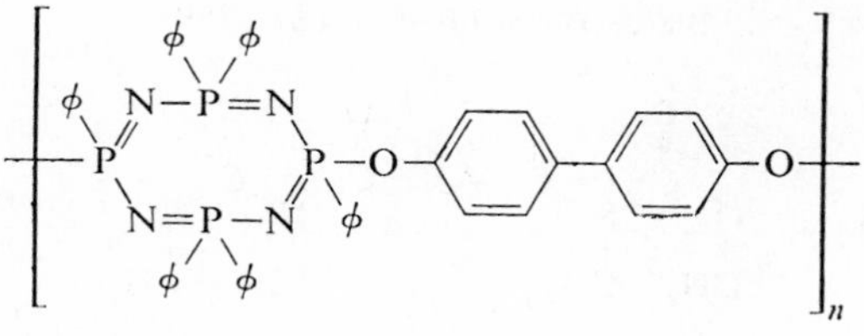

4. Polymers with a Dipropiolate (256)

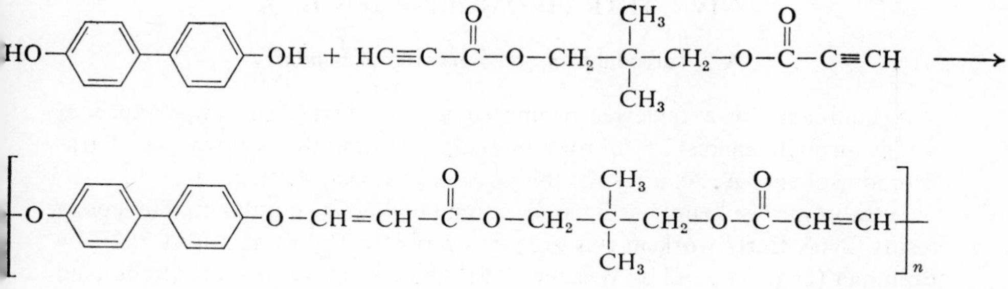

$$\tag{49}$$

5. Poly(aryloxysilanes) (257)

$$HOArOH + R_2Si(NH\phi)_2 \longrightarrow 2\phi NH_2 + \left[OArO-\underset{\underset{R}{|}}{\overset{\overset{R}{|}}{Si}} \right]_n \tag{50}$$

6. POLYMERS WITH CYCLIC SILAZANES (258)

7. POLYETHER-AMIDES BY REACTION WITH N,N'-DISUBSTITUTED BIS(2-OXAZOLIDINONES) (259)

$$\left[OArOCH_2CH_2-NH-\overset{\overset{O}{\|}}{C}-Ar'-\overset{\overset{O}{\|}}{C}-NH-CH_2CH_2 \right]_n + 2CO_2 \quad (52)$$

XIV. TETRABROMOBISPHENOL A

A. Introduction and Survey of Chemistry

A comment by a reviewer prompted us to reassess the importance of 4,4'-isopropylidenebis(2,6-dibromophenol), commonly known as tetra-bromobisphenol A. As a result, this section has been written.

Tetrabromobisphenol A is used commercially in nonflammable epoxy resins (260). Early work in this area was reported by Nametz (261,262), by Bremmer (263,264), and by Wismer (265). Thermoplastic epoxy resins derived from tetrabromobisphenol A were described by Sonnabend (266). Bromine is more effective than chlorine in imparting flame resistance (260,262,265). The polycarbonate (267) of tetrabromobisphenol A has relatively poor mechanical properties (268), but flame-retardant objects with high impact strength can be made from blends of bisphenol A polycarbonate with tetrabromobisphenol A carbonate homopolymer (269) or copolymers (270). The bis(chloroformate) of tetrabromobisphenol A, mp 161–162°C, reported by Stephens (208), has been used in the preparation of polyurethanes (271). Tetrabromobisphenol A is ineffective as a fungicide (272).

The only reported method of preparation of tetrabromobisphenol A is the bromination of bisphenol A, shown in Eq. 53. Early work has been reviewed by Hennis (273).

It is possible to effect the bromination with a mixture of bromine and chlorine (274–276), but bromine alone is more commonly used. An important improvement over early methods involves the introduction of a mixture of water and an alcohol as the reaction medium (273,277).

B. Synthesis

The method for the synthesis of tetrabromobisphenol A described is that of Dietzler (277); it was the basis of a process used in a manufacturing plant (273). The compound made by this method exhibited melting points as high as 181–182°C. More recent workers (278,279), who conducted the bromination in aqueous ethanol, reported melting points of 161–162°C, but products more comparable to Dietzler's have been reported by Losev (280) and particularly by Schnell (281).

> To a solution of bisphenol A (684 g, 3 moles) dissolved in 1050 ml of methanol is added 450 ml of water. The solution is warmed to 42°C and 1920 g (12 moles) of bromine added below the surface of the stirred liquid, while a temperature between 39.5 and 42°C is maintained by cooling. This takes 4.3 hr. After an additional 2 hr at the same temperature, the reaction mixture is cooled to room temperature. After it has been allowed to stand about 60–70 hr, the precipitated solid is filtered, washed eight times with 2-liter portions of water, and dried. There is obtained 1573 g (96.4% yield) of the light yellow tetrabromobisphenol A, mp 173.5–176.5°C. Recrystallization from isopropanol raises the melting point to 181–182°C.

References

1. P. W. Morgan, "Condensation Polymers by Interfacial and Solution Methods," Wiley-Interscience, New York, 1965.
2. Beilsteins Handbuch, **6**:1011, I 493, II 978, III 5459.
3. W. A. Beatty, U.S. Patent 1,225,748 (1917).
4. A. Greth, *Chem.-Z., Chem. Appl.*, **91**:357 (1967).
5. E. Leibnitz and K. Naumann, *Chem. Tech.* (*Berlin*), **3**:5 (1951).
6. W. C. Stoesser and E. H. Sommerfield (to Dow Chemical Co.), U.S. Patent 2,623,908 (1952).

7. H. Schnell and H. Krimm, *Angew. Chem., Int. Ed. Engl.*, **2**:373 (1963).
8. J. I. deJong and F. H. D. Dethmers, *Rec. Trav. Chim. Pays-Bas*, **84**:460 (1965).
9. J. D. MacNaughton and J. Ornstein (to Shawinigan Chemicals Ltd.), U.S. Patent 3,418,378 (1968).
10. W. H. Prahl, S. J. Lederman, and E. I. Lichtblau (to Hooker Chemical Corp), U.S. Patent 3,290,390 (1966).
11. R. P. Perkins and F. Bryner (to Dow Chemical Co.), U.S. Patent 2,359,242 (1944).
12. J. E. Jansen (to B. F. Goodrich Co.), U.S. Patent 2,468,982 (1949).
13. P. K. Ghosh, T. Guha, and A. N. Saha, *J. Appl. Chem. (London)*, **17**:239 (1967).
14. D. B. Luten, Jr. (to Shell Development Co.), U.S. Patent 2,791,616 (1957).
15. L. C. Bostian, K. T. Nilsson, and Z. Oser (to Allied Chemical Corp.), U.S. Patent 3,169,996 (1965).
16. A. G. Farnham, F. N. Apel, and H. L. Bender (to Union Carbide Corp.), U.S. Patent 3,242,219 (1966).
17. F. N. Apel, L. B. Conte, Jr., and H. L. Bender (to Union Carbide Corp.), U.S. Patent 3,153,001 (1964).
18. F. N. Apel, P. Farevaag, and H. L. Bender (to Union Carbide Corp.), U.S. Patent 3,049,569 (1962).
19. A. R. Grover and R. E. Richardson (to Union Carbide Corp.), U.S. Patent 3,221,061 (1965).
20. H. L. Bender, L. B. Conte, Jr., and F. N. Apel (to Union Carbide Corp.), U.S. Patent 2,858,342 (1958).
21. P. H. Deming and H. Dannenberg (to Shell Development Co.), U.S. Patent 2,669,588 (1954).
22. I. M. Bilik, A. M. Serebryanyi, and T. A. Mar'yashkina, *Zh. Prikl. Khim.*, **40**:636 (1967); *J. Appl. Chem. (USSR)*, **40**:617 (1967).
23. T. Kato, *Kôgyô Kagaku Zasshi*, **66**:730 (1963); *Chem. Abstr.*, **59**:13856f (1963).
24. Bergwerkgesellschaft Hibernia AG, British Patent 766,549 (1957).
25. Farbenfabriken Bayer AG, British Patent 963,294 (1964).
26. R. F. Webb and I. Hinton (to CIBA, Ltd.), British Patent 912,288 (1962).
27. A. G. Farnham (to Union Carbide Corp.), U.S. Patent 3,288,864 (1966).
28. D. B. Luten, Jr., S. A. Ballard, and C. G. Schwarzer (to Shell Development Co.), U.S. Patent 2,602,821 (1952).
29. R. I. Hoaglin, C. W. Plummer, and H. C. Schultze (to Union Carbide Corp.), U.S. Patent 2,858,343 (1958).
30. R. T. Holm and C. W. Smith (to Shell Development Co.), U.S. Patent 2,779,800 (1957).
31. C. G. Schwarzer, S. A. Ballard, and D. B. Luten, Jr. (to Shell Development Co.), U.S. Patent 2,602,822 (1952).
32. Badische Anilin- & Soda-Fabrik, Netherlands Application 67,08809 (1967).
33. J. P. Henry, (to Union Carbide Corp.) U.S. Patent 2,884,462 (1959).
34. L. B. Conte and F. N. Apel (to Union Carbide Corp.), German Patent 1,161,284 (1964).
35. D. A. Novokhatka, B. V. Matyushenskii, A. A. Glushkova, and V. I. Seraya, *Khim. Prom*, **42**:175 (1966); *Chem. Abstr.*, **64**:19461f (1966).
36. D. A. Novokhatka, V. I. Seraya, A. S. Nalivaiko, and T. G. Lobko, *Khim. Prom. Ukr. (Russ. Ed.)*, No. 2, 20 (1968); *Chem. Abstr.*, **69**:58896n (1968).
37. Anonymous, *Oil, Paint, Drug Rep.*, March 10, 1969.
38. Union Carbide Corporation, Chemicals and Plastics, "Bulletin F-40900A," June, 1968.

39. W. R. Sorenson and T. W. Campbell, "Preparative Methods of Polymer Chemistry," 2nd ed., Interscience, New York, 1968, p. 140.
40. L. H. Griffin, *Anal. Chem.*, **34**:564 (1962).
41. G. F. Duncan and A. H. Widiger, Jr. (to Dow Chemical Co.), U.S. Patent 3,326,986 (1967).
42. J. I. de Jong (to Koninklijke Zwavelzuurfabrieken v/h Ketjen N. V.), British Patent 949,668 (1964).
43. *Beilsteins Handbuch* **6**:I 503, II 996, III 5646.
44. J. Mleziva, M. Lidarik, and S. Stary, *Plaste Kaut.*, **8**:171 (1961).
45. G. S. Kolesnikov *et al.*, *Plast. Massy*, No. 3, 41 (1966); *ibid.*, No. 9, 20 (1966).
46. K. H. Meyer and H. Schnell (to Farbenfabriken Bayer AG), German Patent 1,031,788 (1958).
47. *Beilsteins Handbuch*, **6**:995, I 488, II 964, III 5412.
48. H. G. Peer, *Rec. Trav. Chim. Pays-Bas*, **79**:825 (1960).
49. A. M. Partansky, *Amer. Chem. Soc., Coatings Plast. Chem., Preprints*, **27**(1):115 (1967).
50. A. G. Farnham and F. P. Klosek (to Union Carbide Corp.), U.S. Patent 2,812,364 (1957).
51. A. J. Conix (to Gevaert Photo-Producten N.V.), U.S. Patent 3,185,664 (1965).
52. R. W. Martin, *J. Org. Chem.*, **17**:342 (1952); (to General Electric Co.), U.S. Patent 2,617,832 (1952).
53. J. Pochwalski and H. Zowall, *Rocz. Chem.*, **33**:661 (1959); *Chem. Abstr.*, **54**:2247b (1960).
54. Y. Ishikawa, T. Ando, and S. Kataoka, Japanese Patent 26,844 (1964); *Chem. Abstr.*, **62**:9064a (1965).
55. G. Zigeuner, K. Jellinek, D. Normann, and K. Elbel, *Monatsh. Chem.*, **90**:473 (1959).
56. M. Levine and S. C. Temin, *J. Org. Chem.*, **22**:85 (1957); (to Industrial Rayon Corp.), U.S. Patent 2,925,444 (1960).
57. H. L. Bender, A. G. Farnham, and J. W. Guyer (to Bakelite Corp.), U.S. Patent 2,464,207 (1949).
58. K. H. Heller, L. Bottenbruch, and H. Schnell (to Farbenfabriken Bayer AG), U.S. Patent 3,277,183 (1966).
59. W. H. Moss, British Patent 491,792 (1938); *Chem. Abstr.*, **33**:1412 (1939).
60. F. Bryner and A. J. Dietzler (to Dow Chemical Co.), U.S. Patent 3,143,575 (1964).
61. F. Bryner (to Dow Chemical Co.), U.S. Patent 3,035,098 (1962).
62. H. Jenkner (to Chemische Fabrik Kalk), German Patent 1,262,283 (1968).
63. H. Schnell and L. Bottenbruch (to Farbenfabriken Bayer AG), German Patent 1,073,504 (1960).
64. E. C. Hurdis and J. F. Petras (to United States Rubber Co.), U.S. Patent 2,891,092 (1959).
65. J. A. Bralley and F. B. Pope (to B. F. Goodrich Co.), British Patent 614,235 (1948).
66. L. Krudenc, Czechoslovak Patent 106,367 (1963); *Chem. Abstr.*, **60**:2834d (1964).
67. Y. Sekine and K. Ikeda (to Rasa Industries Co., Ltd.), Japanese Patent 1011 (1967); *Chem. Abstr.*, **67**:32465e (1967).
68. *Beilsteins Handbuch*, **18**:143, I 373, II 119.
69. L. Molle, *Ing. Chim.* **10**:193 (1926); *Chem. Abstr.*, **21**:2470 (1927).
70. W. Herzog, *Chem.-Z.*, *Chem. Appl.*, **51**:84 (1927).
71. H. R. Gamrath (to Monsanto Chemical Co.), U.S. Patent 2,522,939 (1950).
72. H. R. Gamrath (to Monsanto Chemical Co.), U.S. Patent 2,522,940 (1950).

73. C. Ward, *J. Chem. Soc.*, **119**:850 (1921).
74. R. Pasternack, in "Encyclopedia of Chemical Technology," 1st ed., Vol. 10, R. E. Kirk and D. F. Othmer, Eds., Interscience, New York, 1953, pp. 370–374.
75. American Chemical Society, "Reagent Chemicals," 4th ed., 1968.
76. National Formulary, Am. Pharm. Assoc., **12**:303 (1965).
77. W. Wieniawski, *Acta Pol. Pharm.*, **23**:483 (1966); *Chem. Abstr.*, **66**:68876h (1967).
78. *Beilsteins Handbuch*, **6**:861, II 853, III 4456.
79. J. R. Stack (to Carnegie-Illinois Steel Corp.), U.S. Patents 2,313,371 and 2,313,372 (1943); *Chem. Abstr.*, **37**:4974–4975 (1943).
80. J. W. Andrews (to United States Steel Corp.), U.S. Patent 2,633,450 (1953); *Chem. Abstr.*, **48**:62h (1954).
81. Monsanto Chemical Company, Technical Bulletin No. IN-1, "Dihydroxydiphenylsulfone," 1957.
82. L. E. Hinkel and G. H. R. Summers, *J. Chem. Soc.*, **1949**:2854.
83. M. L. Mausner (to Witco Chemical Co., Inc.), U.S. Patent 3,318,956 (1967).
84. Shell Internationale Research Maatschappij N.V., British Patent 915,768 (1963).
85. J. I. Carr (to E.I. du Pont de Nemours and Co.), U.S. Patent 2,000,061 (1935).
86. J. Huismann (to General Aniline and Film Corp.), U.S. Patent 2,288,282 (1942).
87. M. J. Keogh and A. K. Ingberman (to Union Carbide Corp.), U.S. Patent 3,415,887 (1968).
88. T. Kurano, M. Horiuchi, and M. Fukuda (to Sankyo Chemical Industries Co., Ltd.), Japanese Patent 24,662 (1968); *Chem. Abstr.*, **70**:57419t (1969).
89. F. Dunning, B. Dunning, Jr., and W. E. Drake, *J. Amer. Chem. Soc.*, **53**:3466 (1931).
90. S. Oae, M. Yoshihara, and W. Tagaki, *Bull. Chem. Soc. Jap.*, **40**:951 (1967).
91. P. Carre and D. Liebermann, *Compt. Rend.*, **196**:275 (1933).
92. S. Oae and C. Zalut, *J. Amer. Chem. Soc.*, **82**:5359 (1960).
93. L. N. Nikolenko and N. I. Krizhechkovskaya, *Zh. Obshch. Khim.*, **33**:3731 (1963); *J. Gen. Chem. (USSR)*, **33**:3664 (1963).
94. R. A. Clendinning, Union Carbide Corporation, private communication.
95. *Beilsteins Handbuch*, **6**:796, I 398, II 802, III 4292.
96. Anonymous, *Chem. Week*, **100** (25):75 (1967).
97. Anonymous, *Chem. Eng. News*, **45** (26):13 (1967).
98. H. E. Fierz-David and G. Stamm, *Helv. Chim. Acta*, **25**:364 (1942).
99. H. E. Fierz-David and L. Blangey, "Fundamental Processes of Dye Chemistry," Interscience, New York, 1949.
100. W. R. Cake (to Heyden Newport Chemical Corp.), U.S. Patent 2,856,437 (1958).
101. R. D. Swisher (to Monsanto Chemical Co.), British Patent 679,827 (1952).
102. G. A. Webb (to Koppers Company, Inc.), U.S. Patent 2,736,754 (1956).
103. Directie van de Staatsmijnen in Limburg, British Patent 672,511 (1952).
104. Koppers Company, Inc., French Patent 1,319,454 (1963).
105. E. G. E. Hawkins, D. C. Quin, and F. E. Salt (to Distillers Company, Ltd.), British Patent 641,250 (1950).
106. A. R. Graham, quoted in *Chem. Ind. (London)*, **1967**:2119.
107. J. P. Fortuin, M. J. Waale, and R. P. van Oosten, *Petrol. Refiner*, **38**(6):189 (1959).
108. W. Webster (to Distillers Company, Ltd.), British Patent 727,498 (1955).
109. S. V. Zavgorodnii and I. N. Novikov, *Izv. Vyssh. Ucheb. Zaved., Khim. Khim. Tekhnol.*, **3**:863 (1960); *Chem. Abstr.*, **55**:8340i (1961).
110. R. D. Mair and A. J. Graupner, *Anal. Chem.*, **36**:194 (1964).
111. J. C. Conner, Jr., and V. Verplanck (to Hercules Powder Co.), U.S. Patent 2,856,432 (1958).

112. W. Webster and B. H. M. Thompson (to Distillers Company Ltd.), German Patent 1,004,611 (1957).

113. T. Bewley (to Distillers Company Ltd.), British Patent 748,287 (1956).

114. Hercules Powder Co., French Patent 979,665 (1951).

115. H. Sodomann, B. Hauschulz, and M. Hanke (to Phenolchemie), German Patent 1,136,713 (1962).

116. Distillers Company Ltd., British Patent 982,514 (1965).

117. K. W. Müller and D. Delfs (to Farbenfabriken Bayer AG), German Patent 1,040,563 (1958).

118. A. H. Widiger, Jr. (to Dow Chemical Co.), French Patent 1,487,906 (1967).

119. G. W. Dolman and F. W. Neumann, *J. Amer. Chem. Soc.*, **90**:1601 (1968).

120. C. F. Boehringer and Sons, German Patent 269,544 (1914).

121. E. C. Britton, S. L. Bass, and N. Elliott (to Dow Chemical Co.), U.S. Patent 1,934,656 (1933).

122. S. G. Burroughs (to Pennsylvania Coal Products Co.), U.S. Patent 2,041,592 (1936).

123. K. F. Bursack, H. J. Moltzan, and E. L. Johnston (to Vulcan Materials Co.), U.S. Patent 3,413,341 (1968).

124. J. Kamlet (to Goodyear Tire and Rubber Co.), U.S. Patent 2,835,708 (1958).

125. Koppers Company, Inc., Bulletin CD-2-424, "Resorcinol."

126. F. A. Patty, Ed., "Industrial Hygiene and Toxicology," 2nd ed., Vol. 2, Interscience, New York, 1963.

127. R. Raff and B. V. Ettling, in "Encyclopedia of Chemical Technology," 2nd ed., Vol. 11, R. E. Kirk and D. F. Othmer, Eds., Interscience, New York, 1966, pp. 472–492.

128. N. I. Sax, Ed., "Dangerous Properties of Industrial Materials," 3rd ed., Reinhold, New York, 1968.

129. *Beilsteins Handbuch*, **6**:836, I 413, II 832, III 4374.

130. W. H. Shearon, Jr., L. G. Davy, and H. V. Bramer, *Ind. Eng. Chem.*, **44**:1730 (1952).

131. H. Von Bramer and A. C. Ruggles (to Eastman Kodak Co.), U.S. Patent 2,043,912 (1936).

132. H. Von Bramer and J. W. Zabriskie (to Eastman Kodak Co.), U.S. Patent 1,998,177 (1935).

133. Fuji Shashin Film K.K., French Patent 1,371,138 (1964).

134. Y. Kono, M. Yoshikawa, and S. Matsumoto (to Fuji Shashin Film K.K.), U.S. Patent 3,278,271 (1966).

135. C. F. Gibbs (to B. F. Goodrich Co.), U.S. Patent 2,343,768 (1944).

136. A. D. Lohr and M. A. Taves (to Hercules Powder Co.), U.S. Patent 2,799,715 (1957).

137. W. F. Zimmer, Jr. (to Hooker Chemical Corp.), U.S. Patent 3,028,410 (1962).

138. M. A. Taves (to Hercules Powder Co.), U.S. Patent 2,799,698 (1957).

139. J. D. London, in "Progress in Organic Chemistry," Vol. V, J. W. Cook and W. Carruthers, Eds., Butterworths, Washington, 1961, pp. 46–72.

140. Rhone-Poulenc S.A., French Patent 1,479,354 (1967).

141. Rhone-Poulenc S.A., British Patent 1,099,334 (1968).

142. G. Amedjian, N. Crenne, and J. E. G. Morel (to Rhone-Poulenc S.A.), U.S. Patent 3,376,351 (1968).

143. S. D. Ross, M. Finkelstein, and R. C. Peterson, *J. Amer. Chem. Soc.*, **86**:4139 (1964).

144. H. Palfreeman and N. V. S. Knibbs, U.S. Patent 2,130,151 (1938).

145. S. Horrobin and R. G. A. New (to Imperial Chemical Industries Ltd.), U.S. Patent 2,285,858 (1942).

146. K. Dürkes, German Patents 1,101,436 and 1,102,171 (1961).

147. G. A. Nikiforov and V. V. Ershov, *Izv. Akad. Nauk SSSR, Ser. Khim.*, **1964**:176; *Bull. Acad. Sci. USSR, Div. Chem. Sci.*, **1964**:160.

148. V. V. Kozlov and S. E. Suvorova, *Zh. Obshch, Khim.*, **32**:1882 (1962); *J. Gen. Chem. (USSR)*, **32**:1861 (1962).

149. W. Reppe and H. Vetter, *Ann.*, **582**:133 (1953).

150. W. Reppe and A. Magin (to Badische Anilin- & Soda-Fabrik AG), German Patent 870,698 (1953); *Chem. Abstr.*, **52**:17185f (1958).

151. W. Reppe (to Badische Anilin- & Soda-Fabrik AG), U.S. Patent 2,702,304 (1955).

152. W. Reppe and A. Magin (to Badische Anilin- & Soda-Fabrik AG), U.S. Patent 3,394,193 (1968).

153. B. W. Howk and J. C. Sauer (to E.I. du Pont de Nemours and Co.), U.S. Patent 3,055,949 (1962).

154. Lonza Ltd., British Patents 1,031,877 (1966) and 1,119,520 (1968).

155. P. Pino, G. Braca, G. Sbrana, and A. Cuccuru, *Chem. Ind. (London)*, **1968**:1732.

156. Ajinomoto Co., Inc., French Patent 1,486,666 (1967).

157. E. J. Goethals, *Meded. Vlaam. Chem. Ver.*, **29**:186 (1967); *Chem. Abstr.*, **70**:3360y (1969).

158. Anonymous, *Oil, Paint, Drug Rep.*, May 3, 1965.

159. W. R. Sorenson and T. W. Campbell, "Preparative Methods of Polymer Chemistry," 2nd ed., Interscience, New York, 1968, p. 139.

160. J. Kamlet (to B. F. Goodrich Co.), U.S. Patent 2,614,127 (1952).

161. B. D. Kruzhalov and V. V. Fedorova, *Tr., Nauchn.-Issled. Inst. Sintetich. Spirtov i Organ. Produktov*, **1960**:267; *Chem. Abstr.*, **59**:6298a (1963).

162. E. J. Lorand and J. E. Reese (to Hercules Powder Co.), U.S. Patent 2,548,435 (1951).

163. B. Domencali and G. Canti (to Montecantini), French Patent 1,382,544 (1964).

164. *Beilsteins Handbuch*, **8**:316, I 641, II 355.

165. A. Baeyer and J. B. Burkhardt, *Ber.*, **11**:1299 (1878); *Ann.*, **202**:126 (1880).

166. R. N. Johnson, A. G. Farnham, R. A. Clendinning, W. F. Hale, and C. N. Merriam, *J. Polym. Sci., Part A-1*, **5**:2375 (1967).

167. I. Goodman, J. E. McIntyre, and W. Russell (to Imperial Chemical Industries Ltd.), British Patent 971,227 (1964).

168. N. J. McCarthy, Union Carbide Corporation, personal communication.

169. E. C. Britton (to Dow Chemical Co.), U.S. Patent 1,961,630 (1934).

170. R. Barclay, Jr., Union Carbide Corporation, unpublished data.

171. L. N. Stanley (to General Aniline and Film Corp.), U.S. Patent 3,073,866 (1963).

172. I. A. Kaye, H. C. Klein, and W. J. Burlant, *J. Amer. Chem. Soc.*, **75**:745 (1953).

173. V. Prey, *Ber.*, **74B**:1219 (1941); *ibid.*, **75B**:350 (1942).

174. *Beilsteins Handbuch*, **6**:991, I 485, II 962, III 5389.

175. G. H. Herbst (to Farbwerke Hoechst), British Patent 936,278 (1963).

176. R. L. Jenkins (to Monsanto Chemical Co.), U.S. Patent 2,368,361 (1945).

177. W. L. Carrick and G. L. Karapinka (to Union Carbide Corp.), U.S. Patent 3,322,838 (1967).

178. A. S. Hay (to General Electric Co.), Netherlands Appl. 64,10238 (1965); *Chem. Abstr.*, **63**:5561b (1965).

179. A. S. Hay (to General Electric Co.), Japanese Patent 102 (1967).

180. A. S. Hay, General Electric Co., private communication.

181. J. E. Malowan (to Swann Research Inc.), U.S. Patent 1,951,577 (1934).

182. C. F. Booth (to Swann Research Inc.), U.S. Patent 1,925,367 (1933).

183. V. I. Sevast'yanov, *Zh. Prikl. Khim.*, **30**:1858 (1957); *Chem. Abstr.*, **52**:10991h (1958).

184. A. L. Wilds, C. H. Shunk, and C. H. Hoffman, *J. Amer. Chem. Soc.*, **76**:1733 (1954).
185. A. S. Hay, *J. Org. Chem.*, **34**:1160 (1969).
186. T. Sulzberg and R. J. Cotter, *Macromolecules*, **2**:146 (1969).
187. United States Pipe & Foundry Company, Chemical Division, "Evaluation of Technical Resorcinol," July 24, 1967.
188. W. M. Anderson, G. B. Carter, and A. J. Landua, *Anal. Chem.*, **31**:1214 (1959).
189. G. Challa and P. H. Hermans, *Anal. Chem.*, **32**:778 (1960).
190. N. H. Reinking and A. E. Barnabeo, *Anal. Chem.*, **37**:395 (1965).
191. J. Aurenge, M. Degeorges, and J. Normand, *Bull. Soc. Chim. Fr.*, **1963**:1732.
192. S. Tominaga, *Bunseki Kagaku*, **12**:137 (1963); *Chem. Abstr.*, **58**:11936f (1963)
193. H. H. Gill, *Anal. Chem.*, **36**:1201 (1964).
194. A. Davis and J. H. Golden, *J. Chromatogr.*, **26**:254 (1967).
195. L. E. Brydia, *Anal. Chem.*, **40**:2212 (1968).
196. F. Kanne and K. Stange, *Z. Anal. Chem.*, **189**:261 (1962).
197. T. S. Ma and D. Spiegel, *Microchem. J.*, **10**:61 (1966).
198. E. Pillion, *J. Gas Chromatogr.*, **3**:238 (1965).
199. J. R. Lindsay Smith, R. O. C. Norman, and G. K. Radda, *J. Gas Chromatogr.*, **2**:146 (1964).
200. J. Ratusky and L. Bastar, *Chem. Ind. (London)*, **1964**:579; *Collect. Czech. Chem. Commun.*, **29**:3066 (1964).
201. H. G. Henkel, *J. Chromatogr.*, **20**:596 (1965).
202. E. A. Blakley, *Anal. Biochem.*, **15**:350 (1966).
203. I. V. Adorova, V. Ya. Kovner, and M. I. Siling, *Plast. Massy*, **1968**(1):(60); *Chem. Abstr.*, **68**:68563v (1968).
204. T. Sulzberg, accepted for publication in *Macromol. Syn.*, vol. 4.
205. R. E. Oesper, W. Broker, and W. A. Cook, *J. Amer. Chem. Soc.*, **47**:2609 (1925).
206. R. J. Cotter, M. Matzner, and R. P. Kurkjy, *Chem. Ind. (London)*, **1965**:791.
207. R. P. Kurkjy, M. Matzner, and R. J. Cotter (to Union Carbide Corp.), U.S. Patent 3,255,230 (1966).
208. C. W. Stephens (to E.I. du Pont de Nemours and Co.), U.S. Patent 3,211,774 (1965).
209. C. W. Stephens and W. Sweeny (to E.I. du Pont de Nemours and Co.), U.S. Patent 3,211,775 (1965).
210. C. W. Stephens (to E.I. du Pont de Nemours and Co.), U.S. Patent 3,211,776 (1965).
211. M. Matzner, R. Barclay, Jr., and C. N. Merriam, *J. Appl. Polym. Sci.*, **9**:3337 (1965).
212. L. J. Bellamy, "The Infra-Red Spectra of Complex Molecules," 2nd ed., Wiley, New York, 1958.
213. H. Schnell and L. Bottenbruch, *Makromol. Chem.*, **57**:1 (1962).
214. W. R. Sorenson and T. W. Campbell, "Preparative Methods of Polymer Chemistry," 2nd ed., Interscience, New York, 1968.
215. J. Frados, Ed., "Modern Plastics Encyclopedia," Vol. 45, No. 14A, McGraw-Hill, New York, 1968.
216. A. C. Stewart, "Phenoxy Resins," in J. Frados, Ed., "Modern Plastics Encyclopedia," Vol. 45, No. 1A, McGraw-Hill, New York, 1967, pp. 214–216.
217. N. H. Reinking, A. E. Barnabeo, and W. F. Hale, *J. Appl. Polym. Sci.*, **7**:2135 (1963).
218. H. Schnell, "Chemistry and Physics of Polycarbonates," Interscience, New York, 1964.
219. T. S. Grabowski (to Borg-Warner Corp.), German Patent 1,170,141 (1964); *Chem. Abstr.*, **61**:4556h (1964).
220. H. A. Vaughn, Jr. (to General Electric Co.), U.S. Patent 3,189,662 (1965).

221. A. G. Farnham and R. N. Johnson (to Union Carbide Corp.), U.S. Patent 3,332,909 (1967); *Chem. Abstr.*, **68**:69543a (1968).
222. Atlas Chemical Industries, Inc., *Bulletin LP-29*, 1965; *cf.* P. Kass, U.S. Patent 2,634,251 (1953); *Chem. Abstr.*, **47**:8414i (1953).
223. A. D. F. Toy (to Victor Chemical Works), U.S. Patent 2,435,252 (1948).
224. H. W. Coover, Jr., R. L. McConnell, and M. A. McCall, *Ind. Eng. Chem.*, **52**:409 (1960).
225. V. V. Korshak, *J. Polym. Sci.*, **31**:319 (1958).
226. C. F. Baranauckas and I. Gordon (to Hooker Chemical Corp.), Belgian Patent 623,965 (1963); *Chem. Abstr.*, **59**:9894c (1963).
227. L. Friedman and H. Gould (to Weston Chemical Corp.), U.S. Patent 3,053,878 (1962); *Chem. Abstr.*, **58**:3354a (1963).
228. O. S. Kauder and W. E. Leistner (to Argus Chemical N.V.), Belgian Patent 626,102 (1963); *Chem. Abstr.*, **59**:5325a (1963).
229. Wingfoot Corp., Brit. Patent 585,775 (1947); *Chem. Abstr.*, **42**:2138d (1948).
230. J. A. Bralley and F. B. Pope (to B. F. Goodrich Co.), British Patent 611,529 (1948); *Chem. Abstr.*, **43**:4298g (1949).
231. M. H. Keck (to Goodyear Tire and Rubber Co.), U.S. Patent 3,133,898 (1964).
232. A. J. Conix (to Gevaert Photo-Producten N.V.), U.S. Patents 3,216,970 (1965) and 3,351,624 (1967).
233. I. Goodman and J. A. Rhys, "Polyesters," Iliffe Books, London, 1965.
234. V. V. Korshak and S. V. Vinogradova, "Polyesters," Pergamon Press, New York, 1965.
235. A. S. Carpenter and E. R. Wallsgrove (to Courtaulds Ltd.), British Patent 675,665 (1952).
236. A. S. Carpenter and E. R. Wallsgrove (to Courtaulds Ltd.), British Patent 664,271 (1952).
237. R. W. Martin (to Shell Oil Co.), U.S. Patent 3,378,522 (1968).
238. E. P. Goldberg and F. Scardiglia (to Borg-Warner Corp.), U.S. Patents 3,236,808 and 3,236,809 (1966).
239. Anonymous, *Chem. Week*, **98**:(26):84 (1966).
240. A. J. Conix and U. L. Laridon (to Gevaert Photo-Production N.V.), U.S. Patent 3,227,684 (1966).
241. T. Takekoshi, *Amer. Chem. Soc., Polym. Chem., Preprints*, **10**:103 (1969).
242. R. Barclay, Jr. (to Union Carbide Corp.), U.S. Patent 3,069,386 (1962).
243. A. G. Farnham, R. N. Johnson, and P. E. Sonnet, Union Carbide Corporation, personal communications.
244. E. P. Goldberg (to General Electric Co.), U.S. Patent 3,169,121 (1965).
245. W. W. Moyer, Jr. (to Borg-Warner Corp.), U.S. Patent 3,272,774 (1966).
246. W. E. Bissinger, F. Strain, H. C. Stevens, and W. R. Dial (to Pittsburgh Plate Glass Co.), U.S. Patent 3,215,668 (1965).
247. J. Fontan and O. Laguna, *Chim. Ind., Genie Chim.*, **97**:321 (1967).
248. H. Sawada, *Kôgyô Kagaku Zasshi*, **68**:1279 (1965); *Chem. Abstr.*, **63**:16479h (1965).
249. E. P. Goldberg and F. Scardiglia (to Borg-Warner Corp.), U.S. Patent 3,262,914 (1966).
250. W. T. Jackson, Jr., J. R. Caldwell, and K. P. Perry, *J. Appl. Polym. Sci.*, **12**:1713 (1968).
251. Farbenfabriken Bayer AG, Netherlands Appl. 67, 16729 (1968).
252. J. E. Cantrill (to General Electric Co.), French Patent 1,502,424 (1967); *Chem. Abstr.*, **69**:87635g (1968).

253. D. F. Loncrini, *J. Polym. Sci., Part A-1*, **4**:1531 (1966).
254. R. Audebert and J. Neal, *Compt. Rend.*, **258**:4749 (1964); *Chem. Abstr.*, **61**:4500b (1964).
255. A. J. Bilbo, C. M. Douglas, N. R. Fetter, and D. L. Herring, *J. Polym. Sci., Part A-1*, **6**:1671 (1968).
256. J. M. Butler, L. A. Miller, and G. L. Wesp (to Monsanto Co.), U.S. Patent 3,201,370 (1965).
257. W. R. Dunnavant, R. A. Markle, P. B. Stickney, J. E. Curry, and J. D. Byrd, *J. Polym. Sci., Part A-1*, **5**:707 (1967).
258. R. L. Elliott and L. W. Breed, *Amer. Chem. Soc., Polym. Chem., Preprints*, **5**:587 (1964).
259. S. Nishizaki and A. Fukami, *Kôgyô Kagaku Zasshi*, **70**:1835 (1967); *Chem. Abstr.*, **68**:115059x (1968).
260. H. Lee and K. Neville, in "Encyclopedia of Polymer Science and Technology," Vol. VI, Interscience, New York, 1967, pp. 209–271.
261. R. C. Nametz (to Michigan Chemical Corp.), U.S. Patent 3,058,946 (1962); *Chem. Abstr.*, **58**:3572e (1963).
262. R. C. Nametz (to Michigan Chemical Corp.), U.S. Patent 3,268,619 (1966); *Chem. Abstr.*, **65**:20291h (1966).
263. B. J. Bremmer (to Dow Chemical Co.), Belgian Patent 617,496 (1962); *Chem. Abstr.*, **58**:10357b (1963).
264. B. J. Bremmer (to Dow Chemical Co.), U.S. Patent 3,294,742 (1966); *Chem. Abstr.*, **66**:38487q (1967).
265. M. Wismer (to Pittsburgh Plate Glass Co.), U.S. Patent 3,016,362 (1962); *Chem. Abstr.*, **56**:11797g (1962).
266. L. F. Sonnabend (to Dow Chemical Co.), U.S. Patent 3,277,048 (1966); *Chem. Abstr.*, **66**:19139m (1967).
267. Farbenfabriken Bayer AG, British Patent 857,430 (1960); *Chem. Abstr.*, **55**:21661c (1961).
268. W. J. Jackson, Jr., and J. R. Caldwell, *J. Appl. Polym. Sci.*, **11**:227 (1967).
269. J. K. S. Kim (to General Electric Co.), French Patent 1,386,646 (1965); *Chem. Abstr.*, **63**:13524b (1965).
270. J. K. S. Kim (to General Electric Co.), U.S. Patent 3,334,154 (1967); *Chem. Abstr.*, **67**:74111s (1967).
271. P. W. Morgan (to E.I. du Pont de Nemours and Co.), U.S. Patent 3,391,111 (1968); *Chem. Abstr.*, **69**:44528g (1968).
272. P. B. Marsh, M. L. Butler, and B. S. Clark, *Ind. Eng. Chem.*, **41**:2176 (1949).
273. H. E. Hennis, *Ind. Eng. Chem., Prod. Res. Develop.*, **2**:140 (1963).
274. H. Hahn (to Chemische Fabrik Kalk GmbH), German Patent 1,129,957 (1962); *Chem. Abstr.*, **57**:16490e (1962).
275. Chemische Fabrik Kalk GmbH, Belgian Patent 611,069 (1962); *Chem. Abstr.*, **57**:15002g (1962).
276. J. Nentwig (to Farbenfabriken Bayer AG), German Patent 1,151,811 (1963); *Chem. Abstr.*, **60**:456a (1964).
277. A. J. Dietzler (to Dow Chemical Co.), U.S. Patent 3,029,291 (1962).
278. F. S. Holahan, W. J. Erich, and R. J. Valles, *Makromol. Chem.*, **103**:36 (1967).
279. S. K. Gupta, Y. N. Sharma, and R. T. Thampy, *Makromol. Chem.*, **120**:137 (1968).
280. I. P. Losev, O. V. Smirnova, and E. V. Korovina, *Vysokomol. Soedin.*, **5**:1603 (1963).
281. H. Schnell, *Ind. Eng. Chem.*, **51**:157 (1959).

9. CARBONYL AND THIOCARBONYL COMPOUNDS

W. H. SHARKEY, *Central Research Department, E. I. du Pont de Nemours and Company, Wilmington, Delaware*

Contents

I. INTRODUCTION

The prominence of carbonyl compounds as polymerization monomers is related to the commercial success of poly(formaldehyde) polymers. These polymers, variously called polyoxymethylenes and polyacetals, are the most important members of the general class of acetal resins. Production of acetal resins in 1967 was reported (1) to be 61.4 million lb. They are of two main types (2), namely, Delrin® acetal resin, a homopolymer of formaldehyde manufactured by Du Pont, and Celcon®, a copolymer manufactured by Celanese Corporation of America.

Certain higher aldehydes also form acetal resins. These include acetaldehyde, propionaldehyde, and n- and isobutyraldehyde. The polymerization of formaldehyde and higher aldehydes was reviewed at the 1966 Winter Meeting of the American Chemical Society and published in a book edited by O. Vogl (3). Another higher aldehyde whose polymers have received much attention is acrolein. As would be expected, acrolein can be polymerized through the vinyl group or through the aldehyde group to give a polyacetal. None of the linear polyacetals from the higher aldehydes has been produced commercially.

All the foregoing polymerizable aldehydes are articles of commerce. They are all made in huge amounts by highly developed manufacturing processes; and of their many applications, formation of linear polyacetals, even for formaldehyde, is not the major use. Because of the commercial availability of aldehydes, it is much simpler and more economical to purchase them than to undertake a laboratory synthesis of them. The problem then, as far as their use in polymer synthesis is concerned, is how to purify commercially available aldehydes to the degree needed for polymerization.

The haloaldehydes—chloroacetaldehyde, dichloroacetaldehyde, chloral, and fluoral—comprise another class of polymerizable carbonyl compounds. These compounds have received less attention as polymerization monomers than the other aldehydes.

The importance of carbonyl polymers led inevitably to interest in thio-carbonyl analogs. Thioformaldehyde does not exist in monomeric form, but its polymers have been made by ring-opening polymerization of trithiane. Such polymers have not become important because they decompose when heated above the polymer's melting temperature, which, of course, occurs during molding. In sharp contrast to the behavior of acetone, thioacetone polymerizes very easily; but the polymers obtained are only of academic interest. They decompose easily and give off very malodorous products.

The thiocarbonyl compounds that have given the best polymers are those in the fluorine series. Thiocarbonyl fluoride, $CF_2{=}S$, polymerizes to a tough, highly resilient elastomer. Perfluorothioacetyl fluoride and other fluoro-thioacid fluorides form logy elastomers. Perfluorothioketones also polym-erize, but with difficulty, and the products obtained slowly depolymerize. None of these polymers has been produced commercially.

In order to discuss these monomers and their preparation, purification, and polymerization in a logical fashion, each one is discussed separately. It seems appropriate to start with formaldehyde.

II. FORMALDEHYDE

Formaldehyde was discovered by Butlerov in 1859 (4) and is now a very large-volume chemical. Domestic capacity is about 3.4 million metric tons (7.5 billion lb)/year (5). Free world production in 1965 was reported to be 1.376 million metric tons (6) and at the present time is probably in excess of 3 million metric tons. Formaldehyde is a basic chemical having a large number of uses, the major ones being (7) for the manufacture of phenolic resins, urea and melamine resins, pentaerythritol, hexamethylenetetramine, and fertilizer. Only about 2 % of the formaldehyde produced is used to make acetal resins. An excellent discussion of all aspects of formaldehyde prepara-tion and chemistry is given by Walker (8) in an ACS monograph. A thorough description of the polymerization of formaldehyde is to be found in a review by Brown (9).

A. Synthesis

Formaldehyde is manufactured either by oxidation of hydrocarbons or by oxidation of methanol. By far the major portion is made by methanol oxidation. This was first done by A. W. Hofmann (10), who passed a mixture

$$2CH_3OH + O_2 \xrightarrow{\text{Pt}} 2CH_2O + H_2O + 76 \text{ kcal}$$

of methanol and air over hot platinum. The heat of reaction is sufficient to keep the reaction going without external addition of heat. This reaction is now carried out on a very large scale. A variety of catalysts based on silver and copper (8) have been developed for efficiently carrying out dehydrogenation of methanol in conversions as high as 90.5% to give 87% yields of formaldehyde (6).

The process involves addition of methanol, water, and air to a vaporizing tower; the effluent is passed over a catalyst bed consisting of highly purified silver of a specific particle size. Complete specifications for such a process have not been published. It is interesting to note that water is a necessary ingredient (5). Under optimum conditions of temperature, water content, and throughput, the reaction takes place adiabatically and can be represented by the following three equations:

$$CH_3OH \rightleftharpoons CH_2O + H_2 - 20 \text{ kcal} \tag{1}$$

$$H_2 + \tfrac{1}{2}O_2 \rightarrow H_2O + 58 \text{ kcal} \tag{2}$$

$$CH_3OH + \tfrac{1}{2}O_2 \rightarrow CH_2O + H_2O + 38 \text{ kcal} \tag{3}$$

The main reaction is Eq. 1 and the heat needed to keep it going is supplied by Eqs. 2 and 3. Removal of some of the hydrogen shifts the first reaction to the right and thereby increases formaldehyde conversions. Two side reactions that need to be repressed are:

$$CH_3OH + \tfrac{3}{2}O_2 \rightarrow CO_2 + 2H_2O + 72 \text{ kcal}$$

$$CH_2O + O_2 \rightarrow CO_2 + H_2O + 123 \text{ kcal}$$

Another side reaction is decomposition of formaldehyde to carbon monoxide and hydrogen (Eq. 4). However, this reaction can be controlled because

$$CH_2O \rightleftharpoons CO + H_2 \tag{4}$$

equilibrium is established slowly and is temperature dependent.

Formaldehyde is also made by oxidation of methanol over catalysts composed of molybdenum and iron oxides. A large excess of air is used, and the space-time yield is less than for processes based on dehydrogenation over silver catalysts. A number of methanol oxidation processes are described in a general way in the literature (6,11).

The manufacture of formaldehyde by oxidation of hydrocarbons involves controlled reaction of a gaseous hydrocarbon with air or oxygen. The result is the formation of a variety of products from which formaldehyde is separated (8). This process is economical when the by-products can also be marketed.

A process for obtaining formaldehyde by oxidation of dimethyl ether has also been described (12). It is stated that dimethyl ether is obtained in an amount of 2 wt.% as a by-product in methanol manufacture. The ether is oxidized to formaldehyde (Eq. 5) at 450–530°C over tungsten oxide containing 10% orthophosphoric acid and supported on silicon carbide or α-alumina. The reaction is exothermic, and control of it is exercised by controlling heat removal.

$$(CH_3)_2O + O_2 \rightarrow 2CH_2O + H_2O + 68 \text{ kcal} \tag{5}$$

Formaldehyde is much too reactive to be kept in monomeric form under ordinary conditions. It is sold as an aqueous solution containing 44–52% by weight formaldehyde and less than 1% methanol. The solution must be shipped hot to prevent formation of solid paraformaldehyde. Formaldehyde is also sold as paraformaldehyde, which is a low-molecular-weight polymer capped with hydroxyl groups and containing appreciable amounts of bound water (Eq. 6).

$$nCH_2O + H_2O \rightleftharpoons HO\!\!-\!\!(CH_2O)_n\!H \tag{6}$$

An aqueous solution containing 37% by weight formaldehyde and enough methanol to inhibit formation of solid paraformaldehyde is classified as U.S.P. Formaldehyde. Methanol, in addition to its solvent action, no doubt forms a hemiacetal (Eq. 7) that is also a solvent.

$$CH_2O + CH_3OH \rightarrow CH_3OCH_2OH \tag{7}$$

B. Purification

In order to convert formaldehyde into a high-molecular-weight acetal resin, it is necessary first to obtain the monomer in pure, anhydrous form. Walker (8) describes monomeric formaldehyde as a colorless gas that condenses to a liquid at $-19°C$ and freezes to a crystalline solid at $-118°C$. Since formaldehyde in either the gaseous or liquid state polymerizes readily, the monomer cannot be kept pure for any length of time. Thus, for polymerization purposes, the monomer must be recovered from a commercial source in pure form and used immediately. Although formaldehyde that is substantially anhydrous can be obtained by fractional condensation of vapors, removal of last traces of water in this way is difficult.

A less tedious way of obtaining substantially pure monomeric formaldehyde is pyrolysis of a hemiformal of formaldehyde and an alcohol that boils at least about 95°C at atmospheric pressure (13). An example is pyrolysis of the hemiformal formed by first adding cyclohexanol to an aqueous solution of formaldehyde and then fractionally distilling to remove water. Formaldehyde gas can be purified by passing it over a liquid film of the hemiformal at -15 to $-20°C$.

Probably the easiest way to prepare polymerization-grade monomeric formaldehyde is by decomposition of a suitable formaldehyde polymer. Such a procedure has been described by MacDonald (14) and by Jaacks et al. (15). Paraformaldehyde is first converted to α-polyoxymethylene by the method of Staudinger (16). The purpose of this step is to remove much of the water bound in paraformaldehyde. The α-polyoxymethylene is then thermally decomposed at 190°C and the formaldehyde so generated is passed through several traps cooled to −15°C.

Pure formaldehyde gas does not polymerize if kept dry and held at 80–100°C. Purity is essential, however, since even a trace of water promotes polymerization. Some of the properties (8) of formaldehyde gas are listed in Table 1.

TABLE 1
Properties of Formaldehyde Gas

Property	Value
Boiling point, °C	−19
Heat capacity (C_p°) at 25°C, cal/mole	8.461
Entropy (S°) at 25°C, cal/mole °C	52.261
ΔH_f° at 25°C, kcal/mole	−27.700
ΔF_f° at 25°C, kcal/mole	−26.266
Principal infrared bands, cm^{-1}	3003.3
	2843
	2766.4
	1746.1
	1500.6
	1247.4
	1163.5
Ionization potentials, V	10.8 ± 0.1
	11.8 ± 0.2
	13.1 ± 0.2

Very pure liquid formaldehyde that is kept cold can be characterized. However, after a few hours it will show signs of polymerization. Some of the properties (8) of the liquid are presented in Table 2.

Liquid formaldehyde is miscible with many organic solvents including toluene, chloroform, esters, and ether. It is not very soluble in paraffin hydrocarbons. Below its boiling point it does not react with sodium, sodium hydroxide, or phosphorus pentoxide. Many compounds initiate polymerization, and those with active hydrogen (e.g., alcohols, amines, and acids) form methylol or methylene derivatives.

Formaldehyde is very flammable, and in the gaseous state it forms explosive mixtures with air over the composition range of 7–72 vol.% formaldehyde. It is highly irritating to the eyes, nose, and skin. However, it is its

TABLE 2
Properties of Liquid Formaldehyde

Property	Value
Melting point, °C	−118
Density	
at −80°C	0.9151
at −20°C	0.8153
Mean coefficient of expansion	0.00283
Heat of vaporization at 19.2°C, kcal/gram	5.57
Trouton constant, entropy units	21.9

own warning agent, because it is detectable at 0.8 ppm, causes throat irritation at 5 ppm, and probably is unendurable at 20 ppm. If the compound is handled with reasonable caution, no serious health hazard is involved. Operations with it should be performed in a well-ventilated area.

C. Polymerization

1. POLY(FORMALDEHYDE)

Poly(formaldehyde) can be prepared either by anionic or cationic initiation of highly purified formaldehyde monomer, or by cationic ring-opening polymerization of trioxane. Anionic initiation is easily carried out and gives a high-molecular-weight polyacetal having very good properties. This polymerization requires highly purified monomer, which is prepared as needed and used immediately. The illustration of monomer preparation and anionic polymerization given here uses a procedure that has been described by MacDonald (14).

α-Polyoxymethylene is first prepared from paraformaldehyde. A slurry of 453 g of paraformaldehyde in 670 ml of water heated to 90°C is neutralized to pH 7 by addition of sodium hydroxide. While the cloudy solution is hot, it is filtered with suction into a flask cooled in an ice bath. To the clear filtrate at 40°C is added dropwise a solution of 6.1 g of sodium hydroxide in 12.5 ml of water. A slurry is formed that is stirred at 40°C for 24 hr, then separated on a filter and washed until neutral. The product is washed with acetone and dried in a vacuum oven at 60°C. There is obtained 170 g of α-polyoxymethylene as a white powder.

Pyrolysis of the polyoxymethylene and polymerization of the formaldehyde formed is carried out in sequential steps. The anionic initiator used in this illustration is tri-n-butylamine.

$$CH_2O \xrightarrow{(n-C_4H_9)_3N} \left(CH_2O \right)_x$$

A two-necked 500-ml, round-bottom flask is charged with 50 g of α-polyoxymethylene. The side neck is connected to a nitrogen inlet. The main neck is connected by wide-bore tubing to three U traps in series, each trap being 1 in. in diameter and 12 in. high. The outlet of the traps is led into a 1-liter, three-necked polymerization flask equipped with a high-speed stirrer, a thermometer, and an outlet terminated with a mineral oil bubbler. The entire apparatus is filled with nitrogen by means of the inlet tube, after which 400 ml of highly purified n-heptane containing 0.13 ml of tri-n-butylamine and 0.1 g of diphenylamine are added to the polymerization flask. The traps are cooled to −15°C and the contents of the polymerization flask is stirred vigorously and cooled with a bath kept at 25°C. Then the nitrogen stream is closed off and the polymerization flask is heated to 190°C. Formaldehyde generated by decomposition of the α-polyoxymethylene passes through the traps into the polymerization flask. After about 80% of the α-polyoxymethylene has decomposed, which occurs in 2–3 hr, the pyrolysis is stopped. When the formaldehyde passes into the polymerization flask it polymerizes, after a short induction period. Because of the heat of polymerization the temperature inside the flask increases by 3–5°C. Polymerization is continued for 3 hr, after which the flask contains a slurry of white poly(formaldehyde) in hexane. This polymer is separated on a filter and washed with heptane. It is then washed several times in a Blendor with acetone, the acetone being decanted after each wash. When dried, the polymer weighs 24–30 g. It has a melting point of 178°C and is of high molecular weight, as indicated by an inherent viscosity between 1 and 3 for a 0.5% solution in p-chlorophenol containing 2% of α-pinene.

High-molecular-weight poly(formaldehyde) made as just described can be molded into objects that have good mechanical properties. Its thermal stability is improved by converting its hydroxyl end groups to acyl or alkyl groups. Esterification (17) can be accomplished by mixing 25 g of poly(formaldehyde) with 300 ml of purified acetic anhydride containing 0.12 g of sodium acetate as a catalyst. The mixture is heated under reflux with stirring for about 1 hr and then cooled. The product is recovered on a filter and washed with acetone and then with water. Antioxidant is incorporated by washing with 200 ml of acetone containing 0.135 g of di-β-naphthyl-p-phenylenediamine. After it has been allowed to dry for 4 hr at 65°C in a vacuum oven, the polymer is ready for use.

An alternative procedure in which acetylation takes place in hot dimethylformamide has been described by Jaacks et al. (15).

Formation of poly(formaldehyde) by polymerization of trioxane is accomplished cationically, usually using boron trifluoride as an initiator. The polymerization is characterized by a relatively long induction period.

Kern and Jaacks (18) have suggested that the ionic species **1** and **2** are formed

first. These ionic species then decompose to formaldehyde and a smaller zwitterion until equilibrium concentration is reached. In support of this

$$\text{trioxane} \xrightarrow{\text{BF}_3} \text{polyoxymethylene} \rightleftharpoons \text{formaldehyde}$$

proposal, they have shown that addition of formaldehyde monomer prior to polymerization reduces the induction period.

2. TRIOXANE

Trioxane is a commercially available compound that is prepared by heating paraformaldehyde or a 60–65% formaldehyde solution in the presence of a trace of 2% of sulfuric acid. It is a colorless, crystalline solid that melts at 62–64°C and boils at 115°C. It is purified by fractionation through an efficient column (15). The following procedure for converting trioxane to polymer is one described by Jaacks et al. (15).

A 500-ml flask is flame-dried and then 90 g of pure trioxane and 210 g of pure ethylene dichloride are added to it. The whole system is kept anhydrous and the neck closed with a serum stopper. A solution of 70 mg/7 ml boron trifluoride etherate/ethylene dichloride is injected and then the flask is kept at 45°C, in a thermostat. In a short time precipitation of soluble polymer should start, continuing until the reaction mixture solidifies. After 1 hr, acetone is added to form a slurry from which polymer is recovered. The polymer is then washed a number of times with acetone and dried. Residual initiator is removed by boiling the polymer in ether containing 2% tri-*n*-butylamine. This procedure should give 40–50 g of polyoxymethylene, mp 176–178°C. After stabilization by acetylation of end groups, the inherent viscosity of the polymer (1% in dimethylformamide at 140°C) is 0.5.

Poly(oxymethylene) obtained by cationic polymerization of trioxane is said to be equivalent to that prepared by anionic polymerization of monomeric formaldehyde (19). Mixtures of trioxane and ethylene oxide lead to co-polymers. Price and McAndrew (19) state that a long induction period precedes polymerization, and during this time formaldehyde reacts with ethylene oxide to form 1,3-dioxolane, 1,3,5-trioxepane, and low-molecular-weight linear copolymer. These products then undergo a polymerization reaction that continues until a high-molecular-weight copolymer is produced.

III. ACETALDEHYDE

Acetal resins of type 3 have been obtained from a variety of higher aldehydes (2). Most prominent of these is poly(acetaldehyde) although none

$$+ \underset{\underset{H}{|}}{\overset{\overset{R}{|}}{C}}-O\ \underset{n}{\overbrace{}}$$

3

has become commercially important. The first to prepare polymers of acetaldehyde were Letort (20) and Travers (21). Both obtained amorphous or atactic polymer by low-temperature polymerization of acetaldehyde. Somewhat later, crystalline poly(acetaldehyde) was prepared independently in three different laboratories (22–24). This crystalline polymer was shown by Natta et al. (25) to be isotactic in configuration.

Like formaldehyde, acetaldehyde is manufactured in huge volume. United States capacity is in excess of 1.6 billion lb/year (26). In 1967 43% of the acetaldehyde produced in the U.S. was used in the manufacture of acetic acid and acetic anhydride, which are basic organic chemicals. Most of the remainder served to produce 1-butanol (21%) and 2-ethylhexanol (20%).

A. Chemistry

Acetic acid is made from acetaldehyde by oxidation with air or oxygen (27). Acetic anhydride is prepared by dehydration of acetic acid and is also made directly from acetaldehyde. Ethyl acetate, widely employed as a solvent, is manufactured from acetaldehyde by the Tishchenko reaction (Eq. 8), using aluminum ethoxide as a catalyst.

$$2CH_3CHO \rightarrow CH_3COOCH_2CH_3 \tag{8}$$

The aldol reaction of acetaldehyde is the first step in a series of reactions leading to n-butyraldehyde, 1-butanol, and 2-ethylhexanol. Aldol was also

$$2CH_3CHO \longrightarrow CH_3CHOHCH_2CHO \longrightarrow CH_3CH{=}CHCHO$$

$$CH_3CH{=}CHCHO \xrightarrow{\ H_2\ } CH_3CH_2CH_2CHO + CH_3CH_2CH_2CH_2OH$$

$$2CH_3CH_2CH_2CHO \longrightarrow CH_3CH_2CH_2\underset{\underset{\overset{|}{CHCHO}}{|}}{CHOH}\overset{\overset{CH_3CH_2}{|}}{} \xrightarrow{-H_2O}$$

$$CH_3CH_2CH_2CH{=}\overset{\overset{CH_3CH_2}{|}}{C}-CHO \xrightarrow{\ H_2\ } CH_3CH_2CH_2CH_2\overset{\overset{CH_3CH_2}{|}}{C}HCH_2OH$$

used at one time as a source of butadiene, but this route (Eq. 9) is now obsolete.

$$CH_3CHOHCH_2CHO \rightarrow CH_3CHOHCH_2CH_2OH \rightarrow CH_2\!\!=\!\!CH\!\!-\!\!CH\!\!=\!\!CH_2 \quad (9)$$

Formation of butadiene from ethanol and acetaldehyde (Eq. 10), which was important during World War II, is also obsolete.

$$CH_3CH_2OH + CH_3CHO \rightarrow CH_2\!\!=\!\!CHCH\!\!=\!\!CH_2 + 2H_2O \quad (10)$$

Other important products based on acetaldehyde are vinyl acetate, trimethylolpropane, and pentaerythritol. Vinyl acetate is derived from the aldehyde and acetic anhydride (Eq. 11). This route is now in competition

$$CH_3CHO + (CH_3CO)_2O \longrightarrow CH_3CH(OCOCH_3)_2 \xrightarrow{\Delta}$$
$$CH_2\!\!=\!\!CHOCOCH_3 + CH_3COOH \quad (11)$$

with a synthesis based on oxyacetoxylation of ethylene, which is a cheaper way of preparing vinyl acetate. Trimethylolpropane is obtained by reaction of butyraldehyde, an aldol reaction product derivative, with 3 molecules of formaldehyde. Pentaerythritol is manufactured by reaction of acetaldehyde with formaldehyde, a process that has been in use for many years.

Peracetic acid appears to be developing into an important acetaldehyde derivative. It is used to oxidize cyclohexanone to caprolactone, which is then reacted with ammonia to give caprolactam, the nylon-6 starting material.

Chloral is another large outlet for acetaldehyde, from which it is obtained by chlorination. Chloral is produced in large volume mainly for use in making DDT (4).

$$CCl_3CHO + 2C_6H_5Cl \rightarrow CCl_3CH(p\text{-}C_6H_4Cl)_2 \quad (12)$$

4

B. Synthesis

The oldest acetaldehyde process, and one that is presently widely used, is oxidation of ethanol (Eq. 13). The reaction is catalyzed by copper and silver

$$C_2H_5OH + \tfrac{1}{2}O_2 \rightarrow CH_3CHO + H_2O + 40.55 \text{ kcal} \quad (13)$$

in various forms. Rare-earth oxides and oxides of vanadium and molybdenum are sometimes added, but they are said to make the catalyst too reactive (28). The main catalysts appear to be silver gauze, a 9:1 copper-silver alloy, and copper granules coated with silver. The highest yield is obtained by using 64–77% excess of air over the stoichiometrical amount (27). The maximum yield at high conversion appears to be a little more than 80%. This value has been given for a reaction temperature of 515°C, although temperature is influenced by flow rate.

Control of the reaction is accomplished by dissipation of the heat of reaction. This is done commercially by using aqueous solutions, usually about 50% ethanol, and limiting the conversion per pass to 45–50%. Under such conditions, the yield of acetaldehyde is 94–96%.

Dehydrogenation is another procedure for converting ethanol to acetaldehyde. This was first done by Berthelot (29) in 1886; he passed alcohol through a glass tube heated to 500°C, obtaining ethylene, methane, and carbon monoxide, in addition to acetaldehyde. Sabatier (30) showed that dehydrogenation to acetaldehyde is the sole reaction at temperatures up to 300°C when copper is used as a catalyst. Commercially, copper catalysts activated by chromium have been used (27). By operating at temperatures of 260–290°C and limiting conversions per pass to 30–50%, it is possible to obtain yields of 90–92%. Combinations of dehydrogenation and oxidation are used to achieve an appropriate heat balance. Typically 50–95% aqueous ethanol is vaporized, mixed with preheated air, and passed over a silver gauze catalyst at 375–550°C, with temperature depending on the air/ethanol/steam ratio and the flow rate.

A substantial proportion of the acetaldehyde manufactured abroad is made by hydration of acetylene (Eq. 14). This reaction, discovered in 1881 (31),

$$CH{\equiv}CH + H_2O \xrightarrow[H_2SO_4]{HgSO_4} CH_3CHO + 33 \text{ kcal} \tag{14}$$

is catalyzed by mercury salts and sulfuric acid. A large excess of acetylene is used to sweep out acetaldehyde as fast as it is formed. Miller (27) gives optimum conditions as 20–25% acid, 1–2% mercury, 0.5–1% ferric ion, and a temperature of 70–80°C. Conversions of acetylene as high as 55% per pass and yields of acetaldehyde as high as 93–95% are said to be obtainable.

Nearly 30 years ago, the Celanese Corporation developed a process for preparing acetaldehyde and acetone by oxidation of saturated hydrocarbons. n-Butane and oxygen appear to be the best starting materials for acetaldehyde (32). Isobutane gives acetone as the major product.

The most important recent development in acetaldehyde manufacture is direct oxidation of ethylene (Eq. 15). This reaction was worked out in

$$CH_2{=}CH_2 + \tfrac{1}{2}O_2 \rightarrow CH_3CHO + 58.2 \text{ kcal} \tag{15}$$

Germany in the late 1950s (33). The basic reaction is between ethylene and aqueous palladium chloride (Eq. 16).

$$CH_2{=}CH_2 + PdCl_2 + H_2O \rightarrow CH_3CHO + Pd + 2HCl \tag{16}$$

Apparently the reaction is very complicated and involves a number of palladium complexes. The following sequence is given by Miller (27):

$$CH_2{=}CH_2 + [PdCl_4]^{2-} \longrightarrow [CH_2{=}CH_2{\cdot}PdCl_3]^- \longrightarrow CH_2{=}CH_2{\cdot}PdCl_2(OH)_2 \longrightarrow$$

$$[CH_2{=}CH_2{\cdot}PdCl_2(OH)]^- \longrightarrow [HOCH_2CH_2PdCl_2]^- \longrightarrow \begin{array}{c} CHOH \\ \| \\ CH_2 \end{array}\!\!\text{- - - -}\,PdHCl_2 \longrightarrow$$

$$[CH_3CHOH{\cdot}PdCl_2] \longrightarrow CH_3CHOH{\cdot}O^+H_2 + [PdCl_2]^{2-} \longrightarrow$$

$$CH_3CHO + H_3O^+ + Pd + 2Cl^-$$

Cupric chloride is added to reoxidize palladium to the divalent state. The overall reactions are

$$CH_2{=}CH_2 + 2CuCl_2 + H_2O \xrightarrow{PdCl_2} CH_3CHO + 2CuCl + 2HCl$$

$$2CuCl + 2HCl + \tfrac{1}{2}O_2 \longrightarrow 2CuCl_2 + H_2O$$

One type of commercial process is a single stage in which both reactions are run simultaneously by subjecting a mixture of ethylene, oxygen, cupric chloride, steam, and a catalytic quantity of palladium chloride to 120–130°C at 3 atm of pressure. Two-stage processes are also in use. In the first stage acetaldehyde is obtained by reacting ethylene at 10 atm with aqueous palladium chloride and cupric chloride. The cuprous chloride formed in the first stage is reoxidized to cupric chloride in a separate stage. This reoxidation can be done with air in the two-stage process. The one-stage process requires high-purity oxygen for reoxidation. In either case, acetaldehyde yields are of the order of 95%. It is probable that most future new acetaldehyde plants will use the ethylene process.

Acetaldehyde is colorless and has high fluidity and a pungent odor. Some of its other properties are listed in Table 3. It is soluble in water and most

TABLE 3
Properties of Acetaldehyde

Property	Value
Melting point, °C	−123.5
Boiling point, °C	20.16
Density, d_4^{20}	0.7780
Refractive index, η_D^{20}	1.3311
Heat of fusion, cal/g	17.6
Heat of vaporization, cal/g	139.5
Heat of formation, kcal/mole	−39.55
Free energy of formation, kcal/mole	−32.60
Flash point, closed cup, °C	−38
Ignition temperature in air, °C	165

common solvents. Being flammable, it forms explosive mixtures with air; the limits are said to be 4–57 vol. % acetaldehyde. Although not classed as a highly toxic material, acetaldehyde should be handled with reasonable precautions because exposure to large amounts of it is deleterious to health.

C. Purification

As obtained from commercial sources, acetaldehyde is usually of good quality. It does contain such impurities as acetic acid, water, small amounts of ethanol and paraldehyde, and, in some cases, small amounts of ethyl acetate. Conversion of this material to polymerization-grade monomer is readily accomplished by neutralization of the acid and fractional distillation to remove other impurities. The following procedure is one described by Vogl (34).

Hydrated sodium carbonate is added to acetaldehyde cooled to 0°C, and the mixture is stirred for 1 hr. The aldehyde is decanted from the sodium carbonate and dried by stirring with magnesium sulfate for 30 min. After the acetaldehyde is separated from the magnesium sulfate, 0.1 % of Age-Rite White is added as an antioxidant and the mixture distilled. Acetaldehyde is collected in a receiver cooled to 0°C and is used immediately. A nitrogen atmosphere is employed throughout the distillation to prevent oxidation of the aldehyde by oxygen in the air. After each distillation, the apparatus is rinsed with either soap solution or dilute sodium hydroxide to remove any acid residues that may have collected in the fractionating column.

Pure acetaldehyde can also be prepared by decomposition of purified paraldehyde. This method has been used by Goodman (35) to provide monomer that is polymerized as rapidly as it is generated.

D. Polymerization

As has been mentioned earlier, poly(acetaldehyde) can be prepared either in atactic form, which is elastomeric, or in isotactic form, which is highly crystalline. The crystalline polymer is prepared by low-temperature initiation with such compounds as $(C_2H_5)_3Al$, $(C_4H_9)_2Zn$, $(C_2H_5)_2AlCl$, sec-C_4H_9OLi, C_4H_9Li, and $LiAlH_4$ (Eq. 17). At room temperature, crystalline poly(acetaldehyde) is not soluble to any appreciable extent in common organic solvents.

$$CH_3CHO \xrightarrow[-78°C]{C_4H_9Li} -\!\!\left(\!\!\begin{array}{c} CH_3 \\ | \\ CH\!-\!O \end{array}\!\!\right)_{\!n}$$

(17)

Elastomeric poly(acetaldehyde) is also prepared at low temperatures. Letort's (20) original polymer was formed by polymerization below the crystallization temperature. It has since become apparent that this is a cationic-initiated acetaldehyde polymerization (36), although the situation was not recognized at the time. A more reliable polymerization method is cationic initiation in a liquid olefin solvent at temperatures below $-40°C$ (36). Detailed directions for polymerization of acetaldehyde by initiation with boron trifluoride in liquid ethylene at -40 to $-120°C$ have been published by Vogl (34). High pressure can be used as an alternative for low temperature. Elastomeric poly(acetaldehyde) equivalent to the low-temperature product, as judged by correspondence of infrared spectra, has been prepared by Novak and Whalley (37) by subjecting the monomer to a pressure of 9 kbar for 1–2 days at room temperature in a stainless steel tube. In contrast to its crystalline counterpart, elastomeric poly(acetaldehyde) is soluble in such solvents as methanol, acetone, and ethyl ether.

IV. HIGHER ALDEHYDES

Propionaldehyde was first polymerized by Novak and Whalley (37) by means of the high-pressure method discussed in the previous section. They were following up on work started some years earlier by Bridgman and Conant (38) and Conant and Tongberg (39). They found the same method to be effective for converting n- and isobutyraldehyde, n-valeraldehyde, and n-heptaldehyde (40) into linear polyacetals. Polymers prepared in this manner are amorphous, and they slowly revert to monomer at room temperature. Crystalline polyacetals are obtained by low-temperature, anionic polymerization of propionaldehyde (22–24), n-butyraldehyde (22,25), isobutyraldehyde (24,25), n-valeraldehyde (24), isovaleraldehyde (25), and n-heptaldehyde (24).

A. Synthesis of Aldehyde Monomers

The most important route to propionaldehyde is reaction of ethylene, carbon monoxide, and hydrogen in the oxo process (Eq. 18). In this process

$$CH_2{=}CH_2 + CO + H_2 \rightarrow CH_3CH_2CHO \qquad (18)$$

the reactants are contacted at temperatures of 140–250°C, under pressures of 40–700 atm with an appropriate catalyst (one example is cobalt and thoria on diatomaceous earth). Other commercial routes include reaction of ethylene, carbon monoxide, and hydrogen under pressure over Fischer-Tropsch catalysts, and dehydrogenation of n-propyl alcohol.

n-Butyraldehyde is synthesized in several ways; one method starts from acetaldehyde, as discussed earlier. This aldehyde is also made via the oxo reaction from propylene, carbon monoxide, and hydrogen (Eq. 19). The reaction gives isobutyraldehyde as well as *n*-butyraldehyde. Proportions of each depend on reaction conditions.

$$CH_3CH=CH_2 + CO + H_2 \rightarrow CH_3(CH_2)_2CHO + (CH_3)_2CHCHO \qquad (19)$$

The source of *n*-heptaldehyde is ricinoleic acid, 12-hydroxy-9-octadecenoic acid. This acid occurs as a glyceride in castor oil, from which it is recovered. When ricinoleic acid is heated, part of it is dehydrated and part is split into *n*-heptaldehyde and 10-undecenoic acid (Eq. 20).

$$CH_3(CH_2)_5CH=CH-CH=CH(CH_2)_7CO_2H$$
$$+$$
$$CH_3(CH_2)_4CH=CH-CH_2CH=CH(CH_2)_7CO_2H$$

$$CH_3(CH_2)_5CHOHCH_2CH=CH(CH_2)_7CO_2H \xrightarrow{\Delta}$$

$$CH_3(CH_2)_5CHO + CH_2=CH(CH_2)_8CO_2H \qquad (20)$$

The other aldehydes that have been converted to polyacetals are of less general interest. They can be purchased or they can be prepared by aldehyde synthesis methods; for example, hexanal can be synthesized from an appropriate organometallic compound and orthoester (41).

$$C_5H_{11}MgBr + CH(OC_2H_5)_3 \longrightarrow C_5H_{11}CH(OC_2H_5)_2 + C_2H_5OMgBr$$

$$C_5H_{11}CH(OC_2H_5)_2 + H_2O \xrightarrow{H_2SO_4} C_5H_{11}CHO + 2C_2H_5OH$$

B. Purification

The clean-up procedure described for acetaldehyde is generally applicable to all volatile, aliphatic aldehydes. This involves treatment with hydrated sodium carbonate to remove acids, followed by separation from the carbonate and careful fractionation. The aldehyde must be protected by a nitrogen atmosphere during fractionation to prevent formation of acids by air oxidation. It is advisable to use the aldehyde for polymerization as soon as possible after fractionation.

C. Properties

Physical properties of the higher aliphatic aldehydes appear in Table 4. Although these are flammable compounds, they are less likely to form explosive mixtures with air than formaldehyde and acetaldehyde because they are less volatile. They should be handled with care, and if so treated they pose no serious health problems. They should not be inhaled or otherwise ingested.

TABLE 4
Properties of Aliphatic Aldehydes

	mp, °C	bp, °C	d_4^{20}	η_D^{20}	Heat of vaporization Btu/lb	Heat of combustion, kcal/mole
Propionaldehyde	−81	47.9	0.7970	1.3619	214	7400[a]
n-Butyraldehyde	−91.7	74.8	0.8048	1.3795	104.4[a]	592.4
Isobutyraldehyde	−65.9	64.5	0.7938	1.3730		599.9
n-Valeraldehyde	−92	103	0.8095	1.39436		
Isovaleraldehyde	−51	92.5				
n-Heptaldehyde	−43.3	152.8	0.8495	1.42571		

[a] Cal/g.

V. HALOALDEHYDES

Polyacetals from haloaldehydes have been reviewed by Rosen (42). These include polyacetals from chloral, dichloroacetaldehyde, monochloroacetaldehyde, and trifluoroacetaldehyde. None has found commercial application.

Poly(chloral) (Eq. 21) has been obtained by initiation with the sodium

$$CCl_3CHO \longrightarrow \text{--} (\overset{\overset{\textstyle CCl_3}{|}}{CH} \text{--} O)_{\overline{n}} \tag{21}$$

naphthalene complex or organometallic compounds and by irradiation with ^{60}Co. The polymer is readily cracked back to chloral by heat. Rosen has reported that stability of poly(chloral) is substantially improved by end capping by reaction with acid anhydrides and chlorides. Since the compound is insoluble as well as infusible, measurement of molecular weight by usual methods is not feasible. From end groups obtained by end-capping reactions, Rosen has estimated that polymer obtained by anionic initiation has a molecular weight of 44,000–88,000.

Dichloroacetaldehyde is readily polymerized by initiation with Lewis acids (Eq. 22). The polymer formed is readily end capped by reaction with

$$HCCl_2CHO \xrightarrow{BF_3} \text{--} (\overset{\overset{\textstyle HCCl_2}{|}}{CH} \text{--} O)_{\overline{n}} \tag{22}$$

acetic anhydride, has a molecular weight of about 45,000, and is soluble in many common organic solvents.

Chloroacetaldehyde is converted to a crystalline polyacetal by initiation with diethyl zinc (Eq. 23) or to an amorphous polymer when initiated with

$$CH_2ClCHO \xrightarrow{(C_2H_5)_2Zn} \text{--} (\overset{\overset{\textstyle CH_2Cl}{|}}{CH} \text{--} O)_{\overline{n}} \tag{23}$$

the etherate of boron trifluoride. The chloroacetaldehyde polymers are of high molecular weight and are more stable than poly(dichloroacetaldehyde) or poly(chloral), but less stable than poly(acetaldehyde).

Very little is known about poly(fluoral). It is obtained by spontaneous polymerization (43) of the monomer and is insoluble in usual organic solvents (Eq. 24). The polymer is readily cracked back to monomer by heat.

$$CF_3CHO \longrightarrow -(\overset{\displaystyle CF_2}{\underset{\displaystyle |}{CH}}-O)_{\overline{n}} \tag{24}$$

A. Synthesis of Monomers

Chloraldehydes are formed by chlorination of alcohol or acetaldehyde. The reaction proceeds in steps to give first chloroacetaldehyde, then dichloro-acetaldehyde, and finally chloral. These products are then separated by distillation. Commercial chloral almost always contains small amounts of chloro and dichloroacetaldehydes. These aldehydes are purified by addition of a chemical drying agent, such as phosphorus pentoxide, to remove water; then follows separation of the drying agent and careful fractionation of the dried chloroaldehyde.

Fluoral is more difficult to prepare. The method reported by Schechter and Conrad (43) is nitration of 1,1,1-trifluoropropane. Products are fluoral, which is obtained in 20–24% yield, and 1,1,1-trifluoro-3-nitropropane, which is formed in 16% yields. Purification involves dissolution of the fluoral fraction in water, separation from water-insoluble materials, and then slow addition of aqueous fluoral solution to solid phosphorus pentoxide. Fluoral is evolved as a colorless gas, which is condensed and redistilled, bp -18.8 to $-17.5°C$ at 748 mm.

B. Properties

The chloroacetaldehydes are colorless liquids with sharp, irritating odors. They produce pronounced physiological effects and should not be inhaled or allowed to come in contact with the skin. Chloral in particular is a sedative and hypnotic drug. However, these compounds can be used safely by observing normal laboratory precautions. Physical properties are listed in Table 5.

TABLE 5
Physical Properties of Chloroacetaldehydes

Compound	mp, °C	bp, °C	d_{25}^{25}	η_D^{20}
Chloral	−57	98	1.5060	1.45572
Dichloroacetaldehyde	—	90–91	—	—
Chloroacetaldehyde	—	85	—	—

VI. ACROLEIN

Acrolein, $CH_2{=}CHCHO$, differs in kind from the aldehydes just discussed because under certain conditions it polymerizes through the unsaturated group to give polymers with a high proportion of pendant carbonyl groups, whereas under other conditions it polymerizes through the carbonyl to give polyacetals with a high proportion of unsaturation. Acrolein was first prepared in 1843 by Redtenbacher (44), who also noted that the compound readily polymerizes to a white, infusible, insoluble polymer. This product he termed "disacryl"; a name that is still used for insoluble acrolein polymers. Although much work has been done on acrolein polymerization, there are no commercial poly(acrolein) products. The monomer, however, is manufactured in large quantities for other purposes, notably for conversion to glycerol, 1,2,6-hexanetriol, and methionine. The largest producer is Shell Chemical Company, which makes 1100–1300 tons/month. The preparation and chemistry of acrolein are discussed in a number of review articles (45–47) and in a book by Smith (48).

A. Chemistry

Acrolein is extremely reactive. It is an α,β-unsaturated aldehyde and as a consequence undergoes such reactions as Michael and Diels-Alder additions, as well as standard reactions of olefins and aldehydes. It is not our purpose to give a thorough account of this chemistry because it is described in detail in several good reviews, particularly those by Smith (48) and Guest et al. (46). However, some attention is given to the chemistry that leads to the most important of the products based on acrolein.

Glycerol is obtained by controlled oxidation of acrolein to glyceraldehyde followed by reduction of glyceraldehyde. Reaction with hydrogen peroxide gives glycidaldehyde (5). Hydrolysis of glycidaldehyde gives glyceraldehyde

$$CH_2{=}CHCHO \xrightarrow[\text{pH 8}]{H_2O_2} \begin{array}{c} CH_2\!\!-\!\!CHCHO \\ \diagdown\ \diagup \\ O \end{array}$$

5

(**6**). Glycerol is obtained by hydrogenation of **6**. Smith and Holm (49) have

$$\mathbf{5} \xrightarrow{H_2O} \begin{array}{c} CH_2\!-\!CH\!-\!CHO \\ |\qquad | \\ OH\quad OH \end{array}$$

6

shown that **6** can be made directly by reaction of acrolein with hydrogen peroxide in the presence of osmium oxide.

A dimer (**7**) is readily obtained by Diels-Alder addition of acrolein to itself. This dimer is the starting point for 2-hydroxyadipaldehyde and 1,2,6-

$$
\begin{array}{cc}
\overset{\displaystyle CH_2}{\underset{\displaystyle \|}{}} & \\
CH & CH_2 \\
| & \| \\
CH & CHCHO \\
\underset{\displaystyle O}{\diagdown\!\!\diagdown} &
\end{array}
\quad\longrightarrow\quad
\underset{\textbf{7}}{\text{(pyran)}}\!\!-\!\text{CHO}
$$

hexanetriol. The former is obtained by acid-catalyzed hydrolysis. The latter, which is useful as a softener for cellophane and as a component of alkyd and polyester resins, is obtained by hydrogenation of 2-hydroxyadipaldehyde.

Diels-Alder addition of ethyl vinyl ether to acrolein is the first step in a route to glutaraldehyde (Eq. 25). The pyran formed is easily hydrolyzed under acidic conditions. Glutaraldehyde, of course, is the starting point for

$$
\begin{array}{cc}
\overset{\displaystyle CH_2}{} & \\
CH & CH_2 \\
\| & | \\
CH & CHOC_2H_5 \\
\diagdown O \diagup &
\end{array}
\xrightarrow[\text{H}_2\text{O}]{\text{H}^+}
OCH(CH_2)_3CHO + C_2H_5OH \qquad (25)
$$

other important compounds including glutaric acid, 1,5-pentanediol, and the dehydrogenation product of 1,5-pentanediol, δ-valerolactone.

Additions to the double bond of acrolein are catalyzed by both acids and bases. Examples are additions of either water or alcohols to give 3-hydroxy-propionaldehyde and 3-alkoxypropionaldehydes, respectively. These additions are often competing reactions. Such is the case when acrolein acetals are synthesized. Fortunately, conditions can usually be found that favor the aldehydic reaction when so desired.

Base-catalyzed additions are followed by a Michael condensation of additional acrolein molecules (Eq. 26).

$$
CH_2{=}CHCHO + ROH \xrightarrow{\text{B}^-} ROCH_2CH_2CHO
$$

$$
\Big\downarrow CH_2{=}CHCHO
$$

$$
\overset{\displaystyle CHO \quad CHO}{\underset{\displaystyle ROCH_2-CH(CH_2CH \,\rightarrow_{\!n} CH_2CH_2CHO}{}} \qquad (26)
$$

A variation of this reaction is combination of acrolein and formaldehyde.

$$CH_2\!\!=\!\!CHCHO \xrightarrow[B^-]{H_2O} HOCH_2CH_2CHO \xrightarrow{CH_2=CHCHO}$$

$$\underset{\displaystyle \overset{|}{HOCH_2CH}}{\overset{CHO}{|}}\!\!\!\!\!\!\leftarrow CH_2\underset{\displaystyle \overset{|}{CH}}{\overset{CHO}{|}}\!\!\!\!\!\!\xrightarrow{}_n CH_2CH_2CHO$$

$$\downarrow CH_2O$$

$$HOCH_2\!\!-\!\!\underset{\displaystyle CH_2OH}{\overset{CHO}{|}}\!\!\!\!\!\!\leftarrow CH_2\!\!-\!\!\underset{\displaystyle CH_2OH}{\overset{CHO}{|}}\!\!\!\!\!\!\xrightarrow{}_n CH_2\!\!-\!\!\underset{\displaystyle CH_2OH}{\overset{CH_2OH}{|}}\!\!\!\!\!\!-\!\!CHO$$

The product is a resin used for crease-proofing of textiles. The desirable effect is obtained by interaction of the aldehyde and hydroxyl groups of the resin with each other after the resin is applied to the fiber and probably with reactive sites in the fiber itself.

An important product based on acrolein is *dl*-methionine, which is a chicken feed supplement. Such a synthesis has been described by Gresham and Schweitzer (50).

$$CH_2\!\!=\!\!CHCHO \xrightarrow{CH_3SH} CH_3SCH_2CH_2CHO \xrightarrow{HCN} CH_3SCH_2CH_2\underset{\displaystyle CN}{\overset{OH}{\underset{|}{\overset{|}{CH}}}}$$

$$\downarrow NH_3$$

$$CH_3SCH_2CH_2\underset{\displaystyle CO_2H}{\overset{NH_2}{\underset{}{\overset{|}{CH}}}} \xleftarrow{H_2O} CH_3SCH_2CH_2\underset{\displaystyle CN}{\overset{NH_2}{\underset{|}{\overset{|}{CH}}}}$$

B. Synthesis

As mentioned earlier, acrolein is made in very large volume (45). Shell Chemical employs gas-phase oxidation of propylene. Cuprous oxide is used as a catalyst and the oxidant is air or oxygen (48). Many other oxides are also effective, including a combination of oxides of molybdenum and cobalt, a mixture of bismuth and molybdenum oxides, combinations containing antimony oxide or tungsten oxide, and silver selenite promoted with cuprous oxide (46).

A preheated mixture of propylene and air or oxygen is passed over pellets of the catalyst, which is cuprous oxide supported on alumina, at 300–350°C. Oxidation is exothermic, and the heat given off is used to preheat the gaseous reaction mixture. Because the reaction is catalyzed by cuprous oxide and not by cupric oxide, oxygen content in the reaction mixture must be kept below the level effective for oxidizing the catalyst. The major reactions that occur are:

$$CH_2=CHCH_3 + O_2 \rightarrow CH_2=CHCHO + H_2O$$

$$CH_2=CHCH_3 + \tfrac{9}{2}O_2 \rightarrow 3CO_2 + 3H_2O$$

Minor reactions that also occur lead to formation of small amounts of formaldehyde, acetaldehyde, propionaldehyde, and acetone. These impurities are removed by careful fractionation.

Acrolein is also prepared, particularly in Germany, by cross-condensation of acetaldehyde with formaldehyde (Eq. 27). The reaction is run in the vapor

$$CH_3CHO + HCHO \longrightarrow [CH_3\overset{\overset{\displaystyle OH}{|}}{C}H-CHO] \longrightarrow CH_2=CHCHO + H_2O \quad (27)$$

phase at 300–350°C using such catalysts as lithium phosphate on activated alumina and sodium silicate on silica gel.

A slight excess of acetaldehyde is used to reduce the amount of formaldehyde in the product. This is to assist purification; formaldehyde is more difficult to remove than acetaldehyde. However, excess acetaldehyde introduces significant amounts of crotonaldehyde that must be removed.

$$2CH_3CHO \rightarrow CH_3CH=CHCHO + H_2O$$

C. Purification

The impurities that may be present in acrolein are water, lower alcohols, and aliphatic aldehydes and ketones. Careful fractionation is usually indicated to provide monomer suitable for polymerization. Purity can be determined by gas chromatography, and Peters and Hood (47) have given much useful information on procedures and retention times. They also describe other assay methods.

D. Properties

Physical properties of pure acrolein have been listed by Peters and Hood (47). Critical physical constants are given in Table 6.

TABLE 6
Physical Properties of Acrolein

Property	Value
Boiling point, 760 mm Hg, °C	52.69
Freezing point, °C	−86.95
Refractive index, η_D^{20}	1.4017
Density at 20°C	0.8389

E. Polymerization

It has already been mentioned that acrolein polymerizes adventitiously to a product called "disacryl." It is also possible to bring about polymerization deliberately by use of free radical and ionic initiators. In most cases the structures of the products have not been determined. As a consequence, the nature of these reactions—as a vinyl polymerization, an aldehyde polymerization, or a polycondensation—has not been unequivocally settled.

Adventitious polymerization, ultraviolet-light or γ-ray-initiated polymerizations, and peroxide or azonitrile-initiated polymerizations are all generally considered to involve free radical polymerization of the vinyl group. Because the polymerization of acrolein has a high heat of reaction, control is difficult unless an organic solvent or water is employed as a heat-transfer medium. Polymerizations in water using reduction-oxidation initiators are preferred (45). For monomer concentrations above 20 vol. % (the solubility limit of acrolein in water), an emulsifying agent is usually employed. The polymers obtained are believed to have a carbon backbone on which pendant aldehyde groups are present in acetal form as in structure **8**.

$$\underset{\mathbf{8}}{\begin{array}{c}\text{structure 8}\end{array}}$$

8

Anionic polymerization, such as that brought about by sodium naphthalene in nonaqueous media, is believed to proceed mainly by polymerization through the aldehyde group to give polyacetals (**9**) along with some vinyl polymer groups (**10**).

$$\begin{array}{cc} +\!\!\!\!\left(\text{CH}-\text{O}\right)_{\!\!\overline{n}} & +\!\!\!\!\left(\text{CH}_2-\text{CH}\right)_{\!\!\overline{n}} \\ \big| & \big| \\ \text{CH}=\text{CH}_2 & \text{CHO} \\ \mathbf{9} & \mathbf{10} \end{array}$$

Cationic initiators also bring about polymerization. Little has been said about the structure of the polymers.

Condensation polymerization has already been discussed. Of particular significance here is polymerization by Diels-Alder condensation. The polymer has a ladder structure (**11**), although details concerning its nature have not been determined.

In addition to the foregoing products, acrolein copolymerizes in free radical systems with a number of monomers, including acrylonitrile,

$$O=CH-CH \overset{CH_2}{\underset{}{\parallel}} \quad + \quad \overset{CH_2}{\underset{O}{\searrow}}\overset{CH}{\underset{CH}{\diagup}} \longrightarrow \quad O=CHCH\overset{\overset{H_2}{C}}{\underset{O}{\diagdown}}\overset{CH}{\underset{CH}{\parallel}}$$

$$O=HC-HC\overset{\overset{H_2}{C}}{\underset{O}{\diagup}}\overset{CH}{\underset{CH}{\parallel}} \quad + \quad \overset{CH_2}{\underset{O}{\searrow}}\overset{CH}{\underset{CH}{\diagup}} \longrightarrow \quad O=HC-HC\left[\overset{\overset{H_2}{C}}{\underset{O}{\diagup}}\overset{H}{\underset{H}{C}}\overset{\overset{H_2}{C}}{\underset{O}{\diagdown}}\overset{CH}{\underset{CH}{\parallel}}\right]_n$$

11

methacrylonitrile, acrylamide, vinyl acetate, methyl acrylate, and 2-vinyl-pyridine.

VII. THIOFORMALDEHYDE

Poly(thioformaldehyde) is prepared indirectly because thioformaldehyde is too reactive to have more than a fleeting existence in the monomeric state (51,52). Credali and Russo (52) have reviewed the thioformaldehyde polymer field.

In 1886 a product believed to be a polymer of thioformaldehyde was prepared by Wohl (53) by reaction of hexamethylenetetramine with hydrogen sulfide. Later it was shown (54) that Wohl's product contains substantial quantities of nitrogen. The first successful preparation of poly(thioform-aldehyde) was by Harmon (55), who removed hydrogen sulfide from either methanedithiol or bis(mercaptomethyl)sulfide (Eq. 28).

$$HSCH_2SH \xrightarrow{R_3N} \text{--}(CH_2\text{--}S\text{)}_{\overline{n}} + H_2S$$

$$HSCH_2\text{--}S\text{--}CH_2SH \xrightarrow{R_3N} \text{--}(CH_2\text{--}S\text{)}_{\overline{n}} + H_2S \qquad (28)$$

Ring-opening polymerization of trithiane has also been used as a route to poly(thioformaldehyde). This is accomplished either by the action of Lewis acids on trithiane at elevated temperatures (56–59), or by ^{60}Co γ-irradiation of crystalline trithiane (60–62). Copolymers containing oxymethylene units have been made by the action of Lewis acids on mixtures of trithiane and trioxane (63).

A. Synthesis of Precursors

Trithiane is a common organic chemical easily synthesized by reaction of formaldehyde with hydrogen sulfide in the presence of hydrochloric acid (64). It is a white, crystalline solid melting at 215°C.

Methanedithiol is readily obtained from formaldehyde and hydrogen

$$CH_2=O + H_2S \longrightarrow HSCH_2SH$$

sulfide. This reaction has been discussed by Cairns et al. (65), and the following procedure is based on their work.

A 1-liter, stainless steel rocker bomb is charged with 214 g of methanol-free 37% aqueous formaldehyde and 5 g of sodium dihydrogen phosphate. The bomb is then closed and cooled with solid carbon dioxide, and the air in the free space is displaced with nitrogen and evacuated. Then 450 g of hydrogen sulfide is distilled into the bomb. The bomb is heated to 45°C for 12 hr with its contents kept under autogeneous pressure, after which it is cooled in ice. The bomb is then carefully opened to allow excess hydrogen sulfide to be discharged, and its contents is poured into an acid-washed, separatory funnel. The product separates into two layers; the lower layer is composed of methanedithiol, bis(mercaptomethyl)sulfide, and a small quantity of higher polymer. This lower layer is separated and distilled under nitrogen through a 12-in. Vigreux column. The methanedithiol fraction distills at 40–44°C/40 mm, and the bis(mercaptomethyl)sulfide fraction distills at 65°C/10 mm. The methanedithiol is then further purified by precision distillation.

B. Toxicity

These sulfur-containing materials are toxic. Because methanedithiol has such a bad odor, it is doubtful that very high concentrations could be tolerated. It should be handled under conditions where it can be contained. Poly(thioformaldehyde) has an oral LD_{50} in rats of 365 ± 11 mg/kg (66). It is said to produce dermatitis on human skin and at high doses to give signs of intoxication in animals.

C. Polymerization

Conversion of methanedithiol to poly(thioformaldehyde) requires treatment with a base such as ammonia, an amine, or a phosphine. In a typical preparation, Harmon (55) describes the addition of a small amount of tri-*n*-amylamine to methanedithiol followed by gradual heating to 112°C. Hydrogen sulfide is evolved, and after 30 min the reaction mixture is solid. The temperature is gradually increased to 230°C over 3 hr, and then the reaction product is extracted with benzene to remove low-molecular-weight materials. This procedure leads to a 75% yield of polymer that has a melting point of 230°C.

D. Properties

Melting temperatures reported by various workers are in the range of 220°C–260°C (55,56) Bapseres and Signouret (58) have related melting temperature to the number of thiomethylene groups in the chain (Table 6).

Formation of poly(thioformaldehyde) by solid-state polymerization has involved irradiation of trithiane crystals with γ-rays from ^{60}Co. This has been

TABLE 6
Softening Temperatures of $HS(CH_2S)_nH$ (58)

n	Softening point, °C
4	110–120
14	200–210
20	210–220
25–30	240

done by irradiation at room temperature followed by heating to 180°C (61). Polymerization occurs during heating. A similar result is obtained by irradiating at 180 or 195°C (60). It was noted that higher melting polymers were obtained from larger, better ordered monomer crystallites.

The crystal structure of poly(thioformaldehyde) has been determined (67). The polymer has a hexagonal unit cell with $a = 5.07$ Å and $c = 36.52$ Å. There are 17 monomer units in the unit cell, and the polymer chain is arranged in a helix that has 9 turns in a period of 36.52 Å. The radius of the helix is 0.99 Å, and the internal rotation angle is 65°39′. The carbon–sulfur bond length is 1.815 Å.

VIII. THIOACETONE

Thioacetone was first formed by reaction of acetone with hydrogen sulfide in the presence of an acidic catalyst (68). It was a minor product detected by its most characteristic property, a very disagreeable odor. The major product was trithioacetone (12). It is likely that thioacetone is formed first and that

12

this compound trimerizes to form **12**. Monomeric thioacetone has been prepared by cleavage of acetone diethylacetal with hydrogen sulfide in the presence of an acid catalyst (69). It has been described as an unstable red oil boiling at 80°C/760 mm and having a refractive index η_D^{20} of 1.4690 (70).

The instability of thioacetone is the result of the facility with which it trimerizes and polymerizes. If pure, it gives only a linear polythioacetal. This was shown by Bailey and Chu (71), who prepared thioacetone by pyrolysis of allyl isopropyl sulfide and purified it by gas chromatography or distillation at room temperature under reduced pressure. They found that pure material can be kept at −70°C, but that it polymerizes rapidly at room temperature to linear polymer of at least 2000 molecular weight. Crude material, on the other hand, slowly changes to trithioacetone.

A. Synthesis

The method that appears to be most convenient for the preparation of thioacetone monomer is pyrolysis of trithioacetone (72). This trimer is prepared by hydrochloric a cidor zinc chloride catalysis of the reaction of acetone and hydrogen sulfide, and it contains two isomeric impurities.

$$
\begin{array}{cc}
\underset{\text{C}}{(CH_3)_2} & \underset{\text{C}}{(CH_3)_2} \\
S \qquad S & S \qquad S \quad CH_3 \\
(CH_3)_2C \qquad S & (CH_3)_2C \qquad C \\
\underset{(CH_3)_2}{C} & CH_2 \qquad SH
\end{array}
$$

The mercapto impurity can be removed by stirring with benzoquinone prior to pyrolysis. Pyrolysis is conveniently accomplished by use of a modified ketene lamp, and the pyrolysate is collected in traps cooled with solid carbon dioxide in acetone. The thioacetone freezes to give colored crystals, mp $-40°C$. Polymerization of thioacetone occurs spontaneously when it is allowed to warm above $-20°C$ (Eq. 29). Polymers with molecular weights as high as 14,000 can be obtained in this fashion (72).

$$CH_3CH{=}S \longrightarrow +CH{-}S+_n \qquad (29)$$

with the group CH_3 on the CH.

The following procedure is one described by Burnop and Latham (73) and is a modification of the method first reported by von Ettinghausen and Kendrick (72).

A loose coil made from 175 cm of SWG Chromel A wire is mounted through the top of a 5.5-cm Pyrex tube 30 cm high in a manner that allows it to extend two-thirds of the way down the tube. At a point below the coil, a side arm is sealed to the tube for insertion of an argon bubbler that extends to the bottom of the tube. An outlet for the reaction product is sealed to the side of the tube near the top and connected to a reflux condenser. To the condenser are connected two traps in series; the first is cooled to $0°C$ and the second to $-78°C$. The outlet of the second trap is connected to a vacuum pump.

Trithioacetone that has been purified by recrystallization from methanol (mp $21.8°C$) is placed in the bottom of the tube, melted, and blanketed with argon. The apparatus is evacuated to 1 mm and the trithioacetone heated sufficiently to result in boil-up of 5–6 g/h. A potential of 20–30 V is then applied across the wire coil. As vapors of trithioacetone come into contact with the coil, the trithioacetone is pyrolyzed, and the products formed are carried through the reflux condenser to the traps where they are condensed. Thioacetone is collected in the $-78°C$ trap.

B.   Polymerization

As already indicated, thioacetone polymerizes very readily at any temperature above $-20°C$. Polymerization occurs either neat or in solution. Ether, chloroform, ethylene oxide, propylene oxide, and epichlorohydrin are among the solvents that have been reported as polymerization media. Bailey and Chu found that pure monomer gives only linear polymer; but in the experience of other investigators, polymer and trimer are usually obtained together.

Burnop and Latham state that thioacetone undergoes solid-state polymerization upon exposure to visible light. In this behavior thioacetone is unique. Orange, fernlike crystals that condense as a thin layer on the wall of the trap cooled to $-78°C$ are converted without change of shape to colorless polymer when exposed to daylight. Molecular weights as high as 33,000 are said to be obtained. The polymer, however, is not stable in benzene solution in which measurements are made.

In the absence of light and when kept at $-78°C$, crystalline thioacetone does not polymerize.

Poly(thioacetone) is a white powder, unpleasant in odor, of density 1.21. It melts sharply at $124°C$ to a pink liquid that resolidifies to a brittle, opaque solid when it is cooled. The polymer is very soluble in chloroform and tetrahydrofuran. It is also soluble in carbon tetrachloride and benzene. Nonsolvents include alcohols, acetone, ether, and dimethyl sulfoxide.

C.   Toxicity

Thioacetone is no doubt toxic, but no one could tolerate toxic levels of it because of its disgusting odor. It should be used only in a well-ventilated area, and equipment should be arranged to allow scrubbing of vapors emanating from such equipment through alkaline permanganate. Apparatus should be decontaminated with alkaline permanganate. Nitrogen dioxide has also been used to destroy objectionable vapors.

IX.   THIOCARBONYL FLUORIDE

The most attractive polymers obtained from thiocarbonyl compounds are those prepared from the perfluoro derivatives. Of these, the best is poly(thiocarbonyl fluoride), $+CF_2—S+_n$. This polymer is an unusually resilient elastomer that also has exceptional resistance to degradation by acids and aqueous bases. Although the homopolymer is most efficaciously prepared by anionic polymerization of thiocarbonyl fluoride, $CF_2{=}S$, at low temperature, it can also be obtained by free radical polymerization at low temperature. In free radical systems, thiocarbonyl fluoride also copolymerizes with olefins, vinyl compounds, allyl derivatives, and many acrylates. Poly(thiocarbonyl fluoride) and copolymers of thiocarbonyl fluoride are described in considerable detail elsewhere (74).

Thiocarbonyl fluoride has been prepared by a number of investigators. Yarovenko and Vasil'eva (75) converted thiophosgene to trichloromethyl sulfenyl chloride, fluorinated this product to chlorodifluoromethyl sulfenyl chloride, and then dechlorinated the fluorinated product. The fluorination

$$CCl_2{=}S \xrightarrow{Cl_2} CCl_3SCl \xrightarrow{SbF_3} CF_2ClSCl \xrightarrow{Sn} CF_2{=}S$$

step also gives $CFCl_2SCl$, which they separated and dechlorinated to $CFCl{=}S$.

Downs and Ebsworth (76) prepared thiocarbonyl fluoride from bis(trifluoromethylthio)mercury and iodosilane. Trifluoromethylthiosilane is formed first, and it immediately decomposes to fluorosilane and thiocarbonyl fluoride.

$$(CF_3S)_2Hg + SiH_3I \longrightarrow SiH_3(SCF_3) + HgI_2$$

$$\downarrow$$

$$FSiH_3 + CF_2{=}S$$

Direct reaction of sulfur with tetrafluoroethylene (Eq. 30) also leads to thiocarbonyl fluoride (77). Trifluorothioacetyl fluoride and bis(trifluoromethyl)disulfide are also formed, although in lesser amounts.

$$CF_2{=}CF_2 + S \xrightarrow{500-600°C} CF_2{=}S + CF_3CF{=}S + CF_3SSCF_3 \qquad (30)$$

Very high yields of thiocarbonyl fluoride are formed by reaction of chlorofluoromethane with sulfur (Eq. 31) at high temperatures (78).

$$ClCF_2H + S \xrightarrow{700-900°C} CF_2{=}S + HCl \qquad (31)$$

The most convenient laboratory synthesis of thiocarbonyl fluoride so far developed involves dimerization of thiophosgene to a dithietane, fluorination of the dithietane by reaction with antimony trifluoride, and cracking of the resultant 2,2,4,4-tetrafluoro-1,3-dithietane at high temperatures (77). If highly purified tetrafluorodithietane is added slowly in a current of dry nitrogen to a dry, air-free, hot platinum tube, the thiocarbonyl fluoride obtained is quite pure. After fractionation, which requires a low-temperature still because thiocarbonyl fluoride boils at $-54°C$, it should contain no impurity other than about 15–25 ppm of carbon oxysulfide.

A. Synthesis

Since the dithietane route, although consisting of several steps, requires the least amount of specialized equipment and readily gives thiocarbonyl fluoride of polymerization grade, it is the route recommended. The procedure presented here for synthesizing this compound is from that given by Sharkey et al. (79). The first step (Eq. 32), dimerization of thiophosgene to 2,2,4,4-tetrachloro-1,3-dithietane, has been described by Schonberg and Stephenson (80).

$$CCl_2{=}S \xrightarrow{\text{u.v.}} \underset{S}{\overset{S}{CCl_2 \diagup \diagdown CCl_2}} \tag{32}$$

A 4-in. I.D. tube, 18 in. long, sealed at one end, in a water jacket, is mounted vertically and closed at the top with a rubber stopper. A glass stirrer with a Teflon® paddle is placed in the tube through the stopper. Also attached through the stopper is a Drierite drying tube. Then 4 kg of thiophosgene is placed in the tube and water cooled to just below $20°C$ is circulated through the jacket. Four to six GE H85-C3 lamps are then placed around the tube with suitable reflectors. The mixture is stirred and irradiated for about 46 hr. The stirrer is stopped, the lamps shut off, and the crystals of tetrachlorodithietane allowed to settle. Unreacted thiophosgene is poured off and the solid tetrachlorodithietane removed from the tube. The yield should be 2900–3100 g (72–77%).

The second step (Eq. 33) involves conversion of the tetrachlorodithietane to 2,2,4,4-tetrafluoro-1,3-dithietane.

$$\underset{S}{\overset{S}{Cl_2C \diagup \diagdown CCl_2}} \xrightarrow{\text{SbF}_3} \underset{S}{\overset{S}{F_2C \diagup \diagdown CF_2}} \tag{33}$$

An amount of tetramethylene sulfone equal to three times the weight of tetrachlorodithietane to be fluorinated is placed in a three-necked, 12-liter flask. The flask is equipped with a stirrer, a thermometer, and a reflux condenser. The outlet from the reflux condenser is connected to three traps in series, the first two being cooled with ice and the third with solid carbon dioxide in acetone. Air in the apparatus is displaced with nitrogen, the stirrer is started, and an amount of antimony trifluoride equal to 1.5 times the weight of the tetrachlorodithietane is added. Then the dithietane is added through the condenser. with additional tetramethylene sulfone being used to wash down the material. The flask is heated to 67°C. Bubble formation in the reaction mixture indicates that reaction has started. The heat source is removed and stirring is continued until bubbling decreases. Heat is again applied at a rate required to increase the inside temperature to 180°C over a 1-hr period. The product collects in the traps, and the contents are combined, dried with magnesium sulfate, filtered, and distilled. The fraction boiling at 47.8–49.4°C is collected as 2,2,4,4-tetrafluoro-1,3-dithietane.

Pyrolysis of the tetrafluorodithietane in a platinum tube at 475–500°C (Eq. 34) leads to thiocarbonyl fluoride in nearly quantitative yield (77).

$$F_2C \overset{S}{\underset{S}{\diamond}} CF_2 \xrightarrow{\Delta} 2CF_2{=}S \qquad (34)$$

An unpacked, platinum tube 0.5 in. in diameter and 20 in. long is mounted at a 30° angle and fitted at the upper end with a T tube that has a dropping funnel and a nitrogen source connected to it. The lower end is fitted to a coiled tube trap made of stainless steel and cooled in an ice bath. The outlet from this trap is connected to a glass trap cooled with a solid carbon dioxide in acetone bath. The outlet from the second trap is protected with a drying tube. The platinum tube is heated to 500°C over a 12-in. section, and nitrogen is passed through the tube at a rate of about 100 ml/min, to ensure that all moisture is purged from the apparatus. Then 100 g of 2,2,4,4-tetrafluoro-1,3-dithietane is added in drops over a 5-hr period. Thiocarbonyl fluoride condenses in the trap cooled with solid carbon dioxide and acetone. It is purified (81) by fractionation through a 2 ft × 9 mm vacuum-jacketed column packed with glass helices and fitted with a still-head that is kept at a low temperature with solid carbon dioxide in acetone. Pure monomer boils at −54°C, and the amount obtained is at least 90 g.

This compound must be protected from moisture and should be kept very cold until used. It is not advisable to store it for more than a week, even at low temperatures, because impurities are slowly formed by this very reactive monomer. Toxicity studies have not been made. Thiocarbonyl fluoride is undoubtedly dangerous, however, since it hydrolyzes quickly and completely to hydrogen fluoride and carbon oxysulfide.

B. Polymerization

Thiocarbonyl fluoride polymerizes rapidly at low temperatures when initiated with anions or free radicals (Eq. 35). Both kinds of initiators give

$$CF_2{=}S \xrightarrow[-78°C]{A^\theta \text{ or } R\cdot} \quad +CF_2{-}S +_n \tag{35}$$

tough, high-molecular-weight polymer. Anionic systems are easiest to handle experimentally and are recommended (81). The initiator used is dimethylformamide, which is very effective although not a usual anionic initiator. It probably operates through reaction with thiocarbonyl fluoride to form an anionic species.

> A polymerization flask is fitted with a stirrer and connected through glass tubing with appropriate stopcocks to an evacuation system and a flask containing 100 g of cold, liquid thiocarbonyl fluoride. Then 100 ml of ether that has been dried with sodium is introduced into the polymerization flask through a serum stopper. The flask containing thiocarbonyl fluoride and the polymerization flask are cooled with liquid nitrogen and the system evacuated. The liquid nitrogen bath is removed from the flask containing thiocarbonyl fluoride, and this monomer is distilled into the polymerization flask. Dry nitrogen gas is added to the system to restore internal pressure to 1 atm. Then the liquid nitrogen cooling bath is removed from the polymerization flask. It is replaced with a solid carbon dioxide in acetone bath. As the contents of the polymerization bath warm to $-78°C$, it melts. Stirring is started and 5 drops of dimethylformamide is added from a No. 22 hypodermic syringe. As stirring is continued, polymer forms and separates from the liquid. After 18 hr at $-78°C$, the white, spongy polymer is removed, boiled in water containing 5 ml of 50% nitric acid, and then dried. Smooth films are formed by pressing the polymer in a Carver press at 150°C and 10,000 lb ram pressure for 2 hr. These films are translucent and very elastic. At room temperature they slowly crystallize to a white, opaque, nonelastic crystalline form that can be reconverted to the elastomeric form by warming above 35°C.

As mentioned earlier, thiocarbonyl fluoride also polymerizes in free radical systems. What is most surprising is that it also copolymerizes with many unsaturated compounds (82). The trick here is to carry out the polymerization at very low temperatures. This requires a free radical source that is operable at low temperatures. One such source that is very good for this purpose is the reaction of trialkylboranes with oxygen.

$$R_3B + O_2 \xrightarrow{-78°C} R_2BOOR$$

$$R_2BOOR + 2R_3'B \xrightarrow{-78°C} R_2BOBR_2' + 2R_2'BOR + 2R'\cdot$$

The thiocarbonyl fluoride–propylene copolymer is easily made in high conversion and high molecular weight. It is an elastomer having good flexibility at temperatures as low as $-55°C$, but it is of low strength and is very difficult to cure. Thiocarbonyl fluoride–allyl chloroformate copolymers are also good elastomers, and they are easily cured by heating with zinc oxide (82).

X. HIGHER FLUOROTHIOACID FLUORIDES AND PERFLUOROTHIOKETONES

Addition polymerization of the thiocarbonyl group in perfluoroalkyl compounds is a general reaction, although it becomes more sluggish with increase in size of substituent groups. For example, perfluorothioacetyl fluoride polymerizes somewhat slowly, and its polymers are logy elastomers. Perfluorothioacetone polymerizes only at very low temperatures ($-110°C$), and the polymer obtained is an unstable, logy elastomer. A general route (83,84) to perfluorothioacetyl fluoride and perfluoromethyl alkyl thioketones is

$$CF_2{=}CF_2 + HgF_2 \longrightarrow (CF_3CF_2)_2Hg$$

$$(CF_3CF_2)_2Hg + S \longrightarrow CF_3CF{=}S$$

$$R_fCF{=}CF_2 + HgF_2 \longrightarrow (R_f{-}\underset{\underset{CF_3}{|}}{C}F)_2Hg$$

$$(R_f\underset{\underset{CF_3}{|}}{C}F)_2Hg + S \longrightarrow R_f{-}\underset{\underset{CF_3}{|}}{C}{=}S$$

A. Synthesis

The preparation of a perfluoromethyl alkyl thioketone is illustrated here by preparation of hexafluorothioacetone. Mercuric fluoride is first added to hexafluoroisopropene (Eq. 36).

$$CF_3CF{=}CF_2 + HgF_2 \rightarrow [(CF_3)_2CF]_2Hg \tag{36}$$

The mercurial is then reacted with sulfur (Eq. 37).

$$[(CF_3)_2CF]_2Hg + S \xrightarrow{450°C} CF_3\overset{\overset{S}{\|}}{C}CF_3 \tag{37}$$

This method is applicable to synthesis of perfluorothioacetyl fluoride (84) if mercuric fluoride is added to tetrafluoroethylene in the first step. Since all these fluorine compounds are extremely toxic, they should be handled with great care. They should not be allowed to get into the respiratory system or to come into contact with the skin.

Bis(perfluoroisopropyl)mercury (83) is prepared by placing 240 g of mercuric fluoride in a dry "Hastelloy" bomb. The tube is closed, evacuated, and chilled in a bath of solid carbon dioxide in acetone. Then 70 g of anhydrous hydrogen fluoride and 240 g of hexafluoropropene are admitted to the bomb. The bomb is heated to 110°C for 12 hr and then cooled and vented. The contents is poured into a polyethylene bottle and excess hydrogen fluoride is allowed to escape. Sodium fluoride is added to remove traces of hydrogen fluoride, and the product is distilled. There is obtained about 320 g of bis(perfluoroisopropyl)mercury, bp 115–116°C; mp 20–21°C; n_D^{25} 1.3244, d_4^{25} 2.5301.

A two-necked flask containing 150 g of sulfur is fitted with a dropping funnel and nitrogen inlet to one neck and with an air condenser to the other. The outlet of the air condenser is connected to a trap cooled with a solid carbon dioxide in acetone bath. The flask is heated until the sulfur boils up and refluxes in the air condenser as a slow stream of nitrogen is passed through the flask. Then 108 g (0.2 mole) of bis(perfluoroisopropyl)mercury is added in drops to the vapor of boiling sulfur over a 2-hr period. The product condenses as a deep-blue liquid in the cold trap. The yield of this deep-blue compound, which is hexafluorothioacetone, should be 22 g (60%). It boils at 8°C under atmospheric pressure, but it is best to distill it under reduced pressure to reduce the likelihood of dimerization. At 200 mm perfluorothioacetone boils at −20°C, and these are the recommended distillation conditions.

Perfluorothioacetone dimerizes readily to 2,2,4,4-tetrakis(trifluoromethyl)-1,3-dithietane (Eq. 38), and it is best kept as this derivative. Dimerization proceeds slowly at room temperature (Eq. 38) but takes place rapidly if a drop

$$CF_3\overset{\displaystyle S}{\overset{\|}{C}}CF_3 \longrightarrow (CF_3)_2C\overset{\displaystyle S}{\underset{\displaystyle S}{\diamondsuit}}C(CF_3)_2 \tag{38}$$

of diethyl ether is added. The dimer is a very toxic, white solid that melts near room temperature and boils at 110°C. It is readily reconverted to perfluorothioacetone by pyrolysis at 600°C, using equipment described earlier for preparing thiocarbonyl fluoride from 2,2,4,4-tetrafluoro-1,3-dithietane.

B. Polymerization

As mentioned earlier, higher fluorothioacid fluorides polymerize somewhat sluggishly (81). Polymers of these compounds are formed by a procedure similar to that described for anionic polymerization of thiocarbonyl fluoride. Perfluorothioketones are different. Polymerization of perfluorothioacetone requires a very low temperature (81).

One procedure is to freeze 1 ml of boron trifluoride etherate in a liquid nitrogen bath and then add 5 ml of perfluorothioacetone to it. The bath is then removed and the mixture allowed to warm until it forms a slush. At this point the temperature of the mixture is about $-110°C$ or lower. As polymerization takes place, the blue color disappears. When polymerization is complete, acetone is added to precipitate the polymer, which is then separated and dried. It is a logy elastomer that slowly reverts to hexafluorothioacetone dimer.

C. Properties

Because they are highly reactive and have only been prepared in small quantities, very little information on the physical properties of thiocarbonyl compounds has been reported. Boiling points on a number of these materials are available (77), however, and these are presented in Table 7.

TABLE 7
Boiling Points of Higher Fluorothiocarbonyl Compounds

Compound	bp, °C	Color
$\overset{\text{S}}{\overset{\|}{\text{CF}_3\text{CCF}_3}}$	$\overset{8}{-20}$ (200 mm)	Blue
$\overset{\text{S}}{\overset{\|}{\text{HCF}_2\text{CF}_2\text{CCF}_3}}$	-27 (20 mm)	Blue
$\overset{\text{S}}{\overset{\|}{\text{ClCF}_2\text{CF}_2\text{CCF}_3}}$	-19 (20 mm)	Blue
$\overset{\text{S}}{\overset{\|}{\text{CF}_3\text{CF}}}$	-22	Yellow
$\overset{\text{S}}{\overset{\|}{\text{CF}_3\text{CCl}}}$	$28-29$	Red
$\overset{\text{S}}{\overset{\|}{\text{ClCF}_2\text{CF}}}$	23	Yellow
$\overset{\text{S}}{\overset{\|}{\text{BrCF}_2\text{CF}}}$	$41-42$	Yellow
$\overset{\text{S}}{\overset{\|}{\text{ClCF}_2\text{CCl}}}$	-10 (20 mm) η_D^{24} 1.4465	Red

References

1. Modern Plastics Encyclopedia, McGraw-Hill, New York, 1968, p. 124.
2. C. E. Schweitzer, Ed., "Encyclopedia of Chemical Technology," 2nd ed., Interscience, New York, 1963, Vol. 1, p. 95.
3. "Polyaldehydes," O. Vogl, Ed., Marcel Dekker, New York, 1967.
4. A. Butlerov, Ann., 111:242 (1859).
5. Chem. Week, May 26, 1971, p. 29.
6. Ulrich Gerloff, Hydrocarbon Process., 46(6): 169–172 (1967).
7. Chem. Profiles, April 1, 1966.
8. J. F. Walker, "Formaldehyde," 3rd ed., Reinhold, New York, 1964.
9. Northrup Brown in Ref. 3, p. 9.
10. A. W. Hofmann, Ann., 145:357 (1868); Ber., 2:152 (1869).
11. Hydrocarbon Process., 44(11):215 (1965); 44(11):216 (1965).
12. H. Tadenuma, T. Murakami, and H. Mitsushima, Hydrocarbon Process., 45:195 (1966).
13. D. L. Funck, U.S. Patents 2,848,500 (1958) and 2,943,701 (1960).
14. R. N. MacDonald, in "Macromolecular Syntheses," Vol. 3, Wiley, New York, 1967, p. 58.
15. V. Jaacks, S. Iwabucki, and W. Kern, in Ref. 14, p. 67.
16. H. Staudinger, H. Signer, and O. Schweitzer, Ber., 64:398 (1931).
17. S. H. Jenkins and J. O. Punderson, U.S. Patent 2,964,500 (Dec. 13, 1960).
18. W. Kern and V. Jaacks, J. Polym. Sci., 48:399 (1960).
19. M. B. Price and F. B. McAndrew, in Ref. 3, p. 38.
20. M. Letort, Compt. Rend., 202:767 (1936).
21. M. W. Travers, Trans. Faraday Soc., 32:246 (1936).
22. G. Natta, G. Mazzanti, P. Corradini, and I. W. Bassi, Makromol. Chem., 37:156 (1960).
23. J. Furukawa, T. Saegusa, H. Fujii, A. Kawasaki, H. Imai, and Y. Fujii, Makromol. Chem., 37:149 (1960).
24. O. Vogl, J. Polym. Sci., 46:261 (1960).
25. G. Natta, P. Corradini, and I. W. Bassi, J. Polym. Sci., 51:505 (1961).
26. Chem. Profiles, October 1, 1967.
27. S. A. Miller, Chem. Process Eng., 49:75 (1968).
28. F. R. Lowdermilk and A. R. Day, J. Amer. Chem. Soc., 52:3535 (1930).
29. M. P. Berthelot and E. C. Jungfleisch, "Traite Elementaire de Chemie Organique," 2nd ed., Paris, 1886, p. 256.
30. P. Sabatier and J. B. Senderens, Compt. Rend., 136:738 (1903).
31. M. Kutscherow, Ber., 14:1532, 1540 (1881); 17:13 (1884).
32. M. Sittig, Hydrocarbon Process. Pet. Refining, 41(4):157 (1962).
33. German Patents 1,049,845 (1959), 1,061,767 (1959), 1,080,994 (1960), 1,118,183 (1960); Angew. Chem., 71:176 (1959).
34. O. Vogl, in Ref. 14, p. 71.
35. J. Brandrup and M. Goodman, in Ref. 14, p. 74.
36. O. Vogl, J. Polym. Sci., A2, 1964:4591.
37. A. Novak and E. Whalley, Can. J. Chem., 37:1710 (1959).
38. P. W. Bridgman and J. B. Conant, Proc. Nat. Acad. Sci. (U.S.), 15:680 (1929).
39. J. B. Conant and C. O. Tongberg, J. Amer. Chem. Soc., 52:1659 (1930).
40. A. Novak and E. Whalley, Can. J. Chem., 37:1718 (1959).
41. "Organic Syntheses," Coll. Vol. 2, Wiley, New York, 1943, p. 323.

REFERENCES 687

42. I. Rosen, in Ref. 3, p. 68.
43. H. Schechter and F. Conrad, *J. Amer. Chem. Soc.*, **42**:3371 (1950).
44. J. Redtenbacher, *Ann. Chem.*, **47**:113 (1843).
45. R. C. Schulz, in "Encyclopedia of Polymer Science and Technology," Wiley, New York, Vol. I, 1964, p. 160.
46. H. R. Guest, B. W. Kiff, and H. A. Stansbury, Jr., in "Encyclopedia of Chemical Technology," 2nd ed., Vol. I, R. E. Kirk and D. F. Othmer, Eds., Interscience, New York, 1964, p. 255.
47. E. G. Peters and G. C. Hood, Jr., in "Encyclopedia of Industrial Chemistry Analysis," Snell-Hilton, Vol. 4, 1967, p. 148.
48. C. W. Smith, "Acrolein," Wiley, New York, 1962.
49. C. W. Smith and R. T. Holm, U.S. Patent 2,718,529 (Sept. 20, 1955).
50. W. F. Gresham and C. E. Schweitzer, U.S. Patent 2,485,236 (Oct. 18, 1949).
51. M. Schmidt and K. Blattner, *Z. Angew. Chem.*, **71**:407 (1959).
52. L. Credali and M. Russo, *Polymer*, **8**(9):469 (1967).
53. A. Wohl, *Ber.*, **19**:2344 (1886).
54. R. J. W. Le Fèvre and M. Macleod, *J. Chem. Soc.*, **1931**:474.
55. J. Harmon, U.S. Patent 3,070,580 (Dec. 25, 1962).
56. E. Gipstein, E. Wellisch, and O. J. Sweeting, *J. Polym. Sci.*, *B*, **1**(5):237 (1963).
57. K. Kuellman, E. Fischer, and K. Weissernel, German Patent 1,153,176 (Aug. 22, 1963).
58. P. Bapseres and J. Signouret, French Patent 1,330,819 (June 28, 1963).
59. H. Birkner and F. Stuerzenhofecker, German Patent 1,202,500 (Oct. 7, 1963).
60. J. B. Lando and V. Stannett, *J. Polym. Sci.*, *B*, **2**(4):375 (1964); *A*, **3**(6):2369 (1965).
61. V. Stannett, AEC Accession No. 37176, Rep. No. BNL-874, Arail. OTS, 2 pp. (1964).
62. R. Wakasa, S. Ishida, and H. Ohama, Japanese Patent 16,152 (July 26, 1965).
63. W. J. Roberts and B. B. Jacknow, Belgian Patent 626,040 (March 29, 1963).
64. Ref. 41, p. 610.
65. T. L. Cairns, G. L. Evans, A. W. Larchar, and B. C. McKusick, *J. Amer. Chem. Soc.*, **74**:3982 (1952).
66. R. Fabre, J. Verne, M. Chaigneau, and G. Le Moan, *Compt. Rend.*, **259**(15):2545 (1964).
67. G. Carazzolo, L. Mortillaro, L. Credali, and S. Bezzi, *Chim. Ind. (Milan)*, **46**:1484 (1964); G. Carazzolo and M. Mammi, *J. Polym. Sci.*, *B*, **2**(11):1057 (1964); G. Carazzolo and G. Valle, *Makromol. Chem.*, **90**:66 (1966).
68. E. Baumann and E. Fromm, *Ber.*, **22**:2592 (1889).
69. H. Berthold, Diploma thesis, Technische Universität, Dresden, 1963.
70. R. Mayer, J. Morgenstern, and J. Fabian, *Angew. Chem., Int. Ed.*, **3**:277 (1964).
71. W. J. Bailey and H. Chu, Polymer Preprints, Division of Polymer Chemistry, American Chemical Society, 149th Meeting, Detroit, April 5, 1965.
72. O. G. von Ettingshausen and E. Kendrick, *Polymer*, **7**:469 (1966).
73. V. C. E. Burnop and K. G. Latham, *Polymer*, **8**:589 (1967).
74. W. H. Sharkey, in "Polymer Chemistry of Synthetic Elastomers, Part 2," Joseph P. Kennedy and Erik G. M. Tornqvist, Eds., Interscience, New York, 1969, p. 893.
75. N. N. Yarovenko and A. S. Vasil'eva, *J. Gen. Chem. (USSR)*, **29**:3754 (1959), English transl.
76. A. J. Downs and E. A. V. Ebsworth, *J. Chem. Soc.*, **1960**:3516.
77. W. J. Middleton, E. G. Howard, and W. H. Sharkey, *J. Org. Chem.*, **30**:1375 (1965).
78. D. M. Marquis, U.S. Patent 2,962,529 (Sept. 29, 1960).
79. W. H. Sharkey and H. W. Jacobson, *Macromol. Syn.*, in press.

80. A. Schonberg and A. Stephenson, *Ber.*, **66B**:567 (1933).
81. W. J. Middleton, H. W. Jacobson, R. E. Putnam, H. C. Walter, D. G. Pye, and W. H. Sharkey, *J. Polym. Sci.*, *A3*, **1965**:4115.
82. A. L. Barney, J. M. Bruce, Jr., J. N. Coker, H. W. Jacobson, and W. H. Sharkey, *J. Polym. Sci.*, *A-1*, **4**:2617 (1966).
83. P. E. Aldrich, E. G. Howard, W. J. Linn, W. J. Middleton, and W. H. Sharkey, *J. Org. Chem.*, **28**:184 (1963).
84. C. G. Krespan, *J. Org. Chem.*, **25**:105 (1960).

10. TETRAFUNCTIONAL INTERMEDIATES

J. K. STILLE, M. E. FREEBURGER, W. B. ALSTON, AND E. L. MAINEN
University of Iowa, Iowa City, Iowa

Contents

I. INTRODUCTION

Of the large number of tetrafunctional monomers reported in the literature, only a very few are being used in industry. The majority of the monomers are either too expensive or too difficult to prepare to make their
commercial production feasible. The tetrafunctional monomers that are
used in industrial processes are well characterized with respect to their
physical properties, storage, handling, toxicity, etc., whereas such information is usually not available for the remainder of the monomers.

Tetrafunctional monomers fall into two general classes: (a) the oxygen-
containing compounds, such as the acids, anhydrides, and ketones, and (b)
the nitrogen-containing compounds, such as the amines and hydroxyamines.
The majority of the polymers covered are prepared by copolymerization of
an oxygen group monomer with a nitrogen group monomer.

In this chapter, the monomers are divided into four sections: Section II,
the dianhydrides and tetracids; Section III, the tetraketones, hydroxyquinones, and diglyoxals; Section IV, the tetramines; and Section V, miscellaneous tetrafunctional monomers. In Sections II–IV, a detailed laboratory
preparation is given for several of the monomers, and for the rest a general
method of synthesis is outlined. At the beginning of these sections, an outline
is provided of the general types of polymerizations which the monomers
undergo. Because of the varied nature of the monomers in Section V, no
general type of polymerization can be given, and polymerizations are discussed together with each monomer.

In each section there is a table listing each monomer and references to its preparation and polymerizations. The pertinent physical properties and toxicological data are given for each monomer, when such information is available.

II. TETRACARBOXYLIC ACIDS AND DIANHYDRIDES

The most common tetrafunctional monomer is the tetracarboxylic acid, which is most often employed as the dianhydride. A dianhydride can be polymerized with a diamine to produce a polyimide or with a tetramine to produce a polyimidazopyrrolone. The anhydride ring can be five-, six-, or seven-membered, but the reactions are the same.

When pyromellitic dianhydride (**1**) is polymerized with a diamine such as benzidine, the polyimide **3** is produced in one or two steps. The most common polymerization procedure is to isolate the polyamide acid **2** and convert it to the polyimide in a subsequent step.

When **1** is polymerized with a tetramine, such as 1,2,4,5-tetraminobenzene, the polyamide amino acid **4** is initially formed. In the second stage of the polymerization, **4** is converted first to a mixture of the polyimidazole acid **5** and polyaminoimide **6**, which is then further dehydrated to the polyimidazo-pyrrolone **7** (1).

In general, the dianhydride is used in lieu of the tetracarboxylic acid because a higher molecular weight can be obtained when the dianhydride is employed (2).

A. Pyromellitic Dianhydride

Pyromellitic dianhydride (**1**) is the tetrafunctional monomer that has received the most attention. It is prepared commercially from the oxidation of durene with a vanadium pentoxide catalyst (3). The dianhydride is easily converted to the tetracid (**8**), by hydrolysis (4).

The tetracid **8** can be prepared in the laboratory from the oxidation of charcoal (5). In a 5-liter flask are placed 100 g of pine or spruce charcoal, 650 ml of ca. 85% sulfuric acid, and a drop of mercury. This mixture is heated to 350°C over a period of several hours. When needles of **8** begin to appear on the neck of the flask, 50 ml of sulfuric acid is used to rinse down the walls of the flask and the hot mixture is transferred to a 1-liter distillation flask. The reaction flask is washed with a minimum of water, which is combined with the acid mixture.

The mixture is distilled until the water is removed and 30 g of acid potassium sulfate is added. Distillation of the sulfuric acid is carried out until crystals of **1** begin to appear; then a receiver containing 50 ml of water is attached, and the distillation is completed. The distillate is evaporated to 25 ml, and **8** crystallizes out upon cooling. The distillation flask is washed with 100 ml of water, solids are removed by filtration, and evaporation to 25 ml affords more **8**. The product is recrystallized from water. The reported yield is 6–8 g. The tetracid is converted to the dianhydride by boiling with acetic anhydride.

The dianhydride is a white solid, mp 285–286°C, which must be stored in a moisture-free area to prevent hydrolysis. It is a skin sensitizer and a skin irritant, but the lethal dosage for rats is very high. Care should be taken to avoid inhalation of the dust.

The tetracid is also a white solid, melting at 269–271°C. It is a skin sensitizer, but not a skin irritant. Again, care should be taken to avoid ingestion of the dust (6).

The process recommended for the analysis of pyromellitic-dianhydride–pyromellitic-acid mixtures involves esterification of the anhydride in the mixture with methanol and titration of the remaining acid functions with sodium hydroxide. This result is compared with that obtained by titration of an unesterified sample with sodium hydroxide (6).

Pyromellitic dianhydride can be purified for polymer studies by sublimination (1) or by recrystallization from a variety of solvents (2,7). Pyromellitic dianhydride has been polymerized with numerous diamines and tetramines to produce polyimides and polyimidazopyrrolones, respectively. References to these polymers appear in Table 1.

B. 3,3′,4,4′-Benzophenonetetracardoxylic Dianhydride

The dianhydride (**9**) of 3,3′,4,4′-benzophenonetetracarboxylic acid (**10**) is commercially available. It is prepared by condensation of 2 moles of *o*-xylene with 1 mole of acetaldehyde to afford 1,1-bis(3,4-dimethylphenyl)-ethane (**11**), which can then be oxidized to **10** with nitric acid (8). Alternatively, *o*-xylene can be condensed with pyruvic acid to give 2,2-bis-(3,4-dimethyphenyl)propionic acid (**12**), which is converted to 1,1-bis(3,4-dimethylphenyl)ethene (**13**) by treatment with sulfuric acid. The olefin, **13**, is then oxidized to **10** in two steps (9).

TABLE 1

References Covering the Preparation and Polymerizations of the Principal Dianhydrides

Monomer	Preparation	Polymerizations with		
		Diamines	Tetramines	Other
Pyromellitic Dianhydride	3–5	2,32–56	5,6,57–59	60,61
3,3',4,4'-Benzophenonetetracarboxylic dianhydride	8–10	9,49,50,53–55,66,67	6,7,58	60
3,3',4,4'-Biphenyltetracarboxylic dianhydride	12–14	13,38,49–55,68		
Naphthalene-1,4,5,8-tetracarboxylic dianhydride	16–20	53–55,62,63	64,65	
3,4,9,10-Perylenetetracarboxylic dianhydride	21	49–55	58	
2,3,6,7-Naphthalenetetracarboxylic dianhydride	22–25	49–55		
2,3,5,6-Biphenyltetracarboxylic dianhydride	26,27	27		
2,2',6,6'-Biphenyltetracarboxylic dianhydride	19,28–30a	30b		
2,2',3,3'-Biphenyltetracarboxylic dianhydride	14	49–55		
3,3',4,4'-Diphenylethertetracarboxylic dianhydride	31–33	32,33,49–55	58	
3,3',4,4'-Diphenylsulfonetetracarboxylic dianhydride	34–36	49–55		

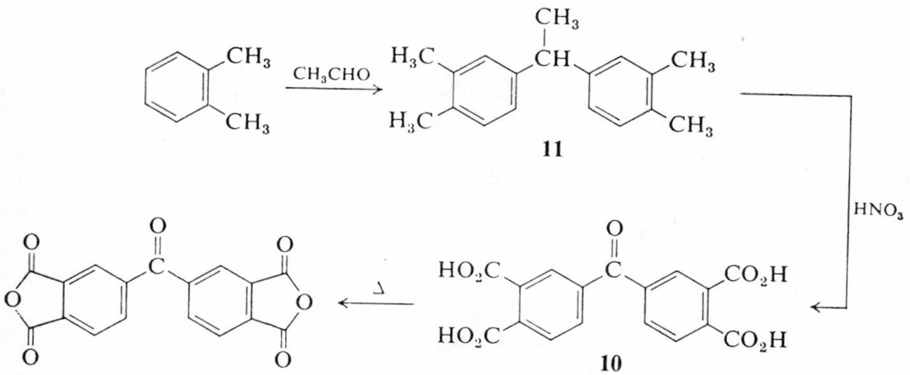

11

10

9

12

13

9 ← Δ — 10 ← 1) Cr₂O₇⁻ 2) MnO₄⁻

CH₃CHO

HNO₃

H_2SO_4

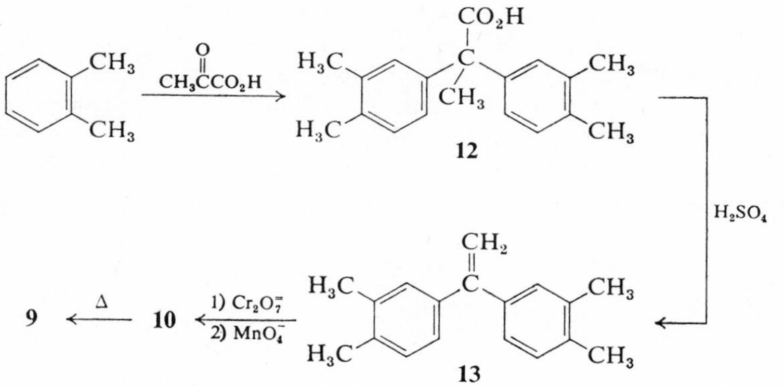

The dianhydride, **9** can be conveniently prepared in the laboratory by placing 238 g of 1,1-bis(3,4-dimethylphenyl)ethane and 2960 g of 30% nitric acid in a 4.5-liter stainless steel autoclave. The temperature is slowly raised to 200°C and held there for 1 hr. The gases are vented intermittently so that the final pressure is 265 psi. The product is cooled to 25°C, removed from the autoclave, and allowed to stand. A solid then precipitates. Evaporation of the filtrate affords more crystals. When the solid has been oven dried, 279 g of 3,3′,4,4′-benzophenonetetracarboxylic acid is obtained (10).

The white dianhydride can be recrystallized from acetone, and it melts at 225.0–226.5°C. Although the dianhydride is rather stable to hydrolysis, it should be stored in a closed container to exclude moisture. The dianhydride is nontoxic to animals and its dust is also nontoxic to the skin, eyes, and respiratory system. The fumes from the molten anhydride are moderately toxic (11). The dianhydride has been polymerized with diamines and tetramines and references to these polymers appear in Table 1.

C. 3,3',4,4'-Biphenyltetracarboxylic Dianhydride

The synthesis of 3,3',4,4'-biphenyltetracarboxylic dianhydride (14) can be accomplished by two methods. The diazotization of *o*-aminotoluene, followed by treatment with cuprous cyanide, affords the coupled product, 4,4'-dicyano-3,3'-dimethylbiphenyl (15). Hydrolysis and oxidation of 15 produces the tetracid, 16, which may be converted to the dianhydride 14 upon heating (12,13).

A suitable laboratory preparation of 14 begins with dimethyl 4-iodophthalate (17). When 17 is heated to 240°C and an equal weight of powdered copper is slowly added with continued stirring, with the resulting mixture being heated at 260°C an additional hour, tetramethyl 3,3',4,4'-biphenyltetracarboxylate (18) is formed. After cooling, the solid is extracted with ether. Concentration of the ether affords crystalline 18, which melts at 106°C upon recrystallization from hot methanol. The ester is hydrolyzed with 2N potassium hydroxide and the tetracid 16 separates upon acidification. The acid is converted to the dianhydride 14 by treatment with excess acetyl chloride (14). No yields are reported for this procedure.

The dianhydride 18 melts at 286°C and has been polymerized with several diamines (Table 1).

D.   Naphthalene-1,4,5,8-tetracarboxylic Dianhydride

Naphthalene-1,4,5,8-tetracarboxylic dianhydride (**19**), another commercially available monomer, is prepared from the oxidation of either 5,6-dialkylacenaphthene or pyrene. If malononitrile is condensed with acenaphthene and the resulting diketimide (**20**) is oxidized with sodium hypochlorite/potassium permanganate or chromic acid (15), or nitric acid (16–18), the tetracid **21** is obtained. Pyrene can be oxidized to **19** with a vanadium pentoxide catalyst (19), with a mixture of sodium dichromate and sulfuric acid followed by sodium hypochlorite and lime (20), or with a two-step oleum-nitric acid oxidation of chlorinated pyrene (20).

The white dianhydride sublimes without melting above 360°C and can be recrystallized from acetic acid. Polymerizations of **19** with diamines and tetramines have been reported (Table 1).

E.   3,4,9,10-Perylenetetracarboxylic Dianhydride

The dianhydride **22** of 3,4,9,10-perylenetetracarboxylic acid is commercially available. Upon heating, naphthalic acid is converted to its anhydride, which forms the corresponding imide when it is treated with ammonium hydroxide. Naphthalimide is fused over potassium hydroxide to produce the diimide of 3,4,9,10-perylenetetracarboxylic acid (**23**). This diimide is converted to the dianhydride **22** upon treatment with sulfuric acid (21). The red-brown dianhydride melts at temperatures above 360°C and has been

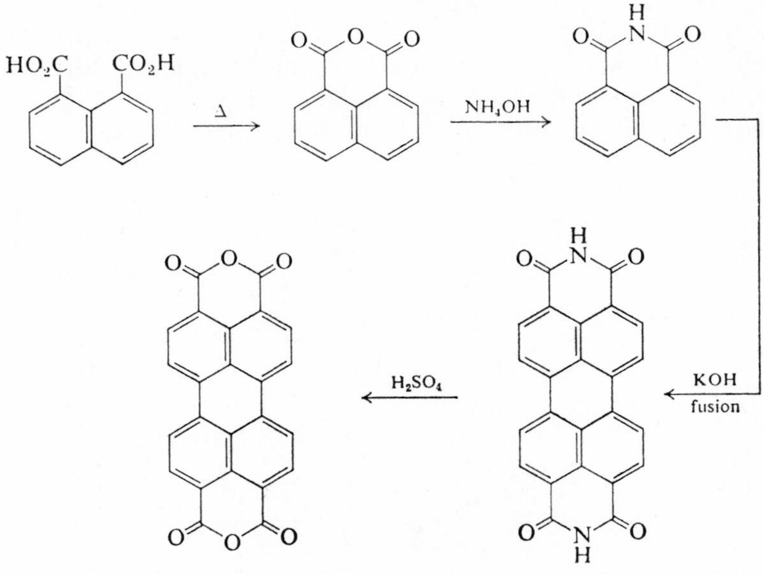

22

23

polymerized with diamines and tetramines (Table 1).

F.   2,3,6,7-Naphthalenetetracarboxylic Dianhydride

The reaction of 2 moles of maleic anhydride with 2 moles of allene produces 1,2,3,4,5,6,7,8-octahydronaphthalene-2,3,6,7-dianhydride (**24**) (22). The same product can be obtained in two steps by isolating the adduct **25** of maleic anhydride with allene dimer and allowing **25** to react with a second mole of maleic anhydride to produce **24** (23). The bromination of the octahydrodianhydride **24** affords an unstable intermediate, which loses hydrogen bromide to produce 2,3,6,7-naphthalenetetracarboxylic dianhydride (**26**) (24,25). The dianhydride, which melts at temperatures above 400°, can be recrystallized from acetic anhydride (25). The use of this dianhydride as a comonomer with diamines is summarized in Table 1.

G. 2,3,5,6-Biphenyltetracarboxylic Dianhydride

The dimerization of phenylpropiolic acid in acetic anhydride and the oxidation of the resulting anhydride (27) with potassium permanganate affords 2,3,5,6-biphenyltetracarboxylic acid (28) (26,27). Heating this tetracid with acetic anhydride converts it to the dianhydride 29. The dianhydride **29** melts at 209–210°C. Table 1 lists the reported polymerizations of **29** with diamines.

H.   2,2′,6,6′-Biphenyltetracarboxylic Dianhydride

The oxidation of pyrene with vanadium pentoxide–silver oxide catalyst produces 2,2′,6,6′-biphenyltetracarboxylic dianhydride (**30**) in low yield (19). Dimethyl 2-iodoisophthalate can be coupled with copper to produce the tetraester, which is converted to the tetracid **31** by saponification (28). The best synthesis of **31** involves the ozonolysis of pyrene to produce the aldehyde acid **32**, followed by oxidation with permanganate to yield **31** (29,30a). Treatment of **31** with acetic anhydride affords the 2:1 mixed anhydride **33**, which is converted to **30** by treatment with hot *o*-dichlorobenzene or nitrobenzene (30b). Reference to the polymerization of **30** with diamines appears in Table 1.

I. 2,2′,3,3′-Biphenyltetracarboxylic Dianhydride

The copper coupling of dimethyl 3-iodophthalate and subsequent saponification of the tetraester produces 2,2′,3,3′-biphenyltetracarboxylic acid (**34**). Treatment of **34** with acetyl chloride or heat affords 2,2′,3,3′-biphenyltetracarboxylic dianhydride (**35**) (14). The dianhydride **35** melts at 267°C

and can be recrystallized from acetic anhydride. References to the polymerizations of **35** with diamines appear in Table 1.

J. 3,3′,4,4′-Diphenylethertetracarboxylic Dianhydride

The dianhydride **36** of 3,3′,4,4′-diphenylethertetracarboxylic acid (**37**) is prepared by the permanganate oxidation of the coupling product (**38**) of

3,4-dimethylphenol and 3,4-dimethylbromobenzene (31–33). The dianhydride, **36**, is purified by recrystallization from acetone/Skellysolve C and melts at 221°C. The polymers of **36** with diamines and tetramines are cited in Table 1.

K.   3,3′,4,4′-Diphenylsulfonetetracarboxylic Dianhydride

The reaction of *o*-xylene with sulfuric acid affords 3,3′,4,4′-tetramethyldiphenylsulfone (**39**) (34). An alternate synthesis of **39** is the transsulfonation of *o*-xylene with *p*-toluenesulfonic acid (35). Oxidation of **39** with nitric acid produces the tetracid **40**, which is converted to 3,3′,4,4′-diphenylsulfonetetracarboxylic dianhydride (**41**) upon heating with acetic anhydride (36).

Both the dianhydride **41** and the tetracid **40** have been employed as comonomers in polymerizations with diamines. These polymers are cited in Table 1.

L.   Miscellaneous Tetracids and Dianhydrides

The literature contains accounts of the preparation of several tetracids and dianhydrides that have received little or no attention as possible monomers. In most cases the preparation of these compounds is too difficult or too expensive to make their manufacture commercially feasible. These compounds, together with references to their preparation and, when available, their polymerizations, appear in Table 2.

III. TETRAKETONES, HYDROXYQUINONES, AND DIGLYOXALS

The use of tetraketones, hydroxyquinones, and diglyoxals as tetrafunctional monomers is mostly limited to research polymers, because of the cost of these monomers, the difficulty of synthesis, or both. However, this has been an active field of research owing to the thermal properties of the resulting polymers.

In most cases, the tetracarbonyl or dihydroxydicarbonyl function is condensed with a tetramine to afford a quinoxaline or dihydroquinoxaline polymer, respectively. Representative polymerizations of each type of monomer are:

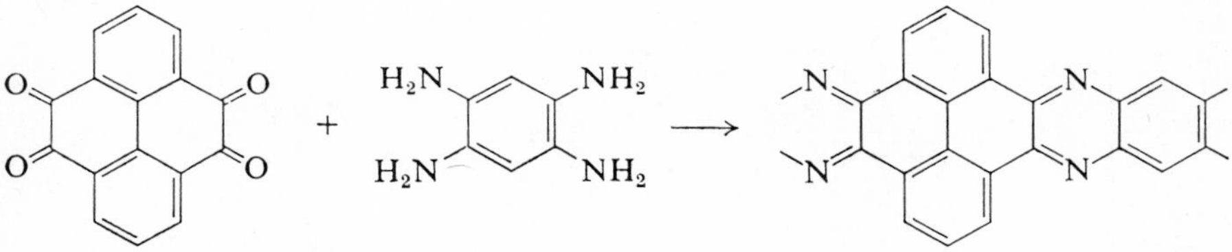

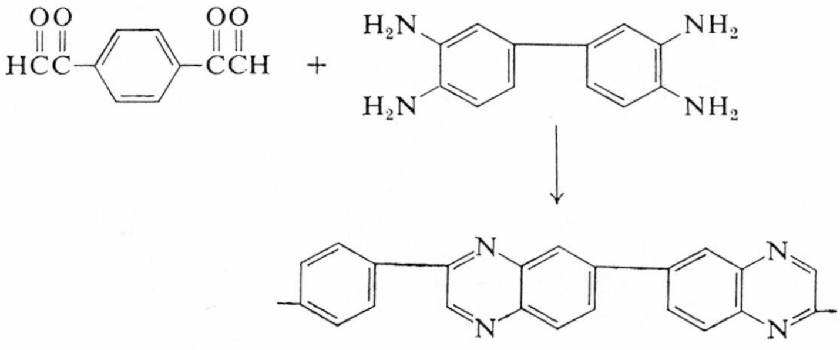

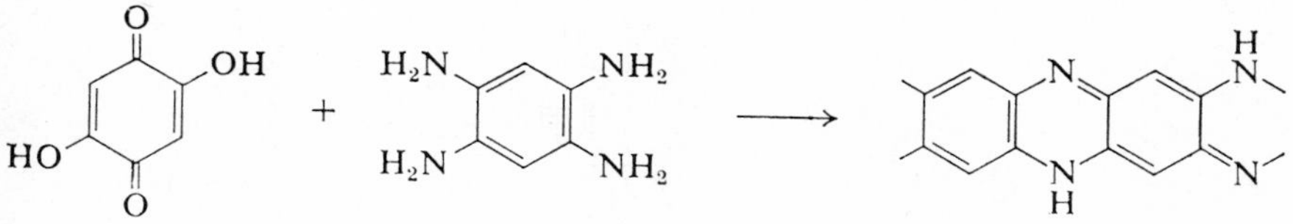

A. Pyrene-4,5,9,10-tetraketone

Pyrene-4,5,9,10-tetraketone (42) is prepared in a low yield by a five-step reaction sequence beginning with pyrene (29,136). A solution of 100 g of pyrene in 500 ml of N,N-dimethylformamide is ozonated with an eightfold excess of ozone, producing the lactone 43. Treatment of 43 with excess phenylhydrazine affords the α-hydroxyhydrazine 44. The hydrazine (44) is dissolved in ca. 800 ml of acetic acid at the reflux temperature, and 378 g of stannous chloride in 250 ml of concentrated hydrochloric acid is added. The solution is heated at the reflux

TABLE 2

References Covering the Preparation and Polymerization of Miscellaneous Tetracids and Dianhydrides

Monomer	Preparation	Polymerizations
1,2,5,6-Naphthalenetetracarboxylic dianhydride	70–72[a]	49–55
2,2-Bis(3,4-dicarboxyphenyl)propane dianhydride	73	49–55
1,2,4,5-Naphthalenetetracarboxylic dianhydride	74	49,50,53–55
2,3,6,7-Tetrachloronaphthalene-1,4,5,8-tetracarboxylic dianhydride	29,75–77[b]	53–55
Phenanthrene-1,8,9,10-tetracarboxylic dianhydride	78–80	53–55
2,2-Bis(2,3-dicarboxyphenyl)propane dianhydride		50,52–55
1,1-Bis(2,3-dicarboxyphenyl)ethane dianhydride		50,52–55
1,1-Bis(3,4-dicarboxyphenyl)ethane dianhydride		49,50,52–55
Bis(2,3-dicarboxyphenyl)methane dianhydride		49,50,52–55
Bis(3,4-dicarboxyphenyl)methane dianhydride		49,50,52–55
Benzene-1,2,3,4-tetracarboxylic dianhydride	81	49,50,53–55
2,2′,3,3′-Benzophenonetetracarboxylic dianhydride		53–55
2,3,3′,4′-Benzophenonetetracarboxylic dianhydride		53–55
Pyrazine-2,3,5,6-tetracarboxylic dianhydride	82–84	49,50,53–55
Thiophene-2,3,4,5-tetracarboxylic dianhydride	85–88	49,50,53–55
1,2,3,4-Cyclopentanetetracarboxylic dianhydride	89–98	51,99–101
2,3,4,5-Pyrrolidinetetracarboxylic dianhydride	102–106[c]	51

Compound		
Ethylenetetracarboxylic dianhydride	107	51
2,2-Bis(3,4-dicarboxyphenyl)hexafluoropropane dianhydride	108,109[d]	50
2,3,5,6-Pyridinetetracarboxylic dianhydride	110–114[e]	114
Azobenzene-3,3',4,4'-tetracarboxylic dianhydride	115	115
p-Benzoquinone-2,3,5,6-tetracarboxylic dianhydride	116–118[f]	
1,1,2,2-Cyclopropanetetracarboxylic acid	119,120[g]	
Anthracene-2,3,6,7-tetracarboxylic dianhydride	121,122	
Furan-2,3,4,5-tetracarboxylic acid	123–125	
Cyclohexane-1,2,3,4-tetracarboxylic dianhydride	93	
1,2,3,4-Butanetetracarboxylic dianhydride	126–131	
1,2,3,4-Cyclobutanetetracarboxylic dianhydride	132–135[h]	

[a] Alkyl-substituted derivatives reported.
[b] 2,6-Dichloro and 2,7-dichloro derivatives reported.
[c] Ref. 103 cites N-methyl and N-ethyl derivatives.
[d] Ref. 108 also cites dichloro tetrafluoro derivative.
[e] The 4-phenyl derivative is described in Refs. 112 and 114, the 4-methyl derivative in 113 and 114.
[f] Refs. 117 and 118 are the hydroquinone derivatives.
[g] Ref. 119 is the 3-methyl derivative; 120 is the 3,3-dimethyl derivative.
[h] Ref. 132 describes alkyl-substituted derivatives.

temperature for 2 hr, and during this time it becomes clear yellow. After cooling to ice bath temperature, the solid amine hydrochloride, **45**, is collected by filtration. A slurry of 10.4 g of **45** in 1 liter of water is made just basic with dilute sodium hydroxide, and dilute sulfuric acid is added until a slurry of the white amine hydrosulfate is formed. Dropwise addition of 13 g of a solution of 50 g of chromium trioxide in 35 ml of water and 35 ml of acetic acid affords the golden yellow diketone, **46**, which is collected by filtration. A solution of 6 g of pyrene-4,5-quinone in 300 ml of acetic acid is prepared by heating the ingredients at the reflux temperature and cooling to 90°C. All the quinone must be in solution at this time. A solution of 15 g of chromium trioxide in 15 ml of water is added in drops and the solution is brought to the reflux temperature for 15 min. After cooling, the golden yellow pyrene-4,5,9,10-tetraketone is collected by filtration and washed with absolute alcohol. The yield is 6–15% based on pyrene.

Pyrene-4,5,9,10-tetraketone melts at 264°C after recrystallization from nitrobenzene, and it is stable indefinitely toward light and air. The specific toxicity of **42** has not been investigated, but the pyrene nucleus is found in many carcinogenic agents and due care should be taken in handling. Polymerizations of **42** with tetramines to produce polyquinoxalines have been reported and are cited in Table 3.

TABLE 3

References Covering the Preparation and Polymerizations of Tetraketones, Hydroxyquinones, and Diglyoxals

Monomer	Preparation	Polymerizations
Pyrene-4,5,9,10-tetraone	29,136	148
3,3,6,6-Tetramethylcyclohexan-1,2,4,5-tetraone	137,138	149
1,2,5,6-Tetraketoanthracene	139–141	148
2,5-Dihydroxy-*p*-benzoquinone	142	148,150
2,5-Dihydroxy-3,6-dichloro-*p*-benzoquinone	142	148,150
2,5-Dihydroxy-3,6-difluoro-*p*-benzoquinone	143	148
1,4-Diglyoxalylbenzene dihydrate	144	144
1,3-Diglyoxalylbenzene dihydrate	145	145
4,4'-Diglyoxalyldiphenylsulfide dihydrate	146	146
4,4'-Diglyoxalyldiphenylsulfone dihydrate	146	146
4,4'-Diglyoxalyldiphenylether dihydrate	145,147	145,147
4,4'-Dibenzil	151	151
4,4'-Oxydibenzil	151	151

B. 3,3,6,6-Tetramethylcyclohexan-1,2,4,5-tetraone

The methylation of 5,5-dimethyl-1,3-cyclohexanedione affords 2,2,5,5-tetramethyl-1,3-cyclohexanedione (**47**) (137), which is then oxidized to 3,3,6,6-tetramethylcyclohexan-1,2,4,5-tetraone (**48**) (138) with selenium dioxide. The light yellow compound **48** melts at 233–235°C, after recrystallization from benzene. Polymerizations of **48** are referenced in Table 3.

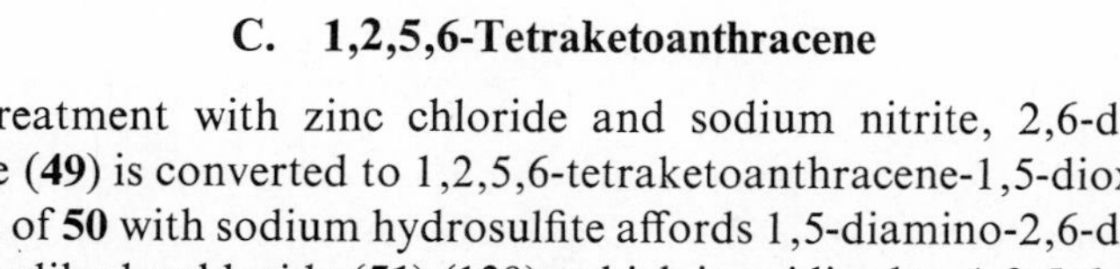

47   48

C. 1,2,5,6-Tetraketoanthracene

Upon treatment with zinc chloride and sodium nitrite, 2,6-dihydroxyanthracene (**49**) is converted to 1,2,5,6-tetraketoanthracene-1,5-dioxime (**50**). Reduction of **50** with sodium hydrosulfite affords 1,5-diamino-2,6-dihydroxyanthracene dihydrochloride (**51**) (139), which is oxidized to 1,2,5,6-tetraketoanthracene (**52**) (140). The dark purple tetraketone, which is purified by recrystallization from dioxane, melts at 320°C with decomposition (140,141). This tetraketone has been polymerized with tetramines (Table 3).

D. 2,5-Dihydroxy-*p*-benzoquinone, 2,5-Dihydroxy-3,6-dichloro-*p*-benzoquinone, and 2,5-Dihydroxy-3,6-difluoro-*p*-benzoquinone

The dihydroxyquinone monomers 2,5-dihydroxy-*p*-benzoquinone (**53**) and 2,5-dihydroxy-3,6-dichloro-*p*-benzoquinone (**54**) are commercially available. The alkaline peroxide oxidation of hydroquinone produces **53** and treatment of 2,3,5,6-tetrachloro-*p*-benzoquinone with sodium hydroxide affords **54** (142).

Commercial 2,5-dihydroxy-*p*-benzoquinone is a yellow-brown powder that decomposes at 210–215°C. It must be sublimed immediately prior to polymerizations. Commercial 2,5-dihydroxy-3,6-dichloro-*p*-benzoquinone sublimes without melting at 300°C and is purified by sublimination. Both of these monomers are skin irritants and are highly toxic to mice when ingested orally. When handling **53** and **54** special care should be taken to avoid contact with the dust (142).

The treatment of 2,3,5,6-tetrafluoro-*p*-benzoquinone with sodium hydroxide affords 2,5-dihydroxy-3,6-difluoro-*p*-benzoquinone (**55**), which decomposes at 230°C. This monomer can be purified by sublimination (143).

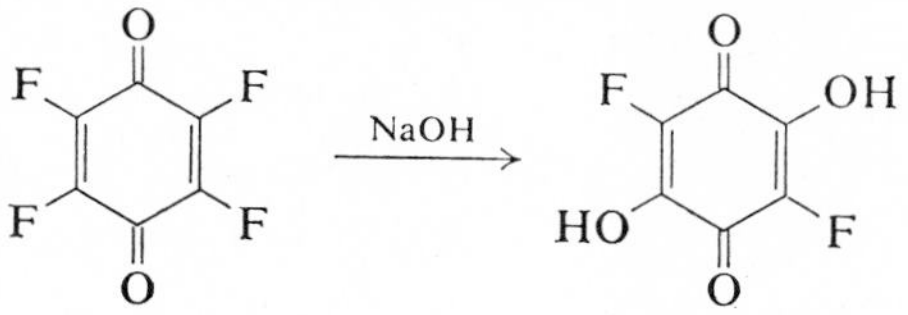

55

For polymerization information about compounds **53–55**, refer to Table 3.

E.   1,4-Diglyoxalylbenzene Dihydrate and 1,3-Diglyoxalylbenzene Dihydrate

To a solution of 22.2 g of selenium dioxide in 100 ml of dioxane and 4 ml of water containing 3 drops of concentrated hydrochloric acid is added 15 g of commercial 1,4-diacetylbenzene. The mixture is heated at the reflux temperature for 4 hr, and during this time selenium metal precipitates. The mixture is filtered hot and the filtrate is cooled to afford yellow crystals of crude 1,4-diglyoxalylbenzene dihydrate (**56**) in 80% yield. Recrystallization from water afforded pure **56**, melting at 140–150°C (dec) (144). The 1,3 isomer, **57**, is prepared by a similar procedure from 1,3-diacetylbenzene and is a yellow oil (145). Polymerizations with a variety of tetramines have been reported for **56** and **57** (Table 3).

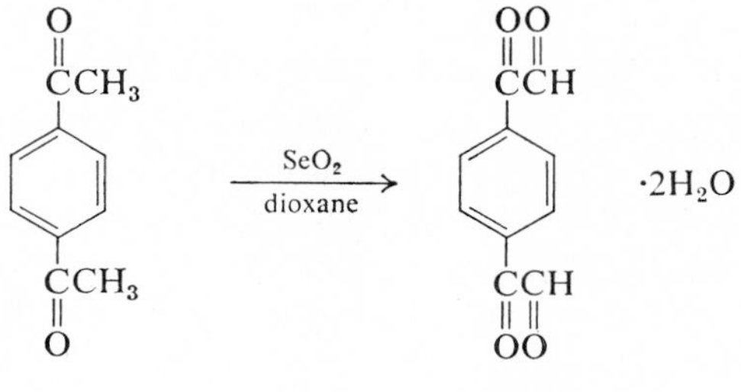

56

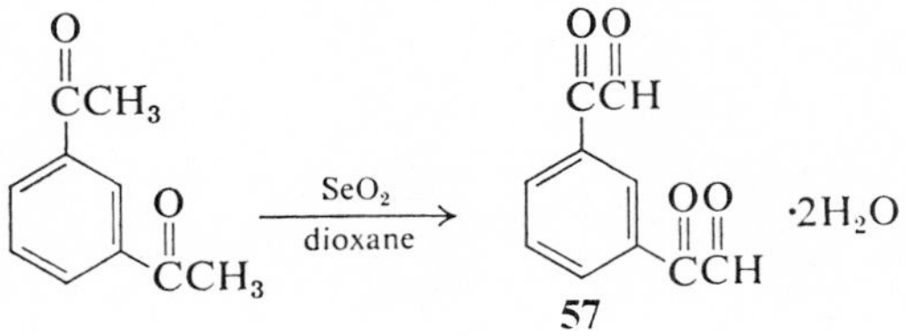

57

F. 4,4′-Diglyoxalyldiphenylsulfide Dihydrate and 4,4′-Diglyoxalyldiphenylsulfone Dihydrate

The acetylation of diphenyl sulfide affords 4,4′-diacetyldiphenylsulfide (58). If 4,4′-diglyoxalyidiphenylsulfone dihydrate (59) is desired, 58 is oxidized to 4,4′-diacetyldiphenylsulfone (60) with hydrogen peroxide. Selenium dioxide oxidation of 58 and 60 produces 4,4′-diglyoxalyldiphenylsulfide dihydrate (61) and 59, respectively (46). The diglyoxalyls are isolated and stored as dihydrates, which are more stable than the unhydrated species. The polymerizations of 59 and 61 with tetramines are referenced in Table 3.

G. 4,4′-Diglyoxalyldiphenylether Dihydrate

Diphenylether can be acetylated to give 4,4′-diacetyldiphenylether (62). The oxidation of 62 with selenium dioxide produces 4,4′-diglyoxalyldiphenylether dihydrate (63) (145,147). The monomer 63 has been employed in a variety of polymerization reactions (Table 3).

H.   4,4′-Dibenzil and 4,4′-Oxydibenzil

The treatment of 4,4′-diglyoxalylbiphenyl and 4,4′-diglyoxalyldiphenyl-ether with aluminum chloride in benzene affords the dibenzoins **64** and **65**, which are oxidized to 4,4′-dibenzil (**66**) and 4,4′-oxydibenzil (**67**). Table 3 cites references to the polymerizations of **66** and **67** with tetramines.

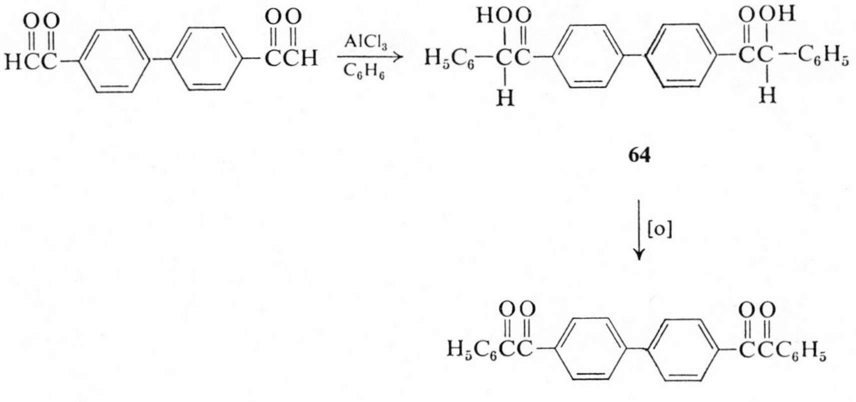

64

66

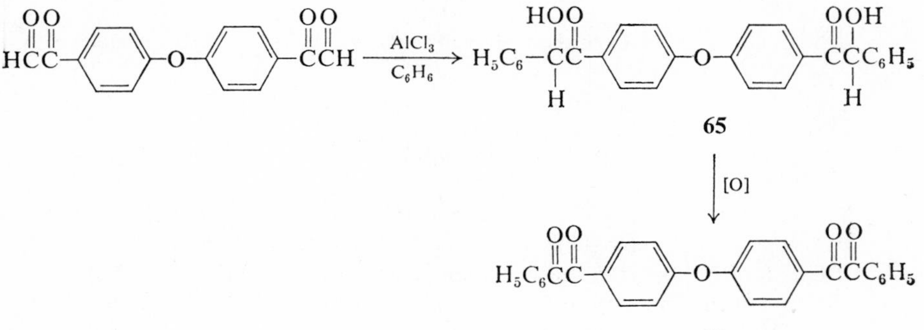

65

67

IV.   TETRAMINES

The tetramine is employed as a comonomer with diacid derivatives, tetracids or dianhydrides, and tetracarbonyl compounds to prepare poly-benzimidazoles, polyimidazopyrrolones, and polyquinoxalines, respectively. A few tetramines are commercially available, but most of the polymerizations are carried out in the course of research, rather than on a commercial scale.

Representative examples of these three kinds of polymers are:

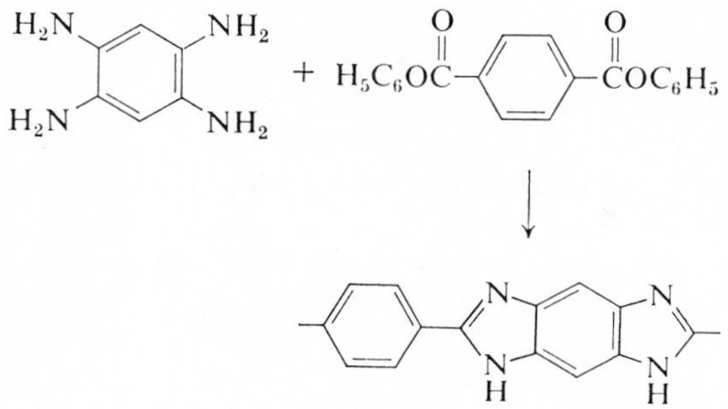

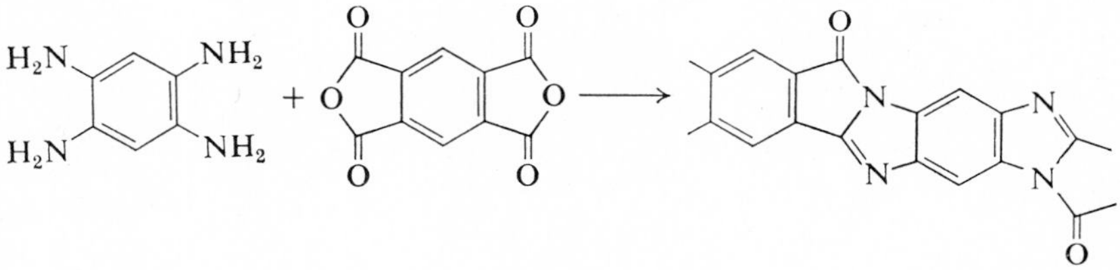

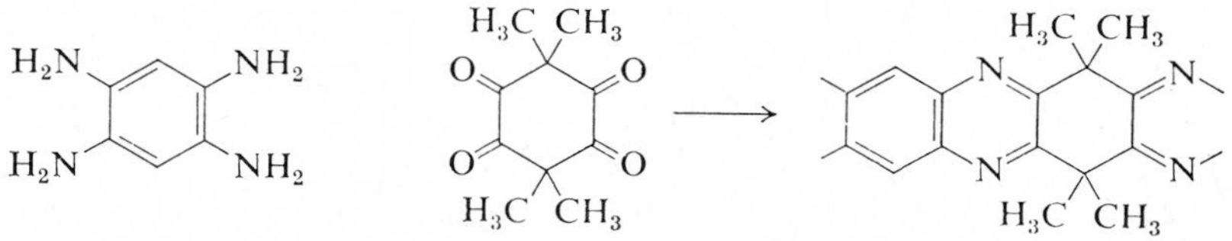

A.   3,3'-Diaminobenzidine

A mixture of 182 g of benzidine and 1500 ml of acetic acid is heated until solution is complete, and 200 ml of acetic anhydride is added. The slurry is stirred at the reflux temperature for 2 hr and allowed to cool to room temperature. The solid, diacetylbenzidine (**68**), is collected by filtration and washed with alcohol. With vigorous stirring, 246 g of **68** is added to 1200 ml of yellow fuming nitric acid (90%), which is cooled to −10°C. The solution is stirred for 30 min and the temperature is allowed to warm to 0°C; at this time it is poured over 6 liters of crushed ice. The precipitated diacetyl-3,3'-dinitrobenzidine (**69**) is collected by filtration and dried. The crude **69** is suspended in 2.5 liters of 95% ethanol and heated to 60°C, and 200 g of potassium hydroxide in 300 ml of water is added. This mixture is heated at reflux for 10 min. After cooling, the solid, 3,3'-dinitrobenzidine (**70**), is collected by filtration, washed with methanol, and dried. To a solution of 1200 g of stannous chloride dihydrate in 2.5 liters of concentrated hydrochloric acid is added 216 g of **70** at a rate that keeps the temperature below 60°C. The slurry is stirred an additional 2 hr at 40°C and cooled to 10°C, and the solid is collected by filtration. The solid, 3,3′-diaminobenzidine tetrahydrochloride is dissolved in 3 liters of water and reprecipitated by the addition of 1.5 liters of concentrated hydrochloric acid. This process is repeated until a white solid, devoid of tin salts, is obtained. This solid is dissolved in 2 liters of water and added to a solution of 200 g of sodium hydroxide in 1800 ml of water. The precipitate is collected by filtration, washed with 200 ml of cold water, and dried. The 3,3′-diaminobenzidine (**71**) can be purified by sublimation or recrystallization from methanol. The overall yield of **71**, which melts at 178–179°C, is 37% (152).

Commercial 3,3′-diaminobenzidine has a melting point of 178–179°C (153,154). It is a light tan (153) or gray (154) crystalline solid that darkens upon exposure to light or air. If the compound is stored in the dark under a nitrogen atmosphere, there is less than 1% decomposition after 90 days at room temperature (153). The tetrahydrochloride salt, which does not melt below 300°C, is a tan powder that darkens in the presence of air (155). The free amine is moderately toxic in mice (154) and the tetrahydrochloride salt is toxic and a probable carcinogen (156). Both the free amine and its salt have been employed as comonomers with dianhydrides, tetracids,

D. 3,3′,4,4′-Tetraminodiphenylsulfide and
3,3′,4,4′-Tetraminodiphenylsulfone

The reaction of ethyl xanthate with *p*-chloronitrobenzene affords 4,4′-dinitrodiphenylsulfide (**76**) (161), which is reduced to 4,4′-diaminodiphenylsulfide (**77**) with stannous chloride and hydrochloric acid (162). Acetylation of **77** yields 4,4′-diaceatmidodiphenylsulfide (**78**) (162). At this point the sulfide, **78**, can be oxidized to 4,4′-diacetamidodiphenylsulfone (**79**) with hydrogen peroxide (163), if the sulfone derivative is desired. The following sequence of reactions is identical for the sulfide and sulfone.

The 4,4′-diacetamidodiphenylsulfide (sulfone) is nitrated to afford 3,3′-dinitro-4,4′-diacetamidodiphenylsulfide (sulfone) (**80 or 81**) (164), which is converted to 3,3′-dinitro-4,4′-diaminodiphenylsulfide hydrochloride (146) (sulfone hydrochloride) (165) (**82 or 83**) by refluxing with hydrochloric acid. Reduction of the nitro groups with stannous chloride and hydrochloric acid and subsequent neutralization of the product yields 3,3′,4,4′-tetraminodiphenylsulfide (146) (**84**) or 3,3′,4,4′-tetraminodiphenylsulfone (163) (**85**).

Both **84** and **85** are white crystalline solids melting at 102.5–103.0°C and 174.0–174.5°C, respectively. Both should be stored in an inert atmosphere and in the absence of light. The tetramines **84** and **85** have been polymerized with a variety of diglyoxals to afford polyquinoxalines (Table 4).

E. 3,3′,4,4′-Tetraminodiphenylether

The acetylation of *p*-oxidianiline affords 4,4′-diacetamidodiphenylether (**86**) (166). The nitration of **86** yields 3,3′-dinitro-4,4′-diacetamidodiphenylether (**87**), which is converted to 3,3′-dinitro-4,4′-diaminodiphenylether (**88**) tetracarbonyls, and diacid derivatives. The polymerizations are referenced in Table 4.

B. 1,2,4,5-Tetraminobenzene

The free amine, 1,2,4,5-tetraminobenzene (72) can be prepared in the laboratory, and 1,2,4,5-tetraminobenzene tetrahydrochloride (73) is commercially available. The nitration of *m*-dichlorobenzene produces 1,5-dinitro-2,4-dichlorobenzene (74), which, upon treatment with ammonia, is converted to 1,5-dinitro-2,4-diaminobenzene (75). The hydrogenation of 75, upon acidification with hydrochloric acid, affords 1,2,4,5-tetraminobenzene tetrahydrochloride (157). Neutralization of the tetrahydrochloride with sodium hydroxide yields the free amine 72, which can be purified by recrystallization or sublimation (158).

The free amine melts at 274–276°C (158) and is sensitive to light and air. The commercial tetrahydrochloride is a white to pale blue, pink, or lavender crystalline solid (159) which does not melt below 350°C (157). Reports of polymerization of 72 and 73 with dianhydrides and tetraketones are found in Table 4.

C. 1,3-Diphenylamine-4,6-diaminobenzene and
1,3-Dimethylamino-4,6-diaminobenzene

If, in the reaction sequence to produce 1,2,4,5-tetraminobenzene (Section IV.B), aniline and methylamine are substituted for ammonia in the nucleophilic displacement of chloride, 1,3-diphenylamino-4,6-diaminobenzene (74) (152) and 1,3-dimethylamino-4,6-diaminobenzene (75) (160) are produced. The polybenzimidazoles obtained when 74 and 75 are polymerized with diacid derivatives are referenced in Table 4.

TABLE 4

References Covering the Preparation and Polymerizations of Tetramines

Monomer	Preparation	Polymerizations with			
		Diacid derivatives	Dianhydrides	Tetracarbonyls	Other
3,3'-Diaminobenzidine	152	152,158,169	6,7,57–59 64,65	144,145,148,151	170,171,177
1,2,4,5-Tetraminobenzene	158	152,158	1,6,7,57,65	148,149	172
1,3-Diphenylamino-4,6-diaminobenzene	161	161			
1,3-Dimethylamino-4,6-diaminobenzene	162	162			
3,3',4,4'-Tetraminodiphenylsulfide	146			146	
3,3',4,4'-Tetraminodiphenylsulfone	163			146	
3,3',4,4'-Tetraminodiphenylether	145,166	166	7	145,151	
1,4,5,8-Tetraminonaphthalene	7,167	167	7		
2,3,6,7-Tetraminodibenzodioxane	148			148,149	

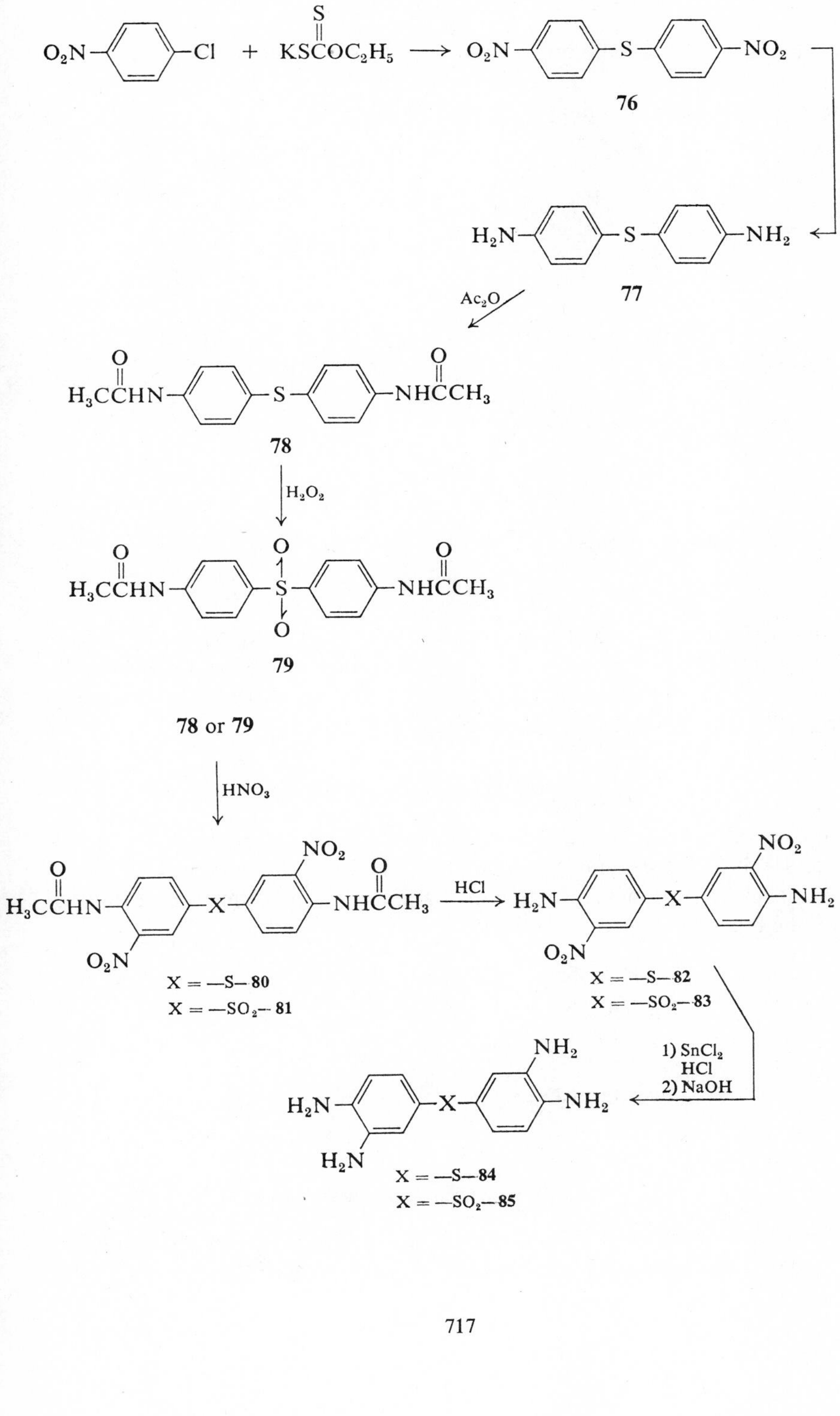

O$_2$N—⟨ ⟩—Cl + KS$\overset{\text{S}}{\text{C}}OC_2H_5$ ⟶ O$_2$N—⟨ ⟩—S—⟨ ⟩—NO$_2$

76

SnCl$_2$
HCl

H$_2$N—⟨ ⟩—S—⟨ ⟩—NH$_2$

77

Ac$_2$O

H$_3$C$\overset{\text{O}}{\text{C}}$HN—⟨ ⟩—S—⟨ ⟩—NH$\overset{\text{O}}{\text{C}}CH_3$

78

H$_2$O$_2$

H$_3$C$\overset{\text{O}}{\text{C}}$HN—⟨ ⟩—$\overset{\text{O}}{\underset{\text{O}}{\text{S}}}$—⟨ ⟩—NH$\overset{\text{O}}{\text{C}}CH_3$

79

78 or **79**

HNO$_3$

H$_3$C$\overset{\text{O}}{\text{C}}$HN—⟨ ⟩—X—⟨ ⟩—NH$\overset{\text{O}}{\text{C}}CH_3$

NO$_2$

O$_2$N

X = —S— **80**
X = —SO$_2$— **81**

HCl ⟶

H$_2$N—⟨ ⟩—X—⟨ ⟩—NH$_2$

NO$_2$

O$_2$N

X = —S— **82**
X = —SO$_2$— **83**

1) SnCl$_2$
    HCl
2) NaOH

H$_2$N—⟨ ⟩—X—⟨ ⟩—NH$_2$

NH$_2$

H$_2$N

X = —S— **84**
X = —SO$_2$— **85**

717

by stirring with methanolic potassium hydroxide. Reduction of **88** with stannous chloride and hydrochloric acid, followed by neutralization, produces 3,3′,4,4′-tetraminodiphenylether (**89**) (146,166).

The tetramine **89**, which melts at 150–151°C, should be stored under nitrogen in the dark. It does not keep well and must be sublimed immediately prior to use in polymerizations. Polymers of **89** with diglyoxals, dibenzils, and diacid derivatives have been reported (Table 4).

F. 1,4,5,8-Tetraminonaphthalene

The unstable 1,4,5,8-tetraminonaphthalene (**90**) is generated *in situ* by the reduction of 1,4,5,8-tetranitronaphthalene (7,167). Polymerizations of **90** with diacid derivatives and dianhydrides are effected without removing **90** from the reduction solvent (Table 4).

G. 2,3,6,7-Tetraminodibenzodioxane

The tetrahydrochloride salt of 2,3,6,7-tetraminodibenzodioxane (**91**) (148) is prepared by the catalytic reduction of 2,3,6,7-tetranitrodibenzodioxane

(168). Purification is accomplished by repeated precipitation from water by the addition of concentrated hydrochloric acid. The light pink tetrahydrochloride must be stored under nitrogen and in the absence of light. Polymerizations of **91** with aromatic and aliphatic tetraketones have been reported (Table 4).

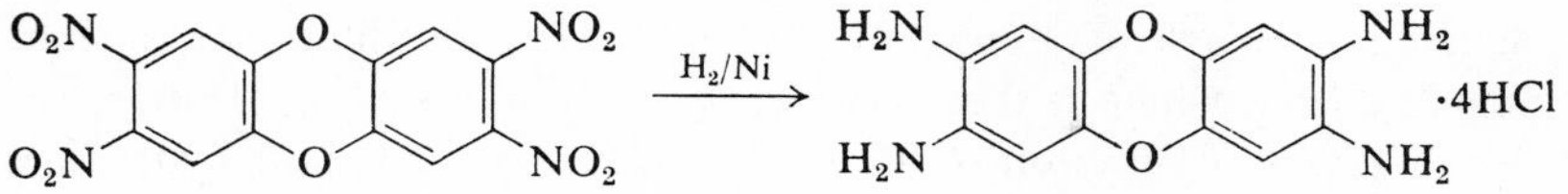

91

V. MISCELLANEOUS TETRAFUNCTIONAL MONOMERS

There are reports in the literature of several tetrafunctional monomers which do not fall into any of the general classes discussed in Sections II–IV. These monomers are disubstituted diamines in which the substituent can be hydroxy, phenoxy, or halide. Since there are no general examples of the polymers formed by these monomers, examples of their polymerizations are included with the discussion of each monomer. A summary of the preparations and polymerizations of these monomers appears in Table 5.

TABLE 5

References Covering the Preparation and Polymerizations of Miscellaneous Tetrafunctional Monomers

Monomer	Preparation	Polymerizations
6,7-Diamino-2,3-diphenoxyquinoxaline	173,174	172
6,7-Diamino-2,3-dihydroxyquinoxaline	173,174	172
2,2′,3,3′-Tetrahydroxy-6,6′-bisquinoxaline	171,172	171
2,2′,3,3′-Tetrachloro-6,6′-bisquinoxaline	171,172	—
2,2′,3,3′-Tetraphenoxy-6,6′-bisquinoxaline	171,172	171
2,3,6,7-Tetrahydroxy-1,4,5,8-tetrazanthracene	171,172	171,172
2,3,6,7-Tetrachloro-1,4,5,8-tetrazanthracene	171,172	172
2,3,6,7-Tetraphenoxy-1,4,5,8-tetrazanthracene	171,172	171,172
4,6-Diaminoresorcinol dihydrochloride	176	150
3,3′-Dimercaptobenzidine dihydrochloride	178	179–181
p-Bis(bromoacetyl)benzene	182,183	182,183
4,4′-Bis(bromoacetyl)biphenyl	183	183
4,4′-Bis(bromoacetyl)diphenyl ether	183	183
4,4′-Bis(bromoacetyl)diphenylmethane	183	183
4,4′-Bis(bromoacetyl)-1,6-diphenylhexane	183	183
3,3′-Diaminobiphenyl-4,4′-dicarboxylic acid		184,185
3,3′-Dihydroxybenzidine	179	179,229
3,3′-Diamino-4,4′-dihydroxybiphenyl	229	229

A.   6,7-Diamino-2,3-diphenoxyquinoxaline and
6,7-Diamino-2,3-dihydroxyquinoxaline Dihydrochloride Hydrate

The monomers 6,7-diamino-2,3-diphenoxyquinoxaline (**92**) and 6,7-diamino-2,3-dihydroxyquinoxaline dihydrochloride hydrate (**93**) are commercially available. A suitable laboratory synthesis of **93** is achieved by condensing *o*-phenylenediamine with oxalic acid to produce 2,3-dihydroxyquinoxaline (**94**), which is then nitrated to afford 6,7-dinitro-2,3-dihydroxyquinoxaline (**95**). Reduction of **95** with stannous chloride and hydrochloric

acid yields 6,7-diamino-2,3-dihydroxyquinoxaline dihydrochloride hydrate (**93**) (173,174). The diphenoxy monomer **92** is prepared by treating **95** with phosphorus pentachloride to produce 2,3-dichloro-6,7-dinitroquinoxaline (**96**), which affords 2,3-diphenoxy-6,7-dinitroquinoxaline (**97**) when reacted with phenol. Catalytic hydrogenation of **97** yields 6,7-diamino-2,3-diphenoxyquinoxaline (**92**) (173,174).

98

The phenoxy derivative **92** is a light tan solid that melts at 208–210°C and may be stored for several months without decomposition. The dihydroxy derivative **93** is also light tan; it does not melt below 300°C, and it may be stored indefinitely under nitrogen (175). Both **92** and **93** undergo self-polymerization, yielding the ladder polyquinoxaline **98** (172).

B.   Tetrahydroxy-, Tetrachloro- and Tetraphenoxy-
Derivatives of 2,2′,3,3′-Tetrasubstituted-6,6′-bisquinoxaline
and of 2,3,6,7-Tetrasubstituted-1,4,5,8-tetrazanthracene

Treatment of 3,3′-diaminobenzidine and 1,2,4,5-tetraminobenzene with oxalic acid affords 2,2′,3,3′-tetrahydroxy-6,6′-bisquinoxaline (**99**) and 2,3,6,7-tetrahydroxy-1,4,5,8-tetrazanthracene (**100**), respectively. The reaction of **99** with phosphorus pentachloride produces 2,2′,3,3′-tetrachloro-6,6′-bisquinoxaline (**101**), and the reaction of **100** with phosphorus oxychloride and antimony trichloride gives 2,3,6,7-tetrachloro-1,4,5,8-tetrazanthracene (**102**). Treatment of **101** and **102** with phenol affords 2,2′,3,3′-tetraphenoxy-6,6′-bisquinoxaline (**103**) and 2,3,6,7-tetraphenoxy-1,4,5,8-tetrazanthracene (**104**), respectively (171,172).

The polymerization of **99** or **103** and the polymerization of **100** or **104** with 3,3′-diaminobenzidine affords the polyquinoxalines **105** and **106**, respectively (171). When **100**, **102**, or **104** is polymerized with 1,2,4,5-tetraminobenzene, the ladder polyquinoxaline **107** is produced (172).

C. 4,6-Diaminoresorcinol Dihydrochloride

The diaminodiphenol, 4,6-diaminoresorcinol dihydrochloride (**108**) is prepared by acetylating resorcinol, affording resorcinol diacetate (**109**), which may then be nitrated to produce 4,6-dinitroresorcinol (**110**). Reduction of **110** with tin and hydrochloric acid produces 4,6-diaminoresorcinol dihydrochloride (**108**) (176).

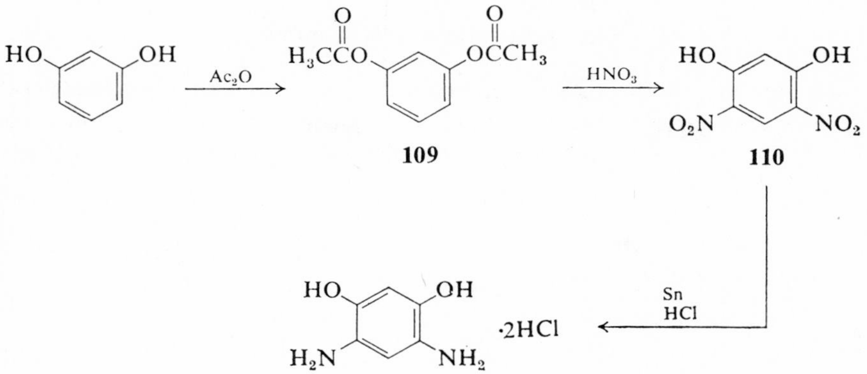

109

110

108

·2HCl

The light purple dihydrochloride, which decomposes at 231°C, can be stored indefinitely under nitrogen. Polymers have been prepared employing **108** and dihydroxyquinones, affording ladder polyphenoxazines such as **111** (150).

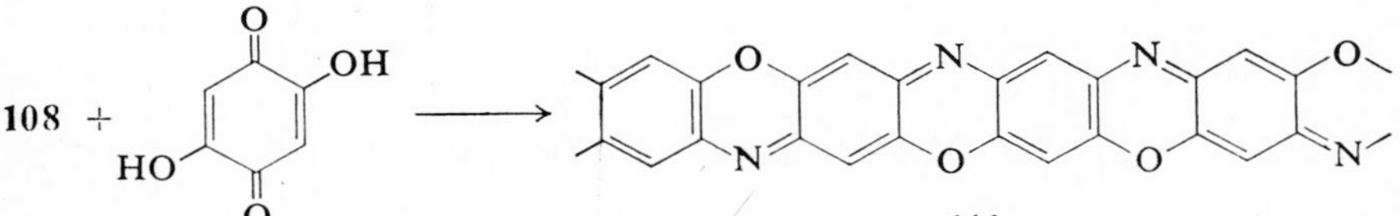

108 +

111

D. 3,3'-Dimercaptobenzidine

This diaminodithiol is prepared from the thiocyanation of benzidine (178). Because of its extreme air sensitivity, it is best handled as the salt, 3,3'-dimercaptobenzidine dihydrochloride (**112**). It has been polymerized with diacid derivatives (179–181) to afford polybenzothiazoles (**113**).

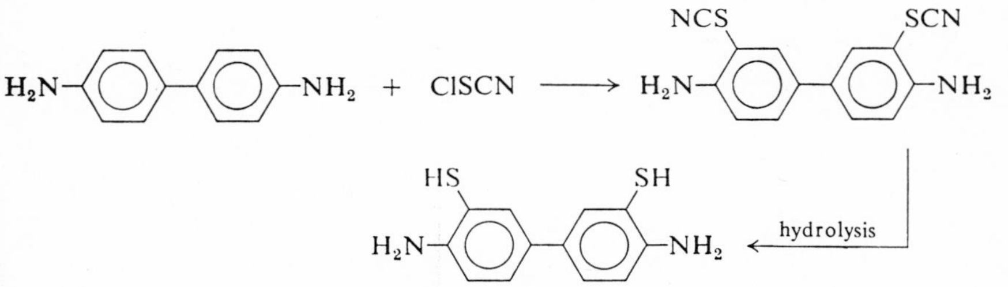

112

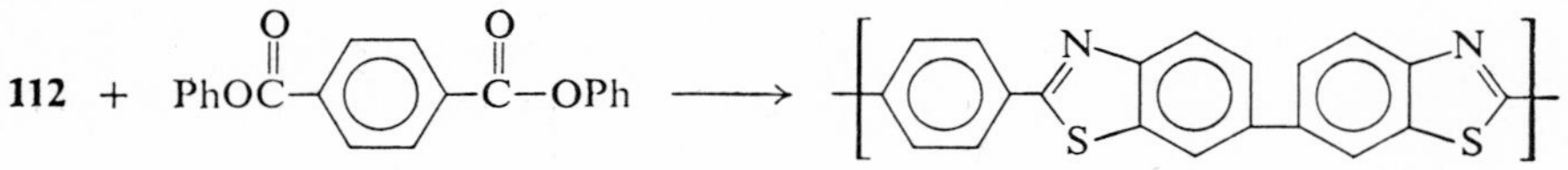

E. *p*-Bis(bromoacetyl)benzene

In addition to the parent compound, *p*-bis(bromoacetyl)benzene (114) (182), a series of bis(bromoacetyl)aromatics has been prepared as monomers (183). They are condensed with dithioamides to form polybithiazoles (115).

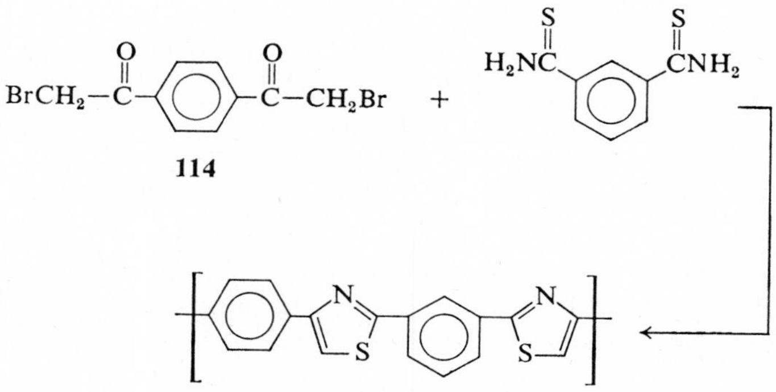

114

115

F. 4,4′-Diaminobiphenyl-3,3′-dicarboxylic Acid

The monomer 4,4′-diaminobiphenyl-3,3′-dicarboxylic acid (116) has been employed in polymerizations, although its preparation and purification has

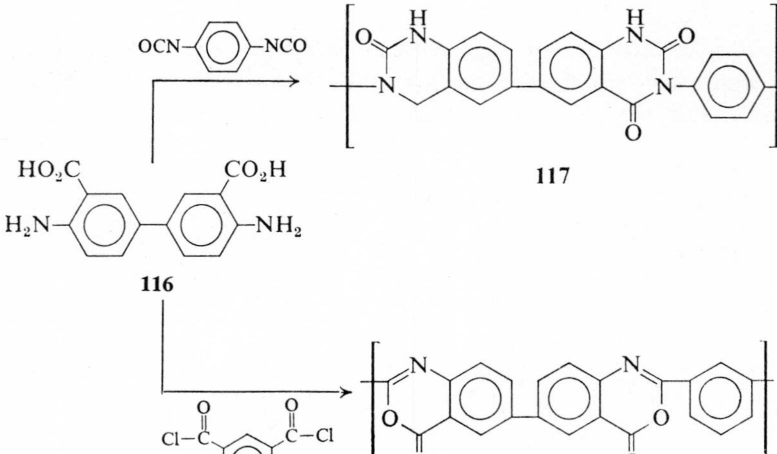

117

116

118

not appeared. When **116** is polymerized with diisocyanates, polyquinazoline-diones (**117**) are formed (184); when polymerized with diacid chlorides,

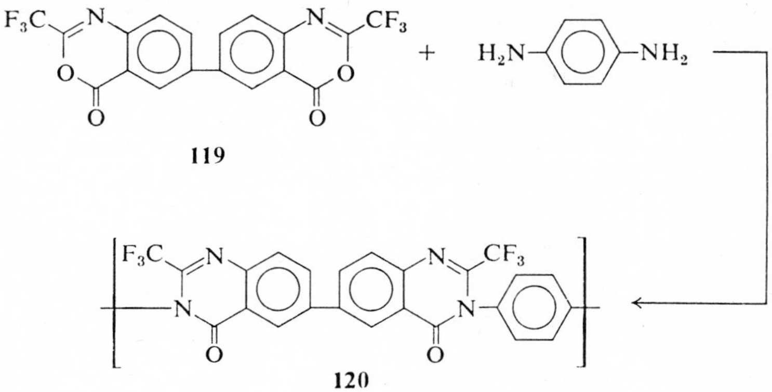

119

120

polybenzoxazinones (**118**) are obtained (185); when its derivative **119** is polymerized with diamines, polyquinazolones (**120**) are produced (222).

G. 3,3'-Dihydroxybenzidine and 3,3'-Diamino-4,4'-dihydroxybiphenyl

The dihydroxydiamines, 3,3'-dihydroxybenzidine (**121**) and its isomer have been polymerized with diacid derivatives to afford polybenzoxazoles (**122**) (179,229).

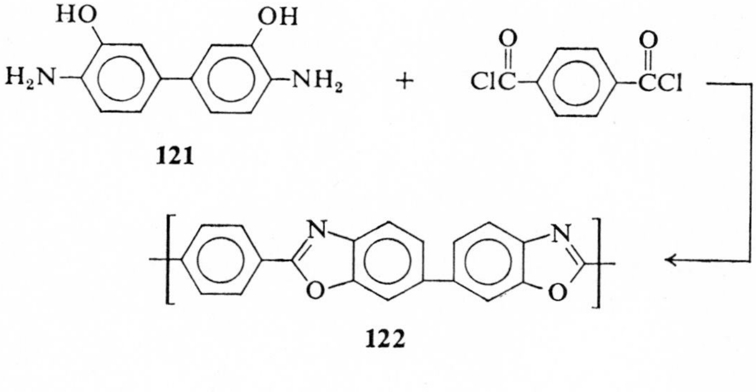

121

122

APPENDIX

Since the completion of this chapter in 1967, a large number of tetrafunctional monomers and their polymerization reactions have been described. Therefore, this appendix attempts to update this chapter through 1970 without imposing the production delay involved in rewriting the chapter.

TABLE 6
Monomers and Polymers Included in the Text

Monomer	Reference
Tetracarboxylic Acids and Dianhydrides	
Pyrazine-2,3,5,6-tetracarboxylic dianhydride	186–188
Naphthalene-1,4,5,8-tetracarboxylic dianhydride	189,210
3,4,9,10-Perylenetetracarboxylic dianhydride	190
Pyromellitic dianhydride	212,223
Tetraketones, Hydroquinones, and Diglyoxals	
4,4'-Dibenzil	191,192,197,198
4,4'-Oxydibenzil	191,192,195–198
1,4-Diglyoxalylbenzene dihydrate	192,193,196
4,4'-Diglyoxalyldiphenylether dihydrate	192,194,195
1,3-Diglyoxalylbenzene dihydrate	193
Tetramines	
3,3'-Diaminobenzidine	191,192,194,199–203
3,3',4,4'-Tetraminodiphenyl ether	191,192,194,199,200
1,2,4,5-Tetraminobenzene	193
3,3',4,4'-Tetraminodiphenyl sulfone	196
Miscellaneous	
6,7-Diamino-2,3-dihydroxyquinoxaline dihydrochloride monohydrate	204
2,2',3,3'-Tetrachloro-6,6'-bisquinoxaline	205,206
2,3,6,7-Tetrachloro-1,4,5,8-tetrazanthracene	205,206
3,3'-Dimercaptobenzidine	202,216,217
4,4'-Diaminobiphenyl-3,3'-dicarboxylic acid	218–221
1,4-Bis(bromoacetyl)benzene	230
4,4'-Bis(bromoacetyl)biphenyl	230
4,4'-Bis(bromoacetyl)diphenyl ether	230

The appendix is divided into two tables. Table 6 lists new syntheses and/or polymerizations of monomers already cited in Tables 1 to 5. Table 7 lists new monomers and/or their polymerizations that have appeared in the period 1968–1970 inclusive. No attempt has been made to separate synthesis from polymerization in these tables.

TABLE 7
Newly Reported Monomers and Polymers

Monomer	Reference
Tetracarboxylic Acids and Dianhydrides	
1,2,5,6-Bis(α,β-dicarboxylpyrazino)anthraquinone dianhydride	203
4,4'-Bis(3,4-dicarboxyphenoxy)biphenyl dianhydride	207
4,4'-Bis(3,4-dicarboxyphenoxy)diphenyl ether dianhydride	207
Tetraketones, Hydroquinones, and Diglyoxals	
5,5'-Biisatyl	200
5,5'-Diisatyl ether	200
Benzene-1,2,4,5-tetraldehyde	208
N,N'-Bis(4-benzilyl)pyromellitimide	201
4,4'-Sulfonyldibenzil	196
Tetramines	
2,3,11,12-Tetraminodiquinoxal[2,3-*e*,2',3'-*l*]pyrene	210
Bis(2,2-aminomethyl)-1,3-diaminopropane	211
1,4-Diamino-1,4-diaminomethylcyclohexane	211
1,2,5,6-Tetraminoanthraquinone	205,212,213
1,2,5,6-Tetraminonaphthalene	214
3,5-Bis(3-amino-4-anilinophenyl)-4-phenyl-4H-1,2,4-triazole	215
3,5-Bis[3-amino-4-(4-phenoxyanilino)phenyl]-4-phenyl-4H-1,2,4-triazole	215
3,3',4,4'-Tetraminobenzophenone	196
Miscellaneous	
5,5'-Bis(2-aminobenzamide)	224
5,5'-Methylenebis(2-aminobenzamide)	223,224
2,2',3,3'-Tetrachloro-6,6'-diquinoxalyl ether	225
1,2,4,5-Tetracyanobenzene	199
4,6-Diaminoisophthalaldehyde	226
2,5-Diaminobenzoquinone	227,228
2,5-Diaminobenzoquinone diimide	227,228
2,5-Diaminohydroquinone	227,228
4,4'-Bis(4-carboxy-3-aminophenoxy)diphenyl sulfone	215
3,3',4,4'-Tetrahydroxybiphenyl	206
1,2,4,5-Tetrahydroxybenzene	206
2,3,5,6-Tetrahydroxy-*p*-benzoquinone	206
2,3,6,7-Tetrahydroxythianthrene	206
1,2,5,6-Tetrahydroxyanthraquinone	206

References

1. G. F. Pezdirtz and V. L. Bell, *J. Polym. Sci. B*, **3**:977 (1965).
2. J. I. Jones, F. W. Ochynski, and F. A. Rackley, *Chem. Ind.* (**London**), **1962**:1686.
3. W. Flavell and V. E. Yarsley, *Chem. Britain*, **3**:375 (1967).
4. "Pyromellitic Dianhydride," E. I. du Pont Technical Products Bull.
5. E. Philippi and R. Thelen, *Org. Syn.*, **Coll. Vol. II.** 551.
6. "Pyromellitic Dianhydride and Pyromellitic Acid," E. I. du Pont Technical Products Bull.

7. C. S. Marvel and F. Dawans, *J. Polym. Sci. A*, **3**:3549 (1965).
8. Gulf Research and Development Co., French Patent No. 1,346,797 (1963).
9. N. A. Adrova, M. M. Koton, and L. P. Ivanova, *Vysokomol. Soedin.*, Ser. *B*, **9**:22 (1967).
10. J. H. McCracken and J. D. Schulz, U.S. Patent 3,078,279 (1963).
11. "Industrial Hygiene and Toxicology Bulletin, Benzophenonetetracarboxylic Dianhydride," Gulf Oil Corporation, Pittsburgh, Pa.
12. R. Loewenherz, *Ber.*, **26**:2486 (1893).
13. N. A. Adrova, M. M. Koton, and E. M. Moskvina, *Dokl. Akad. Nauk SSSR*, **165**:1069 (1965).
14. J. Kenner and A. M. Mathews, *J. Chem. Soc.*, **105**:2471 (1914).
15. I. G. Farben, FAIT Final Rep. 1313, PB 85172, Vol. II, p. 163.
16. H. Greune and W. Eckert, German Patent 566,154 (1931).
17. I. G. Farbenindustrie AG, U.S. Patent 1,803,182 (1931).
18. I. G. Farbenindustrie AG, British Patent 363,044 (1930).
19. J. Arient, J. Slosar, Z. Nerad, and M. Habada, Czechoslovak Patent 111,551 (1964).
20. H. Fierz-David and L. Blangey, "Fundamental Processes of Dye Chemistry," Interscience, New York, 1949, p. 222.
21. A. E. Porai-Koshits and I. S. Pavlushenko, *J. Gen. Chem. USSR*, **17**:1739 (1947).
22. K. Alder and O. Ackermann, *Ber.*, **87**:1567 (1954).
23. A. T. Blomquist and J. A. Verdol, *J. Amer. Chem. Soc.*, **78**:109 (1956).
24. O. W. Webster, U.S. Patent 2,912,442 (1959).
25. O. W. Webster and W. H. Sharkey, *J. Org. Chem.*, **27**:3354 (1962).
26. A. Michael and J. E. Bucher, *J. Amer. Chem. Soc.*, **20**:89 (1898).
27. G. S. Kolesnikov, O. Ya. Fedotova, E. I. Hofbauer, and H. H. Mohamed Ali Al'-Sufi, *Vysokomol. Soedin.*, **8**:1440 (1966).
28. F. Mayer, *Ber.*, **44**:2301 (1911).
29. H. Vollmann, H. Becker, M. Corell, and H. Streeck, *Ann.*, **531**:1 (1937).
30a. P. G. Copeland, British Patent 1,007,012 (1965).
30b. O. K. Goins and R. L. Van Deusen, *Polym. Lett.*, *B*, **6**:821 (1968).
31. C. S. Marvel and J. H. Rassweiler, *J. Amer. Chem. Soc.*, **80**:1197 (1958).
32. G. S. Kolesnikov, O. Ya. Fedolova, E. I. Hofbauer, and V. G. Shelgaeva, *Vysokomol. Soedin.*, Ser. *A*, **9**:612 (1967).
33. W. F. Gresham and M. A. Naylor, Jr., U.S. Patent 2,731,447 (1956).
34. J. A. Jarboe, III, U.S. Patent 2,833,826 (1958).
35. H. Drews, S. Meyerson, and E. K. Fields, *Angew. Chem.*, **72**:493 (1960).
36. O. F. Bennett, M. L. Huber, and R. A. Smiley, U.S. Patent 3,022,320 (1962).
37. E.I. du Pont de Nemours and Co., British Patent 903,272 (1962).
38. M. I. Bessonov, M. M. Koton, E. I. Pokrovskii, and E. E. Fedorova, *Dokl. Akad. Nauk SSSR*, **161**:617 (1965).
39. C. E. Sroog, A. L. Endrey, S. V. Abramo, C. E. Berr, W. M. Edwards, and K. L. Olivier, *J. Polym. Sci.*, *A*, **3**:1373 (1965).
40. W. G. Gall, French Patent 1,365,545 (1964).
41. W. M. Edwards and A. L. Endrey, British Patent 903,271 (1962).
42. F. E. Rogers, Belgian Patent 649,366 (1964).
43. S. V. Vinogradova, V. V. Korshak, and Ya. S. Vygodskii, *Vysokomol. Soedin.*, **8**:809 (1966).
44. V. H. Kuckertz, *Makromol. Chem.*, **98**:101 (1966).
45. Dow Corning Corp., Netherlands Patent Appl. 6,414,419 (1965).
46. W. M. Edwards and I. M. Robinson, U.S. Patent 2,710,853 (1955).

47. W. M. Edwards and I. M. Robinson, U.S. Patent 2,880,230 (1959).
48. J. Preston and W. B. Black, *J. Polym. Sci. A*, **5**:2429 (1967).
49. W. M. Edwards, U.S. Patent 3,179,634 (1965).
50. E. I. du Pont de Nemours and Co., British Patent 1,062,435 (1964).
51. W. M. Edwards, U.S. Patent 3,179,614 (1965).
52. A. L. Endrey, U.S. Patent 3,179,630 (1965).
53. J. A. Kreuz, U.S. Patent 3,271,366 (1966).
54. R. J. Angelo and W. E. Tatum, U.S. Patent 3,316,212 (1967).
55. W. E. Tatum, U.S. Patent 3,261,811 (1967).
56. C. E. Sroog, *J. Polym. Sci. C*, **16**:1191 (1967).
57. J. G. Colson, R. H. Michel, and R. M. Paufler, *J. Polym. Sci. A*, **4**:59 (1966).
58. B. J. Sillion and A. Reboul, *Compt. Rend., Ser. C*, **262**:471 (1966).
59. E. N. Teleshov and A. N. Pravednikov, *Dokl. Akad. Nauk SSSR*, **172**:1347 (1967).
60. M. M. Guy Rabilloud, B. Sillion, and G. De Gaudemaris, *Compt. Rend., Ser. C*, **263**:862 (1966).
61. M. Saga, T. Shono, and K. Shinra, *Bull. Chem. Soc. Japan*, **39**:1795 (1966).
62. Z. Yu. Plonka and W. M. Al'brekht, *Vysokomol. Soedin.*, **7**:2177 (1965).
63. Z. Yu. Plonka and W. M. Al'brekht, *Polym. Sci. USSR*, **7**:2387 (1965).
64. A. A. Berlin, B. I. Liogon'Kii, G. M. Shamraev, and G. V. Belova, *Izv. Akad. Nauk. SSSR, Ser. Khim.*, **1966**:945.
65. R. L. Van Deusen, *J. Polym. Sci. B*, **4**:211 (1966).
66. S. A. Zakoshchikov and G. A. Ruzhentseva, *Vysokomol. Soedin.*, **8**:1231 (1966).
67. J. C. Oromi, Spanish Patent No. 304,410 (1965).
68. N. A. Adrova, M. M. Koton, and E. M. Moskuina, *Dokl. Akad. Nauk SSSR*, **165**:1069 (1965).
69. S. A. Zakoshchikov, K. N. Vlasova, and G. M. Zolotareva, *Vysokomol. Soedin., Ser. B*, **9**:234 (1967).
70. L. W. Butz, A. M. Gaddis, E. W. J. Butz, and R. E. Davis, *J. Org. Chem.*, **5**:379 (1940).
71. L. W. Butz, U.S. Patent 2,329,979 (1944).
72. L. W. Butz, U.S. Patent 2,387,830 (1945).
73. W. F. Gresham and M. A. Naylor, Jr., U.S. Patent 2,712,543 (1955).
74. L. F. Fieser and E. B. Hershberg, *J. Amer Chem. Soc.*, **61**:1272 (1939).
75. H. Reimlinger and G. King, *Ber.*, **95**:1043 (1962).
76. J. M. Straley and C. W. Wayman, U.S. Patent 2,578,759 (1951).
77. J. Arient, J. Dvorak and D. Snobl, *Collect. Czech. Chem. Commun.*, **28**:2479 (1963).
78. A. Zinke, G. Hauswirth, and V. Grimm, *Monatsh.*, **57**:405 (1931).
79. A. Zinke, F. Stimler, and E. Reuss, *Monatsh.*, **64**:415 (1934).
80. B. R. Brown and A. R. Todd, *J. Chem. Soc.*, **1954**:1280.
81. L. I. Smith, "Organic Reactions," Vol. 1, Wiley, New York, 1942, p. 370.
82. F. D. Chattaway and W. G. Humphrey, *J. Chem. Soc.*, **1929**:645.
83. A. R. J. Ramsey, British Patent 565,778 (1944).
84. R. J. Light and C. R. Hauser, *J. Org. Chem.*, **26**:1296 (1961).
85. H. Hopff and J. von der Crone, *Chimia (Switz.)*, **13**:107 (1959).
86. R. D. Vest, U.S. Patent 3,053,853 (1962).
87. H. E. Simmons, R. D. Vest, D. C. Blomstrom, J. R. Roland, and T. L. Cairns, *J. Amer. Chem. Soc.*, **84**:4746 (1962).
88. O. Scherer and F. Kluge, *Ber.*, **99**:1973 (1966).
89. H. Hopff and W. Rapp, U.S. Patent No. 2,203,628 (1940).
90. I. G. Farbenindustrie AG, British Patent 510,638 (1939).

91. K. Alder and S. Schneider, *Ann.*, **524**:189 (1936).
92. K. Alder, F. H. Flock, A. Hausweiler, and R. Reeber, *Ber.*, **87**:1752 (1954).
93. K. Alder, H. H. Möells, and R. Reeber, *Ann.*, **211**:7 (1958).
94. A. W. Carlson and G. C. Schweiker, German Patent 1,132,719 (1962).
95. R. Van Volkenburgh, J. R. Olechowski, and G. C. Royston, U.S. Patent 3,194,816 (1965).
96. R. Van Volkenburgh and J. R. Olechowski, U.S. Patent 3,218,353 (1965).
97. B. L. Moldavskii and V. G. Babel, *Zh. Prikl. Khim.*, **36**:1614 (1963).
98. R. Van Volkenburgh and J. R. Olechowski, U.S. Patent 3,242,206 (1966).
99. A. C. Nawakowski and L. A. Lundberg, U.S. Patent 3,215,553 (1965).
100. G. C. Royston, U.S. Patent 3,247,146 (1966).
101. Hysol Corp., British Patent 1,043,998 (1966).
102. R. A. Nicolaus and L. Mangoni, *Ann. Chim. (Rome)*, **1956**:865.
103. R. A. Nicolaus, *Rass. Med. Sper.*, 7:Suppl. 2 (1960).
104. A. Vitale, M. Piattelli, and R. A. Nicolaus, *Rend. Accad. Sci. Fis. Mat. (Soc. Nazl. Sci. Napoli) Ser. 4*, **26**:267 (1959).
105. M. Piattelli, E. Fattorusso, and S. Magno, *Tetrahedron Lett.*, **1961**:718.
106. L. Chierici and G. C. Artusi, *Ann. Chim. (Rome)*, **53**:1644 (1963).
107. H. R. Snyder and C. W. Kruse, *J. Amer. Chem. Soc.*, **80**:1942 (1958).
108. D. G. Coe, Netherlands Patent 6,406,896 (1964).
109. F. E. Rodgers, Belgian Patent 649,336 (1964).
110. M. P. Oparina, *Zh. Obshch. Khim.*, **19**:1351 (1949).
111. G. Machek, *Monatsh.*, **59**:175 (1932).
112. W. H. Mills, W. H. Palmer, and M. G. Tomkinson, *J. Chem. Soc.*, **125**:2369 (1924).
113. A. Hantzsch, *Ann.*, **215**:57 (1882).
114. S. Hashimoto and V. Nagasuna, *Kobunshi Kagaku*, **24**:633 (1968).
115. Farbenfabriken Bayer AG, Netherlands Patent Appl. 6,515,693 (1966).
116. P. R. Hammond, *Science*, **142**:502 (1963).
117. J. U. Nef, *Ann.*, **237**:32 (1887).
118. J. U. Nef, *Ann.*, **258**:282 (1890).
119. R. P. Mariella and A. J. Roth, *J. Org. Chem.*, **22**:1130 (1957).
120. H. Hart and F. Freeman, *J. Amer. Chem. Soc.*, **85**:1161 (1963).
121. C. Marschalk, *Bull. Soc. Chim., Fr.*, **9**:400 (1942).
122. C. Marschalk, *Bull. Soc. Chim., Fr.*, **1950**:311.
123. T. Reichstein, A. Grüssner, K. Schindler, and E. Hardmeier, *Helv. Chim. Acta*, **16**:276 (1933).
124. H. Sutter, *Ann.*, **499**:47 (1932).
125. H. Gilman, H. Oatfield, and W. H. Kirkpatrick, *Proc. Iowa Acad. Sci.*, **40**:112 (1933).
126. A. P. Dunlop and C. D. Hurd, *J. Org. Chem.*, **15**:1160 (1950).
127. N. P. Buu-Hoi, M. Sy, and N. D. Xuong, *Rec. Trav. Chim. Pays-Bas*, **75**:463 (1956).
128. J. Cley and J. F. Arens, *Rec. Trav. Chim. Pays-Bas*, **78**:929 (1959).
129. E. Buchta and K. Greiner, *Ber.*, **94**:1311 (1961).
130. J. E. Franz and W. S. Knowles, *Chem. Ind. (London)*, **1961**:250.
131. I. G. Farbenindustrie AG, British Patent 510,638 (1939).
132. G. O. Schenck, W. Hartmann, S. P. Mannsfeld, W. Metzner, and C. H. Krauch, *Ber.*, **95**:1642 (1962).
133. E. B. Reid and M. Sack, *J. Amer. Chem. Soc.*, **73**:1985 (1951).
134. G. W. Griffin, J. E. Basinski, and A. F. Vellturo, *Tetrahedron Lett.*, **3**:13 (1960).
135. R. Criegee and W. Funke, *Ber.*, **94**:2358 (1961).
136. R. E. Dessey and M. S. Newman, *Org. Syn.*, **38**:32 (1958).

137. E. J. Meek, J. H. Turnbull, and W. Wilson, *J. Chem. Soc.*, **1953**:811.
138. Y. Gaoni and E. Wenkert, *J. Org. Chem.*, **31**:3809 (1966).
139. M. V. Gorelik, *Zh. Obshch. Khim.*, **34**:2003 (1964).
140. M. V. Gorelik, *J. Org. Chem. USSR*, **1**:584 (1965).
141. P. Boldt, *Ber.*, **99**:2322 (1966).
142. Private communication, Eastman Organic Chemicals Dept., Distillation Products Industries.
143. K. Wallenfels and K. Friedrich, *Ber.*, **93**:3070 (1960).
144. J. K. Stille and J. R. Williamson, *J. Polym. Sci. A*, **2**:3867 (1964).
145. J. K. Stille, J. R. Williamson, and F. E. Arnold, *J. Polym. Sci. A*, **3**:1013 (1964).
146. J. K. Stille and F. E. Arnold, *J. Polym. Sci. A*, **4**:551 (1966).
147. G. De Gaudemaris and B. Sillion, *J. Polym. Sci. B*, **2**:203 (1964).
148. J. K. Stille and E. L. Mainen, *Macromolecules*, **1**:36 (1968).
149. J. K. Stille and M. E. Freeburger, *J. Polym. Sci. B*, **5**:989 (1967).
150. J. K. Stille and M. E. Freeburger, *J. Polym. Sci. A*, **6**:161 (1968).
151. P. M. Hergenrother and H. H. Levine, *J. Polym. Sci. A*, **5**:1453 (1967).
152. H. Vogel and C. S. Marvel, *J. Polym. Sci. A*, **1**:1531 (1963).
153. Technical Bull., "3,3'-Diaminobenzidine," Burdick and Jackson Laboratories, Inc., Muskegon, Mich.
154. Technical Bull., "3,3'-Diaminobenzidine," American Aniline Products, Inc., Paterson, N.J.
155. Analytical Rep., "3,3'-Diaminobenzidine Tetrahydrochloride Hydrate," Aldrich Chemical Co., Inc., Milwaukee, Wis.
156. Private communication, G. Fredrich Smith Chemical Co., Columbus, Ohio.
157. R. Nietzki and A. Schedler, *Ber.*, **30**:1666 (1897).
158. H. A. Vogel and C. S. Marvel, *J. Polym. Sci.*, **50**:511 (1961).
159. Technical Bull., "1,2,4,5-Tetraaminobenzene Tetrahydrochloride," Burdick and Jackson Laboratories, Inc., Muskegon, Mich.
160. K. Mitsuhashi and C. S. Marvel, *J. Polym. Sci., A*, **3**:1661 (1965).
161. C. C. Price and G. W. Stacy, *Org. Syn.*, **Coll. Vol. III.** 667 (1960).
162. G. W. Raiziss, L. W. Clemence, M. Severac, and J. C. Moetsch, *J. Amer. Chem. Soc.*, **61**:2763 (1939).
163. A. M. Van Arendonk and E. C. Kleiderer, *J. Amer. Chem. Soc.*, **62**:3521 (1940).
164. T. Matsukawa, B. Ohta, and T. Imada, *J. Pharm. Soc. Jap.*, **70**:77 (1950).
165. F. Ullman and J. Korselt, *Ber.*, **40**:641 (1907).
166. R. Foster and C. S. Marvel, *J. Polym. Sci. A*, **3**:417 (1965).
167. F. Dawans, B. Reichel, and C. S. Marvel, *J. Polym. Sci. A*, **2**:5005 (1964).
168. H. Gilman and J. Dietrich, *J. Amer. Chem. Soc.*, **80**:366 (1958).
169. J. E. Mulvaney and C. S. Marvel, *J. Polym. Sci.*, **50**:541 (1961).
170. J. E. Mulvaney, J. J. Bloomfield, and C. S. Marvel, *J. Polym. Sci.*, **62**:59 (1962).
171. H. Jadamus, F. De Schryver, W. De Winter, and C. S. Marvel, *J. Polym. Sci. A*, **4**:2831 (1966).
172. F. De Schryver and C. S. Marvel, *J. Polym. Sci. A*, **5**:545 (1967).
173. Private communication, Burdick and Jackson Laboratories, Inc., Muskegon, Mich.
174. W. G. Jackson and W. Schroeder, U.S. Patent 3,326,915 (1967).
175. Technical Bull., 6,7-Diamino-2,3-diphenoxyquinoxaline and 6,7-Diamino-2,3-dihydroxyquinoxaline, Dihydrochloride Hydrate, Burdick and Jackson Laboratories, Inc., Muskegon, Mich.
176. P. G. W. Typke, *Ber.*, **16**:551 (1883).
177. W. Wrasidlo and H. Levine, *J. Polym. Sci. A*, **2**:4795 (1964).

178. Houben-Weyl, "*Methoden der Organischen Chemie*," Verlag Chemie, Weinheim, Germany, 1955, Vol. IX, p. 39.
179. Y. Imai, I. Taoka, K. Uno, and Y. Iwakura, *Makromol. Chem.*, **83**:167 (1965).
180. P. M. Hergenrother, W. Wrasidlo, and H. H. Levine, *J. Polymer Sci.*, *A-1*, **3**:1665 (1965).
181. P. M. Hergenrother and H. H. Levine, *J. Polymer Sci.*, *A-1*, **4**:2341 (1966).
182. J. E. Mulvaney and C. S. Marvel, *J. Org. Chem.*, **26**:95 (1961).
183. D. T. Longone and H. N. Un, *J. Polymer Sci.*, *A-1*, **3**:3117 (1965).
184. N. Yoda, R. Nakanishi, M. Kurihara, Y. Bamba, S. Tohyama, and K. Ikeda, *J. Polymer. Sci.*, *B*, **4**:11 (1965).
185. N. Yoda, M. Kurihara, K. Ikeda, S. Tohyama, and R. Nakanishi, *J. Polymer Sci.*, *B*, **4**:551 (1966).
186. S. S. Hirsch and J. R. Holsten, *Appl. Polymer Symp.*, **9**:187 (1969).
187. S. S. Hirsch, *J. Polymer Sci.*, *A-1*, **7**:15 (1969).
188. S. S. Hirsch, French Patent 1,545,420 (1969).
189. O. K. Goins and R. L. Van Deusen, *J. Polym. Sci. B*, **6**:821 (1968).
190. S. Iwa-shina and J. Aoki, *Bull. Chem. Soc. Japan*, **41**:2789 (1968).
191. P. M. Hergenrother and H. H. Levine, *J. Polym. Sci. A-1*, **5**:1453 (1967).
192. W. Wrasidlo, *J. Polym. Sci. A-1*, **8**:1107 (1907).
193. W. Wrasidlo and J. M. Augl, *J. Polym. Sci. B*, **8**:69 (1970).
194. P. M. Hergenrother and H. H. Levine, *J. Appl. Polym. Sci.*, **14**:1037 (1970).
195. P. M. Hergenrother and D. E. Kiyohara, *Macromolecules*, **3**:387 (1970).
196. W. Wrasidlo and J. M. Augl, *Macromolecules*, **3**:545 (1970).
197. W. Wrasidlo and J. M. Augl, *J. Polym. Sci. B*, **7**:281 (1969).
198. W. Wrasidlo and J. M. Augl, *J. Polym. Sci. A-1*: **7**:3393 (1969).
199. D. I. Packham, J. D. Davies, and H. M. Paisley, *Polymer*, **10**:923 (1969).
200. I. Shopov and N. Popov, *J. Polym. Sci. A-1*, **7**:1803 (1969).
201. J. M. Augl, *J. Polym. Sci. A-1*, **8**:3145 (1970).
202. H. Kokelenberg and C. S. Marvel, *J. Polym. Sci, A-1*, **8**:3199 (1970).
203. P. K. Dutt and C. S. Marvel, *J. Polym. Sci. A-1*, **8**:3225 (1970).
204. R. Liepins, G. S. P. Verma, and C. Walker, *Macromolecules*, **2**:419 (1969).
205. A. Banihashemi and C. S. Marvel, *J. Polym. Sci. A-1*, **8**:3211 (1970).
206. R. Wolf and C. S. Marvel, *J. Polym. Sci. A-1*, **7**:2381 (1969).
207. M. M. Koton and F. S. Florinskii, *Zh. Org. Khim*, **6**:88 (1970).
208. N. Soyer, M. Kerfunto, and J. Vene, *Bull. Soc. Chim. France*, **1967**:1480.
209. W. Wrasidlo and P. M. Hergenrother, *Macromolecules*, **3**:548 (1970).
210. F. E. Arnold, *J. Polym. Sci. A-1*, **8**:2079 (1970).
211. J. Heller, J. H. Hodgkin, and F. J. Martinelli, *J. Polym. Sci. B*, **6**:153 (1968).
212. W. Bracke and C. S. Marvel, *J. Polym. Sci. A-1*, **8**:3177 (1970).
213. R. Pense and C. S. Marvel, *J. Polym. Sci. A-1*, **8**:3189 (1970).
214. J. K. Stille, W. Alston, E. Green, K. Imai, L. Mathias, R. Wilhelms, and J. Wittmann, Air Force Materials Laboratory—Technical Report-70-5, Part II, Jan. 1971.
215. G. Lorenz, M. Gallus, W. Giessler, F. Bodesheim, H. Wieden, and G. N. Nischk, *Makromol. Chem.*, **130**:65 (1969).
216. H. Kokelenberg and C. S. Marvel, *J. Polym. Sci. A-1*, **8**:3235 (1970).
217. M. Okada and C. S. Marvel, *J. Polym. Sci. A-1*, **6**:1744 (1968).
218. M. Kurihara and N. Yoda, *J. Polym. Sci. A-1*, **5**:1765 (1967).
219. N. Yoda, K. Ikeda, M. Kurihara, S. Tohyama, and R. Nakanishi, *J. Polym. Sci. A-1*, **5**:2359 (1967).
220. M. Kurihara and N. Yoda, *Makromol. Chem.*, **107**:112 (1967).

221. M. Kurihara and N. Yoda, *J. Macromol. Sci.*, **A1**:1069 (1967).
222. G. De Gaudemaris, B. Sillion, and J. Preve, *Bull. Soc. Chim. France*, **1965**:171.
223. M. Kurihara and N. Yoda, *J. Polym. Sci. B*, **6**:883 (1968).
224. M. Kurihara, *Macromolecules*, **3**:722 (1970).
225. M. Okada and C. S. Marvel, *J. Polym. Sci. A-1*, **6**:1259 (1968).
226. W. Bracke, *Macromolecules*, **2**:286 (1969).
227. J. Szita and C. S. Marvel, *J. Polym. Sci. A-1*, **9**:413 (1971).
228. J. Szita and C. S. Marvel, *J. Polym. Sci. A-1*, **7**:3203 (1969).
229. T. Kubota and R. Nakanishi, *J. Polym. Sci. B*, **2**:655 (1964).
230. J. Higgins, J. F. Jones, and A. Thornborgh, *Macromolecules*, **2**:558 (1969).

SUBJECT INDEX

DATE DUE

MAY 10			